DYNAMIC BEHAVIOR OF MATERIALS

DYNAMIC BEHAVIOR OF MATERIALS

Marc André Meyers
University of California, San Diego

A WILEY-INTERSCIENCE PUBLICATION

JOHN WILEY & SONS, INC.

New York • Chichester • Brisbane • Toronto • Singapore

A NOTE TO THE READER
This book has been electronically reproduced from digital information stored at John Wiley & Sons, Inc. We are pleased that the use of this new technology will enable us to keep works of enduring scholarly value in print as long as there is a reasonable demand for them. The content of this book is identical to previous printings.

This text is printed on acid-free paper.

Library of Congress Cataloging in Publication Data:

Meyers, Marc A.
Dynamic behavior of materials/Marc A. Meyers.
p. cm.
Includes bibliographical references.
ISBN 0-471-58262-X
1. Deformations (Mechanics) 2. Materials—Mechanical properties. 3. Micromechanics. I. Title.
TA417.6.M49 1994
620.1'123—dc20 93-33109

Lovingly dedicated to the memory
of my father, Henri Meyers.

BLUE WAVES

Soldiers who dream of wars
Mothérs who dream of milk
Lovers their dreams are sweet.
But I, I dream of blue
Blue waves merging thought and skies
Blue waves drowning in the sea
Blue waves in your blue eyes.

Shelly Bustamante
"Chronicles of the Higher Seas"

GOLNAZ OF THE NIGHT

Golnaz of the night
In the curl of your hair
In the curve of your neck
In the scent of your skin
In your eyes full of li(v)es.

And the wave of your tigh
That meets my embrace
In the folds of the night
And my pain and my pride
On the sands of this tide.

The sand is so true
Not mine but the wave's
Not now but for ages
And so are you dear
As the sand by my fingers
You will flow to the wave
Leaving only my longing
And my pain and my pride.

Amdur Hafez
"Poems for a Last Love"

PREFACE

This book has a simple objective: to provide the reader (typically, an engineering student, engineer, or a materials scientist) with a working knowledge of dynamic events in materials. It presents theory, experimentation, and applications in a balanced way. It contains example problems throughout the chapters. It is directed at an engineering/science student with either senior or graduate standing with a working knowledge of calculus; derivations are patiently worked out in a step-by-step manner, and elegance has been sacrificed for ease of comprehension. The sequence of topics progresses from the mechanics of high-strain-rate deformation (primarily, but not exclusively, elastic, plastic, shock, and detonation waves) to the dynamic response of materials (constitutive models, shear instabilities, dynamic fracture). The last chapter addresses the various topics of technological importance in a broad way, developing, where appropriate, the most fundamental quantitative treatment. The topics have not been covered in an exhaustive manner, and a great deal of important work has been left out for the sake of conciseness. The references at the end of the chapters provide some additional information. The study of dynamic processes in materials has been intensively carried out, especially since World War II, at major research laboratories throughout the world. Many thousands of research papers have been published in these past 50 years, and this author was forced to ignore a great fraction of this work. This is primarily a textbook, and no attempt has been made to cover the field exhaustively; thus, important contributions have been overlooked.

This book represents the response to a need clearly felt among students by this author for the past 20 years. Most scientists/engineers entering this field acquire their knowledge through an informal learning process, consisting of seminars, verbal discussions, monographs, and research papers. This is a lengthy and inefficient process, and the development of formal courses in which the topics are presented in a unified and sequential manner presents enormous advantages. This book was developed as a senior level/graduate text, and the material can be comfortably (and integrally) covered in a two-semester sequence. The book was the outgrowth of class notes developed at the Military Institute of Engineering, (Brazil), the New Mexico Institute of Mining and Technology, and the University of California, San Diego (UCSD). It is biased to the author's scientific interests and uses, in a disproportionate manner, examples from the author's research work. Nevertheless, an attempt was made to retain the generality and breadth needed to make it useful to a wide community. It is well suited for short courses to be taught to engineers/scientists in defense/ordnance laboratories.

The writing of this book would not have been possible without the help provided by colleagues and students. I was "baptized" at age 12, when I managed to initiate a detonator and lacerate my hands and face. In the Army, Captain Mendonça continued my education, and I felt the intense rush and ecstasy of TNT explosions. We would, after removing the casing, approach the detonation as much as possible to photograph the event. At the University of Denver, R. N. Orava initiated me into this esoteric science, and I owe him a great deal. He patiently guided me through my initial stumblings in this field. L. E. Murr played a key role in my professional development. He enabled my most fruitful nine-year tenure at New Mexico Tech by attracting me to Socorro. He laid the groundwork for shock-wave research at that institution and I continued his endeavor. The great help given by Jaimin Lee (Taejon, South Korea) should not go unmentioned. He taught me many aspects of wave propagation. Equally helpful was Masatake Yoshida (National Chemical Laboratory for Industry; Japan), whose MYIDL code is an important teaching tool. Drs. N. N. Thadhani; L. H. Leme Louro; A. Ferreira; U. Andrade; S. L. Wang; S. N. Chang; L. H. Yu, J. C. LaSalvia, Mrs. D. A. Hoke, H. C. Chen, and S. S. Shang; and Y. J. Chen, S. Christy, C. Wittman, C. Y. Hsu, and K. C. Hsu were intelligent and dedicated students. N. N. Thadhani taught this course with me in Socorro and provided valuable input. At UCSD, I was fortunate to meet a bright young scholar, G. Ravichandran. He was of great value and his input can be seen throughout this book. Collaborations with D. Benson, A. H. Chokshi, J. Isaacs, S. Nemat-Nasser, J. Starrett, and K. S. Vecchio are gratefully acknowledged. At Sandia National Laboratories, the interactions and constant admonitions of R. Graham (". . . experiments have to be quantitative and reproducible. . . ") were precious and encouraged me to pursue this subject further. While at the Army Research Office, the leadership and guidance provided by George Mayer were essential in helping me to formulate a broader picture of the field; I owe him a great debt of gratitude. The many discussions with K. Iyer have also been, throughout these past years, a source of stimulation.

Many collaborators, from EOD (Explosives and Ordnance Disposal) technicians to graduate students and from secretaries to program managers, are not mentioned here. However, they have all contributed to this endeavor for which I, solely, will be recognized. To all of them, my gratitude. In addition, I asked a group of colleagues to review this manuscript and am very thankful for their criticism and input: D. Benson reviewed and rewrote Section 6.6; R. Graham reviewed Chapter 8; K. H. Oh reviewed Chapter 5; K. Iyer reviewed Chapter 17; R. J. Clifton reviewed Chapter 3; V. F. Nesterenko reviewed the manuscript from cover to cover and improved it considerably.

Last, but not least, I thank the competent typing of Kay Baylor and Tina Casso. The Army Research Office (A. Crowson, E. S. Chen, W. Simmons) and National Science Foundation (R. J. Reynik and B. MacDonald) generously supported me for the past 15 years; 15 more years are needed for completion of the opus.

La Jolla, California
January 1994

CONTENTS

CHAPTER 1

Dynamic Deformation and Waves

1.1 OBJECTIVES AND APPROACH

The dynamic behavior of materials is an area of study at the confluence of many scientific disciplines. The processes that occur when bodies are subjected to rapidly changing loads can differ significantly from those that occur under static or quasi-static situations. A vivid illustration of this is the familiar sand bag that we are accustomed to seeing in old war movies and that is used by soldiers to stop bullets. Soft, free-flowing sand is effective against impacts of a velocity on the order of 1 mile/s (30.06-caliber bullet speed). On the other hand, a simple knife can defeat the sand bag. A solid wood or steel enclosure, on the other hand, if not excessively thick, can easily be perforated by a bullet. A knife would never penetrate it. Dynamic events require special study, and inertia and inner kinetics of materials become an important factor. It is hoped that this book will provide the student with the necessary tools to tackle, on his or her own, specific problems not discussed here. Chart 1.1 presents a schematic description of the scientific fields that are of importance in the study of dynamic processes as well as the broad range of applications in which they are important.

The understanding of the dynamic response of materials is very important. This section will review the most important areas in which the dynamic behavior of materials is involved, and then present the structure of this book. The applications can conveniently be divided into civilian and military. In all applications, the simple equation relating the kinetic energy with the velocity of a mass is of paramount importance:

$$E_c = \tfrac{1}{2} mv^2$$

The kinetic energy of a mass increases with the square of its velocity. The energy delivered by an object of mass m to a target can be expressed as

$$dE = F\, dl$$

where dl is the length over which this force F will act. This kinetic energy is transformed into damage in projectile and target. This is qualitatively and

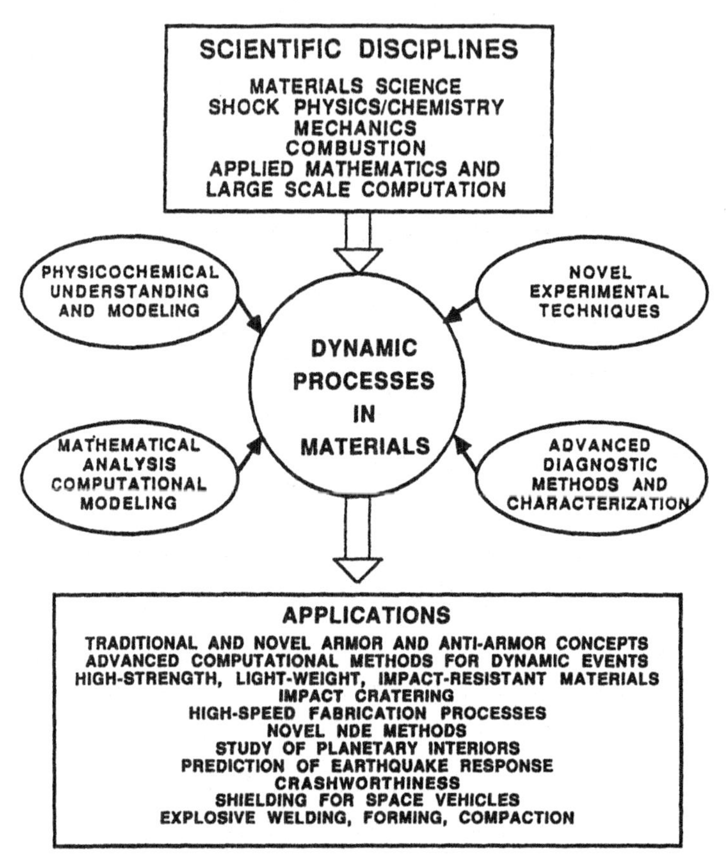

CHART 1.1 Schematic representation of the contributing sciences (disciplines) and principal applications of dynamic processes in materials. (Prepared in collaboration with G. Ravichandran and S. Nemat-Nasser.)

intuitively seen by the blow of a hammer. We all know that the faster the blow, the more penetration it will impart to the nail. All things being equal, the penetration of the nail quadruples, when its velocity is increased from v to $2v$.

Another concept of paramount importance is that there are fundamental differences between static (or quasi-static) and dynamic deformation. In quasi-static deformation we have, at any time, a situation of static equilibrium, that is, any element in the body has a summation of forces acting on it close to zero. When the deformation is imparted from the outside at a very high rate, one portion of the body is stressed while the other portion has not experienced this stress yet. In other words, stress has to travel through the body. Stress (and its associated deformation, or strain) travels in bodies at specified velocities that can be calculated to a good approximation. These are called **waves**, since they have usually well-established velocities. Thus, dynamic deformation often involves wave propagation, whereas quasi-static deformation can be considered as a sequence of states of equilibrium that can be treated by the well-known equations of mechanics of materials (summation of forces equal to zero; summation of moments equal to zero; compatibility of strains; constitutive relations).

Figures 1.1 and 1.2 summarize in a schematic manner common applications where dynamic deformation is involved. One can say that the civilian applications are mostly intended at producing something, whereas the military applications are either intended at defeating or protecting. Explosive metalworking was developed as a technology in the 1960s and 1970s, and some interesting and unique technological applications have resulted. Explosive welding is industrially the most important application; two metal plates are placed on top of each other with a specified spacing. The detonation of an explosive on top of the top plate will propel it against the bottom plate at a velocity V_p. This velocity will produce, at the impact interface, a very high pressure peak, which will induce a jet that will cleanse the two surfaces. The two surfaces so cleansed will metallurgically bond. This technique has been applied to join metals with very different melting points that cannot be welded conventionally. Thus, aluminum and titanium can be welded to steel, as an example. A second application that has been successfully implemented industrially is explosive forming. The detonation of an explosive will produce the energy that drives the metal sheet against an anvil. This can be done in a transmitting medium such as water or through a direct contact operation (explosive in direct contact with metal to be formed). More recent applications that are being intensely researched are shock synthesis and shock consolidation of materials. Shock synthesis has been successfully industrialized by DuPont for the production of diamond powder (from carbon). It is currently being investigated in a number of other material systems. Shock consolidation uses the energy of shock waves to bond fine metal powders (in the micrometer range). The shock wave passes through the powder and densifies it by plastic deformation and/or fracture. The shock wave deposits intense energy at the particle surfaces, often melting them and causing bonding. Metal, polymer, and ceramic powders have been consolidated by this method, which presents a considerable potential for the production of hard compacts of boron nitride and diamond for cutting tool applications. The diagram shown in Figure 1.1 presents one method to shock consolidate powders. In the method sketched in Figure 1.1, the powder is placed in a cannister (tube shaped) surrounded by explosives placed in a cylindrical container that is positioned coaxially with the powder container. Detonation is initiated at one end and will propel the tube walls inward, compressing the powder to very high pressure. This high pressure produces bonding between the powders. Explosives are also used to cut metals, and a linear shaped charge is used for this application. Detonation of the explosive produces a metallic jet that severs the metal under it. Rock blasting is another important application where the dynamic response of rock is important. In rock blasting compressional and tension waves travel through the rock and fragment it. This is very important in both the mining and construction industries.

Explosively generated shaped charges described later in this chapter are used

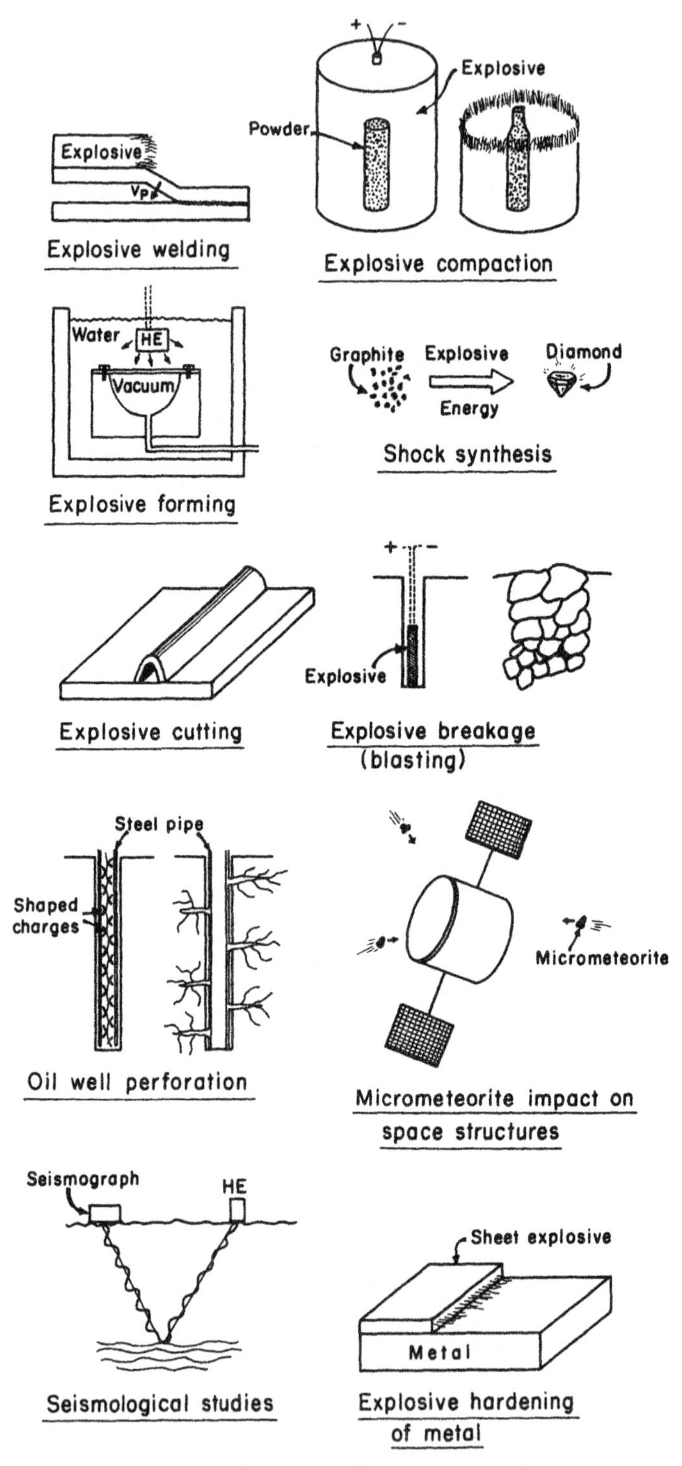

FIGURE 1.1 Civilian applications where high-strain-rate phenomena are important.

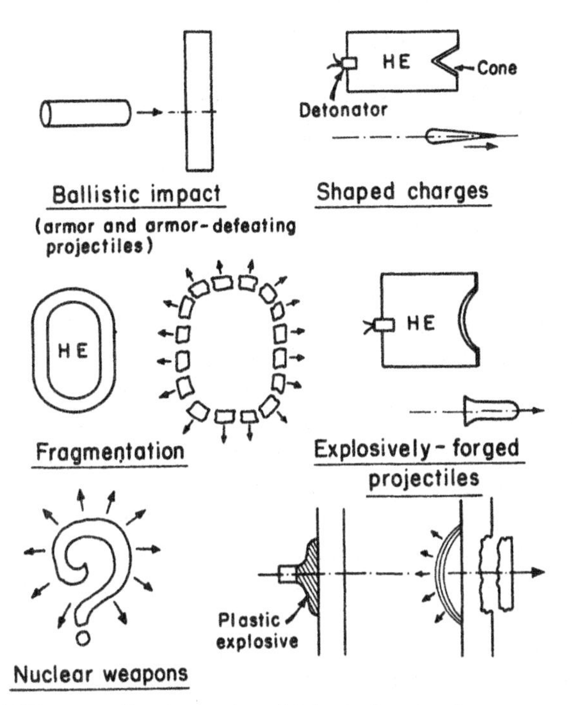

FIGURE 1.2 Military applications where high-strain-rate phenomena are important.

in oil well drilling. The shaped charges are loaded down the hole and are detonated to produce perforations in the rock that will aid in oil recovery. Dynamic deformation by impact at velocities up to 30 km/s occurs when micrometeorites impact space structures and satellites. These impacts can produce a great deal of damage, and the development of shielding mechanisms for space vehicles requires a thorough knowledge of damage mechanics. Seismological studies use reflected and refracted P and S waves from different levels within the earth and rely heavily on a complete understanding of wave propagation in porous media. For instance, underground nuclear detonations in China and Russia can be detected in the United States (and vice versa) with special instrumentation by means of wave propagation through the earth. It has been shown that many metals can be considerably hardened by the passage of a shock wave through them. Hadfield steels (manganese based) are significantly hardened by the detonation of explosives in contact with them, and the hardening of rail frogs constitutes a successful industrial application.

The military applications (Fig. 1.2) in which impact, explosive detonation, and high-strain-rate (dynamic) deformation are used are numerous. Of great importance is the defeat of structures through high-velocity projectiles. The mechanisms by which armor can protect these structures are highly important. Among the projectiles used today, the shaped charge is worth describing. An explosive charge placed in contact with a hollow metal cone detonates and deforms the cone into a long rod that is accelerated to very high velocities (up to 10 km/s). This "jet" has great penetration capability. Other concepts for projectiles are the kinetic energy penetrator, a long cylindrical rod of a high-density metal (tungsten or uranium) traveling at a velocity of approximately 1.5–2 km/s. This kinetic energy penetrator has a penetration capability into a steel target equal to approximately one-half of its length. Other forms of projectiles are the explosively forged projectile, which is similar to the shaped-charge one. The cone is replaced by a disk (shaped like a saucer) that is deformed by the explosive into a long rod. Yet another projectile is the one that has a plastic explosive head that deforms upon impact with an armor and then detonates, producing shock waves that propagate through the armor. These waves reflect at the inner surface of the armor, thereby spalling it. Thus, fragments are emitted from the inside without total perforation. Fragmentation rounds and projectiles are common and well known. An explosive charge located in the core detonates and transmits the energy to the shell, which fragments. These fragments travel at very high velocities. Of particular importance is the detonation of a nuclear weapon effected by a carefully and precisely designed explosive system.

There are additional applications of high-strain-rate deformation, such as machining, accidental impact of vehicles and crashworthiness studies, earthquakes and the associated damage to structures, and explosive devices used in aerospace industry. These applications require a thorough knowledge of (1) the **mechanics of high strain rate deformation** and (2) the **dynamic response of materials**. At the end of this introductory chapter a bibliography to the most important sources is provided. The student should consult these sources for more in-depth knowledge. Five key books are Kolsky's for elastic waves, Rice et al.'s (monograph) for shock waves, Zeldovich and Kompaneets's for detonation waves, Freund's for dynamic fracture, and Rinehart and Pearson's for dynamic deformation.

1.2 STRUCTURE OF BOOK

The sequence of materials covered in this book is the following.

Chapter 1 presents the objectives and approach to the book. The simplest model for a wave is physically introduced and analytically described. A class experiment with a string and with a slinky should be performed.

Chapter 2 describes the equations for longitudinal and shear waves in a continuous body. These derivations will be made for an ideal isotropic material. Elastic waves in bounded media will be introduced, with particular emphasis on elastic waves in a cylindrical rod. The Hopkinson bar, an important experimental tool for the determination of the dynamic strength of materials, will be described. The more advanced treatments, using the indicial notation, and the general solution to the wave equation are developed. Wave interactions (reflection and refraction) are introduced.

Chapter 3 discusses plastic waves. When the amplitude of the wave exceeds the elastic limit of the material, at that strain rate, plastic deformation takes place in the material, if it is ductile. These plastic waves propagate at velocities that are lower than the elastic waves. The treatment given by von Karman and Duwez will be presented. This treatment was developed for tensile stresses in a wire. Taylor's approach was different. He studied compressional plastic deformation in a cylinder. His presentation will also be discussed. Contemporary experimental methods that are used for the study of dynamic plastic deformation will be briefly reviewed. The Taylor test, the Hopkinson split-bar test, and the pressure–shear experiments developed at Brown University will be described. These experiments allow the understanding of the plastic behavior of the material under wave propagation conditions. Plastic shear waves are discussed.

Chapter 4 is about shock waves. When a medium is bound (no lateral flow of material is allowed), the plastic deformation front will have a different configuration. It will become very steep, and the thickness of this front can be as low as a fraction of a micrometer. Extremely high amplitude waves can travel through a material without changing its macroscopic dimensions in this manner, since the material is impeded from flowing laterally and therefore a state of compressive stress close to hydrostatic compression establishes itself. These waves are called shock waves and are treated mathematically by the Rankine–Hugoniot equations and by a material equation of state. These pressures, produced by conventional explosives, can be as high as 50 GPa, or 50 times higher than the flow stress of most materials. In nuclear explosives, pressures higher by orders of magnitude than those of conventional explosives can be achieved. The Rankine–Hugoniot equations will be introduced.

Chapter 5 presents equations of state (EOS) for shock waves. The description of a shock wave traveling in a material requires the incorporation of material-specific parameters. These parameters are experimentally obtained and are usually presented in the form of the linear relationship between the particle and shock velocity ($U_s = C_0 + SU_p$). The shock response of a material can also be obtained, at a more fundamental level, from the Mie–Grüneisen EOS. This equation allows the calculation of the shock-induced temperature as well as the shock response of porous materials.

Chapter 6 discusses the differential form of conservation equations and hydrocodes for shock waves. In order to simulate the behavior of shock waves in computers, special mathematical techniques have to be used. The most common approach is the finite-difference method. This method is based upon the differential form of the conservation equations. These equations will therefore be derived. The frame of reference used in the computational codes is very important, and both Lagrangian and Eulerian frames of reference are used. The frames of reference will be defined and described.

Chapter 7 is a discussion of the attenuation, interaction, and reflection of shock waves. A number of important effects associated with the passage of shock waves through solids are described. Shock waves reflect at interfaces and free surfaces, producing spalling and other effects. The amplitude of shock waves decreases as they travel through a material.

Chapter 8 is a discussion of shock-wave–induced phase transformations and chemical changes. Certain materials undergo phase transformations at specific pressures and temperatures. These transformations change the wave structure. The classic example is the α(BCC)-to-ϵ(HCP) transformation undergone by iron at 13 GPa. The thermodynamics of shock-induced transformation is presented and specific cases are discussed. Other chemical changes, such as reactions, are also discussed.

Chapter 9 discusses explosive-material interactions. In order to produce shock waves in materials, one has to deposit a large amount of energy over a very short duration of time over the surface of the material or inside it. This can be accomplished in many ways, the most common being impact of a projectile against the surface at a high velocity, detonation of an explosive in contact with the surface, and deposition of laser energy at the surface. The fundamental equations predicting the velocities of metal components accelerated by explosives will be derived. The Gurney equations are based on the transformation of chemical energy of the explosive products and metal components. The various forms of the Gurney equation for different geometries will be presented and compared with other predictive models.

Chapter 10 is on detonation, a special wave in which a chemical reaction occurs that transforms a condensed explosive into a gas at a high pressure. The basic conservation equations and EOS for detonation are presented and discussed. Characteristics of different explosives are presented, and the initiation of detonation (both by heat and shock waves) is described.

Chapter 11 presents experimental techniques and diagnostic tools. Measurement techniques (diagnostics) that allow us to probe into short-time events are presented. Cinematography has been successfully used. Streak and framing cameras can be used for very high velocity events. Flash X-rays are a powerful technique to monitor high-velocity events, and the principles of this technique are presented. Gages have been developed to investigate the various ranges of pressure and piezoresistive (manganin and ytterbium) and piezoelectric [quartz and polyvinylidene (PVDF) polymer] gages will be described, together with the required instrumentation. Electromagnetic gages are also important in some applications.

Chapter 12 is a discussion of experimental techniques and methods to produce dynamic deformation. The high-strain-rate mechanical response of materials is discussed. The various techniques that have been developed to investigate the high-strain-rate response are presented. For intermediate strain rates, pneumatic machines, cam plastometers, and flywheel machines are used. At higher strain rates, the Hopkinson (or Kolsky) bar is the standard technique. At still higher rates, impact is used by means of explosively accelerated devices, projectiles propelled in one- and two-state gas guns. Still higher impact velocities are obtained by means of the electromagnetic rail gun and the plasma gun.

Chapter 13 discusses plastic deformation at high strain rates. The flow stress of metals increases as the strain rate increases and as the temperature decreases; it usually increases with plastic strain. These dependencies are incorporated into equations, called constitutive models. Empirical (plain or fancy curve fitting) and physically based models are presented and discussed. The principal agents of plastic deformation are dislocations, and the stress response of their movement determines the flow stress response to strain, strain rate, and temperature. These constitutive models are important components to hydrodynamic codes in which material strength is incorporated.

Chapter 14 discusses plastic deformation in shock waves. At the extreme (maximum) of strain rates one has shock waves. The strain rates at the front are up to 10^8 s^{-1}. The mechanisms by which defects are generated under shock-wave conditions are discussed. The formation of point defects (interstitials and vacancies), line defects (dislocations), interfacial defects (stacking faults and twins), and phase transformation is discussed. Residual microstructures are presented, and the effects of grain size, stacking-fault energy, peak pressure, and pulse duration are discussed.

Chapter 15 is a discussion of shear bands (thermoplastic shear instabilities). The localization of plastic deformation along internal regions having essentially two dimensions (the thickness being of the order of 10-100 μm) is a characteristic of materials subjected to high strain rates. Steels, titanium and its alloys, aluminum alloys, brass, and tungsten and its alloys undergo easy shear localization at high strain rates. These regions of localized plastic deformation in which softening occurs are precursors to failure. The breakup of projectiles, defeat of armor, and fragmentation of bombs are very sensitive to the formation in them of shear bands. These instabilities also play a key role in machining and forging operations. Their microstructure is discussed and constitutive models that predict their formation are presented.

Chapter 16 discusses dynamic fracture, the last phenomenon in the chain of events that occur in deformation. The principal concepts of fracture mechanics are introduced in order to familiarize the neophyte with this subject. Dynamic fracture has unique aspects that are discussed in detail. The limiting velocity for a crack is discussed (Rayleigh wave velocity); the effect of crack propagation velocity on the fracture toughness is presented; spalling, a particular dynamic fracture mode produced by tensile pulses resulting from the reflection of compressive shock waves at a free surface, is presented. At high strain rates, the tendency for materials to fragment into many pieces rather than break into two has been systematically observed; theories addressing this behavior are discussed. The microstructural effects of dynamic fracture are introduced and the experimental techniques used in dynamic fracture are discussed. Fragmentation of brittle materials is discussed.

Chapter 17 focuses on the principal applications of this field, discussed qualitatively in Section 1.1 (Figs. 1.1 and 1.2). These applications are broadly classified into civilian and military, and simple analytical treatments that allow the prediction of the key parameters are developed. Explosive welding, explosive compaction, and explosive forming are presented; explosive breakage of rock (blasting) is succinctly described. The military applications (e.g., shaped charges, explosively forged projectiles, fragmentation of bombs and shells, armor penetration) are also discussed.

This book covers a broad range of topics, and the mathematics is kept very simple. The student should be familiar with calculus and the basic laws of physics and fluid mechanics. Tensors are only introduced where essential; whenever more advanced mathematical analyses are introduced, a preliminary explanation is given. Emphasis is placed on the simple, direct treatments; more advanced treatment of most of the topics covered in this book is available in the literature. It is felt that the most important is to present the physical phenomena and the basic, fundamental formulations.

As an example of the relevance of the material presented in this book, we analyze below the sequence of events that takes place when a shaped charge approaches a target, is activated, and penetrates it. Figure 1.3 shows the sequence of stages. The shaped charge is a very important military weapon as well as having significant civilian applications. The flight of the projectile is not treated in the current book and is in the realm of flight ballistics (or intermediate ballistics). One of the mechanisms of initiation of the explosive is the activation of a piezoelectric sensor at the tip of the ogive. Chapter 11 deals with piezoelectric materials. Initiation of detonation is triggered by a fuse that activates a booster charge that, in turn, creates a shock wave that propagates throughout the explosive. This knowledge is presented in Chapter 10. The detonation wave [Fig. 1.3(c)] interacts with the copper liner, sending a shock

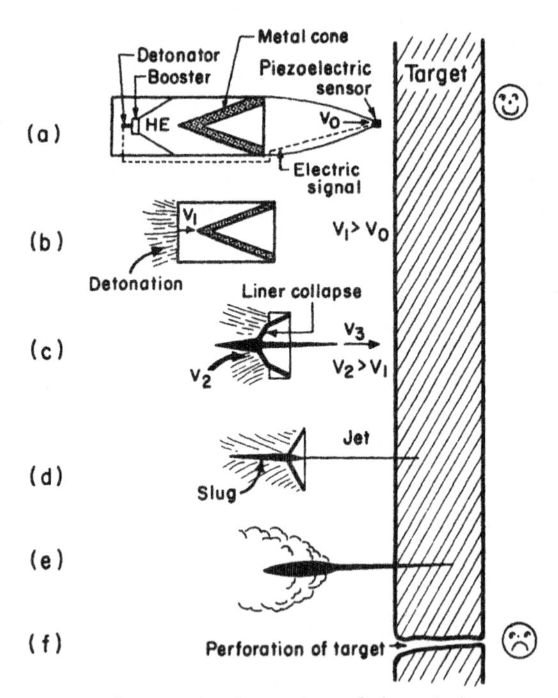

FIGURE 1.3 Sequence of events in detonation of shaped-charge projectile and perforation of target ($v_3 > v_1 > v_2 > v_0$): (a) projectile traveling at velocity v_0; (b) detonation of explosive at v_1; (c) collapse of liner with formation of slug and jet; (d)–(f) perforation of target.

wave through it. This is treated in Chapter 9. After the propagation of this shock pulse, the liner undergoes intense plastic deformation, which requires description by constitutive equations; the processes are presented in Chapter 13. This collapsed metal liner emits a jet (traveling at velocities close to 10 km/s) and a trailing slug (traveling at a lower velocity). Upon impact with the target, very high pressures are produced, which generate shock waves, treated in Chapters 4, 5 and 7. The experimental methods used as diagnostics for these experiments (principally, flash X-rays) are described in Chapter 11. The design of armor and shaped charges uses computer simulations in hydrocodes of increasing complexity; one can significantly reduce the number of experiments by using computer calculations; the foundations for this type of analysis are given in Chapter 6. The illustrative example above shows how the knowledge discussed throughout the chapters of this book comes to bear in the improved understanding of a real problem and how it can help in solving it.

1.3 VIBRATING STRING AND SPRING

The simplest form of a wave to a nonscientific observer is the wave pattern in water when a rock is thrown into it. A circular pattern evolves that spreads radially from the point of impact. Energy and motion are carried outward without any long-range motion of the water molecules. A mechanical wave is a motion transfer mechanism not involving long-range displacements of material. Sound waves, light waves, radio waves, and other electromagnetic waves are other examples of waves. In this book we will concern ourselves exclusively with mechanical waves. They originate when a portion of a body is displaced from its original position and transmits that displacement to adjacent material.

In order to understand waves, we will start by describing them analytically for the most simple situation: that of a wave train traveling through a rope. By taking a long rope and fixing one end of it to a wall and moving swiftly the other end by an up-and-down motion of the hand, we can produce a wave in the rope. Sequential motions of the hand will produce a wave train. We will derive the expression that describes this wave train and its propagation velocity next.

This author vividly remembers the lucid explanation given by Feynman in a TV program just prior to his death. He described the infinity of waves that surround us at any time and how our diagnostics (eyes, ears, skin) just pick up the important waves. In a room, at any time, we have light waves, infrared (heat) waves, sound waves, and radio waves "bouncing" all over the place. Our body judiciously filters in only the waves needed.

This author could add to Feynman's musings by making the (correct) statement that the atoms in an eraser sitting at his desk are all continuously vibrating (at $\sim 10^{13}$ s^{-1}). The fascinating part of this is that these vibrations are not independent, but coupled. Thus, low-amplitude elastic waves (standing waves) are continuously bouncing back and forth through a material. These low am-

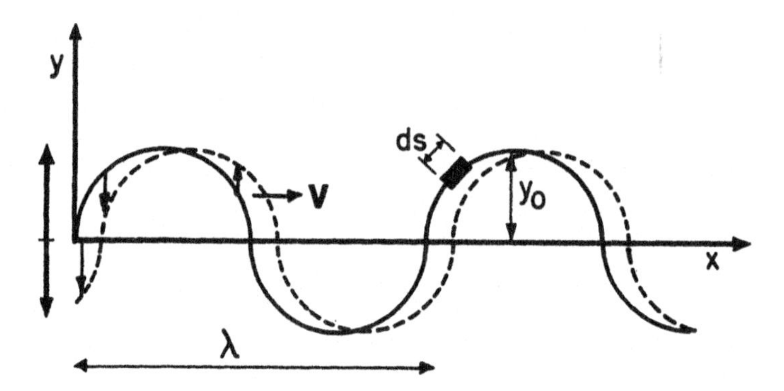

FIGURE 1.4 Wave train produced in a string; particle displacement marked.

plitude waves are called phonons. These waves have propagation velocities approximately equal to the velocities of elastic waves.

Figure 1.4 shows a sinusoidal wave train. This is the simplest shape of wave to study and will therefore be the assumed shape. We have the following:

$$y = 0 \quad \text{at } x = 0, \tfrac{1}{2}\lambda, \lambda, \ldots, t = 0$$

The amplitude is y_0 and the wavelength is λ. This is the appearance of the wave at time $t = 0$. At $t \neq 0$, the end vibrates and the wave consequently propagates. The new position, at time t, is indicated by the interrupted line. The frequency of vibration of the end point ($x = 0$) determines the wavelength and frequency of the wave. For one entire harmonic motion of the end (or hand holding the rope) one full wave unit is produced. Thus, the period of the wave is equal to the period of the vibrating source producing it. This period (τ) is equal to the time required for the wave to propagate a distance equal to the wavelength λ. Since the wave advances at a velocity V, we have

$$V = \frac{x}{t} = \frac{\lambda}{\tau} = \lambda f \tag{1.1}$$

where f is the frequency of the source (equal to $1/\tau$). We can now express the shape of the wave (y) as a function of position (x) and time (t). At $t = 0$:

$$y = y_0 \sin \frac{2\pi x}{\lambda} \tag{1.2}$$

One also has:

$$y = y_0 \sin 2\pi f t \tag{1.3}$$

This equation applies to the harmonic vibration of the source (hand). If we have to express the displacement of y at a position x at the moment t, we have to see that there is a time lag t from the source until this wave reaches that point. At $t = x/V$, we subtract this time lag from the oscillator (hand) equation:

$$y = y_0 \sin\left[2\pi f\left(t - \frac{x}{V}\right)\right] \tag{1.4}$$

We can make a substitution to simplify this equation (from 1.1).

$$y = y_0 \sin\left[2\pi\left(ft - \frac{x}{\lambda}\right)\right] \tag{1.5}$$

This equation describes the displacement at any position x and any time t.

We now proceed to determine the differential form of the wave equation, which will allow us to calculate the wave velocity. For this, we isolate one small segment of the wave, in a general position. Later, in Chapter 2 (Section 2.9) we will develop the solution for a general disturbance.

This segment is shown in Figure 1.5. It is subjected to a line tension T. This segment is in motion, and we will apply Newton's second law to it. We also assume that all motion takes place in the y direction:

$$\begin{aligned} \sum F_x &= 0 \\ \sum F_y &= ma_y = \rho\, ds\, \frac{\partial^2 y}{\partial t^2} \end{aligned} \tag{1.6}$$

The rope has a mass per unit length of ρ and the segment has a length equal to ds. In order to find F_y, we have to take the summation of forces in that direction; we assume a sufficiently thin string and that the tension along it is constant ($\Delta T << T$).

$$F_y = T \sin \theta_B - T \sin \theta_A = T[\sin(\theta_A + d\theta) - \sin \theta_A]$$

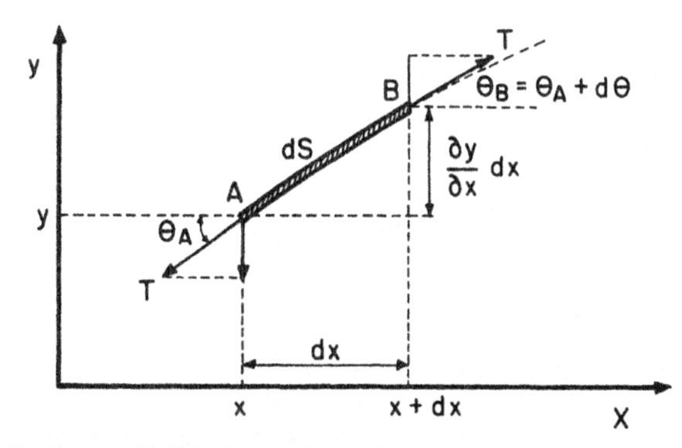

FIGURE 1.5 Segment of vibrating string subjected to a line tension T; this segment is shown in Fig. 1.4.

Multiplying and dividing by $d\theta$ yields

$$T\,d\theta\left[\frac{\sin(\theta_A + d\theta) - \sin\theta_A}{d\theta}\right] = \rho\,ds\,\frac{\partial^2 y}{\partial t^2}$$

But

$$\left[\frac{\sin(\theta_A + d\theta) - \sin\theta_A}{d\theta}\right] = \frac{d}{d\theta}\sin\theta_A = \cos\theta_A$$

Thus

$$T\cos\theta_A\,d\theta = \rho\,ds\,\frac{\partial^2 y}{\partial t^2} \tag{1.7}$$

We also have

$$\cos\theta_A = \frac{dx}{ds} \qquad \therefore\ ds = \frac{dx}{\cos\theta_A} \tag{1.7a}$$

We can eliminate $d\theta$ by the following operations; we see that

$$\tan\theta_A = \frac{y + (\partial y/\partial x)\,dx - y}{x + dx - x} = \frac{\partial y}{\partial x}$$

$$\frac{\partial \tan\theta_A}{\partial x} = \sec^2\theta_A\,\frac{\partial\theta}{\partial x} = \frac{\partial^2 y}{\partial x^2}$$

$$d\theta_A = \frac{\partial^2 y}{\partial x^2}\cos^2\theta_A\,dx \tag{1.7b}$$

Substituting (1.7a) and (1.7b) into (1.7) yields

$$T\cos^4\theta_A\,\frac{\partial^2 y}{\partial x^2} = \rho\,\frac{\partial^2 y}{\partial t^2}$$

Since we are dealing with small oscillations, $\cos\theta \approx 1$ and

$$\rho\,\frac{\partial^2 y}{\partial t^2} = T\,\frac{\partial^2 y}{\partial x^2} \tag{1.8}$$

This is the differential equation describing the wave. It is a second-order hyperbolic differential equation that we will solve in Chapter 2. In order to determine the velocity of the wave, we simply differentiate (1.4) twice with respect to x and t and substitute it into (1.8). It should be noticed that we are applying the general wave equation to a restricted case (harmonic wave):

$$y = y_0 \sin 2\pi f\left(t - \frac{x}{V}\right)$$

$$\frac{\partial y}{\partial x} = -y_0\left(\frac{2\pi f}{V}\right)\cos 2\pi f\left(t - \frac{x}{V}\right)$$

$$\frac{\partial^2 y}{\partial x^2} = -y_0 \frac{4\pi^2 f^2}{V^2} \sin 2\pi f\left(t - \frac{x}{V}\right) \tag{1.9}$$

$$\frac{\partial^2 y}{\partial t^2} = -y_0 4\pi^2 f^2 \sin 2\pi f\left(t - \frac{x}{V}\right) \tag{1.10}$$

Equations (1.9) and (1.10) are then substituted into Eq. (1.8):

$$-y_0\, 4\pi^2 f^2 \sin 2\pi f\left(t - \frac{x}{V}\right) = \frac{T}{\rho}\left[-y_0 \frac{4\pi^2 f^2}{V^2} \sin 2\pi f\left(t - \frac{x}{V}\right)\right]$$

$$V = \pm\left(\frac{T}{\rho}\right)^{1/2} \tag{1.11}$$

Equation (1.11) is a very important relationship that will recur repeatedly, in modified forms, throughout the book. It states that the velocity of the wave is equal to the square root of the ratio between the tension of the rope and its line density. The differential equation (wave equation) can therefore be expressed as

$$\frac{\partial^2 y}{\partial t^2} = V^2 \frac{\partial^2 y}{\partial x^2} \tag{1.12}$$

For the propagation of a disturbance in a helical spring an analogous treatment can be followed. This equation can be derived by considering Figure 1.6. If the spring constant is k and the mass per unit length is ρ, we can readily derive the equation for displacement of material in direction S:

$$\frac{\partial^2 S}{\partial t^2} = \frac{k}{\rho}\frac{\partial^2 S}{\partial x^2} \tag{1.13}$$

We see that $V = (k/\rho)^{1/2}$. Notice that ρ is the mass per unit length.

It will be seen, in Chapter 2, that the two previous cases correspond very closely (if we observe the relationship between the propagation directions of wave and particle) to the propagation of shear and longitudinal waves, respectively. These simple cases help us to understand the more complex behavior of wave propagation in a continuum. In Section 2.9 a general solution for the wave equation will be presented. The second-order partial differential equation will be solved [Eqns. (1.12), (1.13), or a more general form]. It will be shown

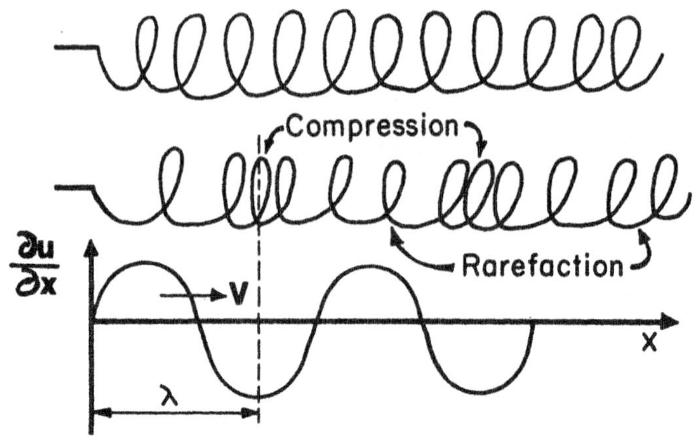

FIGURE 1.6 Compressional (or tension) wave set up in spring by displacing one of its ends (the "slinky" experiment).

that any shape of disturbance propagating at a velocity V in a medium and remaining unchanged is described by the identity

$$\frac{\partial^2 z}{\partial t^2} = V^2 \frac{\partial^2 z}{\partial x^2}$$

This is the most important equation in this book, since it tells us that disturbances travel through materials at characteristic velocities that are a function of the stress response of the material to strain (in the simplified example above, string tension and spring constant) and density (in the example above, linear density).* For elastic waves in bounded and unbounded media, these stress–strain relations are the elastic constants. For plastic waves, they are the work hardening. For shock waves, they are related to the adiabatic compressibility. These are the underpinnings for elastic, plastic, shock, and detonation waves. Oh, student, thy knowledge has grown magnificently! Now, if you think deeply, you will recognize that Eqns. (1.12) and (1.13) are a direct consequence of Newton's second law (conservation of momentum).

Example 1.1 A traveling harmonic wave on a string has a frequency of 30 Hz and a wavelength of 60 cm. Its amplitude is 2 mm. Write the equation for this wave in SI units:

$$f = 30 \text{ Hz} = 30 \text{ s}^{-1} \qquad y = 2 \text{ mm} = 2 \times 10^{-3} \text{ m}$$

$$\lambda = 60 \text{ cm} = 0.6 \text{ m}$$

$$y = 2 \times 10^{-3} \sin\left[2\pi\left(30t - \frac{x}{0.6}\right)\right]$$

*A more general non-linear wave equation, which includes the traditional one as a special case, was proposed by V. F. Nesterenko for longitudinal (High-Rate Deformation of Heterogeneous Materials, Nauka, 1992) and transverse (Fiz. Goren. iVzr.; N.2 (1993) 132) waves.

Example 1.2 A wave along a string has the following equation, in SI units:

$$y = 0.02 \sin (30t - 4x)$$

Determine its amplitude, frequency, speed, and wavelength:

$$y_0 = 0.02 \text{ m}$$
$$y = 0.002 \sin 2\pi(4.7t - 0.63x)$$
$$f = 4.7 \text{ Hz} = 4.7 \text{ s}^{-1}$$
$$\lambda = 1.57 \text{ m}$$

$$V = \lambda f = 7.45 \text{ m/s}$$

Example 1.3 You are bear-hunting in the Sandias. After a careful stalk, you are able to approach a beautiful specimen to within 30 yards. From between the bushes, your heart pounding excitedly, you pull your bow back only 30 cm, to a maximum force of 50 lb. You aim and release the arrow. A furious roar ensues and you find yourself frantically running. Fortunately, you find a tree before the bear finds you and you discover an unexpected ability to climb a tree at high speed and newly developed brachiation skills. You observe the bear under the tree, with the arrow inserted in its flank. The penetration of the tip is only approximately 10 cm. You know that you need approximately 30 cm of penetration to reach a vital organ.

1. What have you done wrong?
2. What was the approximate velocity of your arrow if its weight is 50 g?
3. What velocity would be needed for 30 cm penetration?
4. By how much would you have to pull the bow?

1. You should have stayed home.

2. We need to make some simplifying assumptions to successfully solve this problem. We assume that the force by the bow acting on the arrow varies linearly with displacement. We calculate the potential energy that is stored in the bow when it is pulled back:

$$dE_p = F_1 \, dx$$

$$E_p = \int_0^{x_1} F_1 \, dx$$

$$E_p = \int_0^{x_1} kx \, dx = k\frac{x_1^2}{2}$$

The kinetic energy is

$$E_k = \tfrac{1}{2} m v^2$$

But

$$E_k = E_p$$

$$v_{\text{arrow}} = \left(\frac{k}{m}\right)^{1/2} x$$

We calculate k:

$$k = \frac{F}{x} = \frac{50 \times 4.448}{30 \times 10^{-2}} = 740 \text{ N/m}$$

$$v_{\text{arrow}} = 36.5 \text{ m/s}$$

3. We now calculate a frictional force that the muscle of the bear imparts on the arrow. If we assume that the velocity at impact is equal to the initial velocity and that the friction force, F_2, is concentrated at the arrow tip, then

$$E_{k_2} - E_{k_1} = F_2 l_1 \qquad E_{k_1} = 0$$

$$F_2 = \frac{\tfrac{1}{2} m v_{\text{arrow}}^2}{l_1} = \frac{50 \times 10^{-3} \times (365)^2}{2 \times 10 \times 10^{-2}}$$

$$F_2 = 333 \text{ N}$$

For 30 cm penetration, we have

$$l_2 = 3 l_1$$

$$E_{k_2} = 3E_{k_2}$$

$$v_2^2 = 3 v_1^2$$

$$v_2 = 63.2 \text{ m/s}$$

4. If we assume the same linear behavior as in part 1, we have

$$E_{p_2} = 3E_{p_1}$$

$$k \frac{x_2^2}{2} = 3k \frac{x_1^2}{2} \qquad \therefore\ x_2 = x_1\sqrt{3}$$

$$x_1 = 30 \text{ cm} \qquad x_2 = 52 \text{ cm}$$

BIBLIOGRAPHY

This bibliography is provided to enable the studious reader to expand his or her knowledge. To master this subject, the reader must go to the foundations of knowledge, not once, but many times.

Elastic Waves

T. D. Achenbach, *Wave Propagation in Elastic Solids*, North-Holland, Amsterdam, 1973.

H. Kolsky, *Stress Waves in Solids*, Dover, New York, 1963.

J. S. Rinehart, *Stress Transients in Solids*, Hyperdynamics, Santa Fe, NM, 1975.

A. K. Ghatak and L. S. Kothari, *An Introduction to Lattice Dynamics*, Addison-Wesley, Reading, MA, 1972.

R. J. Wasley, *Stress Wave Propagation in Solids*, Dekker, New York, 1973.

K. F. Graff, *Wave Motion in Elastic Solids*, Ohio State University Press, Columbus, OH, 1975 (also Dover Press).

Plastic Waves

T. von Karman and P. Duwez, *J. Appl. Phys.*, **21** (1959), 987.

A. S. Abou-Sayed, R. J. Clifton, and L. Hermann, *Exptl. Mech.* (1976), 127.

R. J. Clifton, in *Mechanics Today*, Vol. 1, ed. S. Nemat-Nasser, Pergamon, Elmsford, NY, 1972.

R. J. Clifton, *J. Appl. Mech.*, **50** (1983), 941.

High-Strain-Rate Deformation

J. A. Zukas, T. Nicholas, H. F. Swift, L. B. Greszczuk, and D. R. Curran, *Impact Dynamics*, J. Wiley, New York, 1982.

J. A. Zukas, ed., *High Velocity Impact Dynamics*, Wiley, New York, 1991.

J. Duffy, "*The Dynamic Plastic Deformation of Metals: A Review*," AFWAL-TR-4024, Wright-Patterson Air Force Base, Dayton, OH, 1982.

Y. Bai and B. Dodd, *Adiabatic Shear Localization*, Pergamon, Oxford, UK, 1992.

Shock Waves

R. Courant and K. O. Friedrichs, *Supersonic Flow and Shock Waves*, Interscience, New York, 1956.

M. H. Rice, R. G. McQueen, and J. M. Walsh, *Solid State Phys.*, **6** (1958), 1.

L. Davison and R. A. Graham, *Phys. Rep.*, **55** (1979), 257.

R. Kinslow, ed., *Hight-Velocity Impact Phenomena*, Academic, New York 1966.

Y. B. Zeldovich and Y. P. Raizer, *Physics of Shock Wave and High-Temperature Hydrodynamic Phenomena*, Academic, New York, 1966.

L. V. Al'tshuler, *Soviet Phys.-JETP*, **8** (1965), 52.

P. C. Chou and A. K. Hopkins, eds., *Dynamic Response of Metals to Intense Impulse Loading*, Air Force Materials Laboratory, WPAFB, 1972.

J. Cagnoux, P. Chartagnac, P. Hereil, and M. Perez, *Ann. Phys. Fr.*, **12** (1987), 451.

R. A. Graham, *Solids under High-Pressure Shock Compression*, Springer, New York, 1993.

A. V. Bushman, G. I. Kanel, A. L. Ni, and V. E. Fortov, *Intense Dynamic Loading of Condensed Matter,* Taylor and Francis, PA, 1992.

Detonation Waves

I. B. Zeldovich and S. A. Kompaneets, *Theory of Detonation*, Academic, New York, 1960.

L. E. Davison and J. E. Kennedy, eds., "Behavior and Utilization of Explosives in Engineering Design," 12th Annual Symposium, March 1972, University of New Mexico.

W. Fickett and W. C. Davis, *Detonation*, University of California Press, 1979.

C. H. Johansson and P. A. Persson, *Detonics of High Explosives*, Academic, New York, 1970.

R. Cheret, *Detonation of Condensed Explosives*, Springer, New York, 1992.

Materials Effects

R. W. Rohde, B. M. Butcher, J. R. Holland, and C. H. Kearnes, *Metallurgical Effects at High-Strain-Rates*, Plenum, New York, 1973.

M. A. Meyers and L. E. Murr, eds., *Shock Waves and High-Strain-Rate Phenomena in Metals*, New York, 1981.

L. E. Murr, K. P. Staudhammer, and M. A. Meyers, eds., *Metallurgical Effects of Shock Wave and High-Strain-Rate Phenomena*, Dekker, New York, 1986.

Y. M. Gupta, ed., *Shock Waves in Condensed Matter*, Plenum, New York, 1986.

M. A. Mogilevsky and P. E. Newman, *Mechanisms of Deformation under Shock Loading*, Phys. Rep. **97**(6) (1983), 359.

G. E. Duvall and R. A. Graham, *Rev. Modern Phys.*, **49**(3) (1979), 523.

M. A. Meyers, L. E. Murr, and K. P. Staudhammer, eds., *Shock-Wave and High-Strain-Rate Phenomena in Materials*, Dekker, New York, 1992.

J. R. Asay, R. A. Graham, and G. K. Straub, eds., *Shock Waves in Condensed Matter—1983*, North-Holland, Amsterdam, 1984.

S. C. Schmidt and N. C. Holmes, eds., *Shock Waves in Condensed Matter—1987*, North-Holland, Amsterdam, 1988.

S. C. Schmidt, J. N. Johnson, and L. W. Davison, eds., *Shock Compression of Condensed Matter—1989*, North-Holland, Amsterdam, 1990.

C. Y. Chiem, H.-D. Kunze, and L. W. Meyer, eds., *Impact Loading and Dynamic Behavior of Materials*, D. G. M. Verlag, Oberursel, Germany, 1988.

Proceedings of First and Second International Conference on Mechanical and Physical Behavior of Materials under Dynamic Loading, 1988 and 1991, *J. Phys.*, *Colloque C3*, **49** (1988); **1** (1991).

J. Harding, ed., Proceedings of the Oxford Conferences on Mechanical Properties of Materials at High Rates of Strain, Inst. of Physics Conf. Series No, 21, 1974; Inst. of Physics Conf. Series No. 47, 1979; Inst. of Physics Conf. Series No. 70, 1984; Inst. of Physics Conf. Series No. 102, 1989.

J. F. Mescall and V. Weiss, eds., *Material Behavior under High Stress and Ultra-High Loading Rates*, Plenum, New York, 1983.

V. F. Nesterenko, *High-Rate Deformation of Heterogeneous Materials*, Nauka. Sub. Div., Novosibirsk, Russia (in Russian), 1992.

A. B. Sawaoka, ed., *Shock Waves in Materials Science*, Springer, Tokyo, 1993.

Y. Horie and A. B. Sawaoka, *Shock Compression Chemistry of Materials*, Terra, Tokyo, 1993.

S. S. Batsanov, *Effects of Explosions on Materials*, Springer, NY, 1994.

Experimental Techniques

R. A. Graham and J. R. Asay, *High Temperature—High Pressure*, Vol. 10, 1978, p. 355.

P. S. DeCarli and M. A. Meyers, in *Shock Waves and High-Strain-Rate Phenomena in Metals*, eds. M. A. Meyers and L. E. Murr, Plenum, New York, 1981.

L. C. Chabildas and R. A. Graham, *Developments in Measurement Techniques for Shock Loading Solids*, AMD, Vol. 83, ASME, New York 1987.

P. S. Follansbee, The Hopkinson Bar, *Metals Handbook*, Vol. 8, American Society for Metals, Metals Park, OH, 1985, p. 198.

T. Nicholas and S. J. Bless, *Metals Handbook*, Vol. 8, American Society for Metals, Metals Park, OH, 1985, p. 208.

K. A. Hartley, J. Duffy, and R. E. Hawley, *Metals Handbook*, Vol. 8, American Society for Metals, Metals Park, OH, 1985, p. 218.

Dynamic Fracture

M. F. Kanninen and S. N. Atluri, eds., *Eng. Fract. Mech.*, **23** (1986).

M. A. Meyers and C. T. Aimone, *Progr. Mater. Sci.*, **28** (1983), 1.

D. R. Curran, L. Seaman, and D. A. Shockey, *Phys. Rep.*, **147** (1987), 253.

L. B. Freund, *Dynamic Fracture Mechanics*, Cambridge, 1990.

N. Kh. Akmadeyev, *The Dynamic Failure of Solids in Tension Waves*, Ufa, USSR Acad. of Sci., 1988 (in Russian).

L. Davison, D. Grady, and M. Shahinpoor, eds., *Dynamic Fracture and Fragmentation*, Springer, NY, 1994.

Applications

A. A. Ezra, ed., *Principles and Practice of Explosive Metalworking*, Industrial Newspapers, London, 1973.

J. S. Rinehart and J. Pearson, *Explosive Working of Metals*, Pergamon, Elmsford, NY, 1963.

J. S. Rinehart and J. Pearson, *Behavior of Metals under Impulsive Loads*, American Society for Metals, Dover, 1964.

R. A. Graham and A. B. Sawaoka, eds., *High Pressure Explosive Processing of Ceramics*, Trans. Tech. Publ., Switzerland, 1987.

B. Crossland, *Explosive Welding of Metals and Its Application*, Oxford, 1982.

R. Prummer, *Explosivverdichtung Pulvriger Substanzen*, Springer, 1987.

M. E. Backman, "Terminal Ballistics," Report No. AD-A021833, National Technical Information Services, 1976.

W. P. Walters and J. J. Zukas, *Fundamentals of Shaped Charges*, Wiley, New York, 1989.

T. Z. Blazinsky, ed., *Explosive Welding, Forming, and Compaction*, Applied Science Publishers, 1983.

L. E. Murr, ed., *Shock Waves for Industrial Applications*, Noyes, Park Ridge, NJ, 1990.

A. A. Deribas, *Physics of Explosive Hardening and Welding*, Nauka, Novosibirsk, USSR (in Russian) 1980.

C. H. Dowding, *Blast Vibration Monitoring and Control*, 1985, Prentice-Hall, Englewood Cliffs, NJ.

W. Johnson, *Impact Strength of Materials*, Edward Arnold, England, 1972.

W. Goldsmith, *Impact*, Edward Arnold, England, 1960.

P.-A. Persson, R. Holmberg, and J. Lee, *Rock Blasting and Explosives Engineering*, CRC, Boca Raton, 1993.

CHAPTER 2

Elastic Waves

2.1 DYNAMIC PROPAGATION OF DEFORMATION

The application of an external force to a body is, by definition, a dynamic process. However, when the rate of change of the applied forces is low, one can consider the process of deformation as a sequence of steps in which the body can be considered in static equilibrium. Figure 2.1(a) shows how the distance between the atoms changes upon the application of an external force.* For each of the stages of deformation shown in Figs. 2.1(b) and (c), the body can be considered under static equilibrium and one can apply the methods of mechanics of materials to determine the internally-resisting stresses (by the method of sections). Hence, a section made at *AA* or *BB* will yield identical stresses.

However, the internal stresses are not instantaneously transmitted from the force application region to the different regions of the body. The stresses (and strains) are transferred from atom to atom at a certain specific velocity. Figure 2.2 shows the application of a force at a rate dF/dt such that the stresses (and attendant strains) vary from section to section. Section *BB* has not "seen" the application of the force at time t, whereas at section *AA*, the effects of the external force F are already felt.

At an atomistic level, one may envision the wave as a succession of impacts between adjacent atoms. Each atom, upon being accelerated to a certain velocity, transmits all (or part) of its momentum to its neighbor(s). The mass, atom separation, and forces of attraction and repulsion between atoms determine the way in which the stress pulse is carried from one point to the other.

One can perform a simple calculation by considering the individual atoms that comprise a solid and thereby arrive at a fairly realistic value for the velocity of an elastic pulse through it. Individual atoms oscillate continuously about their equilibrium position, and the frequency of this oscillation is approximately 10^{13} oscillations per second. These oscillations, although three dimensional, can be broken down along three directions (x_1, x_2, x_3). Figure 2.3 shows an idealized linear array of atoms. The equilibrium position is shown by the solid

*The atomic motions in Figs. 2.1 and 2.2 are highly idealized; in actual deformation the motions of atoms are determined by crystallographic constraints.

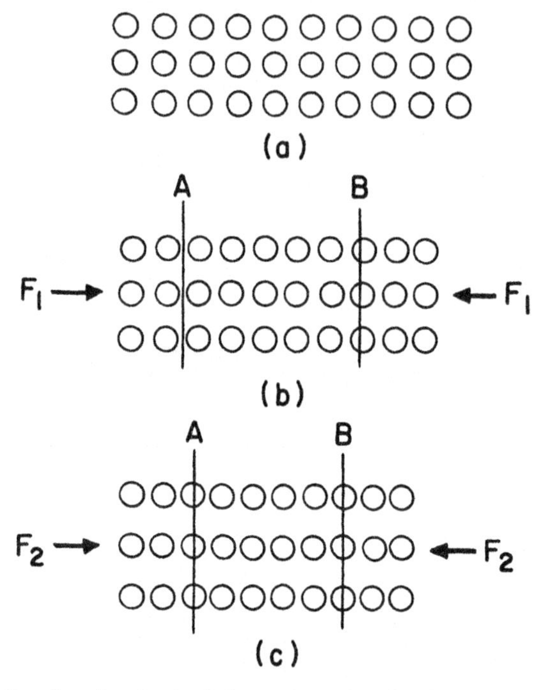

FIGURE 2.1 Quasi-static elastic deformation of a simple two-dimensional array of atoms ($F_2 > F_1$).

lines, and the extreme positions along the three axes are shown by dashed lines. This field of study is called **lattice dynamics**, and the oscillations are actually not independent, but are coupled in waves called **phonons**. However, for our simplified case, all we need to know is that these atoms are vibrating. If we impact the atom on the left, this atom will transmit this impact to its neighbor on the right.

It is interesting to note that all wave propagation equations can be derived by assuming that the material is a continuum (mass uniformly distributed in space), since they are a direct consequence of Newton's second law. Nevertheless, a more correct picture is that these atoms are connected by interatomic

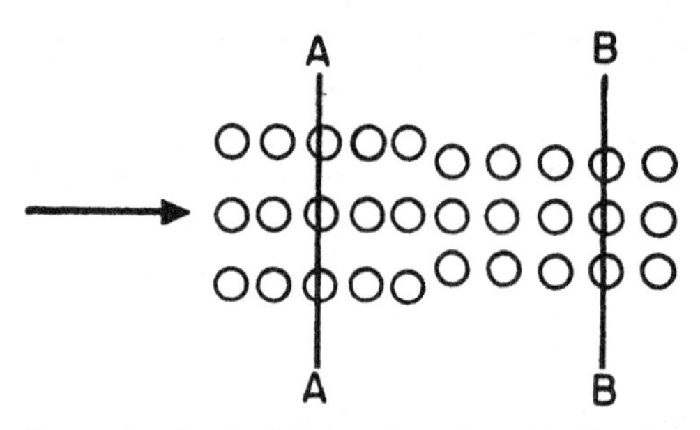

FIGURE 2.2 Dynamic elastic deformation of an idealized array of atoms.

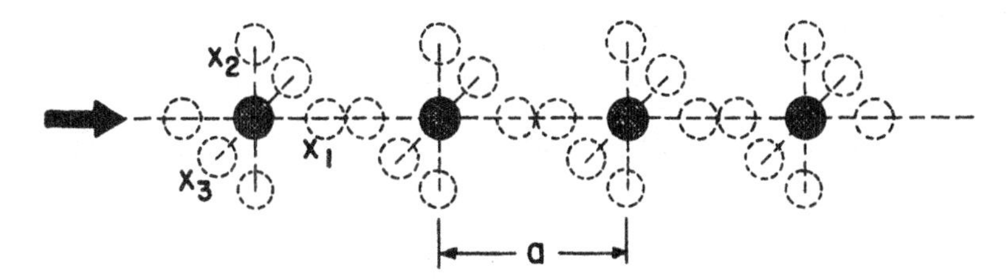

FIGURE 2.3 Transmission of disturbance from atom to atom.

forces that can be envisioned as minute springs connecting spheres (atoms). Momentum is transferred from sphere to sphere (atom to atom) at a rate that provides the velocity of propagation of disturbances.

When the force on the left is applied, there will be a time lag involved in the transmission of this push equal in average to the period of vibration (10^{-13} s). The atoms in the middle will continue the process, transmitting the "push" to the atom on the right. It is a simple matter to calculate the order of magnitude of velocity of propagation of this disturbance if one knows the separation between neighboring atoms. For a solid metal (iron) we can take it approximately equal to 3 A, or 3×10^{-10} m. Thus

$$V = \frac{a}{t} = 3 \times 10^{-10} \times 10^{13} = 3 \times 10^{3} \text{ m/s}$$

This is very close to the velocity of propagation of an elastic wave in iron (3500 m/s). This simple reasoning explains the physical sequence of momentum transfer at the atomic level. In the sections that follow, the material will be considered a continuum. We first derive the field equation for a wave in a slender bar (Section 2.2) and in an infinite (unbounded) material (Sections 2.4–2.5). We also discuss the various types of waves (Section 2.3). In Section 2.6, we derive the equation for surface (Rayleigh) waves. The indicial notation that simplifies considerably the derivations is introduced in Section 2.7. In Section 2.8 we deal with wave reflections, interactions, and refractions. Whereas in Chapter 1 we presented the solution of the wave equations as a harmonic sine function, the general solution does not impose this shape restriction; this general solution is developed in Section 2.9. The method of characteristics, useful for wave propagation studies, is briefly outlined. In Section 2.11 and 2.12 we describe waves in cylindrical bars and spherical waves, respectively. This short description of waves should provide the student with a sufficient background for the subsequent chapters.

2.2 ELASTIC WAVE IN CYLINDRICAL BAR

The calculation of the velocity of propagation of a wave in a thin bar is straightforward and therefore we will start with it. Figure 2.4 shows a striker bar impacting a long cylindrical bar at velocity V. A compressive stress wave is

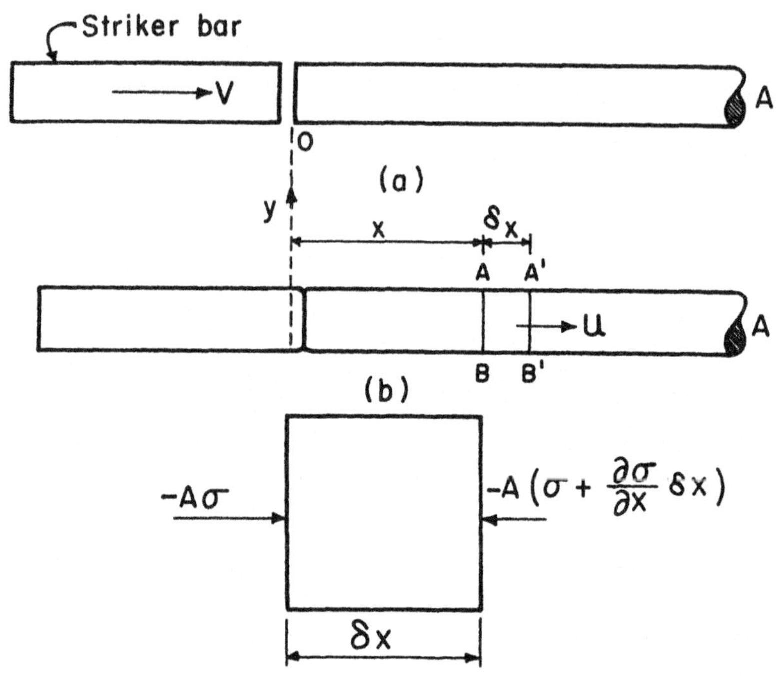

FIGURE 2.4 Propagation of wave in bar produced by impact of striker bar (a) prior to impact and (b) after impact.

produced in the bar that travels from left to right. At time t the front of this disturbance is at x. We will neglect, in the analysis, strains and inertia along the direction transverse to the bar Oy. We consider a section AB and $A'B'$ at the front of the wave at time t. The section $A'B'$ is distant $x + \delta x$ from the origin. Applying Newton's second law to $AA'B'B$ we obtain:

$$F = ma$$

$$-\left[A\sigma - A\left(\sigma + \frac{\partial \sigma}{\partial x}\,\delta x\right)\right] = A\rho\delta x\,\frac{\partial^2 u}{\partial t^2} \tag{2.1}$$

$$\frac{\partial \sigma}{\partial x} = \rho\,\frac{\partial^2 u}{\partial t^2} \tag{2.2}$$

However, the deformation is elastic and we assume Hooke's law to hold:

$$\frac{\sigma}{\varepsilon} = E$$

where ε is the strain, defined as $\partial u/\partial x$. The minus sign is due to the compressive strain. Thus,

$$\frac{\partial}{\partial x}\left[E\,\frac{\partial u}{\partial x}\right] = \rho\,\frac{\partial^2 u}{\partial t^2} \tag{2.3}$$

and

$$\frac{\partial^2 u}{\partial t^2} = \frac{E}{\rho} \frac{\partial^2 u}{\partial x^2} \tag{2.4}$$

This is the differential equation for the wave that we saw before in Chapter 1. We also recall that the velocity of this wave is given by

$$C_0 = \sqrt{\frac{E}{\rho}} \tag{2.5}$$

It will be seen in Section 2.4 that an elastic (dilatational) wave in an unbounded medium travels at a velocity slightly higher than C_0. For a metal with $\nu = 0.3$, this wave velocity is $1.2C_0$.

2.3 TYPES OF ELASTIC WAVES

Different types of elastic waves can propagate in solids, depending on how the motion of the particles of the solid is related to the direction of propagation of the waves themselves and on the boundary conditions. A "particle" can be defined as a small discrete portion of the solid. It is not an atom, which may move in a direction different from the general direction of motion, because of crystallographic requirements. The most common types of elastic waves in solids are

1. longitudinal (or irrotational) waves (in seismology, they are known as push, primary, or P waves); in infinite and semi-infinite media, they are known as "dilatational" waves.
2. distortional (or shear, or transverse, or equivolumal) waves (in seismology, they are known as secondary, shake, or SH and SV waves);
3. surface (or Rayleigh) waves;
4. interfacial (Stoneley) waves;
5. bending (or flexural) waves (in bars and plates).

A brief description of these waves follows:

1. *Longitudinal Waves.* These waves correspond to the motion of the particles back and forth along the direction of wave propagation such that the particle velocity (U_p) is parallel to the wave velocity (U). If the wave is compressive, they have the same sense; if it is tensile, they have opposite senses. A simple example of a longitudinal elastic wave is illustrated in Figure 2.5(a). This wave is produced by the impact of a hammer on the left-hand side of the long cylinder. Equation (2.5) gives the velocity of the wave. Figure 2.5(b)

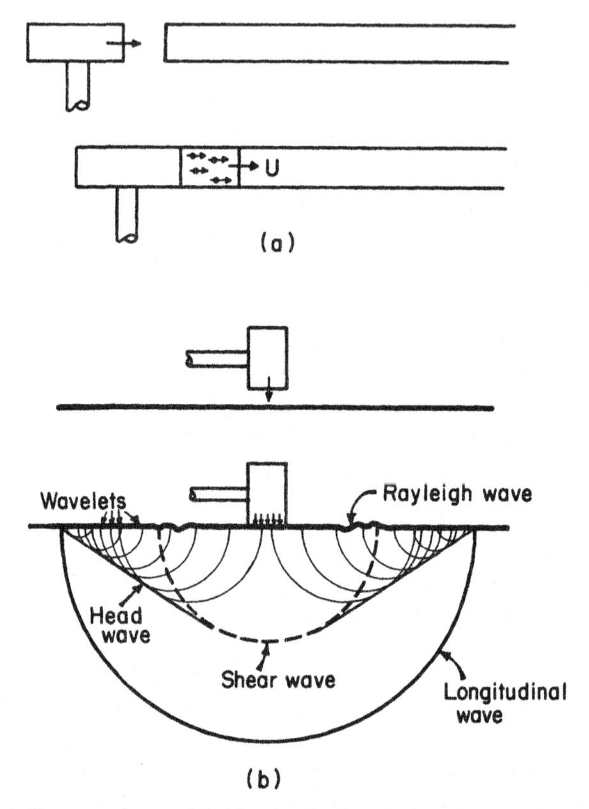

FIGURE 2.5 (a) Hammer impacting slender body; (b) hammer impacting semi-infinite body; notice formation of longitudinal, shear, and Rayleigh waves. Interaction of longitudinal wave with free-surface forms wavelets that comprise headwave.

shows, in schematic manner, the impact of a hammer on a semi-infinite body. In addition to longitudinal waves, distortional and Rayleigh waves are generated. Note that the velocity of a wave in a bounded medium (e.g., cylinder) is different from that in an unbounded medium. This will be discussed further in Section 2.5.

2. *Distortional (or Shear) Waves.* If the motions of the particles conveying the wave are perpendicular to the direction of the propagation of the wave itself, then we have a distortional wave. There is no resulting change in density, and all longitudinal strains ε_{11}, ε_{22}, and ε_{33} are zero. An example of a distortional wave is illustrated in Figure 2.6. When a bar extremity is subjected to a torsion, the region between the end and the clamp will store elastic energy. Upon release of the clamp, a pulse will travel in the bar, toward the right. This is a distortional wave, and the particle displacement is perpendicular to the wave propagation direction (U_p is perpendicular to U).

3. *Surfaces Waves.* Surface waves are analogous to waves on the surface of water. Objects floating on the water act as markers for the particles of water

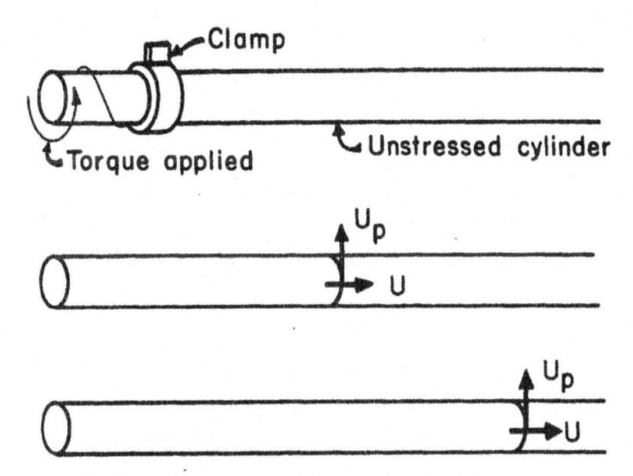

FIGURE 2.6 Wave and particle motion in release of torsional energy; when clamp is released, particle displacement (on cross-sectional plane) travels along longitudinal axis.

and move both up and down and back and forth, tracing out elliptical paths as the water moves by. This type of wave is restricted to the region adjacent to the surface, and "particle" velocity (U_p) decreases very rapidly (exponentially) as one moves away from it. The particles describe elliptical trajectories, as illustrated by the cork in the example of Figure 2.7. Surface waves (called Rayleigh waves in solids) are a particular case of interfacial waves when one of the materials has negligible density and elastic wave velocity.

4. *Interfacial (Stoneley) Waves.* When two semi-infinite media with different properties are in contact, special waves form at their interface.

5. *Waves in Layered Media (or Love waves).* In seismology, these waves are very important. It is known that earthquakes produce waves in which the horizontal component of displacement can be significantly larger than the vertical component, a behavior not consistent with Rayleigh waves. The earth is

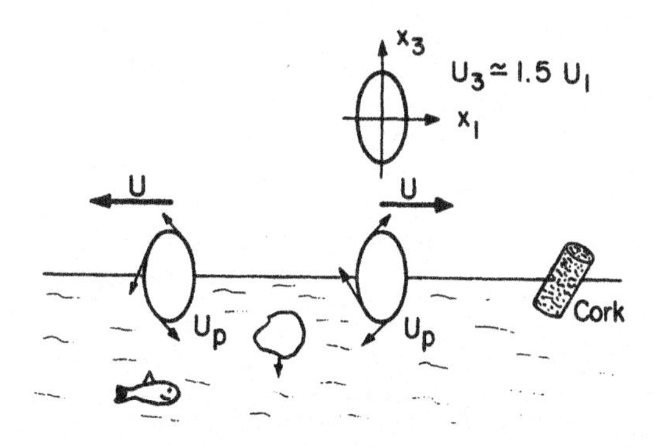

FIGURE 2.7 Surface waves; particle and wave velocities shown; cork would describe elliptical trajectory. Rayleigh waves are analogous to surface (gravity) waves in fluids.

composed of layers, or strata, with different properties, and special wave patterns emerge. Love was the first person to study these waves that are named after him.

6. *Bending (or flexural) Waves (in bars, membranes, plates).* These waves involve propagation of flexure in a one-dimensional (bar) or in two-dimensional configuration. The mathematics of wave propagation in plates and shells is quite formidable and beyond the scope of this book. The reader should consult Graff [1].

The surface waves are the slowest of the three waves; the fastest are longitudinal waves, as will be seen from the derivations that follow. Figure 2.8 (adapted from Woods [2]) shows, in a schematic manner, the amplitudes of the longitudinal, shear, and Rayleigh waves in half-space. The shear wave undergoes a change in particle motion sense ($\pm$); at a point vertically below the load application region this particle velocity and, consequently, the shear-wave amplitude are zero. The horizontal and vertical components of the particle motions as a function of distance from the surface in the Rayleigh wave are also shown. The exponential decay of the horizontal component, on the right side, is clearly seen. The waves decay at different rates. The amplitude of the longitudinal wave decays more readily close to the surface; the width of the band indicates its relative value. The amplitudes of the longitudinal and shear

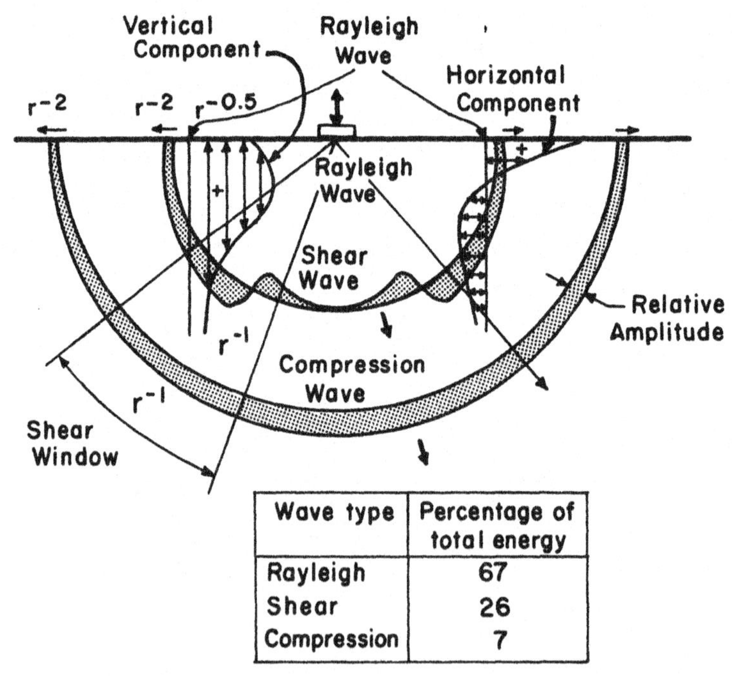

Wave type	Percentage of total energy
Rayleigh	67
Shear	26
Compression	7

FIGURE 2.8 Distribution of displacement and energy in dilatational, shear, and surface waves from a harmonic normal load on a half-space of $\nu = 0.25$.

waves decay as r^{-1} in the region away from the free surface (or free from its effect). Along the surface, they decay faster, as r^{-2}. The bright student will be able to show this decay by strictly energetic arguments. The Rayleigh wave decays much slower, at a rate of $r^{-1/2}$. Thus, it can be picked up at much larger distances. Figure 2.8 also shows how different waves carry different fractions of the energy. Most of the energy is carried by the Rayleigh wave (67% for $\nu = 0.25$). The longitudinal (compression) wave only carries 7% of the energy. Not shown in Figure 2.8 are Love and Stoneley waves. Love waves have particle displacements that are parallel to the wave front (at the surface); these horizontal displacement components are due to differences in elastic constants close to the surface. These waves are observed on the surface of the earth due to the existence of strata with different densities (elastic impedances). Stoneley waves are a more general type of wave that originates at the interface between layers with different impedances. These waves have maximum amplitude at the interface.

2.4 PROPAGATION OF ELASTIC WAVES IN CONTINUUM [1, 3–6]

The stresses acting on the different planes normal to the axes Ox_1, Ox_2, and Ox_3 are σ_{11}, σ_{22}, σ_{33}, σ_{12}, σ_{32}, σ_{13}, σ_{23}, σ_{21}, and σ_{31}, where the first number in the suffix denotes the direction of the stress and the second number defines the plane on which the stress is acting. Under equilibrium,

$$\sum F = 0 \begin{cases} \sum F_{x1} = 0 \\ \sum F_{x2} = 0 \\ \sum F_{x3} = 0 \end{cases} \text{and} \quad \sum M = 0 \begin{cases} \sum M_{x1} = 0 \\ \sum M_{x2} = 0 \\ \sum M_{x3} = 0 \end{cases} \tag{2.6}$$

If the unit cube in Figure 2.9(b) is not in static equilibrium, then the stresses acting on the opposite faces are not equal. In order to obtain the equations for the propagation of elastic waves in continuum, we consider the variation in stress across a small parallelepiped with its sides parallel to a set of rectangular axes, as shown in Figure 2.9(b).

If we consider an infinitesimal rectangular parallelepiped, then the stress acting on the surface element of the solid body has components both normal to the plane and tangential to it. We only need the equation for conservation of momentum to derive the elastic wave equation. It will be shown later (Chapter 4) that the equations for conservation of mass and energy are also needed for shock waves.

Newton's second law can be expressed in relation to the three coordinate axes (Newton's second law expresses the conservation of momentum):

$$F_{x1} = ma_{x1} \qquad F_{x2} = ma_{x2} \qquad F_{x3} = ma_{x3} \tag{2.7}$$

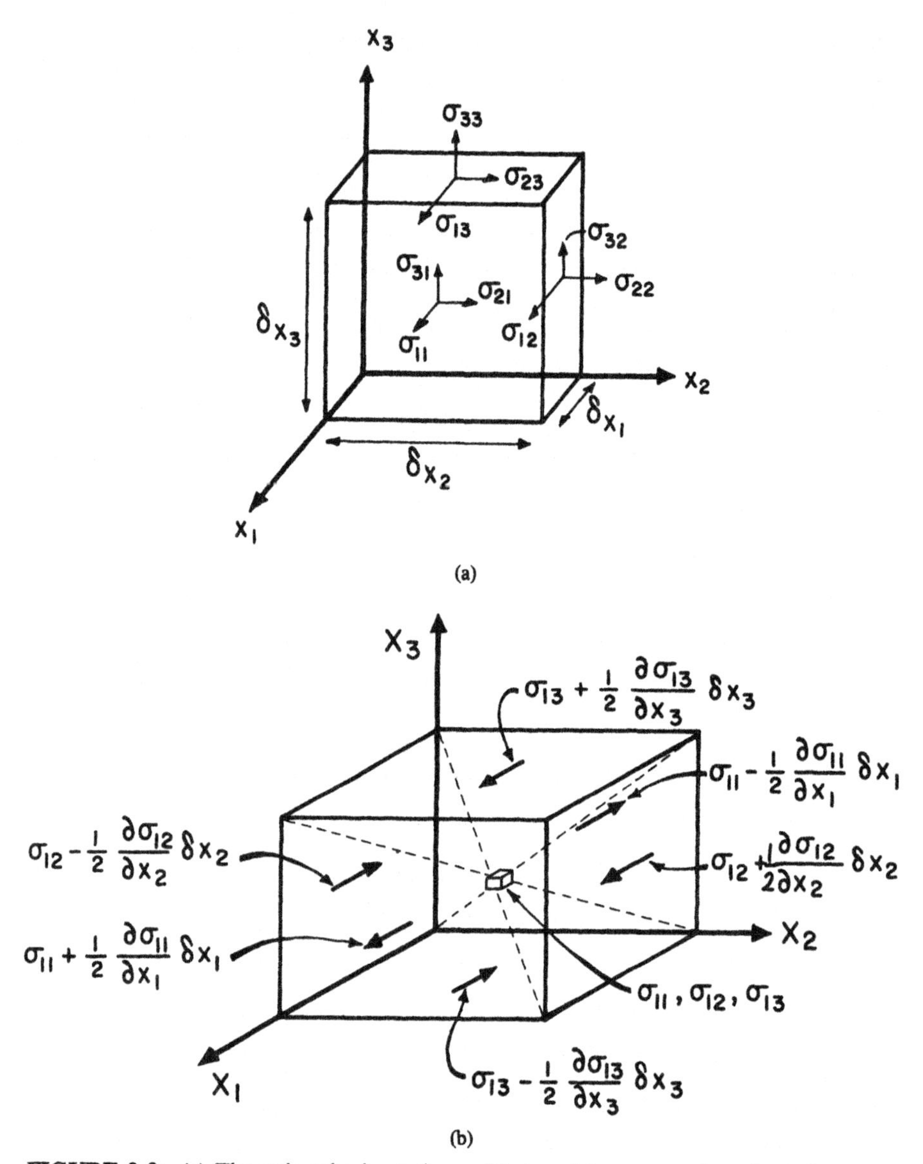

FIGURE 2.9 (a) The unit cube in static equilibrium (stresses on three back faces not shown). (b) The unit cube in dynamic state.

or in the Ox_1 directional only:

$$\sum F_{x_1} = (\rho \, dx_1 \, dx_2 \, dx_3) \frac{\partial^2 u_1}{\partial t^2} \tag{2.8}$$

All stresses acting in the direction Ox_1 are represented in the cube in Figure 2.9. It is considered that at the center of the cube (with dimensions δx_1, δx_2,

δx_3) the stresses have the values σ_{11}, σ_{22}, σ_{33} (normal) and σ_{12}, σ_{13}, σ_{23} (shear). The components of stress on each face of the cube thus are

$$\sigma_{11} + \frac{\partial \sigma_{11}}{\partial x_1} \cdot \tfrac{1}{2}\,\delta x_1 \qquad \sigma_{12} + \frac{\partial \sigma_{12}}{\partial x_2} \cdot \tfrac{1}{2}\,\delta x_2 \qquad \sigma_{13} + \frac{\partial \sigma_{13}}{\partial x_3} \cdot \tfrac{1}{2}\,\delta x_3$$

$$\sigma_{11} - \frac{\partial \sigma_{11}}{\partial x_1} \cdot \tfrac{1}{2}\,\delta x_1 \qquad \sigma_{12} - \frac{\partial \sigma_{12}}{\partial x_2} \cdot \tfrac{1}{2}\,\delta x_2 \qquad \sigma_{13} - \frac{\partial \sigma_{13}}{\partial x_3} \cdot \tfrac{1}{2}\,\delta x_3$$

The term $\pm\,\frac{1}{2}(\partial\sigma_{11}/\partial x_1)\,\delta x_1$ expresses the change in σ_{11}, with respect to Ox_1, as one moves from the center of the cube to one of the faces perpendicular to Ox_1. The other terms also have similar meaning.

Then, to obtain the force acting on each face we take the value of stress at the center of each face times the area of the face. As can be seen from the figure, six separate forces are acting parallel to each axis, and neglecting the effects of gravitational forces and body moments, if one considers the resultant forces acting in the x_1 direction, then

$$\begin{aligned}
&\left(\sigma_{11} + \frac{1}{2}\frac{\partial \sigma_{11}}{\partial x_1} \cdot \delta x_1 - \sigma_{11} + \frac{1}{2}\frac{\partial \sigma_{11}}{\partial x_1} \cdot \delta x_1\right) \delta x_2\, \delta x_3 \\
&\quad + \left(\sigma_{12} + \frac{1}{2}\frac{\partial \sigma_{12}}{\partial x_2} \cdot \delta x_2 - \sigma_{12} + \frac{1}{2}\frac{\partial \sigma_{12}}{\partial x_2} \cdot \delta x_2\right) \delta x_1\, \delta x_3 \\
&\quad + \left(\sigma_{13} + \frac{1}{2}\frac{\partial \sigma_{13}}{\partial x_3} \cdot \delta x_3 - \sigma_{13} + \frac{1}{2}\frac{\partial \sigma_{13}}{\partial x_3} \cdot \delta x_3\right) \delta x_1\, \delta x_2 = \sum F_{x_1} \qquad (2.9)
\end{aligned}$$

Simplifying Eqn. (2.9) and evaluating it with (2.8), one gets

$$\frac{\partial \sigma_{11}}{\partial x_1}\,\delta x_1\,\delta x_2\,\delta x_3 + \frac{\partial \sigma_{12}}{\partial x_2}\,\delta x_1\,\delta x_2\,\delta x_3 + \frac{\partial \sigma_{13}}{\partial x_3}\,\delta x_1\,\delta x_2\,\delta x_3 = \rho\delta x_1\,\delta x_2\,\delta x_3\,\frac{\partial^2 u_1}{\partial t^2}$$

or

$$\frac{\partial \sigma_{11}}{\partial x_1} + \frac{\partial \sigma_{12}}{\partial x_2} + \frac{\partial \sigma_{13}}{\partial x_3} = \rho\,\frac{\partial^2 u_1}{\partial t^2} \qquad (2.10a)$$

Similarly, for other directions,

$$\frac{\partial \sigma_{21}}{\partial x_1} + \frac{\partial \sigma_{22}}{\partial x_2} + \frac{\partial \sigma_{23}}{\partial x_3} = \rho\,\frac{\partial^2 u_2}{\partial t^2} \qquad (2.10b)$$

$$\frac{\partial \sigma_{31}}{\partial x_1} + \frac{\partial \sigma_{32}}{\partial x_2} + \frac{\partial \sigma_{33}}{\partial x_3} = \rho\,\frac{\partial^2 u_3}{\partial t^2} \qquad (2.10c)$$

In generalized indicial notation

$$\frac{\partial \sigma_{ij}}{\partial x_j} = \rho \frac{\partial^2 u_i}{\partial t^2} \tag{2.11}$$

The solutions of this system of differential equations will yield the equations of the wave once the stresses are replaced by the strains. The generalized Hooke's law for an isotropic material in a triaxial state of stress is used. There are six equations relating the stresses to the strains. Here λ and μ are the Lamé constants:

$$\begin{aligned} \sigma_{11} &= \lambda\Delta + 2\mu\varepsilon_{11} & \sigma_{13} &= 2\mu\varepsilon_{13} \\ \sigma_{23} &= 2\mu\varepsilon_{23} & \sigma_{12} &= 2\mu\varepsilon_{12} \\ \sigma_{22} &= \lambda\Delta + 2\mu\varepsilon_{22} & \sigma_{33} &= \lambda\Delta + 2\mu\varepsilon_{33} \end{aligned} \tag{2.12}$$

where Δ is the dilatation, $\Delta = \varepsilon_{11} + \varepsilon_{22} + \varepsilon_{33}$. Replacing the values of σ_{ij} from Eqn. (2.12) into Eqn. (2.11) yields

$$\frac{\partial(\lambda\Delta + 2\mu\varepsilon_{11})}{\partial x_1} + \frac{\partial(2\mu\varepsilon_{12})}{\partial x_2} + \frac{\partial(2\mu\varepsilon_{13})}{\partial x_3} = \rho \frac{\partial^2 u_1}{\partial t^2}$$

or

$$\frac{\partial(\lambda\Delta)}{\partial x_1} + \frac{2\mu\partial\varepsilon_{11}}{\partial x_1} + \frac{2\mu\partial\varepsilon_{12}}{\partial x_2} + \frac{2\mu\partial\varepsilon_{13}}{\partial x_3} = \rho \frac{\partial^2 u_1}{\partial t^2} \tag{2.13}$$

But by definition of strain,

$$\varepsilon_{11} = \frac{\partial u_1}{\partial x_1}, \qquad \varepsilon_{12} = \frac{1}{2}\left(\frac{\partial u_1}{\partial x_2} + \frac{\partial u_2}{\partial x_1}\right) \qquad \varepsilon_{13} = \frac{1}{2}\left(\frac{\partial u_1}{\partial x_3} + \frac{\partial u_3}{\partial x_1}\right)$$

Substituting these expressions for the strains in the above equation, we get

$$\frac{\partial(\lambda\Delta)}{\partial x_1} + \frac{2\mu\partial}{\partial x_1}\left(\frac{\partial u_1}{\partial x_1}\right) + \frac{2\mu\partial}{\partial x_2}\left[\frac{1}{2}\left(\frac{\partial u_1}{\partial x_2} + \frac{\partial u_2}{\partial x_1}\right)\right] + 2\mu\frac{\partial}{\partial x_3}\left[\frac{1}{2}\left(\frac{\partial u_1}{\partial x_3} + \frac{\partial u_3}{\partial x_1}\right)\right]$$
$$= \rho \frac{\partial^2 u_1}{\partial t^2}$$

or

$$\frac{\partial(\lambda\Delta)}{\partial x_1} + 2\mu \frac{\partial^2 u_1}{\partial x_1^2} + \mu \frac{\partial^2 u_1}{\partial x_2^2} + \mu \frac{\partial^2 u_2}{\partial x_1\,\partial x_2} + \mu \frac{\partial^2 u_1}{\partial x_3^2} + \mu \frac{\partial^2 u_3}{\partial x_3\,\partial x_1} = \rho \frac{\partial^2 u_1}{\partial t^2}$$

Let us define an operator ∇^2 as

$$\nabla^2 = \frac{\partial^2}{\partial x_1^2} + \frac{\partial^2}{\partial x_2^2} + \frac{\partial^2}{\partial x_3^2} = \frac{\partial^2}{\partial x_i\, \partial x_i} \tag{2.14}$$

The last term represents the indicial notation. Notice that i appears twice and therefore summation is implied. Simplifying the above equation, we get, since $\partial\lambda/\partial x_1 = 0$

$$\lambda \frac{\partial \Delta}{\partial x_1} + \mu \frac{\partial^2 u_1}{\partial x_1^2} + \mu_1 \left(\frac{\partial^2}{\partial x_1^2} + \frac{\partial^2}{\partial x_2^2} + \frac{\partial^2}{\partial x_3^2} \right) u_1$$

$$+ \mu \frac{\partial^2 u_2}{\partial x_1\, \partial x_2} + \mu \frac{\partial^2 u_3}{\partial x_3\, \partial x_1} = \rho \frac{\partial^2 u_1}{\partial t^2}$$

$$\lambda \frac{\partial \Delta}{\partial x_1} + \mu \frac{\partial^2 u_1}{\partial x_1^2} + \mu\, \nabla^2 u_1 + \mu \frac{\partial^2 u_2}{\partial x_1\, \partial x_2} + \mu \frac{\partial^2 u_3}{\partial x_3\, \partial x_1} = \rho \frac{\partial^2 u_1}{\partial t^2}$$

$$\lambda \frac{\partial \Delta}{\partial x_1} + \mu \frac{\partial}{\partial x_1} \left(\frac{\partial u_1}{\partial x_1} + \frac{\partial u_2}{\partial x_2} + \frac{\partial u_3}{\partial x_3} \right) + \mu\, \nabla^2 u_1 = \rho \frac{\partial^2 u_1}{\partial t^2}$$

$$\lambda \frac{\partial \Delta}{\partial x_1} + \mu \frac{\partial}{\partial x_1} \underbrace{(\varepsilon_{11} + \varepsilon_{22} + \varepsilon_{33})}_{\Delta \text{ (dilatation)}} + \mu\, \nabla^2 u_1 = \frac{\rho \partial^2 u_1}{\partial t^2}$$

$$\lambda \frac{\partial \Delta}{\partial x_1} + \mu \frac{\partial \Delta}{\partial x_1} + \mu \nabla^2 u_1 = \frac{\rho \partial^2 u_1}{\partial t^2}$$

$$(\lambda + \mu) \frac{\partial \Delta}{\partial x_1} + \mu\, \nabla^2 u_1 = \rho \frac{\partial^2 u_1}{\partial t^2} \tag{2.15a}$$

Similarly, for the x_2 and x_3 directions,

$$(\lambda + \mu) \frac{\partial \Delta}{\partial x_2} + \mu\, \nabla^2 u_2 = \rho \frac{\partial^2 u_2}{\partial t^2} \tag{2.15b}$$

$$(\lambda + \mu) \frac{\partial \Delta}{\partial x_3} + \mu\, \nabla^2 u_3 = \rho \frac{\partial^2 u_3}{\partial t^2} \tag{2.15c}$$

These are the equations of motion of an isotropic elastic solid in which the body forces (gravitation and moments) are absent. They are used to develop the equations for the propagation of the two types of elastic waves (longitudinal and distortional).

Thus, if the displacements in the above equations are replaced by strains and the equations are grouped together after taking the derivative of Eqn.

(2.15a) with respect to x_1, Eqn. (2.15b) with respect to x_2, and Eqn. (2.15c) with respect to x_3, we get

$$\frac{\partial}{\partial x_1}\left[(\lambda + \mu)\frac{\partial \Delta}{\partial x_1} + \mu\,\nabla^2 u_1\right] + \frac{\partial}{\partial x_2}\left[(\lambda + \mu)\frac{\partial \Delta}{\partial x_2} + \mu\,\nabla^2 u_2\right]$$

$$+ \frac{\partial}{\partial x_3}\left[(\lambda + \mu)\frac{\partial \Delta}{\partial x_3} + \mu\,\nabla^2 u_3\right]$$

$$= \frac{\rho\partial}{\partial x_1}\left(\frac{\partial^2 u_1}{\partial t^2}\right) + \frac{\rho\partial}{\partial x_2}\left(\frac{\partial^2 u_2}{\partial t^2}\right) + \frac{\rho\partial}{\partial x_3}\left(\frac{\partial^2 u_3}{\partial t^2}\right)$$

or

$$\left[(\lambda + \mu)\frac{\partial^2 \Delta}{\partial x_1^2} + \mu\frac{\partial}{\partial x_1}\nabla^2 u_1\right] + \left[(\lambda + \mu)\frac{\partial^2 \Delta}{\partial x_2^2} + \mu\frac{\partial}{\partial x_2}\nabla^2 u_2\right]$$

$$+ \left[(\lambda + \mu)\frac{\partial^2 \Delta}{\partial x_3^2} + \mu\frac{\partial}{\partial x_3}\nabla^2 u_3\right] = \frac{\rho\,\partial^3 u_1}{\partial t^2\,\partial x_1} + \frac{\rho\,\partial^3 u_2}{\partial t^2\,\partial x_2} + \frac{\rho\,\partial^3 u_3}{\partial t^2\,\partial x_3}$$

or

$$\left[(\lambda + \mu)\frac{\partial^2 \Delta}{\partial x_1^2} + \mu\,\nabla^2\frac{\partial u_1}{\partial x_1}\right] + \left[(\lambda + \mu)\frac{\partial^2 \Delta}{\partial x_2^2} + \mu\,\nabla^2\frac{\partial u_2}{\partial x_2}\right]$$

$$+ \left[(\lambda + \mu)\frac{\partial^2 \Delta}{\partial x_3^2} + \mu\,\nabla^2\frac{\partial u_3}{\partial x_3}\right] = \frac{\rho\,\partial^2}{\partial t^2}\left[\frac{\partial u_1}{\partial x_1} + \frac{\partial u_2}{\partial x_2} + \frac{\partial u_3}{\partial x_3}\right]$$

But

$$\frac{\partial u_1}{\partial x_1} = \varepsilon_{11} \qquad \frac{\partial u_2}{\partial x_2} = \varepsilon_{22} \qquad \frac{\partial u_3}{\partial x_3} = \varepsilon_{33}$$

$$\therefore \left[(\lambda + \mu)\frac{\partial^2 \Delta}{\partial x_1^2} + \mu\,\nabla^2 \varepsilon_{11}\right] + \left[(\lambda + \mu)\frac{\partial^2 \Delta}{\partial x_2^2} + \mu\,\nabla^2 \varepsilon_{22}\right]$$

$$+ \left[(\lambda + \mu)\frac{\partial^2 \Delta}{\partial x_3^2} + \mu\,\nabla^2 \varepsilon_{33}\right] = \frac{\rho\,\partial^2}{\partial t^2}(\varepsilon_{11} + \varepsilon_{22} + \varepsilon_{33})$$

or

$$(\lambda + \mu)\frac{\partial^2 \Delta}{\partial x_1^2} + (\lambda + \mu)\frac{\partial^2 \Delta}{\partial x_2^2} + (\lambda + \mu)\frac{\partial^2 \Delta}{\partial x_3^2} + \mu\,\nabla^2(\varepsilon_{11} + \varepsilon_{22} + \varepsilon_{33})$$

$$= \frac{\rho\,\partial^2}{\partial t^2}(\varepsilon_{11} + \varepsilon_{22} + \varepsilon_{33})$$

But

$$\varepsilon_{11} + \varepsilon_{22} + \varepsilon_{33} = \Delta \quad \text{and} \quad \nabla^2 = \frac{\partial^2}{\partial x_1^2} + \frac{\partial^2}{\partial x_2^2} + \frac{\partial^2}{\partial x_3^2}$$

That is why

$$(\lambda + \mu)\, \nabla^2 \Delta + \mu\, \nabla^2 \Delta = \rho \frac{\partial^2 \Delta}{\partial t^2}$$

or

$$\frac{\partial^2 \Delta}{\partial t^2} = \frac{\lambda + 2\mu}{\rho} \nabla^2 \Delta \tag{2.16}$$

A dimensional analysis shows that the units of the coefficient $(\lambda + 2\mu)/\rho$ are $(\text{distance}/\text{time})^2$. This second-order partial differential equation is analogous to the one that we saw in Chapter 1 [Eqn. (1.8)]. It represents a wave (as will be shown later, in Section 2.9) of a general shape traveling at a velocity

$$V_{\text{long}} = \left(\frac{\lambda + 2\mu}{\rho}\right)^{1/2} \tag{2.17}$$

Equation (2.16) is known as the longitudinal wave equation in an unbounded medium and implies that the dilatation (Δ) is propagated through the medium with velocity V_{long}. It is also called "dilatational" wave. The velocity is often called "bulk sound speed."

From elasticity theory, $\mu = E/[2(1 + \nu)]$ and $\lambda = \nu E/[(1 + \nu)(1 - 2\nu)]$ and one assumes $\nu = 0.3$; then

$$\begin{aligned} V_{\text{long}} &= \left[\frac{(1 - \nu)}{(1 + \nu)(1 - 2\nu)} \frac{E}{\rho}\right]^{1/2} \\ &= \left(\frac{1.346E}{\rho}\right)^{1/2} \end{aligned} \tag{2.18}$$

This is the longitudinal velocity of the wave in an unbounded (or infinite) medium. For a finite medium such as a Hopkinson bar, the longitudinal wave velocity is different, as seen in Section 2.2. It can be wavelength dependent.

Wave velocities vary from 340 m/s for air to quite high values. Beryllium has a low density but a high modulus of elasticity; hence, the longitudinal wave velocity is high. Table 2.1 shows that it is equal to approximately 10,000 m/s.

The similarity between Eqns. (2.5) (uniaxial stress) and (2.17) is better understood if we find the relationship between stress and strain for a uniaxial strain state:

TABLE 2.1 Velocities of Elastic Waves

Material	V_{long} (m/s)	V_{shear} (m/s)
Air	340	—
Aluminum	6,100	3,100
Steel	5,800	3,100
Lead	2,200	700
Beryllium	10,000	—
Glass/window	6,800	3,300
Plexiglas	2,600	1,200
Polystyrene	2,300	1,200
Magnesium	6,400	3,100

Source: From [4], p. 23.

$$\varepsilon_{22} = \varepsilon_{33} = 0 \qquad \varepsilon_{11} \neq 0$$

$$\varepsilon_{11} = \frac{1}{E}[\sigma_{11} - \nu(\sigma_{22} + \sigma_{33})]$$

But

$$\sigma_{22} = \sigma_{33}$$

$$\varepsilon_{11} = \frac{1}{E}(\sigma_{11} - 2\nu\sigma_{22})$$

$$\varepsilon_{22} = 0 = \frac{1}{E}[\sigma_{22} - \nu(\sigma_{11} + \sigma_{22})] \quad \therefore \; \sigma_{22} = \frac{\nu}{1 - \nu}\sigma_{11}$$

Thus:

$$\varepsilon_{11} = \frac{\sigma_{11}(1 + \nu)(1 - 2\nu)}{E(1 - \nu)}$$

$$\frac{\sigma_{11}}{\varepsilon_{11}} = \frac{(1 - \nu)E}{(1 + \nu)(1 - 2\nu)} = \bar{E}$$

This is the uniaxial strain elastic modulus and the corresponding elastic wave velocity is:

$$V_{long} = \sqrt{\frac{\bar{E}}{\rho}} = \left[\frac{(1 - \nu)E}{(1 + \nu)(1 - 2\nu)\rho}\right]^{1/2}$$

This is identical to Eqn. 2.18. For a uniaxial stress state, we have:

$$V_{long} = \sqrt{\frac{E}{\rho}}$$

2.5 CALCULATION OF DISTORTIONAL WAVE VELOCITY

In order to calculate the distortional (shear) wave velocity, we need to eliminate Δ between Eqns. (2.15a) by differentiating it with respect to x_2 and Eqn. (2.15b) with respect to x_1:

$$\frac{\partial}{\partial x_2}[2.15(a)] \Rightarrow (\lambda + \mu)\frac{\partial^2 \Delta}{\partial x_1\, \partial x_2} + \mu \frac{\partial}{\partial x_2}\nabla^2 u_1 = \rho \frac{\partial^3 u_1}{\partial x_2\, \partial t^2} \quad (2.19)$$

and

$$\frac{\partial}{\partial x_1}[2.15(b)] \Rightarrow (\lambda + \mu)\frac{\partial^2 \Delta}{\partial x_1\, \partial x_2} + \mu \frac{\partial}{\partial x_1}\nabla^2 u_2 = \rho \frac{\partial^3 u_2}{\partial x_1\, \partial t^2} \quad (2.20)$$

Taking the difference of Eqns. (2.19) and (2.20), we get

$$\mu\nabla^2\left(\frac{\partial u_1}{\partial x_2} - \frac{\partial u_2}{\partial x_1}\right) = \rho \frac{\partial^2}{\partial t^2}\left(\frac{\partial u_1}{\partial x_2} - \frac{\partial u_2}{\partial x_1}\right) \quad (2.21)$$

But by definition, rigid body rotations are given by

$$\omega_{ij} = \frac{1}{2}\left(\frac{\partial u_i}{\partial x_j} - \frac{\partial u_j}{\partial x_i}\right)$$

Substituting ω_{ij} into Eqn. (2.21), we obtain

$$\mu\, \nabla^2\omega_{12} = \rho \frac{\partial^2 \omega_{12}}{\partial t^2}$$

So, a rotation ω_{12} propagates at a velocity

$$V = \left(\frac{\mu}{\rho}\right)^{1/2}$$

The same can be said for ω_{13} and ω_{23}.

By simply assuming that dilatations are zero, in Eqn. (2.15), one arrives at

$$\mu\, \nabla^2 u_1 = \rho \frac{\partial^2 u_1}{\partial t^2}$$

$$\mu\, \nabla^2 u_2 = \rho \frac{\partial^2 u_2}{\partial t^2}$$

$$\mu\, \nabla^2 u_3 = \rho \frac{\partial^2 u_3}{\partial t^2}$$

Thus, equivoluminal waves (dilatation equal to zero) travel at a velocity

$$V = \left(\frac{\mu}{\rho}\right)^{1/2}$$

TABLE 2.2 Velocities of Longitudinal Elastic Waves in Ceramics

Ceramic	V (km/s)	E (GN/m^2)	ρ (g/cm^3)
Al_2O_3	9.6	365	3.9
BN	6	82.7	2.25
B_4C	3.3	28	2.52
Diamond	17	1000	3.51
WC	5.6	500	14.6
TiC	8.8	380	4.93
SiN	9.9	340	3.44

There are only two possible velocities for waves in elastic isotropic unbounded media:

$$V_{\text{long}} \equiv C_l = \left(\frac{\lambda + 2\mu}{\rho}\right)^{1/2} \tag{2.22}$$

$$V_s \equiv C_s = \left(\frac{\mu}{\rho}\right)^{1/2} \tag{2.23}$$

We will use, from here onward, both V and C for wave velocities. The first wave is an irrotational ($\omega' = 0$), longitudinal, or dilatational wave. The second wave has no dilatation and is both a shear and rotational wave. It should be noted that longitudinal waves have a shear component (in general, $\varepsilon_{11} \neq \varepsilon_{22} \neq \varepsilon_{33}$), whereas shear waves have no dilatational component. Typical velocities of longitudinal and distortional waves are reported in Table 2.1. Table 2.2 lists velocities of longitudinal waves for some ceramics.

2.6 SURFACE (RAYLEIGH) WAVES

The derivation of the Rayleigh (surface) wave equation will not be carried out here, and the reader is referred to Kolsky [3], Wasley [7], or other texts on wave propagation. The Rayleigh wave velocity–shear wave velocity ratio k,

$$k = \frac{C}{C_s}$$

can be obtained from the solution of the cubic equation in k^2:

$$k^6 - 8k^4 + (24 - 16\alpha_1)k^2 + 16(\alpha_1^2 - 1) = 0$$

$$\alpha_1^2 = \frac{1 - 2\nu}{2 - 2\nu}$$

The solution can be approximated by the following expression:

$$C_R = \frac{0.862 + 1.14\nu}{1 + \nu} C_s$$

For steel ($\nu = 0.3$) the Rayleigh wave velocity is 92.6% of the distortional shear velocity.

The Rayleigh wave attenuates exponentially [3] as a function of distance from the surface, and this decay is dependent on its wavelength. For a harmonic (sinusoidal) wave with frequency $p/2\pi$ and wavelength $\Lambda = 2\pi/f$, the velocity is given by

$$C_R = \frac{p}{f}$$

For a material with $\nu = 0.25$, the amplitude u_3 of the vertical (normal to surface) component of particle displacement varies as (x_3 is the distance perpendicular to the boundary surfaces)

$$\frac{u_3}{u_{3s}} = -1.37[\exp(-0.847fx_3) - 1.73\exp(-0.39fx_3)] \qquad (2.25)$$

where u_{3s} is the amplitude at the surface. Figure 2.8 (left side) shows the decay of this amplitude. At first (from $x_3 = 0$ to $x_3 = 0.1\Lambda$), there is an increase in amplitude, followed by an exponential decay. When $x_3 \simeq \Lambda$, the amplitude has been reduced to 20% of the surface amplitude u_{3s}. The horizontal (parallel to bounding surface component decays more rapidly with x_3 and changes sign (see right-hand side of Fig. 2.8). The particle displacements u_1 and u_3 (parallel and perpendicular to bounding surface, respectively) describe an elliptical trajectory. The component perpendicular to the surface is larger than the one parallel to the surface. For $\nu = 0.25$, $u_{35} \simeq 1.5u_{1s}$.

Example 2.1 Determine the velocities of longitudinal (for a slim bar and in an infinite body) and shear elastic waves for the following materials:

Uranium
Copper
Aluminum
Iron
Alumina

Assume $\nu = 0.3$ for the values that you do not find.

$$C_0 = \left(\frac{E}{\rho}\right)^{1/2} \quad C_l = \left(\frac{\lambda + 2\mu}{\rho}\right)^{1/2} \quad C_s = \left(\frac{\mu}{\rho}\right)^{1/2}$$

PROPERTY

Material	E (GPa)	ρ (kg/m^3)	λ (GPa)	μ (GPa)	ν
Uranium	172.0	18,950.0	99.2	66.1	0.3
Copper	129.8	8,930.0	105.6	48.3	0.343
Aluminum	70.3	2,700.0	58.2	26.1	0.345
Iron	211.4	7,850.0	115.7	81.6	0.293
Alumina	365.0	3,900.0	210.6	140.4	0.3

WAVE VELOCITY (m/s)

	Longitudinal Wave		*Shear Wave*
Material	Slim Bar (C_0)	Infinite Body (C_l)	(C_s)
Uranium	3,012.7	3,494.4	1,867.6
Copper	3,812.5	4,758.4	2,325.6
Aluminum	5,102.6	6,394.4	3,109.1
Iron	5,189.4	5,960.6	3,224.1
Alumina	9,674.2	11,225.0	6,000.0

Example 2.2 A small charge is detonated at point A. An observer with a seismograph is located at B. Determine the times at which the various signals from the detonation will reach the seismograph and qualitatively present the nature of the displacement.

We establish the paths of the waves. The sketch below shows the two paths: direct and reflected. We need to calculate the velocities of the longitudinal, shear, and Rayleigh waves along the paths:

$$C_R = kC_s$$

$$k = 0.9422$$

$$C_s = \sqrt{\frac{\mu}{\rho}}$$

$$\lambda = \frac{E\nu}{(1 + \nu)(1 - 2\nu)} = 142.8 \text{ GPa}$$

$$\mu = \frac{E}{2(1 + \nu)} = 35.7 \text{ GPa}$$

$$C_s = 3706 \text{ m/s} \qquad C_l = 9077 \text{ m/s}$$

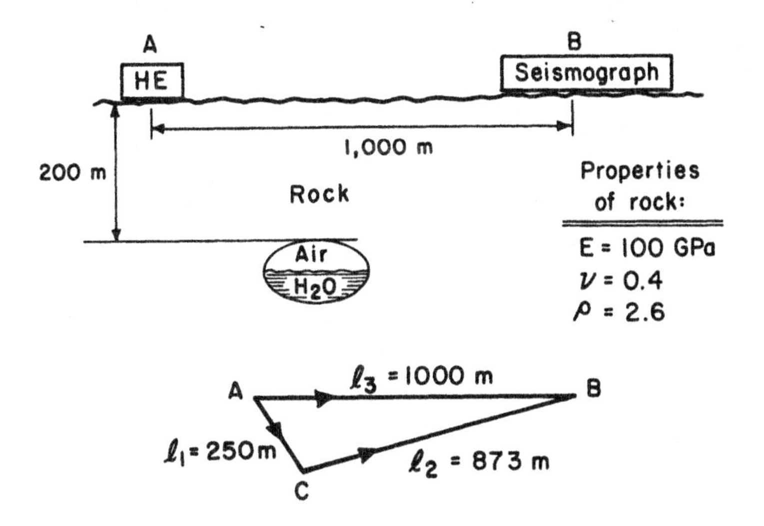

We now compute the times:

Wave	Time (s)
Longitudinal	0.11
Reflected longitudinal	0.12
Shear	0.27
Reflected shear	0.30
Rayleigh	0.29

2.7 ELASTIC WAVES: INDICIAL NOTATION

The indicial notation simplifies the derivation of the wave equation. Stresses σ_{11}, σ_{12}, σ_{13}, σ_{23}, σ_{22}, and σ_{33} are represented by one symbol, σ_{ij}, where $i, j =$ 1, 2, 3. When a subscript (or index) appears twice in the same term, it is implied that summation with respect to this index occurs. For instance, Eqns. (2.10a), (2.10b), and (2.10c) can be represented by Eqn. (2.11):

$$\frac{\partial \sigma_{ij}}{\partial x_j} = \rho \frac{\partial^2 u_i}{\partial t^2} \tag{2.26}$$

The generalized constitutive equations for a linear isotropic material are given by, in indicial notation (see Fung [5, Eqn. 8.5-2]),

$$\sigma_{ij} = \lambda \varepsilon_{kk} \delta_{ij} + 2\mu \varepsilon_{ij} \qquad \delta_{ij} = \begin{cases} 0 & i \neq j \\ 1 & i = j \end{cases} \tag{2.27}$$

$$\varepsilon_{kk} = \varepsilon_{11} + \varepsilon_{22} + \varepsilon_{33} = \frac{\partial u_i}{\partial x_i} \equiv \Delta$$

Then Eqn. (2.26) can be rewritten as

$$\lambda\delta_{ij}\frac{\partial\Delta}{\partial x_j} + 2\mu\frac{\partial\varepsilon_{ij}}{\partial x_j} = \rho\frac{\partial^2 u_i}{\partial t^2}$$

Since $\delta_{ij} = 0$ when $i \neq j$,

$$\lambda\delta_{ij}\frac{\partial\Delta}{\partial x_j} = \lambda\frac{\partial\Delta}{\partial x_i}$$

and since

$$\varepsilon_{ij} = \frac{1}{2}\left(\frac{\partial u_i}{\partial x_j} + \frac{\partial u_j}{\partial x_i}\right)$$

we have

$$\lambda\frac{\partial\Delta}{\partial x_i} + \mu\left(\frac{\partial^2 u_i}{\partial x_j\,\partial x_j} + \frac{\partial^2 u_j}{\partial x_i\,\partial x_j}\right) = \rho\frac{\partial^2 u_i}{\partial t^2} \tag{2.28}$$

$$\lambda\frac{\partial\Delta}{\partial x_i} + \mu\left(\frac{\partial^2 u_i}{\partial x_j\,\partial x_j} + \frac{\partial\Delta}{\partial x_i}\right) = \rho\frac{\partial^2 u_i}{\partial t^2} \tag{2.29}$$

This is the field equation of momentum conservation in indicial notation. Differentiating with respect to x_i and inverting the differentiation order yields

$$\lambda\frac{\partial^2\Delta}{\partial x_i\,\partial x_i} + \mu\frac{\partial^2\Delta}{\partial x_j\,\partial x_j} + \mu\frac{\partial^2\Delta}{\partial x_i\,\partial x_i} = \rho\frac{\partial^3 u_i}{\partial t^2\,\partial x_i} = \rho\frac{\partial^2\Delta}{\partial t^2}$$

Exchanging j for i in the second term yields

$$(\lambda + 2\mu)\frac{\partial^2\Delta}{\partial x_i\,\partial x_i} = \rho\frac{\partial^2\Delta}{\partial t^2}$$

We already know [from Eqn. (2.14)] that

$$\frac{\partial^2}{\partial x_i\,\partial x_i} = \nabla^2$$

Thus

$$(\lambda + 2\mu)\nabla^2\Delta = \rho\frac{\partial^2\Delta}{\partial t^2} \tag{2.30}$$

We recognize this as the wave equation for a dilatation Δ. To obtain the shear wave velocity, we simply make the dilatation equal to zero in the field equation [Eqn. (2.29)] and arrive at

$$\mu\frac{\partial^2 u_i}{\partial x_j\,\partial x_j} = \rho\frac{\partial^2 u_i}{\partial t^2}$$

This is the equation for a disturbance propagating at

$$V_s = \left(\frac{\mu}{\rho}\right)^2$$

Alternatively, we can express the rotations (rigid body)

$$\omega_{ij} = \frac{1}{2}\left(\frac{\partial u_i}{\partial x_j} - \frac{\partial u_j}{\partial x_i}\right)$$

For this, we take the equation for conservation of momentum [Eqn. (2.29)] and differentiate it with respect to x_k:

$$(\lambda + \mu)\frac{\partial^2 \Delta}{\partial x_k\,\partial x_i} + \mu\frac{\partial^3 u_i}{\partial x_j\,\partial x_j\,\partial x_k} = \rho\frac{\partial^3 u_i}{\partial x_k\,\partial t^2}$$

We take the equation for conservation of momentum [Eqn. (2.29)] expressed in terms of u_k and differentiate it with respect to x_i:

$$(\lambda + \mu)\frac{\partial^2 \Delta}{\partial x_i\,\partial x_k} + \mu\frac{\partial^3 u_k}{\partial x_i\,\partial x_j\,\partial x_j} = \rho\frac{\partial^3 u_k}{\partial x_i\,\partial t^2}$$

We subtract one from the other:

$$\mu\frac{\partial^2}{\partial x_j\,\partial x_j}\left(\frac{\partial u_i}{\partial x_k} - \frac{\partial u_k}{\partial x_i}\right) = \rho\frac{\partial}{\partial t^2}\left(\frac{\partial u_i}{\partial x_k} - \frac{\partial u_k}{\partial x_i}\right)$$

$$\mu\frac{\partial^2}{\partial x_j\,\partial x_j}\omega_{kj} = \rho\frac{\partial}{\partial t^2}\omega_{kj} \qquad (2.31)$$

These two velocities, called phase velocities, are obtained much more rapidly using the indicial notation.*

2.8 WAVE REFLECTION, REFRACTION, AND INTERACTION

We will briefly describe the interaction of waves when they encounter a boundary. We will follow a very simple approach in order to better familiarize the reader with the fundamentals. Figure 2.10 shows the longitudinal waves that

*These equations do not include any scale parameters of material structure which are important in real materials. There are attempts, such as the Cosserat model (in "Mechanics of Micropolar Media," O. Brulin and R. K. T. Hsieh. World Scient., Singapore, 1992) which include mesoscale level structure parameters.

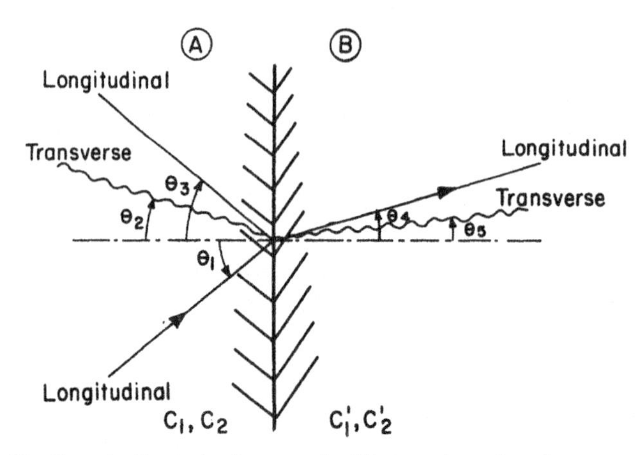

FIGURE 2.10 Longitudinal elastic wave inciding on boundary between media A and B, generating refracted and reflected waves. C_1 and C_2 are the longitudinal and shear wave velocities in medium A; C_1' and C_2' are the respective velocities in medium B.

are reflected and refracted at the boundary as well as the two transverse waves that are generated at the interface. These effects—reflection and refraction—occur when the wave encounters a medium with different sonic impedance. The sonic, or sound, impedance is defined as the product of the medium density by its sound (or elastic) wave velocity. These refraction and reflection angles are given by a simple relationship of the form

$$\frac{\sin \theta_1}{C_1} = \frac{\sin \theta_2}{C_2} = \frac{\sin \theta_3}{C_1} = \frac{\sin \theta_4}{C_1'} = \frac{\sin \theta_5}{C_2'} \tag{2.32}$$

This is identical to Lenz's law in electricity. The amplitudes of these waves are given by, for instance, Rinehart [4].

The interactions of a wave with an interface are much simpler when the incidence is normal ($\theta = 0$). In this case, a longitudinal wave refracts (transmits) and reflects longitudinal waves, and the shear wave refracts (transmits) and reflects shear waves. Figure 2.11(a) shows the front of a wave propagating along a cylinder with cross-sectional area A in a medium in which the wave velocity is C_A. The particle velocity is U_p and the stress is σ. Figure 2.11(b) shows an interface and the forces due to incident, transmitted, and reflected waves; Figure 2.11(c) shows the particle velocities of the incident, transmitted, and reflected waves. We can calculate the amplitudes of the transmitted and reflected waves from the densities and wave velocities of the two media. First, we will derive an expression relating the uniaxial stress σ to particle and wave velocities. The conservation-of-momentum relationship states (a segment of length dx is indicated in Fig. 2.11a)

$$F\,dt = d(mU_p) \quad \text{(impulse = change in momentum)}$$

$$\sigma A\,dt = \rho A\,dx\,U_p$$

$$\sigma = \rho \frac{dx}{dt} U_p$$

$$\sigma = \rho C U_p \tag{2.33}$$

We consider a linear material where the sonic impulses propagate independently. When the longitudinal wave encounters an interface [Figs. 2.11(b) and (c)], we have the formation of a reflected and a transmitted wave. If we look at the interface and consider it to be in equilibrium under the three stresses σ_I (incident), σ_T (transmitted), and σ_R (reflected), we have

$$\sigma_I + \sigma_R = \sigma_T \tag{2.34}$$

For the continuity at the interface (no gaps can be created and matter cannot superimpose itself),

$$U_{pI} + U_{pR} = U_{pT} \tag{2.35}$$

The signs of the stresses and particle velocities in Figure 2.11 were set in such a manner that medium A has a higher impedance than medium B. The particle velocities are obtained from Eqn. (2.33):

$$U_{pI} = \frac{\sigma_I}{\rho_A C_A} \qquad U_{pT} = \frac{\sigma_T}{\rho_B C_B} \qquad U_{pR} = \frac{-\sigma_R}{\rho_A C_A} \tag{2.36}$$

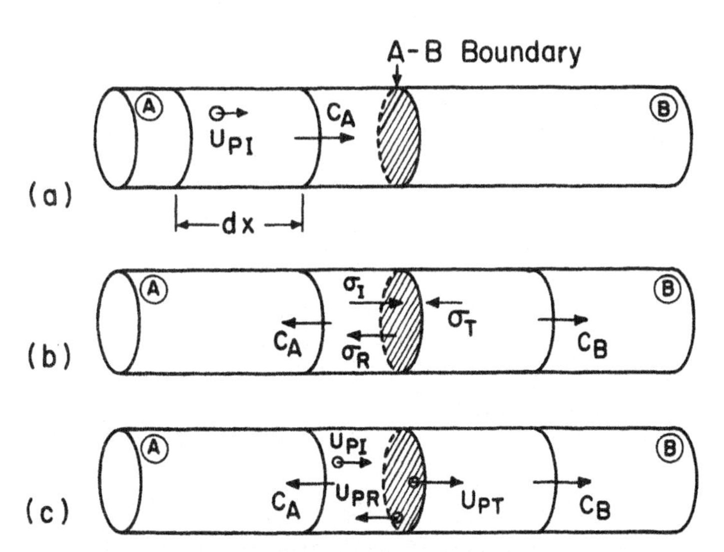

FIGURE 2.11 Longitudinal wave inciding on boundary between two media A and B in normal trajectory: (a) prior to encounter with boundary; (b) forces exerted on boundary (equilibrium condition); (c) particle velocities (continuity). Direction of arrows for reflected wave for case impedance *A* > impedance *B*.

The sign of the reflected particle velocity was inverted because a positive stress causes a negative particle velocity upon reflection. This will become clearer later. Substituting Eqns. (2.36) into Eqn. (2.35) yields

$$\frac{\sigma_I}{\rho_A C_A} - \frac{\sigma_R}{\rho_A C_A} = \frac{\sigma_T}{\rho_B C_B} \tag{2.37}$$

Equations (2.34) and (2.37) provide

$$\frac{\sigma_T}{\sigma_I} = \frac{2\rho_B C_B}{\rho_B C_B + \rho_A C_A}$$

$$\frac{\sigma_R}{\sigma_I} = \frac{\rho_B C_B - \rho_A C_A}{\rho_B C_B + \rho_A C_A} \tag{2.38}$$

It is now clear that the impedances of the materials—ρC product—determine the amplitude of transmitted and reflected pulses. When $\rho_B C_B > \rho_A C_A$, a pulse of the same sign as the incident pulse is reflected. When $\rho_B C_B < \rho_A C_A$, a pulse of opposite sign is reflected. The two limiting conditions, a wave incident on a free surface ($E = 0$) and a wave incident on a rigid boundary ($E = \infty$), will be analyzed below.

Free surface, $\rho_B C_B = 0$:

$$\frac{\sigma_T}{\sigma_I} = 0 \qquad \frac{\sigma_R}{\sigma_I} = -1$$

Rigid boundary ($E = \infty$, leading to $C = \infty$):

$$\frac{\sigma_T}{\sigma_I} = \frac{2}{1 + \dfrac{\rho_A C_A}{\rho_B C_B}} \simeq 2$$

$$\frac{\sigma_R}{\sigma_I} = \frac{1 - \rho_A C_A/\rho_B C_B}{1 + \rho_A C_A/\rho_B C_B} \simeq 1$$

Expressions for the particle velocities (transmitted and reflected) can be similarly obtained by using Eqns. (2.34) and (2.36). One obtains

$$\frac{U_{pR}}{U_{pI}} = \frac{\rho_A C_A - \rho_B C_B}{\rho_A C_A + \rho_B C_B}$$

$$\frac{U_{pT}}{U_{pI}} = \frac{2\rho_A C_A}{\rho_A C_A + \rho_B C_B} \tag{2.39}$$

For a free surface,

$$\frac{U_{pT}}{U_{pI}} = 2 \quad \text{and} \quad \frac{U_{pR}}{U_{pI}} = 1$$

For a rigid boundary

$$\frac{U_{pT}}{U_{pI}} = 0 \qquad \frac{U_{pR}}{U_{pI}} = \frac{\rho_A C_A/\rho_B C_B - 1}{\rho_A C_A/\rho_B C_B + 1} = -1$$

Figure 2.12 shows sequences of stresses and particle velocities for rigid and free boundaries. It is seen that when a compressive wave encounters a free

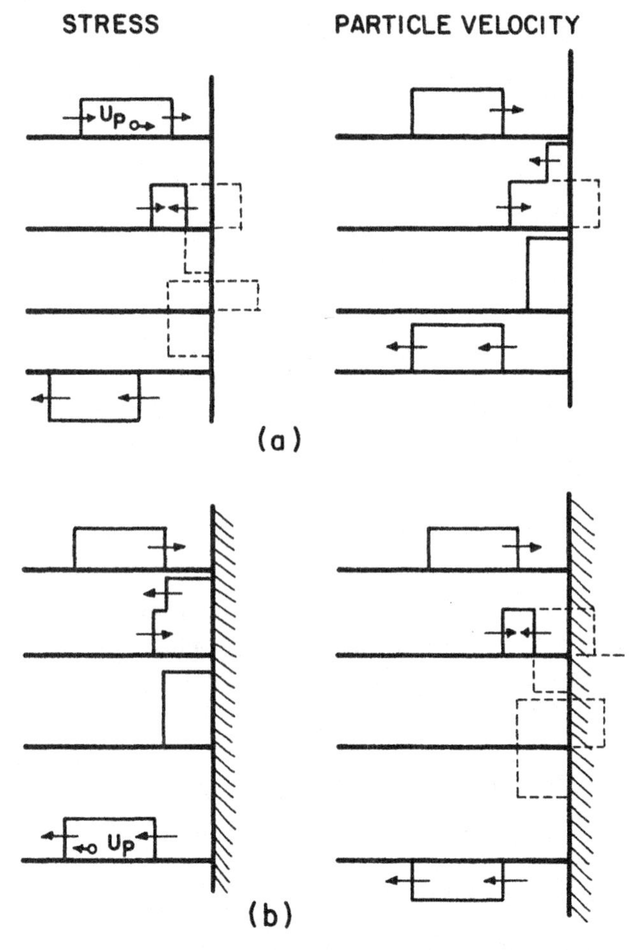

FIGURE 2.12 Reflection of a stress pulse (rectangular) at (a) free surface and (b) rigid boundary.

surface, it reflects back as a tensile wave and vice versa; the stress sign is changed, while the particle velocity is maintained. When a rigid boundary is encountered, the opposite occurs. The stress sign is maintained and the particle velocity direction is reversed. One is reminded that if the particle velocity and wave velocity have the same direction and sense, the stress is compressive; if they have opposite senses, the stress is tensile. The derivation above is the simplest presentation of reflected and refracted waves. We can solve simple shapes, such as rectangular and triangular waves. The general solution of the problem involves the method of characteristics. Prior to the advent of large hydrocodes, the method of characteristics was the major technique to handle complex problems. In Section 2.12 we will give a simple description of this method in order to familiarize the reader with the basic concepts. The wave equation [Eqns. (2.4), (2.16), (2.21)] is a second-order partial differential equation whose solution we are trying to obtain for a wide array of boundary conditions. The solution can be obtained through characteristic curves; the method of characteristics is a way of solving hyperbolic partial differential equations.

2.9 GENERAL SOLUTION OF WAVE EQUATION

The wave equation shown in Eqns. (1.12), (1.13), (2.4), (2.16), (2.21), (2.30), and (2.31) is a second-order partial differential equation. In Chapter 1, we assumed a sine shape, whereas earlier in Chapter 2 we implicitly accepted that the square of the velocity of the pulse was given by the parameter that relates the two second-order partial derivatives. Nevertheless, the equation has a general solution that applies to pulses of general shape propagating, as will be seen from the general solution given below. We will solve the equation for a uniaxial displacement u:

$$\frac{\partial^2 u}{\partial t^2} = C_0^2 \frac{\partial^2 u}{\partial x^2}$$

This is a linear, homogeneous, second-order partial differential equation. Up to now, we have only solved it in Chapter 1, assuming a harmonic waveform. We will now develop a general solution for it. Partial differential equations of the form

$$A \frac{\partial^2}{\partial x^2} + B \frac{\partial^2}{\partial x\, \partial t} + C \frac{\partial^2}{\partial t^2} + \cdots = 0$$

are classified into

$$B^2 - 4AC > 0 \quad \text{(hyperbolic)}$$

$$B^2 - 4AC = 0 \quad \text{(parabolic)}$$

$$B^2 - 4AC < 0 \quad \text{(elliptic)}$$

In our case,

$$C_0^2 \frac{\partial^2 u}{\partial x^2} - \frac{\partial^2 u}{\partial t^2} = 0$$

$$A = C_0^2 \qquad B = 0 \qquad C = -1$$

$$B^2 - 4AC = 4C_0^2 > 0$$

Thus, this is a hyperbolic equation, and there are two principal methods of solution: separation of variables and transformation. The method of separation of variables is more appropriate for standing waves, whereas the method of transformations is better suited for traveling waves. The transformation method (also called the method of characteristics), originally developed by Riemann and Hadamard, transforms the equation to a new set of variables. This transformation is often obtained by experience, although it can also be formally obtained. For the standing-wave solution, the separation of variables for the harmonic wave can be expressed as

$$u(x, t) = u_0 \left(\sin \frac{n\pi x}{l} \cos \frac{n\pi C_0 t}{l} \right)$$

where l is a characteristic length and C_0 is the wave velocity. We can express the trigonometric function as

$$\sin \frac{n\pi x}{l} \cos \frac{n\pi C_0 t}{l} = \frac{1}{2} \left[\sin \frac{n\pi}{l} (x - C_0 t) + \sin \frac{n\pi}{l} (x + C_0 t) \right]$$

Thus, the displacement u can be expressed as

$$u(x, t) = \frac{u_0}{2} \sin \frac{n\pi}{l} (x - C_0 t) + \frac{u_0}{2} \sin \frac{n\pi}{l} (x + C_0 t)$$

Generalizing the above equation, assuming nonharmonic functions F and G, one would have

$$u(x, t) = F(x - C_0 t) + G(x + C_0 t)$$

We will verify whether the variables $x + C_0 t$ and $x - C_0 t$ will provide a correct solution to the equation. We make the following substitution of variables:

$$\xi = x + C_0 t \qquad \eta = x - C_0 t \tag{2.40}$$

We now express $du(\xi, \eta)$ and $\partial u/\partial x$:

$$du = \left(\frac{\partial u}{\partial \xi}\right) d\xi + \left(\frac{\partial u}{\partial \eta}\right) d\eta$$

$$\frac{\partial u}{\partial x} = \left(\frac{\partial u}{\partial \xi}\right) \frac{\partial \xi}{\partial x} + \left(\frac{\partial u}{\partial \eta}\right) \frac{\partial \eta}{\partial x}$$

But

$$\frac{\partial \xi}{\partial x} = \frac{\partial x}{\partial x} + C_0 \frac{\partial t}{\partial x} = 1$$

$$\frac{\partial \eta}{\partial x} = \frac{\partial x}{\partial x} - C_0 \frac{\partial t}{\partial x} = 1$$

Thus

$$\frac{\partial u}{\partial x} = \frac{\partial u}{\partial \xi} + \frac{\partial u}{\partial \eta}$$

$$\frac{\partial^2 u}{\partial x^2} = \frac{\partial}{\partial x}\left(\frac{\partial u}{\partial \xi} + \frac{\partial u}{\partial \eta}\right) = \left(\frac{\partial}{\partial \xi} + \frac{\partial}{\partial \eta}\right)\left(\frac{\partial u}{\partial \xi} + \frac{\partial u}{\partial \eta}\right)$$

$$= \frac{\partial^2 u}{\partial \xi} + 2\frac{\partial^2 u}{\partial \xi\, \partial \eta} + \frac{\partial^2 u}{\partial \eta^2} \tag{2.41}$$

The same is done for $\partial^2 u/\partial t^2$:

$$\frac{\partial u}{\partial t} = \left(\frac{\partial u}{\partial \xi}\right) \frac{\partial \xi}{\partial t} + \left(\frac{\partial u}{\partial \eta}\right) \frac{\partial \eta}{\partial t}$$

From Eqn. (2.40)

$$\frac{\partial \xi}{\partial t} = \frac{\partial x}{\partial t} + C_0 \frac{\partial t}{\partial t} = C_0$$

$$\frac{\partial \eta}{\partial t} = \frac{\partial x}{\partial t} - C_0 \frac{\partial t}{\partial t} = -C_0$$

Thus

$$\frac{\partial u}{\partial t} = C_0 \frac{\partial u}{\partial \xi} - C_0 \frac{\partial u}{\partial \eta}$$

$$\frac{\partial^2 u}{\partial t^2} = C_0 \frac{\partial}{\partial t}\left(\frac{\partial u}{\partial \xi} - \frac{\partial u}{\partial \eta}\right)$$

$$= C_0^2 \left(\frac{\partial}{\partial \xi} - \frac{\partial}{\partial \eta}\right)\left(\frac{\partial u}{\partial \xi} - \frac{\partial u}{\partial \eta}\right)$$

$$\frac{\partial^2 u}{\partial t^2} = C_0^2 \left(\frac{\partial^2 u}{\partial \xi^2} - 2\frac{\partial^2 u}{\partial \xi\, \partial \eta} + \frac{\partial^2 u}{\partial \eta^2}\right) \tag{2.42}$$

Inserting Eqns. (2.41) and (2.42) into the wave equation yields

$$C_0^2 \left(\frac{\partial^2 u}{\partial \xi^2} - 2\frac{\partial^2 u}{\partial \xi\, \partial \eta} + \frac{\partial^2 u}{\partial \eta^2}\right) = \left(\frac{\partial^2 u}{\partial \xi^2} + 2\frac{\partial^2 u}{\partial \xi\, \partial \eta} + \frac{\partial^2 u}{\partial \eta^2}\right) C_0^2$$

$$\frac{\partial^2 u}{\partial \xi\, \partial \eta} = 0 \tag{2.43}$$

The general solution to this equation is

$$u(\eta, \xi) = F(\eta) + G(\xi)$$

where F and G are two functions; one would arrive, by differentiating twice, at Eqn. (2.43):

$$\frac{\partial u}{\partial \eta} = F'(\eta)$$

So

$$\frac{\partial}{\partial \xi}\left(\frac{\partial u}{\partial \eta}\right) = 0$$

Thus, the general solution to the wave equation is

$$u(x, t) = F(x - C_0 t) + G(x + C_0 t) \tag{2.44}$$

The physical meaning of this equation is that we have functions F and G that describe the shape of pulses propagating in the positive and negative directions, along the x axis, respectively, at a velocity C_0. The shapes of these waves are unchanged with time, and they propagate as shown in Figure 2.13. If we only have one arm propagating in the positive x direction, we have $G = 0$. Figure 2.13 shows the wave profiles at two times t_1 and t_2.

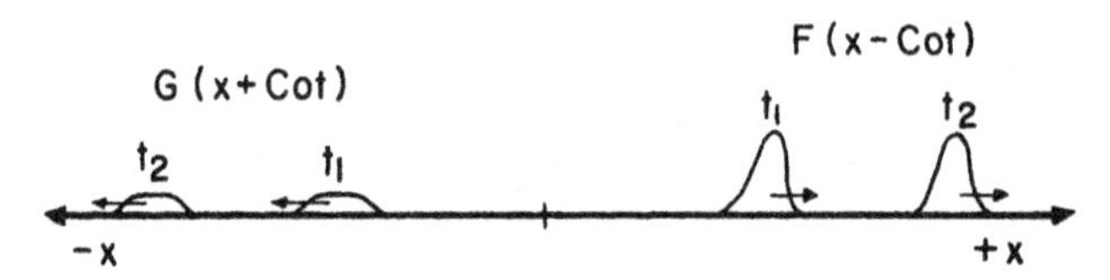

FIGURE 2.13 General solution for wave equation in uniaxial stress; both components traveling in $+x$ and $-x$ directions shown.

2.10 ELASTIC WAVES IN CYLINDRICAL BARS: ADDITIONAL CONSIDERATIONS

It is very important to understand the propagation of elastic waves in cylindrical bars. This is required for the analysis of Hopkinson bar experiments. The simplest analysis, presented in Section 2.2, is not really accurate. When one impacts a cylindrical bar with a cylindrical projectile of length L, one would expect a rectangular pulse of length $2L$ propagating through the bar if the bar and projectile are of the same material. This is shown schematically in Figure 2.14. The impact produces compressive waves propagating at velocities C_0 into projectile and target. As the compressive wave reaches the end of the projectile, it reflects back, as shown in Figure 2.12. This determines the length of the pulse, $\Lambda = 2L$. The velocity of the interface, equal to the particle velocity, can be calculated from the conservation of momentum equation (momentum prior to impact equals momentum after impact):

Prior to impact: $\rho_0 A_0 L V$.
After impact: $\rho_0 A_0 \Lambda U_p = 2\rho_0 A_0 L U_p$.

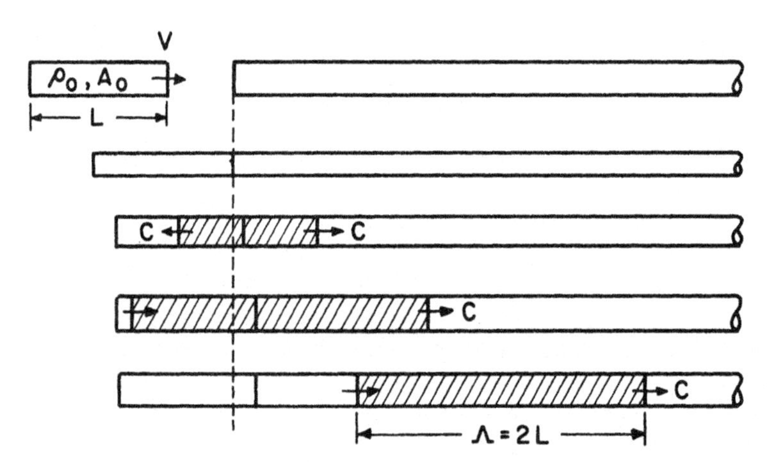

FIGURE 2.14 Cylindrical projectile impacting cylindrical bar.

Thus

$$U_p = \frac{V}{2} \tag{2.45}$$

The stress generated by the impact at a velocity V is given by Eqn. (2.33); substituting in Eqn. (2.45) yields

$$\sigma = \rho C U_p = \tfrac{1}{2} \rho C V \tag{2.46}$$

The expected pulse shape is rectangular, with a duration (at a fixed point) of $2L/C$ [Fig. 2.15(a)]. Gages placed on the cylinder surface allow us to obtain reliable records of the stress pulses. These pulses show significant fluctuations [Fig. 2.15(b)]. The nature of these fluctuations will be discussed. These are called "dispersion" effects; one can also see a decrease in the slope of the rise of the wave. These effects have a significant bearing on the interpretation of split Hopkinson bar experimental results. The equation developed in Section 2.2 [$C_0 = (E/\rho)^{1/2}$] is based on uniaxial stress. In a real case, one has radial inertia and a wave interaction with the external surfaces of the cylinder (free surfaces). Radial inertia is caused by the kinetic energy of the material flowing radially outward as the bar is compressed. The first calculation of the elastic wave velocity in cylindrical bars is due to Pochammer [7] and Chree [8, 9]. Rayleigh [10] also calculated the velocity of a longitudinal wave in cylindrical bars and obtained the expression

$$\frac{C_p}{C_0} = 1 - 3\nu^2\pi^2 \left(\frac{a}{\Lambda}\right)^2 \tag{2.47}$$

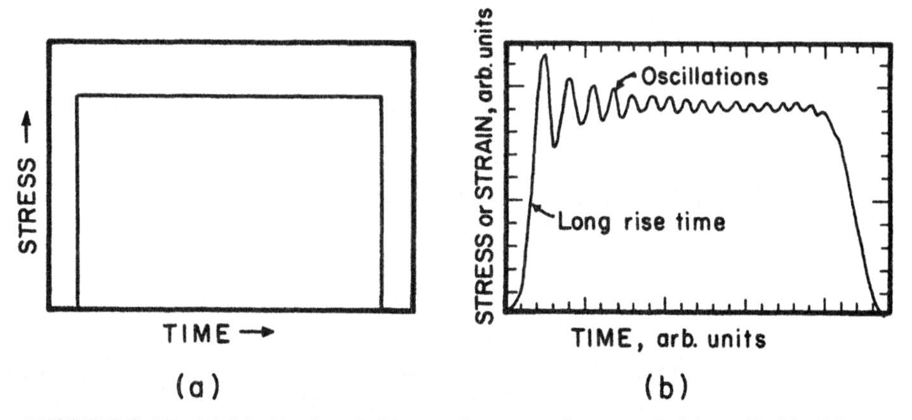

FIGURE 2.15 (a) Idealized and (b) actual stress pulse recorded in cylindrical bar.

where a is the bar radius and C_p is the group velocity. It can be seen that C_p tends to unrealistic values as Λ decreases. This is not a realistic prediction, and the results of the Pochhammer–Chree predictions are better. Figure 2.16 shows the predictions of the two theories for a value of $\nu = 0.29$, as calculated by Davies [11]. It can be seen that $C_p/C_0 = 1$ for $a/\Lambda \to 0$, that is, when the wavelength is long with respect to the bar radius r. For values of $a/\Lambda < 0.1$, the Rayleigh correction is fairly accurate. For larger values of a/Λ, there is a greater and greater deviation from C_0, and the two theories differ markedly. The Pochhammer–Chree theory predicts a plateau for C_p/C_0 at $\sim 0.58C_0$. This is the Rayleigh surface wave velocity C_s, as shown in Figure 2.16.

The fluctuations at the peak of the wave shown in Figure 2.15(b) are easily understood if one looks at Figure 2.17(a), which shows a wave propagating in a cylinder. Release waves are generated at the free surfaces and trail the main wave. These release waves interact continuously and cause fluctuations of particle velocity (and, consequently, of stress and strain) at the surface of the cylinder. When $a/\Lambda < 1$, the displacements and stresses along the cross section of the bar can be assumed to be constant. For $a/\Lambda < 0.1$, the differ-

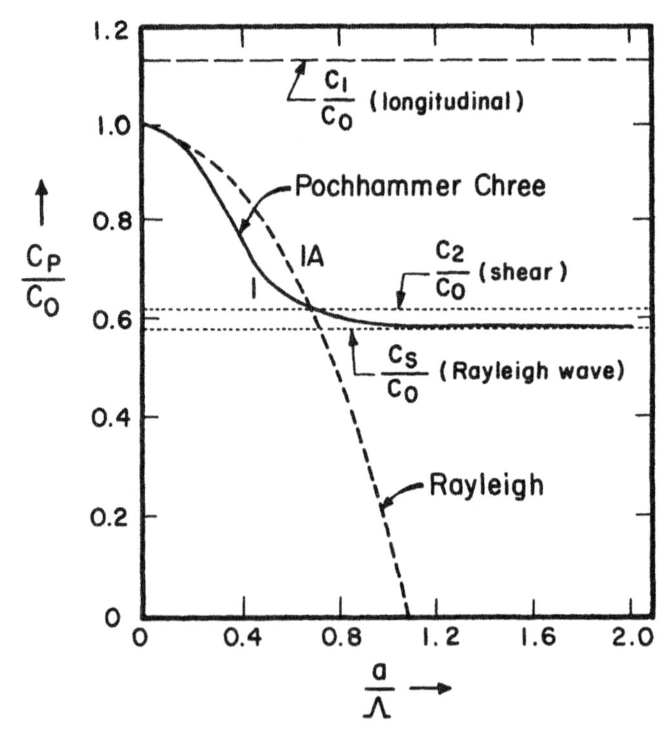

FIGURE 2.16 Phase velocity as a function of the ratio between cylindrical bar diameter a and pulse length Λ. On the right-hand side, the longitudinal, shear, and Rayleigh velocities for unbounded media are shown. (From Kolsky [1], Fig. 14, p. 61. Reprinted with permission of the publisher.)

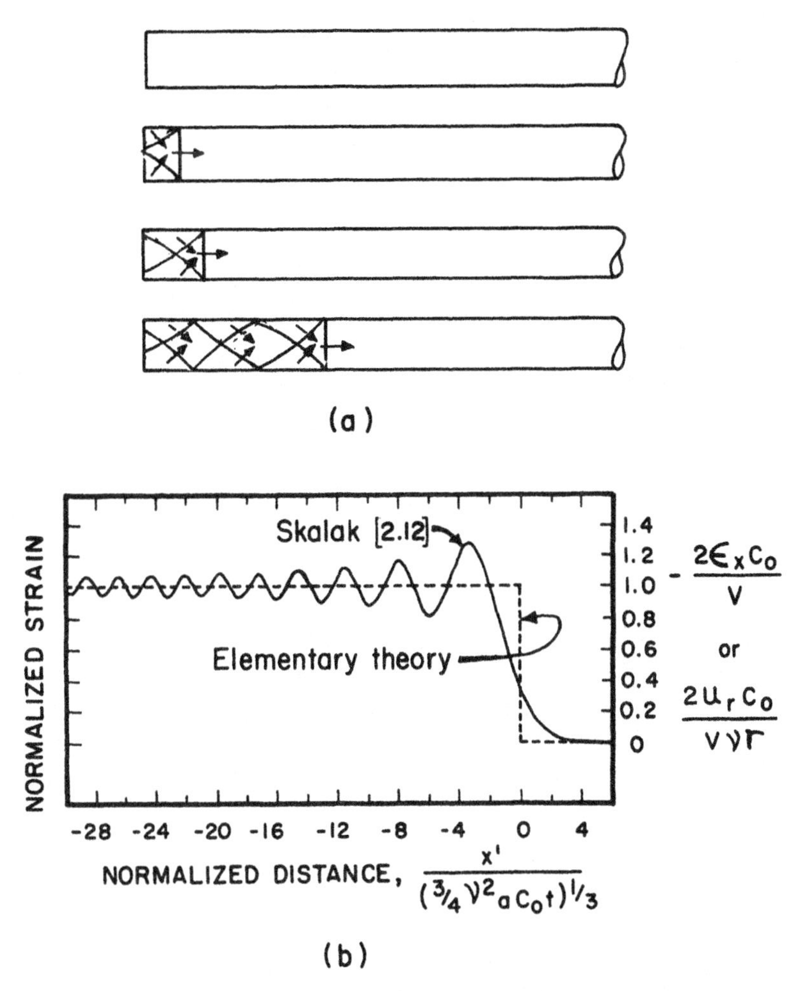

FIGURE 2.17 (a) Longitudinal wave propagation in cylindrical elastic bar showing effects of free boundaries; (b) comparison of idealized and calculated stress pulse shape. (From Skalak [12], p. 63, Fig. 5. Reprinted with permission of the publisher.)

ences are less than 5%. Figure 2.17(b) shows the prediction of the stress pulse produced by impact according to Skalak [12]. Riding on a rectangular pulse are the oscillations; the rise time is also increased. Thus, computations can predict the stress-pulse shape in cylindrical specimens. Skalak [12] used the method of double-integral transforms to solve the problem of the impact of a cylindrical projectile on a cylindrical rod. First, he assumed that elastic deformation in the rod occurred without any lateral strain. This is equivalent to the solution for the wave equation in an unbounded medium, and the velocity is given by Eqn. (2.17). This required the application of lateral restraints (perpendicular to cylinder axis) to retain the dimension of the cylinder. The second part of the problem was to apply tractions exclusively along the lateral surface of the cylinder of a magnitude such as to cancel the restraints from the first part. These tractions started at the impact region and propagated along the surface. The superposition of the two parts provides the general solution, which is schematically shown in Figure 2.17(b). The predominant wavelength of the dominant disturbance is obtained from [12]:

$$x' = x - C_0 t - 4\sqrt[3]{\tfrac{3}{4}\,\nu^2 a C_0 t}$$

The coordinate x' has its origin at the front of the wave, whereas x has origin at the impact interface. The initial amplitude of the oscillation is approximately $\pm 20\%$ of the total stress (or displacement). These oscillations decrease in amplitude and wavelength as the distance from the front increases. Shown in the ordinate are normalized displacements, obtained through Eqn. (2.46):

$$\sigma_x = \rho C_0 U_p = \varepsilon_x E$$

$$\varepsilon_x = \frac{U_p}{C_0} = \frac{V}{2C_0}$$

Thus, $2\varepsilon_x C_0 / V = 1$ is the normalization factor (notice that Skalak [12] used a symmetric impact with each component propagating at a velocity V, whereas results were adapted, in Figure 2.17, to our conditions). The radial displacements are obtained from the longitudinal displacements through

$$U_r = -r\nu\varepsilon_x$$

When the cross-sectional area of a cylindrical bar is changed, one has a geometric impedance difference, although the intrinsic impedance of the material is unaltered. By applying the equations for force and particle velocity balance to the interface, one can arrive easily at expressions predicting the stress and particle velocity in the new cross section. Figure 2.18 shows a cylindrical specimen in which the cross section is reduced from A_1 to A_2. The derivation of the equations is analogous to the ones where one has intrinsic impedance changes. This analysis is important in Hopkinson bars, where we sandwich ceramic and metal specimens between high-strength steel (Maraging

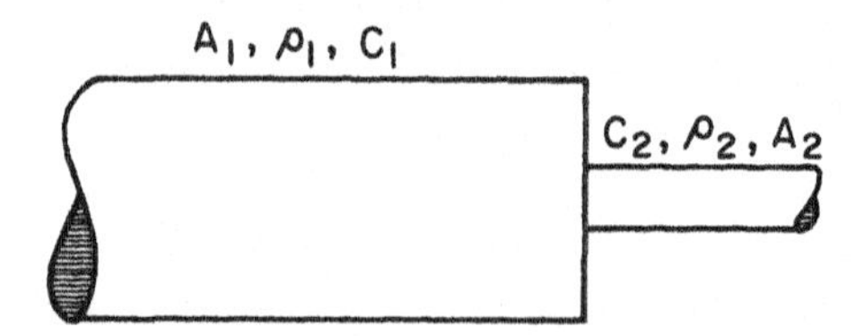

FIGURE 2.18 Change in cross-sectional area in cylindrical bar.

steel) cylindrical bars. The diligent student can easily show that

$$\sigma_T = \frac{2A_1\rho_2C_2}{A_1\rho_1C_1 + A_2\rho_2C_2}\sigma_I \tag{2.48}$$

$$\sigma_R = \frac{A_2\rho_2C_2 - A_1\rho_1C_1}{A_1\rho_1C_1 + A_2\rho_2C_2}\sigma_I \tag{2.49}$$

when the two materials are the same, $\rho_1 = \rho_2$ and $C_1 = C_2$:

$$\sigma_T = \frac{2A_1}{A_1 + A_2}\sigma_I \tag{2.50}$$

$$\sigma_R = \frac{A_2 - A_1}{A_1 + A_2}\sigma_I \tag{2.51}$$

2.11 SPHERICAL STRESS WAVES

Spherical stress waves are important in explosions, when a "point" source creates a spherically expanding stress pulse. Imploding spherical stress waves also have technological importance, and the detonation of nuclear devices may occur by spherical imploding stress waves. In mining and civil engineering, underground explosive charges are detonated to break the rock. These waves produce distinct patterns of rock breakage, and the careful analysis of the stress waves enables understanding and prediction of the fracture pattern. In Chapter 17 we will briefly discuss the damage due to industrial explosives in rock. The underground charges also produce surface waves, which may result in damage to structures.

The basic equation for a spherical wave is obtained from the general wave equation for a disturbance Ψ by changing the coordinates from Cartesian to polar. For a centrosymmetric problem r is the only dimension to consider, and

$$r^2 = x_1^2 + x_2^2 + x_3^2$$

We will transform equation

$$\frac{\partial^2\Psi}{\partial t^2} = C^2\frac{\partial^2\Psi}{\partial x_i\,\partial x_i} = C^2\,\nabla^2\Psi \tag{2.52}$$

where Ψ is a potential function (see Graff [1]). From Cartesian to spherical coordinates, one has the standard conversion (student should consult a handbook)

$$\nabla^2\Psi = \frac{1}{r^2}\frac{\partial}{\partial r}\left(r^2\frac{\partial\Psi}{\partial r}\right) + \frac{1}{r^2\sin^2\theta}\frac{\partial^2\Psi}{\partial\phi^2} + \frac{1}{r^2\sin\theta}\frac{\partial}{\partial\theta}\left(\sin\theta\frac{\partial\Psi}{\partial\theta}\right) \tag{2.53}$$

However, spherical symmetry demands that

$$\frac{\partial \Psi}{\partial \theta} = \frac{\partial \Psi}{\partial \phi} = 0$$

Thus, Eqn. (2.53) becomes

$$\begin{aligned}\nabla^2 \Psi &= \frac{1}{r^2}\left(r^2 \frac{\partial^2 \Psi}{\partial r^2} + \frac{\partial \Psi}{\partial r}\frac{\partial r^2}{\partial r}\right) \\ &= \frac{\partial^2 \Psi}{\partial r^2} + \frac{2}{r}\frac{\partial \Psi}{\partial r}\end{aligned} \tag{2.54}$$

Substituting (2.54) into (2.52) yields

$$\frac{\partial^2 \Psi}{\partial t^2} = C^2\left(\frac{\partial^2 \Psi}{\partial r^2} + \frac{2}{r}\frac{\partial \Psi}{\partial r}\right) \tag{2.55}$$

And since r is an independent variable, it is easily shown that

$$\frac{\partial^2 \Psi}{\partial r^2} + \frac{2}{r}\frac{\partial \Psi}{\partial r} = \frac{1}{r}\frac{\partial^2 r\,\Psi}{\partial r^2}$$

Thus, Eqn. (2.55) becomes

$$r\frac{\partial^2 \Psi}{\partial t^2} = C^2 \frac{\partial^2 r\,\Psi}{\partial r^2}$$

or

$$\frac{\partial^2 r\,\Psi}{\partial t^2} = C^2 \frac{\partial^2 r\,\Psi}{\partial r^2} \tag{2.56}$$

This is the equation for a spherical wave, and the solution is

$$\Psi = \frac{1}{r}F(r - Ct) + \frac{1}{r}G(r + Ct)$$

where the terms F and G correspond to the wave expanding or contracting, respectively.

Cylindrical waves receive a similar treatment, and Graff [1] provides a detailed presentation. The solution of Eqn. (2.56) for different initial and boundary conditions provides the radial and tangential stresses as a function of time and position. Applications of these equations include the detonation of explosives in cylindrical or spherical cavities. Chapter 17 (Section 17.8) touches on the subject. These are the so-called cavity problems. We can apply a step pressure pulse to the internal walls of a cavity of radius a (cylindrical or sperical) located in an infinite medium and express the stress as

$$\sigma_{rr}(a, t) = -P_0 H(t) \qquad \sigma_{r\phi} = \sigma_{r\theta} = 0$$

$H(t)$ is a Heaviside (step) function that is equal to zero for $t < 0$ and unity for $t \geq 0$. The displacement u_r is the derivative of the potential function Ψ with respect to r (by definition):

$$u_r = \frac{\partial \Psi}{\partial r} \tag{2.57}$$

We make $G = 0$ in Eqn. (2.57) to signify that we only have the expanding form. The solution, at distance r from the center, is of the form [1]

$$u(r, t) = \frac{a^3 P_0}{4\mu r^2}\left\{1 - \bar{e}^{\zeta\tau}\left[\left(\frac{2r}{a} - 1\right)\frac{\zeta}{w}\sin w\tau - \cos w\tau\right]\right\} \tag{2.58}$$

where τ is a modified time and

$$\zeta = \frac{2C_s^2}{aC_1}$$

$$w = \zeta\left[\frac{C_1^2}{C_s^2} - 1\right]^{1/2}$$

The important conclusion to be derived from Eqn. (2.58) is that the displacements are oscillatory, not constant, as in the static case; the oscillations are provided by the harmonic terms in the solution. From the displacements, one finds the strains and stresses. The tangential stresses are also oscillatory, and at a distance r, the initial compression is followed by oscillations that lead to tension. These tensile components lead to fracturing of rock. The characteristic pattern of radial cracks surrounding a bore hole is thus clearly explained (see Section 17.8).

2.12 SOLUTION OF WAVE EQUATION BY METHOD OF CHARACTERISTICS

The method of characteristics is very useful in the solution of boundary value problems involving waves. The wave equation, a second-order partial differential equation, is decomposed into first-order partial differential equations, which in turn are transformed into ordinary differential equations. We will briefly describe the methodology used. We start with the equation

$$\frac{\partial^2 u}{\partial t^2} - C^2 \frac{\partial^2 u}{\partial x^2} = 0$$

This equation can be decomposed into

$$\left(\frac{\partial}{\partial t} + C \frac{\partial}{\partial x}\right)\left(\frac{\partial u}{\partial t} - C \frac{\partial u}{\partial x}\right) = 0 \tag{2.59}$$

By performing the operation

$$\frac{\partial^2 u}{\partial t^2} - C \frac{\partial^2 u}{\partial t\, \partial x} + C \frac{\partial^2 u}{\partial x\, \partial t} - C^2 \frac{\partial^2 u}{\partial x^2} = 0 \tag{2.60}$$

one arrives at the wave equation. One can also express the wave equation as

$$\left(\frac{\partial}{\partial t} - C \frac{\partial}{\partial x}\right)\left(\frac{\partial u}{\partial t} + C \frac{\partial u}{\partial x}\right) = 0 \tag{2.61}$$

If we make

$$W = \frac{\partial u}{\partial t} - C \frac{\partial u}{\partial x}$$

$$v = \frac{\partial u}{\partial t} + C \frac{\partial u}{\partial x}$$

we end up with

$$\frac{\partial W}{\partial t} + C \frac{\partial W}{\partial x} = 0$$

$$\frac{\partial v}{\partial t} - C \frac{\partial v}{\partial x} = 0$$

These are called the first-order wave equations. They are solved by transforming them into ordinary differential equations, remembering that

$$dW = \left(\frac{\partial W}{\partial t}\right) dt + \left(\frac{\partial W}{\partial x}\right) dx$$

$$\frac{dW}{dt} = \frac{\partial W}{\partial t} + \frac{dx}{dt}\left(\frac{\partial W}{\partial x}\right) \tag{2.62}$$

The change in the property W with time at a fixed position x is $\partial W/\partial t$. If we want the change of property W with time of a region that is moving at a velocity dx/dt with respect to an observer, we have dW/dt. This difference between Lagrangian (material) and Eulerian (space) referentials will be explained again, in greater detail, in Chapter 3 (Section 3.2) and Chapter 6 (Section 6.4).

In Eqn. (2.62), if $dx/dt = C$, we have

$$\frac{dW}{dt} = \frac{\partial W}{\partial t} + C\frac{\partial W}{\partial x} = 0$$

We now have two ordinary differential equations:

$$\frac{dx}{dt} = C \tag{2.63}$$

and

$$\frac{dW}{dt} = 0 \tag{2.64}$$

We have, by integrating Eqn. (2.63),

$$x = Ct + x_0$$

This equation describes a family of parallel characteristics, shown in Figure 2.19(a). At $t = 0$, $x = x_0$. The property W is constant along this characteristic line. The value of W at (x, t) in Figure 2.19(a) is the same as the one at $(x_0, 0)$.

$$W(x, t) = W(x_0, 0) = P(x_0)$$

We are reminded that, for a one-dimensional wave, $\partial u/\partial x = \varepsilon$ and $\partial u/\partial t$ is the particle velocity U_p. Thus

$$W = U_p - C\varepsilon$$

$$v = U_p + C\varepsilon$$

Thus, the value of W at any point (x, t) can be obtained by proceeding along the characteristic line to $t = 0$. Hence

$$W(x, t) = P(x - Ct)$$

By repeating this procedure for the second first-order partial differential equation, for $x = x_0 - Ct$ (shown in Fig. 2.19(b)), we obtain

$$v(x) = Q(x + Ct)$$

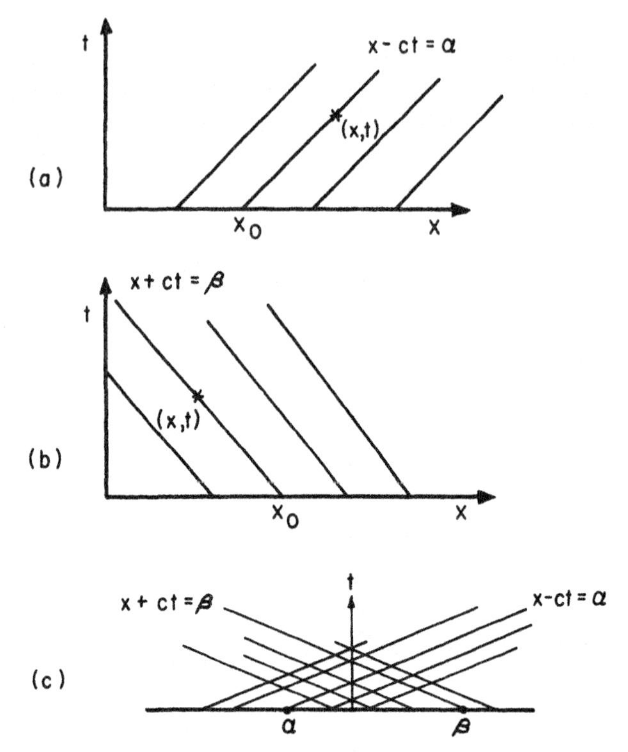

FIGURE 2.19 Characteristics for (a,b) the first-order wave equations; (c) a second-order one-dimensional wave equation.

Adding $W(x)$ and $v(x)$ yields

$$W(x) + v(x) = 2\frac{\partial u}{\partial t} = P(x - Ct) + Q(x + Ct)$$

Integration yields

$$u = F(x - Ct) + G(x + Ct)$$

where F and G are obtained by integrating P and Q, respectively. Thus, we arrive at the same solution as in Section 2.9; this is called the d'Alembert solution. These characteristic curves are very useful in the solution of wave propagation (elastic, plastic, and shock) through solid, liquid, and gaseous media. In the case of nonlinear equations (when C is a function of stress or pressure), the characteristic lines are not parallel, but form a fan (for dispersion) or intersect (for shock waves). The method of characteristics is the basis of one of the first hydrocodes to deal with shock waves, SWAP-7. This code was developed by Barker [13].

REFERENCES

1. K. F. Graff, *Wave Motion in Elastic Solids*, Ohio University Press, Columbus, OH 1975.
2. R. D. Woods, *J. Soil Mech. Founds, Am. Soc., Civil Eng.*, **94** (1968), 115.
3. H. Kolsky, *Stress Waves in Solids*, Dover, New York, 1963.
4. J. S. Rinehart, *Stress Transients in Solids*, Hyperdynamics, Santa Fe, 1975.
5. Y. C. Fung, *Continuum Mechanics*, Prentice-Hall, Englewood Cliffs, NJ, p. 198.
6. R. J. Wasley, *Stress Wave Propagation in Solids*, M. Dekker, NY, 1973.
7. L. Pochhammer, *J. Reine Angew. Math.*, **81** (1876), 324.
8. C. Chree, *Trans. Comb. Phil. Soc.*, **14** (1889), 250.
9. C. Chree, *Q. J. Pure Appl. Math.*, **23** (1889), 335.
10. L. Rayleigh, *Theory of Sound*, Dover, New York, 1894.
11. R. M. Davies, *Phil. Trans. A.*, **240** (1898), 375.
12. R. Skalak, *J. Appl. Mech. Trans. ASME*, **24** (1957), 59.
13. L. M. Barker, "SWAP-7, A Stress Wave Analyzing Program," Report No. SC-RR-67-143, Sandia Natl. Lab., Albuquerque, 1967.

CHAPTER 3

Plastic Waves

3.1 INTRODUCTION

When the stress in a ductile material exceeds the elastic limit, plastic deformation sets in. This occurs both in quasi-static and dynamic deformation. If a pulse is transmitted to a material that has an amplitude exceeding the elastic limit, this pulse will decompose into an elastic and a plastic wave. Three classes of plastic waves are studied in this book:

1. *Plastic Waves in Rods, Wires, and Bars.* The classical example of this situation is the impact of a long plastic rod against a rigid target. If the lateral dimension of the rod (diameter) is small, a state of *uniaxial stress* is established. The study of this type of plastic wave is carried out in this chapter.
2. *Plastic Waves in Semi-infinite Bodies.* When the lateral strains (perpendicular to the direction of propagation of the stress front) are zero, a state of *uniaxial strain* is established and the resulting plastic wave has a very sharp front: it is therefore called a shock wave. Shock waves will be studied in Chapters 4–8.
3. *Plastic Shear Waves.* Torsional waves in bars or shear waves in semi-infinite bodies can, if their amplitude is sufficiently high, generate plastic deformation.

Material strength is strain and strain rate dependent, as exemplified by Figure 3.1. For metals, the stress–strain curve is often represented by a bilinear function [Fig. 3.1(a)] where the first stage is elastic and the second stage is plastic. The stress–strain curve in many metals approaches more closely a power function of the type

$$\sigma = \sigma_0 + k\varepsilon^n \tag{3.1}$$

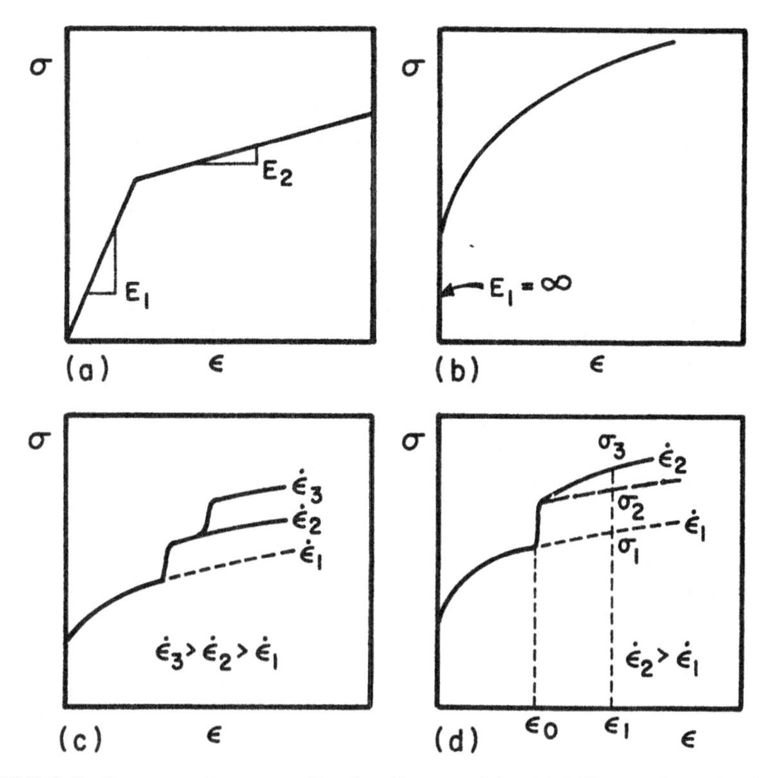

FIGURE 3.1 Stress–strain curves for ductile materials: (a) bilinear elastoplastic; (b) power law work hardening; (c) strain-rate-dependent flow stress; (d) strain rate history dependence of flow stress.

where the exponent $n < 1$ [Fig. 3.1(b)]. The strain rate dependence of the flow stress is schematically illustrated in Fig. 3.1(c): If the strain rate is changed during the test, the flow stress is accordingly changed. This dependence of flow stress on strain rate is often represented by

$$\sigma = \sigma_0 + k\varepsilon^n\dot{\varepsilon}^m \tag{3.2}$$

where m is the strain rate sensitivity and usually varies between zero and unity for metals. In Chapter 13 the effects of temperature, strain, and strain rate on the mechanical behavior of materials will be discussed in greater detail. An additional factor that often has a noticeable effect is the strain rate history. Figure 3.1(d) shows how this effect can change the mechanical response of a material. If the strain rate of a material is changed from $\dot{\varepsilon}_1$ to $\dot{\varepsilon}_2$ at a strain ε_1, the flow stress changes from σ_1 to σ_2. However, if the strain rate change occurred earlier, at a strain of ε_0, the work-hardening rate of the material could have been modified by the higher strain rate. In that case, the material would exhibit a flow stress σ_3 at a strain ε_1 and strain rate $\dot{\varepsilon}_2$. Figure 3.1(d) shows how the work hardening at the strain rate $\dot{\varepsilon}_2$ is higher than at the strain rate $\dot{\varepsilon}_1$. If it were the same, it would follow the dashed line. The treatment of plastic waves increases in complexity as strain rate and strain rate history effects are incorporated. The reader is referred to the review article by Clifton [1] and to the theory for plastic waves in materials with strain rate sensitivity by Malvern [2] and Sokolovsky [3]. In this chapter we will deal only with strain rate-independent behavior, for the sake of simplicity.

In this chapter we will present the derivation of the equations for the propagation of a plastic wave in a wire by von Karman and Duwez [4] (Section 3.2). This derivation will lead to an equation for the velocity of the plastic wave equal to

$$V_p = \left(\frac{d\sigma/d\varepsilon}{\rho}\right)^{1/2} \tag{3.3}$$

where $d\sigma/d\varepsilon$ is the slope of the plastic region of the stress–strain curve. V_p is constant only at a fixed strain ε. If one looks at Eqn. (2.5), which gives the velocity at propagation of an elastic wave in a bar, we clearly see that $d\sigma/d\varepsilon$ is E in the elastic region. This slope is higher in the elastic than in the plastic region:

$$\left(\frac{d\sigma}{d\varepsilon}\right)_{el} > \left(\frac{d\sigma}{d\varepsilon}\right)_{pl}$$

Thus, the velocity of plastic waves is lower than the velocity of elastic waves. A qualitative understanding can be reached from the observation of a stress–strain curve. Because the velocity decreases with decreasing work-hardening rate, the tendency is for the front of the wave to "spread out," or disperse ($V_2 < V_1$ in Fig. 3.2).

After von Karman and Duwez's [4] derivation, we will introduce plastic waves in uniaxial strain (Section 3.3). Under these conditions the plastic wave is usually called a shock wave if the amplitude is sufficiently high. In Section 3.4, we will discuss plastic shear waves. These waves are produced in conjunction with longitudinal waves, and therefore this section is called "Plastic Waves of Combined Stress."

In Section 3.5, we make a brief mention of the Hopkinson–Kolsky pressure bar experiments, emphasizing that plastic waves do not play a role in these experiments.

3.2 PLASTIC WAVES OF UNIAXIAL STRESS

Von Karman and Duwez [4], Taylor [5], and Rakhmatulin [6] independently developed the theory of plastic waves. Von Karman and Duwez's treatment will be presented here. One can use two frames of reference when dealing with

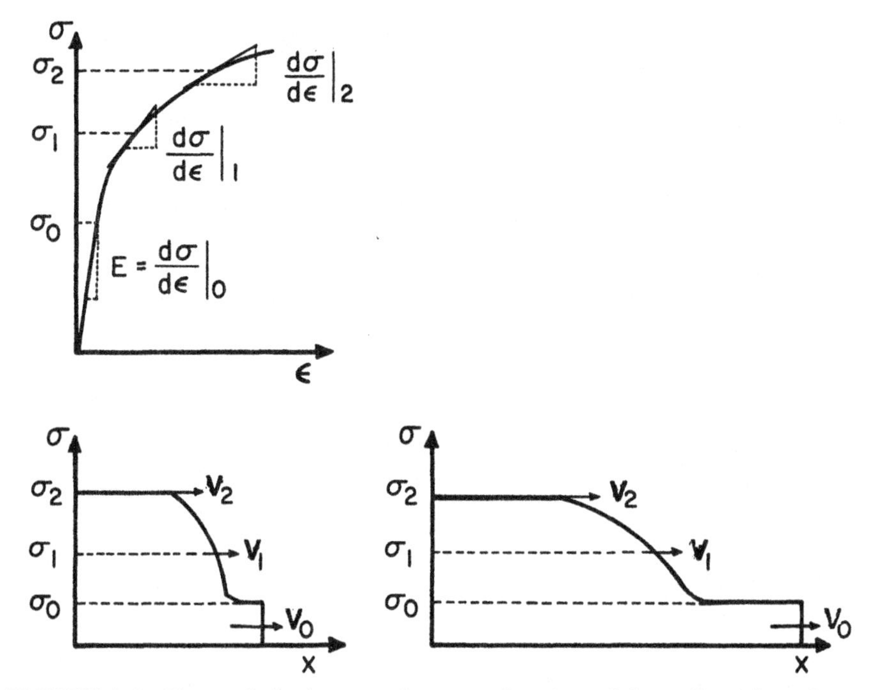

FIGURE 3.2 Shape of plastic wave front as a function of time; dispersion of wave is observed due to decrease in wave velocity with increasing stress.

disturbance propagation problems (or fluid displacement problems): (1) one considers a particle in the material and observes the change of position of this particle with time; (2) one considers a certain region in space and observes the flow of material in and out of it: Approach (1) receives the name of the material, or Lagrangian, coordinates, whereas approach (2) refers to spatial, or Eulerian, coordinates. Hence, a property F, which varies with time, spatial position (X), and particle position (x), can be expressed as

$$F = f(x, t) \qquad \text{or} \qquad F = f(X, t) \tag{3.4}$$

The Lagrangian coordinates were used by von Karman and Duwez and Rakhmatulin whereas Taylor used Eulerian coordinates. Nevertheless, the two approaches yield the same results. These two approaches will be discussed again in Chapter 6, in connection with numerical methods of solution of shock waves.

Von Karman and Duwez considered the simplest possible plastic wave propagation problem, a semi-infinite thin wire being impacted at a certain velocity generating a downward motion at a velocity V_1 (Fig. 3.3). Annealed copper wires were tested and equidistant marks were made along the wire prior to testing. The end of the wire was attached to a rigid plate (P). Figure 3.3(a) shows a schematic of the setup. The weight W was dropped from a certain distance d and was accelerated by rubber bands (R). After impact with the

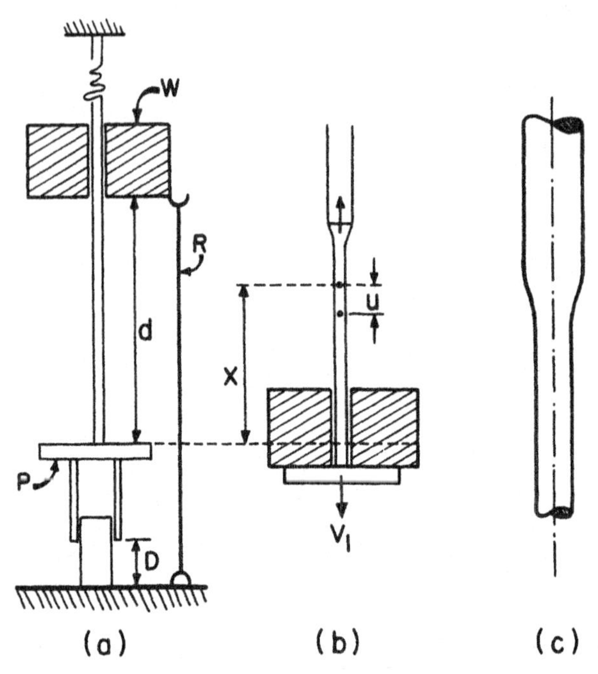

FIGURE 3.3 Wire being impacted by weight in von Karman–Duwez experiments: (a) initial position; (b) position after impact; (c) shape of wire during plastic wave propagation stage. (From von Karman and Duwez [4], Fig. 4, p. 991. Reprinted with permission of the publisher.)

platform the wire was plastically deformed. The total amount of extension was established by the distance D, which could be varied. This distance could be changed by setting the bottom anvil. The hammer was of a sufficient weight not to be decelerated significantly by the plastic deformation in the wire. The initial position of the extremity of the wire is taken as the origin, and we observe the displacement of a particle situated at the position x. At time t, it will be displaced by u. Applying Newton's second law (the acceleration is the second derivative, with respect to time, of the displacement), one has

$$
\begin{aligned}
dF &= dm \frac{\partial^2 u}{\partial t^2} \\
&= \rho_0 \, dV \frac{\partial^2 u}{\partial t^2} \\
&= \rho_0 A_0 \, dx \frac{\partial^2 u}{\partial t^2}
\end{aligned}
$$

Hence

$$\frac{d\sigma}{dx} = \rho_0 \frac{\partial^2 u}{\partial t^2} \tag{3.5}$$

where ρ_0 and A_0 are the initial density and area and σ is the stress acting on it. Since we have a state of plastic deformation and assuming that one has a one-to-one relationship between stress and strain in loading (not in unloading, because of the irreversibility of the process), we can write, for Eqn. (3.5),

$$\rho_0 \frac{\partial^2 u}{\partial t^2} = \frac{d\sigma}{d\varepsilon} \frac{\partial \varepsilon}{\partial x}$$

Notice that d was used instead of ∂ because of the one-to-one relationship.

But $\varepsilon = \partial u/\partial x$, and

$$\frac{\partial^2 u}{\partial t^2} = \frac{d\sigma/d\varepsilon}{\rho_0} \frac{\partial^2 u}{\partial x^2} \tag{3.6}$$

One can see the similarity with Eqn. (2.5). The velocity of the plastic wave at a constant strain ε can be inferred from Eqn. (3.6) and is given by

$$V_p = \left(\frac{d\sigma/d\varepsilon}{\rho_0}\right)^{1/2} \tag{3.7}$$

However, the rigorous proof is given by von Karman and Duwez [4]. In the elastic range,

$$\frac{d\sigma}{d\varepsilon} = E$$

and one has [see Eqn. (2.7)]

$$C_0 = V = \left(\frac{E}{\rho_0}\right)^{1/2}$$

This is exactly the velocity of propagation of elastic waves in rods when the length of the pulse (Λ) is large with respect to the rod radius r ($r/\Lambda < 0.1$). It is slightly lower than the velocity of a longitudinal wave in an unbounded medium; when Poisson's ratio is 0.3, the velocity of the longitudinal wave in an unbounded medium is $1.16C_0$.

The application of the boundary conditions allows the determination of the wave profile. The boundary conditions are

$$u = \begin{cases} V_1 t & \text{at } x_1 = 0 \\ 0 & x = \infty \quad (\text{any } t > 0) \end{cases}$$

1. $x = 0$ to $x = C_1 t$ (C_1 is the velocity of propagation of the plastic wave front); the strain is constant and equal to ε_1.
2. $C_1 t < x < C_0 t$ (C_0 is the velocity of elastic longitudinal waves in bars). In this interval, one has

$$\frac{x}{t} = \left(\frac{d\sigma/d\varepsilon}{\rho_0}\right)^{1/2}$$

3. $x > C_0 t$, $\varepsilon_1 = 0$.

Figure 3.4(a) shows graphically how the strain varies as a function of x/t. As one can see, this plot applies to any time t, and one can have the profile for different times after impact. One has to determine the maximum strain ε_1 as a function of the impact velocity in order to completely solve the problem.

Figure 3.4(b) shows the wave profile at two times, t_1 and t_2. It can be seen that the wave disperses itself, that is, the plastic wave front spreads out as t increases. This is due to the fact that $C_1 < C_0$.

In order to find σ_1 and ε_1 (the maximum stress and plastic strain undergone by the wire) at a specific impact velocity V_1, we will have to establish the displacement of the bottom part of the wire as a function of time. This displacement $u|_{x=0}$ is determined by the velocity V_1 of the falling weight (which will be assumed constant and undisturbed by the wire). Thus,

$$V_1 = \frac{u|_{x=0}}{t}$$

We have to consider that a force increase $d(A\sigma_0)$ that occurs at the stress level σ_0 on the bar travels at a velocity V_{p0} (plastic wave velocity at a stress level σ_0). The time dt taken by an arbitrary length of specimen dx to propagate this plastic wave is

$$V_{p0} = \frac{dx}{dt} \qquad \therefore\ dt = \frac{dx}{V_{p0}}$$

Applying the conservation-of-momentum equation, we have

$$m\, dv = d(A\sigma_0)\, dt$$

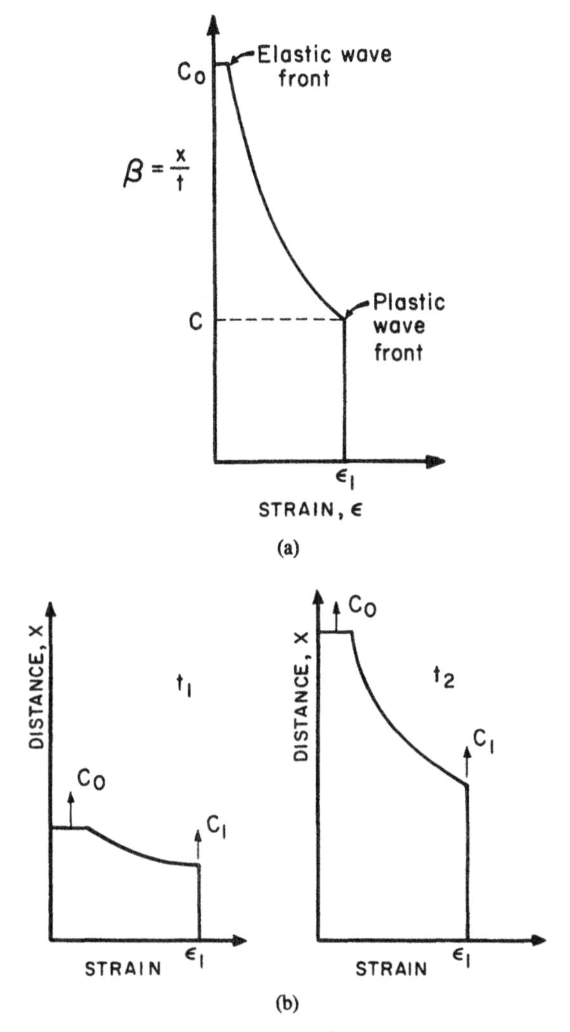

FIGURE 3.4 (a) Strain as a function of $\beta = x/t$. (b) Strain distribution at two different times t_1 and t_2 for an impact producing a total strain ϵ_1.

The mass of the segment dx is $dx\,\rho_0 A$:

$$(dx\,\rho_0 A)\,dv = d(A\sigma_0)\,dt$$

$$= d(A\sigma_0)\,\frac{dx}{V_{p0}}$$

$$dv = \frac{1}{\rho_0}\frac{d\sigma_0}{V_{p0}}$$

The velocity at $x = 0$, V_1, is obtained by integrating the above expression:

$$V_1 = \int dv = \int_0^{\sigma_1} \frac{d\sigma_0}{\rho_0 \sqrt{\dfrac{d\sigma_0/d\varepsilon}{\rho_0}}}$$

By manipulation we arrive at

$$V_1 = \int_0^{\sigma_1} \frac{d\varepsilon^{1/2}}{\rho_0^{1/2}}\, d\sigma_0^{1/2}$$

We can change the integration limits to ε; and by proper manipulation arrive at

$$= \int_0^{\varepsilon_1} \sqrt{\frac{d\sigma_0/d\varepsilon}{\rho_0}}\, d\varepsilon \qquad (3.7a)$$

The integral has to be numerically determined, since the integrand is a variable. Thus, one cannot present a general solution, except if one assumes a functional relationship between σ and ε. For a known impact velocity and a known σ–ε relationship, one can determine ε_1. If we assume a power law relationship that describes well the plastic behavior of most ductile metals,

$$\sigma = k\varepsilon^n$$

$$\frac{d\sigma}{d\varepsilon} = kn\varepsilon^{n-1}$$

$$V_1 = \left(\frac{kn}{\rho_0}\right)^{1/2} \int_0^{\varepsilon_1} \varepsilon^{(n-1)/2}\, d\varepsilon$$

$$V_1 = \left(\frac{kn}{\rho_0}\right)^{1/2} \left(\frac{n-1}{2} + 1\right)^{-1} \varepsilon_1^{(n+1)/2}$$

And

$$\varepsilon_1 = \left[\frac{\rho_0 V_1^2 (n+1)^2}{4kn}\right]^{1/(n+1)} \qquad (3.8)$$

The maximum impact velocity can be easily found by setting it equal to the velocity that will produce necking on the specimen. For a material obeying a power law ($\sigma = k\varepsilon^n$) it is a simple matter to find the stress at which the specimen starts to neck. This is found through Considère's criterion, and this strain is

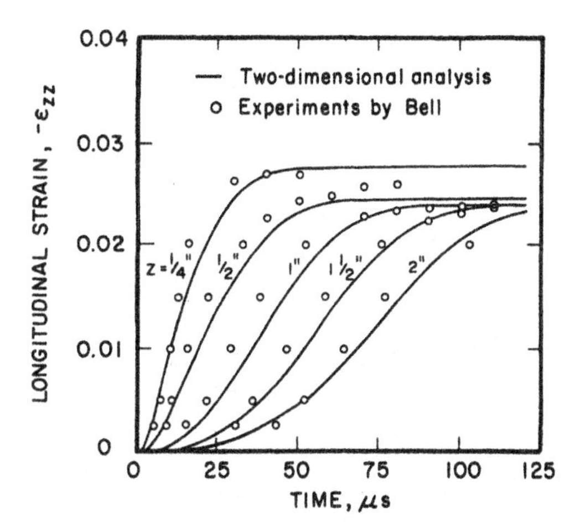

FIGURE 3.5 Analytical predictions of strain–time profiles in aluminum rod compared with experimental results of Bell [7] (Obernhuber et al. [8]); Z is distance from interface.

numerically equal to n:

$$\varepsilon_{\max} = n$$

$$V_1 = \frac{k^{1/2} n^{(n+2)/2}}{\rho_0^{1/2}[\frac{1}{2}(n + 1)]} \tag{3.9}$$

Results of actual experiments conducted by Bell [7] are shown in Figure 3.5; these results are compared with analytical predictions by Obernhuber et al. [8] and are discussed by Nicholas [9]. The results (data points) were produced by a very accurate diffraction grating technique developed by Bell [7]. The experiment consisted of a symmetric impact of two identical rods of commercial purity aluminum with a diameter of 0.99 in. (25 mm). The rate at which strain is accumulated (from 0 to ~0.027) decreases with distance from the impact interface. This is due to dispersion of the plastic wave, as depicted in Figure 3.5. The experimental results match very well the calculations by Obernhuber et al. [8]. Figure 3.6(a), adopted from Goldsmith [10], shows a stress-strain curve for copper. By utilizing Eqn. (3.7b), one can establish the stress σ as a function of impact velocity V_1. This is shown in Figure 3.6(b). The plastic wave velocity is also calculated as a function of velocity for copper.

3.3 PLASTIC WAVES OF UNIAXIAL STRAIN

The theory of von Karman and Duwez [4] is only the first step toward an understanding of plastic waves in solids. A considerable theoretical effort has been devoted, since the 1940s, to developing models for the propagation of

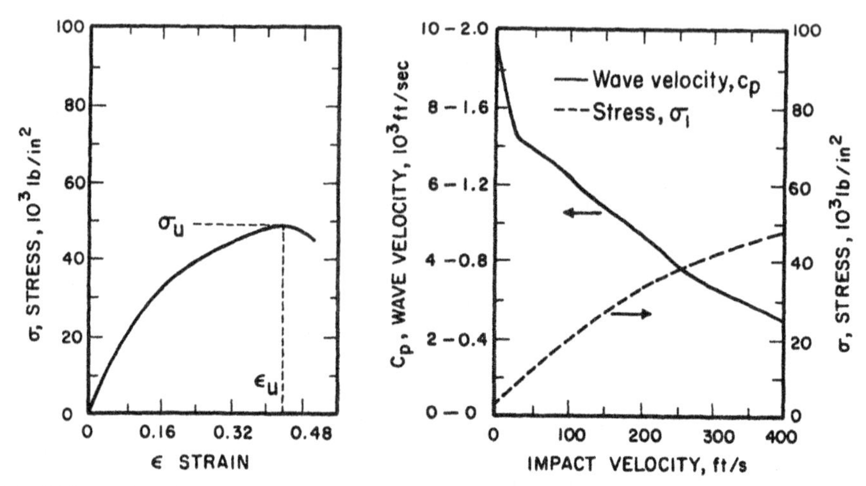

FIGURE 3.6 Plastic wave velocity and peak stress as a function of impact velocity for copper.

plastic waves. A number of mathematical formulations have been proposed, incorporating the material properties by a constitutive equation (the relationship between stress and strain; stress and strain rate; stress, strain, and temperature; etc.). These relationships are, for the most part, nonlinear. Lee [11] developed a theory he called elastoplastic with finite deformation. Herrmann and Nunziato [12] divide the response into several categories, depending on the wave propagation characteristics, and analyze them in terms of linear viscoelastic, infinitesimal elastic–perfectly plastic, nonlinear viscoelastic, and thermoelastic behaviors. Clifton [13] and Herrmann et al. [14] describe the analytical approaches, and a report by the National Materials Advisory Board [15] makes a critical analysis of the state of the art and the areas that need development.

As the lateral dimensions of a system are increased, the boundary conditions are altered and the state of stress at center and surface become different; for the limiting case, the lateral dimensions are infinite and no lateral strains (perpendicular to stress–wave propagation direction) are allowed. The state of uniaxial stress is thus transformed into uniaxial strain, and Eqn. (3.3) is altered because $d\sigma/d\varepsilon$ is changed. This curve actually becomes concave, with the slope $d\sigma/d\varepsilon$ increasing with plastic strain. This leads to a plastic wave front that will "steepen up" as it propagates and becomes therefore a discontinuous jump, called a *shock front*. A simple *hydrodynamic* treatment has been applied to shock waves, in which material strength is ignored. However, it has been realized that the hydrodynamic treatment of shock waves is oversimplified and fails to predict a number of phenomena; even at high pressures strength effects are of importance. The shock front is not an exact discontinuity but possesses a definite structure, due to material effects. We will treat shock waves in Chapters 4–7.

3.4 PLASTIC WAVES OF COMBINED STRESS

One can have plastic shear waves, in an analogous way to elastic shear waves. The particles undergo plastic (permanent) displacement perpendicular to the direction of propagation of the wave. It is very difficult to generate exclusively plastic shear waves. A wave in a rod subjected to a sudden torsion would be a plastic shear wave; however, the amplitude of motion would vary with the distance from the longitudinal axis of the rod. Hence, it is easier to generate a plastic shear wave simultaneously with a longitudinal stress wave. The mathematical treatment of plastic shear waves has been performed by Bleich and Nelson [16] and Ting and Nan [17]. For an elastoplastic material with no rate dependence [$m = 0$ in Eqn. (3.2)], Bleich and Nelson [16] solved the problem shown in Fig. 3.7(a): a half-space subjected to uniform stresses $\sigma_0(+)$ and $\tau_0(+)$ at the surface ($x = 0$). The stresses σ_0 and τ_0 are applied suddenly at $t = 0$ and are kept constant for the simplest case. These stresses exceeded the flow stress of the material. They defined a nondimensional velocity

$$U = \left(\frac{\rho}{\mu}\right)^{1/2} \frac{x}{t} \geq 0$$

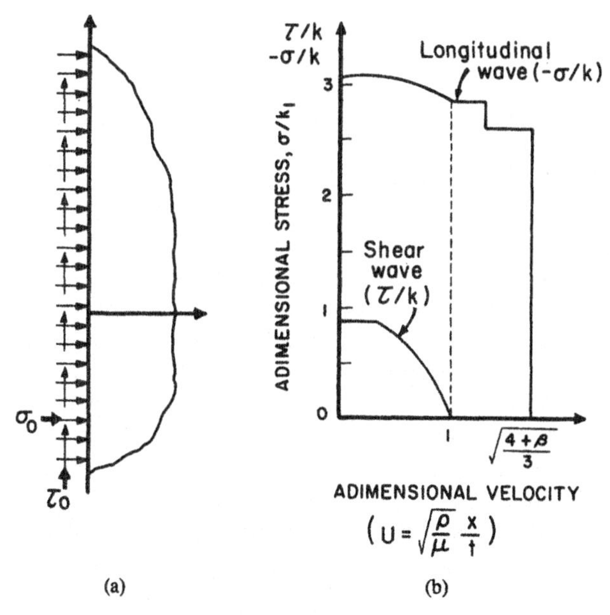

FIGURE 3.7 (a) Half-space loaded simultaneously by normal and shear tractions. (b) Longitudinal and shear plastic wave profiles for $\nu = 0.25$, $\sigma_0 = -3.1k$, $\tau_0 = 0.9k$. Here k is the shear yield strength of material. (From Bleich and Nelson [16], Figs. 1 and 6, pp. 149, 156. Reprinted with permission of the publisher.)

and obtained two solutions for the nondissipative (elastic) fronts:

$$U = \left(\frac{4 + \beta}{3}\right)^{1/3} \qquad U_s = 1$$

where

$$\beta = \frac{2(1 + \nu)}{1 - 2\nu}$$

These are the elastic longitudinal and shear velocities, respectively. For the dissipative (plastic) fronts they obtained the solution shown in Figure 3.7(b) for a material with $\beta > 5$ (if $\nu = 0.3$, $\beta = 6.5$). The elastic longitudinal wave, traveling at the adimensional velocity of $[4 + \beta/3]^{1/2}x/t$, is followed by a plastic wave. Similarly, the elastic shear wave ($U = x/t$) is followed by a plastic shear wave, traveling at a lower velocity. Since the plot is normalized to x/t, the plastic waves are dispersive, that is, they "smear out" with time. The plot in Figure 3.7(b) is drawn for applied stresses:

$$\sigma = -3.1k \qquad \tau = 0.9k$$

where k is the flow stress in pure shear. Notice that the shear stress is only equal to $0.9k$, but the shear strength of the material is exceeded due to the superposition of the normal stress. A flow rule (such as von Mises', or J2) has to be applied.

Lipkin and Clifton [1, 18] studied combined plastic longitudinal and shear waves by means of an apparatus shown schematically in Figure 3.8. A specimen (tubular) is subjected to a torsion by means of a rear mounting assembly. It is then impacted by a striker through a transmitter bar. At the transmitter–specimen interface, elastic reflected longitudinal and shear waves are created (the cross section of the specimen bar is considerably lower than that of the transmitter bar). The tubular specimen was torqued beyond its yield stress prior to impact. The characteristic lines in Figure 3.8(b) show the distribution of stresses (stress constant along characteristic line) and the formation of longitudinal and shear elastic and plastic waves. Both longitudinal and shear waves form "fans." These fans are due to wave dispersion; that is, the wave spreads out with time because a higher stress wave travels at a lower velocity than a lower stress wave. The velocity of the shear plastic wave is established, in analogy with the longitudinal wave, by

$$V_s = \left(\frac{d\tau/d\gamma}{\rho}\right)^{1/2}$$

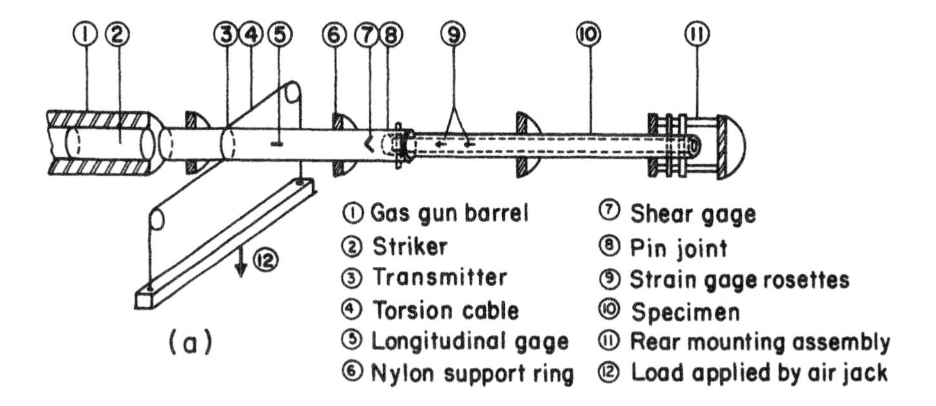

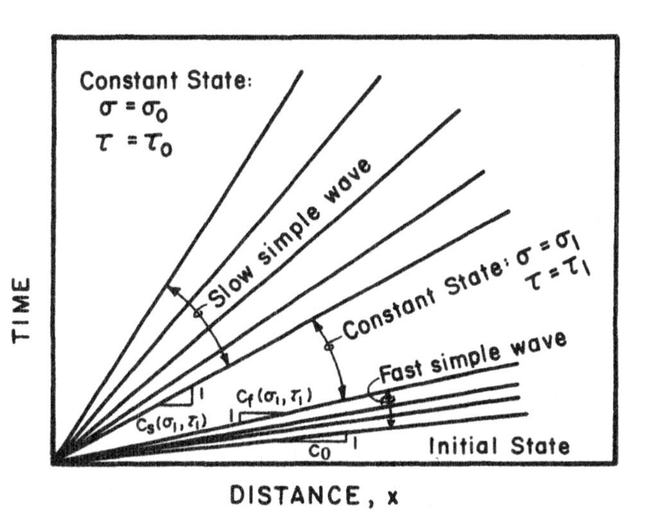

FIGURE 3.8 (a) Experimental setup used by Lipkin and Clifton [13] to produce combined longitudinal and shear (torsional) plastic waves; (b) characteristic fan for longitudinal (fast) and shear (slow) plastic waves. (From Clifton [1], Figs. 13 and 14, pp. 155, 156. Reprinted with permission of the author.)

where τ and γ are shear stress and strain, respectively. The fronts of the longitudinal and shear waves travel with the velocities of the respective elastic waves, C_0 and C_s, respectively.

Plastic waves have also been obtained by the oblique–parallel plate impact method (pressure–shear experiments). This technique was developed by Clifton's group [19–22] and is shown schematically in Figure 3.9(a). A sabot, accelerated by compressed gases in a barrel (see Chapter 12, Section 12.5), carries a flyer plate, at its front. This flyer plate is nonnormal to the direction

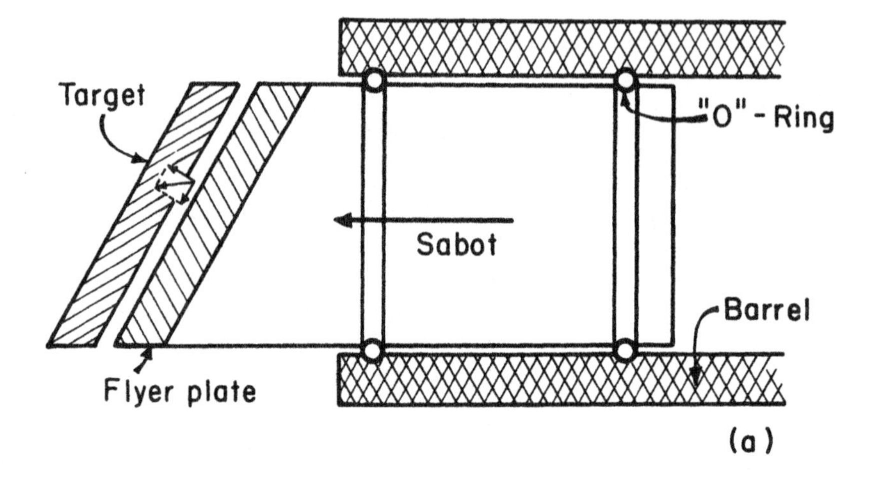

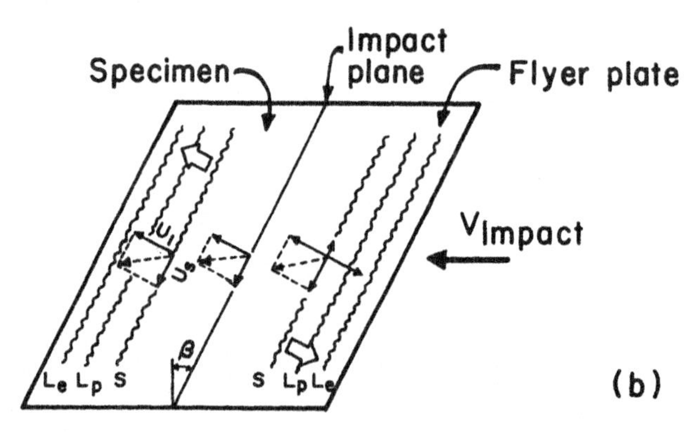

FIGURE 3.9 (a) Schematic representation of the oblique–parallel plate experiment; (b) longitudinal elastic and plastic (L_e and L_p) and shear waves (S) produced by impact.

of motion and impacts a target that is parallel to it. Thus, the impact is oblique and parallel. Longitudinal and shear waves are set up in both flyer plate (projectile) and target, as shown in Figure 3.9(b). Both elastic and plastic longitudinal waves are shown as well as a shear wave. If the impact velocity is sufficiently high, the shear strength of the material is exceeded and the shear wave decomposes into an elastic front and a plastic wave. Mashimo and Nagayama [23] were able to obtain plastic shear waves by using impact velocities of up to 2 km/s. These waves move in directions perpendicular to the interface and have particle velocities that are perpendicular (for longitudinal) or parallel (for shear waves) to the interface. One can establish the amplitude and configuration of these waves by techniques such as electromagnetic gauges (used by Mashimo and Nagayama [23]) and laser interferometric methods. Chapter 12 provides additional information.

3.5 ADDITIONAL CONSIDERATIONS OF PLASTIC WAVES

An experimental technique that has been used often to obtain constitutive relations at high strain rates is the split Hopkinson bar. It is, in its present form, a modified version of the bar originally used by Hopkinson [24], Davies [25], and Kolsky [26], who were responsible for its development. Figure 3.10(a) shows schematically a specimen between two long bars; an elastic pressure pulse is produced in the left-hand end of the bar. Once it reaches the specimen, a complex pattern of reflections is produced, but it passes through it (deforming it plastically) and continues its course in the right-hand bar. The specimen length is chosen "short," so that one can consider the stress to be uniformly distributed in it. In this sense, one does not have a wave propagation configuration, since one is not dealing with a wave front at which plastic deformation is taking place. The pulse length generally exceeds the length of the specimen. Figure 3.11 shows the resultant stress–strain plot obtained from one single experiment. Figure 3.10(b) shows the incident, transmitted, and reflected pulses, duly converted into stresses in specimen. These results, for tantalum, can then be converted into a stress–strain curve. One should not consider the Hopkinson bar as a plastic wave propagation experiment. Nevertheless, it can provide important information (the constitutive relationship) needed in the mathematical models on plastic waves. Hopkinson bars will be described in greater detail in Chapter 13 (Section 13.3.2).

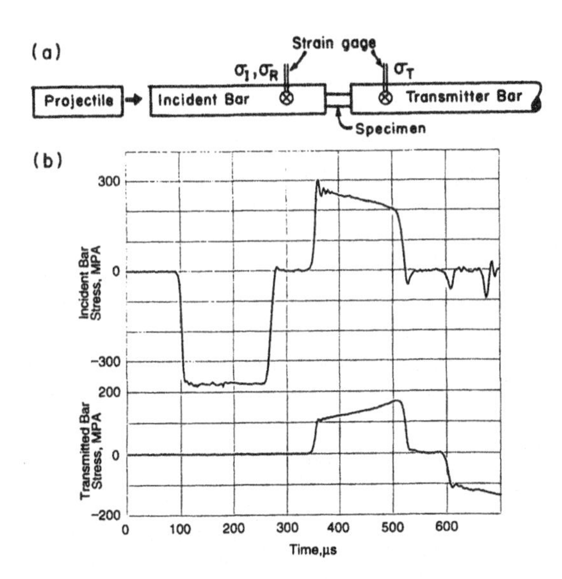

FIGURE 3.10 (a) Schematics of split Hopkinson bar; (b) typical oscilloscope record of a split Hopkinson bar test of tantalum. (Courtesy of J. Isaacs, CEAM, UCSD.)

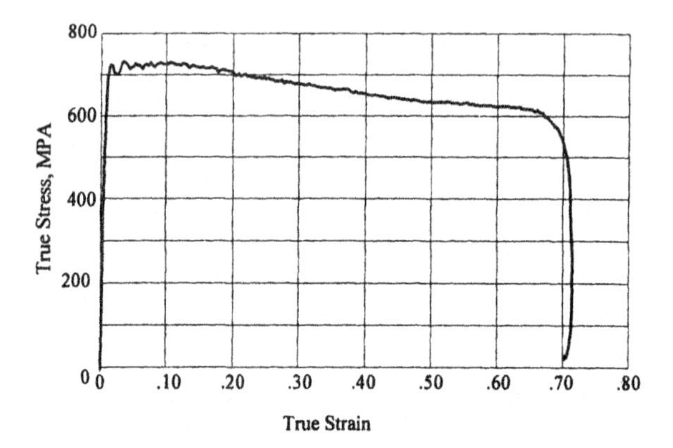

FIGURE 3.11 True stress–true strain curve for tantalum obtained from analysis of incident, reflected, and transmitted pulses of Fig. 3.10(b). (Courtesy of J. Isaacs, CEAM, UCSD.)

3.6 IMPACT OF BARS OF FINITE LENGTH

3.6.1 Taylor's Experiments

During World War II, Taylor [28] conducted an analysis on specimens deformed at very high rates of strain. These experiments involved the propagation of plastic deformation as a wave process. This analysis is considered a classic piece of work, and it will be presented here in a simplified manner. The "Taylor" test has become a standard procedure to verify the constitutive behavior of materials. Figure 3.12a shows a specimen deformed quasi-statistically under compression. It deforms uniformly, so that the cross-sectional area is constant at any time during deformation, if one neglects frictional end effects. On the other hand, if deformation is dynamic, such as the one produced by the impact of the cylinder against a rigid wall, the shape during deformation and after deformation is not uniform. The part of the cylinder undergoing impact undergoes a higher deformation. Taylor depicted the process of deformation as a sequence of elastic and plastic wave propagation into the cylinder. Figure 3.13 shows the sequence of events according to Taylor's simplified approach. A cylindrical projectile of length L impacts a target at a velocity U. At this moment, an elastic wave is faster than the plastic wave and moves at a velocity C. This elastic compressional wave travels until it reaches the back surface of the projectile, reflects there, and then returns as a release wave. When it returns toward the plastic wave, it interacts with it, and this marks the end of the deformation process, because the stress is reduced to zero. Some concepts from wave interaction and reflection that will be discussed later in Chapter 7 are needed here, and the student is referred to that material if he or she cannot understand the concept. The stress within the region that has plastically deformed is assumed to be constant and equal to the yield stress of the material

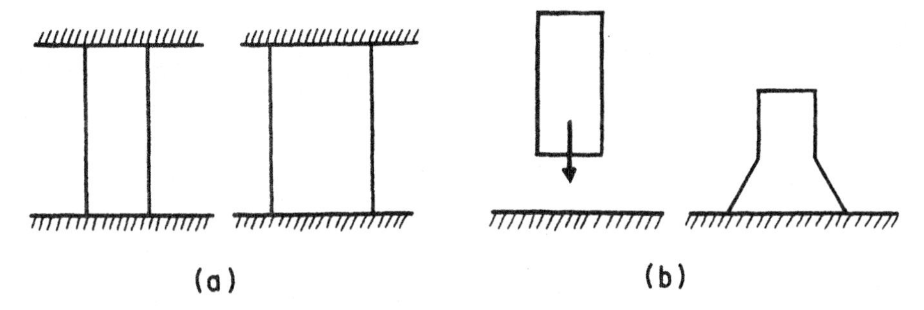

FIGURE 3.12 Quasi-static (a) and dynamic (b) deformation of a cylindrical specimen.

at that strain rate, σ_{yd} (dynamic yield stress). Figure 3.14 shows the two regions, elastic and plastic, and the virgin material moving into the interface I with velocity $U + v$, whereas the material that has been plastically deformed distances itself from that interface with velocity v. Thus, conservation of mass (mass in equals mass out) dictates that

$$A_0(U + v) = Av \qquad \rho_0 \simeq \rho \tag{3.10}$$

Conservation of momentum dictates that the change in momentum equals the impulse. If we assume that the stress on the two sides of the interface is σ_{yd} (the amplitude of the elastic wave), we have*

$$\rho A_0(U + v)U = \sigma_{yd}(A - A_0) \tag{3.11}$$

By eliminating v between (3.10) and (3.11), knowing that $\varepsilon = 1 - L_0/L$, one obtains (notice that the strain was defined as positive in compression); V_p is the volume, assumed constant for plastic deformation.

$$\varepsilon = \frac{L_0 - L}{L_0} = \frac{V_0/A_0 - V_0/A}{V_0/A_0} = 1 - \frac{A_0}{A}$$

$$\frac{\rho U^2}{\sigma_{yd}} = \frac{\varepsilon^2}{1 - \varepsilon} \tag{3.12}$$

*The area A_0 is transformed into A, in Fig. 3.14(a); the material entering A between t and $t + dt$ is:

$$dl = (U + v)\,dt$$

$$dm = \rho A_0 dl = \rho A_0(U + v)\,dt$$

The force is $-(A - A_0)\sigma_{yd}$. From Newton's second law:

$$-(A - A_0)\sigma_{yd} = \gamma dm$$

γ, the acceleration, is equal to change in velocity in time interval dt. Before entering region with area A, the velocity is U. After, it is equal to zero. Thus,

$$-(A - A_0)\sigma_{yd} = \frac{U}{dt}\rho A_0(U + v)\,dt$$

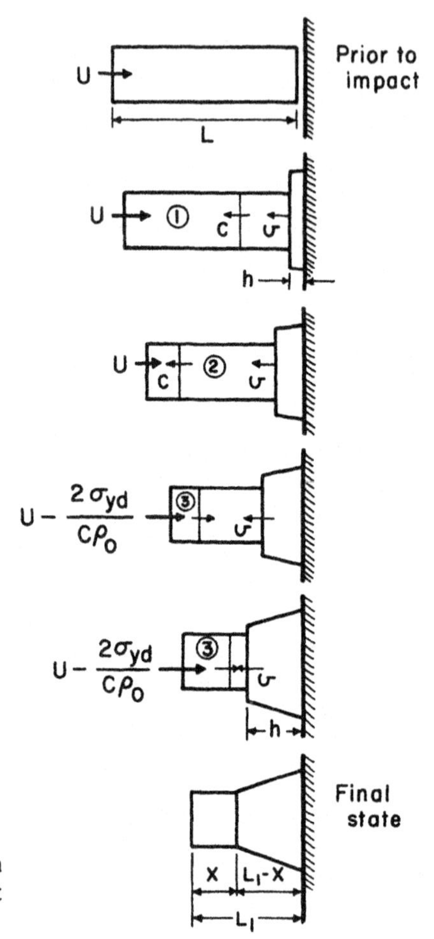

FIGURE 3.13 Sequence of deformation after impact of cylindrical projectile against rigid wall.

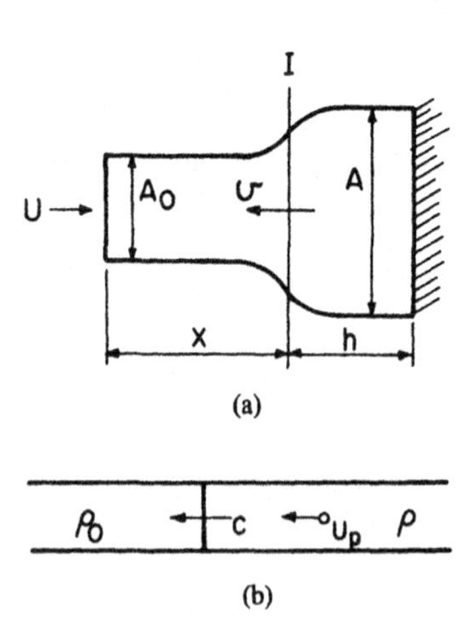

FIGURE 3.14 Detailed view of propagation of (a) plastic and (b) elastic wave front.

We now analyze the velocity at which the particles are moving in the elastic region. We name this velocity u_p. By applying the conservation of mass to the elastic front propagating at a velocity C, one has

$$\rho_0 C = \rho(C - U_p)$$

This is because we assume that the elastic wave compresses the material from an initial density ρ_0 to ρ. The velocity of the particles in the compressed material is

$$U_p = \frac{\rho - \rho_0}{\rho_0} C$$

This is also equal to the strain:

$$\varepsilon_2 = \frac{L_0 - L}{L_0} = \frac{V_0 - V}{V_0} = \frac{\rho - \rho_0}{\rho} \quad \text{(at constant } A_0\text{)}$$

Thus

$$U_p = \varepsilon C = \frac{\sigma C}{E}$$

if the material is elastic and obeys Hooke's law: $\varepsilon = \sigma/E$. From the velocity of elastic waves, $E = \rho_0 C^2$. Thus

$$U_p = \frac{\sigma}{C\rho_0} \tag{3.13}$$

The particles at the free surface will have twice this velocity. In Figure 3.13, we have three regions: 1, 2, and 3. The velocities of the particles are

1. U,
2. $U - \sigma_{yd}/C\rho_0$, and
3. $U - 2\sigma_{yd}/C\rho_0$ (because of reflection).

Referring to Figure 3.14(a), we have, for the duration of the double passage of an elastic wave from the plastic boundary to the free surface and back,

$$dt = \frac{2x}{C}$$

But

$$v = \frac{dh}{dt} \qquad \therefore\ dh = v\,dt = v\,\frac{2x}{C}$$

where h is the thickness of the plastic zone. The change in U, the velocity of the back portion of the projectile during this time, is

$$dU = -\frac{2\sigma_{yd}}{\rho_0 C} \tag{3.13a}$$

We also have $dx = -(U + v)(2x/C)$. (3.13b)

These equations provide the variation in U, h, and x:

$$\frac{dh}{dt} = v \tag{3.14}$$

$$\frac{dx}{dt} = -(U + v) \tag{3.15}$$

$$\frac{dU}{dt} = -\frac{\sigma_{yd}}{\rho_0 x} \tag{3.16}$$

Dividing Eqns. (3.15) and (3.16) (or Eqns. (3.13b) and (3.13a)) yields

$$\frac{dx}{dU} = \frac{(U + v)\rho_0 x}{\sigma_{yd}} \tag{3.17}$$

But

$$v = \frac{A_0 U}{A - A_0} = \frac{A_0/A}{1 - A_0/A} U = \frac{1 - \varepsilon}{\varepsilon} U \tag{3.17a}$$

Substituting Eqn. (3.17a) into Eqn. (3.17) leads to

$$\frac{dx}{dU} = \frac{[U + U(1 - \varepsilon)/\varepsilon]\rho_0 x}{\sigma_{yd}} = \frac{\rho_0 x}{\sigma_{yd}\varepsilon} U$$

$$\frac{dx}{x} = \frac{\rho_0}{\sigma_{yd}\varepsilon} U\, dU$$

From Eqn. (3.12):

$$= \frac{\rho_0}{2\sigma_{yd}\varepsilon} dU^2 = \frac{\rho_0}{2\sigma_{yd}\varepsilon} d\left(\frac{\sigma_{yd}}{\rho_0}\frac{\varepsilon^2}{1 - \varepsilon}\right) = \frac{1}{2\varepsilon} d\left(\frac{\varepsilon^2}{1 - \varepsilon}\right)$$

By integration,

$$\ln x^2 = \int \frac{1}{\varepsilon} d\left(\frac{\varepsilon^2}{1-\varepsilon}\right) = \frac{\varepsilon^2}{\varepsilon(1-\varepsilon)} - \int \frac{\varepsilon^2}{1-\varepsilon} d\left(\frac{1}{\varepsilon}\right)$$

$$= \frac{\varepsilon}{(1-\varepsilon)} + \int \frac{1}{1-\varepsilon} d\varepsilon = \frac{\varepsilon}{1-\varepsilon} - \ln(1-\varepsilon) + K$$

$$= \frac{1}{1-\varepsilon} - \ln(1-\varepsilon) + K'$$

where K' is an integration constant.

At the moment of impact, $x = L$ and $\varepsilon = \varepsilon_1$. This boundary condition will suffice to determine the integration constant: When the projectile is brought to rest, $x = X$ and $\varepsilon = 0$. This means that the plastic region of the projectile will form a truncated cone, as shown in Figure 3.13:

$$\ln\left(\frac{x}{L}\right)^2 = \frac{1}{1-\varepsilon} - \ln(1-\varepsilon) - \frac{1}{1-\varepsilon_1} + \ln(1-\varepsilon_1)$$

For $x = X$, $\varepsilon = 0$,

$$\ln\left(\frac{X}{L}\right)^2 = 1 - \frac{1}{1-\varepsilon_1} + \ln(1-\varepsilon_1) \tag{3.18}$$

But we have (Eqn. (3.12))

$$\frac{\rho U^2}{\sigma_{yd}} = \frac{\varepsilon_1^2}{1-\varepsilon_1} \tag{3.19}$$

By eliminating ε_1, between Eqns. (3.18) and (3.19), we can determine L_1/L as a function of $\rho U^2/\sigma_{yd}$. This is shown in Figure 3.15. In Figure 3.15 L_1/L and $(L_1 - X)/L$ $(=L_1/L - X/L)$ are plotted against $\rho U_2/\sigma_{yd}$. Here L_1/L and $(L_1 - X)/L$ can be obtained if one knows X/L and ε_1. Figure 3.15 shows that L_1/L decreases from 1 to 0.3 when $\rho U_2/\sigma_{yd}$ increases from 0 to 3.5. The shortening of the cylinder (L_1/L) varies with the square of the impact velocity, as predicted from kinetic energy considerations, and with the inverse of the dynamic yield stress. The experimental results shown in Figure 3.15 are for steel projectiles. Taylor noted that the shape of the projectiles obtained experimentally is close to the one shown in Figure 3.12(b) for impact velocities of up to $\rho U^2/\sigma_{yd} = 0.5$. The experimental results were obtained by Whiffin [29].

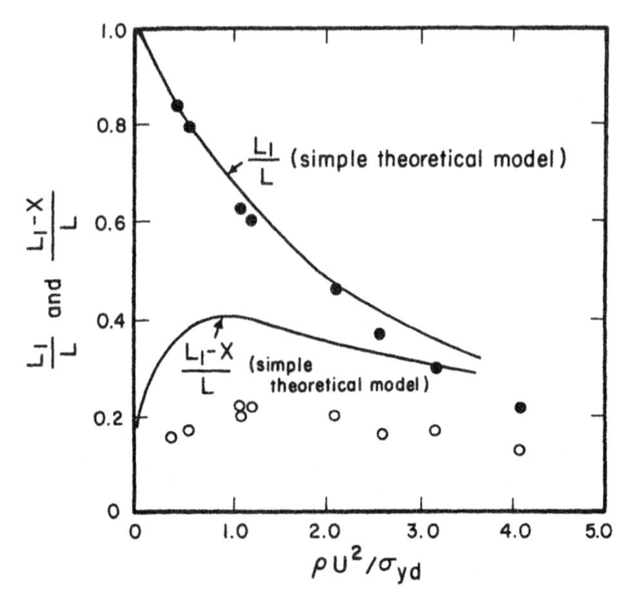

FIGURE 3.15 Theoretical predictions and observed results of L_1/L and $(L_1 - X)/L$ as a function of $\rho U^2/\sigma_{yd}$ for steel. (From Taylor [28], Fig. 2, p. 292. Reprinted with permission of the publisher.)

For steel, this corresponds to a velocity of $\sim$250 m/s. At higher velocities, the projectiles mushroom excessively and deviate from the theoretical analysis.

One very important result from Taylor's analysis is the velocity of propagation of the plastic wave, dh/dt. The change in the distance of the elastoplastic interface h with time gives the velocity of the plastic wave. In order to determine the variation of h with t, Taylor took Eqn. (3.16) and expressed it in terms of t:

$$t = -\int \frac{\rho x}{\sigma_{yd}}\, dU$$

From Eqn. (3.12),

$$U = \sqrt{\frac{\sigma_{yd}}{\rho} \frac{\varepsilon^2}{1 - \varepsilon}}$$

Thus

$$t = \left(\frac{\rho}{\sigma_{yd}}\right)^{1/2} \int_0^{\varepsilon_1} x \frac{1 - 0.5\varepsilon}{(1 - \varepsilon)^{3/2}}\, d\varepsilon$$

But

$$\frac{\rho}{\sigma_{yd}} = \frac{1}{U^2} \frac{\varepsilon_1^2}{1 - \varepsilon_1}$$

and

$$\frac{Ut}{L} = \frac{\varepsilon_1}{(1 - \varepsilon_1)^{1/2}} \int_0^{\varepsilon_1} \frac{x}{L} \frac{1 - 0.5\varepsilon}{(1 - \varepsilon)^{3/2}} \, d\varepsilon$$

Numerical integration of this equation for different values of ε_1 leads to the curves shown in Figure 3.16. It can be seen that h/L is nearly proportional to Ut/L, resulting in a nearly constant velocity of the plastic front (dh/dt). For a strain of 0.5, the ratio between the ordinate and the abscissa is given by

$$\frac{h/L}{Ut/L} = \frac{h}{Ut} = \frac{1}{U}\frac{h}{t} \simeq 1$$

Thus, the velocity of the plastic wave is approximately equal to the impact velocity U for a strain of 0.5. For the strain of 0.7, it is equal to approximately one-half of the impact velocity. It is interesting to note that Taylor's analysis totally ignores the plastic stress–strain curve that establishes the velocity of the plastic wave in von Karman and Duwez's [4] theory.

3.6.2 The Wilkins–Guinan Analysis of the Taylor Test

Figure 3.17 shows a photograph of copper specimens impacted at different velocities. It is readily observed that one has significant deviation from Taylor's

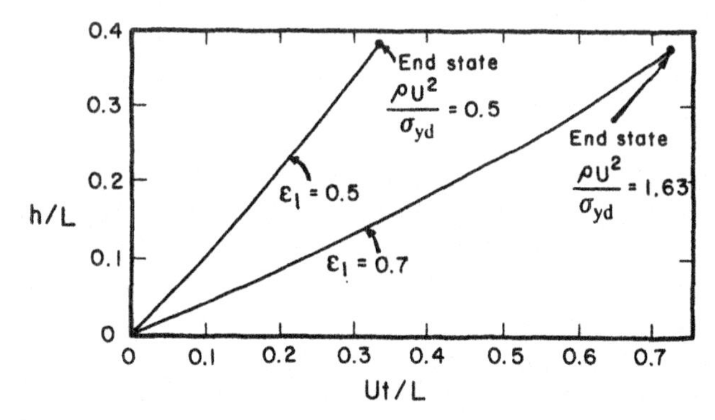

FIGURE 3.16 Propagation of plastic wave front in cylindrical projectile as a function of impact velocity. (From Taylor [28], Fig. 3, p. 294. Reprinted with permission from the publisher.)

FIGURE 3.17 Photograph of copper specimens subjected to normal impact at different velocities. Specimen on the left is original size.

analysis. The main differences are as follows:

1. The deformed part is not conical but presents a "mushroom" at the end that accentuates itself as the velocity of impact increases.
2. The boundary between the plastically deformed and the undeformed regions cannot be easily seen. Thus, it is very difficult to obtain a reliable reading for X (see Fig. 3.13).

Wilkins and Guinan [30] performed a simple mathematical analysis that led to very good predictions of the fractional reductions in length, L_1/L, for a number of materials. They first verified that L_1/L_0 was independent of the original length, L_0. This is shown in Figure 3.18, for 1090 steel specimens with three different lengths: 0.78, 2.35, and 4.70 cm. With this in mind, they developed the following analysis.

They assumed that the change in length with time was equal to the instantaneous velocity U:

$$\frac{dL}{dt} = -U \tag{3.20}$$

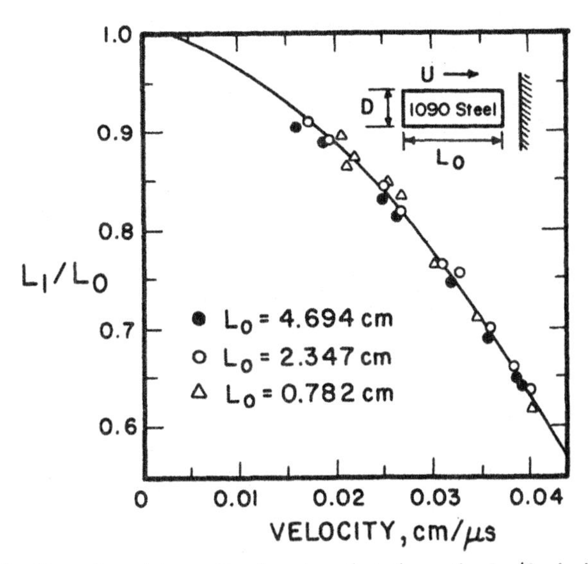

FIGURE 3.18 Experimental results showing that the ratio L_1/L_0 is independent of L_0 for impacts of steel cylinders on a rigid boundary. (From Wilkins and Guinan, [30], Fig. 2, p. 1200. Reprinted with permission of the publisher.)

Applying Newton's second law for the force exerted by the specimen on the rigid wall (equal to the product of the dynamic yield stress σ_{yd} and the cross-sectional area) yields

$$\sigma_{yd}A = -\rho_0 LA\frac{dU}{dt} \tag{3.21}$$

where $\rho_0 LA$ is the mass of the specimen and dU/dt its deceleration. By substituting (3.21) into (3.20), one obtains

$$\frac{dL}{L} = \frac{\rho_0 U}{\sigma_{yd}}\,dU$$

$$\ln\frac{L_1}{L_0} = -\frac{\rho_0 U^2}{2\sigma_{yd}}$$

This equation presents a dependence of L_1/L_0 on $(\rho U^2/\sigma_{yd})$ similar to Taylor's. Wilkins and Guinan [30] introduced a correction into this formulation that rendered the equation applicable to a number of materials. They assumed that a plastic front would form at a distance h from the impact surface and that this plastic front would define a new boundary condition. This plastic front would move in the early stages of plastic deformation to a fixed position from the impact interface. This distance h was found to be independent of velocity and

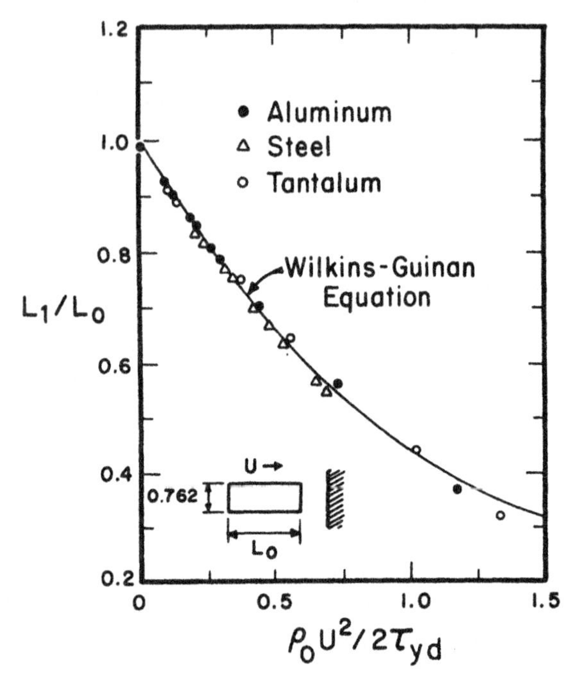

FIGURE 3.19 Comparison between experimental data for aluminum, steel, and tantalum and predictions of Eq. (3.22). (From Wilkins and Guinan, [30]. Reprinted with permission of the publisher.)

proportional to the specimen length L_0; h/L_0 = const. Thus

$$\ln L \Big|_{L_0 - h}^{L_1 - h} = \frac{\rho_0 U^2}{2\sigma_{yd}}\Big|_U^0$$

$$\frac{L_1}{L_0} = \left(1 - \frac{h}{L_0}\right) \exp\left(-\frac{\rho_0 U^2}{2\sigma_{yd}}\right) + \frac{h}{L_0} \qquad (3.22)$$

Figure 3.19 shows the very successful correlation obtained between the experimental data (aluminum, steel, tantalum) and Eqn. (3.22), assuming that L/L_0 = 0.12. This analysis assumes a constant-flow stress, which is inferred from the match between experiments and analysis.

Example 3.1 Copper has the stress–strain curve shown below. Determine (1) the maximum strain ε_1, (2) the stress σ_1, and (3) the plastic wave velocity C_1 as a function of impact velocity for the experimental configuration used by von Karman and Duwez.

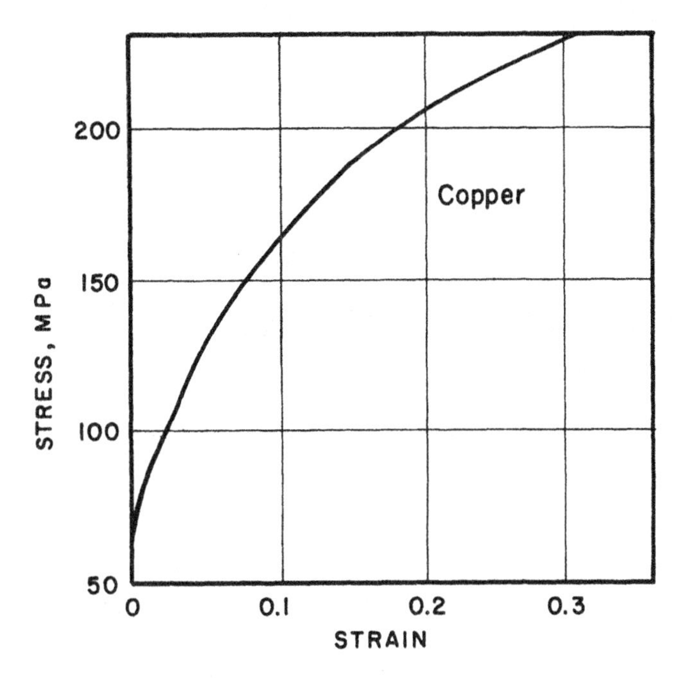

We first have to obtain an equation for the stress–strain curve. We do this by assuming an expression of the form

$$\sigma = k\varepsilon^n$$

By plotting the curve on log paper, we obtain

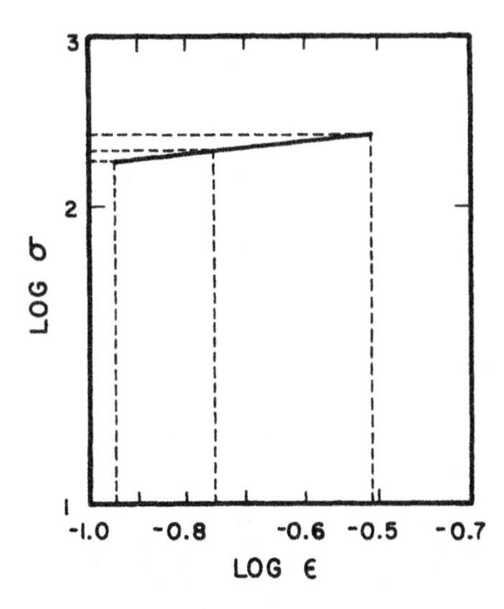

1. The equation $\log \sigma = k + n \log \varepsilon$ provides

$$\begin{array}{ll} K \rightarrow \text{intercept} & k = 2.54 \\ n \rightarrow \text{slope} & n = 0.33 \\ \text{Maximum stress} & \varepsilon_{max} = n = 0.33 \end{array}$$

This gives the maximum wave velocity:

$$C_1 = \left(\frac{d\sigma/d\varepsilon}{\rho_0}\right)^{1/2} = \left(\frac{kn\varepsilon_1^{n-1}}{\rho_0}\right)^{1/2}$$

$$C_0 = \sqrt{\frac{E}{\rho}} = 3812 \text{ m/s} \quad \text{(for wire or slender cylinder)}$$

$$C_1 = 164.1 \text{ m/s}$$

Above this velocity, the wire will simply break.

For strain as a function of impact velocity, we use Eqn. (3.8):

$$\varepsilon_1 = \left[\frac{\rho_0 V_1^2(n + 1)^2}{4kn}\right]^{1/(n+1)} = 4.4 \times 10^{-4} V_1^{1.5}$$

2. For stress as a function of impact velocity,

$$\sigma = k\varepsilon_1^n = 27.09 V_1^{0.5}$$

3. For plastic wave velocity as a function of impact velocity,

$$C_1 = \left[\frac{d\sigma/d\varepsilon}{\rho_0}\right]^{1/2} = \left[\frac{kn\varepsilon_1^{n-1}}{\rho_0}\right]^{1/2}$$

$$= \left[\frac{(kn)^{2/(n-1)} V_1^2(n + 1)^2}{4\rho_0^{2/(n-1)}}\right]^{(n-1)/2(n+1)}$$

Example 3.2 For the material given above, determine the approximate shape of the wave front at times of 0.05 and 0.1 ms after the falling weight impacts the bottom of the wire. Assume that the weight drops from a height of 1 m.

We have to calculate the impact velocity V_1. This velocity is given by

$$V_1 = \sqrt{2gh} = \sqrt{2 \times 9.8 \times 1} = 4.4 \text{ m/s}$$

Inserting the value into the last equation from Example 3.1, one has

$$C_1 = \left[\frac{(346 \times 10^6 \times 0.33)^{2/(0.33-1)}(0.33 + 1)^2 V_1^2}{4 \times (8.9 \times 10^3)^{2/(0.33-1)}}\right]^{(0.33-1)/2(0.33+1)}$$

$$\log C_1 = -0.25[-2.98 \log 12.8 \times 10^3 + \log 8.49]$$

$$C_1 = 660 \text{ m/s}$$

This is higher than the allowable speed. Wire will break. Assuming $C_1 = 164$ m/s, we obtain the following plots:

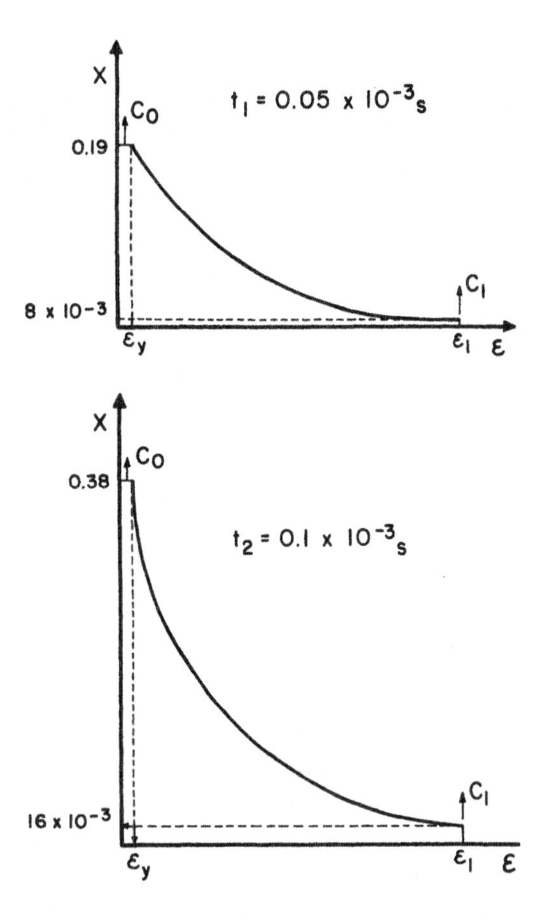

Example 3.3 Make a cylindrical projectile of copper with a length of 10 cm and a diameter of 3 cm and fire it from a gun against a rigid target at velocities of 150 and 400 m/s. Determine the final shape of the projectile using Taylor's analysis. Assume a yield stress that is equal to 100 MPa.

We use Eqn. (3.19) and Figure 3.13:

$$\frac{\rho U^2}{\sigma_{yd}} = \frac{8930 \times 150^2}{100 \times 10^6} = 2$$

$$\frac{\rho U^2}{\sigma_{yd}} = \frac{8930 \times 400^2}{100 \times 10^6} = 14.2$$

Taylor's theory is not valid for 400 m/s. We will only perform the calculation for 150 m/s. From Figure 3.13

$$\frac{L_1 - X}{L} = 0.35 \qquad \frac{L_1}{L} = 0.5$$

From these dimensions, we obtain

$$L_1 = 4.8 \times 10^{-2} \text{ m} \qquad X = 1.28 \times 10^{-2} \text{ m}$$

We also have to find the diameter at the impact interface. This is found by considering the constancy in the volume of the projectile:

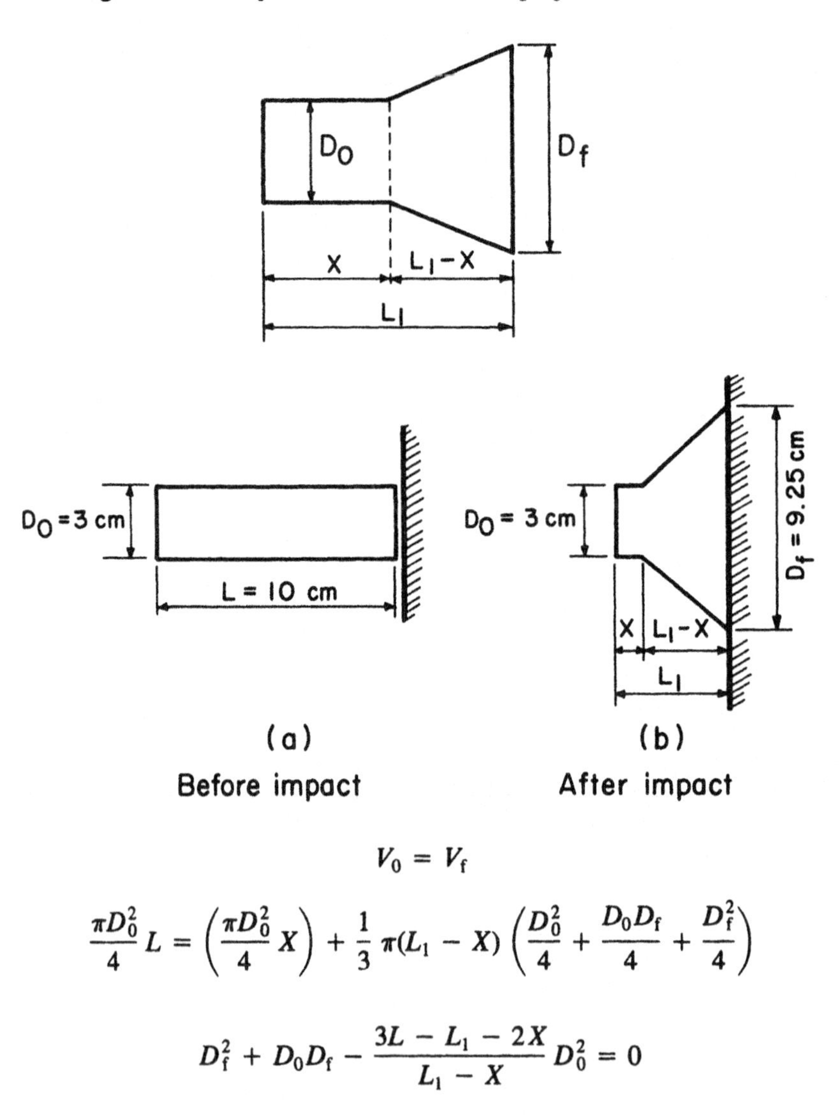

$$V_0 = V_f$$

$$\frac{\pi D_0^2}{4} L = \left(\frac{\pi D_0^2}{4} X\right) + \frac{1}{3}\pi(L_1 - X)\left(\frac{D_0^2}{4} + \frac{D_0 D_f}{4} + \frac{D_f^2}{4}\right)$$

$$D_f^2 + D_0 D_f - \frac{3L - L_1 - 2X}{L_1 - X} D_0^2 = 0$$

REFERENCES

1. R. J. Clifton, in *Mechanics Today*, Vol. 1, ed. S. Nemat-Nasser, Pergamon, Elmsford, NY, 1972, p. 102.
2. L. E. Malvern, *Q. Appl. Math.*, **8** (1951), 405.
3. V. V. Sokolovsky, *Prikl. Math. Mekh.*, **12** (1948), 261.
4. T. von Karman and P. Duwez, *J. Appl. Phys.*, **21** (1950), 987.
5. G. I. Taylor, *J. Inst. Civil Eng.*, **26** (1946), 486.
6. K. A. Rakhmatulin, *Appl. Math. Metch.*, **9**(1) (1945).
7. J. F. Bell, *J. Appl. Phys.*, **31** (1960), 2188.
8. P. Obernhuber, S. R. Bodner, and M. Sagir, *J. Appl. Math. Phys.* (TAMP), **37** (1986), 714.
9. T. Nicholas, in *Impact Dynamics*, ed. J. Zukas et al., Wiley, New York, 1982, p. 277.
10. W. Goldsmith, *Impact*, Edward Arnold, London, 1960, p. 162.
11. E. H. Lee, *J. Appl. Mech.*, **36** (1969), 1.
12. W. Herrmann and J. W. Nunziato, in *Dynamic Response of Materials to Intense Impulsive Loading*, eds. P. C. Chou and A. K. Hopkins, AFML, WPAFB, 1972, p. 123.
13. R. J. Clifton, in *Shock Waves and the Mechanical Properties of Solids*, eds. J. J. Burke and V. Weiss, Syracuse University Press, Syracuse, 1971, p. 3.
14. W. Herrmann, D. L. Hicks, and E. G. Young, cited in V. V. Sokolovsky, *Prikl. Math. Mekh.*, **12** (1948), 261.
15. National Materials Advisory Board, Report No. NMAB-356, 1980, National Academy of Sciences, 1980.
16. H. H. Bleich and I. Nelson, *Trans. ASME, J. Appl. Mech.*, **33** (1966), 149.
17. T. C. T. Ting and N. Nan, *Trans. ASME, J. Appl. Mech.*, **36** (1969), 189.
18. J. Lipkin and R. J. Clifton, *J. Appl. Mech.*, **37** (1970), 1107.
19. A. S. Abou-Sayed, R. J. Clifton, and L. Hermann, *Exptl. Mech.*, **6** (1976), 127.
20. K. S. Kim and R. J. Clifton, *J. Appl. Mech.*, **47** (1980), 11.
21. R. J. Clifton, A. Gilat, and C. H. Li, in *Material Behavior under High Stress and Ultrahigh Loading Rates*, eds. J. Mescall and V. Weiss, Plenum, New York, 1983.
22. R. J. Clifton, *J. Appl. Mech.*, **50** (1983), 941.
23. T. Mashimo and K. Nagayama, Jap., *J. Appl. Phys.*, **25**(Suppl. 25-1) (1986), 103.
24. B. Hopkinson, *Roy. Soc. Phil. Trans.*, **A213** (1914), 437.
25. R. M. Davies, *Roy. Soc. Phil. Trans.*, **A240** (1948), 375.
26. H. Kolsky, *Proc. Roy. Soc. Lond.*, **62B** (1949), 676.
27. R. J. Wasley, *Stress Wave Propagation in Solids*, Dekker, New York, 1973.
28. G. I. Taylor, *Proc. Roy. Soc. Lond.*, **194** (1948), 289.
29. A. C. Whiffin, *Proc. Roy. Soc. Lond.*, **194** (1947), 300.
30. M. L. Wilkins and M. W. Guinan, *J. Appl. Phys.*, **44** (1973), 1200.

CHAPTER 4

Shock Waves

4.1 INTRODUCTION

When the amplitude of stress waves greatly exceeds the dynamic flow strength of a material, one can effectively neglect the shear stresses, in comparison with the compressive hydrostatic component of the stress. One therefore considers a high-pressure state traveling into a material. For an ideal gas, one has the following equation of state for an isentropic process:

$$PV^{\gamma} = K = \text{const}$$

By differentiating,

$$\gamma P V^{\gamma - 1}\, dV + V^{\gamma}\, dP = 0 \qquad \frac{dP}{dV} = -\gamma \frac{P}{V}$$

Since the value P/V increases with pressure, so does $|dP/dV|$. Thus, the compressibility of ideal gases decreases with pressure. We will assume that the disturbance is occurring isentropically. From Chapters 1–3, we know that the velocity of a disturbance is given by $(d\sigma/d\varepsilon/\rho)^{1/2}$. For a gas, in a one-dimensional configuration, this is equivalent to $(dP/dV/\rho)^{1/2}$. The velocity of the disturbance is proportional to $(dP/dV)^{1/2}$. It can be concluded from the above that high-amplitude isentropic disturbances travel faster than low-amplitude ones in gases. The physical reason that the compressibility of solids and liquids decreases with pressure is due to the external electron shells of the atoms being "pushed" against each other and interpenetrating. This is explained in Section 5.2. This is the *sine qua non* requirement for shock waves. A disturbance front will "steepen up" as it travels through the material because the higher amplitude regions of the front travel faster than the lower amplitude regions. This leads to a *shock wave*, which is defined, simply and sweetly, as a discontinuity in pressure, temperature (or internal energy), and density. Hence

$$\left|\frac{d\sigma}{dV}\right| \uparrow \quad \text{as } \sigma \uparrow \quad \rightarrow \text{shock front}$$

$$\left|\frac{d\sigma}{dV}\right| \downarrow \quad \text{as } \sigma \uparrow \quad \rightarrow \text{dispersion of wave}$$

For solids, we have to differentiate between the deviatoric and hydrostatic components of stress. When the former are negligible, we ignore them and apply the treatment for fluids without remorse.

The concept of shock wave propagation and the equations of conservation of mass, momentum, and energy can be very easily understood and derived by means of a simplified conceptual framework, due to Davis [1]. One imagines a cylinder of unit cross-sectional area onto which a piston penetrates. This is shown in Figure 4.1. Initially, the piston is at rest. It then is pushed into the compressible material, initially, at a pressure P_0 and having density ρ_0 at a velocity U_p. After a time t_1, the highly compressed region ahead of the piston has moved forward by a distance equal to $U_s t_1$, where U_s is the velocity of propagation of the disturbance ahead of the piston. During this time interval, the piston has moved by a distance equal to $U_p t_1$. The compressed region has a pressure P and a density ρ. This can be compared to a snow plow moving into snow. If the snow plow moves at a velocity U_p, the packed snow ahead of it moves at a velocity equal to U_p. However, the region separating the packed from the fresh snow will move at a velocity U_s larger than U_p, because snow is being continuously added to the packed region. A shock front can then be visualized as a plane separating "moving" from "stationary" fluid in a cylinder with a moving piston. The velocity of the shock wave is greater than the piston velocity, such that when the piston has traveled a distance $U_p t_1$ in time t_1, the shock wave has traveled a distance $U_s t_1$. We can now look into the conservation of mass, momentum, and energy in the compressed region. In Figure 4.1, two times were used, t_1 and t_2. Conceptually, there is no difference and time t_2 was added only to describe the process sequence. Since the internal cross-sectional area of the cylinder is unity, one has, for the compression length $(U_s - U_p)t_1$,

$$U_s t_1 \rho_0 = \rho(U_s - U_p)t_1$$

The term $U_s t_1 \rho_0$ is the mass of the initial uncompressed material. This equation applies to any time t, and

$$U_s \rho_0 = (U_s - U_p)\rho \tag{4.1}$$

The momentum is defined as the product of mass and velocity. The change of momentum of a system is equal to the impulse given to this system. Initially,

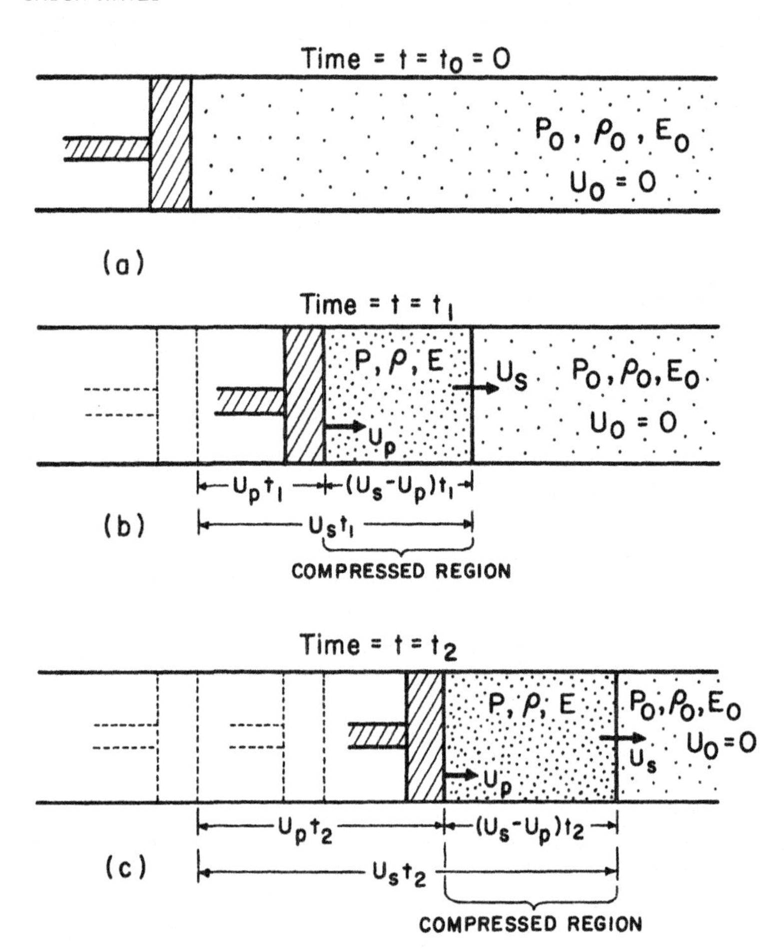

FIGURE 4.1 Successive positions of an idealized piston moving into a cylinder with compressible fluid.

the momentum is zero. The momentum at t_1 is the product of the mass $(U_s - U_p)\rho t$ and the velocity of the material particles in the compressed volume (U_p). Thus

$$\rho(U_s - U_p)U_p t - 0 = (P - P_0)t$$

or

$$\rho(U_s - U_p)U_p = P - P_0 \tag{4.2}$$

The equation for the conservation of energy simply states that the work by the external forces is equal to the change in internal energy plus change of kinetic energy. In the compression region, at time t, the change in internal energy is

$$E_1[\rho(U_s - U_p)t] - E_0[\rho_0 U_s t] = (E_1 - E_0)\rho_0 U_s t$$

The change in kinetic energy is given by $\frac{1}{2}mv^2$:

$$\tfrac{1}{2}\rho(U_s - U_p)U_p^2 t - 0 = \tfrac{1}{2}\rho_0 U_s U_p^2 t$$

For a stationary shock wave, the change in kinetic energy is equal to the change in internal energy.

$$E_1 - E_0 = \tfrac{1}{2}U_p^2 \tag{4.3}$$

Equations (4.1)–(4.3) are, in essence, the Rankine–Hugoniot [2, 3] relationships for a material in which a pressure discontinuity propagates. These equations apply to a piston moving into a compressible medium (gas) and can be extended to a shock wave propagating into gas, liquid, or solid or to a detonation wave. This treatment can be regarded as an analog to the treatment for developing Rankine–Hugoniot conservation equations for shock waves.

4.2 HYDRODYNAMIC TREATMENT

Shock waves are characterized by a steep front and require a state of uniaxial strain (no lateral flow of materials) which allows the buildup of the hydrostatic component of stress to high levels. When this hydrostatic component reaches levels that exceed the dynamic flow stress (i.e., the flow stress at the strain rate established at the front) by several factors, one can, to a first approximation, assume that the solid has no resistance to shear (i.e., $\mu = 0$). The calculation of the shock wave parameters is based, in its simplest form, on the Rankine–Hugoniot [2, 3] conservation equations.

4.2.1 Basic Assumptions

- A shock is a discontinuous surface and has no apparent thickness.
- The shear modulus of the material is assumed to be zero, such that it responds to the wave as a fluid; hence the theory is restricted to higher pressures.
- Body forces (such as gravitational) and heat conduction at the shock front are negligible.
- There is no elastoplastic behavior.
- Material does not undergo phase transformations

The *fundamental requirement* for the establishment of a shock wave is that the velocity of the pulse, U, increases with increasing pressure, that is,

$$\left(\frac{\partial^2 P}{\partial^2 U}\right) > 0 \quad \text{as } P\uparrow,\ U\uparrow$$

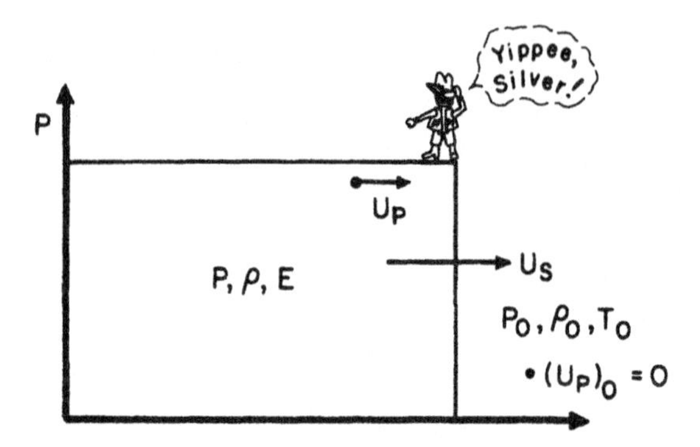

FIGURE 4.2 Schematic of a shock front. Lone Ranger, riding at the front, sees particles moving at him with U_s and moving away at $U_s - U_p$. Lucky guy!

The mathematical treatment of shock waves (discontinuities in pressure, density, and temperature) was originally developed by Rankine and Hugoniot for fluids. The equations can be easily developed by considering regions immediately ahead of and behind the shock front. Figure 4.2 illustrates a shock front. Ahead of the front, the pressure is P_0, the density is ρ_0, and the temperature is T_0; behind it they are P, ρ, and T, respectively. The velocity of the front is U_s; the particles (or atoms) are stationary ahead of the front. At the front and behind it, they are moving at a velocity U_p; this displacement of particles is responsible for the pressure buildup.

If one considers the center of reference as the shock front and moves with it at a velocity U_s into a region of particle velocity $U_p = U_0$ and density ρ_0, then the apparent velocity of the fluid moving toward the center of reference is $U_s - U_0$. At the same time, the material leaving the front or receding from the center of reference is moving at a velocity $U_s - U_p$. With this in mind, we will set up the equations for the conservation of mass, momentum, and energy.

Conservation of Mass.

Mass in = mass out (per unit area)
Mass moving toward the front: $A\rho_0(U_s - U_0)\,dt$
Mass moving away from the front: $A\rho(U_s - U_p)\,dt$

Hence,

$$A\rho_0(U_s - U_0)\,dt = A\rho(U_s - U_p)\,dt$$

or, if $U_0 = 0$

$$\rho_0 U_s = \rho(U_s - U_p) \tag{4.4}$$

This is the equation for the conservation of mass.

Conservation of Momentum. The conservation of momentum requires that the difference in momentum be equal to the impulse per unit cross-sectional area:

$$\text{Momentum} = \text{mass} \times \text{velocity}$$

$$\text{Impulse} = F\,dt$$

$$\text{Difference in momentum} = (\text{momentum})_1 - (\text{momentum})_0$$

$$= \underbrace{\rho A(U_s - U_p)\,dt}_{\text{mass}}\ \underbrace{U_p}_{\text{velocity}} - \underbrace{\rho_0 A(U_s - U_0)\,dt}_{\text{mass}}\ \underbrace{U_0}_{\text{velocity}}$$

$$\text{Impulse} = F\,dt$$

$$= (PA - P_0 A)\,dt$$

Then, equating change in momentum with impulse, we obtain

$$A\rho(U_s - U_p)U_p\,dt - A\rho_0(U_s - U_0)\,U_0\,dt = (P - P_0)\,A\,dt$$

$$\rho_0(U_s - U_0)(U_p - U_0) = P - P_0$$

If $U_0 = 0$

$$(P - P_0) = \rho_0 U_s U_p \tag{4.5}$$

This is the equation for the conservation of momentum, and the quantity $\rho_0 U_s$ is often called the *shock impedance.*

Conservation of Energy. The conservation of energy is obtained by setting up an equation in which the work done by P minus the work done by P_0 is equal to the difference in the total energy (kinetic plus internal) between the two sides of the front. As an exercise, we will change to the original (stationary) reference system. The difference of work done by P and P_0 is given by:

$$\Delta W = \underbrace{(PA)}_{\text{force}}\ \underbrace{(U_p dt)}_{\text{distance}} - \underbrace{(P_0 A)}_{\text{force}}\underbrace{(U_0 dt)}_{\text{distance}}$$

The difference in total energy (kinetic plus internal) per unit mass is equal to the final minus the initial energy:

$$= \tfrac{1}{2}\,[\rho A(U_s - U_p)\,dt]\,U_p^2 + EA\rho(U_s - U_p)\,dt$$

$$- \{\tfrac{1}{2}[\rho_0 A(U_s - U_0)\,dt]\,U_0^2 + E_0 A\rho_0(U_s - U_0)\,dt\}$$

So equating ΔW to ΔE and taking $U_0 = 0$ yields

$$PU_p\, dt\, A = [\tfrac{1}{2}\rho A(U_s - U_p)\, dt]\, U_p^2 + EA\rho(U_s - U_p)\, dt - E_0 A\rho_0(U_s)\, dt$$

or

$$PU_p = \tfrac{1}{2}\rho(U_s - U_p)U_p^2 - E_0\rho_0 U_s + E\rho(U_s - U_p)$$

But from conservation of mass, $\rho(U_s - U_p) = \rho_0 U_s$. Substituting into the above we get

$$PU_p = \tfrac{1}{2}\rho_0 U_s U_p^2 - E_0\rho_0 U_s + E\rho_0 U_s$$

or

$$PU_p = \tfrac{1}{2}\rho_0 U_s U_p^2 + \rho_0 U_s(E - E_0) \tag{4.6}$$

So, the conservation equations are

$$\rho_0 U_s = \rho(U_s - U_p) \quad \text{mass,} \qquad \text{Eqn. (4.4)}$$

$$P - P_0 = \rho_0 U_s U_p \quad \text{momentum,} \qquad \text{Eqn. (4.5)}$$

$$PU_p = \tfrac{1}{2}\rho_0 U_s U_p^2 + \rho_0 U_s(E - E_0) \quad \text{energy,} \qquad \text{Eqn. (4.6)}$$

The conservation-of-energy equation (4.6) can be simplified to obtain a more common form:

$$E - E_0 = \frac{PU_p}{\rho_0 U_s} - \frac{1}{2}\rho_0 \frac{U_s U_p^2}{\rho_0 U_s}$$

But from the conservation of momentum, $U_p = (P - P_0)/\rho_0 U_s$. Substituting for U_p in the above, we obtain

$$E - E_0 = \frac{P(P - P_0)}{\rho_0^2 U_s^2} - \frac{1}{2}\frac{(P - P_0)^2}{\rho_0^2 U_s^2} \tag{4.7}$$

Again, from conservation of mass, $\rho_0 U_s = \rho(U_s - U_p)$, and using Eqn. (4.5):

$$(\rho_0 - \rho)U_s = -\rho U_p = -\frac{\rho(P - P_0)}{\rho_0 U_s}$$

or

$$\rho_0 U_s^2 = -\rho(P - P_0) \frac{1}{(\rho_0 - \rho)}$$

If $1/\rho = V$, then, simplifying, we get

$$\rho_0^2 U_s^2 = \frac{P - P_0}{V_0 - V}$$

Substituting this back into Eqn. (4.7), we get

$$E - E_0 = P(P - P_0) \cdot \frac{V_0 - V}{P - P_0} - \frac{1}{2} \frac{(P - P_0)^2}{P - P_0} (V_0 - V)$$

or

$$E - E_0 = \tfrac{1}{2}(P + P_0)(V_0 - V) \tag{4.8}$$

Equation (4.8) is the more common form of the conservation of energy. In the above conservation equations (4.4)–(4.6), there are five variables: pressure (P), particle velocity (U_p), shock velocity (U_s), specific volume (V), or density (ρ), and energy (E). Hence, an additional equation is needed if one wants to determine all parameters as a function of one of them. This fourth equation, which can be conveniently expressed as the relationship between shock and particle velocities, has to be experimentally determined. A polynomial equation with parameters C_0, S_1, S_2, S_3, . . . empirically describes the relationship between U_s and U_p:

$$U_s = C_0 + S_1 U_p + S_2 U_p^2 + \cdots \tag{4.9}$$

Equation (4.9) is often known as the equation of state (EOS) of a material. Here, S_1 and S_2 are empirical parameters and C_0 is the sound velocity in the material at zero pressure. For most metals, $S_2 = 0$, and Eqn. (4.9) thus reduces to a linear relationship:

$$U_s = C_0 + S_1 U_p \tag{4.10}$$

The linear relationship between U_s and U_p describes fairly well the shock response of materials *not* undergoing phase transitions. Values of C_0 and S_1 are often tabulated in the literature; hence knowing these values and applying Eqns. (4.4)–(4.6), (4.8), and (4.10), one can calculate the P–U_s, P–U_p, P–ρ, P–V/V_0, E–U_s, and other relationships.

Figure 4.3 shows the curve of U_s versus U_p (experimentally measured equation of state) for some typical materials; as expected, the relationship is linear.

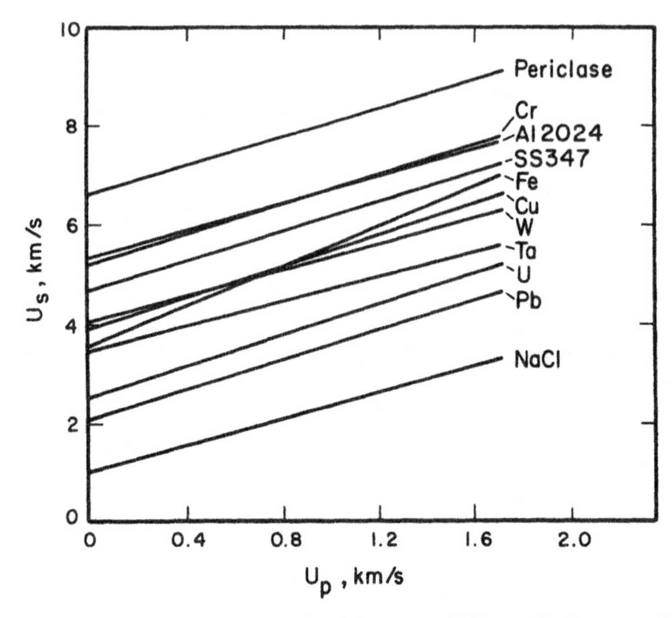

FIGURE 4.3 Experimentally measured EOS curve of U_s vs. U_p for several materials.

Applying the conservation equations, one can then obtain plots of pressure (P) versus specific volume (V/V_0) and pressure (P) versus particle velocity (U_p), as shown in Figures 4.4(a) and (b), respectively. Tables 4.1 and 5.1 list shock wave parameters for some representative materials. A good source of shock wave parameters is Appendix F of Meyers and Murr [4]. These tabulations are based on the extensive work conducted by Walsh and Christian [5], Rice et al. [6], and McQueen et al. [7]. The reader will find, in these sources, the specific values for the equations of state for many materials. See also Section 4.5.

It is important to note that if the material has porosity or undergoes phase transformation, the linear equation of state is no longer applicable and has to be modified. Equation (4.8), derived from the conservation of energy, establishes a relation between P and ρ immediately behind the shock front. This pressure–density relationship is usually known as the Rankine–Hugoniot equation, or simply the "Hugoniot," and in graphic form is shown in Figure 4.5 as a P–V curve. So, a Hugoniot is defined as the locus of all shocked states in a material and essentially describes the material properties. The straight line joining (P_0, V_0) and (P_1, V_1) is known as the Rayleigh line and refers to the shock state at P_1. It is very important to realize that when pressure is increased in a shock front, it does not follow the P–V/V_0 path. Rather, it changes discontinuously from its initial value P_0 to its value P_1. The Rayleigh line is an important line, as will be seen below. If one has a shock pulse of amplitude P_1, one does not reach this point by following the Hugoniot line. There is a discontinuity in pressure and density as defined by the Rankine–Hugoniot relationships. This discontinuity is explained by the slope of the Rayleigh line

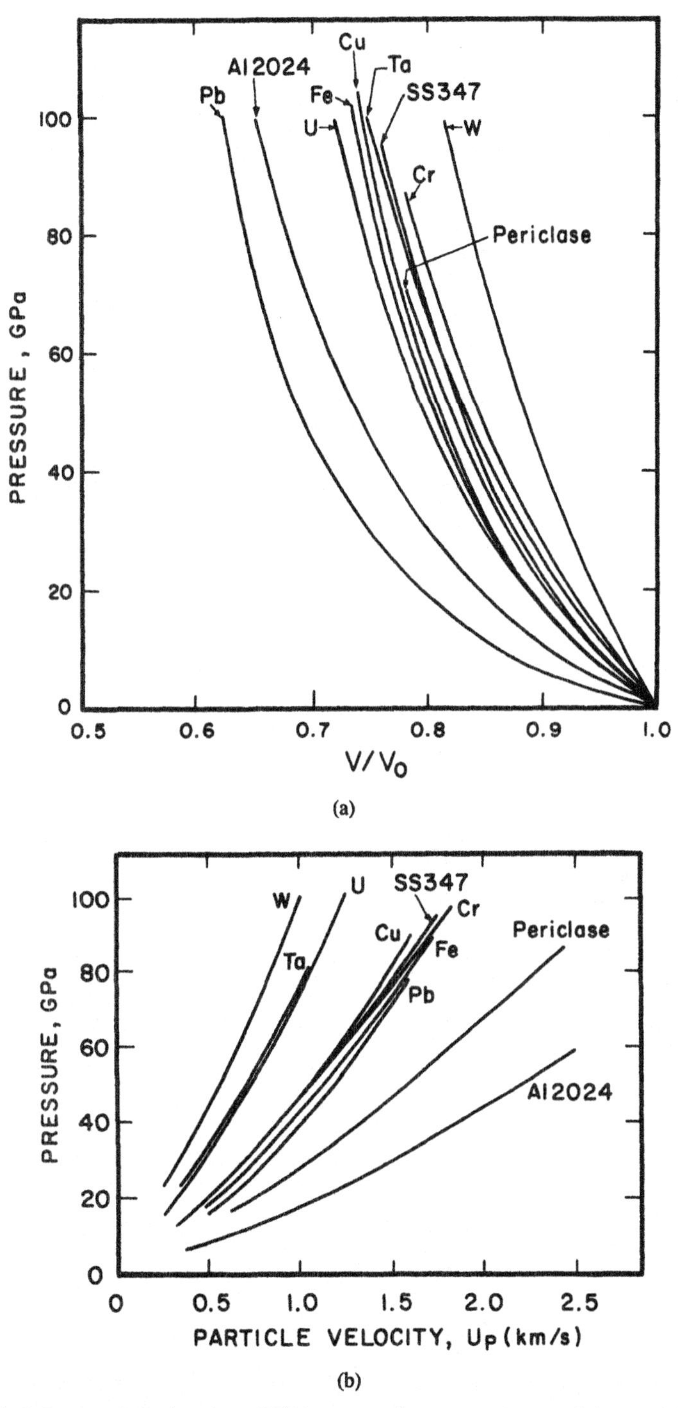

FIGURE 4.4 (a) Calculated P–V/V_0 curve for several materials. (b) Calculated P–U_p curve for several materials.

TABLE 4.1 Shock Wave Parameters for Representative Metals

Metal	Pressure (GPa)	ρ (g/cm^3)	V/V_0	U_s (km/s)	U_p (km/s)	C (km/s)
2024Al	0	2.785	1.	5.328	0	5.328
	10	3.081	0.904	6.114	0.587	6.220
	20	3.306	0.842	6.751	1.064	6.849
	30	3.490	0.798	7.302	1.475	7.350
	40	3.647	0.764	7.694	1.843	7.774
Cu	0	8.930	1.	3.940	0	3.94
	10	9.499	0.940	4.325	0.259	4.425
	20	9.959	0.897	4.656	0.481	4.808
	30	10.349	0.863	4.950	0.679	5.131
	40	10.668	0.835	5.218	0.858	5.415
Fe	0	7.85	1.	3.574	0	3.574
	10	8.479	0.926	4.155	0.306	4.411
	20	8.914	0.881	4.610	0.550	4.054
	30	9.258	0.848	4.993	0.759	5.602
	40	9.543	0.823	5.329	0.945	6.092
Ni	0	8.874	1.	4.581	0	4.581
	10	9.308	0.953	4.916	0.229	5.005
	20	9.679	0.917	5.213	0.432	5.357
	30	9.998	0.888	5.483	0.617	5.661
	40	10.285	0.863	5.732	0.786	5.933
304SS	0	7.896	1.	4.569	0	4.569
	10	8.326	0.948	4.950	0.256	5.051
	20	8.684	0.909	5.283	0.479	5.439
	30	8.992	0.878	5.583	0.681	5.770
	40	9.264	0.852	5.858	0.865	6.061
Ti	0	4.528	1.	5.220	0.	5.220
	10	4.881	0.928	5.527	0.4	5.420
	20	5.211	0.869	5.804	0.761	5.578
	30	5.525	0.820	6.059	1.094	5.708
	40	4.826	0.777	6.296	1.403	5.815
W	0	19.224	1.	4.029	0.	4.029
	10	19.813	0.970	4.183	0.124	4.207
	20	20.355	0.944	4.326	0.240	4.365
	30	20.849	0.922	4.462	0.350	4.508
	40	21.331	0.901	4.590	0.453	4.638

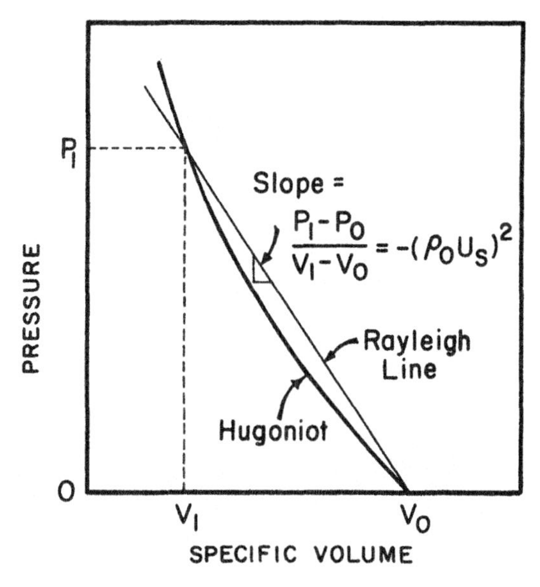

FIGURE 4.5 Characteristic Hugoniot (P–V) curve showing Rayleigh line.

that is proportional to the square of the velocity U_s of the shock wave. One can see this from the following derivation:

$$P - P_0 = \rho_0 U_s U_p$$

or

$$\frac{P - P_0}{U_p} = \rho_0 U_s$$

But

$$\rho_0 U_s = \rho(U_s - U_p)$$

$$= \rho U_s - \rho U_p$$

$$\therefore U_p = U_s \frac{\rho - \rho_0}{\rho}$$

Thus

$$\frac{P - P_0}{(\rho - \rho_0)/\rho} = \rho_0 U_s^2 = \frac{P - P_0}{1 - \rho_0/\rho}$$

But

$$\frac{\rho_0}{\rho} = \frac{V}{V_0}$$

So

$$\frac{P - P_0}{1 - V/V_0} = \rho_0 U_s^2 = \frac{P - P_0}{(V_0 - V)/V_0}$$

This is equivalent to

$$\frac{P - P_0}{V - V_0} = -\rho_0^2 U_s^2 \tag{4.11}$$

$(P - P_0)/(V - V_0)$ is the slope of the Rayleigh line in Figure 4.5. One sees clearly that the higher the pressure, the higher the magnitude of the slope and the higher the velocity of the wave.

In a P–U_p plot, the Rayleigh lines have a slope that is simply given by the equation for the conservation of momentum:

$$\frac{P - P_0}{U_p} = \rho_0 U_s$$

This rationale explains the concave shapes of the P–V and P–U_p plots.

4.3 IMPACT

Planar, normal, parallel, impact [8] is the simplest situation encountered and the method of production of shock waves most commonly used. Planar impact implies that two flat surfaces are involved. Parallel impact means that the two surfaces are parallel, so that contact of these surfaces occurs simultaneously, that is, all points of the two surfaces establish contact at the same time. Normal impact (in contrast to inclined impact) implies that the direction of motion of the projectile is perpendicular to its surface. Thus, one has the situation schematically depicted in Figure 4.6. Prior to impact, projectile 1 is traveling at velocity V, whereas the target is at rest. After impact, two compressive shock waves are created: one travels into the target, with velocity U_{s2}, and one travels into the projectile, with velocity U_{s1}. The uncompressed portion of the projectile still moves at a velocity V, whereas the uncompressed portion of the target is at rest. There are two conditions that will be used to establish the equations that will predict the pressure in both target and projectile. The material has to be continuous across the impact interface, that is, the same velocity exists in the compressed region (hatched in Fig. 4.6). If particle velocities were different on the two sides, either voids or regions of superhigh density would form. The second condition is that the pressure has to be the same. This is easy to understand because otherwise we would have the formation of another pulse. The

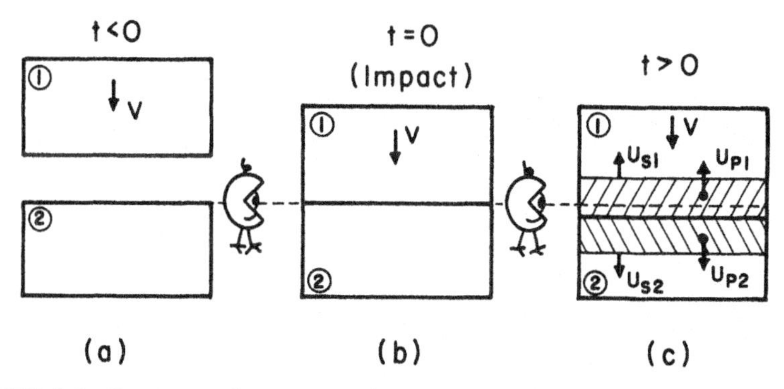

FIGURE 4.6 Sequence of events on impact: (a) projectile of material 1 flying at velocity V; (b) position at instant of impact; (c) position after impact during wave propagation stage. Notice how interface moves with respect to external observer.

analog of two gases in a cylinder separated by a mobile membrane helps us understand this concept. The central membrane will move until pressure is equilibrated. Thus,

$$P_1 = P_2 \tag{4.12}$$

In order to establish the equality of the particle velocities, we have to be careful with the reference system. We are setting a fixed referential in space. This is indicated by the schematic observer in Figure 4.6, looking at the impact from the outside. Prior to impact, all particles in the projectile have the same velocity V. Upon impact, the particle velocity in the compressed region of the projectile is reduced by a value U_p (the upward motion of the particles) so that the resultant particle velocity is $V - U_{p1}$. In the target, the particle velocity (in the compressed region) is U_{p2}. Thus

$$V - U_{p1} = U_{p2} \tag{4.13}$$

and

$$U_{p1} + U_{p2} = V$$

It should be noted that U_{p1} is the velocity of the particles in the projectile with respect to moving a referential (impact interface, called a Lagrangian referential). In order to determine the pressure, we make use of the equation for conservation of momentum.

For the target

$$P_2 = \rho_{02} U_{s2} U_{p2} \tag{4.14}$$

For the projectile

$$P_1 = \rho_{01} U_{s1} U_{p1} \tag{4.15}$$

The EOS for the two materials are

$$U_{s1} = C_1 + S_1 U_{p1}$$

$$U_{s2} = C_2 + S_2 U_{p2}$$

Substituting this into Eqns. (4.14) and (4.15) yields

$$P_1 = \rho_{01}(C_1 + S_1 U_{p1}) U_{p1} = \rho_{01} C_1 U_{p1} + \rho_{01} S_1 U_{p1}^2 \tag{4.16}$$

$$P_2 = \rho_{02}(C_2 + S_2 U_{p2}) U_{p2} \tag{4.17}$$

We now set $P_1 = P_2$ and perform a transformation of axes on Eqn. (4.16) (the projectile). We will express U_{p1} as a function of U_{p2}, substituting $V - U_{p2}$ for U_{p1}. This will lead to an equation with only one unknown, U_{p2}:

$$P_1 = \rho_{01} C_1(V - U_{p2}) + \rho_{01} S_1(V - U_{p2})^2 \tag{4.17a}$$

Then if $P_1 = P_2$, we can apply Eqns. (4.17) and (4.17a)

$$U_{p2}^2(\rho_{02} S_2 - \rho_{01} S_1) + U_{p2}(\rho_{02} C_2 + \rho_{01} C_1 + 2\rho_{01} S_1 V) - \rho_{01}(C_1 V + S_1 V^2) = 0 \tag{4.18}$$

Taking the roots $[(-b \pm \sqrt{b^2 - 4ac})/2a]$ yields

$$U_{p2} = \frac{-(\rho_{02} C_2 + \rho_{01} C_1 + 2\rho_{01} S_1 V) \pm (\Delta)^{1/2}}{2(\rho_{02} S_2 - \rho_{01} S_2)}$$

where $\Delta = (\rho_{02} C_2 + \rho_{01} C_1 + 2\rho_0 S_1 V)^2 - 4(-\rho_{01})(\rho_{02} S_2 - \rho_{01} S_1)$

From the above equation, one can readily determine $P_2 = P_1 = \rho_{02}(C_2 + S_2 U_{p2}) U_{p2}$ using Eqn. (4.17). When target and projectile are of the same material, Eqn. (4.18) becomes a first-order equation because $(\rho_{02} S_2 - \rho_{01} S_1) = 0$. Thus

$$U_p = \frac{\rho_{01}(C_1 V_1 + S_1 V^2)}{\rho_{02} C_2 + \rho_{01} C_1 + 2\rho_{01} S_1 V}$$

Since $\rho_{01} = \rho_{02} = \rho_0$, $C_1 = C_2 = C$, and $S_1 = S_2 = S$,

$$U_p = \tfrac{1}{2}V_1 \tag{4.20}$$

This means that the particle velocity is equal to one-half the impact velocity for symmetric impact. The particles in the projectile transfer half of their momentum to the target.

The above calculation can be done more rapidly by a graphic solution, usually called the impedance matching technique. It just consists of expressing Eqns. (4.17) and (4.17a) graphically. Figure 4.7 shows the pressure-particle velocity curves for target (2) and projectile (1). These curves are available for all materials whose EOS are known. With these equations one can determine the pressure by making use of Eqns. (4.16) and (4.17). To express Eqn. (4.17) in the P–U_p plane, one has to make a change in coordinates. The origin is changed from 0 to V and the curve is inverted (the change of U_p to $-U_p$). This

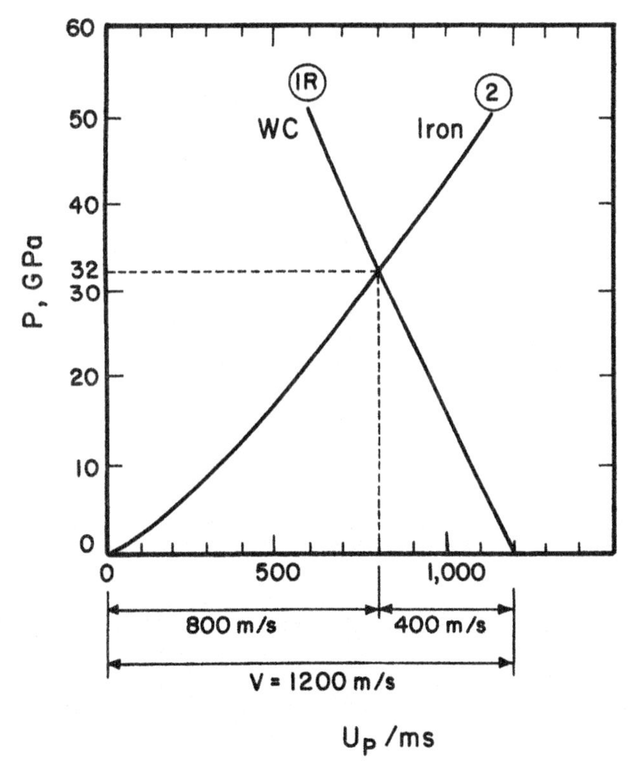

FIGURE 4.7 Graphical solution for 1200 m/s velocity impact at WC projectile on steel target.

is indicated in Figure 4.7 by $1R$. Curve $1R$ and curve 2 correspond to Eqns. (4.17) and (4.17a). The solution is simply given by their intersection. The pressure P and particle velocities U_{p1} and U_{p2} are thus found.

This procedure is very simple to carry out graphically. One traces the P–U_p curves for the different materials on transparent papers. By overlaying the direct curve (for the target) with the inverted curve (for the projectile), setting its origin at V_1, one easily finds P, U_{p1}, and U_{p2}. From these values one can obtain the other shock parameters by using the Rankine–Hugoniot equations.

Example 4.1 Calculate the pressure generated by the impact of a copper projectile against a copper target at 500 m/s (Data in Table 5.1, p. 133)

$$C_0 = 3.94 \text{ mm}/\mu\text{s}$$

$$= 3.94 \times 10^3 \text{ m/s}$$

$$S_1 = S_2 = S = 1.489 \cong 1.49$$

$$\rho_{01} = \rho_{02} = \rho_0 = 8.92$$

The particle velocity is determined directly from Eqn. (4.20):

$$U_p = \tfrac{1}{2}V = 250 \text{ m/s}$$

From Eqn. (4.17):

$$P = \rho_0(C_2 + SU_p)U_p$$

$$= 8.92 \times 10^3(3.94 \times 10^3 + 1.49 \times 250) \times 250$$

$$= 8.4 \times 10^9 \text{ N/m}^2$$

$$= 8.4 \text{ GPa}$$

Note that the density had to be converted from g/cm^3 to kg/m^3.

Example 4.2 A tungsten carbide projectile is impacting a steel target at a velocity of 1200 m/s. Determine the pressure in the target and projectile.

Analytical Solution. From the appendix of Kinslow [18],

$$\text{WC} \rightarrow U_s = 4.920 + 1.339U_p \qquad \rho_{01} = 15 \text{ g/cm}^3$$

$$\text{Steel} \rightarrow U_s = 3.57 + 1.92U_p - 0.068U_p^2 \qquad \rho_{02} = 7.85 \text{ g/cm}^3$$

For the analytical solution we will neglect the quadratic term, which complicates the solution. It is due to a phase transformation, as will be explained in Chapter 8. Applying Eqn. (4.19) yields

$$U_p = \frac{-(7.85 \times 3.57 + 15 \times 4.92 + 2 \times 15 \times 1.34 \times 1.200) \pm \sqrt{\Delta}}{2(7.85 \times 1.92 - 15 \times 1.34)}$$

where

$$\sqrt{\Delta} = [(7.85 \times 3.57 + 15 \times 4.92 + 2 \times 15 \times 1.34 \times 1.2)^2 + 4(15)(7.85 \times 1.92 - 15 \times 1.34)(4.92 \times 1.2 + 1.34 \times 1.2)^2]^{1/2}$$

$$U_p = \frac{-(150.06) \pm [22.509 + 60 \times (-5.073) \times (7.83)]^{1/2}}{-10.026}$$

$$= \frac{-(150.03) \pm 141.96}{-10.026}$$

Only positive solutions are acceptable:

$$U_p = 0.805 \text{ km/s (or mm}/\mu\text{s)}$$

$$U_p = 29.02 \text{ km/s (or mm}/\mu\text{s)}$$

The particle velocity has to be lower than the impact velocity. Let us calculate the pressure:

$$P_2 = \rho_{02}(3.57 + 1.92U_p)U_p$$

$$= 7.85 \times 10^3(3.57 \times 10^3 + 1.92 \times 0.805 \times 10^3)(0.805 \times 10^3)$$

$$= 32.3 \times 10^9 = 32.3 \times 10^9 \text{ N/m}^2$$

For the projectile,

$$P_1 = 15 \times 10^3 \times 4920(1200 - 805) + 15 \times 10^3 \times 1.34(1200 - 805)^2$$

$$= 32.5 \times 10^9 \text{ N/m}^2$$

The solution is 32.5 GPa.

Graphical Solution. The pressure–particle velocity plots of Figure 4.7 provide the graphical solution (impedance matching).

TABLE 4.2 Relationships between Shock Parameters ($\Delta E = E - E_0$)

$V_1 - U_s$	$V_1 = V_0\left(1 - \frac{1}{S} + \frac{C_0}{SU_s}\right)$	1
$U_s - V_1$	$U_s = \frac{C_0 V_0}{V_0 - S(V_0 - V_1)}$	2
$P - U_s$	$P = \frac{\rho_0}{S}(U_s^2 - C_0 U_s)$	3
$U_s - P$	$U_s = \frac{C_0}{2}\left(1 + \sqrt{1 + \frac{4V_0 S}{C_0^2}P}\right)$	4
$\Delta E - U_s$	$\Delta E = \frac{(U_s - C_0)^2}{2S^2}$	5
$U_s - \Delta E$	$U_s = C_0 + S\sqrt{2\Delta E}$	6
$U_p - V_1$	$U_p = \frac{C_0(V_0 - V_1)}{V_0 - S(V_0 - V_1)}$	7
$V_1 - U_p$	$V_1 = V_0\left(1 - \frac{U_p}{C_0 + SU_p}\right)$	8
$P - V_1$	$P = \frac{C_0^2(V_0 - V_1)}{[V_0 - S(V_0 - V_1)]^2}$	9
$V_1 - P$	$V_1 = \frac{C_0^2}{2S^2P}\left[\sqrt{1 + \frac{4SV_0}{C_0^2}}P + \frac{2S(S-1)V_0}{C_0}P - 1\right]$	10
$\Delta E - V_1$	$\Delta E = \frac{1}{2}\frac{C_0^2(V_0 - V_1)^2}{[V_0 - S(V_0 - V_1)]^2}$	11
$V_1 - \Delta E$	$V_1 = 1 - \frac{\sqrt{2\Delta E}}{\sqrt{C_0} + S\sqrt{2\Delta E}}$	12
$P - U_p$	$P = \rho_0(C_0 U_p + SU_p^2)$	13
$U_p - P$	$U_p = \frac{C_0}{2S}\left(\sqrt{1 + \frac{4S}{\rho_0 C_0^2}}P - 1\right)$	14
$\Delta E - P$	$E = \frac{1}{2}PV_0 - \frac{C_0}{4S^2}\left[\sqrt{1 + \frac{4SV_0}{C_0}}P + \frac{2S(S-1)V_0}{C_0}P - 1\right]$	15
$P - E$	$P = 2\Delta E\frac{\sqrt{C_0} + S\sqrt{2\Delta E}}{\sqrt{2E} + (\sqrt{C_0} + S\sqrt{2\Delta E})(V_0 - 1)}$	16
$\Delta E - U_p$	$\Delta E = \frac{1}{2}U_p^2$	17

TABLE 4.2 (*Continued*)

$U_p - \Delta E$	$U_p = \sqrt{2\Delta E}$	18
$U_s - U_p$	$U_s = C_0 + SU_p$	19
	$U_p = \dfrac{U_s - C_0}{S}$	20

4.4 RELATIONSHIPS BETWEEN SHOCK PARAMETERS

The pioneering work conducted at Los Alamos National Laboratory by Rice, McQueen, Walsh, and co-workers [6, 7] on shock Hugoniots resulted in the tabulation of shock parameters for a large number of materials. This work is collected in the *LASL Shock Hugoniot Data* handbook [9]. There are also compilations prepared by Sandia National Laboratories [10] and Lawrence Livermore National Laboratory [11]. It is important to note that one needs only two shock parameters to determine the remaining ones once the constants in the EOS are known. One should remember that there are five shock parameters with which we deal: P, E, ρ (or V), U_s, and U_p. With the use of EOS, these relationships can be separated into 10 pairs. These 10 pairs provide, in turn, 20 equations. In order to illustrate this, we have the equations in Table 4.2, kindly derived by K. H. Oh (the only scientist who had a cereal named after him). These relationships can be derived by the student as an exercise.

4.5 REAL SHOCK WAVE PROFILES

Life is complicated, beyond the strictures of religious or military discipline, and so are shock waves. While an ideal shock wave profile would predict a discontinuity at the front, a plateau at the top, and a gradual return to zero pressure [Fig. 4.8(a)], real shock waves exhibit a number of peculiarities that are material and pressure dependent. A few of them are discussed below. A generic profile of the interface velocity (that represents, after appropriate conversion of units, the shock wave pressure) is shown in Fig. 4.8(b); it is due to Grady [12]. Experimental techniques used to obtain such a pulse profile are discussed in Chapter 11. The specific profile is obtained by the VISAR technique. The pressure–volume curve for a real material is not identical to the Hugoniot (hydrostatic) curve because of the deviatoric component of stress. This is shown in Figure 4.9. The rate of rise of stress with volume is much higher in the elastic range. When the elastic limit under the imposed stress and strain rate conditions is reached [this is called the Hugoniot elastic limit (HEL)], the pressure–volume curve shows a change in slope. In Chapter 5 (Section 5.2) the difference between this dynamic compressibility curve (uniaxial strain) and

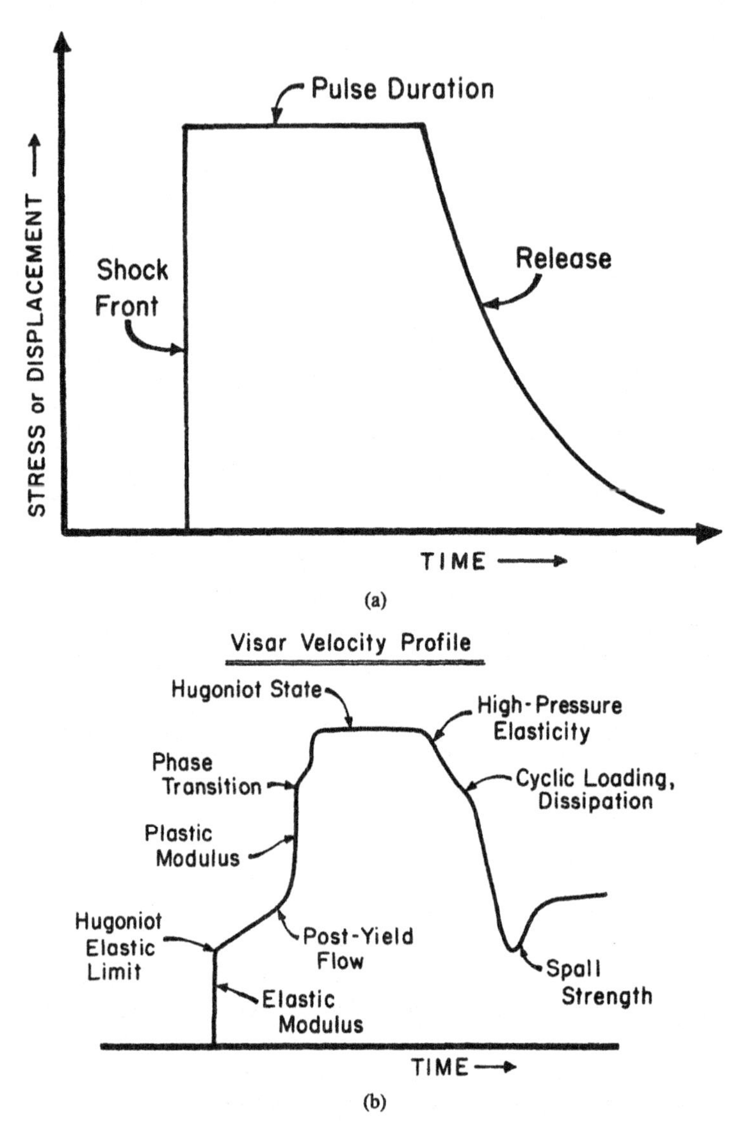

FIGURE 4.8 (a) Idealized and (b) "generic" realistic shock wave profile. (From Grady [12], Fig. 2, p. 36. Reprinted with permission.)

the Hugoniot curve (hydrostatic) will be explained. The three arrows indicate the flow stress of the material at the imposed strains. Three cases are shown in Figure 4.9:

Curve 1: Flow strength of material is independent of pressure.

Curve 2: Flow strength of material decreases with pressure: softening.

Curve 3: Flow strength in material increases with pressure: hardening.

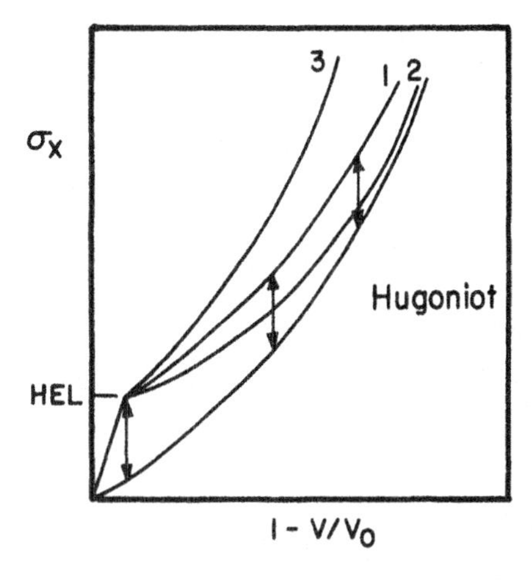

FIGURE 4.9 Schematic representation of shock response of different materials.

In metals, the flow stress (HEL) is fairly low, and these effects are reasonably unimportant. However, in ceramics such is not the case. For instance, the HEL of sapphire is close to 20 GPa, whereas that of alumina is around 6–8 GPa. In Figure 4.8(b) the elastic portion of the wave is separated from the plastic portion. This elastic portion, below the HEL, travels at a velocity higher than the plastic wave.

Thus, the effects observed in Figure 4.8 can be explained. After an initial steep rise in pressure (or particle velocity, measured by VISAR), the HEL is reached. Beyond the HEL, the pressure rises continuously to the top (actually, we do not have a discontinuity). The rate of rise of this pressure is dictated by the constitutive behavior of the material, and this will be discussed further later. If there is a phase transition (transformation), there can be a clear signal in the wave profile. The wave may separate into two waves. This is discussed in greater detail in Chapter 8. At the top of the plot of Figure 4.8(b) we have the pulse duration plateau. When unloading starts, this occurs initially elastically and then plastically. This elastoplastic transition in unloading leaves a signal, in an analogous manner to the HEL on loading. Since the free surface velocity is being measured by VISAR, the wave reflects and can fracture the material. The signal produced by spalling is discussed again in Section 16.8.2, and wave interactions leading to spalling are described in Section 7.3.

Actual shock wave profiles for real materials are shown in Figure 4.10. One metal (beryllium) and two ceramics are represented. The separation between the elastic and plastic portions of the loading is evident. For silicon carbide and boron carbide, two shock wave amplitudes are shown. The profiles for boron carbide are irregular (jagged) whereas those for silicon carbide are smooth. These irregularities could be due to fracturing of the ceramic under shock compression, but the exact source is not known. It can be seen in Figures 4.10(b) and (c) that the shock rise becomes steeper at the pressure increases.

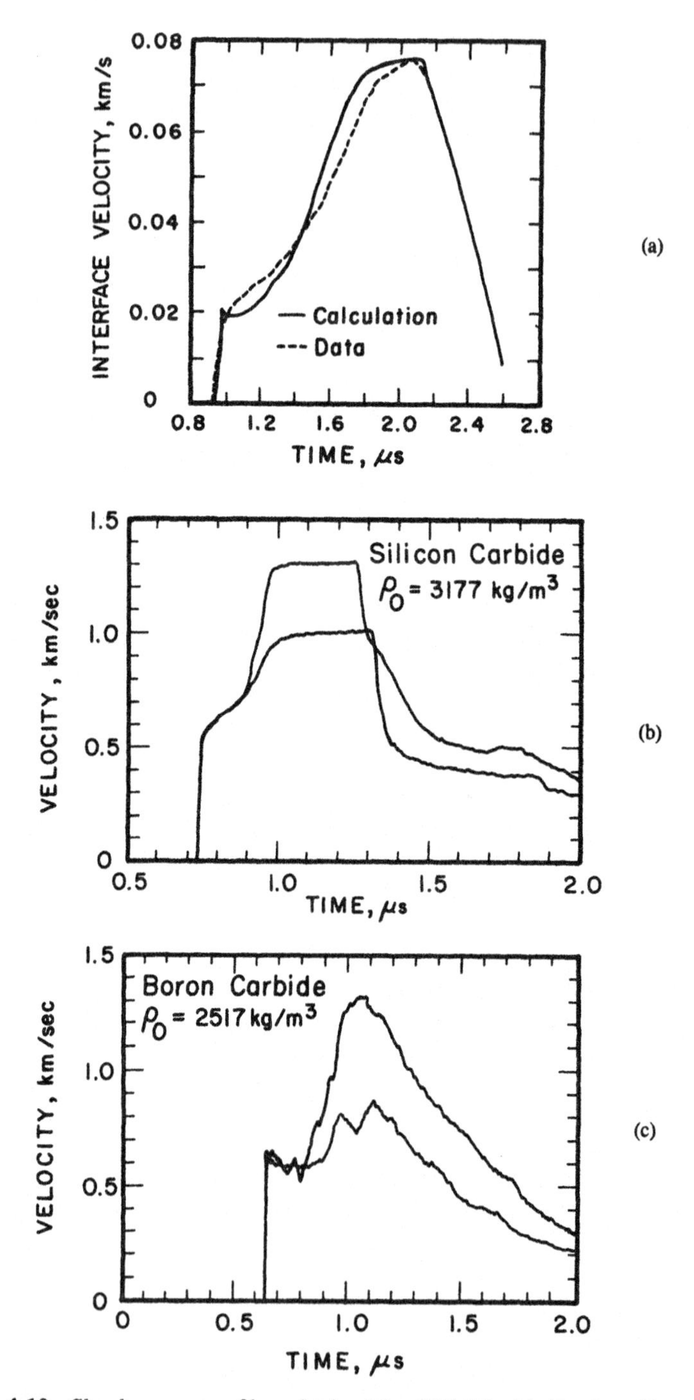

FIGURE 4.10 Shock wave profiles obtained by VISAR. (a) For beryllium. (From Steinberg [11], Fig. 3, p. 310. Reprinted with permission of the publisher.) (b) For silicon carbide. (c) For boron carbide. (From Grady [12], Fig. 7, p. 41. Reprinted with permission.)

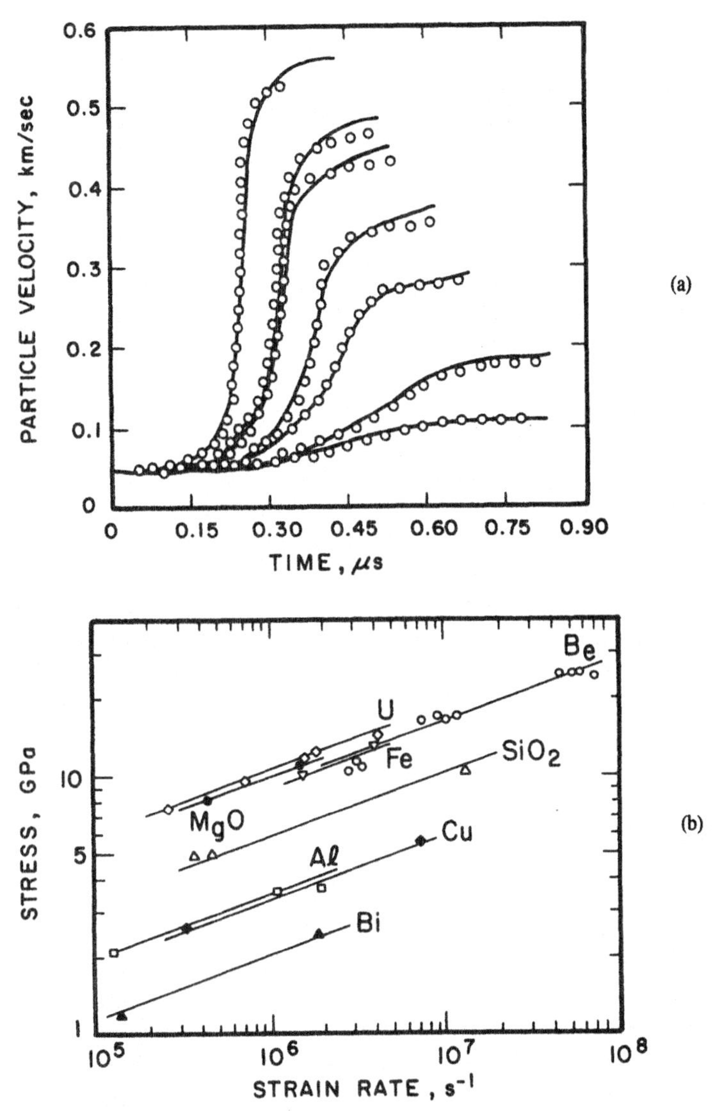

FIGURE 4.11 (a) Comparison of experimental (dotted line) and calculated (solid line) shock profiles for uranium at stress levels of 3.3, 5.7, 8.7, 11.2, 13.2, 14.4, and 16.7 GPa. (b) Relationship between stress and strain rate for a number of materials. (From Swegle and Grady [13], Figs. 3 and 1, pp. 356, 354. Reprinted with permission of the publisher.)

This phenomenon has been experimentally observed for a number of materials and is shown with more detail in Figure 4.11. These plots (for uranium) show the increase in slope with increasing particle velocity (or pressure); the dotted lines represent the experimental points. Swegle and Grady [13, 14] plotted the linear portion of this, which they converted into a strain rate, as a function of pressure (or stress) for a number of materials. Their results are shown in Figure

4.11(b). These plots provide linear relationships in a log–log plot. The astute student will immediately reason that most phenomena are linear on a log–log plot where one of the axes varies over one cycle. Nevertheless, the relation empirically obtained by Swegle and Grady [13, 14] indicated that, for most materials,

$$\dot{\varepsilon} \propto \Delta\sigma^4$$

This expression states that the stress rise (or strain rise) rate is proportional to the peak stress to the power 4. This relationship has some bearing on the constitutive models discussed in Chapter 13. The strain rates in Figure 4.11(b) range from 10^5 to 10^8 s^{-1}. The upper limit (10^8 s^{-1}) represents some of the highest strain rates achieved to date. The fundamental underpinning for this behavior is not completely understood yet. Nevertheless, it is a manifestation of the strength dependence on strain rate, discussed in Section 3.1 in a preliminary manner and in Chapter 13, in greater depth. By incorporating a strain rate dependence of material strength into proper computational models, it is possible to reproduce the experimentally determined plots of Figures 4.10 and 4.11(a). Swegle and Grady [13, 14] postulated a viscosity term, while Steinberg and co-workers [15–17] used a constitutive model developed over the years at Lawrence Livermore National Laboratory. The solid line in Figure 4.10(a) represents the calculated profile based on a constitutive model developed by Steinberg and others [16, 17].

REFERENCES

1. W. C. Davis, *Sci. Am.*, **255**(5) (1987), 105.
2. W. J. M. Rankine, *Phil. Trans. Roy. Soc. Lond.*, **160** (1870), 270.
3. H. J. Hugoniot, *J. L'Ecole Polytech.*, **58** (1989), 3.
4. M. A. Meyers and L. E. Murr, eds., *Shock-Waves and High-Strain-Rate Phenomena in Metals*, Plenum, New York, 1981, p. 1059.
5. J. M. Walsh and R. H. Christian, *Phys. Rev.*, **97** (1955), 1554.
6. M. H. Rice, R. G. McQueen, and J. M. Walsh, *Sol. State Phys.*, **6** (1958), 9.
7. R. G. McQueen, S. P. Marsh, J. W. Taylor, J. N. Fritz, and W. J. Carter, in *High Velocity Impact Phenomena*, ed. R. Kinslow, Academic, New York, 1970, p. 299.
8. P. S. DeCarli and M. A. Meyers, in *Shock Waves and High-Strain-Rate Phenomena in Metals: Concepts and Applications*, Plenum, eds. M. A. Meyers and L. E. Murr, New York, 1981, p. 341.
9. S. P. Marsh, ed., *LASL Shock Hugoniot Data*, University of California Press, Berkeley, 1980.
10. C. E. Anderson, J. S. Wilbeck, J. C. Hokanson, J. R. Asay, D. E. Grady, R. A. Graham, and M. E. Kipp, in *Shock Waves in Condensed Matter—1985*, ed. Y. M. Gupta, Plenum, New York, 1986, p. 185.

11. D. J. Steinberg, "Equation of State and Strength Properties of Selected Materials," Report No. UCRL-MA-106439, Lawrence Livermore National Laboratory, Livermore, CA, 1991.
12. D. E. Grady, "Dynamic Material Properties of Armor Ceramics," Report No. SAND 91-0147.4C-704, Sandia National Laboratory, 1991.
13. J. W. Swegle and D. E. Grady, in *Shock Waves in Condensed Matter—1985*, eds. Y. M. Gupta, Plenum, New York, 1986, p. 353.
14. J. W. Swegle and D. E. Grady, *J. Appl. Phys.*, **58** (1985), 692.
15. D. J. Steinberg, in *Shock Compression of Condensed Matter—1989*, eds. S. C. Schmidt, J. H. Johnson, and D. W. Davison, Elsevier, 1990, p. 309.
16. D. Steinberg and C. Lund, *J. Appl. Phys.*, **65** (1989), 1528.
17. D. Steinberg, S. Cochran, and M. Guinan, *J. Appl. Phys.*, **51** (1980), 1498.
18. R. Kinslow, ed., *High Velocity Impact Phenomena*, Academic, New York, 1970.

CHAPTER 5

Shock Waves: Equations of State

5.1 EXPERIMENTAL METHODS FOR OBTAINING EOS DATA

A large volume of shock EOS (or Hugoniot) data has been accumulated pertaining to the behavior of metals, polymers, liquids, and ceramics, subjected to shock loading (see Chapter 4). Hugoniot curves have been determined from empirical data obtained by performing measurements during impact experiments. Equations of state are essential to describe the shock response of materials. Equations of state can be experimentally determined and the linear equation $U_s = C_0 + SU_p$ is the simplest form for the shock Hugoniot. If the parameters C_0 and S are known, one can determine all other shock parameters. Several methods have been used to determine the EOS of materials of importance.

The early experiments used explosives in contact with the material. These explosives were detonated by means of a plane-wave generator (explosive lens) that provided the planar shock wave traveling through the metal. Figure 5.1(a) shows the experimental configuration. The shock and particle velocities were measured by either pin contactors or a flash-gap technique. The free-surface velocity can be used to obtain the particle velocity, since it can be shown that the free-surface velocity U_{fs} is calculated as

$$U_{fs} = U_p + U_r \simeq 2U_p$$

where U_r is the particle velocity of the reflected wave, which in essence is equal to the particle velocity of the shock pulse. This will be discussed in greater detail in Chapter 7. From different pairs of U_s–U_p values the EOS can be established. For pressure values higher than the ones obtained by direct detonation, a flyer plate assembly was used, also accelerated by explosives. The free-surface velocities are determined by pins or a flash technique. These diagnostics are placed at the back of the material. In Figure 5.1(a) one can see four different positions where these particle velocity measurements are made.

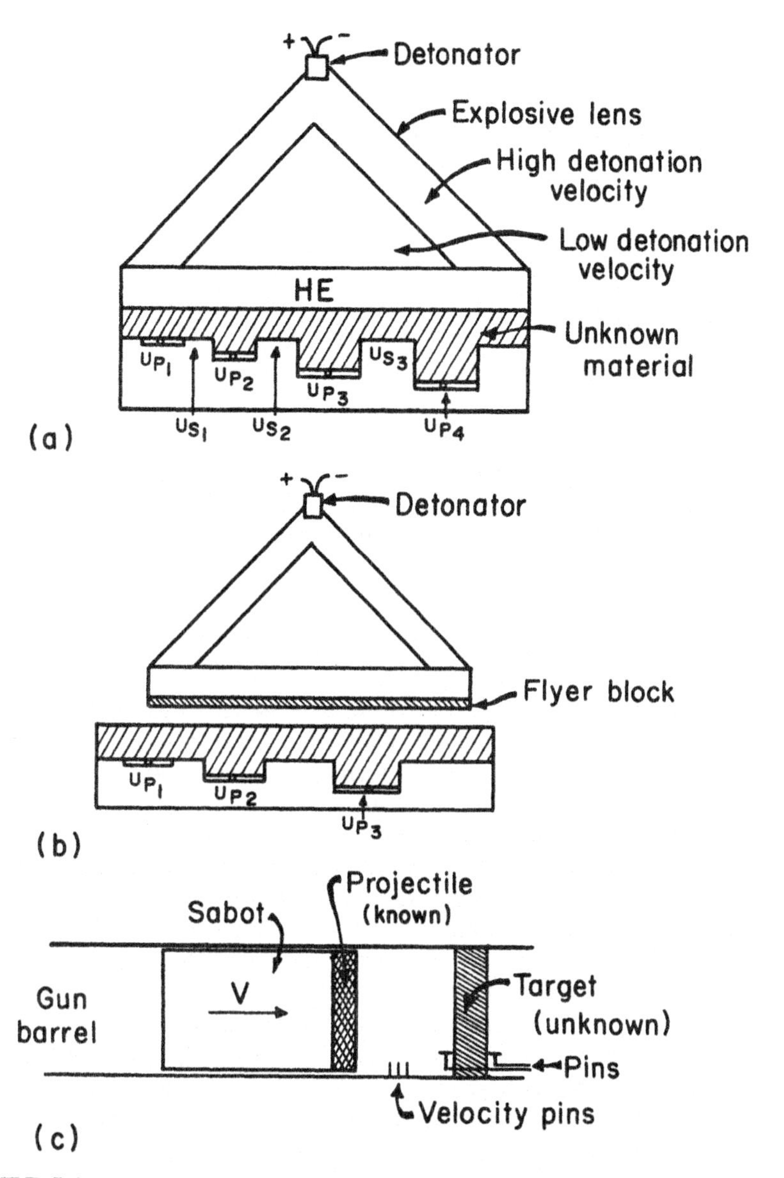

FIGURE 5.1 Experimental configurations for EOS determination: (a) explosives in direct contact with material; (b) flyer plate driven by explosive detonation; (c) gas gun impact.

The shock wave velocity is given by the transit times of the wave between pins (or flash gaps). These diagnostic techniques will be described in greater detail in Chapter 11.

However, gas guns provide a much greater control of the impact velocity and planarity of impact, and one can measure the velocity of the flyer plate

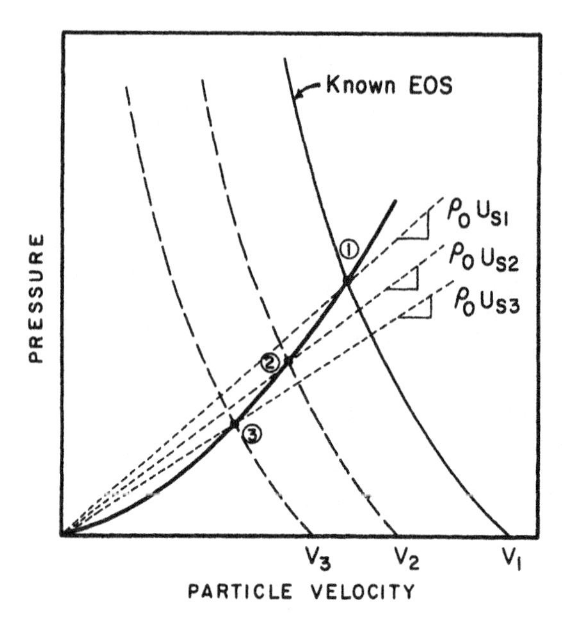

FIGURE 5.2 Impedance matching technique for determination of EOS.

(projectile) much more reliably. Therefore, they are currently used more often. The impedance matching technique is then used, and one uses a projectile with a known EOS impacting an unknown target at a fixed and experimentally determined velocity. By measuring the shock velocity or particle velocity (from free-surface velocity measurements) in the back of the target, one can determine the impact pressure. By repeating this procedure for different impact velocities, one can construct the P–U_p curve for the unknown material. The principal technique presently used to determine the free-surface velocity is laser interferometry. Figure 5.1(c) shows a gas gun experiment. Three pins in the barrel establish the velocity of the projectile. Figure 5.2 shows the inverted shock Hugoniot for the projectile (known) starting at the known impact velocity V_1. By measuring U_s (and knowing the density ρ_0 of the unknown material), one draws the Rayleigh line with slope $\rho_0 U_s$. It intersects the projectile shock Hugoniot at point 1. By repeating the procedure for different impact velocities, one determines points 1, 2, and 3 on the shock Hugoniot of the unknown material.

There are other techniques for the determination of the EOS for materials (solids, porous materials, etc). They are discussed by Rice et al. [1] and McQueen et al. [2].

5.2 THEORETICAL CALCULATIONS OF EOS

The experimental determination of EOS is important but should not preclude theoretical analyses. In an ideal world, we would like to be able to determine the shock parameters of materials from first principles. This is currently only possible in a few cases and using very advanced quantum-mechanical computations. If we compress together atoms, the electronic shells interpenetrate and we have very high short-range forces. The Condon–Morse curve, which describes approximately the energy and force between two atoms as a function of their separation, provides a good idea of the forces involved. When the electronic shells start interacting, strong repulsions occur. This is shown in the first plot of Figure 5.3. For ionic materials, the repulsive forces vary roughly with r^4 while the attractive forces vary with r^2 (coulombic). The dip in the energy curve provides a minimum that is the distance at which atomic radii are at an equilibrium distance. From the force–distance plot one could calculate a pressure–volume plot. This can be done by normalizing the force and by converting the separation between atoms into a volume. One would result with an isothermal (0 K) compressibility curve, which shows an increasing slope (in magnitude) with increasing pressure. This shape of the compressibility curve is very important and makes possible the existence of shock waves. From this

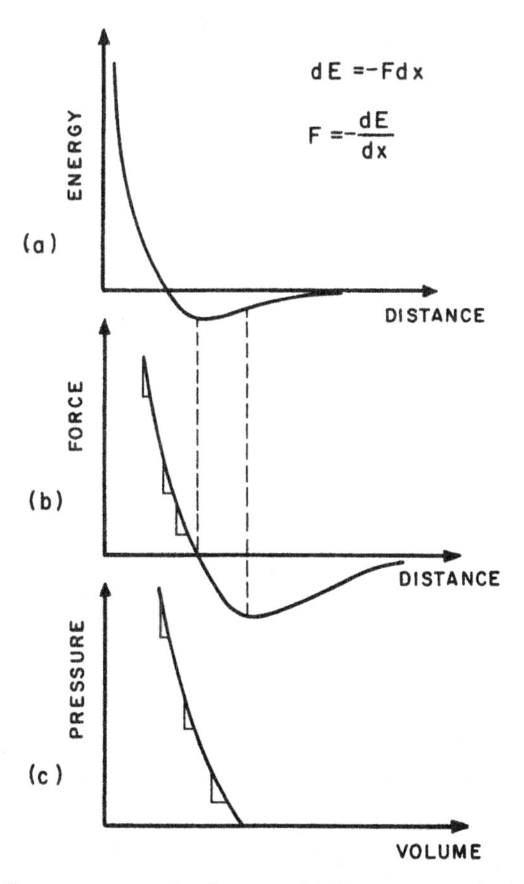

FIGURE 5.3 (a) Energy vs. atomic distance; (b) force vs. atomic distance; (c) pressure vs. volume.

"cold compression" curve one could, as a next stage, calculate the shock Hugoniot curve.

Early efforts centered around obtaining the shock compression curve from the isothermal compression curve. Bridgman [3] has established compressibility curves by static high pressures for a large number of materials. However, these static pressure–volume curves were not available at sufficiently high pressures, and experimental determinations of the shock Hugoniots was necessary for the high pressures not achievable statically ($P > 10$ GPa). In this respect, the use of the Mie–Grüneisen EOS has been very successful and useful. Additionally, the Mie–Grüneisen EOS is very important in the determination shock and residual temperatures and for predicting the shock response of porous materials. We will present here a brief account of the Mie–Grüneisen EOS. The Grüneisen constant comes from statistical mechanics. Statistical mechanics treats the energies of individual atoms and arrives at expressions that are equal to thermodynamics.

Statistical mechanics is microscopic, whereas thermodynamics is macroscopic. Atoms are considered to be oscillators in statistical mechanics. We are reminded of Figure 2.3. Atoms are considered as quantized oscillators, each having three directions of vibration. Figure 5.4 shows an array of oscillators. The energy of a quantized oscillator (nth level) is given by $nh\nu$, where h is Planck's constant and ν is the frequency of vibration. This does not include the ground level $\frac{1}{2}h\nu$.

The total mean vibrational energy of a crystal is equal to the sum of the mean energies of the individual oscillators $\bar{\varepsilon}_j$,

$$\bar{E} = \sum_{j=1}^{3N} n_j h\nu_j = \sum_{j=1}^{3N} \bar{\varepsilon}_j$$

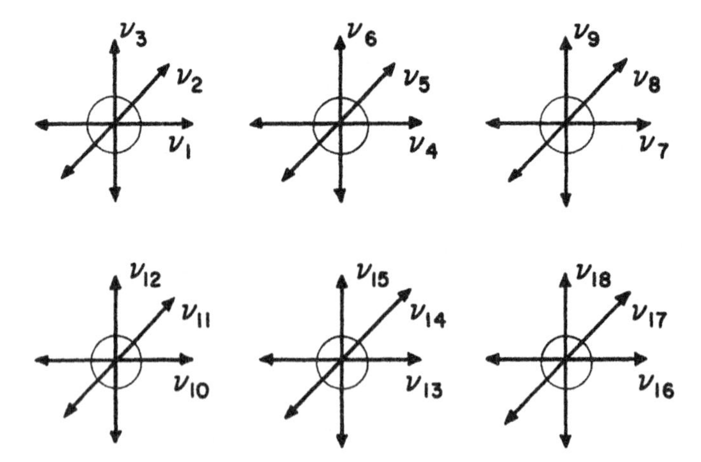

FIGURE 5.4 Assemblage of atoms acting as independent oscillators (three oscillators per atom).

We will now proceed to calculate the mean energy of an oscillator. It is derived by assuming discrete energy levels in accordance with quantum theory.

The relative probability of finding in its ith level a system that has levels of energy ε_i and degeneracy g is

$$P_i = g_i e^{-\varepsilon_i/kT} \tag{5.1}$$

where k is Boltzmann's constant.

The absolute probability P_i is found by dividing P_i by the total number of events:

$$P_i = \frac{g_i e^{-\varepsilon_i/kT}}{\sum_j g_j e^{-\varepsilon_j/kT}} \tag{5.2}$$

where j is summed over all levels. The mean energy $\bar{\varepsilon}$ of the system obviously is

$$\bar{\varepsilon} = \sum_i P_i \varepsilon_i = \frac{\sum_i \varepsilon_i g_i e^{-\varepsilon_i/kT}}{\sum_j g_j e^{-\varepsilon_j/kT}} \tag{5.3}$$

$$= -\frac{d}{d(1/kT)} \log \left(\sum_i g_i e^{-\varepsilon_i/kT}\right) \tag{5.4}$$

We call the sum

$$\sum_i g_i e^{-\varepsilon_i/kT} \tag{5.5}$$

the partition function and designate it by f.

We now evaluate Eqn. (5.4) for a harmonic oscillator of which the energy levels, in accordance with quantum theory, are given by

$$\varepsilon = nh\nu \tag{5.6}$$

where n takes all integer values and ν is the natural frequency of the oscillator. The levels are not degenerate in this case, where g_i is unity.

Thus, the partition function for the system is

$$f = \sum_{n=0}^{\infty} e^{-nh\nu/kT} = \sum_{n=0}^{\infty} (e^{-h\nu/kT})^n = 1 + e^{-h\nu/kT} + (e^{-h\nu/kT})^2 + \cdots \tag{5.7}$$

This series converges to

$$f = \frac{1}{1 - e^{-h\nu/kT}} \tag{5.7}$$

According to Eqn. (5.4) the mean energy is

$$\bar{\varepsilon} = \frac{d}{d(1/kT)} \log (1 - e^{-h\nu/kT})$$
$$= \frac{h\nu e^{-h\nu/kT}}{1 - e^{-h\nu/kT}} = \frac{h\nu}{e^{h\nu/kT} - 1} \tag{5.8}$$

The mean total energy of an assembly of $3N$ oscillators of different frequencies ν_j $(j = 1, 2, \cdots, 3N)$ is

$$\bar{E} = \sum_{j=1}^{3N} \frac{h\nu_j}{e^{h\nu_j/kT} - 1} \tag{5.9}$$

The total energy of one gram-atom of material is given by the sum of the potential energy of the atoms (energy in the absence of vibration), ϕ, and the vibration energy (note that the ground state $\frac{1}{2}h\nu$ is now added)

$$E = \phi(\nu) + \sum_{j=1}^{3N} \tfrac{1}{2}h\nu_j + \sum_{j=1}^{3N} n_j h\nu_j$$
$$= \phi(\nu) + \sum_{j=1}^{3N} \left[\frac{1}{2} h\nu_j + \frac{h\nu_j}{e^{h\nu_j/kT} - 1}\right]$$

The Helmholtz free energy is:

$$A = -kT \ln \sum \exp(-E_j/kT) = \phi(\nu) + \sum_{j=1}^{3N} \tfrac{1}{2}h\nu_j + kT \sum_{j=1}^{3N} \ln (1 - e^{-h\nu_j/kT})$$

One obtains the pressure by taking the differential of the Helmholtz free energy with respect to volume, at constant temperature (by definition). Thus,

$$P = -\left(\frac{\partial A}{\partial V}\right)_T = -\frac{d\phi}{dV} + \frac{1}{V} \sum_{j=1}^{3N} \gamma_j \left[\frac{1}{2} h\nu_j + \frac{h\nu_j}{e^{h\nu_j/kT} - 1}\right] \tag{5.10}$$

At this point, a bridge between statistical mechanics and thermodynamics was built, and we have a relationship between pressure and volume. The term γ_j is defined as

$$\gamma_j = -\frac{V}{\nu_j}\left(\frac{\partial \nu_j}{\partial V}\right)_T = -\left(\frac{\partial \ln \nu_j}{\partial \ln V}\right)_T$$

Since the resistance of a crystal to compression increases with pressure, we have an increase in the vibrational frequencies ν_j. If ν_j increases with decreasing volume, $\gamma > 0$. Grüneisen simplified the overall analysis by assuming that all oscillators had the same γ, called the Grüneisen constant. Hence

$$\gamma = -\left(\frac{\partial \ln \nu}{\partial \ln V}\right)_T \tag{5.11}$$

A value of $\gamma = 1$ corresponds to changes in frequencies being inversely proportional to changes in volume. For instance, decreasing the volume from V to 0.5 V would result in a doubling of the vibrational frequency. Making all the γ equal implies that they can be taken out of the summation term in Eqn. (5.10):

$$P = \frac{-d\phi}{dV} + \frac{\gamma}{V} \underbrace{\sum_{j=1}^{3N} \left[\frac{1}{2} h\nu_j + \frac{h\nu_j}{e^{h\nu_j/kT} - 1}\right]}_{E_{\text{VIB}}} \tag{5.12}$$

$$= -\frac{d\phi}{dV} + \frac{\gamma}{V} E_{\text{VIB}}$$

Applying this equation to 0 K leads to

$$P_{0K} = -\frac{d\phi}{dV} + \frac{\gamma}{V} E_{0K} \tag{5.13}$$

Subtracting Eqn. (5.13) from (5.12) yields

$$P - P_{0K} = \frac{\gamma}{V}(E - E_{0K}) \tag{5.14}$$

This is the Mie–Grüneisen EOS. It relates a (P, V, E) state to the pressure and internal energy at 0 K. It can also be related to another reference state, like a point on the Hugoniot plot. In that case

$$P - P_{\text{H}} = \frac{\gamma}{V}(E - E_{\text{H}}) \tag{5.15}$$

that is, the pressure and internal energy at a point off the Hugoniot are related to the pressure and internal energy in the Hugoniot by the equation *at the same volume*. The condition of constancy of volume is important, as will become clear when we calculate EOS for porous materials.

The Grüneisen constant can be expressed from Eqn. (5.15):

$$dP = \frac{\gamma}{V} dE$$

Since we have V as a constant, this equation is equivalent to

$$\gamma = V\left(\frac{\partial P}{\partial E}\right)_V = V\left(\frac{\partial P}{\partial T}\right)_V\left(\frac{\partial T}{\partial E}\right)_V$$

$$= \frac{V}{C_v}\left(\frac{\partial P}{\partial T}\right)_V = -\frac{V}{C_v}\left(\frac{\partial P}{\partial V}\right)_T\left(\frac{\partial V}{\partial T}\right)_P$$

Because

$$\left(\frac{\partial P}{\partial T}\right)_V = -\left(\frac{\partial P}{\partial V}\right)_T\left(\frac{\partial V}{\partial T}\right)_P$$

$(1/V)(\partial V/\partial T)_P$ is the volumetric thermal expansion (3α), and $-(1/V)(\partial V/\partial P)_T$ is the isothermal compressibility (K). Thus,

$$\frac{\gamma}{V} = \frac{3\alpha}{C_v K} \tag{5.16}$$

Table 5.1 presents calculated Mie–Grüneisen constants for a number of materials. It is interesting to note that we have the following practical approximation:

$$\gamma_0 \cong 2S - 1 \tag{5.17}$$

This is a very useful relationship that can help us obtain approximate values for unknown materials. There is another approximate model of considerable importance. The ratio γ/V is assumed to be constant. This implies that

$$\frac{\gamma}{V} = \frac{\gamma_0}{V_0} = \text{const} \tag{5.18}$$

where γ_0 and V_0 are the Grüneisen constant and specific volume at zero pressure, respectively.

At this point it would be helpful to note that γ is approximated as a function of volume, which is correct only for the harmonic solid we assumed in deriving Eqn. (5.11). By the thermodynamic laws (first and second) a thermodynamic function is a function of two variables. Thus the $\gamma \sim \gamma(V)$ is an approximation, and equally, the simple model given by Eqn. (5.18) may be applicable up to a few hundred gigapascals in pressure for usual solids and for porous materials whose initial density is above one-third of its solid density.

Example 5.1. With known values of thermal expansion, specific heat, and bulk modulus, calculate the Grüneisen constant for copper.

TABLE 5.1 Shock and Thermodynamic Properties for Different Materials ($U_s = C_0 + SU_p$)

Material[a]	ρ_0 (g/cm^3)	C_0 (mm/μs)	S	C_p (J/g K)	γ
Ag	10.49	3.23	1.60	0.24	2.5
Au	19.24	3.06	1.57	0.13	3.1
Be	1.85	8.00	1.12	0.18	1.2
Bi	9.84	1.83	1.47	0.12	1.1
Ca	1.55	3.60	0.95	0.66	1.1
Cr	7.12	5.17	1.47	0.45	1.5
Cs	1.83	1.05	1.04	0.24	1.5
Cu	8.93	3.94	1.49	0.40	2.0
Fe[b]	7.85	3.57	1.92	0.45	1.8
Hg	13.54	1.49	2.05	0.14	3.0
K	0.86	1.97	1.18	0.76	1.4
Li	0.53	4.65	1.13	3.41	0.9
Mg	1.74	4.49	1.24	1.02	1.6
Mo	10.21	5.12	1.23	0.25	1.7
Na	0.97	2.58	1.24	1.23	1.3
Nb	8.59	4.44	1.21	0.27	1.7
Ni	8.87	4.60	1.44	0.44	2.0
Pb	11.35	2.05	1.46	0.13	2.8
Pd	11.99	3.95	1.59	0.24	2.5
Pt	21.42	3.60	1.54	0.13	2.9
Rb	1.53	1.13	1.27	0.36	1.9
Sn	7.29	2.61	1.49	0.22	2.3
Ta	16.65	3.41	1.20	0.14	1.8
U	18.95	2.49	2.20	0.12	2.1
W	19.22	4.03	1.24	0.13	1.8
Zn	7.14	3.01	1.58	0.39	2.1
KCl[b]	1.99	2.15	1.54	0.68	1.3
LiF	2.64	5.15	1.35	1.50	2.0
NaCl[c]	2.16	3.53	1.34	0.87	1.6
Al-2024	2.79	5.33	1.34	0.89	2.0
Al-6061	2.70	5.35	1.34	0.89	2.0
SS-304	7.90	4.57	1.49	0.44	2.2
Brass	8.45	3.73	1.43	0.38	2.0
Water	1.00	1.65	1.92	4.19	0.1
Teflon	2.15	1.84	1.71	1.02	0.6
PMMA	1.19	2.60	1.52	1.20	1.0
PE	0.92	2.90	1.48	2.30	1.6
PS	1.04	2.75	1.32	1.20	1.2

Source: Courtesy of the American Physics Society Topical Group on Shock Compression.

[a]PMMA, polymethylmethacrylate; PE, polyethylene; PS, polystyrene.

[b]Above phase transition.

[c]Below phase transition.

The definition of γ_0 is

$$\gamma_0 = -\frac{V_0}{C_v}\left(\frac{\partial P}{\partial V}\right)_T\left(\frac{\partial V}{\partial T}\right)_P$$

which is equivalent to

$$\gamma_0 = -\frac{V_0}{C_v}\left(\frac{\partial P}{\partial(V/V_0)}\right)\left(\frac{\partial(V/V_0)}{\partial T}\right)_{\substack{P=0\\T=298}} \tag{1}$$

From tabulated data we obtain values for the heat capacity at constant volume (C_v), thermal expansion coefficient ($\partial(V/V_0/\partial T)$, and isothermal bulk modulus $[-\partial P/\partial(V/V_0)]$.

From Gschneider [4] we have

$$C_v = 376\,\frac{J}{\text{kg } K} \tag{2}$$

For

$$\partial P/\partial(V/V_0)|_{\substack{P=0\\T=298}}$$

we use data from the equation

$$\frac{\Delta V}{V} = -7.49 \times 10^{-8}P + 2.018 \times 10^{-12}P^2 \ (P \text{ in kgf/cm}^2)$$

For $V = V_0$

$$\frac{\Delta V}{V_0} = -7.49 \times 10^{-8}P$$

$$\left.\frac{\partial P}{\partial(V/V_0)}\right|_{\substack{P=0\\T=298}} = \frac{-1}{7.49 \times 10^{-8}}\left(\frac{\text{kgf}}{\text{cm}^2}\right) \tag{3}$$

For

$$\left.\frac{\partial(V/V_0)}{\partial T}\right|_{\substack{P=0\\T=298}}$$

we use the linear expansion coefficient, which must be multiplied by 3:

$$\frac{\Delta V}{V_0} = 3\frac{\Delta l}{l_0} = 3 \times 16.7 \times 10^{-6} T \qquad \therefore \frac{\partial(V/V_0)}{\partial T} = \frac{50.1}{10^6} \tag{4}$$

We also have

$$\rho_0 = 8.93 \text{ g/cm}^3 \qquad V_0 = 1/\rho_0 \tag{5}$$

Substituting (2), (3), (4), and (5) into (1) yields

$$\gamma_0 = \left(\frac{1}{8.93}\right)\left(\frac{-1}{0.376}\right)\left(\frac{\text{g}\,K}{J}\right)\left(\frac{\text{cm}^3}{\text{g}}\right)\left(-\frac{1}{7.5 \times 10^{-8}}\right)\left(\frac{\text{kgf}}{\text{cm}^2}\right) \times \frac{50.1}{10^6}\left(\frac{1}{K}\right)$$
$$= 2.005 \simeq 2$$

5.3 EQUATION OF STATE FOR ALLOYS AND MIXTURES

In real-life situations, we rarely have the opportunity to deal with pure metals, for which the EOS are known. Although the EOS for the most important alloys have been established (e.g., 304 stainless steel, Al alloys, U–Mo alloys, U–Rh alloys), we are very often confronted with experiments with materials for which we do not have the EOS. The theoretical analysis of the EOS for alloys and mixtures is very complex and has not been carried out yet. What we have is a number of interpolation methods with varying degrees of rigorousness. A single-phase alloy is structurally different from a phase mixture or a mechanical mixture between different materials. In a single-phase alloy the material is microscopically and macroscopically homogeneous, whereas in a mixture we have regions that are structurally different. Figure 5.5 shows the mixtures of these different phases. In a single-phase alloy the internal energy cannot be just obtained from an interpolation of the internal energies of the constituents. There is a mixing free energy and a configurational free energy. Thus, the EOS for the alloy cannot be directly obtained. For two-phase (or more) mixtures, we have an equilibrium taking place between the pressures in the two phases during the passage of the pulse. This Hugoniot pressure that equilibrates produces, by virtue of the different EOS for the constituents (γ_0 and V_0 are different), different internal energies. This is obvious if we apply the Mie–Grüneisen EOS for the two materials, A and B:

$$\text{A:} \qquad (P_{\text{HA}} - P) = \frac{\gamma_{0\text{A}}}{V_{0\text{A}}}(E_{\text{HA}} - E_{\text{A}})$$

$$\text{B:} \qquad (P_{\text{HB}} - P) = \frac{\gamma_{0\text{B}}}{V_{0\text{B}}}(E_{\text{HB}} - E_{\text{B}})$$

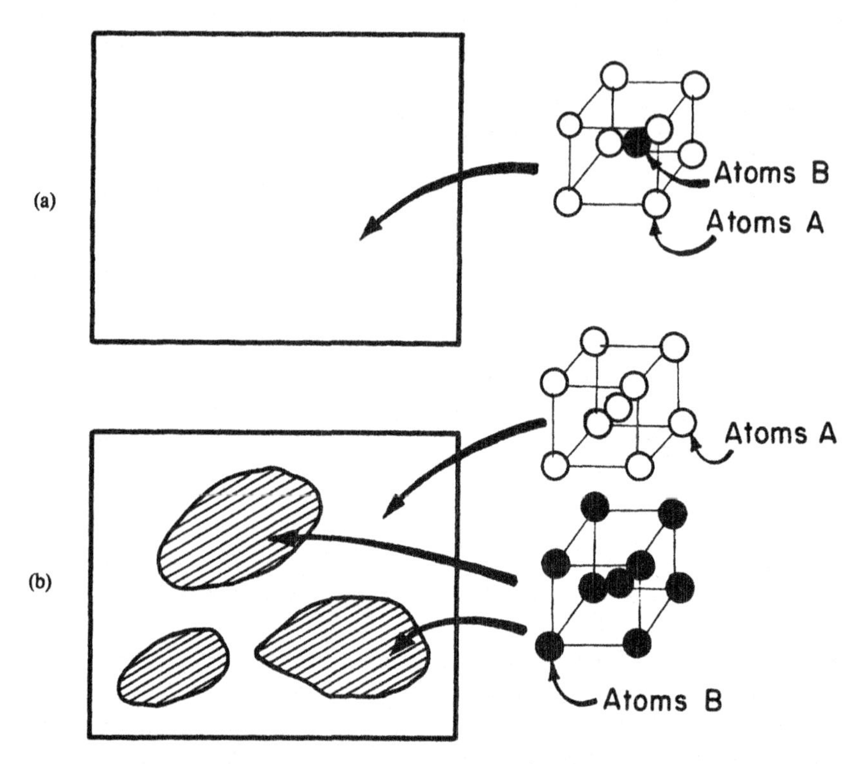

FIGURE 5.5 (a) Alloy composed of atoms A and B in one single phase and (b) a mixture of phases A and B.

This, in turn, leads to differences in temperature between the two constituents (see Section 5.5). One manner in which we can circumvent this problem is to calculate the 0 K isotherm for the two constituents. At this point, the temperature effect disappears. We therefore apply this procedure, which we call method A, which considers the mixture to be in thermal equilibrium.

Method A

1. Construct 0 K isotherms from the various component elements from the shock Hugoniots. This is done by using

$$(P_{0K} - P_H) = \frac{\gamma_0}{V_0}(E_{0K} - E_H)$$

   At each value of V we find the P_{0K}–E_{0K} pair.

2. Mix the 0 K isotherm on a mass fraction basis (mass fraction of ith component is m_i) and obtain the isotherm for the mixture alloy. A constant value of C_v is assumed:

$$V = \sum m_i V_i \qquad \frac{V}{\gamma} = \sum m_i \left(\frac{V}{\gamma}\right)_i \qquad E_k = \sum m_i E_i$$

3. From the 0 K isotherm for the mixture/alloy obtain the shock Hugoniot for the alloy using the same equation:

$$(P_H - P_{0K}) = \frac{\gamma_0}{V_0}(E_H - E_{0K})$$

In this equation, we use mass averages of γ_0 and V_0 for the mixture/alloy.

This is a long procedure that is best performed using a computer program. Yoshida [5] developed such a code during his stay at CETR-New Mexico Tech.

Method B. This is a much simpler procedure and is based on the interpolation of the C_0 and S values in the EOS by mass averaging. Thus, for

$$C_0 = \sum m_i C_{0i} \qquad S = \sum m_i S_i$$

one has

$$\rho_0 = \sum m_i \rho_{0i}$$

(this value can also be directly measured for the alloy/mixture). Since these are approximate methods and since real alloys undergo phase changes not encountered in the component elements, the latter method is considered satisfactory.

Example 5.2. Determine the EOS for an alloy with the following composition: 83 wt % Ti, 5 wt % Al, 2 wt % Sn, 2 wt % Zr, 4 wt % Mo, Cr.

We will first solve this problem by method B. Then, we will approach it by method A, indicating the method of solution.

METHOD B

Element	wt %	C_0 (m/s)	S	ρ_0 (kg/m^3)	γ
Ti	83	5220	0.761	4528	1.09
Al	5	5328	1.338	2750	2.00
Sn	2	2668	1.428	7287	2.03
Zr	2	3757	1.018	6506	1.09
Mo	4	5124	1.232	10206	1.52
Cr	4	5173	1.473	7117	1.19

$$C_0 = \sum m_i C_{0i} = 5.139 \qquad \rho_0 = \sum m_i \rho_{0i} = 4.86 \text{ g/cm}^3$$

$$S = \sum m_i S_{0i} = 0.855 \qquad U_s = 5.139 + 0.855 U_p$$

METHOD A

This is the rigorous method recommended by McQueen et al. [2].

One establishes the 0 K pressure–volume plot from the Hugoniot, for each element, by means of the equation that will be derived below:

$$P = f(V_1 E)$$

$$\left(\frac{dP}{dV}\right)_H = \left(\frac{\partial P}{\partial V}\right)_E + \left(\frac{\partial P}{\partial E}\right)_V \left(\frac{\partial E}{\partial V}\right)_H$$

Along 0 K isotherm:

$$\left(\frac{dP}{dV}\right)_{T_0} = \left(\frac{\partial P}{\partial V}\right)_E + \left(\frac{\partial P}{\partial E}\right)_V \left(\frac{\partial E}{\partial V}\right)_{T_0}$$

Subtracting these equations yields

$$\left(\frac{dP}{dV}\right)_H - \left(\frac{dP}{dV}\right)_{T_0} = \left(\frac{\partial P}{\partial E}\right)_V \left[\left(\frac{dE}{dV}\right)_H - \left(\frac{dE}{dV}\right)_{T_0}\right]$$

But

$$E_H - E_0 = \tfrac{1}{2} P_H (V_0 - V)$$

$$\left(\frac{dE}{dV}\right)_H = \frac{\gamma}{V} = \frac{\gamma_0}{V_0}$$

Along the 0 K isotherm,

$$dE = T\,dS - P\,dV = -P\,dV$$

$$\left(\frac{dE}{dV}\right)_{T_0} = -P_{T_0}$$

Substituting these expressions, we arrive at

$$\left(\frac{dP}{dV}\right)_{T_0} + \frac{\gamma_0}{V_0} P_{T_0} = \frac{\gamma_0}{2V_0}\left[P_H + \left(\frac{V_0}{\gamma_0} + V - V_0\right)\left(\frac{dP}{dV}\right)_H\right] \tag{5.19}$$

All elements on the right-hand side are known. By solving this differential equation, one finds the pressure–volume 0 K isotherm. This can be done by computer, and Yoshida's [5] MIXTURE program addresses this. It can also be done analytically if we solve the equation.

From

$$P_H = \frac{C^2(V_0 - V)}{[V_0 - S(V_0 - V)]^2}$$

(one of the equations in Table 4.2), one can find

$$\frac{dP_H}{dV} = \left(\frac{dP}{dV}\right)_H = \frac{C^2 + 2S[V_0 - S(V_0 - V)]}{[V_0 - S(V_0 - V)]^4}$$

and

$$\left(\frac{dP}{dV}\right)_{T_0} + \frac{\gamma_0}{V_0} P_{T_0} = \frac{\gamma_0}{2V_0}\left[P_H + \left(\frac{V_0}{\gamma_0} + V - V_0\right) \cdot \frac{\{C^2 + 2S[V_0 - S(V_0 - V)]\}}{[V_0 - S(V_0 - V)]^4}\right]$$

By numerical integration, these values were found, for the 0 K isotherm, by K. H. Oh (South Korean Defense Laboratory, Taejon); they are shown in Table 5.2.

One now averages all the values based on the mass fraction. The γ_0 values are assumed to be unchanged with temperature. From the method described

TABLE 5.2 Zero K Isotherm for Alloy Components

	V (cm^3/g)						
P (GPa)	Ti[a]	Al[b]	Sn[c]	Mo[d]	Zr[e]	Cr[f]	Alloy[g]
2.0	0.21803	0.352	0.136	0.08976	0.152	0.1396	0.21381
4.0	0.21527	0.346	0.135	0.09723	0.150	0.1389	0.21112
6.0	0.21258	0.339	0.134	0.0969	0.148	0.1381	0.02843
8.0	0.20994	0.335	0.1325	0.0965	0.1472	0.1373	0.20595
10.0	0.20736	0.328	0.1314	0.0961	0.1457	0.1365	0.20335
12.0	0.20483	0.323	0.1303	0.0958	0.1442	0.1359	0.20092
14.0	0.20235	0.318	0.1293	0.0955	0.1428	0.1351	0.19852
16.0	0.19992	0.313	0.1283	0.0951	0.1414	0.1344	0.19616
18.0	0.19753	0.308	0.1275	0.0948	0.14	0.1337	0.19384

Source: Courtesy of K. H. Oh.

[a]Mass fraction 0.83, $V_0 = 0.22085$.
[b]Mass fraction 0.05, $V_0 = 0.35907$.
[c]Mass fraction 0.02, $V_0 = 0.13736$.
[d]Mass fraction 0.04, $V_0 = 0.09798$.
[e]Mass fraction 0.02, $V_0 = 0.15373$.
[f]Mass fraction 0.04, $V_0 = 0.14051$.
[g]Mass fraction 1.0, $V_0 = \Sigma\, m_i v_{0i} = 0.21662$, $V = \Sigma\, m_i v_i$.

before,

$$\left(\frac{V_0}{\gamma_0}\right)_{\text{alloy}} = \sum m_i \left(\frac{V_0}{\gamma_0}\right)_i = 0.1883$$

$$\left(\frac{V_0}{\gamma_0}\right)_{\text{alloy}} = 5.31$$

With this value one can now convert the 0 K isotherm into the shock Hugoniot for the alloy. This is also done by computer, using, for the simplest solution, a quadratic fit for the 0 K pressure–volume curve, listed in the previous table:

$$P_{0\text{K}} = 329 - 2333V + 3757V^2$$

or

$$\left(\frac{dP}{dV}\right)_{0\text{K}} = -2333 + 7514V$$

The inverse differential equation to be solved is

$$\left(\frac{dP}{dV}\right)_{\text{H}} + \frac{P_{\text{H}}}{2V_0/\gamma_0 + V - V_0} = \frac{(2V_0/\gamma_0)(dP/dV)_{0\text{K}} + 2P_{0\text{K}}}{2V_0/\gamma_0 + V - V_0} \tag{5.20}$$

Notice that this equation applies to the alloy. By solving this equation one obtains the shock Hugoniot pressure–volume. It can be seen that this procedure is much more complicated than the former one (method B).

5.4 THE EOS FOR POROUS AND DISTENDED MATERIALS

In order to determine exactly the EOS for a powder, instrumented gas gun experiments have to be performed. However, reliable calculation procedures have been developed and will be described here. McQueen et al. [6], Al'tshuler [7], Mader [8], and Herrmann [9] used calculational procedures for the Hugoniots of shock powders based on the Mie–Grüneisen equation, given by

$$P = P_{\text{H}} + \frac{\gamma}{V}(E - E_{\text{H}}) \tag{5.21}$$

where P_{H} and E_{H} are the pressure and specific internal energy along the solid Hugoniot line in thermodynamic space and are functions of the specific volume; P and E are the pressure and specific energy for the porous material to be compacted; $1/V$ is the density of the material. The three equations for the conservation of mass, momentum, and energy (Rankine–Hugoniot relationships) are then applied to both the powder and the solid. For the solid, it is assumed that the particle velocity U_{pH} is linearly related to the shock velocity U_{pH}. In the equations below, the subscript H will be used when the parameter refers to the solid:

$$\rho(U_{sH} - U_{pH}) = \rho_0 U_{sH} \tag{5.22}$$

$$P_H = \rho_0 U_{sH} U_{pH} \tag{5.23}$$

$$E_H = \tfrac{1}{2} P_H(V_0 - V) \tag{5.24}$$

$$U_{sH} = C + SU_{pH} \tag{5.25}$$

For the powder

$$\rho(U_s - U_p) = \rho_{00} U_s \tag{5.26}$$

$$P = \rho_{00} U_s U_p \tag{5.27}$$

$$E = \tfrac{1}{2} P(V_{00} - V) \tag{5.28}$$

Equation (5.23) can be expressed as

$$P_H = \frac{C^2(V_0 - V)}{[V_0 - S(V_0 - V)]^2} \tag{5.29}$$

Inserting (5.24), (5.28), and (5.29) into (5.21) leads to relationship between P and V for the powder:

$$P = \frac{[2V - \gamma(V_0 - V)]C^2(V_0 - V)}{[2V - \gamma(V_{00} - V)]\,[V_0 - S(V_0 - V)]^2} \tag{5.30}$$

For an equation expressing pressure as a function of particle velocity, V is obtained as a function of U_p from Eq. (5.26) and (5.27) and P is incorporated into Eqn. (5.31)

$$V = V_{00} - \frac{U_p^2}{P} \tag{5.31}$$

For a relationship between shock and particle velocities the simplest procedure is to solve Eq. (5.30) and find P–V pairs, applying them to Eq. (5.31), which is a combination of Eqs. (5.26) and (5.27):

$$U_s = V_{00}\left(\frac{P}{V_{00} - V}\right)^{1/2} \tag{5.32}$$

For obtaining particle velocities, one uses the equation

$$U_p = [P(V_{00} - V)]^{1/2} \tag{5.33}$$

The constants C and S in the empirical equation $U_H = C + SU_H$ can be estimated from the experimentally determined values for pure elements [1, 2]. The simplest approach is to find C and S from the weighted average of weight fractions of principal constituents Method B, p. 137. For In 718, one obtains C and S equal to 4.535 and 1.530, respectively [10]. This is an alloy containing 53% Ni, 18% Cr, 18% Fe, 5% Nb + Ta, 3% Mo, 1% Ti, and a balance of other elements.

The pressure-specific volume curves obtained by applying Eqn. (5.30) for five different distentions are shown in Figure 5.6.

It is worth noticing that the Mie-Grüneisen EOS predicts volume increases at all pressures for a porosity of 50%. This aspect is discussed by Altschuler [7].

The energy deposited by the shock wave in the powder is an important parameter; it can be converted into an equilibrium shock temperature. Although this value does not describe the microscopic deformation phenomena at the particles it is a good estimate of the intensity of work done on the particles by the pressure pulse.

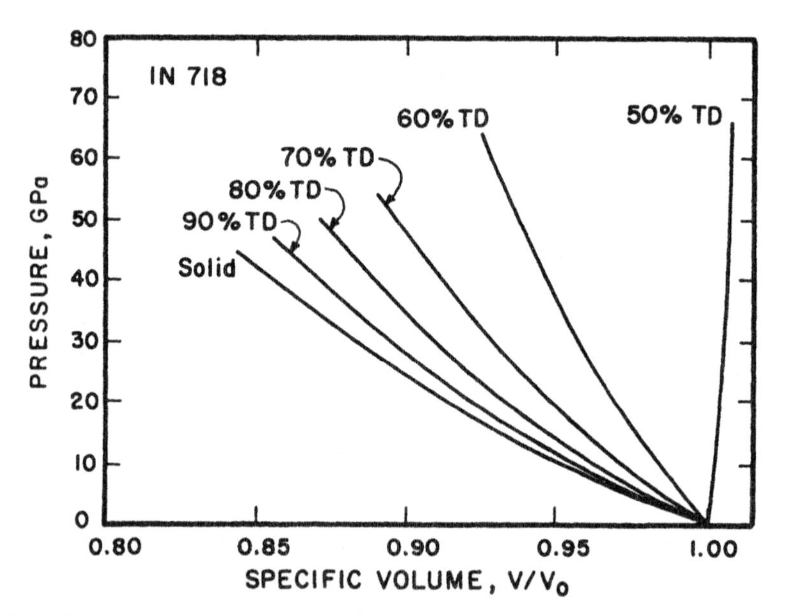

FIGURE 5.6 Calculated pressure-specific volume curves for Inconel 718 powders of varying initial densities. (Reprinted from *Acta Met.*, vol. 36, M. A. Meyers and S. L. Wang, Fig. 2(a), p. 928), Copyright 1988, with permission from Pergamon Press, Ltd.)

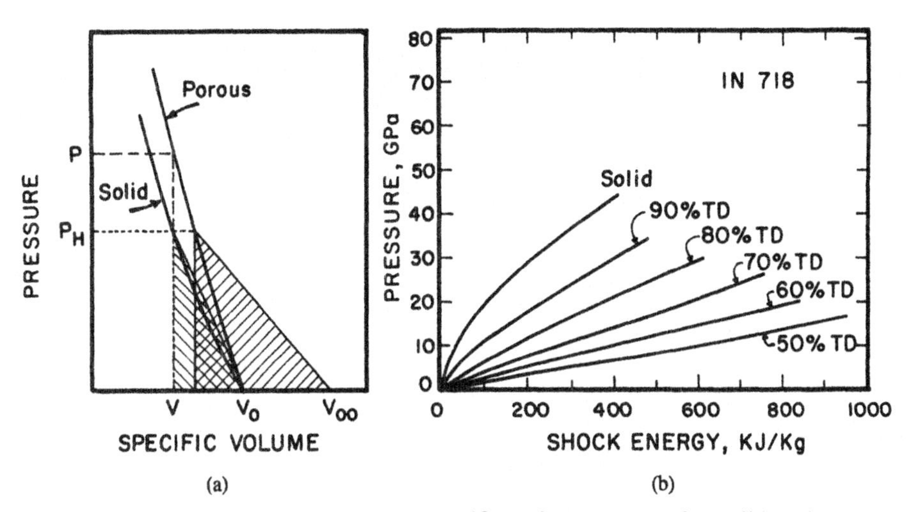

FIGURE 5.7 (a) Schematic pressure-specific volume curves for solid and porous material. (b) Calculated pressure-shock energy for IN 718 powder at various distentions. (Reprinted from *Acta Met.*, vol. 36, M. A. Meyers and S. L. Wang, Fig. 3, p. 929), Copyright 1988, with permission from Pergamon Press, Ltd.)

Figure 5.7(a) shows schematically how this energy is determined. From the definition $E = 1/2P(V - V_0)$, the energy at a pressure level P_H is equal to the shaded areas for the solid and porous Hugoniot curves. This energy is larger for the powder than for the solid at the same pressure. These energies were determined as a function of pressure and porosity and are plotted in Figure 5.7(b). The Mie–Grüneisen EOS converts the state (P_H, V) on the solid material into the state (P, V) on the porous material. The procedure is based on the same volume. Point (P, V) is marked in Figure 5.7(a).

Boade [11] used quartz sensors to study the shock response of porous copper and reported both high-pressure and low-pressure data. Figure 5.8 shows the calculated results that match the theoretical Mie–Grüneisen Hugoniots well at higher pressures. For lower pressures, in general, substantial deviations are observed that are due to material strength effects.

Figure 5.9 shows the EOS for porous nickel at an initial density of 60% of the theoretical value; this curve was calculated using the Mie–Grüneisen EOS, which neglects the strength of the powder—thus the question mark (?) on plot. For higher pressures, the Mie–Grüneisen predictions are good. Oh and Persson [12] developed a more advanced formulation, in which the Grüneisen parameter γ is a function of the energy and not the volume. This formulation was successfully used for the determination of the shock response of porous materials. The Grüneisen treatment ignores the strength of the material, and Herrmann [9] developed an alternative treatment in which the material strength is incorporated. At lower pressures, the constitutive model describes the compaction of a material with a flow stress σ_f, and at higher pressures the model merges with the Hugoniot.

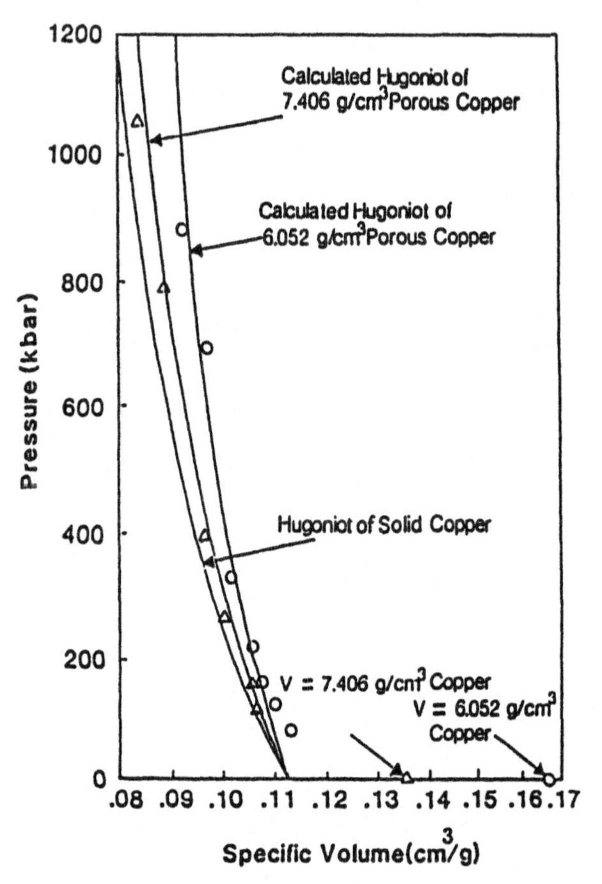

FIGURE 5.8 Measured (circles and triangles) and calculated Hugoniots for copper with two initial densities (6.052 and 7.406 g/cm^3). (From Boade [11], Fig. 4, p. 5696. Reprinted with permission of the publisher.)

Thouvenin [13] proposed a model in which the porous material is assumed to consist of successive plates of solid material and gaps. The solid material is accelerated into the gaps until it impacts the next plate. A succession of impacts represents the wave traveling through the powder. Thus, the phase velocity of the shock wave, $U*$, in a porous material is represented as

$$\frac{1}{U*} = \frac{\rho_{00}/\rho_0}{U_s} + \frac{(\rho_0 - \rho_{00})/\rho_0}{U_p}$$

where U_s is the shock velocity in the solid material, U_p is the particle velocity, ρ_0 is the initial density for the solid material, and ρ_{00} is the initial density for the porous material. This model assumes that the solid plate is accelerated to a velocity U_p into the gap.

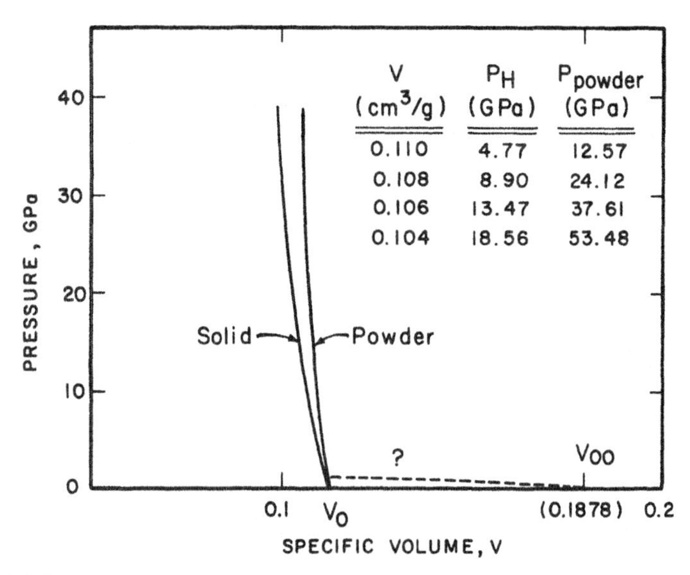

FIGURE 5.9 Pressure vs. specific volume for nickel powder at 60% of theoretical density.

The interest in the passage of shock waves through porous material (powders, foams) is both scientific and technological. The technological interest stems from the fact that high-amplitude shock waves can be used to compact powders and produce monoliths. This is described in Chapter 17. An additional application is the use of porous materials to "damp" shock waves. Since the shock wave deposits a great deal of energy in the material (the area of the hatched triangle in Fig. 5.7), it is very rapidly attenuated. We all know that the sound is attenuated by porous materials and the same occurs with the shock wave. Thus, powders can be used in devices to trap shock waves. A third, scientific area of interest is that the passage of a shock wave through a powder enables us to reach very high internal energy states at relatively low pressures. This internal energy, given by the area in the triangle defining initial, shock, and final states in P–V spall, can be orders of magnitude higher in a porous than in a solid material. This has been used by scientists to access high-energy states.

Example 5.3 Determine the pressure–volume curve for nickel powder with a density equal to 60% of the theoretical density, subjected to a shock wave.

We first start with the EOS for solid nickel. It is given by

$$U_s = 4.667 + 1.41U_p$$

The initial density of nickel (ρ_0) is 8.874, yielding $V_0 = 0.1126$. The calculation of V_{00}, the specific volume of the powder, is sometimes tricky, and one has to be careful. In this case,

$$\rho_{00} = 0.65\rho_0 = 5.768 = 5768 \text{ kg/m}^3$$

$$V_{00} = V_0/0.65 = 0.17336 = 0.173 \times 10^{-3} \text{ m}^3/\text{kg}$$

The Grüneisen parameter is taken as approximately

$$\gamma_0 = 2S - 1 = 2.82 - 1 = 1.82$$

We now define the ratio

$$\frac{\gamma_0}{V_0} = \frac{\gamma}{V} = \text{const}$$

We have all parameters needed as input for Eqn. (5.30). The EOS for the powder is established by setting values for V and calculating the corresponding values of P:

$$P = \frac{[2V - (\gamma_0/V_0)(V_0V - V^2)]C^2(V_0 - V)}{[2V - (\gamma_0/V_0)(V_{00}V - V^2)][V_0 - S(V_0 - V)]^2}$$

Equation (5.30) was slightly modified to incorporate the constant ratio γ_0/V_0.

5.5 TEMPERATURE RISE ASSOCIATED WITH SHOCK WAVES

Shock waves compress the material, and as a consequence the temperature rises. The thermodynamic process at the shock front is assumed to be adiabatic. The release from the shock state to the initial state is usually assumed to be isentropic. In solid materials the shock Hugoniot and release isentrope are fairly close. Figure 5.10 shows schematically a material that was shocked to pressure P_1 from an initial state at atmosphere pressure. The point P_1, V_1 is on the shock Hugoniot. When the pressure is released, unloading follows the release isentrope to point 2. It can be seen that V_2 is different from V_0, because T_2 is higher than T_0. This irreversibility of the process produces lost energy (shown by the hatched area). Indicated in the same plot are three temperatures: T_0, T_1, and T_2. By means of the Grüneisen EOS and thermodynamic relationships, it is possible to calculate the temperatures T_1 and T_2. This will be done next.

For the process 0 → 1. From the first law of thermodynamics: $dE = \delta Q - \delta W$, when all work is against volume, we obtain

$$\delta W = P\,dV$$

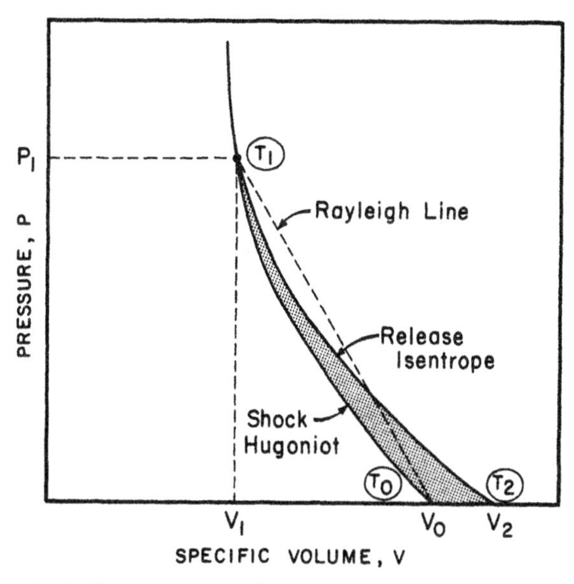

FIGURE 5.10 Shock Hugoniot and release isentrope leading to calculation of temperatures T_1 and T_2.

and

$$\frac{\delta Q}{T} = dS$$

Thus

$$dE = T\,dS - P\,dV \tag{5.34}$$

We will now obtain a thermodynamic expression for $T\,dS$:

$$S = f(T, V)$$

$$dS = \left(\frac{\partial S}{\partial T}\right)_V dT + \left(\frac{\partial S}{\partial V}\right)_T dV$$

$$T\,dS = T\left(\frac{\partial S}{\partial T}\right)_V dT + T\left(\frac{\partial S}{\partial V}\right)_T dV \tag{5.35}$$

$$\text{But } C_v = \left(\frac{\partial E}{\partial T}\right)_V = T\left(\frac{\partial S}{\partial T}\right)_V \tag{5.36}$$

There are four Maxwell relationships. From

$$dA = -P\,dV - S\,dT$$

it can be seen that

$$\left(\frac{\partial P}{\partial T}\right)_V = \left(\frac{\partial S}{\partial V}\right)_T \tag{5.37}$$

Substituting (5.36) and (5.37) into (5.35) yields

$$T\,dS = C_v\,\partial T + T\left(\frac{\partial P}{\partial T}\right) dV \tag{5.38}$$

We now use the following identity:

$$\left(\frac{\partial P}{\partial T}\right)_V = \left(\frac{\partial P}{\partial E}\right)_V \left(\frac{\partial E}{\partial T}\right)_V$$

$$= \frac{\gamma}{V} C_v \tag{5.39}$$

We have applied the Grüneisen equation

$$\frac{\gamma}{V} = \left(\frac{\partial P}{\partial E}\right)_V$$

We can now substitute (5.38) and (5.39) into (5.34):

$$dE = C_v\,dT + T\frac{\gamma}{V} C_v\,dV - P\,dV \tag{5.40}$$

However, for a Hugoniot shock process, we have

$$\Delta E = (E_1 - E_0) = \tfrac{1}{2}(P_1 + P_0)(V_0 - V_1) \tag{5.41}$$

If we express changes of internal energy with volume along the Hugoniot, we have, for Eqns. (5.40) and (5.41),

$$\left(\frac{dE}{dV}\right)_{\mathrm{H}} = C_v\left(\frac{dT}{dV}\right)_{\mathrm{H}} + \frac{\gamma T}{V} C_v - P \tag{5.42}$$

$$\left(\frac{dE}{dV}\right)_{\mathrm{H}} = \frac{1}{2}\left(\frac{dP}{dV}\right)_{\mathrm{H}} (V_0 - V) - \frac{P}{2} \tag{5.43}$$

By substituting (5.43) into (5.42), we obtain

$$C_v \left(\frac{dT}{dV}\right)_H + \frac{\gamma T C_v}{V} = \frac{1}{2}\left(\frac{dP}{dV}\right)_H (V_0 - V) + \frac{P}{2}$$

This is a differential equation of the form

$$Ay' + By = F(V)$$

We know $(dP/dV)_H$ from the Rankine–Hugoniot relationships. By solving this differential equation, we can determine the temperature as a function of specific volume on any point along the Hugoniot.

The standard solution is of the form

$$T = T_0 \exp\left[\left(\frac{\gamma_0}{V_0}\right)(V_0 - V)\right] = \frac{(V_0 - V)}{2C_v} P + \frac{\exp\left[(-\gamma_0/V_0)V\right]}{2C_v} \int_{V_0}^{V} P \cdot \exp\left[(\gamma_0/V_0)V\right]\left[2 - \left(\frac{\gamma_0}{V_0}\right)(V_0 - V)\right] dV \qquad (5.44)$$

The integral has to be numerically calculated.

Once we know the temperature T_1 on the shock Hugoniot, we can determine the temperature T_2 after the pressure has returned to zero. This is done by following an isentropic path from 1 to 2. The thermodynamic equation

$$T\,dS = C_v\,dT + T\frac{\gamma}{V}C_v\,dV$$

$$dS = 0$$

$$\frac{dT}{T} = -\frac{\gamma}{V}\,dV$$

Integrating this from 1 to 2 yields

$$\ln\frac{T_2}{T_1} = -\int_{V_1}^{V_2} \frac{\gamma}{V}\,dV$$

$$T_2 = T_1 \exp\left(-\int_{V_1}^{V_2} \frac{\gamma}{V}\,dV\right)$$

If we use the approximation

$$\frac{\gamma_0}{V_0} = \frac{\gamma}{V}\ \text{const}$$

then

$$T_2 = T_1 \exp\left[\frac{\gamma_0}{V_0}(V_1 - V_2)\right] \tag{5.45}$$

Table 5.3 presents some representative temperatures at three shock pressure levels (30, 50, and 100 GPa)

Example 5.4 Calculate the temperature rise produced by the 90-GPa shock pressure in AISI 304 stainless steel:

$$U_s = 4.569 + 1.49U_p \qquad \rho_0 = 7.896 \qquad \gamma_0 = 2.170$$

$$C_v = 0.1055 \text{ cal/g K} \qquad T_0 = 300 \text{ K}$$

At 90 GPa, we calculate, from the Rankine–Hugoniot equations,

$$\rho_0 = 10.293 \qquad U_p = 1.63$$

We now estimate the three terms in Eqn. (4.65):

$$T_0 \exp\left[\frac{\gamma_0}{V_0}(V_0 - V)\right] = 486 \text{ K}$$

$$\frac{P(V_0 - V)}{2C_v} = 3010 \text{ K}$$

$$\frac{\exp[(-\gamma_0/V_0)V]}{2C_v}\int_{V_0}^{V} P \exp\left[\frac{\gamma_0}{V_0}V\right]\left[2 - \frac{\gamma_0}{V_0}(V_0 - V)\right] dV = -2156 \text{ K}$$

$$T_1 = 1340 \text{ K}$$

TABLE 5.3 Selected Adiabatic (T_H) and Residual (T_R) Temperature Rises (°C)

	Metal	30 GPa	50 GPa	100 GPa
Copper				
	T_H (= T_1)	179	425	1462
	T_R (= T_2)	67	194	657
Lead				
	T_H (= T_1)	1050	2430	8925
	T_R (= T_2)	307	604	1520
Titanium				
	T_H (= T_i)	242	644	2095
	T_R (= T_2)	134	389.5	1134

Source: From McQueen and Marsh [14].

Note: Initial temperature 20°C.

The integral has to be numerically calculated. The residual temperature is calculated from the release isentrope:

$$T_2 = T_1 \exp\left[\frac{-\gamma_0}{V_0}(V_2 - V_1)\right]$$

$$T_2 = 1340 \exp\left[-2.17 \times 7.896\left(\frac{1}{7.896} - \frac{1}{10.293}\right)\right]$$

$$T_2 = 808.4 \text{ K}$$

REFERENCES

1. M. H. Rice, R. G. McQueen, and J. M. Walsh, *Solid State Phys.*, **6** (1958), 1.
2. G. McQueen, S. P. Marsh, J. W. Taylor, J. N. Fritz, and W. J. Carter, in *The Equation of State of Solids from Shock Wave Studies, High Velocity Impact Phenomena*, ed. R. Kinslow, Academic, New York, 1970, p. 230.
3. P. W. Bridgman, *Proc. Am. Acad. Arts Sci.*, **77** (1949), 189.
4. K. Gschneider, *Solid State Phys.*, **16.**
5. M. Yoshida, "Program MIXTURE," CETR Report, New Mexico Institute of Mining and Technology, 1986.
6. R. G. McQueen, S. P. Marsh, J. W. Taylor, and W. J. Carter, in *Symposium on High Dynamic Pressure*, IUTAM, Paris, Gordon and Breach, New York, 1968, p. 67.
7. L. V. Al'tshuler, *Soviet Phys.*, **8** (1965), 52.
8. C. L. Mader, Report No. LA 4381 Los Alamos Scient. Lab., Los Alamos, NM, 1970.
9. W. Herrmann, *J. Appl. Phys.*, **40** (1969), 2490.
10. M. A. Meyers and S. L. Wang, *Acta Met.*, **36** (1988), 925.
11. R. R. Boade, *J. Appl. Phys.*, **39** (1968), 693.
12. K. H. Oh and P. A. Persson, *J. Appl. Phys.*, **65** (1989), 3853.
13. J. Thouvenin, in *Proc. Fourth Symposium on Detonation*, Naval Ordinance Laboratory, Silver Spring, MD, 1965, p. 258.
14. R. G. McQueen and S. P. Marsh, *J. Appl. Phys.*, **31** (1960), 1253.

CHAPTER 6

Differential Form of Conservation Equations and Numerical Solutions to More Complex Problems

6.1 INTRODUCTION

In Chapter 5 we introduced the Rankine–Hugoniot conservation equations and the EOS for different materials (solids, powders, mixtures, and alloys). With the use of these equations one can describe mathematically the propagation of shock waves in a one-dimensional configuration: a planar shock front in which the particle and shock velocities are parallel. In many situations the geometry is more complex, and this simple set of equations is not sufficient. We have curved shock fronts, spherically (radially) expanding shock fronts, and complex interactions between shock waves and materials that require numerical solutions only possible in computers. Computer calculations are used routinely to tackle dynamic deformation and shock wave propagation and detonation problems. An example of such a complex problem is shown in Figure 6.1. An explosively forged projectile (EFP) is formed from the detonation of an explosive charge placed behind it. This EFP is propelled forward at a velocity V that can be calculated. The shape of the EFP can also be predicted from numerical computations, since it results from gradients in velocities imparted to the initial dish. The EFP eventually hits a target and produces a shock wave traveling through it. This is followed by the penetration process, which involves erosion in both projectile and target.

This and many other dynamic deformation problems cannot be analytically tackled if realistic answers are sought. The design of the shape of the EFP, the explosive charge, the prediction of the crater diameter and depth, the optimization of the EFP material (and shape and armor materials), and the configuration can currently be done using computer codes. These codes are commonly referred to as **hydrocodes**. They are very complex and involved, and their study and utilization constitutes a specialized field of knowledge. This is much beyond the scope of this book. Nevertheless the *fundamental aspects* of hydrocodes will be presented here. In order to understand hydrocodes we have to derive the conservation equations of fluid mechanics in their differential form. These differential equations are then transformed into difference equa-

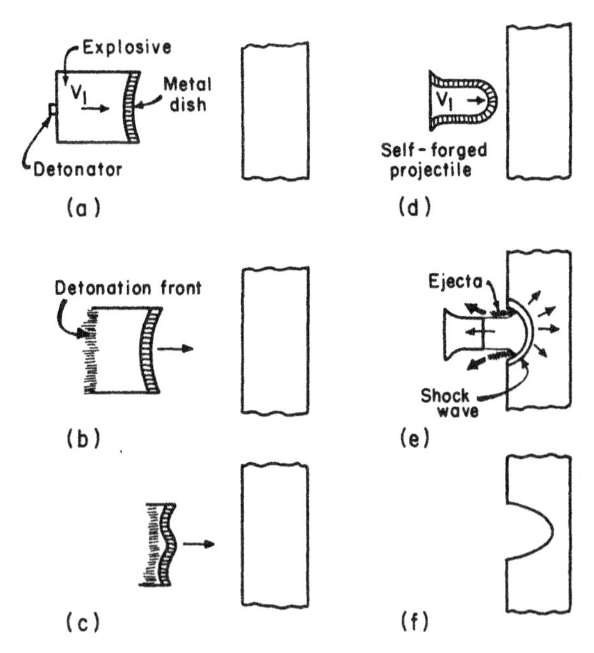

FIGURE 6.1 Schematic sequence of events involved in explosively forged fragment formation and impact with target.

tions that are tractable by the computer. These are the finite-difference codes. Another class of codes, finite-element programs, use a different scheme of converting the differential conservation equations. In order to accommodate the discontinuity at the shock front, a smearing-out technique is introduced, by which the stress rise at the front is spread out over a number of cells. This is done by introducing an "artificial viscosity" term.

Some knowledge of vector calculus is needed in order to derive the conservation equations in their differential form. Thus, a brief review is presented in the next section.

6.2 MATHEMATICAL REVIEW [1, 2]

Figure 6.2 shows two vectors **A** and **B** in a Cartesian coordinate system; we have

$$\mathbf{A} = A_x\mathbf{i} + A_y\mathbf{j} + A_z\mathbf{k}$$

$$\mathbf{B} = B_x\mathbf{i} + B_y\mathbf{j} + B_z\mathbf{k}$$

The vectors **i**, **j**, **k** are unit vectors along the coordinate axes Ox, Oy, and Oz. The terms A_x, A_y, A_z and B_x, B_y, B_z are the components of vectors **A** and **B** with respect to the Cartesian coordinate system. We will now define the scalar

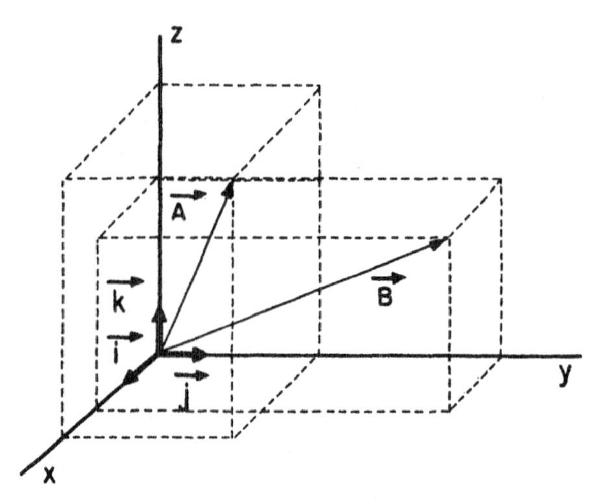

FIGURE 6.2 Vectors decomposed in an orthogonal Cartesian system of reference with unit vectors **i**, **j**, **k**. Scalar product (product is a scalar).

and vectorial products of these two vectors, also called "dot" and "cross" products:

$$\mathbf{AB} = \mathbf{A} \cdot \mathbf{B} = |\mathbf{A}|\,|\mathbf{B}| \cos\theta$$

$$= A_xB_x + A_yB_y + A_zB_z$$

A vector product (product is a vector) is defined as

$$\mathbf{A} \times \mathbf{B} = \begin{vmatrix} \mathbf{i} & \mathbf{j} & \mathbf{k} \\ A_x & A_y & A_z \\ B_x & B_y & B_z \end{vmatrix}$$

$$= \mathbf{i}\begin{vmatrix} A_y & A_z \\ B_y & B_z \end{vmatrix} - \mathbf{j}\begin{vmatrix} A_x & A_z \\ B_x & B_z \end{vmatrix} + \mathbf{k}\begin{vmatrix} A_x & A_y \\ B_x & B_y \end{vmatrix} \tag{6.1}$$

The vectorial product $\mathbf{A} \times \mathbf{B}$ is a vector having magnitude $|\mathbf{A}|\,|\mathbf{B}| \sin\theta$ and is perpendicular to both **A** and **B**. The sense of this vector is such that if **A** is aligned with $0x$ and **B** with $0y$, then $\mathbf{A} \times \mathbf{B}$ is along $0z$.

Two perpendicular vectors have a scalar product equal to zero. Thus

$$\mathbf{i} \cdot \mathbf{j} = 0 \qquad \mathbf{i} \cdot \mathbf{k} = 0 \qquad \mathbf{j} \cdot \mathbf{k} = 0$$

However, their vectorial product has the same magnitude, since sin θ = sin $\pi/2 = 1$:

$$\mathbf{i} \times \mathbf{j} = \mathbf{k} \qquad \mathbf{i} \times \mathbf{k} = -\mathbf{j} \qquad \mathbf{j} \times \mathbf{k} = \mathbf{i}$$

$$\mathbf{i} \times \mathbf{i} = 0 \qquad \mathbf{j} \times \mathbf{j} = 0 \qquad \mathbf{k} \times \mathbf{k} = 0$$

If the arrowtip of a vector **OP** denotes a particle moving in space and the position is a function of time (see Figure 6.3) we have

$$P(t) = x(t)\mathbf{i} + y(t)\mathbf{j} + z(t)\mathbf{k}$$

The velocity of this particle at any time t is defined as

$$\mathbf{u}(t) = \frac{dx}{dt}\mathbf{i} + \frac{dy}{dt}\mathbf{j} + \frac{dz}{dt}\mathbf{k}$$

$$\mathbf{u}(t) = u_x\mathbf{i} + u_y\mathbf{j} + u_z\mathbf{k}$$

This is also shown in Figure 6.3. The velocity vector $u(t)$ is tangential to the trajectory at all times. The acceleration is given by the second derivative:

$$a(t) = \frac{d\mathbf{u}}{dt} = \frac{d^2x}{dt^2}\mathbf{i} + \frac{d^2y}{dt^2}\mathbf{j} + \frac{d^2z}{dt^2}\mathbf{k}$$

We now define a differential operator:

$$\nabla = \frac{\partial}{\partial x}\mathbf{i} + \frac{\partial}{\partial y}\mathbf{j} + \frac{\partial}{\partial z}\mathbf{k} \tag{6.2}$$

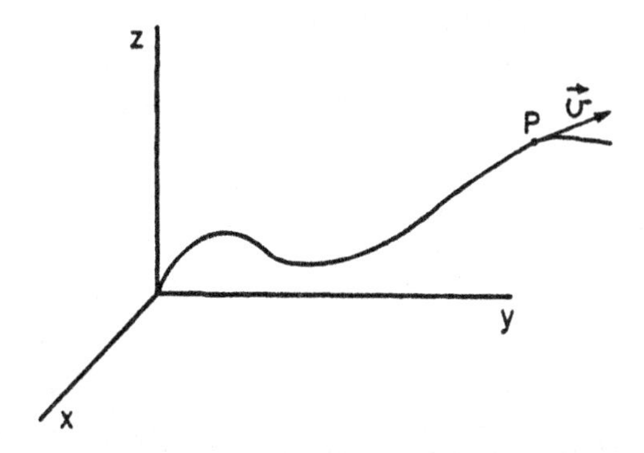

FIGURE 6.3 Point P in coordinate system and its trajectory.

This operator is denoted as "del." By applying it to a scalar field ϕ, where ϕ is a function of x, y, z, we have

$$\nabla\phi = \frac{\partial\phi}{\partial x}\mathbf{i} + \frac{\partial\phi}{\partial y}\mathbf{j} + \frac{\partial\phi}{\partial z}\mathbf{k} \tag{6.3}$$

This is called the gradient of the scalar field ϕ. This gradient has the following properties:

1. $\nabla\phi$ points in the direction in which ϕ increases at its greatest rate.
2. $\nabla\phi(P_0)$ has magnitude equal to the maximum rate of increase of ϕ per unit distance at P_0.
3. If $\nabla\phi \neq 0$, $\nabla\phi$ is normal to the surface $\phi(x, y, z) = \text{const}$.

Figure 6.4 shows a surface $\phi(x, y, z) = k$ and the gradient of the function.

Example 6.1 Determine the gradient of

$$\phi = 2xy + e^z - x^2z$$

at point $P_0(1, 2, -1)$:

$$\nabla\phi = (2y - 2xz)\mathbf{i} + 2x\mathbf{j} + (e^z - x^2)\mathbf{k}$$

$$\nabla\phi(1, 2, -1) = 6\mathbf{i} + 2\mathbf{j} + (e^{-1} - 1)\mathbf{k}$$

By substituting the coordinates at P_0 into ϕ, we find this value:

$$\phi(P_0) = 4 + e^{-1} + 1 = 5 + e^{-1}$$

Thus $\nabla\phi(1, 2, -1)$ is the normal to surface $2xy + e^z - x^2z = 5 + e^{-1}$.

If we multiply the gradient of a surface by a unit vector, we obtain the directional derivative along that surface. We now define the divergence of a

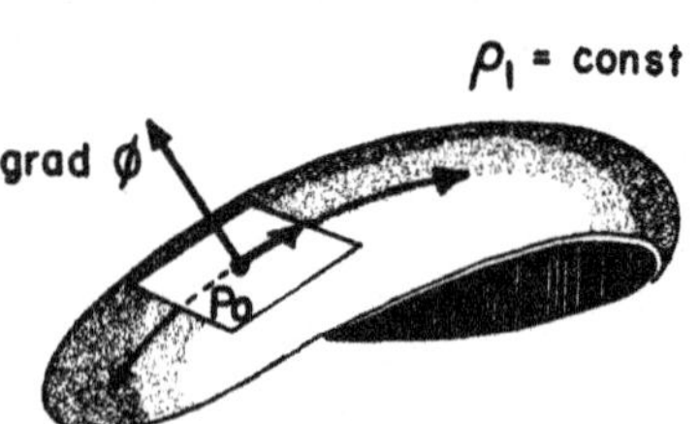

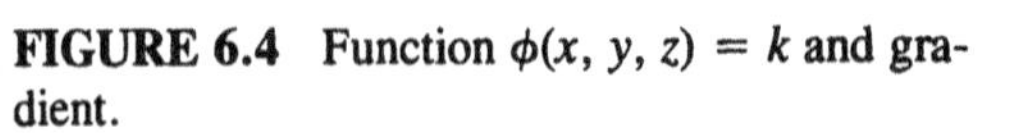

FIGURE 6.4 Function $\phi(x, y, z) = k$ and gradient.

vector field. Whereas the gradient acts on a scalar, the divergence acts on a vector:

$$\nabla \cdot \boldsymbol{\varphi} = \text{div}\, \boldsymbol{\varphi} = \left(\frac{\partial}{\partial x}\mathbf{i} + \frac{\partial}{\partial y}\mathbf{j} + \frac{\partial}{\partial z}\mathbf{k}\right)(\phi_x\mathbf{i} + \phi_y\mathbf{j} + \phi_z\mathbf{k})$$

$$= \frac{\partial \phi_x}{\partial x} + \frac{\partial \phi_y}{\partial y} + \frac{\partial \phi_z}{\partial z} \tag{6.4}$$

The curl is the vector product of the operator ∇ with $\boldsymbol{\varphi}$:

$$\nabla \times \boldsymbol{\varphi} = \begin{vmatrix} i & j & k \\ \frac{\partial}{\partial x} & \frac{\partial}{\partial y} & \frac{\partial}{\partial z} \\ \phi_x & \phi_y & \phi_z \end{vmatrix}$$

$$\nabla \times \boldsymbol{\varphi} = \mathbf{i}\begin{vmatrix} \frac{\partial}{\partial y} & \frac{\partial}{\partial z} \\ \phi_y & \phi_z \end{vmatrix} - \mathbf{j}\begin{vmatrix} \frac{\partial}{\partial x} & \frac{\partial}{\partial z} \\ \phi_x & \phi_z \end{vmatrix} + \mathbf{k}\begin{vmatrix} \frac{\partial}{\partial x} & \frac{\partial}{\partial y} \\ \phi_x & \phi_y \end{vmatrix}$$

$$= \left(\frac{\partial \phi_z}{\partial y} - \frac{\partial \phi_y}{\partial z}\right)\mathbf{i} - \left(\frac{\partial \phi_z}{\partial x} - \frac{\partial \phi_x}{\partial z}\right)\mathbf{j} + \left(\frac{\partial \phi_y}{\partial x} - \frac{\partial \phi_x}{\partial y}\right)\mathbf{k} \tag{6.5}$$

6.3 FLUID FLOW

We will now see the physical significance of the divergence and curl of a vector field. Figure 6.5 shows a fluid flowing in a referential system (x, y, z). We define an elemental parallelepiped in the fluid. The velocity of the fluid is given by

$$\mathbf{U}(x, y, z, t) = u_1(x, y, z, t)\mathbf{i} + u_2(x, y, z, t)\mathbf{j} + u_3(x, y, z, t)\mathbf{k}$$

where $\mathbf{U}$ is the velocity of the fluid at (x, y, z) and time t. We will measure the rate per unit volume of fluid flow out of the box, $\delta x \cdot \delta y \cdot \delta z$. We will first look at faces I and II (back and front faces) on plane yz. The normal to the faces is $\mathbf{i}$.

This represents the flux out of the box in face II. The components u_2 and u_3 do not contribute anything because they are parallel to the face. Similarly, the flux out of face I is

$$\mathbf{u}(x, y, z, t) \cdot (-\mathbf{i})\, \delta y\, \delta z = -u_1(x, y, z, t)\, \delta y\, \delta z$$

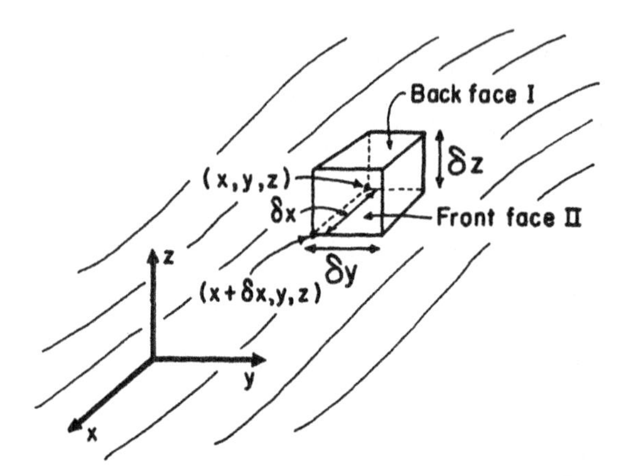

FIGURE 6.5 Element (volume δx, δy δz) of a parallelepiped geometry within fluid that is flowing.

by considering the other faces,

$$[u_1(x + \delta x, y, z, t) - u_1(x, y, z, t)]\,\delta y\,\delta z + [u_2(x, y + \delta y, z, t) - u_2(x, y, z, t)]\,\delta x\,\delta z + [u_3(x, y, z + \delta z, t) - u_3(x, y, z, t)]\,\delta x\,\delta y$$

The outward flux per unit volume is obtained by dividing the above expression by the volume of the box, $\delta x \cdot \delta y \cdot \delta z$:

$$\frac{u_1(x + \delta x, y, z, t) - u_1(x, y, z, t)}{\delta x} + \frac{u_2(x, y + \delta y, z, t) - u_2(x, y, z, t)}{\delta y} + \frac{u_3(x, y, z + \delta z, t) - u_3(x, y, z, t)}{\delta z} = \nabla \cdot u$$

This expression is equal to the divergence of u when the size of the box tends to zero. Thus, the divergence of u represents the expansion, or the divergence away from the point, of the fluid. If the fluid were incompressible, its "flow out of P" would be zero, and

$$\nabla \cdot u = 0 \quad \text{for incompressible fluid}$$

By the same reasoning it can be shown that the case $\nabla \times \mathbf{u}$ represents the degree to which a fluid *swirls*, or rotates, at a point P. We can obtain the continuity equation for the fluid motion from the considerations above. The flow of mass through the box can be represented by similar equations. We can express the conservation of mass by similar equations. If ρ is the density of the fluid, then we can express the mass flow by

$$\mathbf{m} = \rho\mathbf{u} = m_1\mathbf{i} + m_2\mathbf{j} + m_3\mathbf{k}$$

If we define a unit area, the mass crossing it per unit time is equal to the volume of material flowing through it ($\mathbf{u}$) multiplied by the density. We now express the mass entering the faces of the parallelepiped. In face II, the mass exiting per unit time Δt is

$$\rho(x + \delta x, y, z, t)u_1(x + \delta x, y, z, t)\ \delta x\ \delta y\ \delta t$$

At face I, the mass entering is

$$\rho(x, y, z, t)u_1(x, y, z, t)\ \delta x\ \delta t\ \delta t$$

By repeating this procedure for the six faces, the change in mass is $\nabla \cdot \mathbf{m}$. This change in mass is equal to the time rate of the change of density. This is the conservation-of-mass equation:

$$\nabla \cdot \mathbf{m} = -\frac{\partial \rho}{\partial t}$$

$$\nabla \cdot \rho\ \mathbf{u} + \frac{\partial \rho}{\partial t} = 0 \tag{6.6}$$

In steady state, $\partial\rho/\partial t$ and

$$\nabla \cdot \rho\ \mathbf{u} = 0$$

If ρ is constant, $\nabla \cdot \mathbf{u} = 0$.

We now define *total* and partial derivatives for a function ϕ:

$$\phi = \phi(x, y, z, t)$$

$$d\phi = \frac{\partial \phi}{\partial t}\,dt + \frac{\partial \phi}{\partial x}\,dx + \frac{\partial \phi}{\partial y}\,dy + \frac{\partial \phi}{\partial z}\,dz$$

$$\frac{d\phi}{dt} = \left(\frac{\partial \phi}{\partial t}\right) + \left(\frac{\partial \phi}{\partial x}\right)\left(\frac{dx}{dt}\right) + \left(\frac{\partial \phi}{\partial y}\right)\left(\frac{dy}{dt}\right) + \left(\frac{\partial \phi}{\partial z}\right)\left(\frac{dz}{dt}\right)$$

But these are the velocities along the three directions:

$$u_x = \frac{dx}{dt} \qquad u_y = \frac{dy}{dt} \qquad u_z = \frac{dz}{dt}$$

$$\frac{d\phi}{dt} = \left(\frac{\partial \phi}{\partial t}\right) + \left(\frac{\partial \phi}{\partial x}\right)u_x + \left(\frac{\partial \phi}{\partial y}\right)u_y + \left(\frac{\partial \phi}{\partial z}\right)u_z$$

$$= \left(\frac{\partial \phi}{\partial t}\right) + \mathbf{u}\nabla\phi$$

We will hereafter call d/dt the total derivative and denote it by D/Dt, to avoid confusion:

$$\frac{D}{Dt} = \left(\frac{\partial}{\partial t} + \mathbf{u} \cdot \nabla\right) \tag{6.7}$$

This is a very important equation. It will be used in a new definition of coordinate systems.

6.4 EULERIAN AND LAGRANGIAN REFERENTIALS

We will now consider the two referentials: Eulerian and Lagrangian. In order to accomplish this, we will again look at the flow of a fluid, represented in Figure 6.6. We define a property ϕ that is dependent on position (x) and time (t). We have reduced the problem to one dimension to simplify it. In Figure 6.6, we have

$$x_2 - x_1 = \Delta x \qquad t_2 - t_1 = \Delta t$$

$$u = \frac{\Delta x}{\Delta t} = \frac{x_2 - x_1}{t_2 - t_1}$$

We define the property ϕ as, for example, $\phi(x, t) = at^2 + bx^2$.

In a Lagrangian (*material*) referential, we look at the variation of a property ϕ as a function of time. We follow the same particle in the flow (at t_1 it is at x_1; at t_2 it is at x_2):

$$\left(\frac{D\phi}{Dt}\right)_L = \frac{\phi_2 - \phi_1}{\Delta t} = \frac{at_2^2 + bx_2^2 - (at_1^2 + bx_1^2)}{\Delta t}$$

$$= \frac{a(t_2 - t_1)(t_2 + t_1)}{t_2 - t_1} + \frac{b(x_2 - x_1)(x_2 + x_1)}{t_2 - t_1}$$

$$= a(t_2 + t_1) + bu(x_2 + x_1)$$

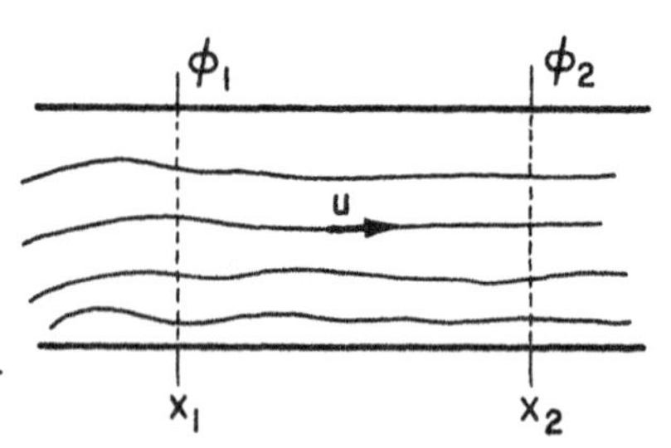

FIGURE 6.6 Definition of Eulerian and Lagrangian referentials.

In a *Eulerian* referential, we follow the flow of the fluid through the same region. Thus, this referential is called a *space* referential. The change in ϕ expressed in this referential is

$$\left(\frac{\partial \phi}{\partial t}\right)_E = \frac{\phi(t_2) - \phi(t_1)}{t_2 - t_1} = \frac{at_2^2 + bx_1^2 - (at_1^2 + bx_1^2)}{t_2 - t_1} = a(t_2 + t_1)$$

Let us now determine $\mathbf{u}\nabla\phi$:

$$\nabla\phi = \frac{\partial}{\partial x}(\phi)_{t=t_2} = \frac{at_2^2 + bx_2^2 - (at_2^2 + bx_1^2)}{x_2 - x_1}$$

$$\mathbf{u}\nabla\phi = ub(x_2 + x_1)$$

If we now apply Eqn. (6.7), we have

$$a(t_2 + t_1) + bu(x_2 + x_1) = a(t_2 + t_1) + ub(x_2 + x_1)$$

Indeed the identity is proven, and Eqn. (6.8) provides the *bridge* between Eulerian and Lagrangian referentials. Thus

Lagrangian referential: $\dfrac{D}{Dt}$

Eulerian referential: $\dfrac{\partial}{\partial t}$

$$\left(\frac{D}{Dt}\right)_L = \left(\frac{\partial}{\partial t}\right)_E + \mathbf{u}\nabla \tag{6.8}$$

6.5 DIFFERENTIAL FORM OF CONSERVATION EQUATIONS

Differential conservation equations have been developed for fluid dynamics and are the fundamental basis of shock-wave physics. These equations are the well-known equations for the conservation of mass, momentum, and energy. They take somewhat different configurations for liquids (assumed to be incompressible) and gases (where there are large density changes). If we use the velocity of sound (C) as a basis and relate the velocity of the material (u) to C, we have

$u \ll C$ for solids and liquids

$u \sim C$ for gases

We also have

$$\Delta\rho \ll \rho_0 \quad \text{for solids and liquids}$$

$$\Delta\rho \sim \rho_0 \quad \text{for gases}$$

For explosives we use the equations for gas dynamics, because we have an expanding gas. We review the concepts of partial and total derivatives:

$\partial/\partial t$ refers to a partial derivative in time at a given point in space (Eulerian).
D/Dt refers to a total derivative, describing the time change in any property, following a moving fluid particle.

An analogy that helps the young investigator to understand the differences between the two referentials is given below. Let us imagine Feynman sitting at the beach in Copacabana. Bikini-clad girls pass by. If he stares fixedly in front of him, he sees different girls passing by. This would be a rigid Germanic (Euler!) approach. On the other hand, if he locks his eyesight on the one girl and follows her, he would be setting a Gallic (Lagrangian) referential. In Eulerian coordinates, the objects move in a fixed and rigid referential. In the Lagrangian referential, individual objects are "tracked."

We set in an orthogonal Cartesian coordinate system a parallelepiped with dimensions δx, δy, and δz. This parallelepiped has coordinates x, y, z (Fig. 6.5). We can now proceed to the equations.

6.5.1 Conservation of Mass

We use the basic equation

$$\text{Input} - \text{output} = \text{accumulation} \tag{6.9}$$

This equation is expressed as a rate:

$$\text{Accumulation} = \frac{\partial}{\partial t}(\rho\ \delta x\ \delta y\ \delta z)$$

$$= \delta x\ \delta y\ \delta z\ \frac{\partial \rho}{\partial t} \tag{6.10}$$

$$\text{Input} = \rho_x u_x\ \delta y\ \delta z + \rho_y u_y\ \delta x\ \delta z + \rho_z u_z\ \delta x\ \delta z \tag{6.11}$$

$$\text{Output} = \rho_{x+\delta x} u_{x+\delta x}\ \delta y\ \delta z + \rho_{y+\delta y} u_{y+\delta y}\ \delta x\ \delta z + \rho_{z+\delta z} u_{z+\delta z}\ \delta x\ \delta y \tag{6.12}$$

We assumed that input occurs in the back faces of the parallelepiped, whereas output occurs in the frontal faces (the ones with coordinates $x + \delta x$, $y + \delta y$, $z + \delta z$):

$$\frac{\partial \rho}{\partial t} = \frac{(\rho_x u_x - \rho_{z+\delta x} u_{x+\delta x})}{\delta x} + \frac{(\rho_y u_y - \rho_{y+\delta y} u_{y+\delta y})}{\delta y} + \frac{(\rho_z u_z - \rho_{z+\delta z} u_{z+\delta z})}{\delta z} \tag{6.13}$$

When δx, δy, $\delta z \rightarrow 0$, Eqn. (6.13) becomes

$$\frac{\partial \rho}{\partial t} = -\left[\frac{\partial}{\partial x}(\rho_x u_x) + \frac{\partial}{\partial y}(\rho_y u_y) + \frac{\partial}{\partial z}(\rho_z u_z)\right]$$

$$= -\nabla(\rho \mathbf{u})$$

$$\frac{\partial \rho}{\partial t} + \nabla \cdot (\rho \mathbf{u}) = 0 \tag{6.14}$$

This is the Eulerian form of the equation for the conservation of mass. The Lagrangian form is found by effecting a transformation. We know

$$\nabla \cdot (\rho \mathbf{u}) = \rho \nabla \cdot \mathbf{u} + \mathbf{u} \nabla \rho$$

Substituting into Eqn. (6.14) yields

$$\frac{\partial \rho}{\partial t} + \rho \nabla \cdot \mathbf{u} + \mathbf{u} \nabla \rho = 0$$

But Eqn. (6.8), applied to the density, gives

$$\frac{D\rho}{Dt} = \frac{\partial \rho}{\partial t} + \mathbf{u} \nabla \rho$$

Thus

$$\frac{D\rho}{Dt} + \rho \nabla \cdot \mathbf{u} = 0 \tag{6.15}$$

Equation (6.15) is the Lagrangian form of the conservation equation.

6.5.2 Conservation of Momentum

We again set the accumulation rate equal to input rate minus output rate. The accumulation rate is

$$\frac{\partial}{\partial t}(\rho \mathbf{u}\ \delta x\ \delta y\ \delta z) \tag{6.16}$$

We define the stresses acting on the different faces of the parallelepiped as P_x, P_y, P_z, $P_x + \delta_x$, $P_y + \delta_y$, $P_z + \delta_z$.

The input rate is

$$[(\rho_x u_x\ \delta y\ \delta z)u_x + P_x\ \delta y\ \delta z]\mathbf{i} + [(\rho_y u_y\ \delta x\ \delta z)u_y + P_y\ \delta x\ \delta z]\mathbf{j} + [(\rho_z u_z\ \delta x\ \delta y)u_z + P_z\ \delta x\ \delta y]\mathbf{k} \tag{6.17}$$

The output rate is

$$[(\rho_{x+\delta x} u_{x+\delta x}\ \delta y\ \delta z)u_{(x+\delta x)} + P_{x+\delta x}\ \delta y\ \delta z]\mathbf{i} + [(\rho_{y+\delta y} u_{y+\delta y}\ \delta x\ \delta z)u_{(y+\delta y)} + P_{y+\delta y}\ \delta x\ \delta z]\mathbf{j} + [(\rho_{z+\delta z} u_{z+\delta z}\ \delta x\ \delta y)u_{(z+\delta z)} + P_{z+\delta z}\ \delta x\ \delta y]\mathbf{k} \tag{6.18}$$

Setting (6.17) − (6.18) = 6.16 yields

$$\frac{\partial}{\partial t}(\rho\mathbf{u}) = \frac{\rho_x u_x^2 - (\rho_{x+\delta x})u_{x+\delta x}^2 + P_x - P_{x+\delta x}}{\delta x}\mathbf{i} + \frac{\rho_y \mathrm{u}_y^2 - (\rho_{y+\delta y})u_{y+\delta y}^2 + P_y - P_{y+\delta y}}{\delta y}\mathbf{j} + \frac{\rho_z \mathrm{u}_z^2 - (\rho_{z+\delta z})u_{z+\delta z}^2 + P_z - P_{z+\delta z}}{\delta z}\mathbf{k}$$

when the size of the elemental parallelepiped is reduced to zero, Eqn. (6.19) becomes

$$\frac{\partial}{\partial t}\rho\mathbf{u} = -\left[\frac{\partial\rho_x u_x^2}{\partial x}\mathbf{i} + \frac{\partial\rho_y \mathrm{u}_y^2}{\partial y}\mathbf{j} + \frac{\partial\rho_z \mathrm{u}_z^2}{\partial z}\mathbf{k}\right] - \left[\frac{\partial P_x}{\partial x}\mathbf{i} + \frac{\partial P_y}{\partial y}\mathbf{j} + \frac{\partial P_z}{\partial z}\mathbf{k}\right]$$

But

$$\rho_x = \rho_y = \rho_z \quad \text{and} \quad P_x = P_y = P_z$$

$$\frac{\partial\rho\ u_x^2}{\partial x} = u_x\frac{\partial\rho\ u_x}{\partial x} + \rho u_x\frac{\partial u_x}{\partial x}$$

$$\frac{\partial\rho\ u_y^2}{\partial y} = u_y\frac{\partial\rho\ u_y}{\partial y} + \rho u_y\frac{\partial u_y}{\partial y}$$

$$\frac{\partial\rho\ u_z^2}{\partial z} = u_z\frac{\partial\rho\ u_z}{\partial z} + \rho u_z\frac{\partial u_z}{\partial z}$$

$$\frac{\partial}{\partial t}\rho\mathbf{u} = -[\mathbf{u}\nabla(\rho\mathbf{u}) + \rho\mathbf{u}\ \nabla\mathbf{u}] - \nabla P$$

$$\frac{\partial}{\partial t}\rho\mathbf{u} + \mathbf{u}\nabla(\rho\mathbf{u}) + \rho\mathbf{u}\ \nabla\mathbf{u} = -\nabla P$$

$$\rho \frac{\partial \mathbf{u}}{\partial t} + \mathbf{u} \frac{\partial \rho}{\partial t} + \mathbf{u}\nabla(\rho \mathbf{u}) + \rho \mathbf{u} \nabla \mathbf{u} = -\nabla P$$

$$\rho \frac{\partial \mathbf{u}}{\partial t} + \mathbf{u} \left[\frac{\partial \rho}{\partial t} + \nabla(\rho \mathbf{u}) \right] + \rho \mathbf{u} \nabla \mathbf{u} = -\nabla P$$

From Eqn. (6.14) (conservation of mass) we obtain

$$\rho \left(\frac{\partial \mathbf{u}}{\partial t} + \mathbf{u} \nabla \mathbf{u} \right) = -\nabla P \tag{6.19}$$

This is the Eulerian form of the equation for the conservation of momentum. In order to obtain the Lagrangian form, we just use Eqn. (6.8):

$$\frac{\rho\, D\mathbf{u}}{Dt} = -\nabla P \tag{6.20}$$

6.5.3 Conservation of Energy

The same basic accumulation equation is used. We consider the internal energy E (potential energy) and the kinetic energy, and the accumulation is the change in this total energy in the elemental parallelepiped:

$$\text{Accumulation rate} = \frac{\partial}{\partial t} (E + \tfrac{1}{2}u^2)\underbrace{\rho\, \delta x\, \delta y\, \delta z}_{\text{mass}}$$

where E is the internal energy per unit mass. This accumulated energy is equal to the energy entering the elemental cube (kinetic, potential, and chemical) minus the energy leaving this parallelepiped. Again, we have to look at the different sides of the parallelepiped to determine the incoming and outflowing energies (Fig. 6.5):

$$\begin{aligned}\text{Energy in} = {} & (E_x + \tfrac{1}{2}u_x^2)\rho u_x\, \delta y\, \delta z + (E_y + \tfrac{1}{2}u_y^2)\rho u_y\, \delta x\, \delta z \\ & + (E_z + \tfrac{1}{2}u_z^2)\rho u_z\, \delta x\, \delta y\end{aligned}$$

(back sides)

$$\begin{aligned}\text{Energy out} = {} & (E_{x+\delta x} + \tfrac{1}{2}u_{x+\delta x}^2)\rho u_{x+\delta x}\, \delta y\, \delta z \\ & + (E_{y+\delta y} + \tfrac{1}{2}u_{y+\delta y}^2)\rho u_{y+\delta y}\, \delta x\, \delta z \\ & + (E_{z+\delta z} + \tfrac{1}{2}u_{z+\delta z}^2)\rho u_{z+\delta z}\, \delta x\, \delta y\end{aligned}$$

(front sides)

$$\text{Energy in due to work} = P_x u_x \, \delta y \, \delta z + P_y u_y \, \delta z \, \delta x + P_z u_z \, \delta x \, \delta y$$

$$\begin{aligned}\text{Energy out due to work} = {} & P_{x+\delta x} u_{x+\delta x} \, \delta y \, \delta z + P_{y+\delta y} u_{y+\delta y} \, \delta z \, \delta x \\ & + P_{z+\delta z} u_{z+\delta z} \, \delta x \, \delta y\end{aligned}$$

The chemical energy being produced (in the case of a chemical reaction such as combustion or detonation) can also be incorporated into the process. This energy is equal to $\rho Q \, \delta x \cdot \delta y \cdot \delta z$, where Q is the chemical energy per unit mass. Putting these equations together and expressing the differences as differentials yields

$$\begin{aligned}\frac{\partial}{\partial t}(E + \tfrac{1}{2}u^2)\rho = {} & -\frac{\partial}{\partial x}(E_x + \tfrac{1}{2}u_x^2)\rho u_x - \frac{\partial}{\partial y}(E_y + \tfrac{1}{2}u_y^2)\rho u_y \\ & - \frac{\partial}{\partial z}(E_z + \tfrac{1}{2}u_z^2)\rho u_z - \frac{\partial P u_x}{\partial x} - \frac{\partial P u_y}{\partial y} \\ & - \frac{\partial P u_z}{\partial z} + \rho Q \\ = {} & -\nabla \cdot [(E + \tfrac{1}{2}\mathbf{u}^2)\mathbf{u}\rho + P\mathbf{u}] + \rho Q\end{aligned}$$

For a process not involving a chemical reaction (such as a shock wave),

$$\frac{\partial}{\partial t}(E + \tfrac{1}{2}u^2)\rho + \nabla \cdot [\rho(E + \tfrac{1}{2}u^2)\mathbf{u} + P\mathbf{u}] = 0 \tag{6.21}$$

In order to express the equation in Lagrangian coordinates, we can make the appropriate substitutions. An alternative method is to use the first law of thermodynamics and to assume that we have an adiabatic process:

$$dE = T\,dS - P\,dV$$

$$\frac{dE}{dt} = T\frac{dS}{dt} - P\frac{dV}{dt}$$

Here $dS/dt = 0$ for a reversible adiabatic process, because

$$\delta Q = 0$$

$$\delta Q_{\text{rev}} = 0 \quad \text{(isentropic)}$$

$$dS = \frac{\delta Q_{\text{rev}}}{T}$$

Thus

$$\frac{dE}{dt} + P\frac{dV}{dt} = 0 \tag{6.22}$$

This is the Lagrangian form.

6.6 FINITE DIFFERENCES AND ARTIFICIAL VISCOSITY [3, 4]

The equations for the conservation of mass, momentum, and energy have to be expressed as differences in order to be processed by the computer. We will explain briefly the method used in solving this problem. We know the solution to a problem at t^n. A set of simultaneous equations is available that may be solved to obtain the values at $t + \Delta t(t \rightarrow t + \Delta t)$. Thus, $t^n \rightarrow t^{n+1}$. This procedure is repeated in time and space (x). In Lagrangian coordinates, we select a grid with spacing Δx and set the time intervals at Δt. There are three finite-difference approximations depending on where we take our differences: forward, central, and backward.

We will now apply this methodology to the conservation equation in one dimension. Another change has to be imparted to the conservation equations in Lagrangian form. The Eulerian coordinate x has to be transformed into the Lagrangian coordinate $X(\rho_0\, \partial X = \rho\, \partial x)$.

We start with the conservation of mass [Eqn. (6.15)]:

$$\frac{D\rho}{Dt} + \rho \nabla \mathbf{u} = 0$$

In one-dimensional treatment,

$$\frac{D\rho}{Dt} + \rho\frac{\partial u}{\partial x} = 0$$

Multiplying and dividing by ρ yields

$$\frac{D\rho}{Dt} = -\frac{\rho^2}{\rho}\frac{\partial u}{\partial x} = \frac{\rho^2}{\rho_0}\frac{\partial u}{\partial X}$$

$$\rho = \frac{1}{V}$$

$$\frac{D(1/V)}{Dt} = \frac{1}{\rho_0 V^2}\frac{\partial u}{dX}$$

$$\frac{\partial V}{\partial t} = \frac{1}{\rho_0}\frac{\partial u}{\partial X} \tag{6.23}$$

We now express this form of the equation as finite differences:

$$\frac{V_l^{n+1/2} - V_l^{n-1/2}}{\Delta t} = \frac{u_{l+1}^{n+1/2} - u_l^{n+1/2}}{\rho_0 \, \Delta X} \tag{6.24}$$

We will explain the meaning of the subscripts and superscripts. The upper subscript (n) denotes time, while the lower subscript (l) denotes position. Thus, in the derivation with respect to time the superscript changes. In the derivative with respect to X the subscript ($l + 1$; l) changes, whereas the superscript is constant. Figure 6.7 shows the time position grid where we see the intervals. These intervals are expressed as $l - 1, l, l + 1$. However, points sitting at the middle of these intervals are also used ($l + \frac{1}{2}, n + \frac{1}{2}, \ldots$).

For the conservation of momentum, we repeat the procedure. In a one-dimensional situation, Eqn. (6.19) reduces itself to

$$\rho \frac{du}{dt} + \frac{\partial P}{\partial x} = 0 \tag{6.25}$$

We replace the Eulerian coordinate x with X ($\rho \, \partial x = \rho_0 \, \partial X$):

$$\frac{du}{dt} = -\frac{\partial P}{\rho \, \partial x} = -\frac{\partial P}{\rho_0 \, \partial X}$$

$$\frac{u_l^{n+1/2} - u_l^{n-1/2}}{\Delta t} = -\frac{P_{l+1/2}^n - P_{l-1/2}^n}{\rho_0 \, \Delta X} \tag{6.26}$$

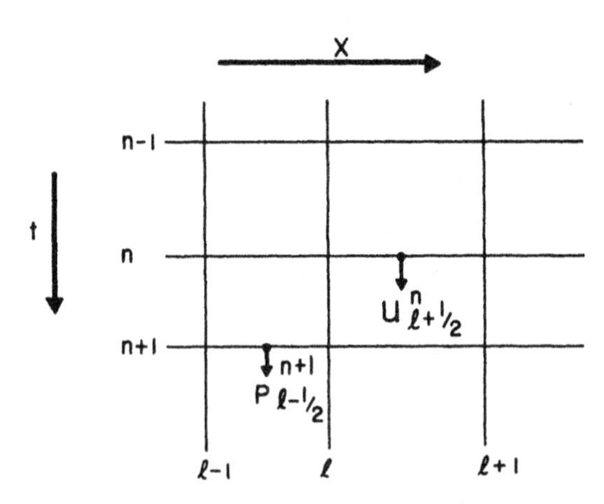

FIGURE 6.7 Time position (Lagrangian) grid.

The conservation-of-energy equation only has time derivatives and is therefore already in full Lagrangian form:

$$\frac{dE}{dt} = -P\frac{dV}{dt} \tag{6.27}$$

$$E^{n+1}_{l+1/2} - E^n_{l+1/2} = -(P^n_{l+1/2})(V^{n+1}_{l+1/2} - V^n_{l+1/2}) \tag{6.28}$$

A more accurate equation for the energy is

$$E^{n+1}_{l+1/2} - E^n_{l+1/2} = -P^{n+1/2}_{l+1/2}\,\Delta V^{n+1/2}_{l+1/2} \tag{6.29}$$

Note that Eqn. (6.29) requires the pressure at time $t^{n+1/2}$ in order for the expression to be second order accurate. We can approximate the pressure at the midpoint of the time step as the average of the pressure at times n and $n + 1$ and solve for the energy at $n + 1$:

$$E^{n+1}_{l+1/2} = E^n_{l+1/2} - \tfrac{1}{2}(P^n_{l+1/2} + P^{n+1}_{l+1/2})\,\Delta V^{n+1/2}_{l+1/2} \tag{6.30}$$

Unfortunately, Eqn. (6.30) for the energy at $n + 1$ involves the pressure at $n + 1$, which is a function of the energy at $n + 1$! The equation is therefore implicit and must be solved numerically for an arbitrary EOS. Most EOS are linear in the energy, which makes Eqn. (6.30) a linear equation in $E^{n+1}_{l+1/2}$, which can then be solved trivially.

For general EOS that are not linear in the internal energy, the pressure at $n + \frac{1}{2}$ can be approximated by $P^n_{l+1/2}$. This approximation is only first order accurate and corresponds to the forward Euler method of integration. A better method, which is almost as efficient from a computational point-of-view, is successive substitution. Let P_i be the ith iteration for $P^n_{l+1/2}$, and P_0 be $P(\rho^{n+1/2}_{l+1/2}, E^n_{l+1/2}/M_{l+1/2})$. M is the element mass. At each iteration, P_i is substituted into Eqn. (6.30) to generate a corresponding E_i. The new value of the energy is used to evaluate the next pressure:

$$P_{i+1} = P(\rho^n_{l+1/2}, E_i/M_{l+1/2}) \tag{6.31}$$

The iterations are continued until the changes in P_i are small. For many EOS, only one to three iterations are necessary to achieve second order accuracy. The algorithms given above for transforming the differential into difference equations [Eqns. (6.24), (6.26), (6.28), (6.29), (6.30)] are illustrative of algorithms used in actual hydrocodes. Wilkins [5] and Benson [6] provide more details of these finite-differencing methods. By applying these equations to a shock front, unfortunately, one does not obtain a correct solution. Rather, as shown in Figure 6.8, one obtains a spurious high-frequency signal that dominates the shock region. These large errors are due to the discontinuity in the

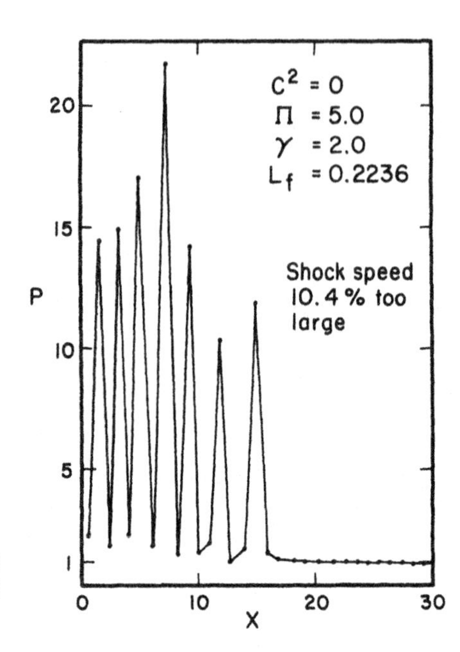

FIGURE 6.8 Solution of difference equations for shock front without artificial viscosity ($c = 0$). ([7], Fig. 26, p. 226.)

shock front. The calculation shown in Figure 6.8 is from Richtmyer and Morton [7]. At the shock front, there is a discontinuity in P and E, and the method breaks down. The most successful method to overcome this deficiency is the introduction of an artificial viscosity by von Neumann and Richtmyer [8]. This consists of adding a term q to the pressure, so that the pressure change is spread out over a few cells, instead of occurring discontinuously. This viscosity term has the form

$$q = -\frac{(c\,\Delta x)^2}{V} \cdot \frac{\partial u}{\partial x}\left|\frac{\partial u}{\partial x}\right| \tag{6.32}$$

where c is a dimensionless constant. We see that this term has the units of

$$\frac{\text{m}^2}{\text{m}^3/\text{kg}} \times \frac{\text{m}^2/\text{s}^2}{\text{m}^2} = \frac{\text{kg} \cdot \text{m}^4}{\text{m}^2\ \text{m}^3 \cdot \text{s}^2} = \frac{\text{N}}{\text{m}^2}$$

These are the same units as pressure. When there is a sharp change in u, q is large; when u is constant, q vanishes. Thus, q acts only on the shock front. The addition of the artificial viscosity term changes the equations for the conservation of mass and energy [from Eqns. (6.25) and (6.27)]:

$$\frac{\partial u}{\partial t} = -\frac{1}{\rho_0}\frac{\partial}{\partial X}(P + q) \tag{6.33}$$

$$\frac{\partial E}{\partial t} = -(P + q)\frac{dV}{dt} \tag{6.34}$$

In finite-difference form, one has

$$\frac{u_l^{n+1/2} - u_l^{n-1/2}}{\Delta t} = -\frac{P_{l+1/2}^n + q_{l+1/2}^{n-1/2} - P_{l-1/2}^n - q_{l-1/2}^{n-1/2}}{\rho_0 \, \Delta X} \tag{6.35}$$

$$E_{l+1/2}^{n+1} - E_{l+1/2}^n = -(P_{l+1/2}^n + q_{l+1/2}^{n+1/2})(V_{l+1/2}^{n+1} - V_{l+1/2}^n) \tag{6.36}$$

Figure 6.9, based on calculations by Richtmyer and Morton [7], shows how the shock front is modified by the application of the artificial viscosity. The constant c^2 was given values of 0.6, 1.7, and 4. At increasing values, the shock front has an increasingly smooth appearance. One can also see that the shock front is spread over a larger and larger region (slope of front decreases with increasing c).

Whereas the original viscosity of von Neumann and Richtmyer [8] had only a quadratic term [Eqn. (6.32)], most viscosities now use an additional linear term. This viscosity is expressed as

$$q = -\left[\frac{(c_Q \, \Delta x)^2}{V} \frac{\partial u}{\partial x} \left|\frac{\partial u}{\partial x}\right| + \frac{c_L \, \Delta x}{V} \frac{\partial u}{\partial x}\right]$$

The constants c_Q and c_L are specified by the user and typical values for them are (1→2) and (0.05–0.25), respectively. Expressed in a central finite-difference scheme, the artificial viscosity takes the form

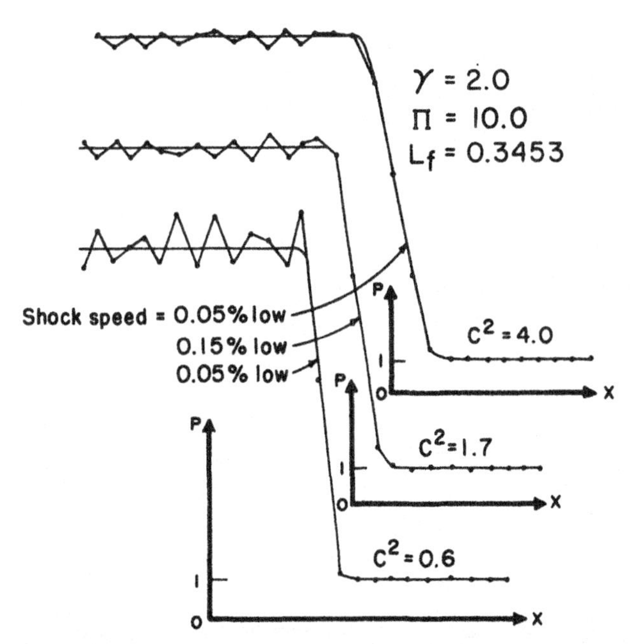

FIGURE 6.9 Solution of difference equations for shock front with different values of artificial viscosity. ([7], Fig. 25, p. 225.)

the artificial viscosity takes the form

$$q_{l+1/2}^{n+1/2} = -(c_Q \rho_{l+1/2}^{n+1/2} |\Delta u_{l+1/2}^{n+1/2}| + c_L c \rho_{l+1/2}^{n+1/2}) \, \Delta u_{l+1/2}^{n+1/2} \tag{6.37}$$

where $\Delta u_{l+1/2}^{n+1/2} = u_{l+1}^{n+1/2} - u_l^{n+1/2}$. The shock viscosity spreads the shock over three to five zones. A narrower shock can be obtained by improving the estimate of the velocity jump across the shock [9]. Some hydrocodes have an option that allows the user to set the viscosity in zones that are expanding to zero. Spurious oscillations are not as much of a problem in expanding regions, and setting the viscosity to zero eliminates unnecessary viscous dissipation.

6.7 HYDROCODES

Large-scale numerical computations can provide very detailed information on the dynamic deformation processes. The code prediction of the problem seen in Figure 6.1 is presented in Figure 6.10 [10]. We see the initial arrangement of the EFP in Figure 6.10(a); Figure 6.10(b) shows the deformation sequence of the dish and its final velocity of 1890 m/s. This is an axisymmetric problem (has a rotational symmetry axis). A perspective view of the initial configuration and intermediate configurations (including the initiation of the folding of the flaps to form fins) is shown in Figure 6.10(c). These results are very impressive and illustrate the capability of modern codes. Although referred commonly to as hydrocodes, these computer codes use more complex material models incorporating strength that is strain, strain rate, and temperature dependent. They also incorporate failure.

These codes contain, in general, the following components:

1. *Conservation Equations*

   Mass: Eqns. (6.14) (Eulerian) and (6.15) (Lagrangian)

   Momentum: Eqns. (6.19) (Eulerian) and (6.20) (Lagrangian)

   Energy: Eqns. (6.21) (Eulerian) and (6.22) (Lagrangian)

2. *Constitutive Equations.* These equations describe the material behavior in elastic, plastic, and shock (hydrodynamic) regimes. These equations are treated in detail in Chapters 5 and 13. Examples are as follows:

   Plastic regime: Johnson–Cook, Steinberg–Guinan, Zerilli–Armstrong, MTS

   Shock response: Mie–Grüneisen, gamma law (explosives), Tillotson.

3. *Failure Models.* These deal with fracture, spalling, and shear band formation. This type is described in Chapters 15 and 16. A number of formulations are presented. Examples are:

   Fracture (Chapter 15): Cochran–Banner, Davison–Stevens, NAG-FRAG (Curran, Seaman, and Shockey)

   Shear bands (Chapter 16): Wright (BRL), Shear (SRI International), Bai, Recht, Clifton

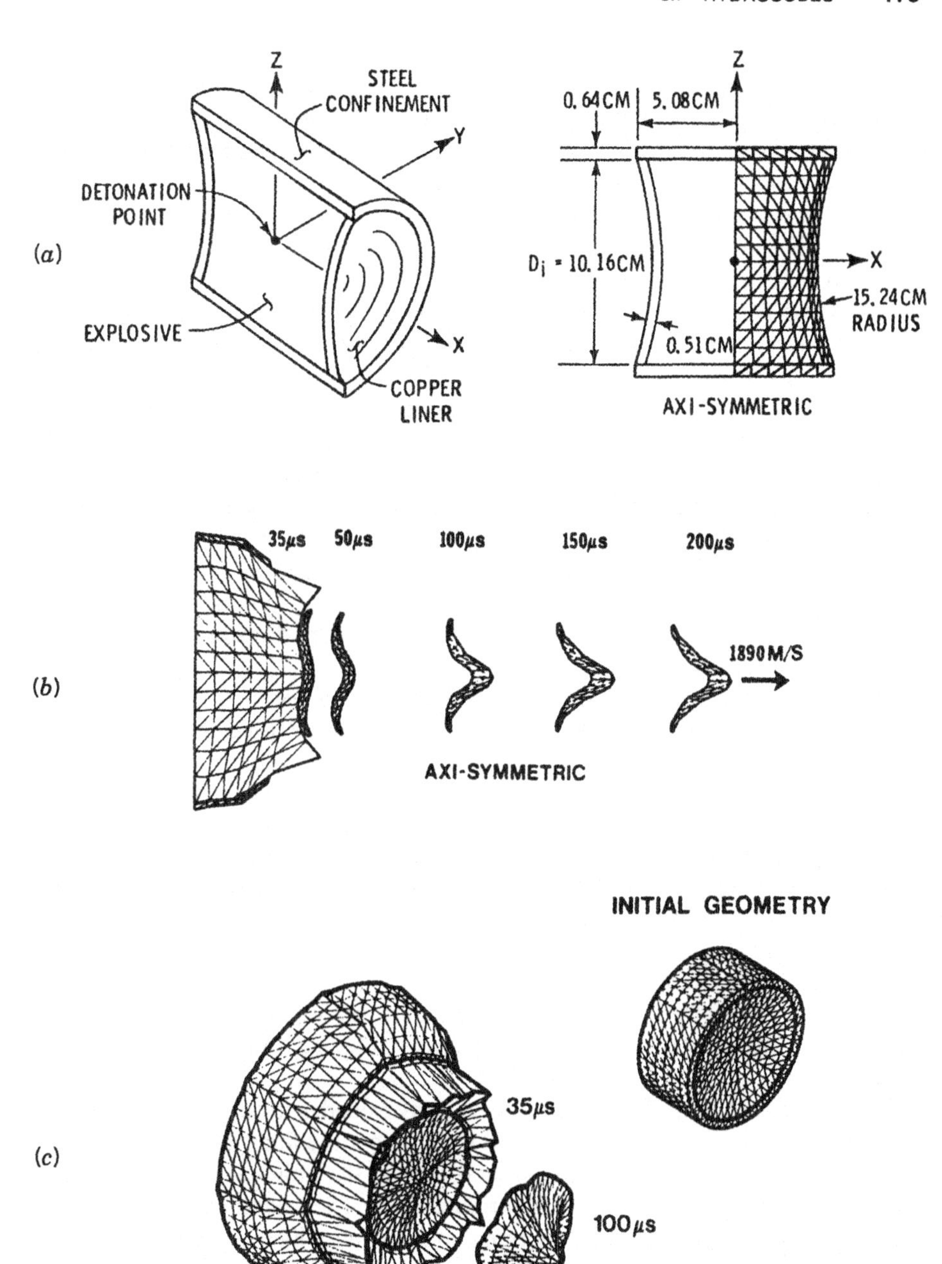

FIGURE 6.10 Calculation of EFP formation using EPIC-3 Lagrangian three-dimensional code. (From Zukas et al. [10], Figs. 19–21, pp. 402–404.) (a) EFP configuration. (b) Progressive distortion and acceleration of dish. (c) Three-dimensional view of (b).

These three ingredients provide the basic physics of the hydrocodes. The constitutive equations and failure models are based on actual experiments and vary from purely phenomenological (i.e., based on experimental results) to microstructurally based predictions based on a physical understanding of the underlying mechanisms.

These codes are usually classified into two major groups: Eulerian and Lagrangian. Figure 6.11 shows a sphere impinging on a target in Lagrangian and Eulerian referentials. As seen before, in a Lagrangian referential the computational grid deforms with the material whereas in a Eulerian referential it is fixed in space. The Lagrangian codes are more straightforward conceptually. However, for problems in which large deformations take place and in which mixing of the materials takes place, the Eulerian approach is necessary.

These codes are also classified as one, two, or three dimensional, according to their dimensional capability. Plane-shock-wave propagation problems (uniaxial strain, plane shock front) are run in one-dimensional codes. Two-dimensional axisymmetric problems (normal impact of cylinders, rods, spheres, axisymmetric projectiles on targets) are successfully treated in two-dimensional codes. More complex problems (e.g., inclined impact) require three-dimensional codes. Table 6.1 shows a listing of some common codes with some of their most prominent features.

The pioneering work at Los Alamos National Laboratory (LANL) and Lawrence Livermore National Laboratory (LLNL) [5, 11, 12] led to the development of hydrocodes and their successful application to a number of important dynamic problems, ranging from nuclear weapons to conventional impact and penetration to shock consolidation of materials. Wilkins [5] incorporated plasticity into the codes through the J2 flow criterion and Drucker's normality postulate. Algorithms were specifically developed to render his finite-difference code HEMP (Hydrodynamic Elasto Magneto Plastic, the acronym a tribute to the 1960s) operational.* These first hydrocodes have been followed by a large number of finite-difference codes (see Table 6.1). Since the 1970s, finite-element codes originally developed for quasi-state problems have been applied with increasing success to dynamic problems, including shock wave propagation situations.

The numerical algorithms that are used in hydrocodes to solve the governing partial differential equations are derived using either the finite-element or finite-difference methods. Under certain circumstances, the finite-element and finite-difference methods give identical algorithms [13]. The Lagrangian form of the conservation equations leads to Lagrangian codes in which the mesh is determined by the material and deforms with it, whereas the Eulerian mesh is fixed. Figure 6.11 shows these two referentials. Whereas Lagrangian codes are much more efficient to run (less computational time spent), the mesh becomes excessively distorted after a critical plastic strain and the predictions lose their accuracy. On the other hand, Eulerian codes can handle large deformations very well but present unique problems. As an example, if we have two materials with different constitutive equations, the same element will have different ma-

*The radial return algorithm introduced by M. L. Wilkins enables the treatment of unloading of the material when it is central-loaded triaxially.

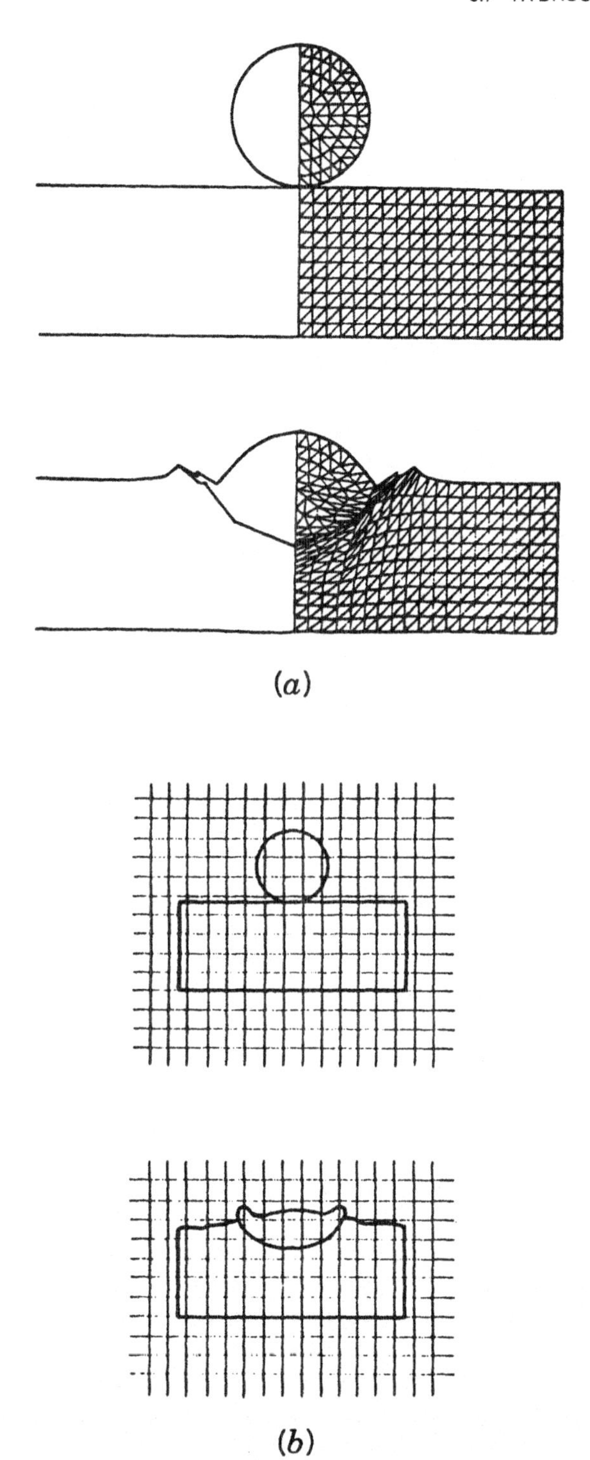

FIGURE 6.11 (a) Lagrangian and (b) Eulerian computational grids. (From Zukas et al. [10], Figs. 2 and 3, pp. 373, 376.)

TABLE 6.1 Principal Hydrocodes Used in the United States[a]

Code	Origin*	Numerical Method	Dimensional Capability	Numerical Coordinate Scheme
SWAP		MC	1-D	Lagrangian
WONDY		FD	1-D	Lagrangian
TOUDY	SNL	FD	2-D	Lagrangian
DUFF		FD	1-D	Lagrangian
HEMP	LLNL	FD	1, 2, 3-D	Lagrangian
STEALTH	SAIC	FD	1, 2, 3-D	Lagrangian
PRONTO	SNL	FD	2, 3-D	Lagrangian
MESA	LANL	FD	2, 3-D	Eulerian
PAGOSA	LANL	FD	3-D	Eulerian
JOY	LLNL	FD	3-D	Eulerian
DYNA	LLNL	FE	2, 3-D	Lagrangian
CALE	LLNL	FD	2-D	Lagrangian
CAVEAT	LANL	FD	2, 3-D	
CTH	SNL	FD	2, 3	Eulerian
PICES	Phys. Intl.	FD	2, 3-D	Coupled Lagrangian/ Eulerian
CRALE		FD	1, 2-D	Arbitrary Lagrangian/ Eulerian
AFTON		FD	1-D	
CSQ II	SNL	FD	2-D	Eulerian
EPIC-2	Honeywell	FE	2-D	Lagrangian
EPIC-3	Honeywell	FE	3-D	Lagrangian
NIKE-2D, 3D	LLNL	FE	2, 3-D	
Codes for personal computers				
ZEUS	Segletis/Zukas			Lagrangian
AUTODYN		FE	2-D	
TDL MADER	C. Mader	FD	2-DD	Lagrangian

[a]Abbreviations: MC, method of characteristics; FD, finite differences; FE, finite elements.
*SNL: Sandia National Labs; LLNL: Lawrence Livermore National Labs.; SAIC: Science Appl. Inc.; Phs. Intl: Physics International.

terials at different times, and it is difficult to track the material interfaces. The Eulerian codes also have a more limited spatial resolution. One solution to this is to stop the computation in the Lagrangian referential at a certain deformation and to rezone the mesh. New approaches that combine Lagrangian and Eulerian methodologies have been developed. The arbitrary Lagrangian–Eulerian (ALE) and the simple arbitrary Lagrangian–Eulerian (SALE) formulations are successful approaches; they are described by Benson [6, 9]. With these formulations, problems involving very large deformations (>10) such as shaped charges can be successfully treated. Figure 6.12 demonstrates this.

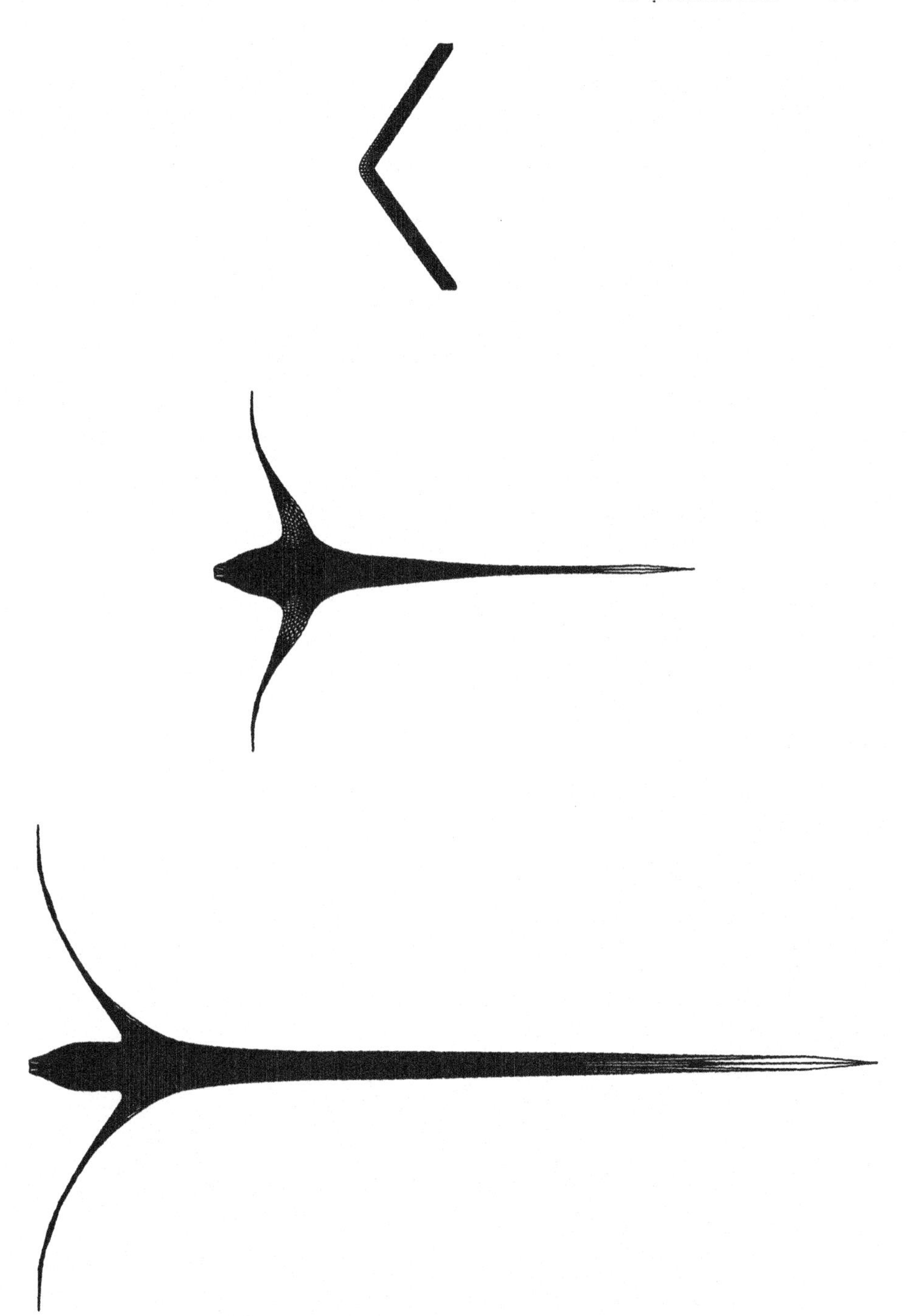

FIGURE 6.12 Computer simulation of the development of a shaped-charge jet using DYNA-2D Lagrangian code with SALE (simple arbitrary Lagrangian–Eulerian) formulation, which enables large deformations. (Different scales from top to bottom.) (From Benson [17], Fig. 14, p. 345. Reprinted with permission of the publisher.)

Detailed descriptions of large-scale computations applied to dynamic problems can be found in Benson [6], Zukas [14], Belytschko and Hughes [15], and Walters and Zukas [16]. The reader is referred to these sources for more detailed study.

REFERENCES

1. E. Kreyszig, *Advanced Engineering Mathematics*, 3rd ed., Wiley, New York, 1972.
2. P. V. O'Neill, *Advanced Engineering Mathematics*, Wadworth, 1987.
3. R. Courant and K. O. Friedrichs, *Supersonic Flow and Shock Waves*, Interscience, New York, 1956.
4. P. C. Chou and A. K. Hoskins, eds., *Dynamic Response of Materials to Intense Impulsive Loading*, Air Force Materials Lab., Dayton, Ohio, 1972.
5. M. Wilkins, in *Methods in Computational Physics*, Vol. 3, *Fundamental Methods in Hydrodynamics*, Academic, New York, 1964, p. 211.
6. D. J. Benson, *Computer Meth. Appl. Mech. Eng.*, **99** (1992), 235.
7. R. D. Richtmyer and K. W. Morton, *Difference Methods for Initial-Value Problems*, Interscience, New York, 1957.
8. J. von Neumann and R. D. Richtmyer, *J. Appl. Phys.*, **21** (1950), 232.
9. D. J. Benson, *Computer Meth. Appl. Mech. Eng.*, **93** (1991), 39.
10. J. A. Zukas, T. Nicholas, H. F. Swift, L. B. Greszezuk, and D. R. Curran, *Impact Dynamics*, Wiley, New York, 1982, pp. 373, 376.
11. W. F. Noh, in *Methods in Computational Physics*, Vol. 3, *Fundamental Methods in Hydrodynamics*, Academic, New York, 1964, p. 117.
12. W. F. Noh, Report No. UCRL-52112, Lawrence Livermore Laboratory, Livermore, CA, 1976.
13. T. B. Belytschko, J. M. Kennedy, and D. F. Schoeberle, *Proc. Computational Methods in Nuclear Engineering*, April 1975, Charleston, SC, American Nuclear Society, 1975, p. IV-39.
14. J. A. Zukas, in *High Velocity Impact Dynamics*, ed. J. A. Zukas, Wiley, New York, 1990, p. 593.
15. T. Belytschko and T. J. R. Hughes, *Computational Methods for Transient Analysis*, North-Holland, Amsterdam, 1983.
16. W. Walters and J. A. Zukas, *Fundamentals of Shaped Charges*, Interscience, New York, 1989.
17. D. J. Benson, *Computer Meth. Appl. Mech. Eng.*, **72** (1989), 305.

CHAPTER 7

Shock Wave Attenuation, Interaction, and Reflection

7.1 INTRODUCTION

A shock wave consists, ideally, of a shock front, a flat top, and a release part. Figure 7.1 shows a simplified configuration for a shock wave. It is preceded by an elastic precursor (1) that has an amplitude equal to the elastic limit of the material, usually called the *Hugoniot elastic limit* (HEL). At sufficiently high pressures the shock wave overcomes the elastic precursor, since the velocity U_s increases with pressure P. Two configurations are shown in Figure 7.1.

The shock wave produced by plate impact has initially a square shape. This shape has a flat top (3) that has a length determined by the time required for the wave to travel through the projectile. For shock waves generated by explosives in direct contact with the material or by laser pulses, the shape of the pulse is triangular. This shape will be better explained in Chapter 10. The portion of the wave in which the pressure returns to zero is called the "release" portion of the wave (4). This is also called, incorrectly, "rarefaction." As the wave propagates through the material, the rate of release decreases, that is, the slope of the back portion decreases. This is explained in greater detail in Section 7.2.

The rate of release of a shock wave is determined primarily by three factors: (1) the material used as the projectile, (2) the distance the pulse has traveled through the material, and (3) the material in which the pulse is traveling.

The distance–time plots are very helpful in understanding the sequence of events that occur as a shock pulse propagates through a target. Figure 7.2 shows a distance–time (commonly called x–t) plot for a projectile of thickness d_0 impacting a semi-infinite target. The slopes of the lines are the inverses of the velocities. It is assumed that the elastic waves have velocities higher than the shock waves and that both target and projectile are of the same material. The flyer plate is propagating at a velocity V_p and at time t_0 impacts the target.

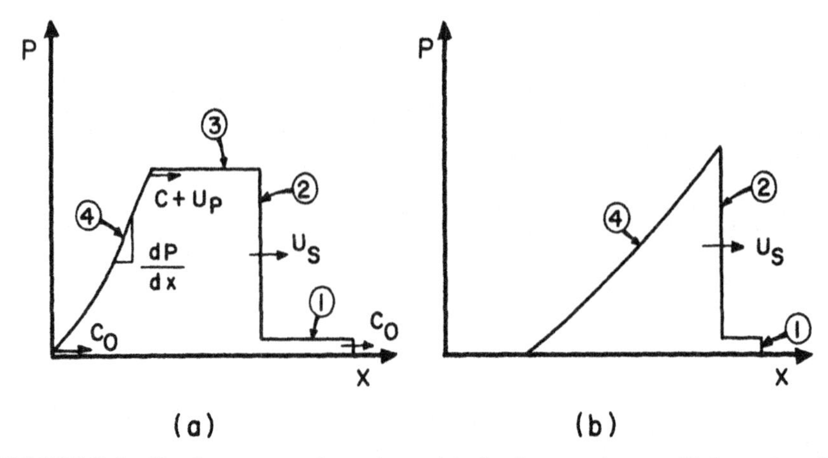

FIGURE 7.1 Shock wave configurations: (a) shock wave (trapezoidal) produced by plate impact; (b) shock wave (triangular) produced by explosive detonation or pulsed laser.

Its velocity is reduced to V, as seen in Chapter 4. Elastic waves with velocities C_0 and shock waves with velocities U_s are emitted into the target and projectile. The slopes of the elastic waves are lower than those of the plastic waves. The waves are reflected at the back surface of the projectile and return into the target as release waves. This process is explained in more detail in Section 7.3. These release waves penetrate into the target. It is the relative position of these various waves at different times that establishes the shape of the stress pulse. These x–t plots are very convenient in determining wave propagation patterns, wave interactions, wave reflection at free surfaces, and phase transformations. The initial duration of the peak shock pulse is approximately equal to twice the travel time of the shock wave through the projectile, as seen from Figure 7.2. Thus,

$$t_p \simeq \frac{2d_0}{U_s} \tag{7.1}$$

This is an approximate expression only.

The state of stress at the shock front is considered in hydrodynamic theory as hydrostatic compression. Thus, in a Cartesian coordinate system one would have $\sigma_x = \sigma_y = \sigma_z = -P$. However, in a real material with a nonnegligible strength, we have to consider the nonhydrodynamic components. In this sense, the HEL is the stress at which a metal deforms plastically. For ceramics, there is considerable debate as to the meaning of the HEL. The plot in Figure 7.3 shows the HEL. We have a state of stress of uniaxial strain; that is, the shock wave does not alter the dimensions of the material perpendicular to the shock propagation direction. If the shock propagation direction is $0X$, then $\epsilon_y = \epsilon_z =$

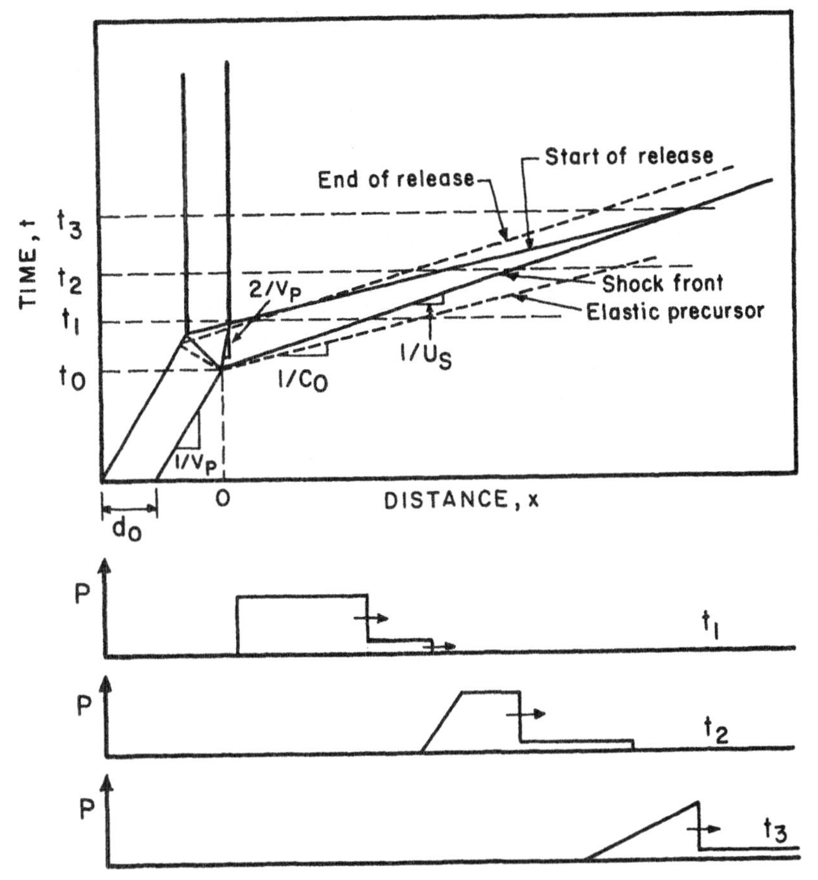

FIGURE 7.2 (a) Impact of flyer plate of thickness d_0 on semi-infinite target; distance x–time t plots are very helpful in visualizing sequence of events; (b) pressure–distance x profiles at three times t_1, t_2, and t_3.

0 and the engineering strain is (Fig. 7.3(b))

$$\varepsilon_x = \frac{l_0 - l}{l_0} = \frac{l_0 - l}{l_0} \times \frac{l_0^2}{l_0^2} = \frac{V_0 - V}{V_0} = 1 - \frac{V}{V_0} \tag{7.2}$$

The true strain is

$$d\varepsilon_x = \frac{dl}{l} \qquad \therefore \ \varepsilon_x = \ln \frac{l}{l_0} = \ln \frac{V}{V_0}$$

In Figure 7.3 the Rankine–Hugoniot curve is drawn for an ideal hydrodynamic material. The stress σ_x was substituted for P and the strain ϵ_x was substituted

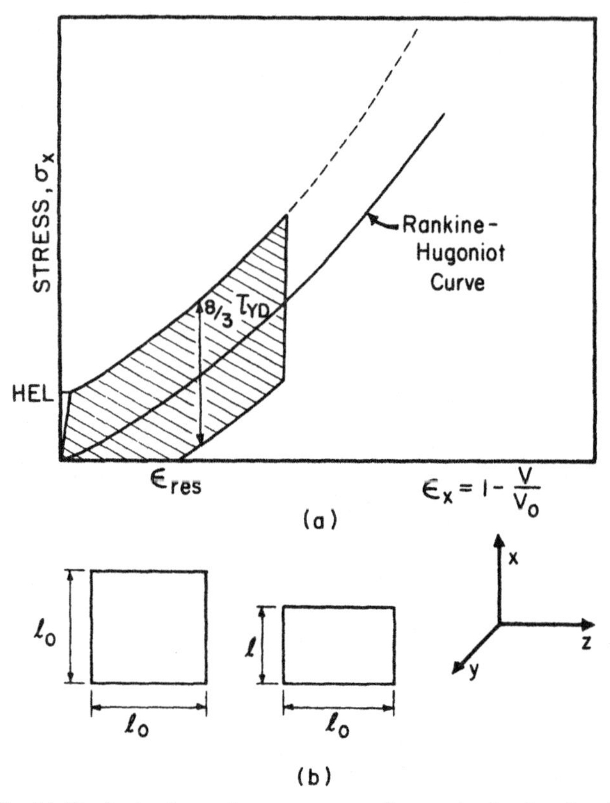

FIGURE 7.3 (a) Nonhydrodynamic component of stress in shock pulse due to strength effects; Rankine–Hugoniot curve shown in same plot. (b) Uniaxial strain deformation produced by shock wave.

for $1 - V/V_0$. For a material having a nonnegligible HEL, loading occurs first elastically. At the HEL, the loading occurs plastically, with permanent deformation in the material. In terms of material strength, a metal deforms plastically when the shear stress equals to the dynamic yield shear stress. This value is shown in the plot in Figure 7.3(a). We have a loading curve that is displaced from the Rankine–Hugoniot curve by $\frac{4}{3}\tau_{YD}$, where τ_{YD} is the yield stress in shear. This stress conversion is demonstrated below:

$$P = \tfrac{1}{3}(\sigma_x + \sigma_y + \sigma_z) = \tfrac{1}{3}(\sigma_x + 2\sigma_y)$$

$$= \tfrac{1}{3}[3\sigma_x - 2(\sigma_x - \sigma_y)] = \sigma_x - \tfrac{2}{3}(\sigma_x - \sigma_y)$$

But

$$\tau_{max} = \tfrac{1}{2}(\sigma_x - \sigma_y)$$

TABLE 7.1 Hugoniot Elastic Limit for Selected Materials

Materials	HEL (*EPa*)
Al_2O_3 (sapphire)	12–21
Al_2O_3 (polycrystalline)	9
Fused quartz	9.8
WC	4.5
2024 Al	0.6
Cu cold worked	0.6
Fe	1–1.5
Ni	1.0
Ti	1.9

So

$$P = \sigma_x - \tfrac{4}{3}\tau_{\max} \quad \text{and} \quad \sigma_x = P + \tfrac{4}{3}\tau_{\max} \tag{7.3}$$

The release path is different from the loading because of hysteresis effects. Unloading will require a stress of $\frac{8}{3}\tau_{YD}$ at the maximum strain. At this point, a curve "parallel" to the Rankine–Hugoniot curve brings the material to zero stress. Notice the irreversibility of the process produced by this elastoplastic effect, indicated by the hatched area in Figure 7.3(a). This energy dissipation contributes to a more rapid decay (or attenuation) of the shock wave. This is discussed at the end of Section 7.2. Another aspect to notice is that the shock wave passage will produce residual strain ϵ_{res}.

As the pressure increases, the relative effect of the HEL vanishes, and the three curves in Figure 7.3(a) collapse essentially into one: the Rankine–Hugoniot curve. Table 7.1 gives examples of HELs for some materials.

7.2 ATTENUATION OF SHOCK WAVES

The rarefaction—or release—part of the shock wave is the region beyond the peak pressure, where the pressure returns to zero. The attenuation of a wave, on the other hand, is the decay of the pressure pulse as it travels through the material. Figure 7.4 shows schematically how a shock wave is changed as it progresses into the material. At t_1, the wave has a definite peak pressure, pulse duration, and rarefaction rate (mean slope of the back of the wave). The inherent irreversibility of the process is such that the energy carried by the shock pulse continuously decreases. This is reflected by a change of shape of the pulse. If one assumes a simple hydrodynamic response of the material, the change of shape of the pulse can be simply seen as the effect of the differences between the velocities of the shock and rarefaction part of the wave. It can be seen in Figure 7.4 that the rarefaction portion of the wave has a velocity

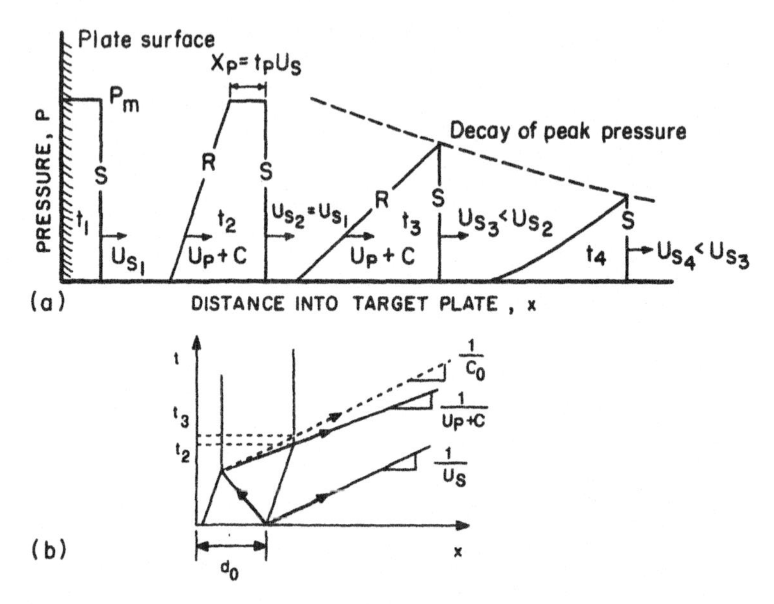

FIGURE 7.4 (a) Progress of a shock pulse through material, rarefaction front steadily overtakes shock front; (b) x–t plot of (a).

$U_p + C$, where U_p is the particle velocity and C the sound velocity at a particular pressure. As the wave progresses, the release part of the wave overtakes the front, because $U_p + C > U_s$. This will reduce the pulse duration (at maximum pressure) to zero. After it is zero, the peak pressure starts to decrease. As this peak pressure decreases, so does the velocity of the shock front: $U_{s4} < U_{s3} < U_{s2} = U_{s1}$. This can be easily seen by analyzing the data in Table 4.1. By appropriate computational procedures one can actually calculate the change in pulse shape based on the velocities of the shock and rarefaction portion of the wave. This is done below, in a very simplified manner. In order to calculate the rarefaction rate of the pressure pulse as it enters the material, it is best to use Figure 7.1. Although the curve is concave upward, we can, as a first approximation, assume it to be a straight line and calculate an average dP/dt. The average rarefaction rate can be obtained by dividing the peak pressure by the difference between the time taken for the head and the tail of the rarefaction wave to pass through a certain point. The head and tail of the rarefactions are shown and travel with velocities $U_p + C$ and C_0, respectively. If one wants to determine the rarefaction rate at the collision interface, one has to find the difference $t_3 - t_2$ in Figure 7.4(b). One should notice that when the flyer plate is under compression, its thickness is $(\rho_0/\rho)d_0$. Hence, one sees, from Figure 7.4(b),

$$t_3 - t_2 = d_0\left(\frac{\rho_0}{\rho}\right)\frac{1}{U_p + C} - d_0\left(\frac{1}{C_0}\right)$$

$$\dot{P} = \frac{dP}{dt} = P_m \left[d_0 \left(\frac{\rho_0}{\rho(U_p + C)} - \frac{1}{C_0} \right) \right]^{-1}$$

Again, when a protective cover plate of thickness t^c is used, then

$$\dot{P}_R = \frac{P_m}{d_0(\rho_0/\rho(U_p + C) - 1/C_0) + t^c(\rho_0^c/\rho^c(U_p + C) - 1/C_0^c)} \quad (7.4)$$

The rate of rarefaction is very sensitive to pressure. So, if either the impact velocity or the flyer plate thickness is changed, for the same target–projectile system, different rarefaction rates will result.

The attenuation rate (or decay rate) measures the rate at which the pressure pulse dissipates itself as it travels through the material. The energy carried by the pressure pulse is dissipated as heat, defects generated, and other processes. Figure 7.4 shows schematically how the impulse carried by the wave decreases as it travels from the front to the back face of the target. Up to a certain point the pressure remains constant; it can be seen that at t_3 the pressure has already decreased from its initial value and that at t_4 it is still lower. The greater the initial duration of a pulse, the greater will be the impulse carried by it and, consequently, its ability to travel throughout the material. The simplest approach to calculating the decay rate of a pulse is to assume the hydrodynamic response of the material. As illustrated in Figure 7.4, the relative velocities of the shock and release waves will determine the attenuation. The head of the release wave travels at a velocity $U_p + C$; the distance that the peak pressure is maintained, S, is given by the difference between the shock velocity U_s and $U_p + C$:

$$U_s = \frac{S}{t}$$

$$U_p + C = \frac{S + X_p}{t}$$

$$\frac{U_s}{S} = \frac{U_p + C}{S + X_p}$$

$$S(U_p + C) = U_s(S + X_p) = U_s(S + t_p U_s)$$

Hence

$$S(U_p + C - U_s) = U_s^2 t_p$$

$$S = \frac{U_s^2 t_p}{U_p + C - U_s} \quad (7.5)$$

X_p and t_p are the initial thickness and duration of pulse (Fig. 7.4(a)).

Beyond this point numerical techniques have to be used to compute the pressure decay. This can be done in an approximate way by drawing the pulse shape at fixed intervals, assuming the shock wave velocity constant in each of them.

The above calculations for the rate of attenuation and rarefaction are based on the hypothesis of hydrodynamic response of the material; they are extremely oversimplified and not too realistic, as will be shown by some experimental results presented below. The attenuation of a pressure pulse in copper produced by normal detonation of explosive (composition B, a military explosive based on RDX and TNT) in direct contact with the metal is shown in Figure 7.5. The attenuation of the pulse starts immediately, because the pulse duration is zero. This will be shown in Chapter 10. The ordinate on the left shows the particle velocity, and that on the right shows the shock velocity. The abscissa represents the ratio between the distance into the metal plate and the explosive thickness. It can be seen that the correlation between Drummond's calculations [4] and the experimental results is quite satisfactory using either the Walsh or the Bridgman EOS. The term X/X^* is the ratio of metal thickness to explosive thickness. This ratio can be decreased or increased at will without affecting the results. The method of characteristics was used in solving the problem.

The first report on nonhydrodynamic attenuation was made by Al'tshuler et al. [2].

The attenuation of pressure pulses with a finite pulse duration produced by

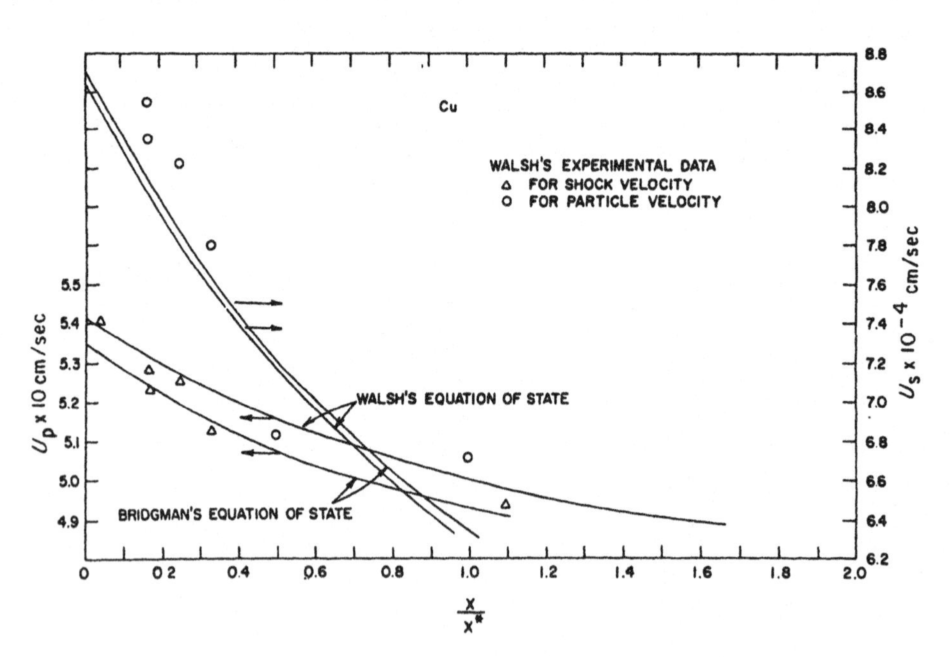

FIGURE 7.5 Attenuation of shock wave in copper produced by direct contact, normal detonation of explosives; X/X^* is the ratio between metal thickness to explosive thickness. (From Drummond [1], Fig. 4, p. 1440. Reprinted with permission of the publisher.)

plate impact has also received some attention, and it has been found that the predictions of the hydrodynamic theory are not realistic. Indeed, the actual attenuation rate of the wave is considerably higher. This is shown in Figure 7.6 for an aluminum target impacted by an aluminum projectile. The hydrodynamic theory predicts a region (up to $R = 16$) in which the peak pressure is constant. The actual results show that the peak pressure starts to drop substantially much earlier ($R \simeq 4$). An elastoplastic model is used to describe the attenuation in a more realistic way. It uses an artificial viscosity term. Sophisticated mathematical methods have been developed and applied to the simulation of shock waves over the past years, for example, finite-difference method, method of characteristics, and artificial viscosity. The work by Curran [3] and Erkman and co-workers [4–6] in this regard is very important. A detailed treatment is provided by Chou and Hopkins [7]. A number of computer codes have evolved to simulate shock waves traveling through solid media (see Chapter 6).

Hsu et al. [8] conducted a number of experiments to determine whether the attenuation of a shock wave in nickel was dependent on the metallurgical state, such as the grain size or the deformation structure (presence or absence of dislocations). They found no such effect. They measured the peak pressure close to the impact interface and at a distance of 100 mm from the impact interface. These experimental results are reproduced in Figure 7.7. Hsu et al. used a plate impact assembly and conducted experiments at two initial peak pressures: 10 and 25 GPa. The experiments were conducted at an initial pulse

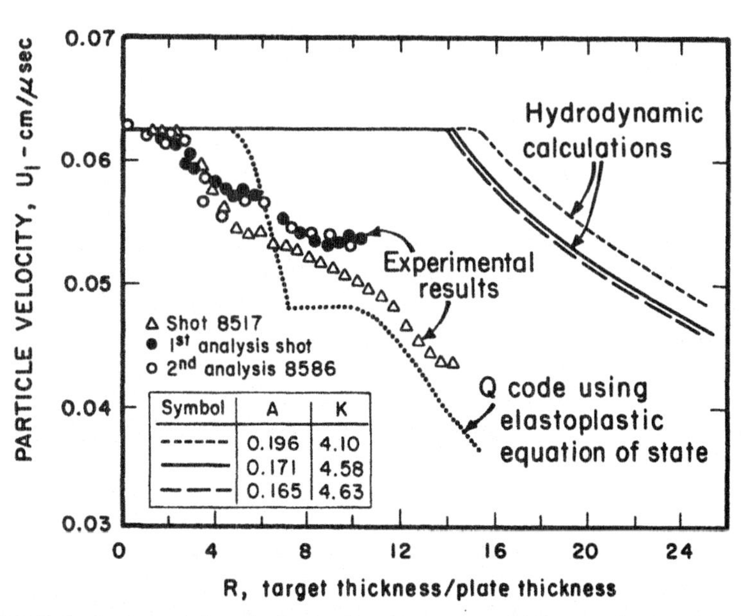

FIGURE 7.6 Peak particle velocity in aluminum target hit by an aluminum projectile at 3 mm/μs. (From Erkman et al. [4], Fig. 16, p. 43.)

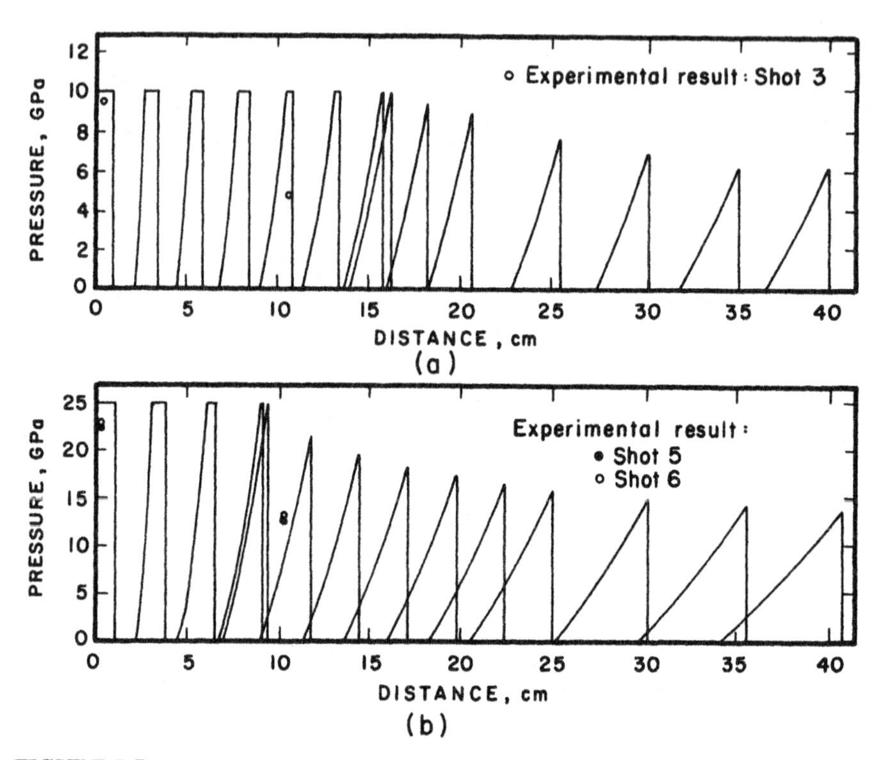

FIGURE 7.7 Attenuation of shock waves in nickel calculated by hydrodynamic theory (continuous lines) and obtained from experiments (data points). From Hsu et al. [8], Fig. 5, p. 441. Reprinted with permissions of the publisher.)

duration of 2 μs. The calculations of pulse attenuation assuming hydrodynamic behavior are shown by the sequence of pulses in Figure 7.7. The pulses are initially square and then take a triangular appearance. The peak stresses were measured by manganin piezoresistive gages. The measured stresses are shown in Figure 7.7. They were made close to the impact interface and at a distance of 10 cm from the impact interface. The agreement between calculated and measured pressures close to the impact interface is quite good.

At a depth of 10 cm, the measured pressure is considerably lower than the one predicted from hydrodynamic theory, in accordance with the earlier results of Al'tshuler et al. [2], Curran [3], and Erkman et al. [4–6]. Thus, additional *dissipative* mechanisms of importance take place. These will be discussed in Chapter 13.

7.3 SHOCK WAVE INTERACTION AND REFLECTION

When a shock wave propagating in a medium A enters a medium B, changes in pressure, wave velocity, density, and so on, occur. It is convenient to define the shock impedance as the product of the initial density ρ_0 and the shock wave velocity U_s. This can be approximated as the product of the initial density ρ_0 and the sonic wave velocity C_0 (see Section 2.8). The impedance is highest for materials with high sonic velocities and high densities. It presents a measure of the pressure that will be generated at a certain value of the particle velocity. In high-impedance materials, the best way to treat the transfer of the wave from medium A to medium B is by means of the impedance matching technique. Continuity at the boundary dictates that the particle velocity and pressure will be the same in both materials. We will treat two cases:

1. *Transmission of Shock Wave from Material A with Low Impedance to Material B with High Impedance.* Figure 7.8 shows the pressure–particle velocity curves for materials A and B. The slope of the dashed line (Fig. 7.8(a)) at pressure P_1, is the shock impedance $\rho_0 U_s$. At the interface, the pressure P_1 will change in order for equilibrium to be reached. This is done by means of the impedance matching method, shown schematically in Figure 7.8(a). One passes the reflected curve (AR) through the pressure P_1. This reflected curve intersects curve B at P_2. This is the pressure in medium B. Figure 7.8(b) shows the sequence of pressure profiles. As the shock front reaches the interface, the pressure rises to P_2. A pressure

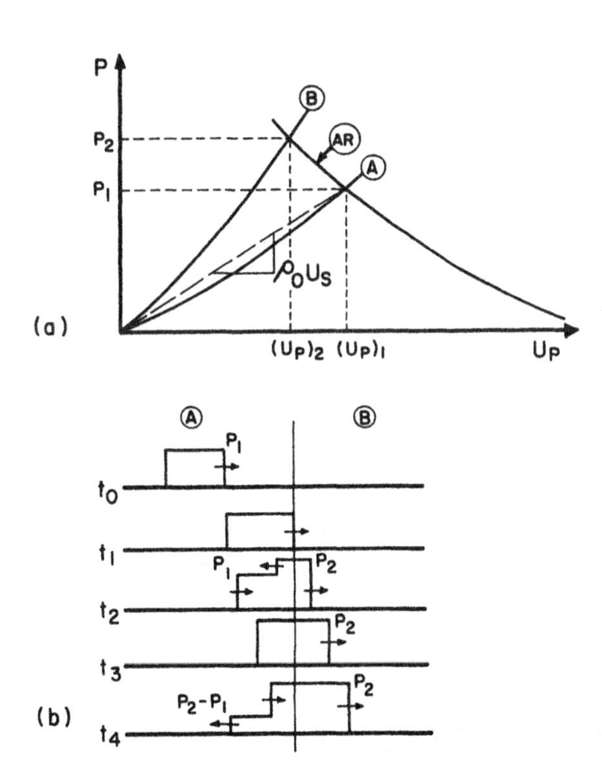

FIGURE 7.8 Transmission of shock wave from medium with low to medium with high shock impedance: (a) pressure–particle velocity plots; (b) stress profiles.

front is propagated into A and another into B. Between t_3 and t_4 this pressure front encounters the release portion of the initial shock wave and the pressure drops to $P_2 - P_1$. This pulse continues to propagate to the left. The particle velocity within the high-pressure region will be U_{p2} in both A and B. Thus, continuity of pressure and particle velocity is assured.

2. *Transmission of Shock Wave from Material A with High Impedance to Material B with Low Impedance.* We have the inverse situation. The graphic solution is shown in Figure 7.9. We invert the curve for A at the pressure P_1 (mirror reflection around P_1). This curve will intersect the curve for material B at P_2. The pressure P_2 is lower than P_1. This pressure will produce a release pulse to be sent through material A. We assume that both A and B are semi-infinite and that the release pulse can travel freely until it encounters the release portion of the primary pulse. At this time (t_4) a tensile pulse will form that will propagate in both directions. If this tensile pulse has sufficient amplitude, a spall will be formed. It should be noticed that the release portion of the wave is not a shock wave but has a slope that decreases with increasing propagation velocity.

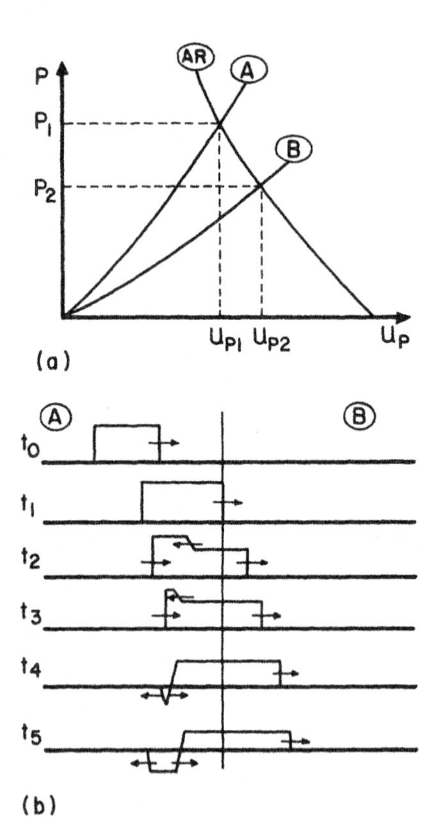

FIGURE 7.9 Transmission of shock wave from medium with high to medium with low impedance: (a) pressure–particle velocity plots; (b) stress profiles.

Example 7.1. Calculate the pressure generated when a 30-GPa pulse traveling in metal A enters metal B:

1. A → Cu
   B → Al
2. A → Al
   B → Cu

The answer is graphically presented in Figure 7.10. The plots for Figure 7.10 were made using data from Table 4.1. For case 1, we have

$$P_2' = 17.6 \text{ GPa}$$

For case 2, we have $P_2 = 43.5$ GPa.

More complex examples follow that involve the impact of a flyer plate onto a material consisting of two layers (pp. 194–200).

We now study the superposition of two shock waves. This is a situation that ideally occurs when two stress waves of opposite senses encounter each other. When the two waves have parallel wave fronts and particle velocities with the

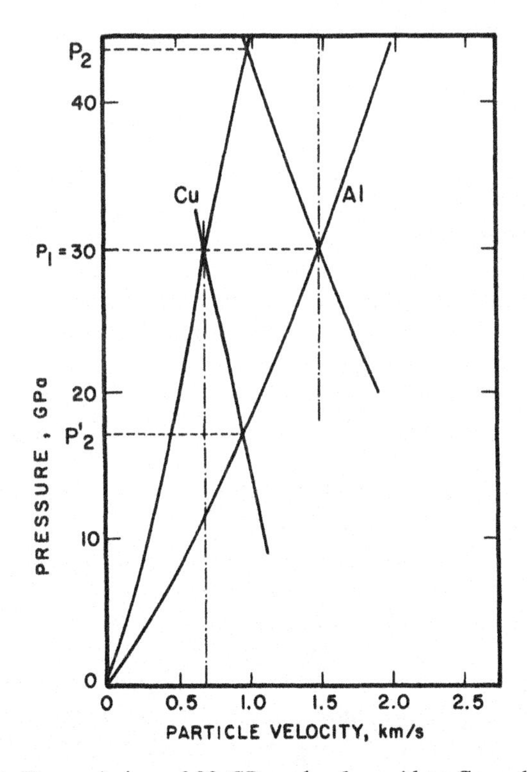

FIGURE 7.10 Transmission of 30-GPa pulse from Al to Cu and from Cu to Al.

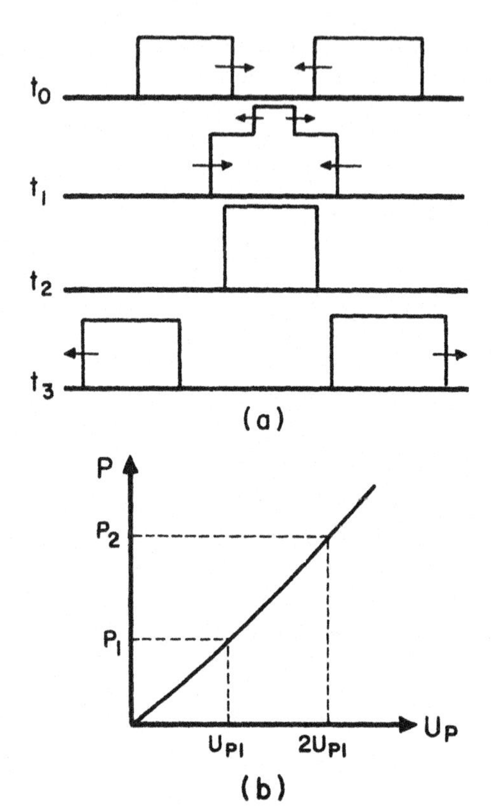

FIGURE 7.11 (a) Shock waves in collision course, time sequence showing profiles at different times; (b) calculation of pressure generated during superposition.

same direction and opposite senses, we solve the problem simply by adding the particle velocities and computing the equivalent pressure. This corresponds to a change in referential. Figure 7.11 shows the wave configuration on superposition and the schematic approach to calculate the peak pressure on a P-U_p plot. If the two waves have the same particle velocity, we just double it to obtain P_2. Thus, the pressure P_2 is not twice P_1. Rather, it is the pressure determined by the doubling of the particle velocity.

A problem of considerable practical importance is the reflection of a shock wave at a free surface. We shall consider two situations: a square pulse and a triangular pulse.

The interaction of a wave with a free surface is a particular case of the transmission of a wave from medium A to medium B when we assume that medium B has a shock impedance equal to zero. Figure 7.12 shows the propagation of a square wave that encounters a free surface. The particle velocity at the interface is obtained by taking the reflected curve (AR) and passing it through the point (P_1, U_{p1}). This curve intersects a P–U_p curve for a material with zero impedance at (0, $2U_{p1}$). This is the particle velocity at the interface. Thus, this will cause a stress pulse to propagate into material A, with pressure equal to zero (the pressures have to become equal on the two sides of the

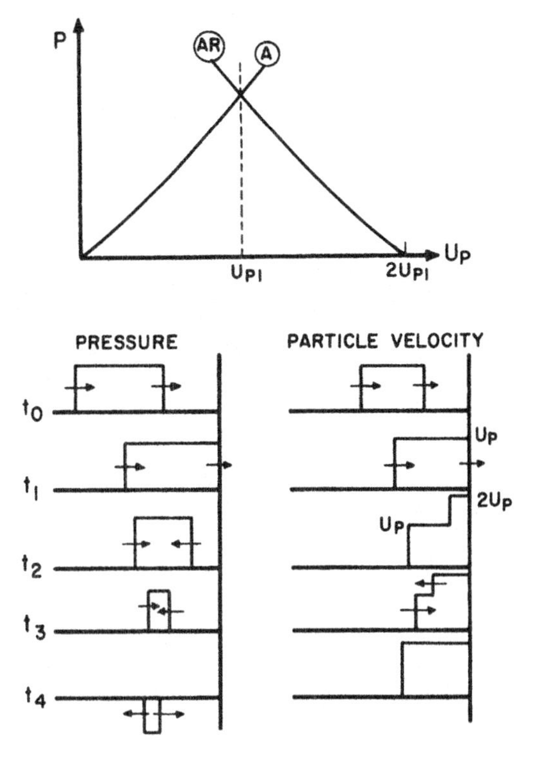

FIGURE 7.12 (a) Reflection of shock wave at free surface (medium with zero shock impedance); (b) pressure–distance profiles at t_1, t_2, t_3, and t_4; (c) partial velocity–distance profiles.

interface; this is the criterion for pressure continuity). Figures 7.12(b) and (c) show a sequence of pressure and particle velocity profiles. At t_4 the release portions of the two waves have encountered, generating tension.

A simple way of visualizing reflection at a free surface and spalling is by means of an x–t plot, shown in Figure 7.13 in a simplified manner. A flyer plate of thickness d_0 impacts a target. Neglecting all secondary effects (attenuation, elastic precursor, etc.), the initial tension will occur at the position marked in the figure, which should be at a distance from the back surface approximately equal to d_0. This can be verified by looking at the configuration of the waves. Thus, one concludes that no tension takes place for a perfectly symmetrical impact when the thicknesses of projectile and target are the same (and both are of the same material).

For triangularly shaped pressure pulses the situation is slightly more complex. The possibility exists for formation of multiple spalls. The number and spacing of the spalls will depend on the tensile strength of the material. The tensile stress builds up gradually from the back surface; as the material spalls

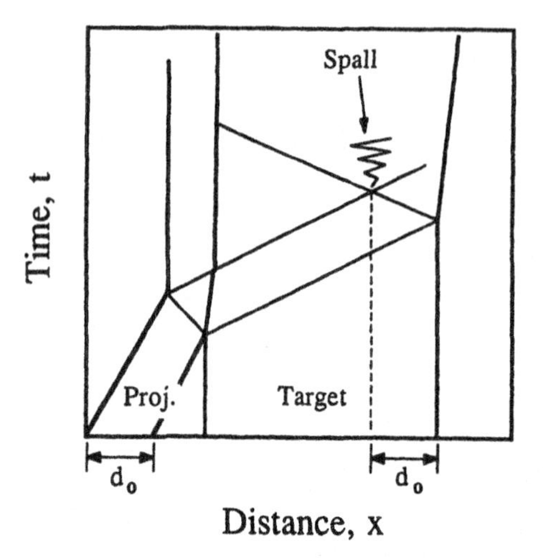

FIGURE 7.13 Position–time plot for case where reflection occurs at back of target.

and separates, new free surfaces are formed, which can generate a continuation of the process.

In the early 1950s, Rinehart [9] reported the results of experiments on steel, brass, copper, and an aluminum alloy using a modification of Hopkinson's technique. He found that there was a critical value of the normal tensile stress (σ_c) required to produce spalling, and this value was a characteristic of the material. He also observed and correctly explained the phenomenon of multiple spalling produced when a triangularly shaped pulse is reflected and has an amplitude substantially higher than σ_c. The shape of the pulse and the spall strength of the material (σ_c) determine the number and spacing of the spall regions. In Chapter 16 (Section 16.8.1) a more detailed analysis is given. Figure 16.24 shows the spall configuration.

We will discuss next the propagation of a shock wave through laminar materials consisting of layers of different materials. This is an important technological problem, because armor often consists of layers of different materials, and we can use laminar materials (composites) to "trap" shock waves. Examples 7.2, 7.3, and 7.4 consider different situations. It is important to recognize, as was demonstrated by Nesterenko et al. [10], that the attenuation of shock waves cannot be considered as a result of the interaction of the head wave alone with the interfaces. The interactions are strongly dependent on the scale of events, i.e., on the relative thickness of the layers in the laminate material and on their relationship with the initial pulse width.

Example 7.2 Determine the shock wave interactions when an iron projectile impacts a setup consisting of a thin Al layer and a thick Plexiglas backing:

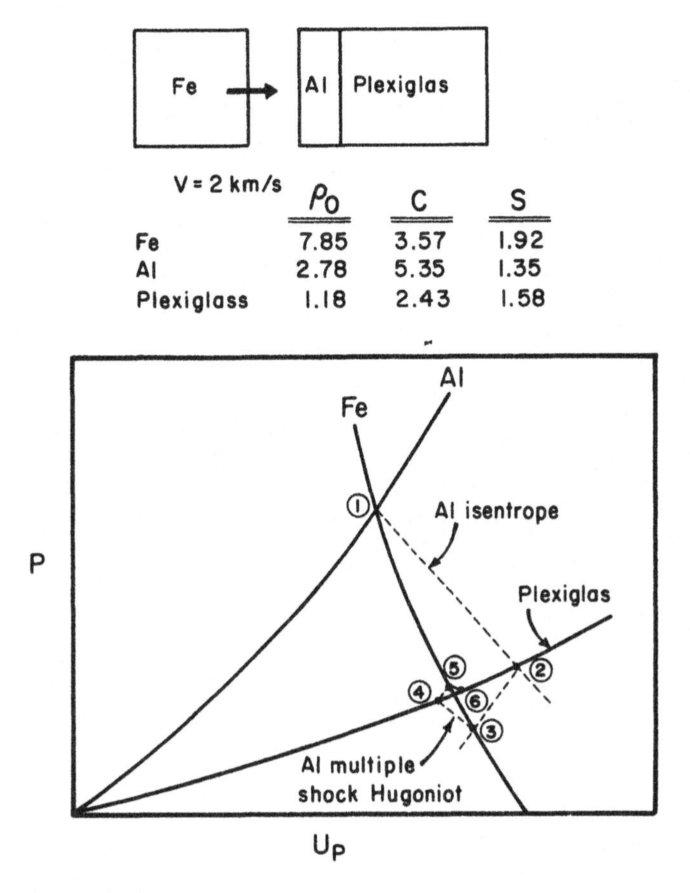

V = 2 km/s

	ρ_0	C	S
Fe	7.85	3.57	1.92
Al	2.78	5.35	1.35
Plexiglass	1.18	2.43	1.58

Assumptions:

1. The thickness of the Al plate is negligible compared to the other plates.
2. No attenuation in the plate.
3. Isentropes are parallel to Hugoniots.

At the impact interface

$$P_{Fe} = P_{Al} \qquad U_{PFe} = U_{PAl}$$

1. *First Point ~ 0.* For Al

$$P_{Al} = \rho_{0Al}(C_{Al} + S_{Al}U_p)U_p$$

For Fe

$$P_{Fe} = \rho_{0Fe}[C_{Fe} + S_{Fe}(2 - U_p)](2 - U_p)$$

From $P_1 = P_2$

$$P_{\text{Al}} = P_{\text{Fe}}$$

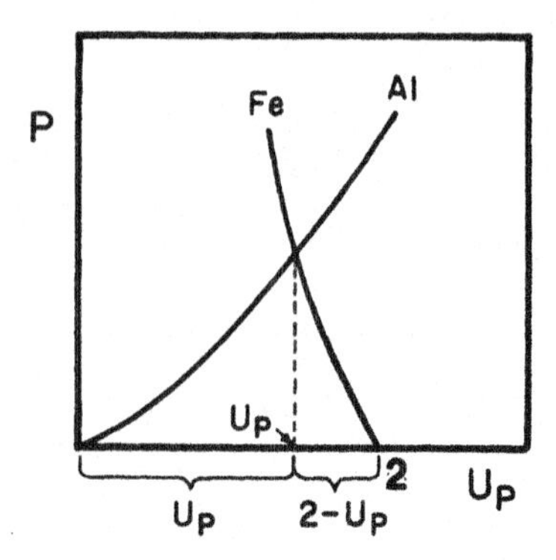

By rearranging,

$$\alpha U_p^2 + \beta U_p + \gamma = 0$$

$$\alpha = \rho_{0\text{Al}} S_{\text{Al}} - \rho_{0\text{Fe}} S_{\text{Fe}}$$

$$\beta = \rho_{0\text{Al}} C_{\text{Al}} + \rho_{0\text{Fe}} C_{\text{Fe}} + 4\rho_{0\text{Fe}} S_{\text{Fe}}$$

$$\gamma = -4\rho_{0\text{Fe}} S_{\text{Fe}} - 2\rho_{0\text{Fe}} - C_{\text{Fe}}$$

By substituting all values,

$$\alpha = -11.3123 \qquad \beta = 103.2437 \qquad \gamma = -116.3998$$

or

$$U_p^2 - 9.1267 U_p + 10.2897 = 0$$

$$U_p = \frac{9.1267 \pm \sqrt{9.1267^2 - 4(10.2897)}}{2}$$

$$U_p = \underline{1.3177} \quad \text{or } 7.8091 \text{ (km/s)}$$

$$\therefore U_p = 1.3177$$

$$P = 2.785[5.35 + (1.35)(1.3177)](1.3177)$$

$$= 26.1067 \text{ GPa}$$

Therefore the first point is

$$U_p = 1.318 \text{ km/s} \qquad P = 26.161 \text{ GPa}$$

2. *Second Point.* For Plexiglas-aluminum interface

$$P_{\text{Pl}} = \rho_{\text{0pl}}(C_{\text{pl}} + S_{\text{pl}}U_{\text{p}})U_{\text{p}}$$

$$P_{\text{Al}} = \rho_{\text{0Al}}[C_{\text{Al}} + S_{\text{Al}}(2 - U_{\text{p}})](2 - U_{\text{p}})$$

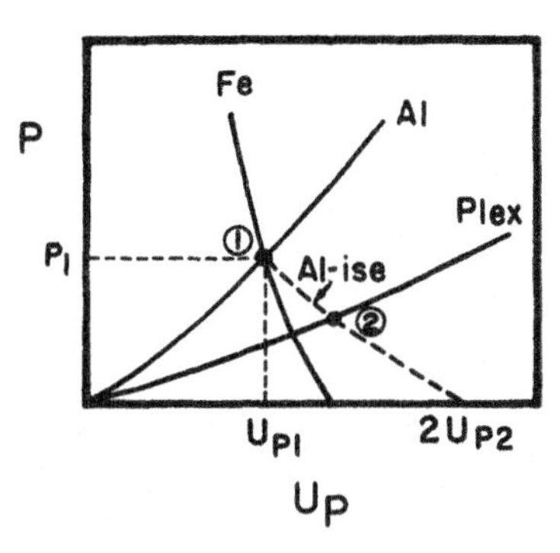

By setting $P_{\text{Pl}} = P_{\text{Al}}$

$$U_{\text{p}}^2 - 19.8182U_{\text{p}} + 34.427 = 0$$

$$U_{\text{p}} = 1.9267 \text{ km/s} \qquad P = 12.446 \text{ GPa}$$

3. *Third Point.* For aluminum-iron interface

For Fe_1

$$P_{\text{Fe}} = \rho_{\text{0Fe}}[C_{\text{Fe}} + S_{\text{Fe}}(V_0 - U_{\text{p}})](V_0 - U_{\text{p}})$$

For Al,

$$P_{\text{Al}} = \rho_{\text{0Al}}[C_{\text{Al}} + S_{\text{Al}}(2U_{\text{P2}} - 2U_{\text{P1}} - U_{\text{p}})](2U_{\text{p2}} - 2U_{\text{p1}} - U_{\text{p}})$$

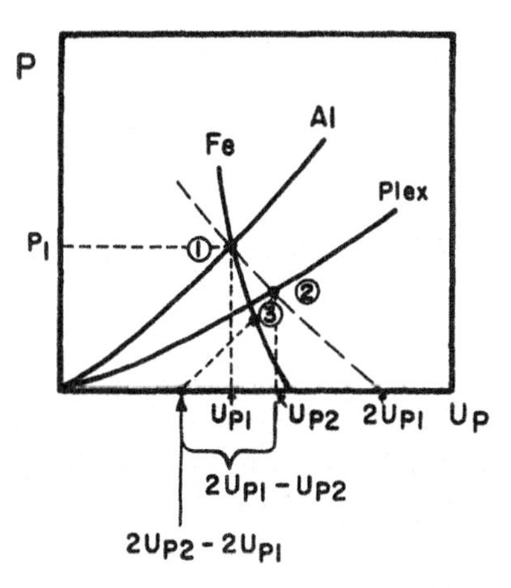

By setting $P_{Fe} = P_{Al}$,

$$U_p^2 - 8.3170U_p + 11.4010 = 0$$

$$U_p = 1.731 \text{ km/s} \qquad P = 8.632 \text{ GPa}$$

For iron/Plexiglas, by calculation,

$$U_p = 1.6894 \text{ km/s} \qquad P = 10.1664 \text{ GPa}$$

From a hydrocode calculation,

$$U_p = 1.689 \text{ km/s} \qquad P = 10.18 \text{ GPa}$$

Thus, if Al is fairly thin, the state in the Al plate is the same as that at the interface iron/Plexiglas. After the "ringing" of the shock wave (1-2-3-4-5-6 · · ·) it will equilibrate at the Fe-Plexiglas intersection (P ~ 10 GPa).

The hydrocode predictions are given below:

1. IRON-ALUMINUM-PLEXIGLAS

	U_p	P
1	1.317	26.23
2	1.946	12.06
3	1.749	7.90
4	1.666	9.71
5	1.690	10.24
6	1.700	10.03
7	1.695	9.94
8	1.692	10.00
9	1.694	10.04
⋮		
	1.689	10.18

Example 7.3 Calculate the shock wave interactions when an Al flyer plate impacts a setup consisting of a thin Fe plate backed by Plexiglas.

We repeat the procedure delineated in Example 7.2.

Using a calculator leads to

$$U_p = 1.496 \text{ km/s} \qquad P = 8.464 \text{ GPa}$$

From the hydrocode

	U_p	P
1	0.683	26.06
2	1.176	5.77
3	1.342	11.49
4	1.440	8.13
5	1.469	9.06
6	1.482	8.56
7	1.487	8.72
⋮		
	1.496	8.525

$$U_p = 1.496 \text{ km/s} \quad P = 8.525 \text{ GPa}$$

The pressure will, after "ringing" (1-2-3-4-5-6 · · ·) reach the Al-Plexiglas intersection (P ~ 8.5 GPa).

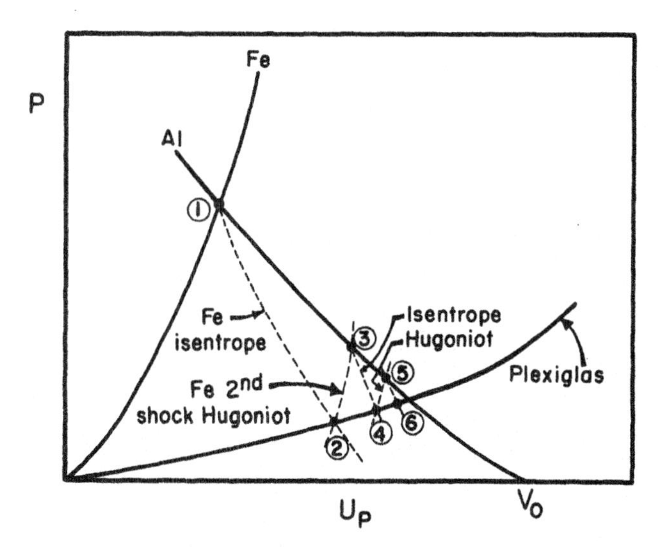

Example 7.4 Determine the shock wave interactions when an Al projectile impacts a setup consisting of a thin Plexiglas layer backed by iron.

We repeat the procedure delineated in Example 7.2. For Al/Fe

$$U_p = 0.682 \text{ km/s} \qquad P = 26.16 \text{ GPa}$$

This is the final pressure.

From the hydrocode we obtain the successive "ringup" values:

ALUMINUM–PLEXIGLAS–IRON
(aluminum moving at 2 km/s)

	U_p	P
1	1.496053	8.463522
2	0.5167098	18.52082
3	0.8323479	22.52384
4	0.6565066	24.91493
5	0.7051487	25.59671
6	0.6784914	25.97409
7	0.6857176	26.07684
8	0.6817663	26.13311
9	0.6828337	26.14832
10	0.6822503	26.15664
11	0.6824083	26.15889
12	0.682322	26.16013
13	0.6823454	26.16045

$$U_p = 0.683 \text{ km/s} \qquad P = 26.18 \text{ GPa}$$

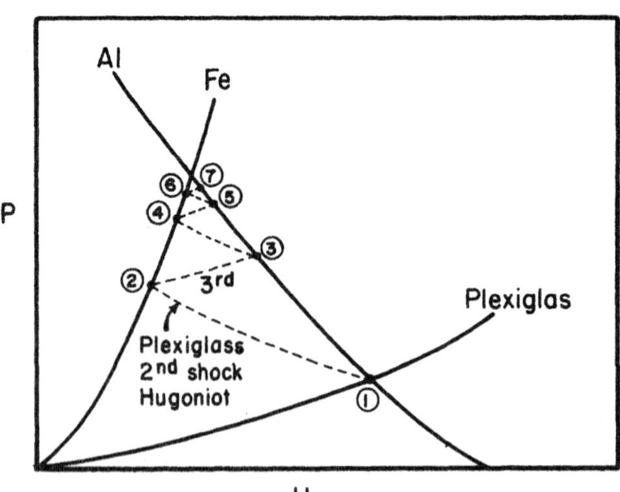

REFERENCES

1. W. E. Drummond, *J.A.P.*, **29** (1957), 1437.
2. L. V. Al'tschuler, S. B. Kormer, M. L. Brazhnik, L. A. Vladimov, et al., *Soviet Phys., JETP,* **11** (1960), 766.
3. D. R. Curran, *J.A.P.* (1963), 2677.
4. J. O. Erkman, A. B. Christensen, and G. R. Fowles, "Attenuation of Shock Waves in Solids," Technical Report No. AFWL-TR-66-72, Stanford Research Institute, Air Force Weapons Laboratory, May 1966.
5. J. R. Rempel, D. N. Schmidt, J. O. Erkman, and W. M. Isbell, "Shock Attenuation in Solid and Distended Materials," Technical Report No. WL-TR-65-119, Stanford Research Institute, Air Force Weapons Laboratory, February 1966.
6. J. O. Erkman and A. B. Christensen, *J.A.P.*, **38** (1967), 5395.
7. P. C. Chou and A. K. Hopkins, eds., *Dynamic Response of Materials to Intense Impulsive Loading*, Air Force Materials Laboratory, Wright-Patterson Air Force Base, OH, 1972.
8. C. Y. Hsu, K. C. Hsu, M. A. Meyers, and L. E. Murr, in *Shock Waves and High-Strain-Rate Phenomena in Metals*, eds. M. A. Meyers and L. E. Murr, Plenum, New York, 1981, p. 433.
9. J. S. Rinehart, *J.A.P.*, **23** (1952), 1229.
10. V. F. Nesterenko, V. M. Fomin, P. A. Cheskidov, "Attenuation of strong shock waves in laminate materials," in *Nonlinear Deformation Waves*, Springer-Verlag, Berlin; N.Y., 1983, p. 191–197.

CHAPTER 8

Shock Wave–Induced Phase Transformations and Chemical Changes

8.1 INTRODUCTION

The landmark paper by Dremin and Breusov [1], published in 1968, summarizes the early work on the process occurring in materials under the action of shock waves. The energy of the shock wave is dissipated within the material as it travels through it. This intense deposition of energy can lead to a number of physical and chemical changes that Dremin and Breusov [1] classified into

1. polymorphic phase transformations,
2. chemical decomposition processes,
3. chemical synthesis processes,
4. polymerization of monomers, and
5. defect formation (point defects, line defects, twins, etc.).

We have seen in Chapters 4–6 that both pressure and temperature are very elevated. Additionally, high shear stresses and strains are generated, either locally or in the entire body. When the material is porous or consists of a mixture of powders of different elements or compounds, these effects are even more pronounced. Local gradients of temperature form as the voids collapse by high-velocity flow of material. These events can produce melting of interfaces. Relative motion of the powders produces intense frictional effects. Some of the effects are summarized by Graham and others [2–4], who coined the CONMAH acronym:

CON	→	*Configuration change:* morphology and porosity are changed during shock wave passage
M	→	*Mass mixing:* material from neighboring particles is forced together and forced to undergo relative motion by high pressures and shear stresses
A	→	*Activation:* a high density of defects (point, line,

interfacial defects) may lead to an enhanced reactivity of the powders

H → *Heating:* intense temperature fluctuations are created in the shocked medium

It is important to realize that during the application of a shock wave—typical duration of 1–10 μs—there is virtually no time for significant diffusion and different mechanisms are involved in the formation or dissociation of phases or compounds. Indeed, in general, the pressure decreases the diffusion coefficients, since it "pushes" atoms together and it becomes more difficult for a foreign atom to move within a "crowded" lattice.

Thus, the significant physicochemical changes observed in materials after the passage of shock waves have been the object of great scientific curiosity. The first report of a shock-induced phase transformation is due to Minshall [5], in 1954. He reported a polymorphic transformation during the shock loading of iron, when the 13 GPa pressure was reached. Initially, the transformation was identified from α(BCC) (body-centered-cubic) to γ(FCC) (face-centered-cubic), but this was later corrected as α(BCC) to ε(HCP) (hexagonal-close-packed). In 1961, De Carli and Jamieson [6] reported the transformation of carbon to diamond at high shock pressures. This latter discovery was indeed very significant, since the scientific community was convinced, at the time, that there was not sufficient time for the graphite-to-diamond transformation for such small time pulses. This is yet another example of how science (or nature) outsmarts scientists! We will discuss this transformation further in Section 8.7. Very fine diamonds (0.1–60 μm) can be produced by this process, and this is indeed an industrial production method used by Dupont [7]. In Russia, ultrafine diamonds (nanocrystalline) have been synthesized directly from detonation of explosives and from explosive–graphite mixtures [9, 10].

In this chapter, we will review the basic thermodynamic laws governing phase transformations (Section 8.2) and the effects of shock pressure (Section 8.3) and describe the principal polymorphic shock-induced transformations (Section 8.4). Melting and solidification are presented in Section 8.5. In Section 8.6 we introduce phase transformations that are produced by tensile pulses generated by the reflection of shock waves at free surfaces; thus, these are transformations induced by negative stress pulses. Section 8.7 reviews, succinctly, the broad field of shock synthesis and shock-induced reactions.

8.2 THERMODYNAMICS OF PHASE TRANSFORMATIONS

This and the following sections rely heavily on the excellent monograph written by Duvall and Graham, published in 1977 and entitled "Phase Transitions under Shock Waves Loading" [11]. This article is strongly recommended. Although more recent work is available in the literature, it contains the foundation for the field.

Solids exist in many structures, and the change from one structure to the other is governed by thermodynamics and kinetics. Phase stability is governed by external factors such as pressure and temperature and internal factors such as composition and internal stresses due to defects (dislocations, point defects, interfaces). Shock waves (and reflected tensile waves) produce sudden changes of pressure and temperature that may result in the production of new phases.

Phase transformations can be classified into diffusional (or reconstructive) and diffusionless (or displacive). This classification encompasses the following:

Diffusional Transformations

1. Nucleation and growth (precipitates)
2. Spinodal decomposition
3. Cellular transformations

Diffusionless Phase Transformations

1. Displacive transformations
2. Massive transformations
3. Melting/solidification
4. Order–disorder transitions
5. Vaporization/condensation
6. Sublimation

Of the above transformations, the diffusional transformations are of no great importance in shock wave propagation, because of the limited time for diffusion (on the order of microseconds). We will restrict ourselves to transformations occurring in the shock pulse. Residual heating after the passage of the shock wave has been known to induce precipitation in some alloys, but these are postshock effects that are not of concern here.

Transformations have further been classified into first order and second order; this classification is reviewed in Figure 8.1. Phase stability is given, under conditions of constant pressure and temperature, by the Gibbs free energy. Of two possible phases, the one with the lowest Gibbs free energy is the stable form. Figure 8.1(a) shows the change in G for two phases α and β as P is changed. At P_T one has equilibrium. Below P_T, β is the stable phase; above P_T, α is the stable phase. In Figure 8.1(b) one has two curves that have the same slope at the transformation pressure P_T. From the definition of Gibbs free energy, one has

$$G = H - TS \qquad G = E + PV - TS$$

$$dG = dE + P\,dV + V\,dP - T\,dS - S\,dT \tag{8.1}$$

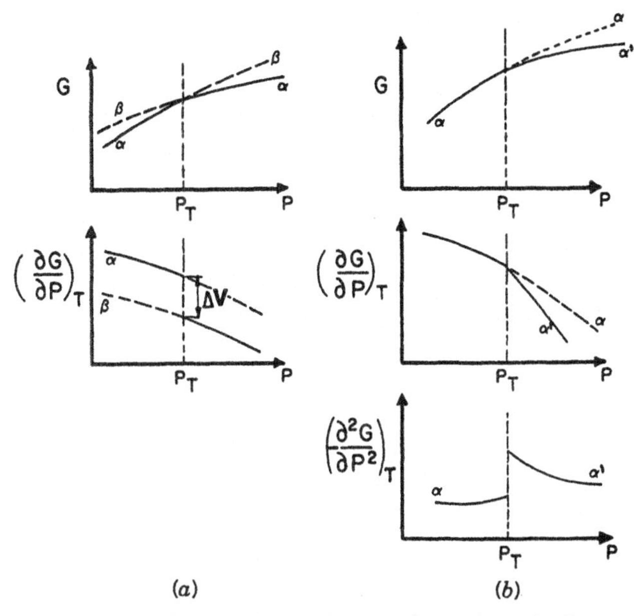

FIGURE 8.1 (a) First- and (b) second-order transformations. In first-order transformation, $\partial G/\partial P$ is discontinuous at transformation pressure (or temperature). In second-order transformation, $\partial^2 G/\partial P^2$ is discontinuous at transformation pressure (or temperature).

But

$$dE = \delta Q - P\,dV \quad \text{and} \quad \frac{\delta Q}{T} = dS$$

$$dG = V\,dP - S\,dT$$

One arrives at

$$\left(\frac{\partial G}{\partial P}\right)_T = V \qquad \left(\frac{\partial G}{\partial T}\right)_P = -S \tag{8.2}$$

Thus, the first derivative with respect to pressure, shown in Figure 8.1, gives the change in volume accompanying the transformation. In first-order transformations, there is a discontinuity in $\partial G/\partial P$ and a volume change ΔV involved in the transformation. In second-order transformations, the first derivative $(\partial G/\partial P)_T$ is continuous (no volume change) but there is a discontinuity in the second derivative of the free energy with respect to pressure $(\partial^2 G/\partial P^2)_T$. Examples of first-order transformations are the martensitic transformation (usually, a 4% volume increase), melting (usually, a volume increase), and solidification (usually, volume decrease). Second-order phase transformations are not as obvious structurally. Magnetic transitions and order–disorder transformations fall in this domain.

It should be noticed that the plots given in Figure 8.1 are sections at a constant temperature. In reality, the free energies can be represented as surfaces when both P and T are included. One can also make isobaric sections along these surfaces and see that first-order transformations result in a net change in entropy $[S = -(\partial G/\partial T)_P]$. The curve of intersection of two surfaces $G_1(P, T)$ and $G_2(P, T)$ provides the Clausius–Clapeyron equation, which gives the change in transformation temperature as a function of pressure. This equation can be obtained from the equations for the surfaces:

$$G = f(P, T)$$

$$dG = \left(\frac{\partial G}{\partial P}\right)_T dP + \left(\frac{\partial G}{\partial T}\right)_P dT \tag{8.3}$$

For two surfaces 1 and 2 using Eqn. (8.3),

$$dG_1 - dG_2 = \left[\left(\frac{\partial G}{\partial P}\right)_T^1 - \left(\frac{\partial G}{\partial P}\right)_T^2\right] dP + \left[\left(\frac{\partial G}{\partial T}\right)_P^1 - \left(\frac{\partial G}{\partial T}\right)_P^2\right] dT$$

At equilibrium, $dG_1 = dG_2$, and from Eqn. (8.2),

$$0 = (V_1 - V_2)\, dP + (S_2 - S_1)\, dT$$

Thus

$$\frac{dP}{dT} = \frac{\Delta S}{\Delta V} \quad \text{or} \quad \frac{dP}{dT} = \frac{\Delta H}{T\,\Delta V} \tag{8.4}$$

Inspection of Eqn. (8.4) provides some insight into the effects of pressure on the transformation temperature.

8.3 PHASE TRANSFORMATIONS AND THE RANKINE–HUGONIOT CURVES

In shock loading we often do not have thermodynamic equilibrium because of the short times involved. Therefore non-equilibrium processes are very important. Nevertheless, we will first treat all processes as equilibrium processes.

Figure 8.2 shows the transformation in an isothermal plot for a change in volume $\Delta V < 0$. In the interval V_1–V_2 the two phases coexist. In shock loading we do not have an isotherm, and we have to develop the Rankine–Hugoniot curve for a family of isotherms. However, we will first inspect Eqn. (8.4). If $\Delta V < 0$ and $\Delta S < 0$, $dP/dT > 0$. Thus, the transformation pressure increases with temperature. Figure 8.3 shows three isotherms at temperatures $T_1 < T_2$

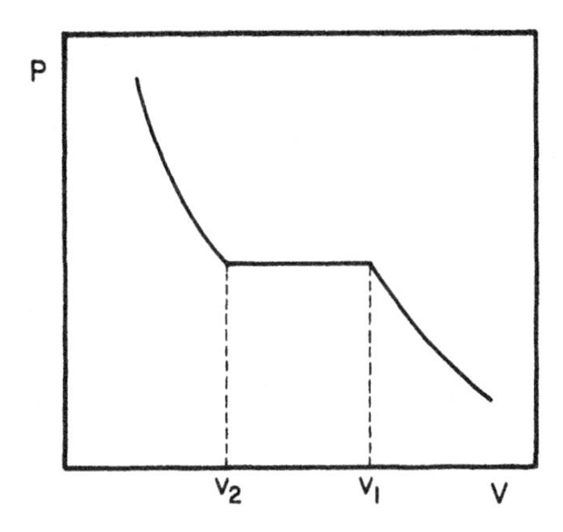

FIGURE 8.2 Pressure–volume isotherm for a first-order phase transformation.

$< T_3$. It can be seen that the transformation pressure increases with increasing temperature on the isotherms. The Rankine–Hugoniot curve, on the other hand, is usually approximated as an adiabat and can be determined from the isothermal compression curves by means of the Grüneisen EOS (Chapter 5). It is schematically shown in Figure 8.3. Its origin ($P = 0$) is set at 0 on the T_1 isotherm. It traverses the two-phase region at $Q'H$, then rises again to J. Figure 8.4 shows the Rankine–Hugoniot curve expressed in a simple P–V plane, as presented in Chapter 5. The principal difference between this curve and the curves seen in Chapter 4 [Fig. 4.4(a)] is that there is a discontinuity in the slope of the curve. This discontinuity marks the onset of transformation.

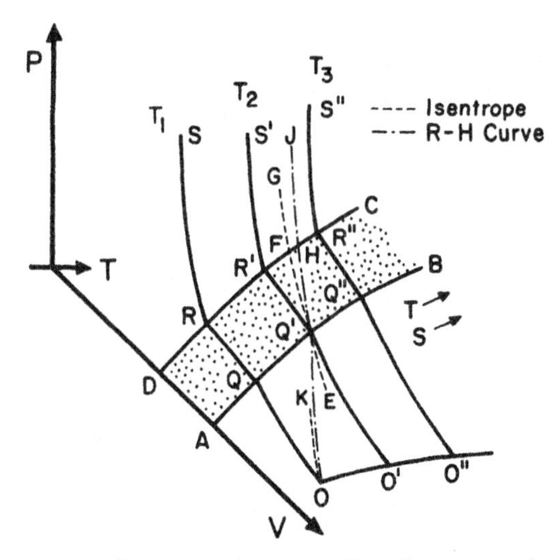

FIGURE 8.3 Pressure–volume–temperature surface for a normal polymorphic transition; $\Delta S < 0$, $\Delta V < 0$, $dP/dT > 0$; OQRS, O′Q′R′S′, O″Q″R″S″ are isotherms at temperatures $T_1 < T_2 < T_3$, respectively; OQ′HJ is a Rankine–Hugoniot curve. (From Duvall and Graham [11], Fig. 9, p. 529. Reprinted with permission of the publisher.)

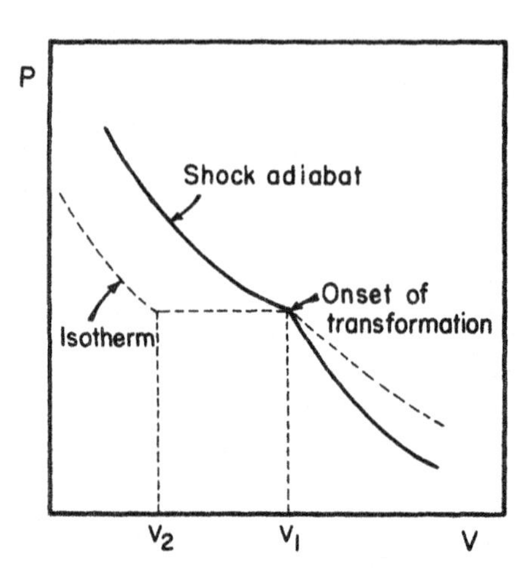

FIGURE 8.4 Rankine–Hugoniot (shock adiabat) of material undergoing phase transformation of the $\Delta V < 0$, $\Delta S < 0$ type.

A second type of situation occurs when $\Delta V < 0$ and $\Delta S > 0$. In this case, $dP/dT < 0$ and the Clausius–Clapeyron relationship predicts a behavior schematized by Figure 8.5. The dashed lines show the isentropic ($S = S_0$) and adiabatic compressions ($S_1 > S_0$). The transformation pressure decreases with temperature and the resulting shock adiabat has a different shape. There is a decrease in temperature as the transformation takes place. The complete transformation will occur at a temperature $T_0 < T_1$. Thus, the slope of the Rankine–Hugoniot curve after transformation shows a second discontinuity. Figure 8.5(b) shows this configuration. This situation is less common than the one in which $\Delta S < 0$.

We will now see how the shape of the P–V curve in Figure 8.4 affects the shape of the shock wave. For this we will refer ourselves to Figure 8.6, which shows a two-shock-wave structure. Shock wave S_1 is traveling at a velocity U_{s1} and with a particle velocity U_{p1}. With respect to the material behind it (between S_1 and S_2) the shock velocity is $U_{s1} - U_{p1}$. Shock wave S_2 "rides on" the material moving at U_{p1} and has a velocity U_{s2} in a Eulerian referential. With respect to the material between S_1 and S_2, its velocity is $U_{s2} - U_{p1}$. If $(U_{s2} - U_{p1}) < (U_{s1} - U_{p1})$ (or, simply speaking, $U_{s2} < U_{s1}$), the second wave falls continuously behind the first wave and the two-wave structure is stable. If, however, $U_{s2} > U_{s1}$, the second wave overtakes the first one, and a single-wave structure ensues. By applying the Rankine–Hugoniot equations, one obtains the Rayleigh slope [Eqn. (4.11)]:

$$(U_{s1} - U_{p0})^2 = V_0^2 \left(\frac{P_1 - P_0}{V_0 - V_1} \right) \tag{8.5}$$

We also have, from Eqn. (4.4) (conservation of mass),

$$U_{s1} - U_{p1} = \frac{V_1}{V_0} (U_{s1} - U_{p0}) \tag{8.6}$$

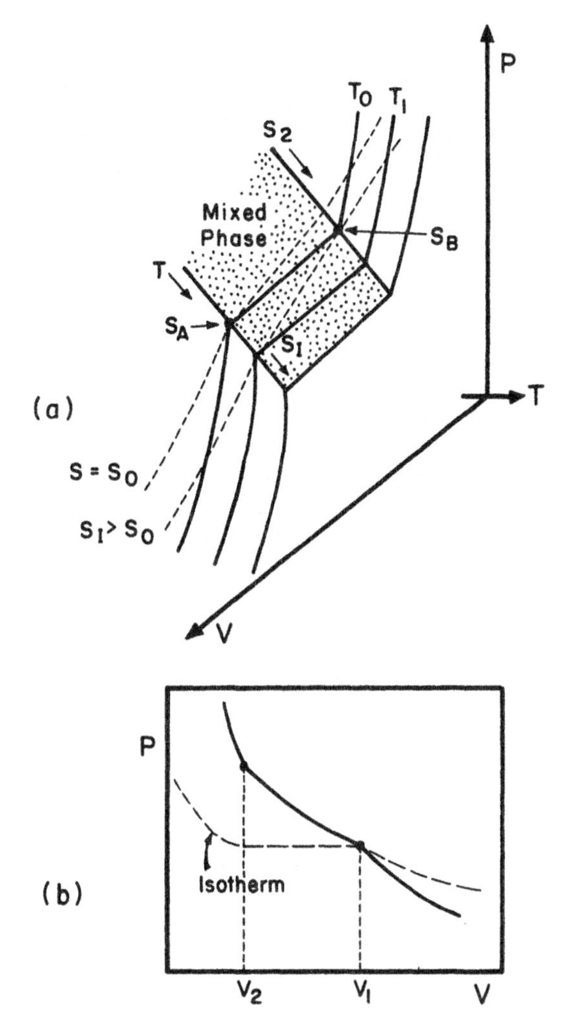

FIGURE 8.5 (a) Pressure–volume–temperature surface for phase transformation with $\Delta V < 0$, $\Delta S > 0$, and $dP/dT < 0$. (b) Rankine–Hugoniot of materials shown in (a). (From Duvall and Graham [11], Fig. 11, p. 529. Reprinted with permission of the publisher.)

For the pressure P_2, we reapply these expressions, taking P_1 as a reference. Notice, in Figure 8.6(a), that the second shock is traveling into material with initial particle velocity U_{p1}, initial specific volume V_1, and initial pressure P_1. Thus

$$(U_{s2} - U_{p1})^2 = V_1^2 \frac{(P_2 - P_1)}{V_1 - V_2} \tag{8.7}$$

$$U_{s2} - U_{p2} = \frac{V_2}{V_1}(U_{s2} - U_{p1}) \tag{8.8}$$

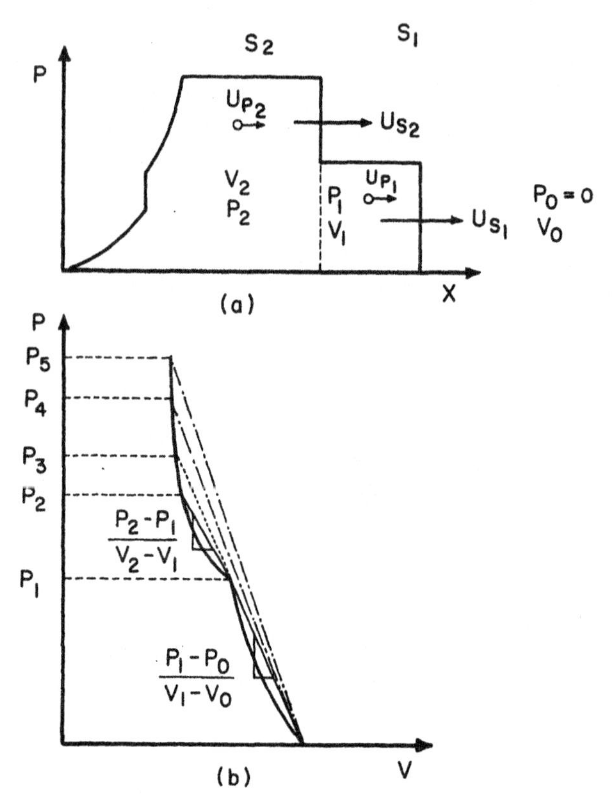

FIGURE 8.6 (a) Two-wave structure; (b) Rankine–Hugoniot curve with discontinuity in slope.

From (8.5) and (8.7),

$$\frac{P_1 - P_0}{V_1 - V_0} = -\left(\frac{U_{s1} - U_{p0}}{V_0}\right)^2 \tag{8.9}$$

$$\frac{P_2 - P_1}{V_2 - V_1} = -\left(\frac{U_{s2} - U_{p1}}{V_1}\right)^2 \tag{8.10}$$

Substituting (8.6) into (8.9) yields

$$\frac{P_1 - P_0}{V_1 - V_0} = -\left(\frac{U_{s1} - U_{p1}}{V_1}\right)^2 \tag{8.11}$$

From the condition previously established for the stability of the two-wave shock front,

$$U_{s2} - U_{p1} < U_{s1} - U_{p1}$$

We have, from Eqns. (8.10) and (8.11),

$$\frac{P_2 - P_1}{V_2 - V_1} > \frac{P_1 - P_0}{V_1 - V_0} \tag{8.12}$$

In Figure 8.6(b) these two values are the slopes of the two Rayleigh lines. These slopes provide the velocities for the two waves S_2 and S_1. When these values become equal (at a peak pressure P_3), the two-wave structure becomes unstable and a single-wave structure is recreated. At P_4, in Figure 8.6(b), a single wave with Rayleigh slope $(P_4 - P_0)/(V_4 - V_0)$ exists. The two-wave structure is stable in the pressure range $P_1 \rightarrow P_3$.

Example 8.1

1. Determine the shock wave velocities for iron shock loaded at a pressure of 20 GPa ($\rho_0 = 7.9$ g/cm^3).

2. Draw the shock wave profile for iron after the wave has traveled a distance of 10 mm, assuming an initial pulse duration of microseconds.

From Figure 8.7, we obtain the two slopes:

$$-\frac{P_1 - P_0}{V_1 - V_0} = \frac{13 \times 10^9}{0.935 - 1} \times \frac{1}{V_0} = 200 \times 10^9 \rho_0 = 1580 \times 10^{12}$$

From Figure 8.7 we see that $V_1/V_0 = 0.935$ and that $P_1 = 8.45$:

$$-\frac{P_2 - P_1}{V_2 - V_1} = \frac{7 \times 10^9}{0.87 - 0.935} \times \frac{1}{V_0} = 107 \times 10^9 \rho_0 = 850 \times 10^{12}$$

These values are equal to the Rayleigh slopes:

$$(\rho_0 U_{s1})^2 = 1580 \times 10^{12} \qquad \therefore\ \rho_0 U_{s1} = 39.7 \times 10^6$$

$$U_{s1} = 5.03 \times 10^3 \text{ m/s}$$

$$(\rho_1 U_{s2})^2 = 850 \times 10^{12} \qquad \therefore\ \rho_1 U_{s2} = 2.92 \times 10^6$$

$$U_{s2} = 3.45 \times 10^3 \text{ m/s}$$

The wave profile is drawn on the next page. The back (release) portion of the wave was drawn assuming that the reverse transformation occurs at 9.8 GPa. A rarefaction shock was arbitrarily drawn. For the release portion, the velocities $C = 5.04$ mm/μs and $C_0 = 3.57$ mm/μs were used. The particle velocity at 20 GPa is 0.550 mm/μs (from Appendix F of Meyers and Murr, *Shock Waves and High-Strain-Rate Phenomena in Metals*, Dekker, New York, 1981, p. 1065).

IRON, $\gamma_0 = 1.69$

P (GPa)	ρ (g/cm^3)	V/V_0	U_s (km/s)	U_p (km/s)	C (km/s)
0.00	7.850	1.000	3.574	0.000	3.574
2.50	8.033	0.977	3.737	0.085	3.813
5.00	8.195	0.958	3.887	0.164	4.029
7.50	8.341	0.941	4.025	0.237	4.227
10.00	8.474	0.926	4.155	0.306	4.411
12.50	8.596	0.913	4.278	0.371	4.584
15.00	8.710	0.901	4.394	0.434	4.748
17.50	8.815	0.891	4.504	0.493	4.905
20.00	8.914	0.881	4.610	0.550	5.054
22.50	9.007	0.872	4.711	0.605	5.198
25.00	9.095	0.863	4.808	0.658	5.337
27.50	9.179	0.855	4.902	0.710	5.471
30.00	9.258	0.848	4.993	0.759	5.602
32.50	9.334	0.841	5.081	0.808	5.729
35.00	9.407	0.835	5.166	0.855	5.853
37.50	9.476	0.828	5.248	0.901	5.974
40.00	9.543	0.823	5.329	0.945	6.092
42.50	9.608	0.817	5.407	0.989	6.208
45.00	9.670	0.812	5.483	1.032	6.322
47.50	9.730	0.807	5.557	1.074	6.434
50.00	9.788	0.802	5.630	1.115	6.544
52.50	9.844	0.797	5.701	1.155	6.652
55.00	9.899	0.793	5.770	1.194	6.759
57.50	9.952	0.789	5.838	1.233	6.865
60.00	10.003	0.785	5.904	1.271	6.969

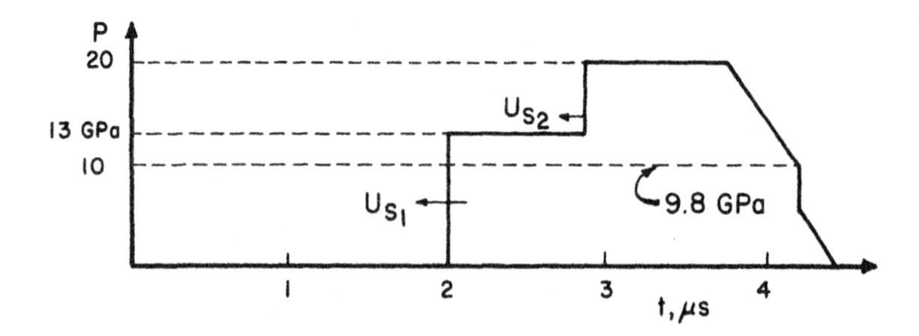

8.4 MATERIALS UNDERGOING SHOCK-INDUCED PHASE TRANSFORMATIONS

Of all shock-induced phase transformations, the $\alpha \rightarrow \varepsilon$ transformation undergone by iron at 12.8 GPa is the best known. Figure 8.7(a) shows the regions of stability of the three phases, α, γ, and ϵ for iron in a P–V diagram. These regimes of stability obey Gibbs' rule ($P + F = C + 2$), where the number of phases is P, the number of components is C, and the number of degrees of freedom is F. For a one-component system (pure iron), one has $P + F = 3$. For three phases ($P = 3$) there is no degree of freedom. This is the triple point, where the phases coexist. For two phases, one has one degree of freedom; by fixing, for instance, pressure, the volume is also fixed. And there are two degrees of freedom (e.g., pressure and temperature) in the one-phase regions. The α(BCC) $\rightarrow$ ε(HCP) transformation was first detected by Bancroft et al. [12] in shock-loading experiments. It was later confirmed by static high-pressure experiments. This phase was identified as having an HCP structure by X-ray diffraction in static high-pressure experiments by Jamieson and Lawson [13]. There is a 6.5% volume reduction associated with the transformation ($\Delta V/V_0 = 0.065$). The plot in Figure 8.7(a) shows that the pressure for the $\alpha \rightarrow \varepsilon$ transition decreases with increasing temperature, leading us to believe that the entropy increases with the transformation ($\Delta V < 0$, $\Delta S > 0$, $dP/dT < 0$). Figure 8.7(b) shows the shock Hugoniot for iron, with the expected discontinuity in slope observed at 12.8 GPa. From this plot, one can determine the wave structure. This is done as an exercise in Example 8.1. Barker and Hollenbach [14], using VISAR interferometry, obtained the unloading curve and found that the reverse transformation is complete at 9.86 GPa. Thus, there is a hysteresis in the process. The ε phase is totally retransformed to α after the passage of the shock pulse. However, a considerable amount of transformation "debris" is left behind, altering the etching characteristics and residual hardness of iron. In low-carbon steels, this is very clearly seen, and Figure 8.8 shows an optical micrograph of the interface between untransformed and transformed regions and the microhardness profile. This profile of hardness was obtained by detonating a plastic explosive (at a grazing incidence) at the surface of a steel block by Meyers et al. [15]. At the surface, the pressure exceeded 13 GPa, and therefore the hardness is close to $R_A = 60$. The thickness of the transformed region is $\sim$4 mm and the drastic drop in hardness is clearly evident. In the optical micrograph, the transformed region is dark because of the profuse transformation "debris," whereas the untransformed region is light, with occasional deformation twins. The microhardness in regions that underwent the phase transformation is considerably higher.

Other important shock-induced phase transformations are graphite to diamond and the transformation from the graphitic to the wurtzite and cubic structures of BN (boron nitride) under shock pressure. These transformations can be retained after the passage of the shock pulse and have important technological applications. Figure 8.9 shows the pressure–volume phase diagram for carbon. The region in which diamond is stable is seen in Figure 8.9. DeCarli and Jamieson [6] were the first to synthesize diamond from graphite. This phase seems to form at pressures starting at 2.3 GPa. The diamond crystals formed are very small because of the limited time available for their growth. Diamond particles with sizes ranging from 500 Å to 100 μm diameter are commercially produced by DuPont. This process is described by Bergmann [7] and Bergman and Bailey [8]. One of the requirements is rapid cooling of the diamond crystals

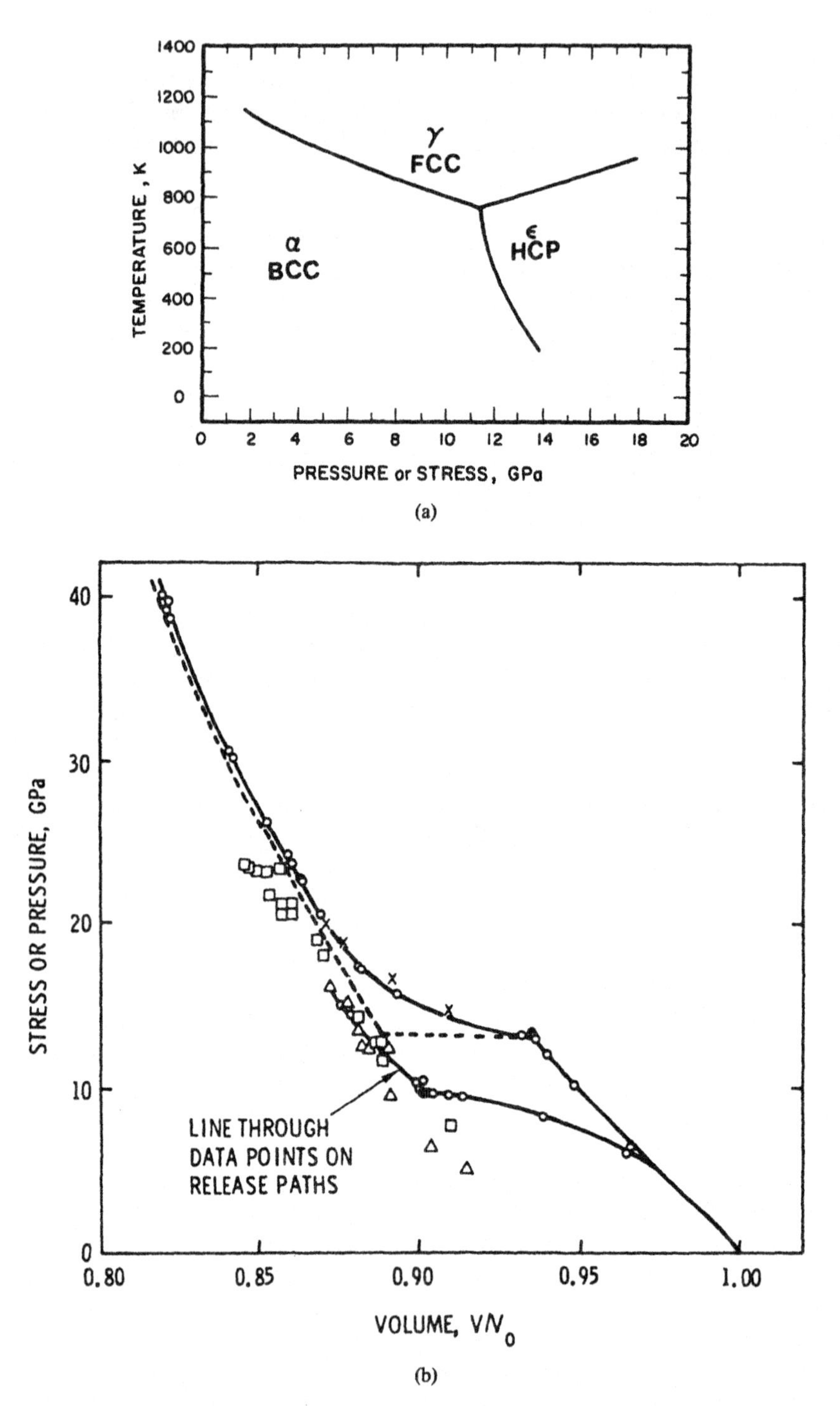

FIGURE 8.7 (a) Temperature–pressure phase diagram for iron. (b) Pressure–volume shock Hugoniot for iron (loading and unloading). (From Duvall and Graham [11], Fig. 18, p. 542. Reprinted with permission of the publisher.)

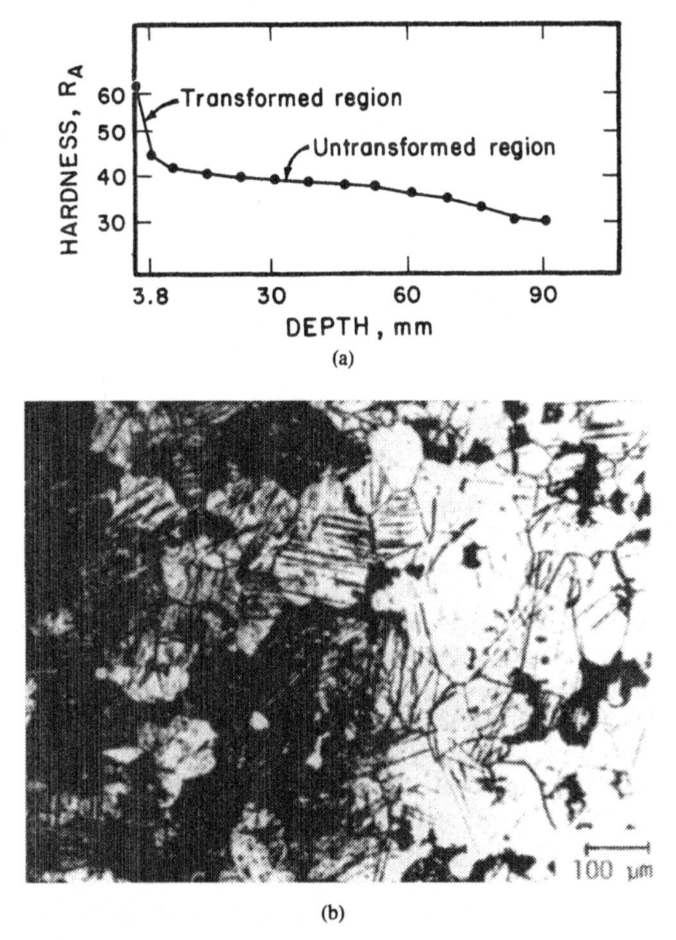

FIGURE 8.8 (a) Microhardness profile and (b) optical micrograph in boundary of transformation region. (From Meyers et al. [15], Figs. 5(b) and 6(c), p. 1742. Reprinted with permission of the publisher.)

after the shock pulse has passed, to avoid graphitization. This is accomplished in the commercial process by mixing copper powder with graphite. The copper provides the quenching medium. The use of graphite powder as compared to solid graphite provides the temperature necessary for the transformation. The graphite–diamond phase transformation is probably martensitic, and the diamond is actually hexagonal (such as the naturally occurring carbonado diamond) and not cubic. The particles are composed of a large number of crystallites (~100 Å diameter) and are therefore polycrystalline. These polycrystalline diamond particles have found a good application as polishing agents.

Table 8.1 shows the other polymorphic phase transformations. It can be seen that a number of materials undergo these changes. Erskine and Nellis [16] shock compressed pyrolytic graphite normal to the graphite basal plane and

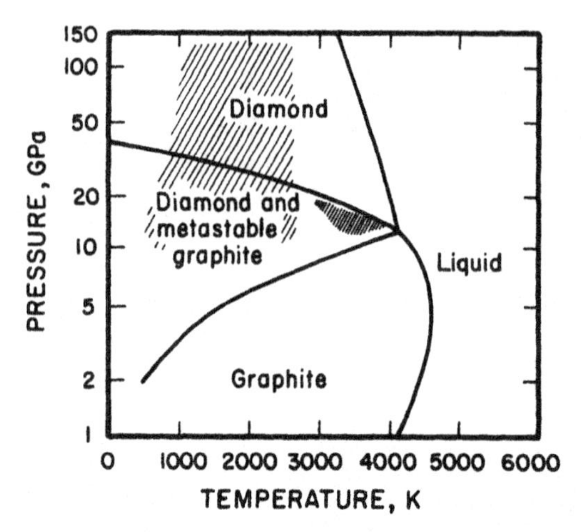

FIGURE 8.9 Pressure–temperature phase diagram for carbon. Diamond range is indicated by hatching.

obtained, by VISAR, stress wave profiles. The graphite-to-diamond transformation was clearly visible by the splitting of the shock front into two waves. The onset of the transformation was 20 GPa, and at 40 GPa the wave structure became a single again because the high-pressure shock velocity exceeds the transformation wave velocity. This effect is shown in Figure 8.6(b). The pressure P_3 corresponds to equal Rayleigh slopes for the two wave fronts. The results obtained by Erskine and Nellis [16] are shown in Figure 8.10. Graphite

TABLE 8.1 Summary of Shock-induced Polymorphic Phase Transformations

Material	Condition	Transition Conditions Stress (GPa)	Remarks
Iron	α(BCC)	12.86	6.5
Fe–0.5 wt % C		13	6.4
Fe–0.2 wt % C		14.7	5.8
Bismuth		~2.5	6.5
Carbon (pressed graphite)		23	28
Graphite		23	28
Titanium		9.4	~12
Zirconium		~23	~16
Uranium		~50	20
Plutonium	δ Phase	0.6	
SiO_2	α Crystal	14.5	16.5
Antimony	Rhombohedral cubic	7	cubic
		8.2	HCP

Source: Summarized from Duvall and Graham [11].

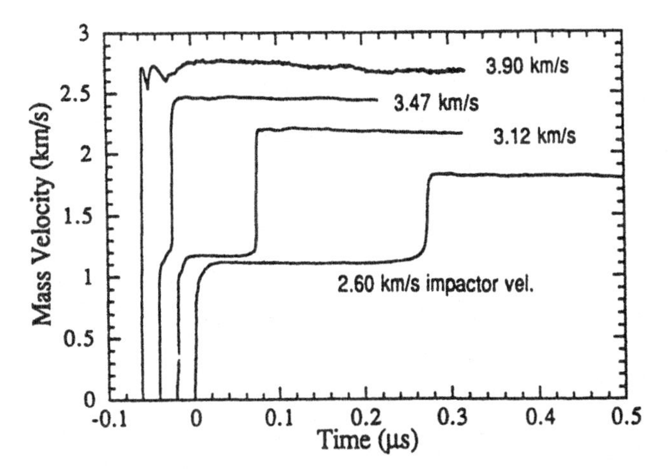

FIGURE 8.10 Wave profiles obtained by VISAR for highly oriented graphite impacted by copper flyer plate at different velocities, marked in plot; notice two-wave structure. (From Erskine and Nellis [16], Fig. 3, p. 4883. Reprinted with permission of the publisher.)

specimens were impacted by copper flyer plate at the four velocities indicated in Figure 8.10: 2.60, 3.12, 3.47, and 3.90 km/s. The 2.60 km/s impact velocity produces a pressure above 20 GPa and a clear two-wave structure. At 3.90 km/s the pressure is approximately 50 GPa, and the second shock overtakes the first (transformation) front. The student can, as an exercise, convert the ordinate axis into pressures by using the 20 GPa transformation pressure as a normalization fraction.

8.5 SHOCK-INDUCED MELTING, SOLIDIFICATION, AND VAPORIZATION

Melting is usually associated with a volume and entropy increase, and therefore the melting point increases with pressure. A notable exception is ice, which has a lower density than liquid water. Thus, an increase in pressure should produce a lowering in the melting point of ice. This phenomenon is indeed observed. The skates apply a high pressure on ice, with localized melting, which decreases the friction and increases the speed of the skater. Less known are the melting of plutonium and silicon, accompanied by a significant contraction.

In materials in which $dP/dT > 0$ ($\Delta V > 0$; $\Delta S > 0$), one can still obtain melting at sufficiently high temperatures. Shock-induced melting has been observed by means of changes in the shock response. An inverse phenomenon can also be produced. Since $dP/dT > 0$, one can produce solidification by starting from the molten material. The different possibilities are illustrated in Figure 8.11. The mixed phase region (solid to liquid) is shown by the shaded

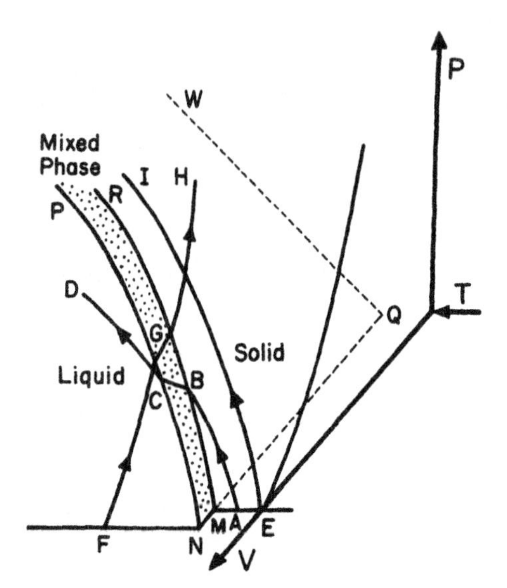

FIGURE 8.11 Pressure–volume–temperature surface for a "normal" ($\Delta V > 0$, $\Delta S > 0$, $dP/dT > 0$) liquid. Note that the melting pressure increases with temperature. (From Duvall and Graham [11], Fig. 14, p. 534. Reprinted with permission of the publisher.)

strip. If shock compression occurs at a low temperature, the Rankine–Hugoniot adiabat does not intersect the mixed phase region. This is the *EI* trajectory shown in Figure 8.11. At a higher temperature, the shock adiabat penetrates into the liquid–solid region, with melting. This is the *ABCD* trajectory shown in Figure 8.11. The solidification induced by the pressure release is qualitatively different from conventional processes. The loading rate (10^9 K/s) is determined by the release time of the wave, which is of the order of $\sim 10^{-7}$ s. At even higher temperatures the material is molten at atmospheric pressure, and shocking can produce solidification. This is indicated by the trajectory *FGH*. Table 8.2 lists instances of shock-induced melting. Shock-induced melting is often experimentally determined by a break in the U_s–U_p line. For lead, for which the melting temperature under shock loading conditions is available, the melting point is approximately three times as high as the one at ambient pressure.

Example 8.2. Calculate the melting temperature for aluminum under shock loading conditions, knowing that the pressure required for melting is 100 GPa.

TABLE 8.2 Melting Pressures for Some Metals

Materials	Melting Pressure (GPa)	Temperature (K)
Pb	28	1210
Zn	44	
Sn	28	
Al	105–102	
Fe	>184	
Ce	43	3600

Shock temperatures can be calculated by means of the Grüneisen EOS (Section 5.5) or simply determined from shock tables. From Kinslow [41, p. 530], we have

$$1 \text{ megabar (MB)} \rightarrow T = 3540 \text{ K}$$

Thus, this is the answer.

8.6 TRANSFORMATIONS INDUCED BY TENSILE STRESS PULSES

The martensitic transformation in Fe–Ni, Fe–Ni–C, and Fe–Mn–Ni alloys involves a positive volume change of approximately 4%. This expansion led to the suggestion that, although compressive shock stresses would hinder transformation, tensile stresses would favor it. This is indeed the case, and the γ(FCC) $\rightarrow$ α(BCC) transformation was produced by tensile waves produced by reflection of compression shock waves at a free surface. Figure 8.12(a) shows schematically how an impact of a projectile of thickness t on a target of thickness $2t$ produces tension in the center of the target. Figure 8.12(b) shows the cross section of an Fe–31% Ni–0.035% C disk after impact. The dark regions are produced by chemical etching and denote the martensite. Systematic experiments were conducted by Thadhani and Meyers [17] and Chang and Meyers [18] and led to a good understanding of this phenomenon. In the opposite experiments, [19, 20] in an alloy of similar composition that had been transformed to α(BCC) previous to shocking by thermally cooling it, shock waves (compressive) produced the α(BCC) $\rightarrow$ γ(FCC) transformation, which has a $\Delta V = -0.04$ (volume decrease). In Chapter 14 (Section 14.5) these transformations will be discussed further.

8.7 SHOCK-INDUCED CHEMICAL REACTIONS

This is a fertile field that is receiving considerable attention. The early work, following DeCarli and Jamieson's [6] discovery of shock synthesis of diamonds, was mostly conducted in the former Soviet Union with notable contributions by Adadurov et al. [21], Dremin et al. [22], Batsanov [23], and Batsanov and Deribas [24]. Under shock conditions, mixtures of powders of different compositions react and produce new compounds. Usually exothermically reacting materials (in powder form) are preferred. The use of powders produces higher temperatures at the interfaces between the reactants. Titanium, nickel, niobium aluminides, silicides, and a variety of other compounds can be synthesized by shock waves. The pioneering work by Ryabinin (1956) [26], DeCarli and Jamieson (1961) [6], and Nomura and Horiguchi (1963) [25], in Russia, the United States, and Japan, respectively, was followed by a great deal of activity. In the United States Graham and co-workers [27, 28] have carried long-ranging and systematic efforts. The shock-induced chemical reac-

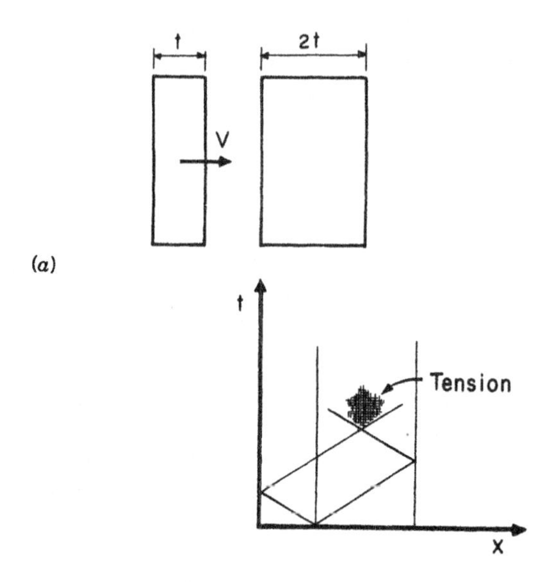

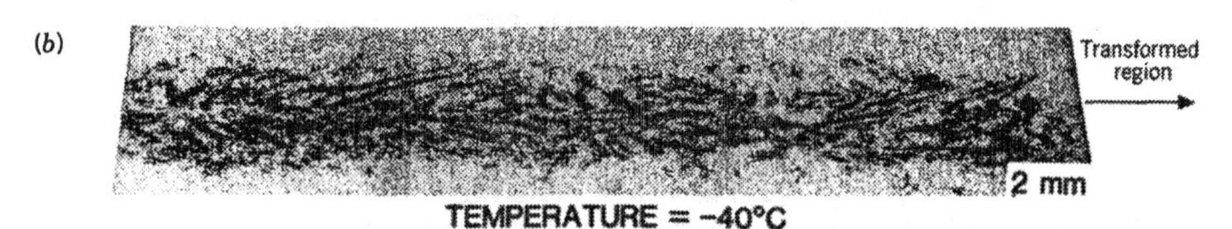

FIGURE 8.12 (a) Experiment designed to generate tensile pulse in target; (b) martensitic transformation [(FCC) → (BCC) with $\Delta V = 0.04$] produced in Fe–30 wt % Ni–0.35 wt % C alloy by tensile pulse technique; note that specimen is rotated 90° from sketch in (a). Reprinted from *Acta Met.*, vol. 34, N. N. Thadhani and M. A. Meyers, Fig. 4, p. 1631, Copyright 1986, with permission from Pergamon Press, Ltd.)

tions can be classified into two groups:

1. Synthesis: $xA + yB \rightarrow A_xB_y$.
2. Decomposition: $A_xB_y \rightarrow xA + yB$.

where A and B can be elements or compounds.

Many exothermic reactions can be triggered by shock waves; a few examples are

$$\text{Ti} + \text{C} \rightarrow \text{TiC}$$

$$\text{Ti} + 2\text{B} \rightarrow \text{TiB}_2$$

$$x\,\text{Ni} + y\,\text{Al} \rightarrow \text{Ni}_x\text{Al}_y$$

$$x\,\text{Nb} + y\,\text{Al} \rightarrow \text{Nb}_x\text{Al}_y$$

$$x\,\text{Nb} + y\,\text{Si} \rightarrow \text{Nb}_x\text{Si}_y$$

We use the letters x and y to specify the unknown values; different compounds can be produced for the same two elements. For example, NiAl, $NiAl_3$, Ni_2Al_3, and Ni_3Al can be produced from the Ni + Al reaction. A systematic overview of the synthesis is provided by Thadhani [29].

Decomposition reactions, on the other hand, have also been observed; Dremin and Breusov [1] note the following:

$$3CuO \rightarrow Cu + Cu_2O_3$$

$$PbO_2 \rightarrow PbO + \tfrac{1}{2}\,O_2$$

$$4FeO \rightarrow Fe + Fe_3O_4$$

$$Al_2SiO_5 \rightarrow Al_2O_3 + SiO_2$$

A very important question is whether these reactions are only initiated by the shock wave or whether they proceed to their conclusion. After the passage of the shock wave the material is in a heated state due to shock and reaction heating, and this could lead to the continuation of the reaction. Figure 8.13 shows, in a schematic fashion, the propagation of a shock wave through a porous reactive medium. The dark region represents reacted products. They increase in size with time. Three possible situations are depicted:

1. Material is fully reacted within the duration of the pulse.
2. Material is partially reacted during duration of the pulse; reaction is stopped by release of pressure.
3. Material is partially reacted during duration of the pulse; reaction continues after pulse passage.

At the microstructural level, these reactions are very complex and depend on a number of factors such as the porosity of material, size and shape of particles, shock pressure, and initial temperature. Meyers et al. [30] and Vecchio et al. [31] have addressed some of these issues in the development of a mechanism. Figure 8.14 shows a Nb–Si mixture. The two photomicrographs show unreacted [Fig. 8.14(a)] and partially reacted regions [Fig. 8.14(b)]. The initial mixture was porous with separate Nb and Si powders. The reaction produces [Fig. 8.14(b)] small spherical $NbSi_2$ particles; these particles are formed at the Nb–Si interface and "float" away, into silicon. It is thought that silicon melts and that a reaction between solid Nb and molten Si takes place under shock compression. Meyers et al. [30] developed an analytical framework based on these observations. (See Section 17.7.5).

The use of porous materials helps in initiating reactions because the energy level that can be reached is much higher. In Chapter 4 we learned that the energy is given by

$$E - E_0 = \tfrac{1}{2}\,(P + P_0)(V_{00} - V)$$

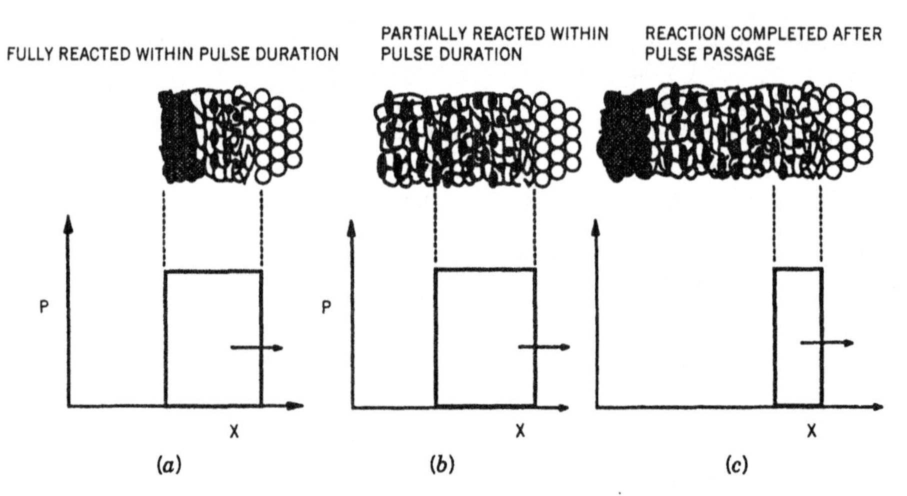

FIGURE 8.13 Propagation of shock wave through a reactive powder mixture with three possible situations schematically shown (dark regions represent reacted material: (a) material fully reacted within duration of pulse; (b) material partially reacted; (c) material fully reacted, with reaction initiated during shock pulse and terminating after this. (From L. H. Yu, Ph.D. dissertation, New Mexico Tech., 1992, p. 149.)

When V_{00}, the specific value, is high (due to the high porosity, $V = 1/\rho$), the energy level at a prescribed pressure is much higher than for the solid. Thus, high temperatures and interparticle melting can be reached, and these effects can initiate the reaction.

Boslough [32] and Yu and Meyers [33] developed a simple thermodynamic treatment for the shock compression of reactive materials. They added the energy of reaction to the conservation-of-energy equation:

$$E - E_0 = \tfrac{1}{2}(P + P_0)(V_{00} - V) + \Delta Q$$

For a porous Nb–Si mixture initially at a density equal to 60% of the theoretical density, the calculations can be easily conducted and the results are shown in Figure 8.15. The Hugoniot curve for the solid material is on the left-hand side; the one for the porous material is in the middle. The Hugoniot for the porous material was calculated as described in Chapter 5 (Section 5.4). The addition of the energy of reaction shifts the curve of the porous material to the right, and this effect is analogous to what occurs in explosives; this will be studied in Chapter 10 (Section 10.2).

Geological materials have also received considerable attention, and shock compression has been used to simulate the high pressures during the impact of meteorites and in the center of the earth. Ahrens and co-workers [35–38] and Syono [34] have identified a number of significant phase transitions. However, due to the focus of this book on materials, these will not be discussed here.

(a)

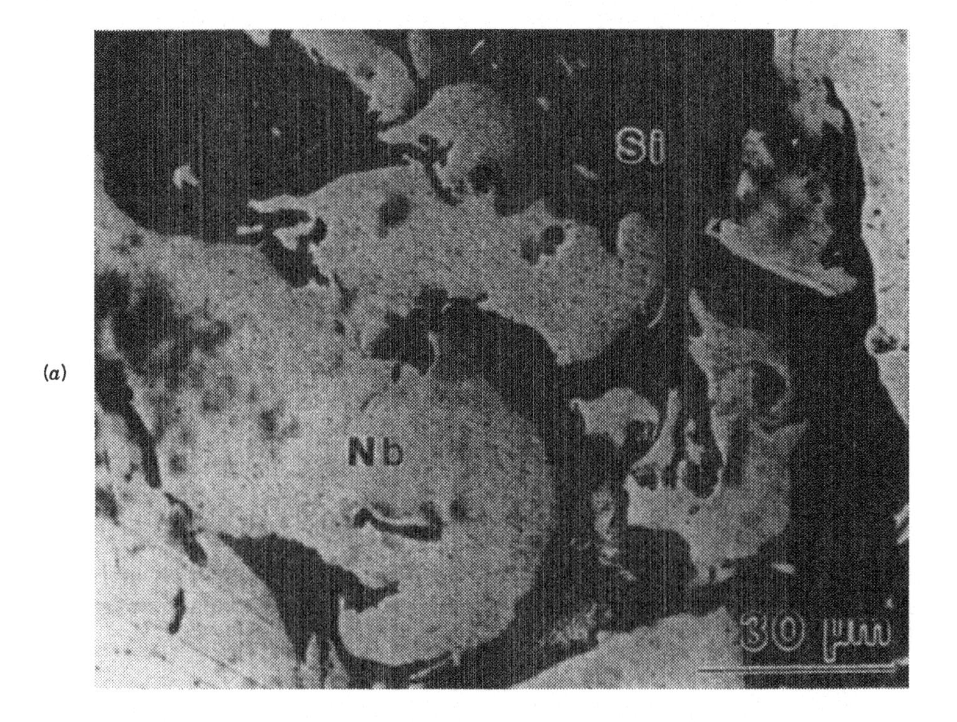

(b)

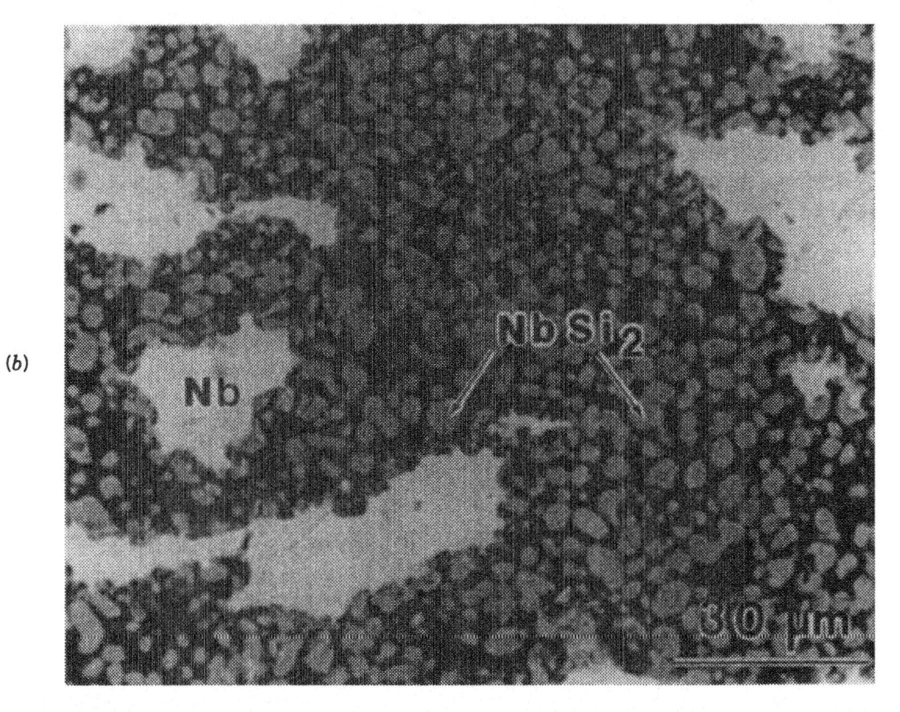

FIGURE 8.14 Niobium–silicon powder mixture after shock wave passage: (a) unreacted material (Nb, light; Si, dark); (b) partially reacted material (small $NbSi_2$ spherules seen as grey in silicon phase). (From L. H. Yu, Ph.D. dissertation, New Mexico Tech., 1992, p. 14.)

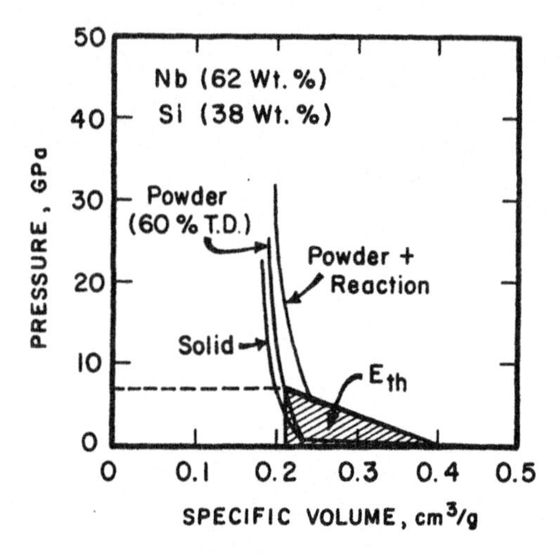

FIGURE 8.15 Hugoniot curves for Nb–Si mixture; from left to right are the curves for solid material, inert powder mixture, and powder mixture with reaction. (From Yu and Meyers [33], Fig. 9, p. 606. Reprinted with permission of the publisher.)

Suffice it to say that most oxides and silicates undergo phase transitions under shock compression. Chapter 17 (Section 17.8) will briefly deal with these changes. Examples are the pressure-induced change undergone by SiO_2. Under shock compression, Chhabildas [39] found that the quartz–stishovite phase transformation commences at 15 GPa and is completed at 40 GPa. The product phase is Stischovite, named after the Russian scientist Stischov, who identified it for the first time. A pressure-induced decomposition studied by Lange and Ahrens [40] is

$$CaCO_3 \rightarrow CaO + CO_2$$

Example 8.3. Using the Clausius–Clapeyron equation, estimate the pressure at which copper would start to melt under shock loading conditions:

$$T_m\ (p = 1\ \text{atm}) = 1083\ \text{K}$$

$$H_m = 49\ \text{cal/g}$$

$$\rho\ (\text{liquid at } T_m) = 7.99 \times 10^3\ \text{kg/m}^3$$

$$\rho\ (P = 1\ \text{atm};\ T = 258\ \text{K}) = 8960\ \text{kg/m}^3$$

Thermal expansion L_T (in degrees Fahrenheit) is calculated as

$$L_T = L_0(1 + 16.73T + 0.002626T^2 + 9.1 \times 10^{-7}T^3) \times 10^{-1}$$

The Clausius–Clapeyron equation will be applied with the assumption that ΔV is independent of pressure. Let us first determine ΔV:

$$\Delta V = V_{\text{liquid}} - V_{\text{solid}}|_{\text{MP}}$$

$$\Delta V = \frac{1}{7.99} - \frac{1}{8.96}\left(1 + 3\,\frac{L_T(T_m) - L_0}{L_0}\right)$$

We use the factor 3 in order to convert the linear expansion coefficient into a volumetric expansion coefficient:

$$T_m = 1490°\text{F}$$

$$\frac{L_T(T_m) - L_0}{L_0} = 0.0338$$

$$\Delta V = 2.2 \times 10^{-6}\ \text{m}^3/\text{kg}$$

Applying the Clausius–Clapeyron equation yields

$$dP/dT = \Delta S/\Delta V = \Delta H/T\,\Delta V = \frac{49\ \text{cal/g} \times 4.2\ \text{J/cal}}{1083\ \text{K} \times 2.2 \times 10^{-6}}$$

$$= 0.9 \times 10^8\ N/\text{m}^2\ \text{K}$$

We can plot the shock temperature versus pressure as obtained from the data below (from the Appendix in R. Kinslow, *High Velocity Impact Phenomena*, Academic, New York):

Parameters on the Hugoniot (*Continued*)

P (Mbar)	ρ (g/cm^3)	T (K)	S	C (km/s)	U_s (km/s)	U_p (km/sec)
0.00	8.930	293	5.16	3.94	3.94	0.00
.10	9.499	336	5.22	4.42	4.33	.26
.20	9.959	395	5.51	4.81	4.66	.48
.30	10.349	479	5.98	5.13	4.95	.68
.40	10.688	589	6.57	5.41	5.22	.86
.50	10.990	726	7.19	5.67	5.47	1.02
.60	11.262	888	7.82	5.90	5.70	1.18
.70	11.510	1072	8.42	6.12	5.91	1.33
.80	11.738	1279	8.99	6.32	6.12	1.46
.90	11.949	1505	9.52	6.51	6.32	1.60
1.00	12.145	1751	10.01	6.69	6.50	1.72
1.10	12.329	2014	10.48	6.86	6.68	1.84

Parameters on the Hugoniot (*Continued*)

P (Mbar)	ρ (g/cm^3)	T (K)	S	C (km/s)	U_s (km/s)	U_p (km/sec)
1.20	12.502	2294	10.91	7.03	6.86	1.96
1.30	12.666	2589	11.31	7.19	7.03	2.07
1.40	12.820	2900	11.69	7.34	7.19	2.18
1.50	12.967	3224	12.04	7.49	7.35	2.29
1.60	13.107	3561	12.37	7.63	7.50	2.39
1.70	13.241	3910	12.69	7.77	7.65	2.49
1.80	13.368	4271	12.98	7.91	7.79	2.59
1.90	13.491	4643	13.26	8.04	7.93	2.68
2.00	13.608	5026	13.53	8.17	8.07	2.77
2.10	13.721	5419	13.78	8.30	8.21	2.87
2.20	13.829	5821	14.02	8.42	8.34	2.95
2.30	13.934	6232	14.25	8.54	8.47	3.04
2.40	14.035	6653	14.47	8.66	8.60	3.13

The third column represents the shock temperatures. We now plot the shock temperature and melting point as a function of pressure. Their intersection marks the melting pressure.

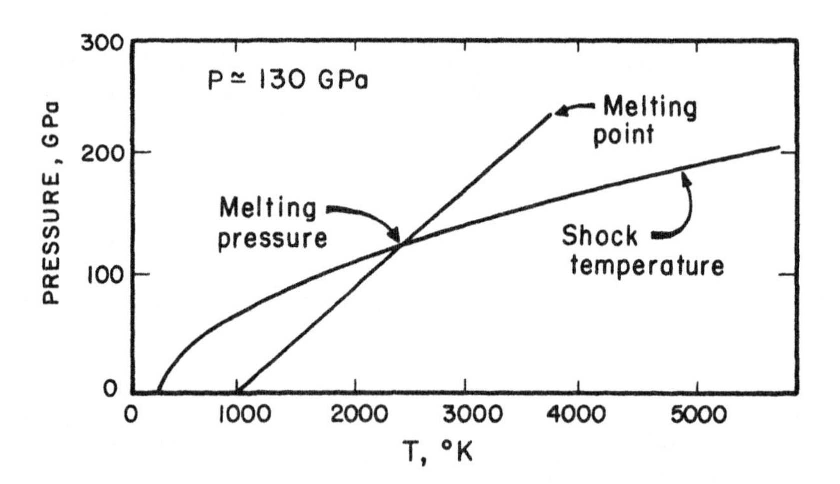

REFERENCES

1. A. N. Dremin and O. N. Breusov, *Russian Chem. Rev.*, **37** (1968), 392.
2. R. A. Graham, B. Morosin, E. L. Venturini, and M. J. Carr, *Ann. Rev. Mat. Sci.*, **16** (1986), 315.

3. R. A. Graham, *Proc. Symp. on High Dynamic Pressures*, La Grande Motte, France, 1989.
4. R. A. Graham, *Solids under High Pressure Shock Compression: Mechanics, Physics, and Chemistry*, Springer-Verlag, New York, 1993.
5. F. S. Minshall, *Bull. APS,* **29** (1954), 23.
6. P. S. DeCarli and J. C. Jamieson, *Science,* **133** (1961), 1821.
7. O. R. Bergmann, in *Shock Waves in Condensed Matter—1983*, eds. J. R. Asay, R. A. Graham, and G. K. Straub, North-Holland, Amsterdam, 1984, p. 429.
8. O. R. Bergmann and N. F. Bailey, in *High Pressure Explosive Processing of Ceramics*, eds., R. A. Graham and A. Sawaoka, Trans. Tech. Publications, 1987, Basel, Switzerland, p. 65.
9. A. N. Dremin and A. Staver, eds., *Proc. Fifth All-Union Meeting on Detonation*, Krasnoyarsk, Russia, 1991.
10. V. M. Titov, *Proc. Intl. Symp. on Intense Dynamic Loading and its Effects*, Chengdu, China, June 1992, p. 85.
11. G. E. Duvall and R. A. Graham, *Rev. Mod. Phys.*, **49** (1977), 523.
12. D. Bancroft, E. L. Peterson, and S. Minshall, *J. Appl. Phys.*, **27** (1956), 291.
13. J. C. Jamieson and A. W. Lawson, *J. Appl. Phys.*, **33** (1962).
14. L. Barker and Hollenbach, *J. Appl. Phys.*, **45** (1974), 4872.
15. M. A. Meyers, C. Sarzeto, and C. Y. Hsu, *Met. Trans. A.*, **11A** (1980), 1737.
16. D. J. Erskine and W. J. Nellis, *J. Appl. Phys.*, **71** (1992), 4882.
17. N. N. Thadhani and M. A. Meyers, *Acta Met.*, **34** (1986), 1625.
18. S. N. Chang and M. A. Meyers, *Acta Met.*, **36** (1988), 1085.
19. R. W. Rohde, J. R. Holland, and R. A. Graham, *Trans. AIME,* **242** (1968), 2017.
20. R. W. Rohde and R. A. Graham, *Trans. AIME,* **245** (1969), 2441.
21. G. A. Adadurov, A. N. Dremin, and G. I. Kanel, *Combust. Expl. Shock Waves,* **6** (1970), 456.
22. A. N. Dremin, S. V. Pershin, and V. F. Porgorelov, *Combust. Expl. Shock Waves,* **1** (1965), 1.
23. S. S. Batsanov, in *Behavior and Utilization of Dense Media under High Dynamic Pressures*, Gordon and Breach, New York, 1968, p. 371.
24. S. S. Batsanov and A. A. Deribas, *Mauch. Teka. Probl. Goreniya Vzryua,* **1** (1965), 103.
25. Y. Nomura and Y. Horiguchi, *Bull. Chem. Soc. Jpn.*, **36** (1963), 486.
26. Y. N. Ryabinin, *Zhur. Tekh. Fiz.*, **26** (1956), 2661.
27. R. A. Graham, B. Morosin, E. L. Venturini, and M. J. Carr, *Ann. Rev. Mater. Sci.*, **16** (1986), 315.
28. Y. Horie, R. A. Graham, and I. K. Simonsen, *Mater. Lett.*, **3** (1985), 354.
29. N. N. Thadhani, *Prog. Mater. Sci.*, **37** (1993), 117.
30. M. A. Meyers, L. H. Yu, and K. S. Vecchio, *Acta Met. et Mat.*, **42** (1994) 701.
31. K. S. Vecchio, L. H. Yu, and M. A. Meyers, *Acta Met. et Mat.*, **42** (1994) 715.
32. M. B. Boslough, *J. Chem. Phys.*, **92** (1990), 1939.
33. L. H. Yu and M. A. Meyers, *J. Mater. Sci.* **26** (1991), 601.

34. Y. Syono, in *Materials Science of the Earth's Interior*, ed. I. Sunnagawa, Terra Sci. Comp., Tokyo, 1984, p. 395.
35. T. J. Ahrens, D. L. Anderson, and A. E. Ringwood, *Rev. Geophys.*, **7** (1969), 667.
36. T. J. Ahrens, *Science*, **207** (1980), 1035.
37. R. Jeanloz, T. J. Ahrens, J. S. Lally, G. L. Nord, J. M. Christie, and A. H. Hamer, *Science*, **206** (1977), 457.
38. R. Jeanloz, T. J. Ahrens, H. K. Mao, and P. M. Bell, *Science*, **206** (1979), 829.
39. L. C. Chhabildas, in *Shock Waves in Condensed Matter*, ed. Y. M. Gupta, Plenum, New York, 1986, p. 601.
40. M. A. Lange and T. J. Ahrens, *J. Earth Plan. Sci. Lett.*, **77** (1986), 409.
41. R. Kinslow, *High-Velocity Impact Phenomena*, Academic, New York, 1970.

CHAPTER 9

Explosive–Material Interactions

9.1 INTRODUCTION

The interaction of a detonating explosive with a material in contact with it or in close proximity is extremely complex, since it involves detonation waves, shock waves, expanding gases, and their interrelationships. For a simple one-dimensional geometry the sequence of events is qualitatively depicted in Figure 9.1. Let us imagine a small segment of an infinite explosive slab placed adjacent to a metal plate [Fig. 9.1(a)]. Initiation is starting simultaneously over the entire surface of the explosive [Fig. 9.1(b)], producing a pressure pulse that results in detonation. Detonation velocity and peak pressure (Chapman–Jouguet pressure) are given by the formulations developed in Section 10.5. As the detonation propagates, the greater and greater amount of detonation products accumulating in the left-hand side results in a gradual increase of the duration of the pressure pulse, whereas the peak pressure remains constant (Chapman–Jouguet pressure). This is shown in Figures 9.1(b) and (c). When the detonation front encounters the metal, an interaction will occur, and a pressure pulse is transferred to the metal. The peak pressure of this pulse is determined, using the impedance matching method, from the intersection of the pressure–particle velocity curves for explosive and metal (Chapter 8). Let us assume that $P_2 > P_1$. At the same time, a reflected wave is transmitted into the explosive detonation products. This is shown in Figure 9.1(e). When the shock wave in metal encounters the free surface, it accelerates it at a velocity of $2U_{p1}$ and reflects back as a release wave [Fig. 9.1(e)]. This reflected wave will encounter the back face of metal (explosive–metal interface) and produce, by interaction, a pressure change [Fig. 9.1(f)]. A new shock wave is sent through the metal [Fig. 9.1(g)] that will, in turn, drive the free surface to a velocity that is increased by $2U_{p2}$. The process continues, with successive reflections. One has $U_{p1} > U_{p2} > U_{p3}$. As the successive reflections take place, the explosive gases continue their expansion, and attenuation of the shock wave in the metal takes place. The situation depicted above is one possible sequence. Different wave configurations can occur, depending on the Chapman–Jouguet pressure, the metal shock impedance, the thicknesses of metal and explosive, and the presence of a gap or an "attenuator" between explosive and metal. Figure 9.2 shows the free-surface velocity of the metal plate as a function of time. The

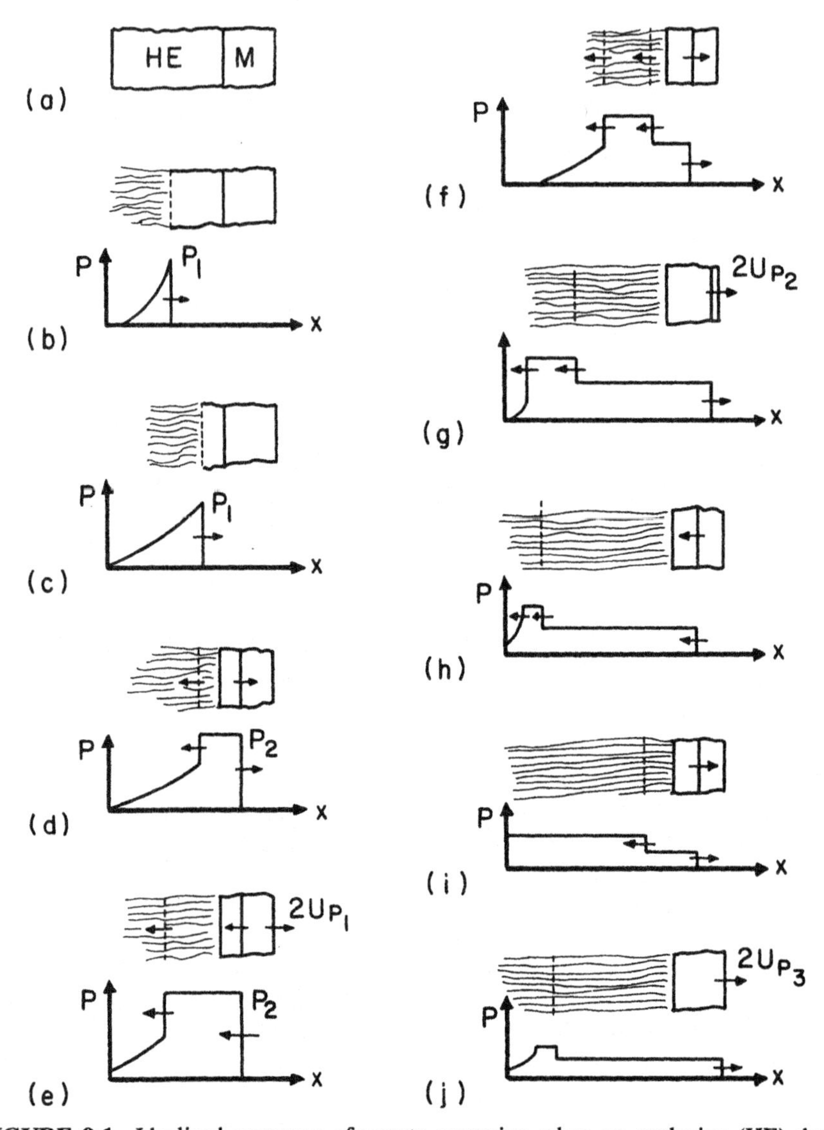

FIGURE 9.1 Idealized sequence of events occurring when an explosive (HE) detonates in contact with material (M); pressure–distance plots given below each figure.

velocity increases in a discontinuous fashion until the steady-state velocity V_p is reached. The first increment is $2U_{p1}$, the second $2U_{p2}$, the third, $2U_{p3}$, and so on. The times t_1, t_2, t_3 are approximately given by the transit time of the shock wave through the metal plate:

$$t_1 - t_0 \approx \frac{2S}{U_s} \tag{9.1}$$

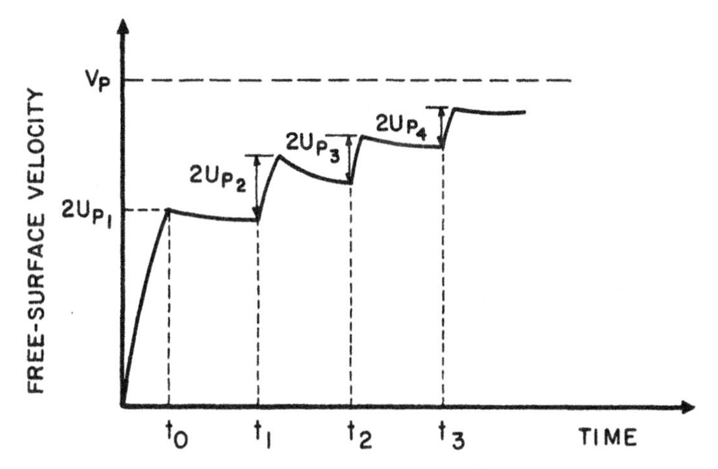

FIGURE 9.2 Velocity of front face (free surface of projectile) as a function of time.

As the pressure pulse attenuates, U_s decreases and one should see a slight increase in t. Between t_0 and t_1 the free-surface velocity decreases once $2U_{p1}$ is reached. This is due to the triangular shape of the pulse, "pumping into" metal a shock with a decreasing amplitude.

The simplified picture described above can be calculated more rigorously, but this requires specialized techniques. One approach is to calculate the acceleration of the metal by using gas dynamics calculations. In this case no shock waves or detonation waves are involved. Aziz et al. [1] used this procedure, and the results are shown in Figure 9.3. The results are plotted in dimensionless coordinates. The projectile accelerates in a continuous fashion from rest to its steady-state velocity V_p. Figure 9.3(b) shows the scaled time, t' (Dt/L, where L is the explosive thickness and D is the detonation velocity). The velocity increases at a decreasing rate until the steady state is achieved. This calculation applies for a ratio between explosive and metal mass equal to 0.341. We can assume that the gas-dynamic computation is equivalent to the curve from shock reflections by means of a smoothing out of the discontinuities.

Aziz et al. [1] were able to calculate the final plate velocity V_p for the case $\gamma = 3$ (polytropic gas constant for explosive; see Chapter 10). They obtained

$$V_p = \frac{z-1}{z+1} \quad \text{where } z = \left(1 + \frac{32}{37}\frac{C}{M}\right)$$

and M is the metal mass and C is the explosive mass.

Numerical computations using finite-element, finite-difference, or method-of-characteristics codes provide a more realistic representation of the sequence of events. The great advantage of these calculational methods is that they can be extended to two- and three-dimensional problems that are encountered in

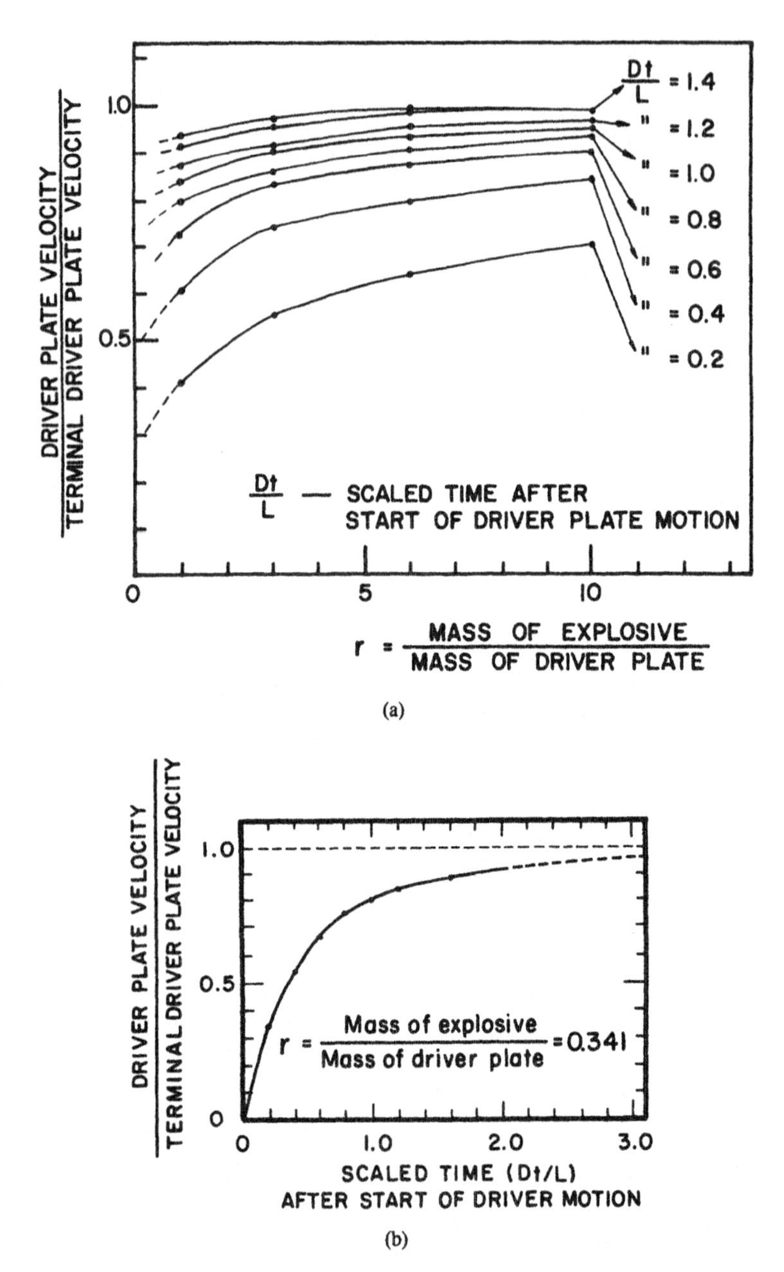

FIGURE 9.3 (a) Master plot for acceleration of flyer plate from explosive detonation; (b) scaled time vs. scaled velocity, according to gas dynamic calculations; wave interactions not considered (t = time, L = thickness of driver plate, D = detonation velocity). (From Aziz et al. [1]. Reprinted with permission of the publisher.)

applications. An example of a calculation using a method-of-characteristics code named NIP (for Normal Initiation Program) is shown in Figure 9.4. The same configuration as in Figure 9.1 is calculated. The x–t diagram with the characteristics network is shown in Figure 9.4(b). One can see that the successive reflections at the metal plate are represented, together with the rarefac-

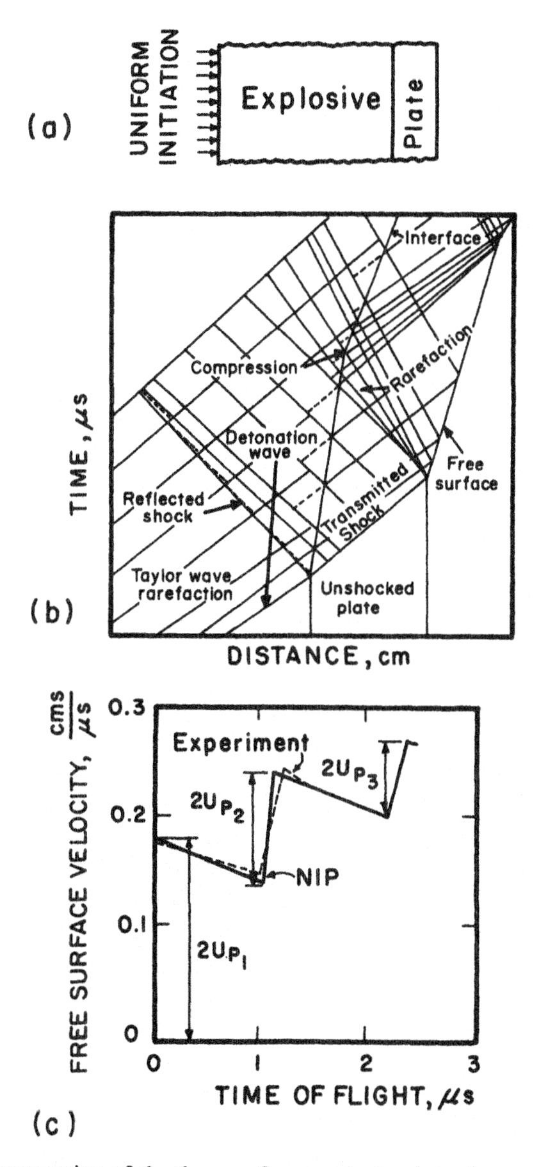

FIGURE 9.4 Propagation of shock wave from explosive into plate and its acceleration toward a steady-state velocity; successive reflections produce velocity discontinuities. (From Karpp and Chou [2].)

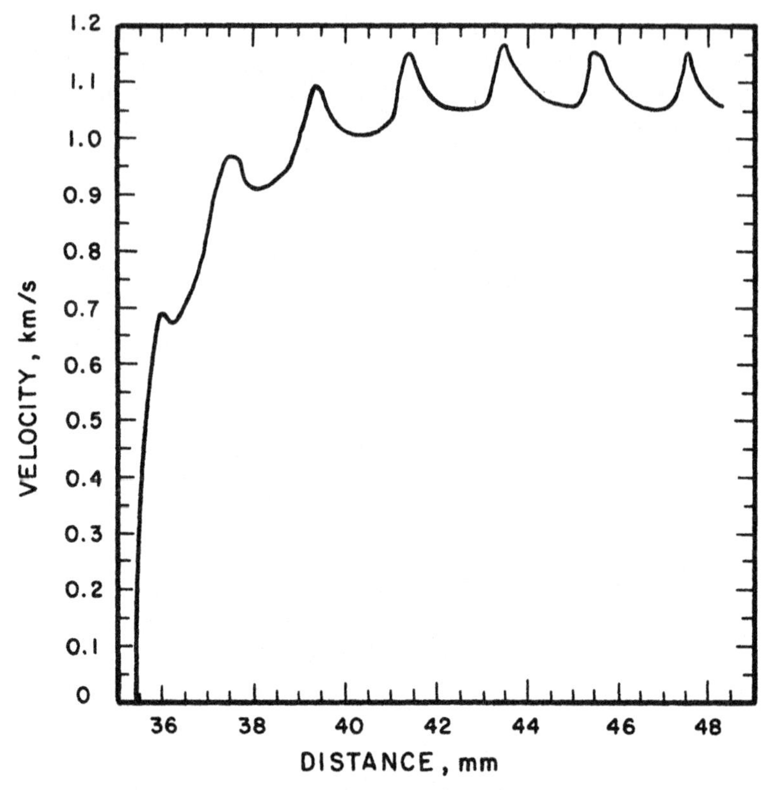

FIGURE 9.5 Velocity–time profile of mild steel plate driven by baratol. (From Yoshida, MY1DL handbook [3]. Reprinted with permission.)

tion fans. Figure 9.4(c) shows the progress of the free-surface velocity. Three velocity increments are seen, and they quantitatively predict the values qualitatively shown in Figure 9.2. One can see that $2U_{p1} = 1.8\ \text{mm}/\mu\text{s}$, $2U_{p2} = 1\ \text{mm}/\mu\text{s}$, and $2U_{p3} = 0.7\ \text{mm}/\mu\text{s}$. This decay in the velocity increments is also qualitatively shown in Figure 9.2.

Predictions of plate acceleration by the code MY1DL can also be made, and the resultant velocity profile for a 1-in. Baratol explosive accelerating an 8-mm mild steel plate is shown in Figure 9.5. The terminal velocity is approximately 1.15 km/s. This computation was done assuming hydrodynamic behavior for the mild steel. The same type of acceleration response of the flyer plate can be observed.

9.2 THE GURNEY EQUATION

In 1943, Gurney [4] proposed a simple approach that has been very successful in predicting the terminal velocity of explosively accelerated devices. By determining the velocities of fragments from bombs weighing as much as 3000 pounds and grenades containing as little as 1.5 ounces of a high explosive, he concluded that the governing factor was the ratio between the mass of the fragments (M) and the mass of the explosive (C). He made the simple assumption that the chemical energy of the explosive was transformed into kinetic energy of explosive products and metal fragments and arrived at expressions relating the velocity of fragments to the ratio C/M. He conducted his calculations for both spherical and cylindrical bombs. For the latter, he assumed that the longitudinal "spray" of fragments was negligible and that all fragments travel along radial trajectories. We will now derive Gurney's equation for a cylindrical geometry.

9.2.1 Cylindrical Geometry

The velocity of the gases that result from the detonation is assumed to vary linearly from the core to the outside of the explosive cylinder. The initial velocity of the fragments that we are attempting to calculate is V_0. We ignore wave propagation effects, the energy consumed in deformation and fracture, and the fact that detonation starts at one end of the cylinder and "runs down" the cylinder. Figure 9.6(a) shows the cylinder with an internal radius a. For a general radius r, the velocity is given by Figure 9.6(b):

$$V = V_0 \left(\frac{r}{a} \right) \tag{9.2}$$

The total kinetic energy (KE) is equal to the kinetic energy of the shell casing and the kinetic energy of the gases:

$$\text{KE} = \frac{1}{2} \sum m_i V_0^2 + \frac{1}{2} \int V^2 \, dm_g \tag{9.3}$$

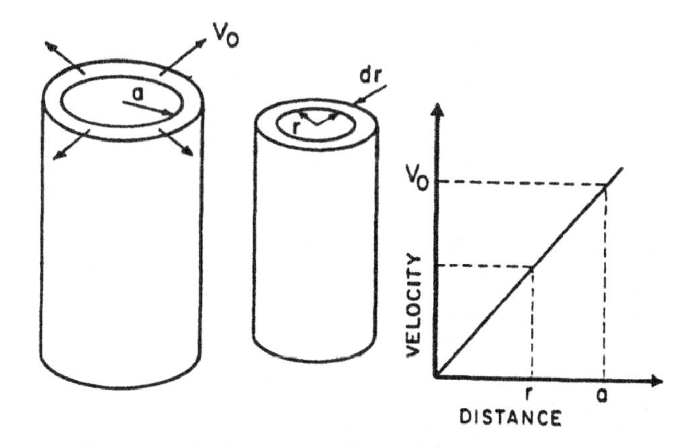

FIGURE 9.6 (a) Cylindrical shell filled with high explosive; (b) element of gas; (c) velocity–distance plot for gases.

We assumed fragments with mass m_i. For the gases, we take a tube-shaped element (Fig. 9.6b). Thus, the element of mass dm_g is equal to

$$dm_g = \rho\, dV = 2\pi r \rho\, dr \tag{9.4}$$

where ρ is the explosive density. Assuming that all fragments have the same velocity and substituting (9.2) and (9.4) into (9.3) yields

$$\text{KE} = \frac{1}{2} MV_0^2 + \frac{1}{2}\int_0^a V_0^2 \frac{r^2}{a^2} 2\pi r \rho\, dr$$

where M is the steel mass and C is the explosive mass:

$$\text{KE} = \tfrac{1}{2}MV_0^2 + \tfrac{1}{4}\pi V_0^2 a^2 \rho$$

But

$$\frac{C}{\pi a^2} = \rho$$

$$\text{KE} = \tfrac{1}{2}MV_0^2 + \tfrac{1}{4}CV_0^2$$

We equate this kinetic energy to a chemical energy (E) of the explosive. The exact nature of this chemical energy will be described later. This represents the portion of the chemical energy of the explosive, per unit mass, that is converted into kinetic energy. Thus, V_0 can be found:

$$V_0^2 = \frac{2EC}{M + \frac{1}{2}C}$$

or

$$V_0 = \sqrt{2E}\left(\frac{M}{C} + \frac{1}{2}\right)^{-1/2} \tag{9.5}$$

The Gurney equations are usually of the above form. They express the fragment velocities as a function of C/M (or M/C) ratios. The term $\sqrt{2E}$ always appears and actually has the units of velocity.

9.2.2 Spherical Geometry

We repeat the above derivation for a spherical shell with internal radius a. We equate the kinetic energy of fragments and gases to the chemical energy of the explosive:

$$E = \frac{1}{2} M V_0^2 + \frac{1}{2} \int V^2 \, dm_g$$

But now we use a spherical shell of thickness dr as the element of mass. Thus

$$dm_g = 4\pi r^2 \rho \, dr$$

$$E = \frac{1}{2} M V_0^2 + \frac{1}{2} \int_0^a V_0^2 \frac{r^2}{a^2} 4\pi r^2 \rho \, dr$$

$$= \frac{1}{2} M V_0^2 + \frac{4\pi\rho V_0^2}{2a^2} \int_0^a r^4 \, dr$$

$$= \frac{1}{2} M V_0^2 + \frac{4\pi\rho a^3}{2 \times 5} V_0^2$$

But

$$V = \tfrac{4}{3}\pi a^3 \quad \text{and} \quad 4\pi\rho a^3 = 3V\rho = 3C$$

$$V_0 = \sqrt{2E} \left(\frac{M}{C} + \frac{3}{5} \right)^{-1/2}$$

9.2.3 Asymmetric Plate Geometry

The asymmetric plate geometry is commonly encountered in plate impact experiments (shock hardening, shock compaction, explosive welding). An explosive is placed on top of a plate and detonated at the top or at one end. Since this is an asymmetrical situation, the conservation-of-energy equation is not sufficient to arrive at a final expression. Kennedy [5] has a detailed derivation of this case. We have two situations: open-faced and closed-faced sandwich. In the latter, we have a metal–explosive–metal assemblage. In the former case, we just have an explosive–metal assemblage. Figure 9.7 shows the velocity distribution for an open-faced sandwich. The velocity distribution can be expressed as (linear variation)

$$V_{gas}(y) = (V_0 + V) \frac{y}{y_0} - V_0$$

where $V_{gas}(y)$ is the gas velocity at position y, the velocity of the metal plate is V_0, and the velocity (maximum) of the gas on the opposite face is V. The

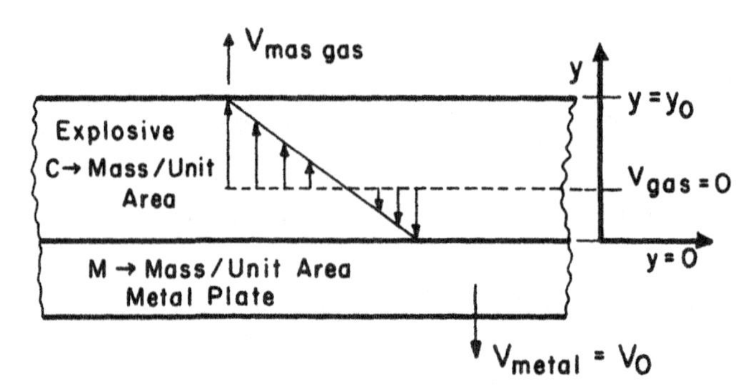

FIGURE 9.7 Velocity distribution for an open-faced sandwich.

conservation-of-energy equation gives

$$E = \frac{1}{2} MV_0^2 + \frac{1}{2} \int_0^{y_0} \rho V_{\text{gas}}^2 |_y \, dy$$

$$= \frac{1}{2} MV_0^2 + \frac{1}{2} \rho_e \int_0^{y_0} \left[(V_0 + V) \frac{y}{y_0} - V_0 \right]^2 dy$$

The equation for the conservation of momentum (difference of forces equals impulse) gives

$$0 = -MV + \rho_e \int_0^{y_0} \left[(V_0 + V) \frac{y}{y_0} - V_0 \right] dy$$

The total charge is $C = \rho y_0$. Integration leads to

$$V_0 = \sqrt{2E} \left[\frac{(1 + 2M/C)^3 + 1}{b(1 + M/C)} + \frac{M}{C} \right]^{-1/2}$$

An identical expression is

$$V_0 = \sqrt{2E} \left[\frac{3}{1 + 5(M/C) + 4(M/C)^2} \right]^{1/2}$$

If one knows the thicknesses and densities of explosive and plate, the following expression gives the ratio M/C:

$$\frac{M}{C} = \frac{\rho_m}{\rho_e} \cdot \frac{t_m}{t_e}$$

9.2.4 Gurney Energies

Figure 9.8 gives the Gurney equations for a few simple geometries. A careful analysis of the Gurney equation is given by Henry [6].

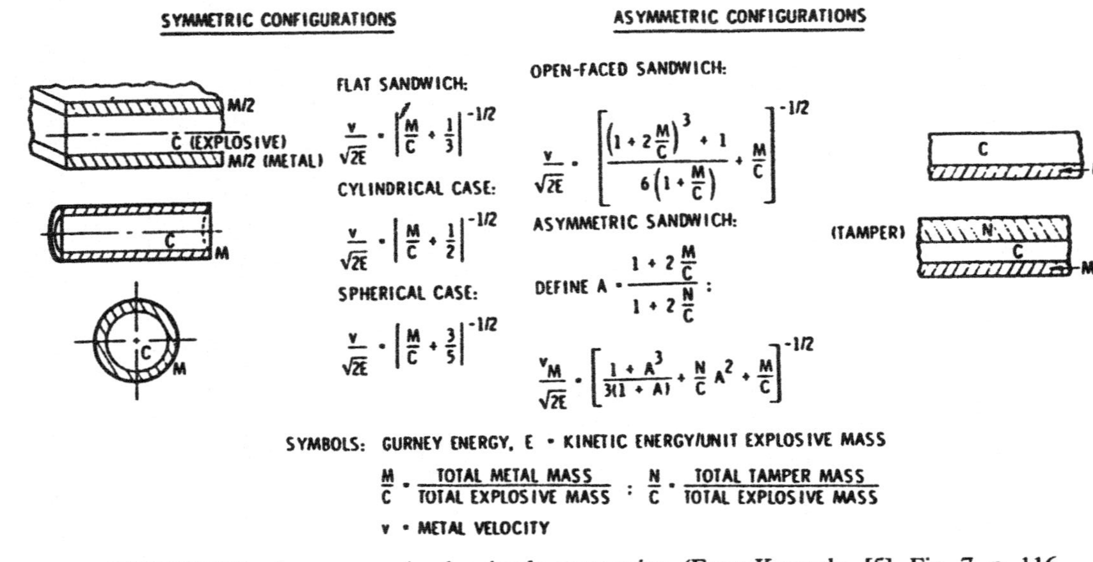

FIGURE 9.8 Gurney energies for simple geometries. (From Kennedy, [5], Fig. 7, p. 116. Reprinted with permission of the publisher.)

TABLE 9.1 Gurney Energies ($\sqrt{2E}$) and Specific Impulses (I_{sp}) of Explosives

Density (g/cm^3)	Explosive	$\sqrt{2E}$ (mm/μs)	I_{sp} (ktaps/g explosive cm^2)	E (kcal/g)	ΔH_d (kcal/g)	$E/\Delta H_d$
1.77	RDX	2.93	254	1.03	1.51	0.68
		2.83	245	0.96		0.64
1.60	Composition C-3	2.68	232	0.86	—	—
1.63	TNT	2.37	205	0.67	1.09	0.61
		2.44	211	0.71		0.65
1.72	Tritonal[a]					
	(TNT/Al = 80:20)	2.32	201	0.64	1.77	0.36
1.72	Composition B	2.71	235	0.87	1.20	0.72
	(RDX/TNT)	2.77	240	0.91		0.76
		2.70	234	0.87		0.72
		2.68	232	0.86		0.72
1.89	HMX	2.97	257	1.06	1.48	0.72
1.84	PBX-9404	2.90	251	1.01	1.37	0.74
1.62	Tetryl	2.50	217	0.75	1.16	0.65
1.61	TACOT	2.12	184	0.54	0.98	0.55
1.14	Nitromethane	2.41	209	0.69	1.23	0.56
1.76	PETN	2.93	254	1.03	1.49	0.69
	duPont Sheet					
1.46	EL506D	2.28	197	0.62	—	—
		2.50	217	0.75	—	—
1.56	EL506L	2.20	191	0.58	—	—
1.1	Trimonite No. 1[a]	1.04	90	0.13	1.26	0.10

Source: From Kennedy [5].)

[a]Detonates nonideally.

Values for $\sqrt{2E}$ are given in Table 9.1. These values were experimentally determined by Kennedy [5] by measuring terminal velocities of metals. Gas leakage between metal fragments was minimized in the early stages of acceleration. If considerable gas leakage occurs, this value ($\sqrt{2E}$) can be decreased by as much as 20%. The enthalpy of detonation (ΔH_d) is given in Table 9.1. It can be seen that the ratio $E/\Delta H_d$ varies between 0.76 and 0.5 for most explosives.

In spite of the fact that more complex formulations have been proposed for the velocities of fragments and plates, the original Gurney equation has been proven to be very reliable [5, 6]. In systematic investigation conducted at Stanford Research Institute, Roth [7] concludes that the simple Gurney theory is as good as the more sophisticated ones; its only limitation is that it is not applicable to situations before the terminal conditions are achieved and when M/C tends to zero. Figure 9.9 shows the comparison between experimental and theoretical plate velocities. It can be seen that the Gurney equation compares well with experimental results for M/C values above 0.1. The parameter

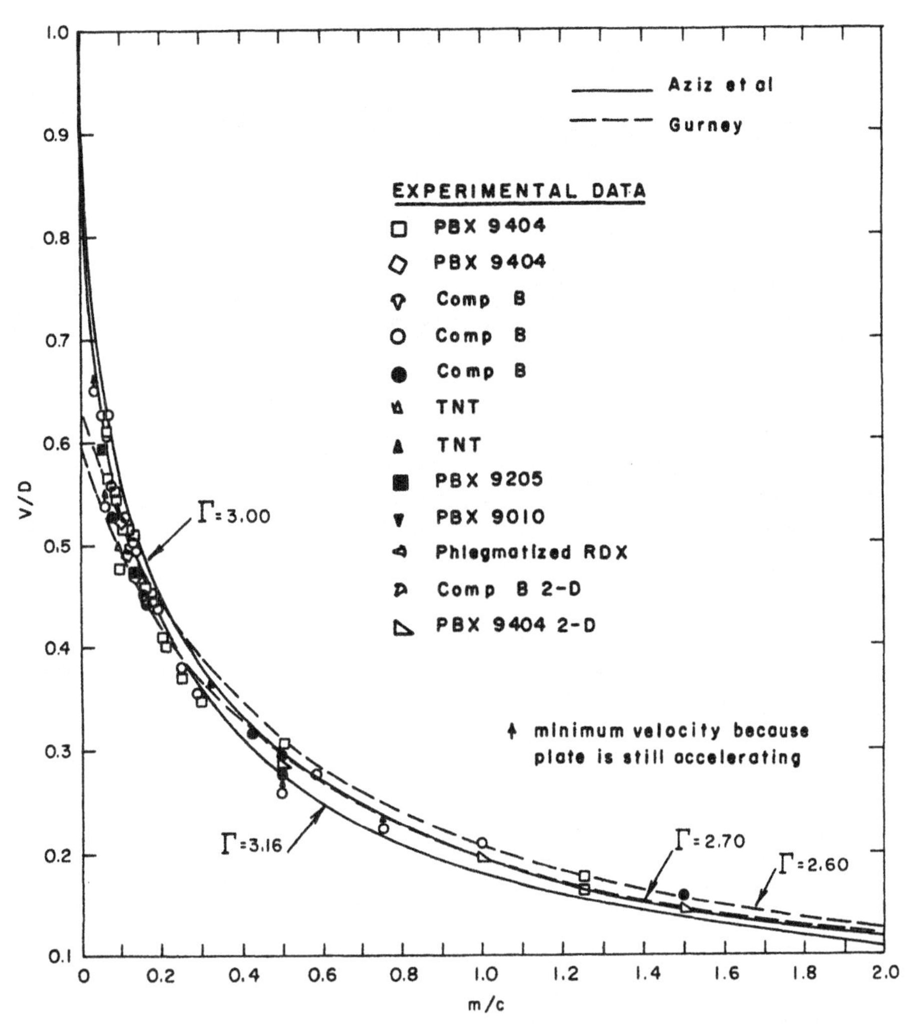

FIGURE 9.9 Comparison of the prediction of Gurney equation (------) for flyer plate velocities at various M/C (mass of material/mass of charge) values with experimental results and more complex equation that includes gas dynamics (——) (Γ is polytropic gas constant). (From Roth [7].)

γ is related to the EOS of the detonation products and is included in the tabulated Gurney energies (see Chapter 10).

It is very important to notice that the Gurney equation only predicts the terminal velocity of the plate; actually, it accelerates from its initial rest position. Enough stand-off distance has to be left between the flyer plate and the target.

For explosive charges of limited lateral dimensions, accelerating plates, the lateral effects are important and have to be taken into account. Kennedy [5]

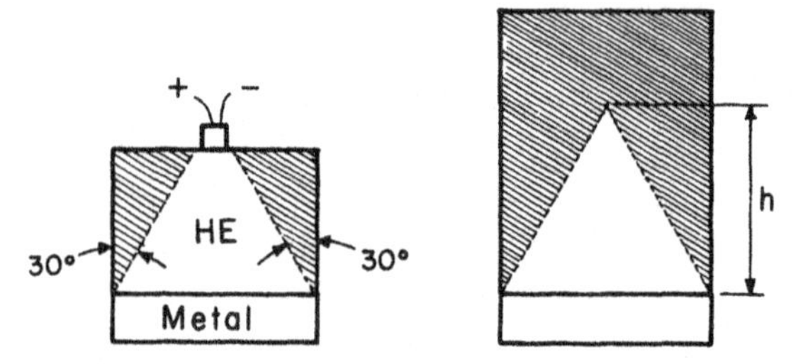

FIGURE 9.10 Discounting of sides of explosive to account for lateral release of explosive pressure; mass from hatched region is not included.

recommends the procedure shown in Figure 9.10. The lateral edges of the explosive have to be subtracted from the total mass C, and an angle of 30° seems to produce the best correlation with experimental results. Thus, there exists a maximum height (h) of explosive (apex of cone) that will drive the metal plate to a top velocity that cannot be increased by increasing the explosive height. The student is now ready to design the most lethal contraptions.

Example 9.1. Calculate the velocity produced by the detonation of a slab of PBX 9404 of 25.4 mm places on top of a 3.2-mm steel plate. Assume simultaneous initiation over the entirety of the top surface:

PBX 9404: $\rho_c = 1.84 \text{ g/cm}^3$
Steel: $\rho_m = 7.89 \text{ g/cm}^3$

$$\sqrt{2E} = 2.90 \text{ mm}/\mu\text{s}$$

$$\frac{M}{C} = 0.54$$

$$V_p = \sqrt{2E}\left[\frac{3}{1 + 5(M/C) + 4(M/C)^2}\right]^{1/2}$$

$$= 2.27 \text{ mm}/\mu\text{s}$$

$$= 2.27 \text{ km/s}$$

REFERENCES

1. A. K. Aziz, H. Hurwitz, and H. M. Sternberg, *J. Appl. Phys.*, **4** (1961), 380.
2. R. Karpp and P. C. Chou, in *Dynamic Response of Materials to Intense Impulse Loading*, eds. P. C. Chou and A. H. Hopkins, Air Force Materials Laboratory, WPAFB, Dayton, Ohio, 1972.
3. M. Yoshida, Program MY1DL, CETR report, 1986.

4. R. Gurney, "The Initial Velocities of Fragments from Bombs, Shells, and Grenades," Report No. 405, Ballistic Research Laboratory, Aberdeen, MD, September 1943, AII-36218.
5. J. E. Kennedy, "Behavior and Utilization of Explosives in Engineering Design," 12th Annual Symposium, ASME, UNM, Albuquerque, NM 1972 (copies can be procured from R. L. Henderson, 2917 Dakota Street NE, Albuquerque, NM 87110).
6. I. G. Henry, "The Gurney Formula and Related Approximations for High-Explosive PUB-189," Hughes Aircraft Co., Culver City, CA, 1967, AD 813398.
7. J. Roth, "Correlation of the Gurney Formulae for the Velocity of Explosively-Driven Fragment with Detonation Parameters of the Driver Plate," Technical Report No. 001.71, Poulter Laboratory, Stanford Research Institute, Menlo Park, CA, 1971.

CHAPTER 10

Detonation

10.1 INTRODUCTION

The release of energy from a detonation is not inordinately high. This statement might appear contradictory to our common sense, but it is actually true. However, there are two unique aspects in detonation:

1. The *rate* at which the energy is released is extremely high.
2. The detonation products are gases in a highly compressed state, and the expansion of these gases performs a great amount of work.

The energy released by detonation is of the order of 1 kcal/g (4–5 MJ/kg). As a comparison, we list, in Table 10.1, enthalpies of reaction for some explosives and some of the other exothermic reactions.

Thus, it is the differences between the rate of release of energy that separates the explosives from other reactions. Popolato [1] provides the following values for explosives:

1. Power output: explosives, 10^9 W/cm^2; fuel oil–air burn, 10 W/cm^2).
2. Energy output: explosives, 1 kcal/g (4–5 MJ/kg); fuel oil–air burn, 10 kcal/g (4–5 MJ/kg).

Fast chemical reactions involving the production of gases can be classified into (in order of growing release rate of energy)

- Burning
- Explosion/deflagration
- Detonation

In burning, ordinary physicochemical mechanisms are involved and propagation is controlled by conventional transport phenomena. If burning is confined, a high pressure builds up, leading to what is known as an explosion. The first explosive mixture dates back to the ninth century [1]. It consisted of a mixture of sulfur, saltpeter, and charcoal. Although it does not detonate, its burning rate, when confined, is sufficient to provide considerable power. Detonation

TABLE 10.1 Heats of Reaction for Selected Materials (1 cal = 4.19 J)

Reactant	Product	H (cal/g)
Composition B	Gases	1,240
TNT	Gases	1,000
PETN	Gases	1,400
Datasheet	Gases	1,000
Ni + Al	NiAl (solid/liquid)	330
Ti + B	TiB_2 (solid/liquid)	1,150
Fuel oil–air	Gases	10,000

requires the propagation of a shock wave. The high pressure is required for the explosive decomposition, because the activation energy for the process is very high. This activation energy is of the order of 150 kJ/mol. Trinitrotoluene [TNT, $C_7H_5(NO_2)_5$] has a molecular weight of 227. This yields an activation energy of 660 J/g, or 158 cal/g. The position at which the reaction is complete is called the Chapman–Jouguet point.

The pressure pulse in the explosive is generated by the highly exothermic chemical reaction behind it, producing detonation products in a state of high pressure and temperature. Figure 10.1 shows the activation energy. The process of detonation is shown in a schematic way in Figure 10.2. A high-pressure shock wave travels into the explosive and initiates the reaction. A detonation front is moving into the virgin material at a velocity D (detonation velocity). Behind the front, we have a reaction zone of thickness t. This transforms the solid into gaseous detonation products at very high pressures. The interface at which the reaction is completed is called the Chapman–Jouguet (C–J) interface. This interface moves at a velocity $U_p + C$ with respect to the unshocked material. This velocity is given by the sound velocity in the compressed gas, C, plus the particle velocity U_p. For this process to be steady state, it is necessary that

$$C_{CJ} + (U_p)_{CJ} = D \tag{10.1}$$

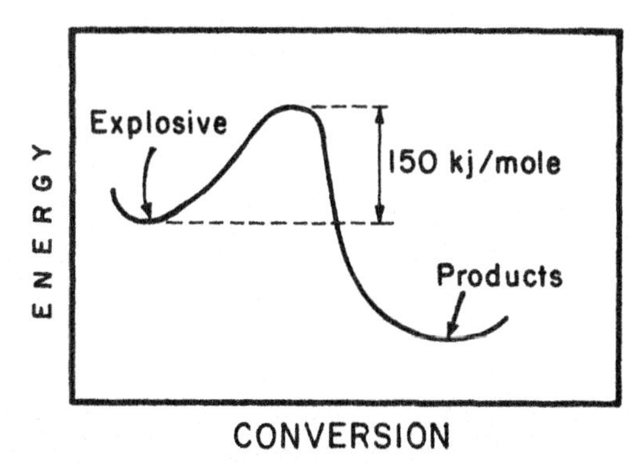

FIGURE 10.1 Schematic of energy involved in detonation.

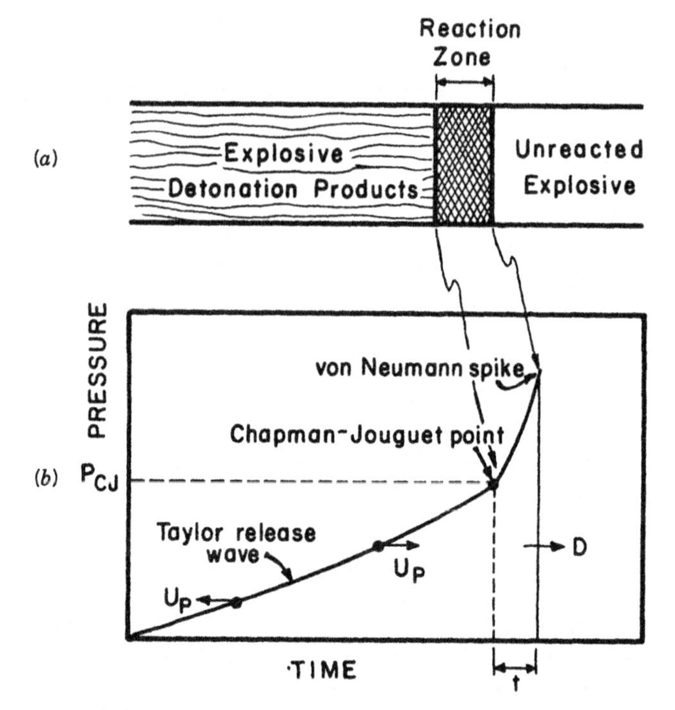

FIGURE 10.2 (a) Structure of detonation wave; (b) pressure profile.

This establishes the constancy in the thickness of the reaction zone. An important feature of this interface is that any signal behind it is *subsonic* with respect to the front, whereas any disturbance ahead of it is *supersonic* with respect to the front.

The rarefaction (or release) wave propagates downstream into the products immediately behind the detonation wave. This wave is called a Taylor wave, after the same notable individual who postulated dislocations. Ahead of the chemical reaction there is a pressure spike, called the von Neumann spike. Figure 10.2 shows a schematic pressure pulse. The von Neumann spike is very narrow and is attenuated almost immediately upon entering a metal. The rate of attenuation of the Taylor wave is much lower, and its highest pressure, indicated as P_{CJ} and particle velocity $(U_p)_{CJ}$ are used to characterize the explosive. The experimental results of Duff and Houston [2] illustrate the above. They found, for an explosive mixture consisting of 63% RDX and 37% TNT (composition B), that the von Neumann spike had a peak pressure of 38.6 GPa, whereas the Chapman–Jouguet pressure was 27.2 GPa. However, the von Neumann spike had completely disappeared beyond a depth of 1.25 mm into an aluminum plate in contact with the explosive, whereas the Taylor wave propagated to much greater depths. At the C–J interface the product gases have a particle velocity that has the same direction and sense as the detonation direction; as the gases expand behind the C–J interface, the particle velocity sense is inverted. The determination of the state variables in a detonating explosive is done by applying the three equations of conservation of mass, momentum, and energy; the same approach is used in the treatment of shock waves.

10.2 CONSERVATION EQUATIONS

The conservation equations are identical to the ones derived in Chapter 4, with a change in symbol for the shock velocity from U_s to D:

$$\rho_0 D = \rho(D - U) \quad \text{(conservation of mass)} \tag{10.2}$$

$$P - P_0 = \rho_0 UD \quad \text{(conservation of momentum)} \tag{10.3}$$

For the conservation of energy, we have to introduce the chemical energy released by the detonation into the equation. If this energy is Q, per unit mass, we have [the student should derive Eqn. (10.4) as an exercise]

$$E_1 - E_0 = \tfrac{1}{2}U^2 \tag{10.4}$$

where E_1 refers to the internal energy per unit mass at the C–J point, and E_0 to the internal energy per unit mass in the totally expanded state; for the gases,

$$E_0 = E_{00} + Q$$

where E_{00} is the internal energy of the solid explosive. Thus

$$E_1 - E_{00} = \tfrac{1}{2}U^2 + Q \tag{10.5}$$

The equations for the conservation of mass, momentum, and energy together with the Chapman–Jouguet condition ($C_{CJ} + (U_p)_{CJ} = D$) and the EOS for the detonation products fully describe the detonation process. The EOS will be discussed in the next section.

10.3 EQUATIONS OF STATE

There is a considerable number of EOS that describe the behavior of the gaseous detonation products. In Chapter 4, we defined the following equation for solids:

$$U_s = C_0 + S_1 U_p + \cdots \tag{10.6}$$

and

$$\rho_0 U_p U_s = \rho_0 (C_0 U_p + S_1 U_p^2) \tag{10.7}$$

The above EOS describes the propagation of a wave through a material and, in particular, through a solid explosive. If this explosive reacts and produces gases, the EOS for gases has to be used. Equation (10.6) does not describe well the behavior of the detonation products for two reasons: (1) the products are *gaseous*, not solids or liquids and (2) there is an energy release at the detonation front. Thus, we have to develop other EOS. Johansson and Persson [3] present a description of the EOS. A brief summary of their presentation is given here. Since the gases are at a highly compressed state, initially the volume occupied by the molecules is a substantial portion of the total volume. This has been incorporated by Abel (as described by Johansson and Persson [3]) who subtracted from the total volume of the gas a fraction corresponding to the molecule fraction. Figure 10.3 shows schematically how the covolume (fraction of volume occupied by molecules) is equal to zero at low pressures and increases with the fraction of total volume occupied by the molecules. The expression proposed by Abel is simply the ideal gas law with a corrected volume:

$$P(V - b) = nRT \tag{10.8}$$

Van der Waals and Cook (as described by Johansson and Persson [3]) proposed equations in which the covolume depends on the specific volume (V), or density (ρ):

$$P = \frac{\rho nRT}{1 - b\rho} - a\rho^2 \tag{10.9}$$

$$P = \frac{\rho nRT}{1 - \rho \exp(-0.4\rho)} \tag{10.10}$$

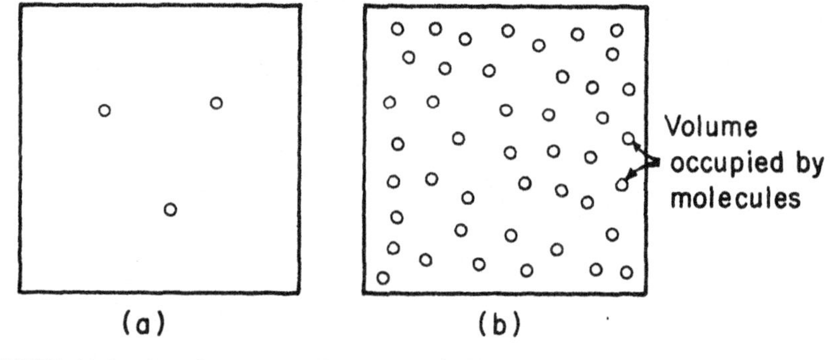

FIGURE 10.3 Covolume, or volume occupied by molecules at (a) low pressure; (b) high pressure. The covolume b is defined as the volume fraction occupied by the molecules.

Another form was introduced by Taylor [5]. He expressed the variation of ρ as a polynomial (virial expansion):

$$P = \rho nRT(1 + b\rho + 0.625b^2\rho^2 + 0.287b^3\rho^3 + 0.193b^4\rho^4 + \cdots) \tag{10.11}$$

Kistiakowski and Wilson [6] expressed the polynomial series as an exponential, where

$$P = \rho nRT[1 + \rho K(T + \theta)^{-\alpha} \exp \beta\rho K(T + \theta)^{-\alpha}] \tag{10.12}$$

This equation is known as the BKW EOS. Mader [7] applied this equation successfully to RDX and TNT, with the values given in Table 10.2. However, it is perhaps the polytropic gas law (equation describing isentropic expansion of an ideal gas) that is the most *convenient* and *easy-to-use* EOS:

$$P\rho^{-\gamma} = PV^{\gamma} = K \tag{10.13}$$

From Eqn. (10.13),

$$\ln P + \ln V^{\gamma} = \text{const} \quad \text{(for isentropic process)}$$

$$d[\ln P + \gamma \ln V] = 0$$

$$\gamma = -\left.\frac{\partial \ln P}{\partial \ln V}\right|_s \tag{10.14}$$

where γ is the polytropic gas constant and should *not* be confused with the Grüneisen coefficient for solids and liquids. Figure 10.4 shows schematically how the exponent γ determines the release of pressure from a high-pressure state (e.g., Chapman–Jouguet point).

The value of γ is found, for most explosives, to vary between 1.3 and 3. From Eqn. (10.13), one finds the expression for the internal energy due to the compression of the gases.

TABLE 10.2 Parameters in Kistiakowsky–Wilson EOS

BKW Parameter	RDX	TNT
ρ	1.8	1.64
k	10.90	12.69
β	0.16	0.09
θ	400	400
α	0.5	0.5

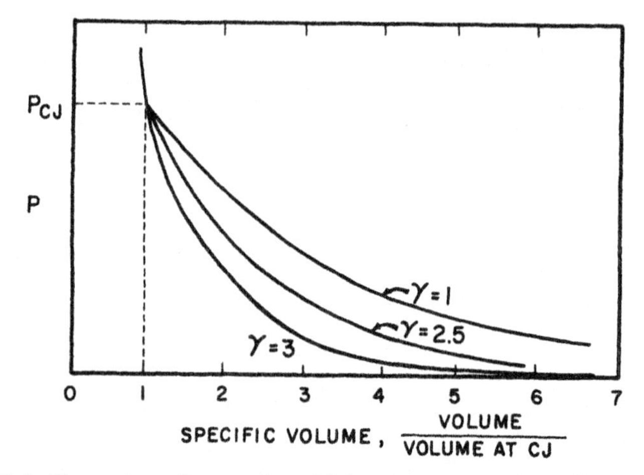

FIGURE 10.4 Expansion of gases from high-pressure state for different values of γ (polytropic gas constant).

From $dE = \delta Q - P\,dV$ (first law),

$$dE = -P\,dV$$

$$E = -\int_{V_0}^{V} P\,dV \tag{10.15}$$

From Eqn. (10.13),

$$P = KV^{-\gamma} \tag{10.16}$$

Substituting Eqn. (10.16) into Eqn. (10.15) yields

$$E = -K \int_{V_0}^{V} V^{-\gamma}\,dV = -K \left. \frac{V^{-\gamma+1}}{-\gamma+1} \right|_{V_0}^{V}$$

$$= -K \left. \frac{PV}{K(1-\gamma)} \right|_{V_0}^{V}$$

For V_0, we set $P = 0$. Thus

$$E = -\frac{PV}{1-\gamma} = \frac{P}{\rho(\gamma-1)} \tag{10.17}$$

This equation can be used to obtain the pressure as a function of particle velocity and as a function of detonation velocity. From Eqn. (10.5),

$$\frac{P}{\rho(\gamma-1)} = \tfrac{1}{2}U^2 + Q \tag{10.18}$$

Notice that the reference energy level E_{00} was made equal to zero. Now, using ρ from Eqn. (10.2) and substituting it into Eqn. (10.18) and using D from Eqn. (10.3) and substituting it into Eqn. (10.18) yield

$$P(D - U) = \tfrac{1}{2}\rho_0 D U^2(\gamma - 1) + \rho_0 D Q(\gamma - 1)$$

$$D = \frac{P}{\rho_0 U}$$

$$P\left(\frac{P}{\rho_0 U} - U\right) = \tfrac{1}{2}\rho_0 \frac{P}{\rho_0 U} U^2(\gamma - 1) + \frac{\rho_0 P}{\rho_0 U} Q(\gamma - 1)$$

$$\frac{P}{\rho_0 U} - U = \tfrac{1}{2}U(\gamma - 1) + \frac{Q}{U}(\gamma - 1)$$

$$\frac{P}{\rho_0 U} = \tfrac{1}{2}U(\gamma + 1) + \frac{Q}{U}(\gamma - 1)$$

$$P = \tfrac{1}{2}\rho_0 U^2(\gamma + 1) + Q\rho_0(\gamma - 1) \qquad (10.18a)$$

We proceed to calculate the pressure as a function of detonation velocity. By repeating the above procedure, using

$$U = \frac{P}{\rho_0 D}$$

we find

$$P = \frac{\rho_0 D^2}{\gamma + 1} \qquad (10.19)$$

Equation (10.19) is plotted in Figure 10.5(b). In Figure 10.5(a) a material obeying the Rankine–Hugoniot equations of conservation is shown. The Rayleigh line with a slope equal to $\rho_0 U_s$ gives the velocity of a wave with pressure P_1. For the detonation wave, the pressure does not pass through zero, and the Rayleigh line, indicated by a dashed line, does not uniquely define a state. It intersects the curve at P_1 and P_2. The only line that defines uniquely one state is the one with lowest slope, indicated by a full line. This state thus defined is the Chapman–Jouguet state. The pressure P_{CJ} and particle velocity $(U_p)_{CJ}$ are the point at which the curve is tangent to the Rayleigh line passing through the origin. From Eqn. (10.3), one has

$$\frac{P_{CJ}}{(U_p)_{CJ}} = \rho_0 D \qquad (10.20)$$

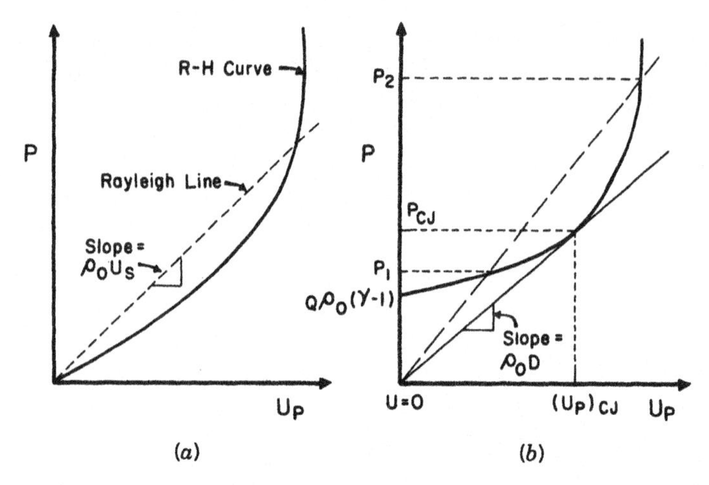

FIGURE 10.5 Pressure vs. particle velocity for (a) solid or liquid material obeying the Rankine–Hugoniot equations and (b) a detonation wave.

Thus, the slope of the line is $\rho_0 D$. Figure 10.5(b) is an *extremely important plot*. It allows us to determine the detonation velocity, the Chapman–Jouguet pressure, and the particle velocity if we know the chemical energy released by the explosive, the polytropic gas constant, and the initial density.

Example 10.1. Determine the detonation velocity and the Chapman–Jouguet pressure for explosive RDX if

$$\rho_0 = 1.77 \text{ g/cm}^3 = 1.77 \times 10^3 \text{ kg/m}^3$$

$$Q = 1500 \text{ cal/g}$$

$$\gamma = 3$$

The procedure to follow is to plot P versus U_p for RDX and to determine the Rayleigh line by using the tangent. This is shown in Figure 10.6. We have to convert the energy into joules per gram:

$$\begin{aligned} Q &= 1500 \times 4.18 \text{ J/g} \\ &= 6.270 \text{ J/g} = 6.27 \times 10^6 \text{ J/kg} \\ Q\rho_0(\gamma - 1) &= (6.27 \times 10^6)(1.77 \times 10^3)(2) \\ &= 22.2 \text{ GPa} \end{aligned}$$

The values of pressure for three detonation velocities calculated using Eqn. (10.19) are shown in Figure 10.6. These values were calculated for particle velocities of 1000, 2000, and 3000 m/s. The Rayleigh line is tangent to this

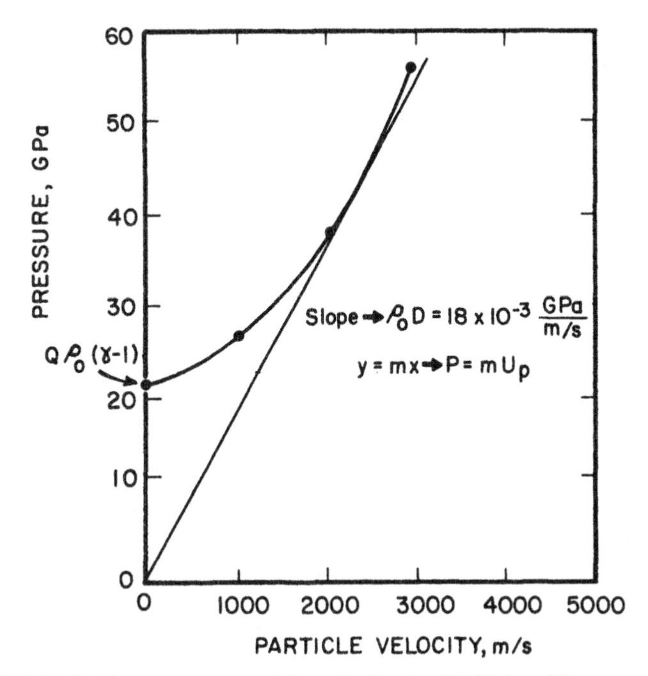

FIGURE 10.6 Pressure vs. particle velocity for RDX (a military explosive).

curve at a pressure of approximately 43 GPa. Thus, P_{CJ} = 43 GPa. The corresponding particle velocity is approximately 2500 m/s. The detonation velocity can be calculated from the slope at the Rayleigh line.

It should be noticed that the "γ-law" EOS allows a rapid and straightforward calculation of detonation velocity and Chapman–Jouguet parameters. The student should repeat this calculation for other explosives. Table 10.3 lists important properties of a number of explosives. One should be somewhat careful with the use of this table, since the values of the energies are the Gurney energies, which are not exactly the chemical energies.

10.4 VON NEUMANN SPIKE [11] AND CHARGE DIAMETER EFFECTS

Figure 10.2 shows the detonation wave profile. The von Neumann "spike" precedes the Chapman–Jouguet point [8, 9]. The region between the two is the reaction region. The condition established by Chapman and Jouguet, that $(U_p)_{CJ} + (C)_{CJ} = D$, is easily understood if one assumes a steady state and reaction region with constant thickness. Thus, all signals reach the C–J point ahead of the von Neumann front (supersonic with respect to it), whereas all signals behind it are subsonic with respect to it.

The reaction rate therefore determines the pressure at the front (P_{VN}) and

TABLE 10.3 Properties of Important Explosives

Explosive	Composition[a]	Heat of Explosion E (cal/g)	Detonation Velocity U_D (m/s)	Density ρ (g/cm^3)	$\sqrt{2E}$ (m/s)	$\gamma = \sqrt{U_D^2/2E + 1}$
EL-506D	PETN/75, Other 25	870	7100	1.40	2700	2.80
Composition B	RDX/60, TNT/40	1240	7840	1.68	3220	2.63
Composition C-2	RDX/79, TNT/5, DNT/12, Other/4	1120[b]	7660	1.57	3050	2.70[c]
Composition C-3	RDX/77, Tetryl/3, TNT/4, DNT/10 MNT/5, NC/1	1100[b]	7630	1.60	3040	2.70[c]
Composition C-4	RDX/91, Nonexplosive plasticizer/9	1230[b]	8040	1.59	3200	2.70[c]
RDX		1280	8180	1.65	3270	2.70
HMX (beta)		1360	9120	1.84	3370	2.89
PETN		1390	8300	1.70	3410	2.63
Tetryl		1100	7850	1.71	3040	2.37
Cyclotol	RDX/75, TNT/25	1230	8000	1.70	3200	2.69
Pentolite	PETN/50, TNT/50	1220	7470	1.66	3200	2.54
TNT		1080	6700	1.56	3000	2.44
Nitroglycerin		1600	7700	1.6	3660	2.33
Nitroguanidine		720	7650	1.55	2680	3.27
Picric acid		1000	7350	1.71	2890	2.73
Ammonium picrate		800	6850	1.55	2590	2.83

Nitrocellulose	N/14.14	1060	7300	1.20	2980	2.65
Low-velocity dynamite (Picatinny Arsenal)	TNT/68	625	4400	0.9	2290	2.17
Detasheet C	PETN/63, Nitroc./8, plast.	990	7200	1.45	2270	2.70
Medium-velocity dynamite (Hercules)	RDX/75, TNT/15	940	6000	1.1	2800	2.36
Minol-2	AN/20, TNT/40, Al/20	1620	5820	1.68	3680	1.87
Torpex	RDX/42, TNT/40, Al/18	1800	7500	1.81	3880	2.18
Tritonol	TNT/80, Al/20	1770	6700	1.72	3950	2.01
DBX	TNT/40, AN/21, RDX/21 Al/18	1700	6600	1.65	3770	2.02
Lead azide		370	4070	2.0	1750	2.53
			4630	3.0	—	2.83
			5180	4.0	—	3.12
Lead styphnate		460	5200	2.9	1950	2.85
Mercury fulminate		430	3500	2.0	1890	2.10
			4250	3.0	—	2.46
			5000	4.0	—	2.83

Source: From Meyers and Murr [10].

Data in columns 2–4 from Ordnance Corp Pamphlet 20-177, "Properties of Explosives of Military Interest."

[a]Abbreviations: N, nitrogen; AN, ammonium nitrate; Al, aluminum.

[b]Based on detonation rate and assumed value of 2.70 for γ.

[c]Taken the same as for cyclotol.

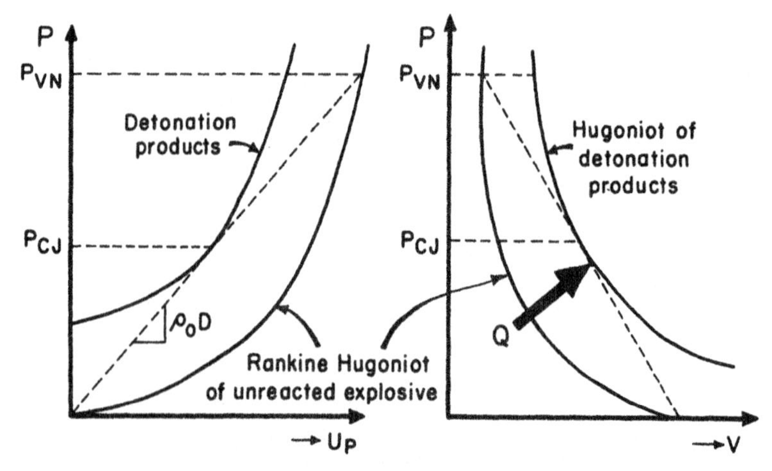

FIGURE 10.7 Determination of Chapman–Jouguet and von Neuman pressures from Rayleigh line in P–V space.

the detonation velocity. It is clear that the von Neumann shock wave travels into a virgin explosive. Figure 10.7 shows the relationship between the reacted and unreacted explosive and between the two states (von Neumann and Chapman–Jouguet). The solid explosive has P–U_p and P–V curves analogous to the ones described in Chapter 4. The slope of the Rayleigh line determines the detonation velocity D. Thus, the tangency line, which defines unequivocally and uniquely the Chapman–Jouguet state, also defines the von Neumann pressure. The von Neumann point is given by the intersection of the Rayleigh line with the Rankine–Hugoniot plot for the unreacted explosive. The curve for the reaction products in Figure 10.7(b) is translated to the right of the curve for the reactants; the amount of translation is given by the reaction heat Q.

Example 10.2. If the P–V curve for the unreacted nitroglycerin and for the detonation products is given, calculate the von Neumann and Chapman-Jouguet pressures and the detonation velocity.

Figure 10.8 shows the P–V plots. We simply take the line starting at V_0 (specific volume for nitroglycerin) and tangent to the curve for the detonation products. It can be seen from Table 10.3 that the density of nitroglycerin is 1.6. Thus, the specific volume is 0.625 cm^3/g. The slope of the line drawn through point V_0 and tangent to the curve for the detonation products is equal to -148 GPa/(cm^3/g). This slope is (see Fig. 4.5) equal to $-(\rho_0 D)^2$:

$$S = -148 \frac{\text{GPa} \times \text{g}}{\text{cm}^3} = (\rho_0 D)^2$$

$$D = \left(148 \times 10^9 \frac{\text{N} \times \text{g}}{\text{m}^2 \times 10^{-6}\,\text{m}^3}\right)^{1/2} \times \frac{\text{m}^3}{1.6 \times 10^3\,\text{kg}}$$

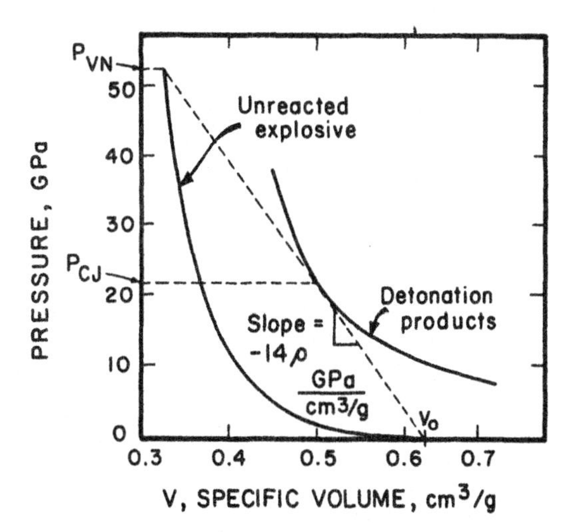

FIGURE 10.8 Pressure-specific volume curves for unreacted and detonated nitroglycerin. (From Johansson and Persson [3], Fig. 1.2.1, p. 22. Reprinted with permission of the publisher.)

$$= \left(148 \times 10^{15} \times 10^{-3} \frac{\text{kg} \times \text{m} \times \text{kg}}{\text{s}^2 \times \text{m}^5}\right)^{1/2} \times \frac{\text{m}^3}{1.6 \times 10^3 \text{ kg}}$$

$$= \frac{12.16 \times 10^6}{1.6 \times 10^3} \frac{\text{kg}}{\text{s m}^2} \times \frac{\text{m}^3}{\text{kg}}$$

$$= 7.6 \times 10^3 \text{ m/s}$$

This value is very close to the one given in Table 10.3 (7700 m/s). The von Neumann and Chapman–Jouguet pressures are read directly from the plot:

$$P_{VN} = 52 \text{ GPa} \qquad P_{CJ} = 21.5 \text{ GPa}$$

The above calculations (in this and the previous sections) apply to *ideal* detonations. The areas of detonation of physics and chemistry are very complex, and it is not the objective of this book to provide the student with an in-depth knowledge. This can be acquired if the diligent student reads the books in the field: Zeldovich and Kompaneets [12], Fickett and Davis [13], and Fickett [14]. However, a brief introduction to the complexities of detonation theory will be given here.

First, detonation velocity varies with the diameter of the charge. If a detonation wave is propagating in a cylindrical explosive, with the direction of propagation parallel to the cylinder axis, its velocity will decrease below a certain threshold diameter. Jaimin Lee [15] incorporated this behavior into an analytical framework. Figure 10.9 exemplifies this behavior. It can be clearly seen that D decreases for radii smaller than 10 cm. The ideal detonation velocity

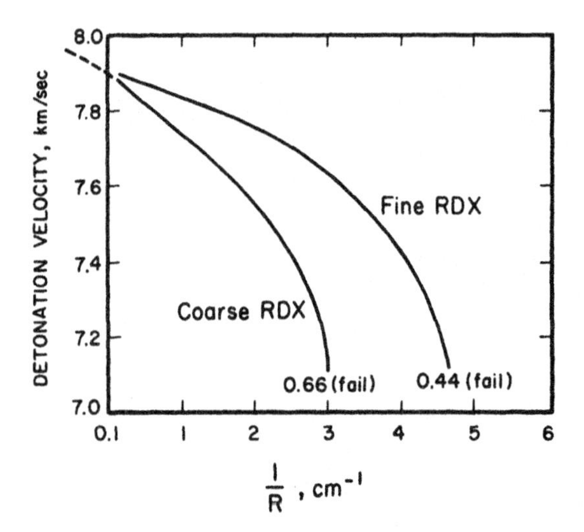

FIGURE 10.9 Variation of detonation velocity with charge diameter for coarse and fine RDX. (From Popolato [1], Fig. 3, p. 22.)

for RDX, given in Table 10.2, is 8180 m/s. This is due to the fact that the detonation gases can escape laterally, and this decreases the overall pressure. Thus, we conclude that the behavior deviates from ideality for small diameters. This behavior is observed for most explosives, as a rule, and there is also a minimum diameter below which detonation does not propagate. This is called the *critical* diameter or *failure* diameter. In Figure 10.9 the failure diameters are 0.66 and 0.44 cm for the coarse and fine RDX, respectively.

Second, the detonation velocity also varies with the density of the explosive. For instance, powder TNT has a lower detonation velocity than solid TNT. This is relatively simple to understand, since the time required for detonation to propagate from one "hot spot" to another is larger. Figure 10.10 illustrates the variation in detonation velocity with density for TNT and RDX. If one assumes that the detonation process is still ideal (Chapman–Jouguet) at the lower densities, one can use Eqns. (10.19) and (10.20) to determine the $P_{CJ} - (U_p)_{CJ}$ state for an explosive having a density lower than the full density; one just uses the density and detonation velocities for the actual porous (powders, flakes, etc.) explosive.

Example 10.3. Determine the Chapman–Jouguet pressure and particle velocity for solid (cast) TNT and powder TNT with a density of 1 g/cm^3. Assume ideal detonation.

From Table 10.3, one has, for TNT,

$$\rho_0 = 1.56 \text{ g/cm}^3$$

$$\gamma = 2.44$$

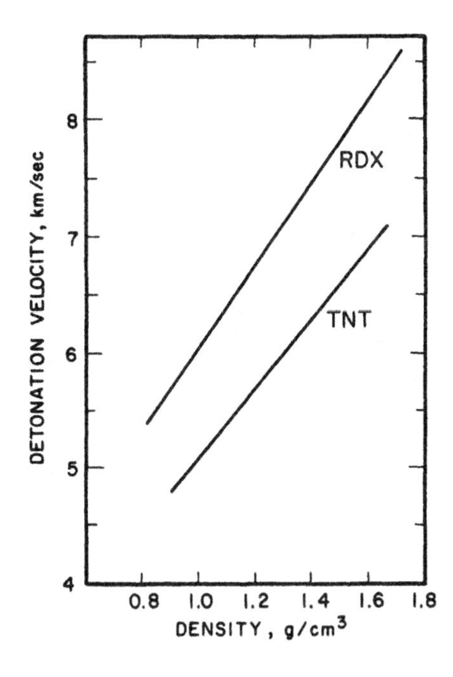

FIGURE 10.10 Variation of detonation velocity with density for TNT and RDX. (From data of Urizar et al. [19].

Assuming a polytropic EOS for the detonation products, using Eqns. (10.19) and (10.20) and the detonation velocities from Figure 10.10 yields

$$D_1 = 6.750 \text{ km/s} \qquad P = \frac{\rho_0 D^2}{\gamma + 1}$$

$$D_2 = 5.10 \text{ km/s} \qquad (U_p)_{CJ} = \frac{p_{CJ}}{\rho_0 D} = \frac{\rho_0 D^2}{\gamma + 1} \cdot \frac{1}{\rho_0 D} = \frac{D}{\gamma + 1}$$

For $\rho = 1$,

$$P_{CJ} = \frac{1 \times 10^3 \times 5.1^2 \times 10^6}{3.44} = 7.56 \text{ GPa}$$

$$(U_p)_{CJ} = \frac{D}{\gamma + 1} = \frac{5.1 \times 10^3}{3.44} = 1.48 \times 10^3 \text{ m/s}$$

For $\rho = 1.56$,

$$P_{CJ} = \frac{1.56 \times 10^3 \times 6.75^2 \times 10^6}{3.44} = 20.6 \text{ GPa}$$

$$(U_p)_{CJ} = \frac{P_{CJ}}{\rho_0 D} = \frac{D}{\gamma + 1} = \frac{6.75 \times 10^3}{3.44} = 1.96 \times 10^3 \text{ m/s}$$

So:

For Solid TNT ($\rho = 1.56$)	For Powder TNT ($\rho = 1$)
$P_{CJ} = 20.6$ GPa	$P_{CJ} = 7.56$ GPa
$U_{pCJ} = 1.96$ km/s	$U_p = 1.48$ km/s

It should be noticed that we used the same γ for both cases. The polytropic gas constant is a function of gas composition but should not significantly depend on initial explosive density. The above calculations illustrate the strong effect of density on the Chapman–Jouguet parameters. The pressure decreased by a factor of 3 with the decrease in density from 1.56 to 1.

It can be seen that *variation in density* and the incorporation of inert additives are ways in which we can *adjust the detonation velocity* of an explosive.

10.5 EXPLOSIVE–MATERIAL INTERACTION

We saw in Chapter 9 how a detonating explosive transfers its energy to a material in contact with it. A series of shock and release waves propagate through the material as it accelerates discontinuously to its steady-state velocity. Figure 9.1 shows qualitatively the sequence of events leading to the acceleration of a plate to its steady-state velocity V_p (shown in Fig. 7.2). We will, in this section, learn to calculate the pressure generated in a material by an explosive detonating in contact with it. For all interests, we neglect the von Neumann pressure spike, because the thickness of the reaction region is very small and the high-pressure "spike" attenuates very readily in the material. Of importance is the Chapman–Jouguet pressure and the release in pressure behind it. We use exactly the same procedure of "impedance matching" described in Chapters 4 (Section 4.3 and 7 (Section 7.3). We have pressure–particle velocity curves for the explosive and for the material, shown in Figure 10.11. We have a pressure in the explosive equal to the Chapman–Jouguet pressure. At the explosive–material interface we invert the pressure–particle velocity curve for the explosive (curve R in Fig. 10.11).

This curve intersects the curve for the material at P_1, U_{p1}. These are the pressure and particle velocity generated in the material. Figure 10.12 schematically shows the sequence. A shock wave is sent into the explosive gases by the higher pressure P_1. As the pressure in the explosive drops, so does the pressure in the material. Thus, a triangular (or sawtooth) shaped pressure pulse is generated in the material. The peak pressure of this sawtooth-shaped pulse starts attenuating immediately. Thus, the pressures are marked P_2 and P_1.

One can also encounter the situation where the Chapman–Jouguet pressure for the explosive is at the left of the material curve in pressure–particle velocity space. Such a situation is shown in Figure 10.13. In this case, a release wave

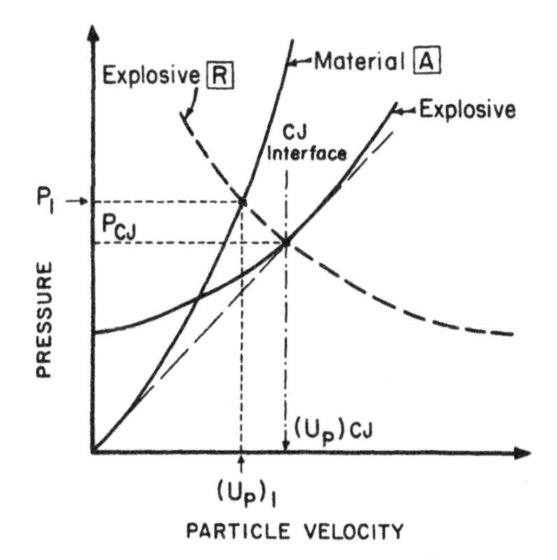

FIGURE 10.11 Impedance matching technique applied to an explosive transferring a shock wave to a material A.

is sent into the explosive from the interface. This happens for materials with very low impedance, such as water and styrofoam.

It is interesting to note that this impedance matching technique is actually used to determine the Hugoniot curves for explosive. This is described by Davis [16]; the technique is due to Deal, at LANL. One uses materials with

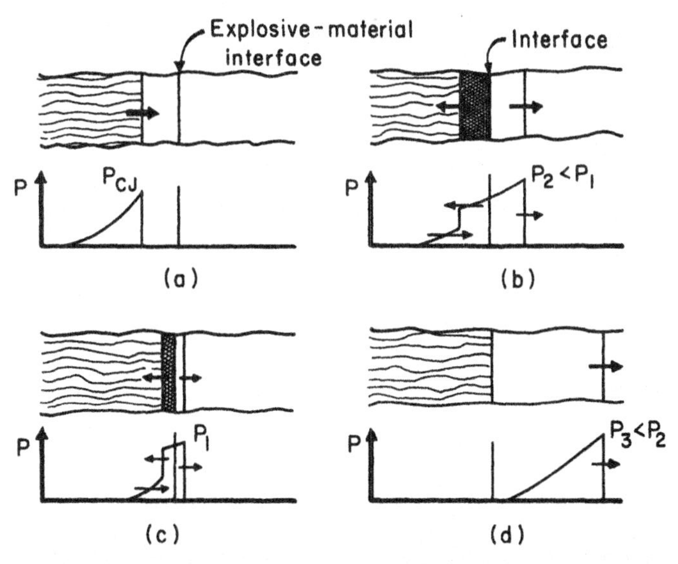

FIGURE 10.12 Sequence of events for situation depicted in Fig. 10.11; pressure profiles below each sketch.

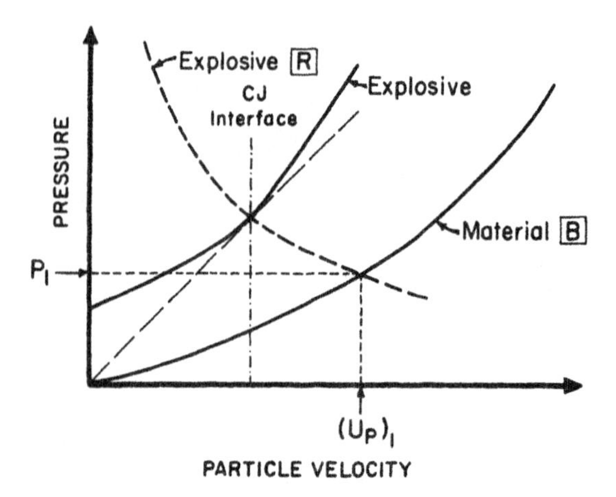

FIGURE 10.13 Impedance matching technique for case where pressure in material (P_{I}) is lower than Chapman–Jouguet pressure.

known EOS and measures the free-surface velocities at the back of the material. Since the pressure P attenuates with distance, as shown in Figure 10.12, by using successively thinner material, one should obtain higher and higher free-surface velocities. Figure 10.14(a) shows a series of experiments for 2024 aluminum alloy, which has a known EOS. The explosive composition B was placed in direct contact with the aluminum alloy and different experiments with different thicknesses were conducted. The free-surface velocity, corresponding to zero thickness is 3.4 mm/μs. The particle velocity can be approximated as (see Chapter 4)

$$U_{\mathrm{p}} = \tfrac{1}{2}U_{\mathrm{fs}} = \tfrac{1}{2}(3.4) = 1.7 \text{ mm}/\mu\text{s}$$

The EOS for 2024 aluminum is [16]

$$U_{\mathrm{s}} = 5.328 + 1.338U_{\mathrm{p}}$$

The initial density is $\rho_0 = 2.785$. From these values one obtains the pressure, by using Eqn. (4.5) (conservation of momentum),

$$\begin{aligned} P &= \rho_0 U_{\mathrm{s}} U_{\mathrm{p}} \\ &= 2785 \times 10^3 \times 1.7 \times 10^3 \times (5.328 \times 10^3 + 1.338 \times 1.7 \times 10^3) \\ &= 36 \times 10^9 \text{ Pa} \end{aligned}$$

By repeating this procedure, we obtain P–U_{p} pairs. Figure 10.14(b) shows several of these points for uranium, brass, magnesium, water, and stafoam A and B. In order to obtain the Chapman–Jouguet point, one should experimen-

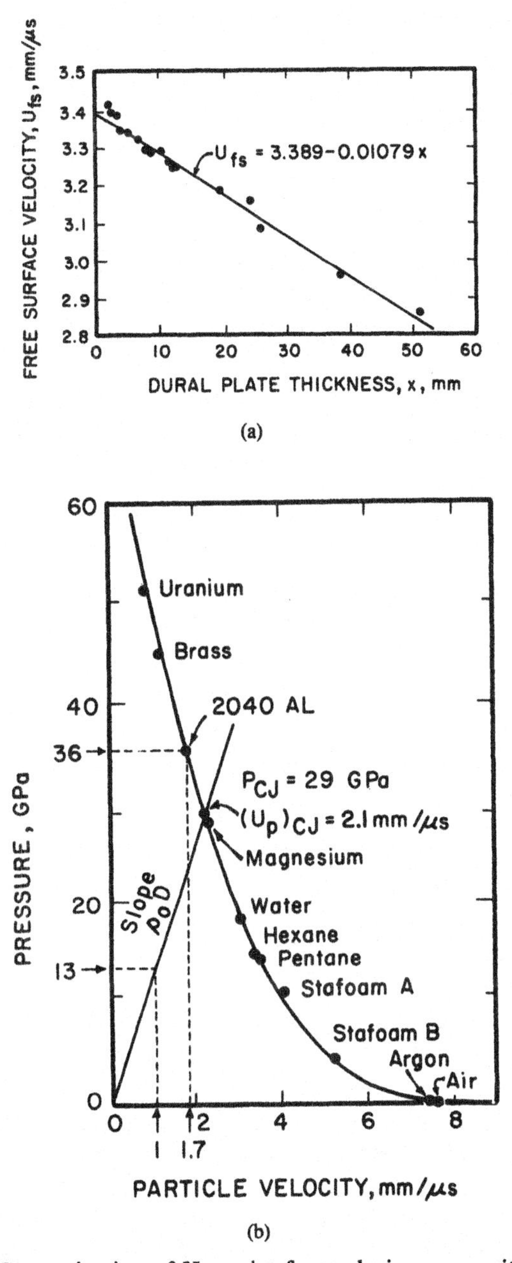

FIGURE 10.14 Determination of Hugoniot for explosive composition B. (a) Experimentally determined free-surface velocities in 2024 aluminum of various thicknesses when composition B is detonated in contact with it. (b) Pressure–particle velocity points obtained by procedure (a) and known EOS for different materials. (From Davis [16], Figs. 10 and 11, pp. 57, 58. Reprinted courtesy of Los Alamos National Laboratory.)

tally determine the detonation velocity D and establish the Rayleigh line with slope $\rho_0 D$.

For composition B one has, from Table 10.3,

$$D = 7840 \text{ m/s}$$

$$\rho_0 = 1.68 \text{ g/cm}^3$$

$$\rho_0 D = 13{,}171{,}200 \frac{\text{kg}}{\text{m}^2 \text{ s}} = 13{,}171{,}200 \frac{\text{kg} \cdot \text{m}}{\text{m}^2 \text{ s}^2} \times \frac{\text{s}}{\text{m}}$$

$$= 13 \times 10^9 \text{ Pa/(m/}\mu\text{s)}$$

In Figure 10.14(b) the numbers 13 and 1 delineate the slope. One obtains

$$P_{\text{CJ}} = 29 \text{ GPa}$$

$$(U_{\text{p}})_{\text{CJ}} = 2.1 \text{ mm/}\mu\text{s}$$

We will, as an exercise, calculate how these values compare with values determined from the polytropic EOS [Eqns. (10.19) and (10.20)] and data from Table 10.3.

One has

$$P_{\text{CJ}} = \frac{\rho_0 D^2}{\gamma + 1} = \frac{1.68 \times 10^3 \times 7840^2}{2.63 + 1} = 28.5 \text{ GPa}$$

$$(U_{\text{p}})_{\text{CJ}} = \frac{D}{\gamma + 1} = \frac{7840}{2.63 + 1} = 2.16 \text{ mm/}\mu\text{s}$$

These values are indeed very close.

10.6 PROPAGATION OF DETONATION IN SOLID AND LIQUID EXPLOSIVES

Detonation involves a chemical reaction producing gases. The explosive contains both an oxidizer and a reducing agent, either intimately mixed or built into the structure. A classic example is a mixture of ammonium nitrate and fuel oil, where ammonium nitrate is the oxidizer and fuel oil is the reducing agent. Thus, oxygen from the air is not required to fuel the reaction.

The reaction that produces the gaseous products and releases energy is responsible for the formation of the detonation front. In liquid and single-crystal solid explosives the microstructure is homogeneous and one has to envisage a uniform reaction zone behind the front. However, this is an oversimplified picture, to which most solid explosives do not adhere. It is well known that the propagation of detonation in most solid polycrystalline explosives involves the formation of "hot spots" [17]. These are small areas at the shock front where the temperature is much higher than the mean bulk temperature under shock conditions. Reactions follow the well-known Arrhenius relationship, and the presence of these hot spots is extremely important to provide rapid decomposition and pressure buildup. The microstructural phenomena that generate high temperatures have been classified by Davis into the categories given in Figure 10.15. High temperatures can be generated between particles by (a) jetting, (b) the collapse of voids (impact of inner walls of void), (c) the heating due to plastic deformation around a void [Fig. 10.15(c)], (d) shock interaction (reinforcement) around particles with high impedance, (e) frictional heating between particles, and (f) adiabatic shear bands. These bands are regions of

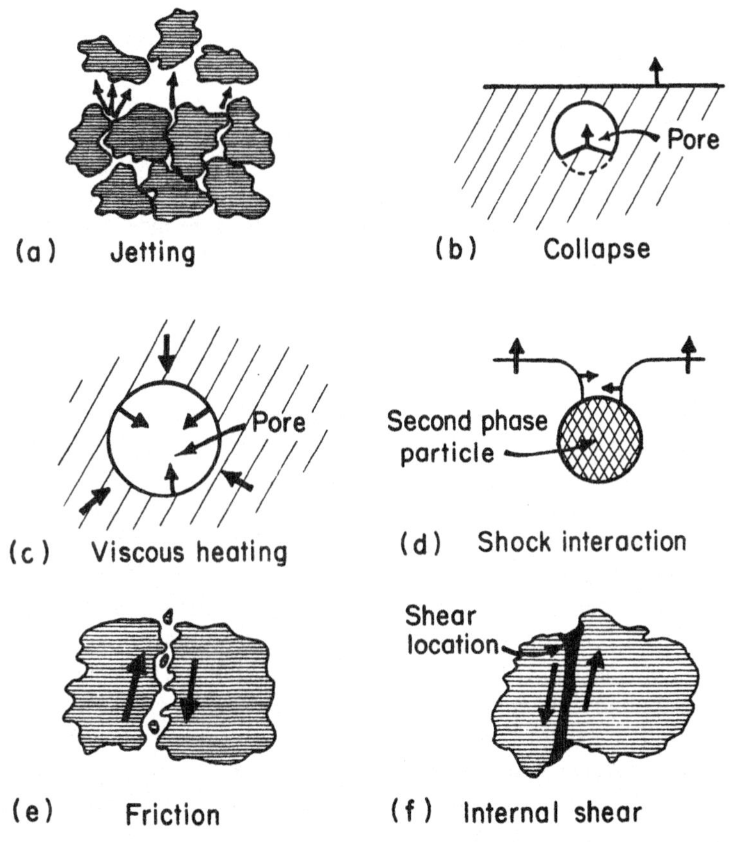

FIGURE 10.15 Six ways in which energy in explosives can be concentrated in localized regions (hot spots): (a) material jetting between explosive grains; (b) viscous heating from void collapse; (c) friction between particles; (d) void collapse; (e) shock wave interaction around high-impedance region; (f) shear instabilities inside individual particles. (From Davis [16], Fig. 25, p. 70. Reprinted courtesy of Los Alamos National Laboratory.)

intense plastic deformation where the heat generated elevates the temperature and softens the material, leading to a decrease in strength.

The importance of hot spots can be clearly assessed by removing their sources. For instance, highly pressed TNT, in which all the voids have been eliminated, is much more difficult to detonate than porous TNT. Similarly, the addition of microballoons (small, hollow glass spheres) to emulsion explosives makes it possible for them to develop detonation waves. Shells containing explosives with large voids sometimes detonate during the acceleration in the gun barrel because of localized heating at the voids. Thus, nondestructive testing by penetration X-rays is required. Explosives that have been excessively pressed (dead-pressed) to the point where all voids have been eliminated become very insensitive and cannot be detonated. Sometimes the accidental passage of a stress wave through an explosive will collapse all its voids, desensitizing it.

10.7 THERMAL AND SHOCK INITIATION OF EXPLOSIVES

The initiation of detonation is usually carried out by passing a shock wave through an explosive. If the shock wave has an amplitude higher than the Chapman–Jouguet pressure, initiation automatically takes place. This initiation is done by a "booster" charge that is in turn initiated by a primary explosive. Thus, the primary explosive detonates, producing a shock wave that initiates the booster. The booster, in turn, detonates, producing initiation in the main charge.

Popolato [1] studied the effect of the pressure and duration shock of pulses on their ability to initiate an explosive when the pressure is below the Chapman–Jouguet pressure. He observed that the required pressure decreased as the duration of the shock pulse increased. This is shown in Figure 10.16; these plots are known as POP plots, after Popolato. The ordinate axis shows the distance to detonation. This is the distance that the shock wave will travel before detonation sets in. The duration of the shock pulse should be such that the distance to detonation is reached at the required pressure.

Figure 10.16 also shows that the explosive PETN of lower density (1 g/cm^3) requires significantly lower pressures to detonate than the one of higher density (1.7 g/cm^3). The explanation resides on the greater presence of potential hot spots for the explosive with lower density as well as the higher shock temperature reached in the porous material (see, e.g., Chapter 5, Section 5.4). The plot of Figure 10.16 also shows that PBX 9404 is less sensitive than PETN.

Thermal initiation is a complex process that we usually want to avoid. Examples of accidental detonation involving thermal initiation include dropping the explosive, localized rubbing, and so on. Very small localized regions that heat up rapidly can give rise to detonation if conditions are right. Thermal decomposition can give rise to a buildup of pressure if the heat transfer is not

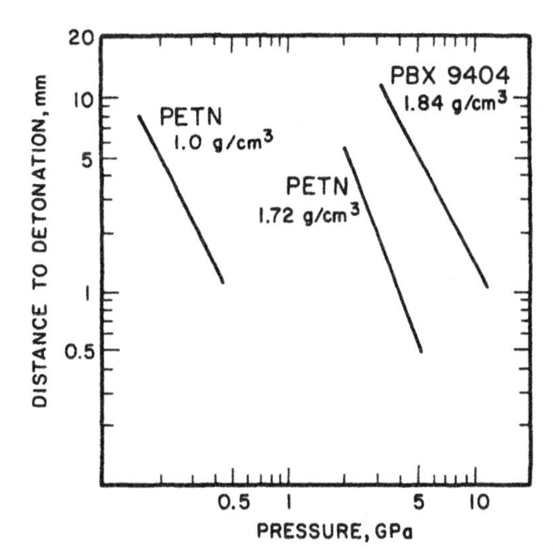

FIGURE 10.16 Effect of pressure on distance to initiation in explosives (POP plot). (From Popolato [1], Fig. 4, p. 23.)

sufficiently rapid. Frank-Komanetsky [18] successfully calculated critical temperatures for explosion.

Primary explosives are initiated by rapid heating of a wire. For the uninitiated, primary explosives are very sensitive explosives used to initiate the detonation in a detonator. The detonator produces a "point initiation," from which detonation propagates either directly into the main charge or into a booster and then into the main charge. This sequence is called explosive train. Explosives of decreasing sensitivity and increasing output (energetic) are sequenced. Hot-wire detonators are the common old-fashioned detonators that can be initiated by a relatively low current. These conventional detonators are composed of three layers, as shown in Figure 10.17. The resistor heats up and ignites the ignitor charge (lead styphnate), which initiates the primary explosive (lead azide or mercury fulminate), which then initiates a small amount of PETN. These old-fashioned electric detonators are very sensitive to static electrical charges in the air and can be easily detonated by the heat of a match. This author once experienced detonation of one of these and was one of the lucky ones in that he did not lose his fingers; he escaped with profuse lacerations to his fingers and face. The power of a detonator is sufficient to cause substantial damage.

Electric detonators are being replaced by EBW (electric bridge wire) detonators, shown in Figure 10.17(b). These detonators require the vaporization of a wire by means of the discharge of a capacitor. The explosive train does not include any primary explosive that can be thermally initiated. The vaporization of the bridge wire produces a shock wave in the secondary explosive, initiating it. Thus, static electricity and electric currents induced in the circuits cannot initiate these detonators, and they are therefore much safer.

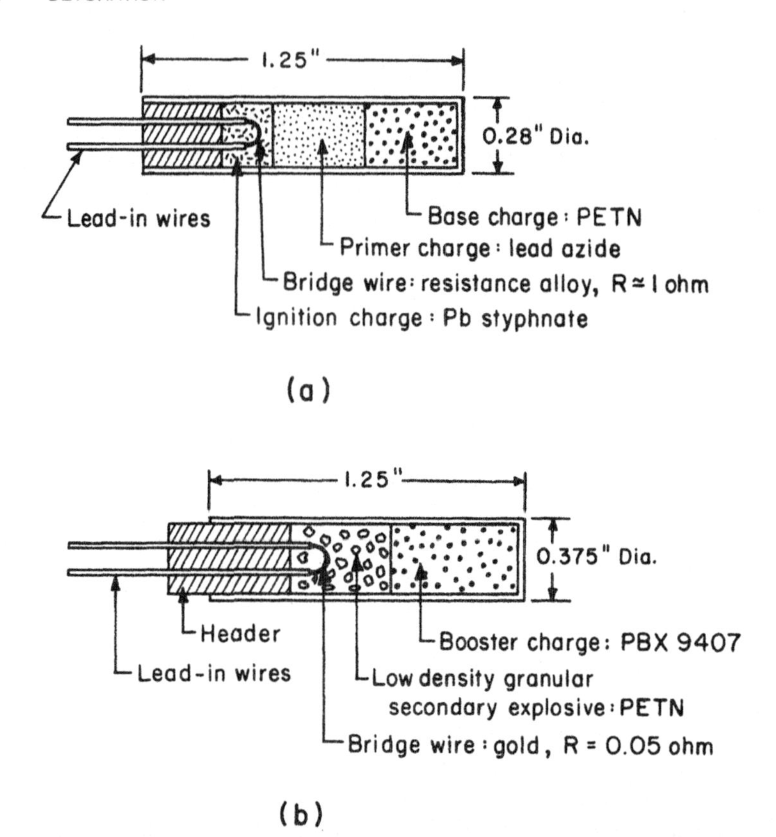

FIGURE 10.17 Longitudinal sections of (a) hot-wire and (b) EBW detonators. (Courtesy of N. N. Thadhani, Georgia Institute of Technology.)

The sensitivity of explosives to thermal, friction, and impact initiation is studied in standardized laboratory tests. Four common tests are:

Drop Weight Test. A weight is dropped from increasing heights on a small explosive sample placed on sand paper.

Skid Test. Explosive is dropped on a ramp covered with high-friction material such as sand paper.

Susan (the author's ex-wife) Test. Explosive is placed in front of a projectile.

Aquarium Test. Explosive is heated at a constant rate in liquid bath until explosion occurs.

The burning-to-deflagration-to-detonation sequence is not too well understood, but this author experienced its effects in Japan; while a visiting scholar at a research institute, 2.2 kg of RDX was placed in a furnace by one of the researchers. The heat controller of the furnace malfunctioned. The furnace was strategically placed in the office building. When the temperature reached a critical level, the entire building was rattled by a powerful detonation. Meyers vividly remembers the tiles that flew off the bathroom walls. (Could it be a spalling phenomenon?) This room adjoined the room where the furnance exploded. A steel door in the hallway flew off 20 yards. Thus, we leave the following reminder: leave heat treatments to metals.

Example 10.4. TNT explosive in powdered form is used to drive a metal plate (titanium). The TNT density is 0.8 and the detonation velocity was measured to be 4.2 km/s. Determine the pressure generated in titanium.

We will use Figure 10.18, which is a compilation of pressure–particle velocity data for a number of explosives and materials. Obviously the TNT used does not fit the TNT curve in the plot, which applies to ideal detonation. The point on the explosive curves marks the C–J state.

For TNT, it is

$$P_{CJ} = 19.5 \text{ GPa}$$

$$(U_p)_{CJ} = 1.65 \text{ mm}/\mu\text{s}$$

In the slowly detonating explosive both P_{CJ} and $(U_p)_{CJ}$ will be different, and we have to calculate these corrected values. Using Eqns. (10.19) and (10.20)

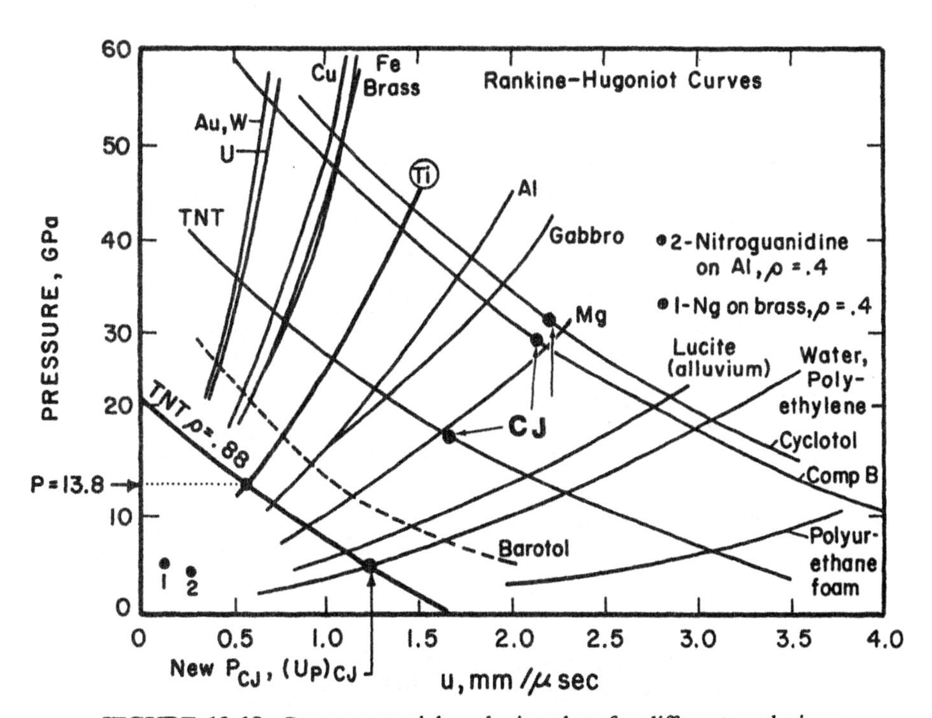

FIGURE 10.18 Pressure–particle velocity plots for different explosives.

and assuming that the polytropic gas constant is independent of detonation velocity ($\gamma = 2.33$ in Table 10.3), we have

$$P_{CJ} = (0.8 \times 10^3 \times 4.2^2 \times 10^6)/3.33 = 4.2 \text{ GPa}$$

$$(U_p)_{CJ} = \frac{D}{\gamma + 1} = \frac{4.2}{2.33 + 1} = 1.27 \times 10^3 \text{ mm}/\mu\text{s}$$

We simply draw a curve parallel to the ideal detonation, full-density TNT curve and passing through the $(P, U_p)_{CJ}$ new point. It should be noted that all P–U_p curves for the explosives are approximately parallel and that drawing these "parallel" curves is fairly accurate. However, more exact answers can be obtained through the MY1DL code, which handles explosives:

$$P = 15.8 \text{ GPa}$$

REFERENCES

1. A. Popolato, in *Behavior and Utilization of Explosives in Engineering Design*, ed. R. L. Henderson, ASME Albuquerque, New Mexico 1972.
2. R. E. Duff and E. Houston, *J. Chem. Phys.*, **23** (1955), 1268.
3. C. H. Johansson and P. A. Persson, *Detonics of High Explosives*, Academic, New York, 1970.
4. M. A. Cook, *The Science of High Explosives*, Reinhold, New York, 1958.
5. J. Taylor, *Detonation in Condensed Explosives*, Clarendon, Oxford, 1952.
6. G. B. Kistiakowski and E. B. Wilson, "The Hydrodynamic Theory of Detonation and Shock Waves," Report No. 114, OSRD, 1941.
7. C. L. Mader, "Detonation Properties of Condensed Explosives Using the BKW Equation of State," Report No. LA-2900, Los Alamos Scientific Laboratory, 1963.
8. D. L. Chapman, *Lond. Edinb. Dubl. Phil. Mag.*, **47** (1899), 90.
9. E. Jouguet, *J. Math. Pure Appl.*, **60** (1905), 347; **61** (1906), 1.
10. M. A. Meyers and L. E. Murr, eds., *Shock Waves and High-Strain-Rate Phenomena in Metals*, Plenum, New York, 1981, Appendix A, p. 1033.
11. T. von Neumann, "Progress Report on the Theory of Detonation Waves," Report No. 549, OSRD, 1942.
12. Y. B. Zeldovich and S. A. Kompaneets, *Theory of Detonations*, Academic, New York, 1970.
13. W. Fickett and W. C. Davis, *Detonation*, University of California Press, 1979.
14. W. Fickett, *Introduction to Detonation*, University of California Press, 1987.
15. Jaimin Lee, Ph.D. Thesis, New Mexico Tech., 1989.
16. W. C. Davis, *Los Alamos Sci.*, **2** (1) (1981), 48.
17. F. P. Bowden, *Proc. Roy. Soc. Lond.*, **246** (1958), 146.
18. D. A. Frank-Komanetski, *Acta Phys. Chim. USSR*, **10** (1939), 365.
19. M. J. Urizar, E. James, and L. C. Smith, *Phys. Fluids*, **4** (2) (1961).

CHAPTER 11

Experimental Techniques: Diagnostic Tools

11.1 INTRODUCTION

Diagnostics are extremely important in high-strain-rate and shock propagation events. They provide the essential dynamic parameters of materials (EOS, strain rate dependence of flow stress, etc.) that can be compared to theoretically calculated parameters. These parameters, which describe the dynamic response of materials, are then used as input in the solution of more complex problems that closely parallel actual real-life situations. These diagnostic tools are an integral part of many high-strain-rate experiments described in Chapter 12.

The range of instruments, gages, and techniques is continuously increasing. We will describe in this chapter the principal tools available to the modern researcher. More details on these techniques can be found in specialized papers published in the literature. The most important sources of information on which this chapter is based are the review articles by Graham and Asay [1], Chhabildas and Graham [2], Chhabildas [3], and Cagnoux et al. [4]. The student should read these papers, which contain references to the specific techniques. Graham and Asay [1] classified detectors as follows:

1. Wave or particle arrival time: In these detectors, the arrival of a signal (wave or moving object) triggers a pulse. Several of these detectors sequentially spaced allow the determination of velocity. Electrically charged pins, flash gaps, and fiber-optic charged pins are the prime example.
2. Discrete particle displacement versus time.
3. Continuous particle displacement versus time: Laser interferometry in the displacement mode is one such device.
4. Stress versus time: Piezoresistive (carbon, manganin, ytterbium) and piezoelectric [X-cut quartz, lithium niobate, the polymer-gage PVDF (polyvinylidene fluoride)] detectors record directly voltage or current, which can be converted to stress.

In addition to the classification by Graham and Asay [1], we can add high-speed cinematrography, flash X-rays, holography, and Doppler radar. Graham and Asay [1] point out that in shock wave propagation events, mechanical devices cannot be used because of the rapidity of events. Thus, electromagnetic and optical devices are preferred. In the following sections, the various detectors will be presented in some detail.

11.2 ARRIVAL TIME DETECTORS

Electrically charged pins and flash gaps have been used for a number of years and are the earliest devices. They record the arrival time of an object (its free surface). There are numerous designs for charged pins, but a simple design is shown in Figure 11.1. Chapter 5 presents a description of arrival time detectors used to determine the shock wave velocity (Fig. 5.1). Pins are set at different spacings from the target plate (Fig. 11.1). These pins are charged so that when contact is established, a small curent ensues that is recorded in a high-speed oscilloscope (or high-speed counter); when the flyer plate is conducting, single pins can be used. When the flyer plate is an insulator, a design similar to the one shown on the right-hand side should be used. These are coaxial pins, and electrical contact is established by the collapse of the central wire, which protrudes from a small metallic tube. A difference of voltage is established between the tube and the central wire by the charging of a capacitor. The collapse of the wire discharges the capacitor. A possible recording is shown in the plot. If we know the distance between pins (d_1 and d_2) and the time intervals (t_1 and t_2), we readily obtain V, the plate velocity.

Since the traveling flyer plate compresses the air ahead of it, a region of high pressure might precede the plate. The air can become ionized and con-

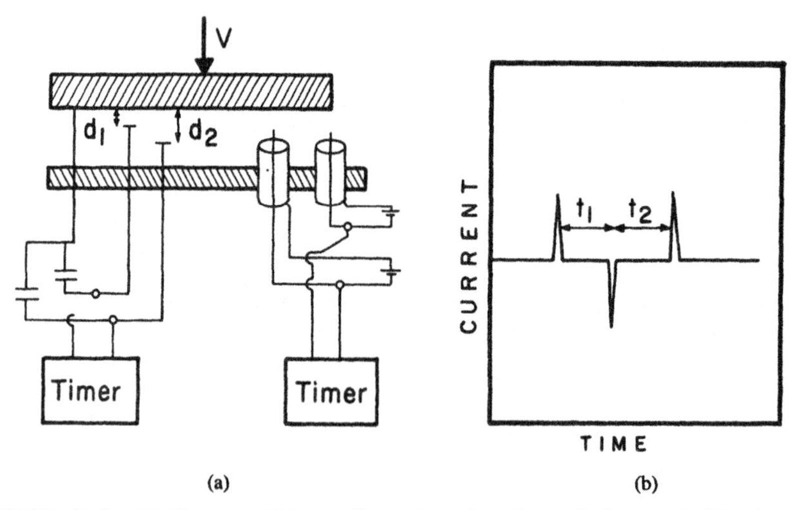

FIGURE 11.1 (a) Two possible configurations for charged pins; and (b) schematic showing recorded signals from three pins.

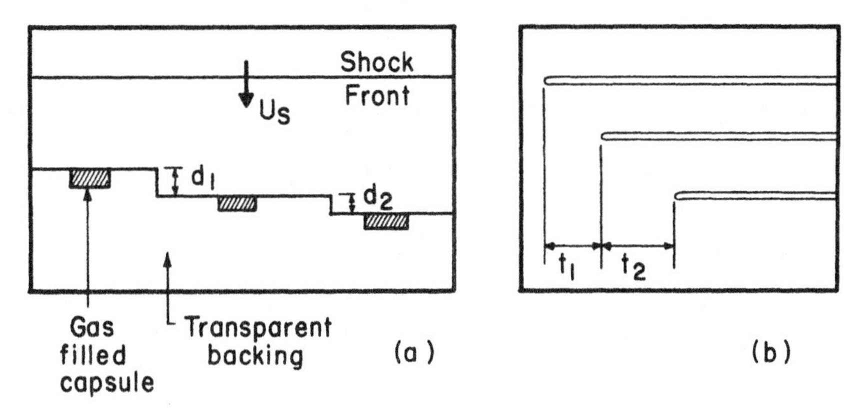

FIGURE 11.2 (a) Flash gap device; and (b) schematic recording from a streak camera.

ductive, leading to an early "shorting" of the pins. This is undesirable, and for this reason it is recommended to fill the gap between the target and flyer plate with a nonionizing gas (such as methane) or to evacuate this region, if possible.

The flash-gap technique uses a small container with a gas that ionizes when subjected to the compression resulting from the impact by the free surface. These flashes are recorded in a high-speed streak or framing camera. Figure 11.2 shows the back face of a material whose shock wave velocity we want to establish. It is stepped, with spacings d_1 and d_2. The gas will ionize when the shock wave reaches it. By this method it is possible to monitor the progression of the shock front and, consequently, to determine the shock wave velocity.

A more modern version of the flash-gap technique is the optical pin. In this technique a glass microballoon (filled with a gas that emits a flash upon being impacted) is placed at the extremity of an optical fiber. Figure 11.3 shows the schematic diagram. Glass microballoons are very small hollow glass spheres (less than 1 mm diameter) that can be produced with different gas fills. Good gases for flash are argon and xenon. The optical signal is carried by the optical fiber and can be processed in a photomultiplier or recorded in a streak camera. The advantage of these fiber-optic pins over flash gap is that they are very small.

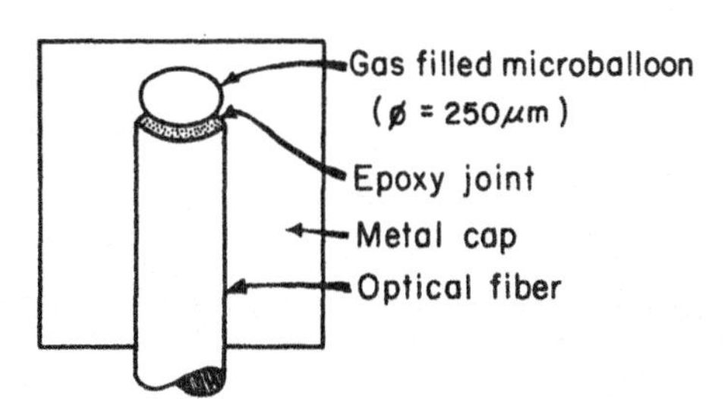

FIGURE 11.3 Fiber-optical pin. (Reprinted from *Int. J. Impact Eng.*, vol. 5, L. E. Chhabildas, p. 205, Copyright 1987, with permission from Pergamon Press Ltd.)

Example 11.1. Calculate the velocity of a flyer plate impacting a system containing capsules with powders to be shock processed (see Fig. 17.38) if the different pins, positioned at preestablished distances, give the readings shown below.

We measure the time between different pin readings. The polarity of pins 2 and 4 was inverted to make the identification of the pins easier:

$$t_{1\text{-}2} = 0.65\ \mu\text{s}$$

$$t_{2\text{-}3} = 0.52\ \mu\text{s}$$

$$t_{3\text{-}4} = 0.69\ \mu\text{s}$$

$$t_{4\text{-}5} = 0.65\ \mu\text{s}$$

Since the pins are positioned 1 mm apart (see the illustration below), the velocities are

$$v_{1\text{-}2} = 1.54\ \text{km/s}$$

$$v_{2\text{-}3} = 1.92\ \text{km/s}$$

$$v_{3\text{-}4} = 1.45\ \text{km/s}$$

$$v_{4\text{-}5} = 1.54\ \text{km/s}$$

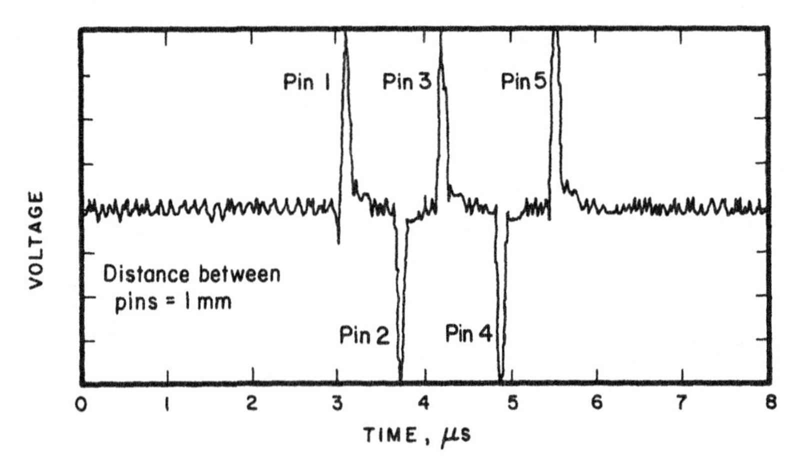

EXAMPLE 11.1 Pin signals for flyer plate impact geometry; pins are placed 1 mm apart, starting at the target surface (Courtesy of N. N. Thadhani, Georgia Institute of Technology.)

11.3 LASER INTERFEROMETRY

Laser interferometry has been an extremely successful technique; it was developed in the 1960s. In 1965 Barker and Hollenbach [5] described the first displacement laser interferometer. The more sophisticated VISAR (Velocity Interferometer System for Any Reflector) was developed by Barker and Hollenbach [6] in 1972; the latter is a velocity interferometer.

Laser interferometry is based on interference fringes that appear when different laser beams interact. Laser is a highly coherent (phased-in) monochromatic light beam. If two beams either are offset (having the same wavelength) or have slightly different wavelengths, interference patterns will occur. We will describe two types of interference. The wavelength of lasers is determined by the source and is of the order of 500 nm (0.5 μm). The displacement interferometer, which uses the Michelson interferometer principle, is explained in Figure 11.4(a). When the free surface (reflecting) moves to the right, the reflected beam is displaced. This reflected beam is given by the discontinuous line. The reflected beam is later juxtaposed to a reference beam that remains unchanged. At every displacement of the mirror by λ/x, the crest of the re-

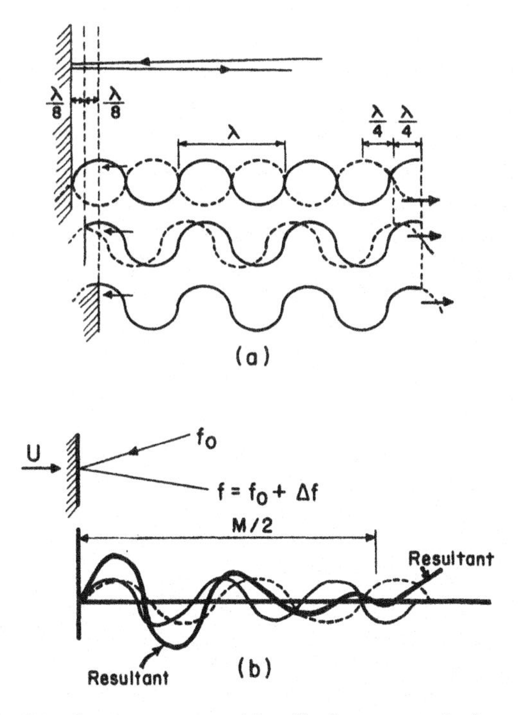

FIGURE 11.4 (a) Interference generated by displacement of mirror; (b) interference generated by Doppler shift.

flected wave is displaced by $2\lambda/x$. When the reflected wave is displaced by $\lambda/2$ (mirror displaced by $\lambda/4$), the reference and reflected beams cancel each other totally, and an interference fringe is generated.

For the velocity interferometer, another principle is used, that of Doppler-shifted frequency. Figure 11.4(b) shows a laser beam with frequency f_0 incident on a reflecting surface that is moving at a velocity U. The frequency of the reflected beam is different from that of the incident beam because of the movement of the mirror. Doppler shifts are readily understood by the classic "train" experiment. A trumpet player is on a train moving at a velocity U with respect to the observer at the train station. The sound will have a different pitch when the train is approaching and distancing itself from the observer. This is because the frequency of the sound is shifted because of the moving source (velocity U). The same phenomenon occurs when the mirror moves to the right in Figure 11.4(b). The frequency of the reflected beam is

$$f' = f_0 \frac{1 + 2U/C}{\sqrt{1 - (U/C)^2}} \tag{11.1}$$

where U is the velocity of the mirror and C is the velocity of light (3×10^5 km/s); f' and f_0 are actually fairly close, and the interference beat (distance at which extinction occurs) is very large.

Next we describe the basic features of the displacement interferometer. Figure 11.5 shows a simple sketch. The laser source emits the beam that is divided (split) at A. One half of the beam is reflected to the back surface of the specimen, whereas the other half goes through the mirror and is then reflected back. This second beam is the reference beam, which remains unchanged during the experiment. The two beams are then brought together and enter a photomultiplier. The displacement of the specimen will cause an interference pattern, and the velocity $U(t)$ of the back surface can be determined from the equation

$$U(t) = \frac{\lambda}{2} \frac{dF}{dt} \tag{11.2}$$

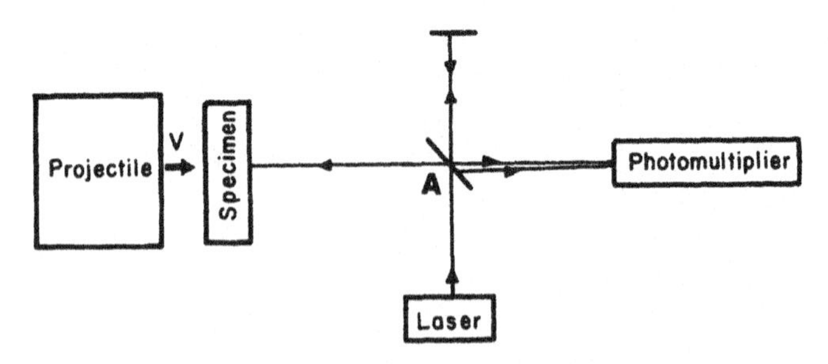

FIGURE 11.5 Displacement interferometer.

where λ is the beam wavelength and dF/dt is the fringe frequency (change in fringe spacing with time). Often a window is used between the back surface of the specimen and the beam. This window avoids jetting and irregular particle deformation at the back surface of the specimen and helps to render the reading an "in-material reading," especially if it is made of the material with shock impedance close to that of the specimen. Common window materials (with their pressure range) are

Polymethylmethacrylate (PMMA)
Fused silica ($\sigma < 8.7$ GPa)
Z-cut sapphire ($\sigma < 30$ GPa)
Lithium fluoride ($\sigma < 160$ GPa)

The maximum stress is dictated by the retention of transparency of the window. The window has to be transparent even after the passage of the stress front in order for the laser to continue monitoring the interface displacement. When a window is used, a correction has to be introduced into Eqn. (11.2) to account for the effect of the change in refractive index:

$$U(t) = \frac{\lambda}{2(1 + \Delta V/V)} \frac{dF}{dt} \tag{11.3}$$

where $\Delta V/V$ is a correction factor accounting for the change of refractive index of the window.

In the displacement interferometer the change in fringe spacing with time provides the velocity. In velocity interferometers the fringe spacing provides the velocity directly. In velocity interferometers one leg of the beam is delayed with respect to the other by a few nanoseconds. The direct and delayed beams will thus have close (but not equal) frequencies, and an interference pattern such as the one shown in Figure 11.4 will be produced. The delay is produced by passing the beam through a glass "etalon" or by a lens system with a delay leg. The beat frequency is given by Eqn. (11.1). Equation (11.4) gives the expression for the velocity:

$$U\left(t - \frac{\tau}{2}\right) = \frac{\lambda}{2\tau} F(t) \tag{11.4}$$

where τ is the optical delay leg time, $U(t - \tau/2)$ is the velocity at time $t - \tau/2$, λ is the wavelength of the laser, and $F(t)$ is the fringe count. Figure 11.6 shows a velocity interferometer with an optical delay leg. If a window is used, the correction term is used, and we have

$$U\left(t - \frac{\tau}{2}\right) = \frac{\lambda}{2\tau} \frac{1}{1 + \Delta V/V} F(t) \tag{11.5}$$

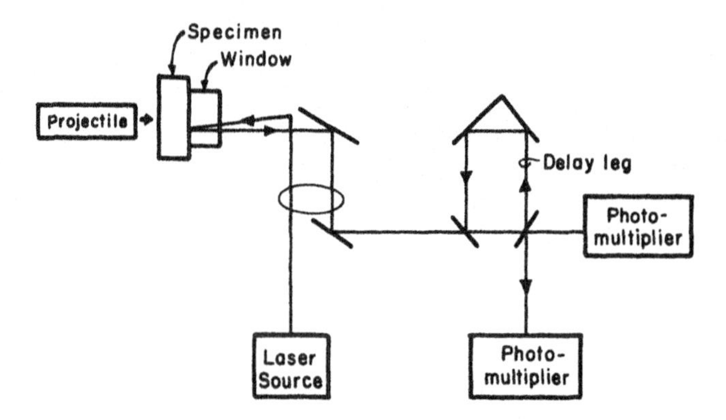

FIGURE 11.6 Velocity interferometer with delay leg.

Two types of velocity interferometers are most commonly used. The VISAR system developed at Sandia and the Fabry–Perot system. The VISAR system has the great advantage that it can be used on irregular surfaces (porous materials, geological materials, composites). It has excellent accuracy and resolution and measures velocities with an accuracy of $\pm 1\%$ with a time resolution of approximately 2 ns. Figure 11.7 shows a VISAR signal from the back surface of an aluminum target impacted by a steel flyer plate at a velocity of 400 m/s. The free-surface velocity can be converted into particle velocity by dividing it by 2. Thus, the peak particle velocity is ~350 m/s. This corresponds to a pressure of 6.1 GPa, close to the one determined by the impedance matching method. The shape of the pulse can be analyzed in terms of spalling taking place at the back surface. If no spalling occurs, the pulse will return to zero, as shown by the dotted line. Spalling produces a "pull-back" signal (full line)

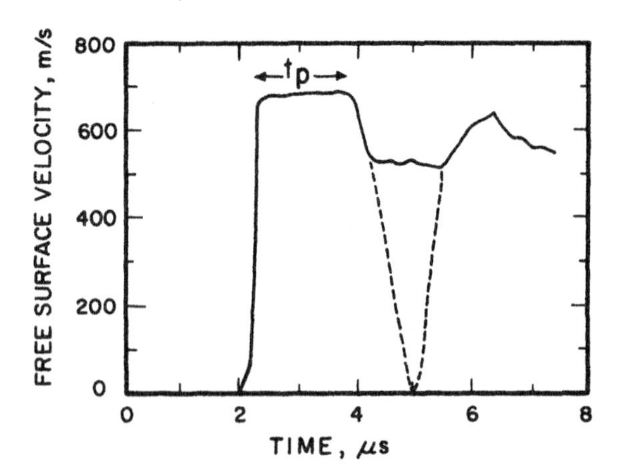

FIGURE 11.7 VISAR signal from steel projectile impacting Al target at 400 m/s. (From Cagnoux et al. [4], Fig. 26, p. 494. Reprinted with permission of the publisher.)

and the difference between the two lines gives the Spall strength (more details in Section 16.8.2). The duration of the pulse is marked t_p and is approximately 1.5 μs. The Fabry–Perot interferometer uses two parallel plates separated by a distance L and that are normal to the beam. The reflection and refraction at these plates cause an interference pattern. The Fabry–Perot interferometer requires a streak camera to record the fringe pattern. The photomultiplier tubes are not suitable. There are also transverse displacement interferometers that are used to monitor transverse displacements produced by shear waves. These interferometers have been extensively used by the Brown University group [7].

11.4 PIEZORESISTIVE GAGES

Piezoresistors are materials whose electrical resistivity changes with pressure. These materials have been used to record pressures in quasi-static experiments. As early as 1911, Bridgman used Manganin coils immersed in pressurized fluids for accurate determination of pressure. Manganin is an alloy with 24 wt % Cu, 12 wt % Mn, and 4 wt % Ni. It displays very important properties: Its electrical resistance is *sensitive* to pressure but *insensitive* to temperature at atmospheric pressure. Thus, the effects of shock heating are minimal. The relationship between the stress σ_x (in the shock propagation direction) and the resistance is given by

$$\sigma_x = \frac{1}{K}\frac{\Delta R}{R_0} \tag{11.6}$$

where R_0 is the initial resistance and K is the piezoresistance coefficient. Figure 11.8 shows the effect of σ_x on the resistivity of this alloy. The resistivity doubles

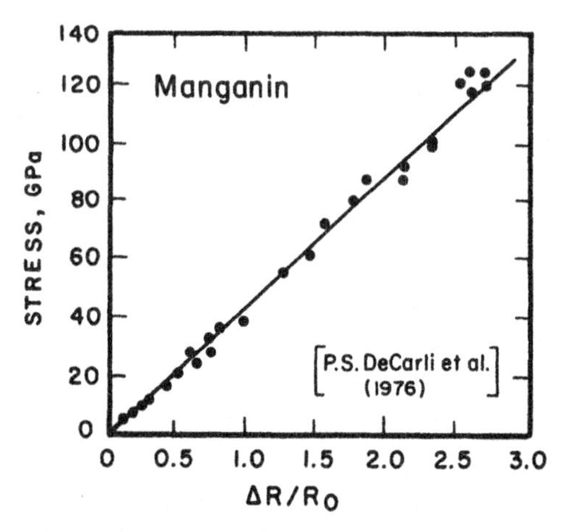

FIGURE 11.8 Fractional change in resistance as a function of shock stress for Manganin. (From DeCarli et al. [8]. Reprinted with permission of the publisher.)

at a stress of 40 GPa ($\Delta R/R_0 = 1$). This gives the following value of K:

$$K = 2.5 \times 10^{-2}\ \text{GPa}^{-1}$$

De Carli et al. [8] recommend the use of the more accurate equations

$$P = (42\ \text{GPa})(\Delta R/R_0) - (120\ \text{GPa})(\Delta R/R_0)^2 \qquad (P < 2\ \text{GPa})$$

$$= (35\ \text{GPa})(\Delta R/R_0) \qquad 2\ \text{GPa} \leq P \leq 15\ \text{GPa}$$

$$= (48\ \text{GPa})(\Delta R/R_0) - 5.6\ \text{GPa} \qquad P \geq 15\ \text{GPa}$$

This value is dependent on the composition of the alloy and the gages should be calibrated. The Manganin gages are successfully used up to stresses of 30 GPa but have been used up to 100 GPa. The time resolution is not as good as with lasers and is of the order of 50–100 ns. Manganin gages are best suited for pressures above 5 GPa. For lower pressures one uses carbon (up to 2 GPa) and ytterbium (up to 4 GPa). These gages have a higher pressure dependence of electrical resistivity, as can be seen in Figure 11.9. The carbon gage deviates from linearity at stresses higher than 2 GPa, and ytterbium undergoes a phase transformation at a slightly higher pressure.

The gages are embedded in the material and are protected from the surrounding by insulating materials, such as Kapton, mica, or Teflon. One has to be careful with the material because some polymers become conductive at high pressures. Manganin gage packages as small as a few millimeters and as large as 2 m have been used, from applications ranging from careful gas gun experiments to underground nuclear tests. Either foil or wire can be used, and a common "dog-bone" design is shown in Figure 11.10. It is made from thin foil that is stamped out in the desired shape. The resistance of gages is ap-

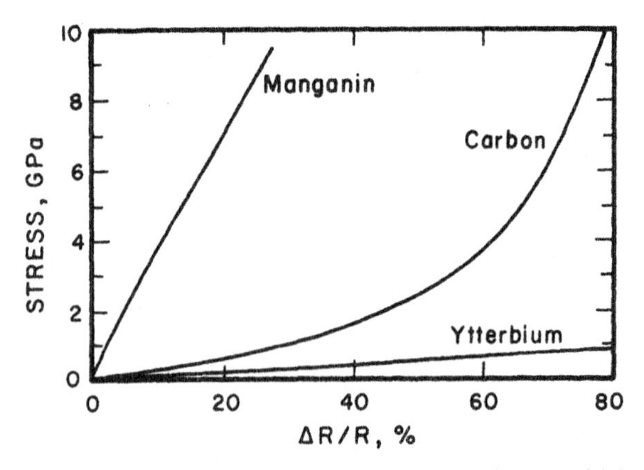

FIGURE 11.9 Comparison of response of carbon, ytterbium, and Manganin gages. (From Chhabildas and Graham [2], p. 11.)

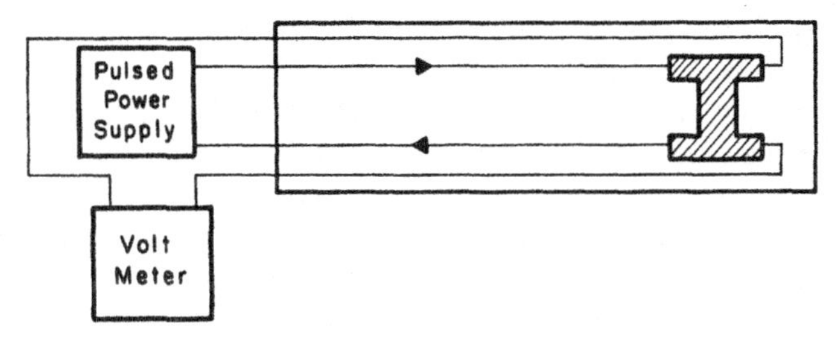

FIGURE 11.10 Schematic of piezoresistive pressure gage.

proximately 50 Ω, although gages with much lower resistance can be used. A pulsed power supply provides the current through the gage prior to the test. The voltage change is monitored across the gage by an oscilloscope. The pulse power supply can only be activated a few milliseconds prior to the passage of the shock wave because resistive heating would prematurely "burn" the gage. By using a high-impedance circuit the change in resistance in the gage due to the shock pressure does not affect the current i and the voltage change across the gage truly represents the change in resistance:

$$\frac{\Delta V}{V_0} = \frac{\Delta R}{R_0} \tag{11.7}$$

An important question is: what pressure does the gage experience? Neglecting secondary effects such as plastic deformation of the gage and lateral stretching, it can be easily seen that the pressure experienced by the gage depends on the nature of the stress pulse. Figure 11.11 shows a gage package (consisting, probably, of two Kapton layers, copper leads, Manganin foil, and epoxy glue) with thickness t_1. The pulse traveling through it has a thickness $\Delta x >> t_1$. The stress sequence experienced by the gage is shown in the P–U_p diagram. The pressure in the bulk of material is P_1. We assume that the shock impedance

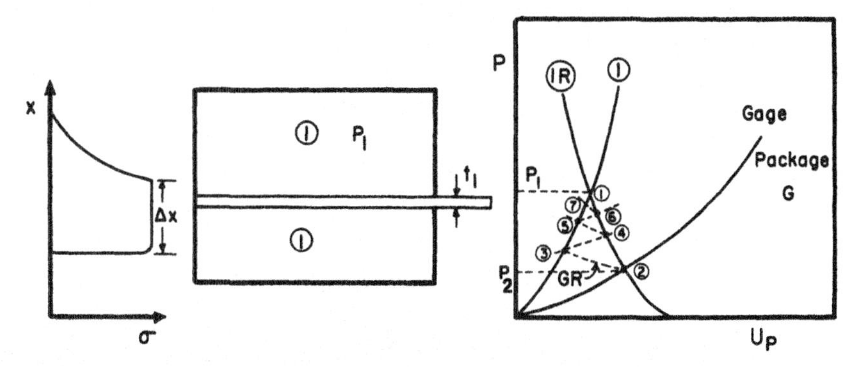

FIGURE 11.11 Thin gage (thickness t_1) subjected to shock pulse with duration Δx.

of the gage package is lower than I. The plot shows this. When the shock pulse enters G, it reflects back into I. Thus, one has IR. The intersection of IR with G gives the pressure P_2 in the gage, corresponding to 2. After traveling through the gage, the pressure pulse reenters material I (bottom side of pulse gage). At this point, the shock pulse is reflected from this interface into the gage. This curve intersects the curve for material 1 at 3. This pressure pulse travels upward, toward the top interface of the gage. At this point, it is re-reflected into the gage, and we have to use, once more, the G Hugoniot passing through P_3. It intersects the material IR curve at 4. Successive reflections lead to P_5, P_6, P_7, which approach, ever more closely, the initial pressure P_1. These successive reverberations have the effect of equilibrating the gage pressure to that of the surrounding medium as long as the gage thickness is much lower than the pulse width. Thus, the pressure in the gage is the same as that in the surrounding material (more details on this analysis are given in Chapter 7, Examples 7.2 and 7.3).

11.5 PIEZOELECTRIC GAGES

Piezoelectric materials generate electricity when stressed, and this behavior, encountered in quartz, lithium niobate, and some polymers, is successfully used to produce gages. The advantage of these gages is that they do not require an external power source; the current is directly produced by the gage. As early as 1921, Keys [9] made piezoelectric measurements. These gages are used in two different geometries that produce electrical currents that are analyzed along two different routes: thin-element and thick-element configurations. In the thin-element configuration, the gage thickness is much lower than the thickness of the pulse traveling through it. This is a configuration similar to the one shown in Figure 11.1 for piezoresistive gages. In the thick-gage configuration the entire shock width occupies only one fraction of the gage thickness. The thin-element configuration was the first to be used. After the discovery by Benedick and Neilson [10] that thick gages emitted special pulses, this geometry became more prevalent. Recently, the introduction of the PVDF polymer gage marked the return of the thin-element configuration. The PVDF gage was developed by Bauer [11] in France and is a superb diagnostic tool, much simpler to use than piezoresistive gages. The PVDF gages are very thin (25 μm) and act as material gages, totally embedded in the material.

The signals in the thick- and thin-element configuration will be described next. Piezoelectricity is generated between two electrodes when there is a pressure gradient between them. The expression for the stress is

$$\sigma(0, t) = \frac{fA}{t_0} i(t)_j \qquad 0 < t < t_0 \tag{11.8}$$

where $\sigma(0, t)$ is the stress at electrode 0 and time t, f is the piezoelectric constant, A is the area of the piezoelectric element, t_0 is the wave transit time

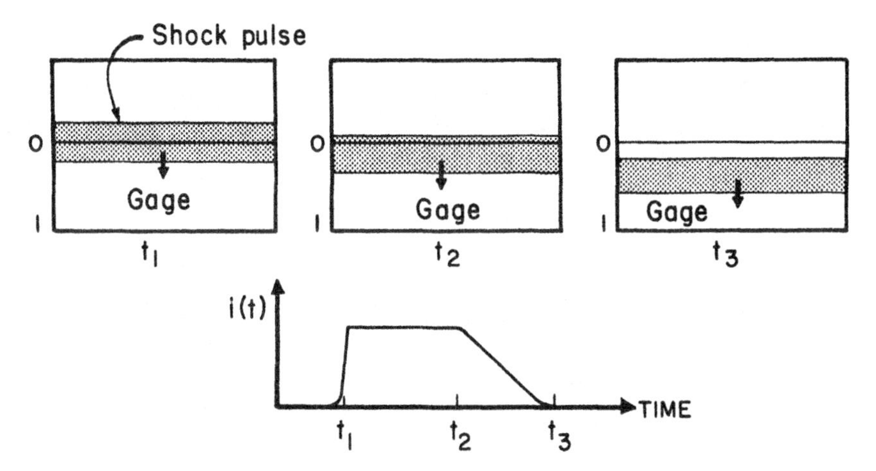

FIGURE 11.12 Progression of stress through thick-element gage.

through the piezoelectric gage. Figure 11.12 shows the thick-element gage placed at the end of the material. Two electrodes are attached at the ends of the piezoelectric gage and a current is generated by the passage of the stress pulse. This pulse is indicated by hatching at three times, t_1, t_2, and t_3. At t_3 none of the electrodes is under stress and therefore the current is zero. Equation (11.8) shows that the current is proportional to the stress in the electrode as long as electrode 1 is not loaded ($t < t_0$). Graham and co-workers [12] conducted very important research on the quartz piezoelectric gage, calibrating it and developing the optimum physical configuration for it. The piezoelectric constant f is actually a parameter that is stress dependent. For quartz (X-cut), Graham [13] showed that it was dependent on stress by the following equation:

$$f = 2.004 + 0.0965\sigma \quad (\text{in mC m}^{-2}\ \text{GPa}^{-1})$$

Thus, at 10 GPa the correction is approximately 5%. For compressional waves, quartz electrodes are cut along [100]; this orientation is commonly known as X-cut. The highest achievable stress with quartz gages is 4 GPa; for lithium niobate gages, the maximum stress is 1.8 GPa. Quartz and lithium niobate gages are most often used in the thick-element configuration.

In the thin-element configuration, the transit time of the pulse through the gage is much lower than the pulse duration. The gage will, in a similar manner, generate a current that is proportional to the difference in stress level between the two electrodes. Figure 11.13 shows a thin-element gage. At t_1, the stresses acting on the two sides are σ_1 and σ_2. Thus,

$$i(t) \propto \frac{\sigma_1 - \sigma_2}{t_0} \tag{11.9}$$

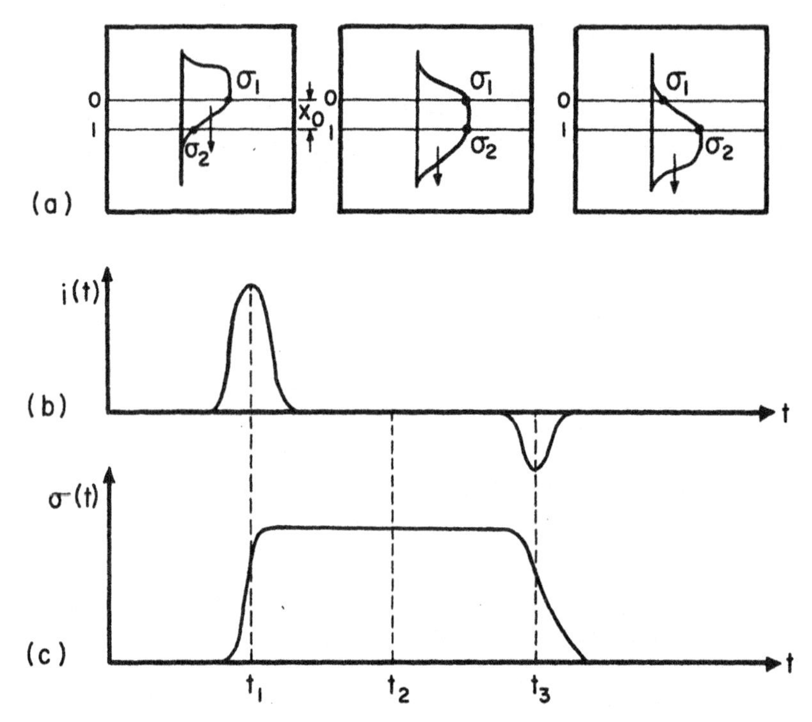

FIGURE 11.13 Thin-element gage and signal obtained from it: (a) pulse position at three times t_1, t_2, and t_3; (b) current as a function of time; (c) stress profile obtained from (b) and Eq. (11.9).

The transit time t_0 is given by

$$t_0 = \frac{U}{x_0}$$

where x_0 is the thickness of the gage and U is the pulse velocity. In Figure 11.13 the pulse is entering the gage at t_1. Thus, $\sigma_1 - \sigma_2 > 0$ and there is a net current. At t_2, the top of the pulse is flat and $\sigma_1 = \sigma_2$; therefore, the current is zero. At t_3, the pulse is exiting the gage, and the current is reversed. The corresponding current–time profile is plotted in Figure 11.13(b). From Eqn. (11.9) one obtains, by integration,

$$\sigma \propto \int i(t)\, dt$$

The profile shown in Figure 11.13(c) can thus be obtained. Figure 11.14 shows how an actual thin-element gage is built. This is a PVDF (or PVF2) gage and the aluminum leads are deposited on the two sides of the gage. The active area

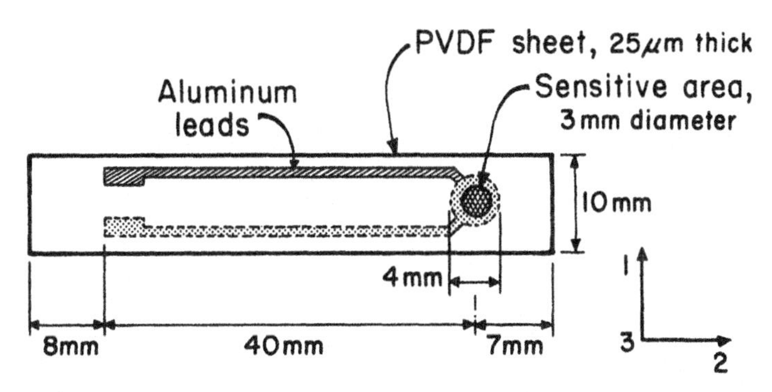

FIGURE 11.14 PVDF piezoelectric gage. (From Cagnoux et al. [4], Fig. 17, p. 482. Reprinted with permission of the publisher.)

of the gage is between the two concentric electrodes plated on the opposite surfaces of the gage. These PVDF gages have been calibrated between 0.3 and 25 GPa and the output is of excellent quality. It is, however, important that the material be manufactured in a reproducible manner in order for the output to be the same from batch to batch. Graham [14] critically calibrated these gages and verified their use to 20 GPa stress.

11.6 ELECTROMAGNETIC VELOCITY GAGES

Zaitsev et al. [15], in 1960, and Dremin et al. [16], in 1964, introduced electromagnetic velocity (EMV) gages.* They are based on Faraday's law of induction, which simply states that if a conductor is moving in a magnetic field, an electromotive force (EMF) is generated. This is the basis for electric motors and generators. The vectorial form of the law is

$$\mathbf{E} = l(\mathbf{U} \times \mathbf{B})$$

where $\mathbf{B}$ is the magnetic field, $\mathbf{U}$ is the velocity of motion of the conductor, l is its length, and $\mathbf{E}$ is the EMF. If the motion is made perpendicular to the magnetic field, and by taking l parallel to $\mathbf{U} \times \mathbf{B}$, one ends up with the simplified expression

$$E = lUB$$

From this equation, one can easily obtain the velocity $\mathbf{U}$ if l and $\mathbf{B}$ are known and $\mathbf{E}$ is measured. This principle is used in EMV gages. They lend themselves very well to determine particle velocities in dielectric materials (rocks, poly-

*According to L. V. Al'tshuler (Soviet Physics–JETP, **8**(1965)52), E. Zavoiski was the first to introduce this method.

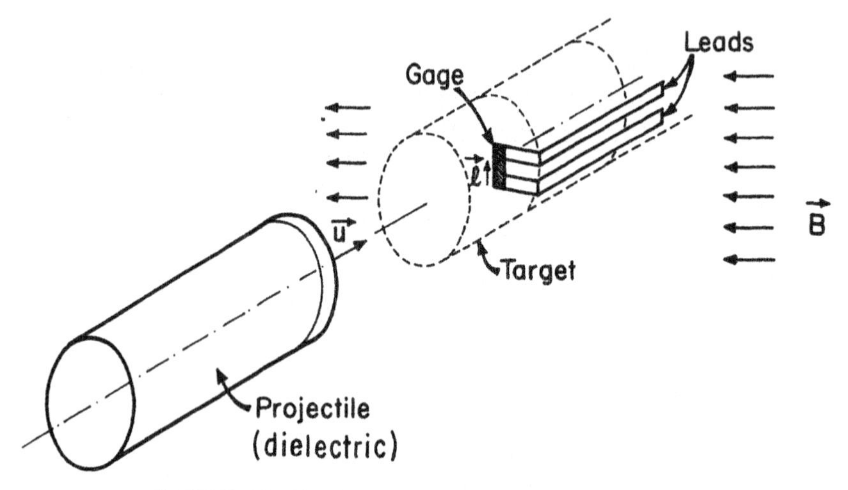

FIGURE 11.15 Electromagnetic particle velocity gage.

mers, ceramics). These gages are embedded in the material and a magnetic field is externally applied. By measuring the EMF induced, one obtains the particle velocity. Figure 11.15 shows a possible geometry for the gage. The magnetic field is set parallel to the impact surface and the conductor l is placed parallel to the impact surface and perpendicular to the vector **B**. The particle velocity **U** is perpendicular to **B**. Other geometries are also possible. Figure 11.16 shows the output of two EMV gages presented by Cagnoux et al. [4].

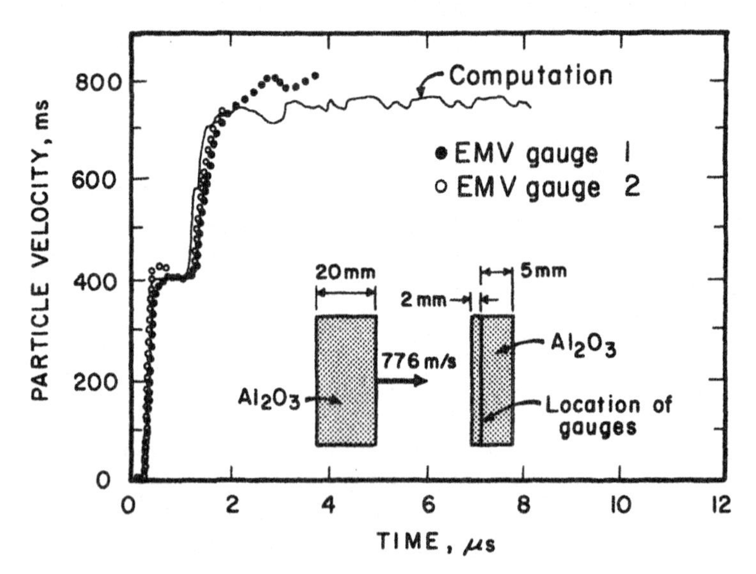

FIGURE 11.16 EMV gage output for alumina. (From Cagnoux et al. [4], Fig. 23, p. 490. Reprinted with permission of the publisher.)

The two gages were placed 2 mm from the impact interface. It can be seen that the main pulse is preceded by an elastic precursor pulse with a particle velocity of 400 m/s. The amplitude of this elastic precursor can be obtained from the corresponding particle velocity.

11.7 CINEMATOGRAPHIC AND FLASH X-RAY TECHNIQUES

For most dynamic events the naked eye is of little value, and we may see a flash or hear a bang. The sequence of events in dynamic processes can be visually recorded by a number of cinematographic methods (high-speed photographs) or flash radiographs. These methods are extensively used, and we will briefly review the most common techniques. For greater details, the reader is referred to the chapters by Swift [17] and Jameson [18]. Jamet and Thomer [19] authored a book on flash radiography. While human beings (and our four-legged cousins) can resolve times of 5×10^{-2} s visually without aid, the fastest cameras available have resolution times approaching 10^{-13} s. A wide range of specialized cameras and techniques have been developed.

11.7.1 Shadowgraphs

In this method, only the shadow of the object is seen, and the image is produced on the film by a rapid flash of light, creating a shadow of the object. Spark gap generators can produce exposure times as short as 5×10^{-9} s, whereas simple spark circuits easily generate times of 10^{-6} s. Pulsed lasers can produce times in the femtosecond range. The Cranz–Schardin camera uses this principle with an array of spark gap sources and an array of lenses. The light sources are sequentially activated, creating a cinematographic effect. The Cranz–Schardin camera is very simple to operate and the framing rate can be adjusted independently from the exposure time for each picture; Figure 11.17 shows a Cranz–Schardin camera; the placement of object, light sources (1 → 2 → 3 → 4), and images is shown.

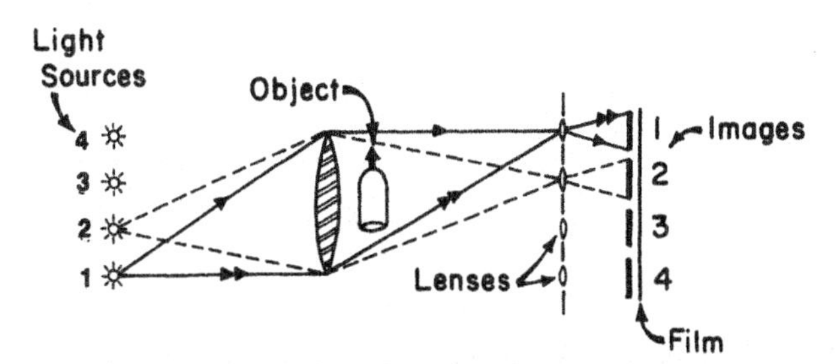

FIGURE 11.17 Schematic of Cranz–Schardin spark-shadowgraph camera showing sequential activation of four light sources and formation of four images.

11.7.2 Rotating-Mirror Camera

The most common rotating-mirror camera uses a mirror rotating at a high velocity by means of a gas-driven turbine. The image is recorded in a large drum in either a framing or streak mode. Figure 11.18 shows schematically such a camera. The objective lens forms an image of the object on the face of mirror. As the mirror rotates, driven by the turbine, it forms successive images of the object on a long film strip placed on a drum concentric with the rotating-mirror axis. Framing speeds of up to 10 million frames per second can be obtained. When the camera is used in the streak mode, a slit is placed between the objective and the relay lens. This streak mode enables continuous monitoring of an object, such as the motion of a surface, but does not reproduce the image. Figure 11.18(b) shows a streak camera. Streak cameras can produce excellent quantitative data.

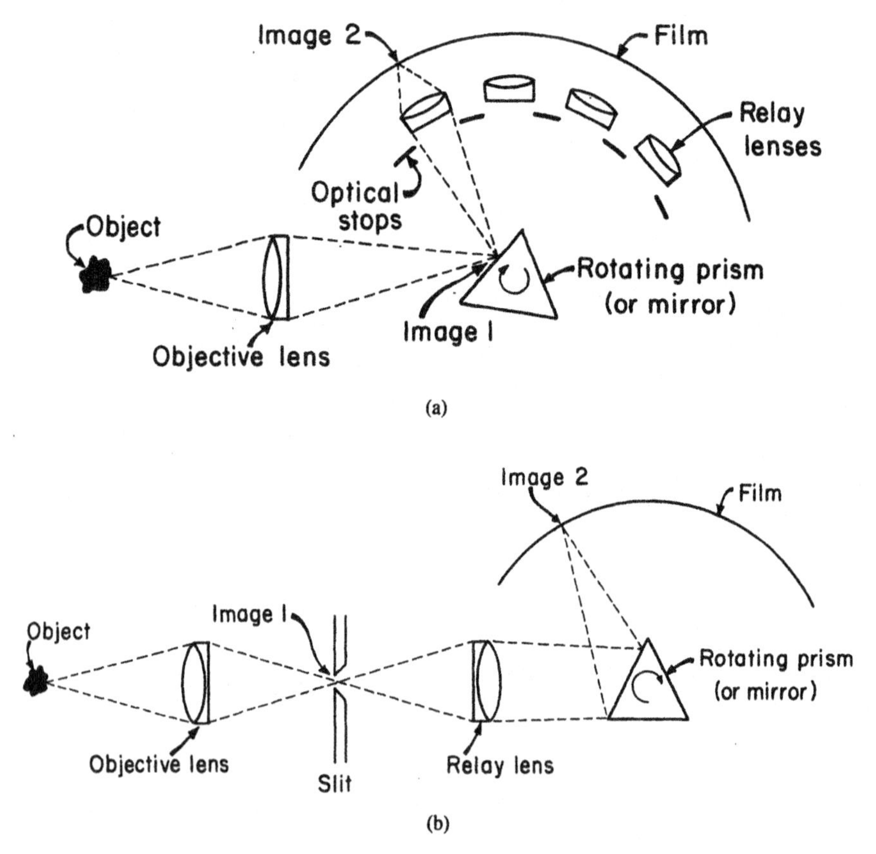

FIGURE 11.18 Rotating-mirror cameras: (a) rotating-mirror framing camera; (b) rotating-mirror streak camera. (Adapted from Swift [17], pp. 263 and 271.)

11.7.3 Image Converter, Electronic, and High-Speed Television Cameras

A wide range of novel systems have been developed using electronics, fiber optics, and microchannel plate technology. Although conventional optical instruments are limited in speed by the exposure time necessary to create an image on film, electronic cameras can use devices that magnify enormously the light-gathering capability. Figure 11.19 shows a sequence of image converter camera photographs taken from a cylindrical copper specimen impacting a rigid target. The velocity of the projectile can be determined from the change in the width of the gap between photographs 1–2 and 2–3 (photographs at 20-μs intervals). The progression of plastic deformation in the specimen (Taylor anvil) can be followed. Image converter cameras are very useful research instruments.

High-speed television cameras are built with speeds of up to 2000 frames per second.

Example 11.2. Calculate the impact velocity and radial strain rate in the Taylor impact specimen shown in Figure 11.19.

The impact velocity is calculated by measuring the gaps at sequential photographs:

$$d_1 - d_2 = 2.8 \text{ mm}$$

$$d_2 - d_3 = 2.8 \text{ mm}$$

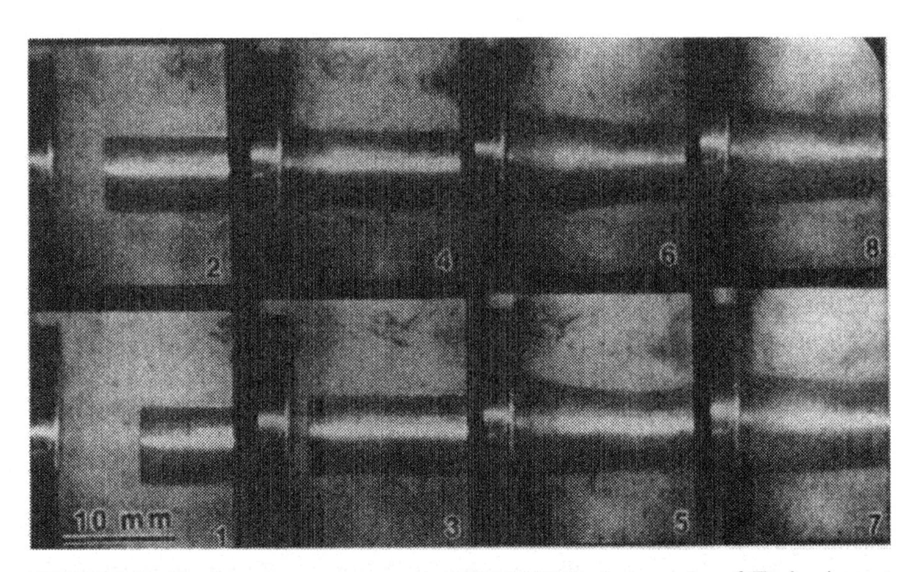

FIGURE 11.19 Image converter camera (IMACON) photographs of Taylor impact specimen (described in Section 3.5) impacting a rigid target; sequential photographs (1–8) taken at 20-μs intervals. (Courtesy of J. Isaacs, D. Lischer, and J. Starrett, University of California, San Diego.)

The interval between sequential photographs is 20 μs. Thus

$$U = \frac{2.8 \text{ mm}}{20 \ \mu\text{s}} = 140 \text{ m/s}$$

The radial strain rate at the contact surface can be established by finding the rate of change in the radius:

$$\varepsilon_r = \frac{dr}{r}$$

$$\dot{\varepsilon}_r = \frac{1}{r}\frac{dr}{dt}$$

This can be done between frames 4 and 5:

$$2\,\Delta r = 2 \text{ mm}$$

$$2r \cong \tfrac{1}{2}(r_4 + r_5) = 9.1 \text{ mm}$$

$$\dot{\varepsilon}_r = \frac{1}{4.55} \times \frac{1}{20 \times 10^{-6}}$$

$$= 1.1 \times 10^4 \text{ s}^{-1}$$

11.7.4 Flash Radiography

Flash radiography has been and is the work horse in many dynamic deformation experiments. It is especially useful in experiments involving explosive detonation, because the gaseous detonation products mask the view in optical cameras. Ballistic penetration experiments, shaped-charge detonation experiments, exploding shells, and hypervelocity impacts against space bumper shields are a few examples of experiments where flash X-rays are routinely used. In these applications X-rays are used in a penetration mode, similar to how we use them in the medical profession. The shadow created by the object reveals the details of the experiment. Flash X-rays have also been used, in a limited way, in the diffraction mode, and this aspect will be discussed at the end of this section. X-ray sources for flash radiography produce a very short duration flash (approximately 10^{-7} s^{-1}). This is in contrast with conventional sources, where the dosage rate is much lower and high times are needed to achieve the same dose. For example, a projectile traveling at 10 km/s will move 0.1 mm in the same time of 10^{-7} s^{-1}. This will produce a rather insignificant smearing. Figure 11.20 shows an experimental setup used by the author and Wittman at New

FIGURE 11.20 Flash X-ray setup based at New Mexico Tech. for projectile velocity measurements, showing the gun tube mounted on a V-block mount, target, grid, and two X-ray tubes (one above and one on the side of the target). Surrealistic frame product of Socorro "artists."

Mexico Tech. [20]. X-ray sources were placed vertically and horizontally along the path of a projectile, fired from a propellant gun. On the opposing sides, horizontal and vertical cassettes containing film backed by an intensifying screen were placed. The flash X-ray units (100 kV each) were triggered by the breakup of wires placed on a screen in the trajectory of the projectile. The yaw and pitch (rotation of projectile from its horizontal and vertical axes perpendicular to cylinder axis) could be determined from these two images. Sequential triggering of the two-flash X-ray units forming the vertical image enabled the determination of the projectile velocity. Figure 11.21 shows the horizontal and vertical images of these flash X-ray cassettes. The steel projectile and gun muzzle are whiter because they screen the X-rays. Reference wires are shown. Some of these wires were sectioned by the projectile; these are the triggering wires. The vertical image shows two projectiles, one from each pulse. There is a delay between the two pulses, and the distance between the two images divided by the time delay provides the projectile velocity. Additional examples of the use of flash X-rays are shown in Chapter 17; shaped-charge jets are shown as they travel (Figs. 17.3 and 17.6). The acceleration voltage for the electrons in the X-ray sources varies from 150 to 2000 kV for most commercial units. The PHERMEX [21] facility, at Los Alamos National Laboratory, is unique in that it has a beam with an energy of 27 MeV that can produce a 200-ns (2×10^{-9}-s) flash. Large explosive–metal systems can be radiographed with a very high resolution. The propagation of detonation waves in complex systems, such as simulated nuclear weapons detonation devices, can be followed with great precision. The higher the energy of the X-rays, the higher the penetration capability and the thicker the systems that can be observed. On the other hand, the image contrast seems to suffer at very high voltages.

A very creative and fundamentally significant application of X-rays in dynamic deformation resides in the use of diffraction. X-ray diffraction provides information on the structure of materials and interatomic spacings through the well-known Bragg law:

$$\lambda = 2d \sin \theta$$

where λ is the wavelength of the X-ray beam, d is the spacing between general (hkl) planes, and θ is the incidence angle. This method has been used to establish that the shaped-charge jets are at least partially crystalline (i.e., solid) in shaped charges by Green [22] and Jamet [23, 24]. Figure 11.22(a) shows a shaped-charge jet and a flash X-ray beam (molybdenum $K\alpha$ radiation). The X-ray beam is diffracted by the crystalline material in the jet and a characteristic pattern is produced on the intensifying screen. This technique enables the determination of the texture and crystallinity of a jet.

Another example of the use of flash X-ray diffraction is to obtain fundamental information on the nature of the shock front. This is done by obtaining a flash X-ray diffraction pattern from it. Wark et al. [25] and Zaretsky [26] applied this method to the shock wave. One can observe the distortion of the lattice planes as the shock wave traverses them. Synchronization between shock wave arrival of the free surface and the flash X-ray is critical. Figure 11.22(b)

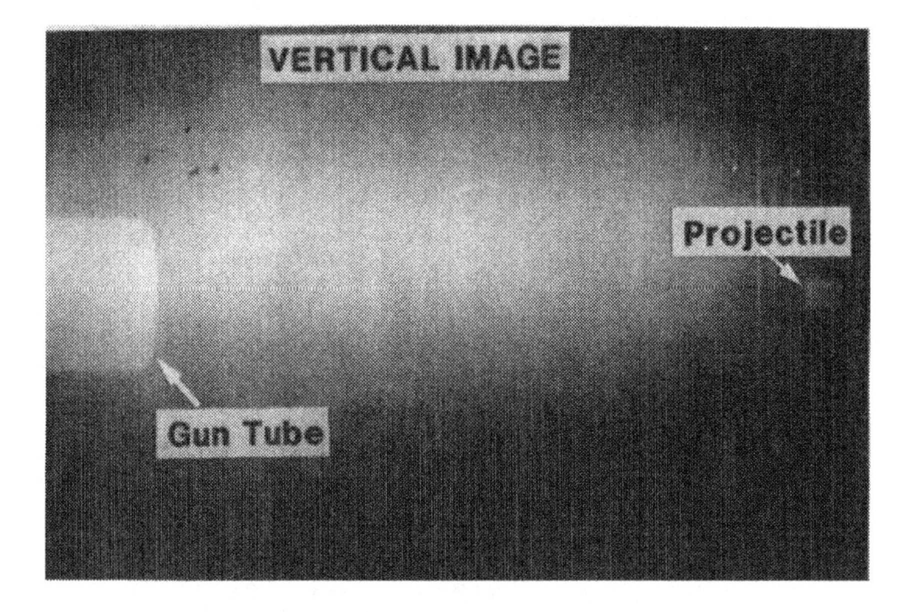

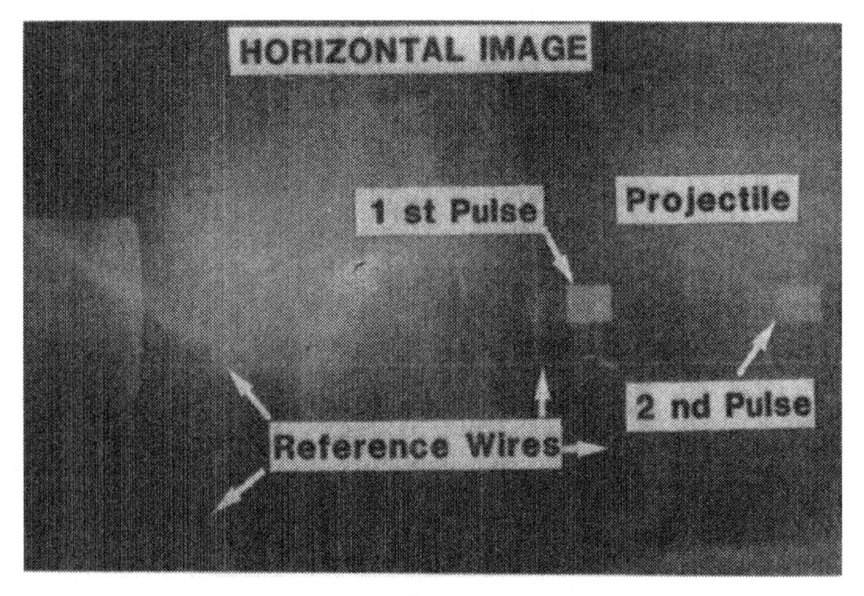

FIGURE 11.21 Flash X-ray image of the projectile in flight; (a) horizontal image; (b) vertical image.

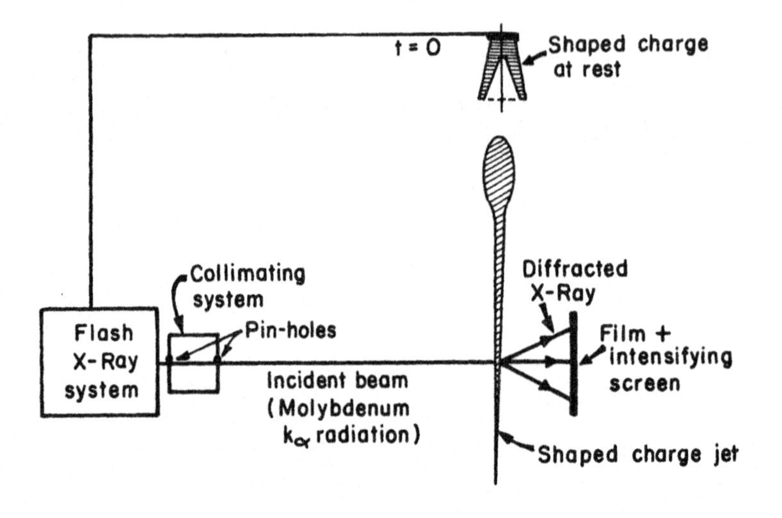

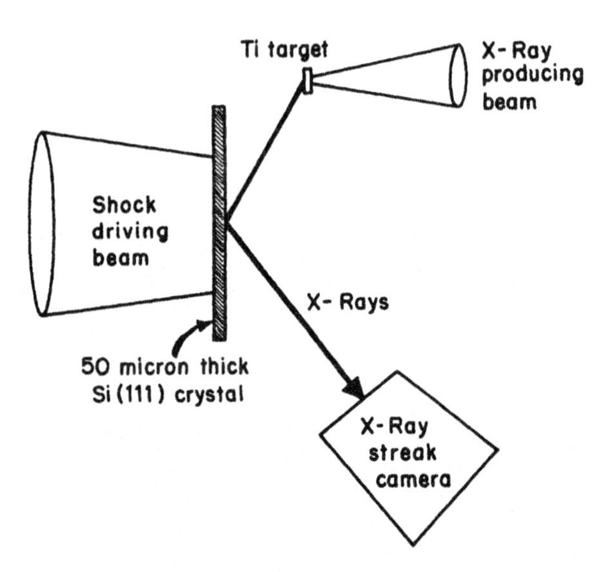

FIGURE 11.22 (a) Schematic illustration of X-ray diffraction; form measurement of shaped-charge structure. (From Jamet [23], Fig. 5. Reprinted with permission of the author and the French-German Research Institute, St. Louis.) (b) Flash X-ray diffraction apparatus for determination of lattice distortions in silicon produced by shock pulse. (From Wark et al. [25], Fig. 1, p. 4532. Reprinted with permission of the publisher.)

shows the setup used by Wark et al. [25]. A shock wave is produced in a 50 μm Si wafer. Flash X-rays are generated at a Ti target. These X-rays are made to incide at the back surface. An X-ray streak camera records the variation in the 2θ angle.

REFERENCES

1. R. A. Graham and J. R. Asay, *High Temp. High Press.*, **10** (1978), 355.
2. L. C. Chhabildas and R. A. Graham, *Techniques and Theory of Stress Measurements for Shock-Wave Applications*, Vol. 83, Appl. Mech. Div., Am. Soc. Mech. Eng.
3. L. E. Chhabildas, *Intl. J. Impact Eng.*, **5** (1987), 205.
4. J. Cagnoux, P. Chartagnac, P. Hereil, and M. Perez, *Ann. Phys. Fr.*, **12** (1987), 451.
5. L. M. Barker and R. E. Hollenbach, *J. Scient. Instr.*, **36** (1965), 1617.
6. L. M. Barker and R. E. Hollenbach, *J. Appl. Phys.*, **43** (1972), 4669.
7. R. J. Clifton, Brown University, personal communication, 1987.
8. P. S. DeCarli, D. C. Erlich, L. B. Hall, R. G. Bly, A. L. Whitson, D. D. Keough, and D. Curran, "Stress-Gage System for the Megabar (100 GPa) Range," Report No. DNA 4066F, Defense Nuclear Energy, SRI Intl., Palo Alto, CA, 1976.
9. D. A. Keys, *Phil. Mag.*, **42** (1921), 473.
10. W. B. Benedick and R. W. Neilson, as reported by R. A. Graham and J. R. Asay, *High Temp. High Press.*, **10** (1978), 379.
11. F. Bauer, in *Shock Waves in Condensed Matter*, eds. W. J. Nellis, L. Seaman, and R. A. Graham, American Physical Society, 1981, p. 251.
12. R. A. Graham, F. W. Neilson, and W. B. Benedick, *J. Appl. Phys.*, **36** (1965), 1775.
13. R. A. Graham, *Phys. Rev. B*, **6** (1972), 4779.
14. R. A. Graham, in *Shock Waves in Condensed Matter*, eds. S. C. Schmidt and N. C. Holmes, Elsevier, Amsterdam, 1988, p. 11.
15. V. M. Zaitsev, P. F. Pokhil, and K. K. Shvedov, *Proc. Acad. Sci. USSR*, **132** (1960), 529.
16. A. N. Dremin, S. V. Pershin, and V. F. Pogorelev, *Comb. Expl. Shock Waves*, **1** (1965), 1.
17. H. F. Swift, in *Impact Dynamics*, eds. J. A. Zukas et al., Wiley, New York, 1982, p. 241.
18. R. L. Jameson, in *High Velocity Impact Dynamics*, ed. J. A. Zukas, Wiley, New York, 1990, p. 831.
19. F. Jamet and G. Thomer, *Flash Radiography*, Elsevier, New York, 1976.
20. M. A. Meyers and C. L. Wittman, *Met. Trans.*, **21A** (1990), 3153.
21. C. L. Mader, *LASL PHERMEX Data*, Vols. I–III, University of California Press, Berkeley, 1976.
22. R. Green, *Rev. Sci. Instr.*, **46** (1975), 1261.
23. F. Jamet, "La Diffraction Instanee," Report No. CO 227/84, Institut St. Louis, France, August 1984.
24. F. Jamet, "Methodes d'Analyse de l'Etat Physique d'un Jet de Charge Creuse," Report No. CO 227/82, Institut St. Louis, France, December 1982.
25. J. S. Wark, D. Riley, N. C. Woolsey, G. Kuhn, and R. R. Whitlock, *J. Appl. Phys.*, **68** (1990), 4531.
26. E. Zaretsky, private communication, 1990.

CHAPTER 12

Experimental Techniques: Methods to Produce Dynamic Deformation

12.1 INTRODUCTION

This chapter will present the various experimental methods that are used to produce dynamic deformation. In Chapter 2 a brief description of the Taylor test, the Hopkinson bar test, and the pressure–shear test was given. In this chapter, other techniques will be presented, producing strain rates ranging from 10 to 10^8 s^{-1}. The definition and description of strain rate will be given first, and the entire range of methods available to the experimentalist will then be discussed.

We start with the important statement that *strain rate*, and not *velocity of deformation*, is the critical parameter. Strain rate is the rate of change of strain with time. Its unit is the reciprocal second:

$$\dot{\varepsilon} = \frac{d\varepsilon}{dt} \tag{12.1}$$

Example 12.1. Calculate the strain rates for the three deformations given below:

1. A tensile specimen with a length of 10 cm is extended in a testing machine at a velocity of 1 m/s [Fig. 12.1(a)]:

$$\dot{\varepsilon} = \frac{\Delta\varepsilon}{\Delta t} = \frac{\Delta l v_0}{l_0 \Delta l} = \frac{v_0}{l_0} = \frac{100}{10}$$

$$\dot{\varepsilon} = 10\ s^{-1}$$

2. A projectile (cylindrical) with a length of 5 cm is impacting a rigid target at a velocity of 1000 m/s. If we assume that the projectile will decelerate linearly to rest when the length is reduced to 2.5 cm (truncated cone), we can establish an approximate strain rate [Fig. 12.1(b)]:

$$\dot{\varepsilon} = \frac{\Delta l}{l_0 t} = \frac{2.5}{5t}$$

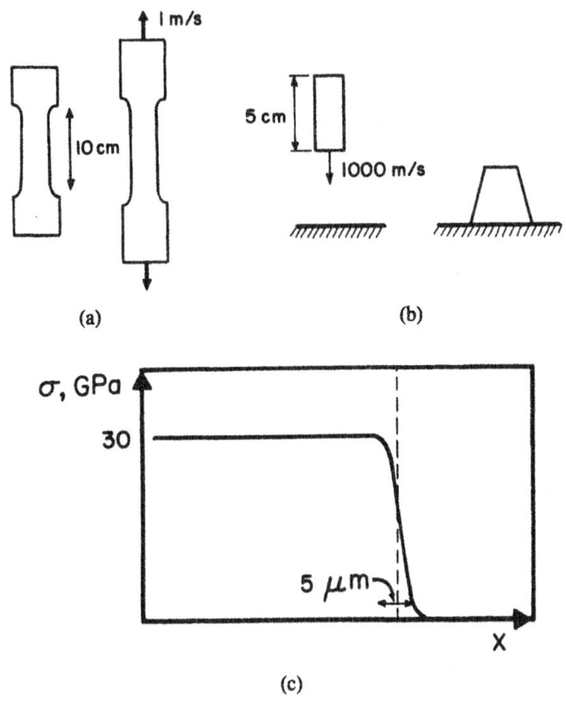

FIGURE 12.1 Examples of three different strain rates: (a) specimen pulled in tension; (b) projectile rigid target; (c) shock wave propagating in copper.

The time, assuming linear velocity change, is

$$S = 2.5 \text{ cm} \qquad v = \frac{2S}{t} \qquad t = \frac{2.5 \times 2 \times 10^{-2}}{v}$$

$$t = \frac{5 \times 10^{-2}}{1000} = 5 \times 10^{-5} \qquad \varepsilon = \frac{2.5}{5 \times 5 \times 10^{-5}} = \frac{1}{10^{-4}} = 10^{4} \text{ s}^{-1}$$

3. The strain rate experienced by a copper specimen traversed by a shock wave with an amplitude of 30 GPa. The plastic deformation takes place at the shock front, during the rise of the shock wave. Thus, it is the rise time of the shock wave (or thickness of the shock front), and the reduction in volume (V/V_0) that determine the strain rate. Let us calculate the strain rate for an assumed shock front width of 5 μm. We have, from Table 4.1,

$$U_s = 4.95 \text{ km/s}$$

$$\frac{V}{V_0} = 0.863$$

$$S = 5\ \mu\text{m}$$

$$\varepsilon = \frac{V - V_0}{V_0} = \frac{L - L_0}{L_0} = 0.137$$

The loading time (to maximum pressure) is:

$$t = \frac{S}{U_s} = \frac{5 \times 10^{-6}}{4.95 \times 10^3} = 10^{-9}\ \text{s}$$

$$\dot{\varepsilon} = \frac{0.137}{10^{-9}} = 1.4 \times 10^8\ \text{s}^{-1}$$

The three examples above produced strain rates from 10 to $10^8\ \text{s}^{-1}$. On the other side of the spectrum, creep tests produce strain rates that can be lower than $10^{-7}\ \text{s}^{-1}$. Figure 12.2 shows the range of strain rates of concern to Modern Man (MM). We do not know whether this a limiting strain rate, and the strain rates produced on nuclear detonation are certainly higher than $10^8\ \text{s}^{-1}$. The shock front thickness decreases with increasing pressure, and V/V_0 increases. Thus, one can predict strain rates of the order of $10^9\ \text{s}^{-1}$ and higher. However, the response of the materials is not known. The strain rates in Figure 12.2 have been grouped into two classes: lower than $5\ \text{s}^{-1}$ and higher than $5\ \text{s}^{-1}$. For lower strain rates the inertial forces described in Chapter 2 are negligible, and we always have equilibrium. At higher strain rates we start having an increasing influence of inertial forces due to wave propagation effects. For relatively low strain rates, elastic wave propagation in specimen and machine becomes important. The machines used are hydraulic, servo-hydraulic, and pneumatic. At higher strain rates, the expanding ring, the Hopkinson bar, and the Taylor test are used. The strain rate range for these tests is around 10^3–$10^5\ \text{s}^{-1}$. In the range 10^5–$10^8\ \text{s}^{-1}$, shear wave and shock wave propagation are involved and one uses means of very rapid deposition of energy at the surface of the material. This can be accomplished by impact (normal or inclined), by the detonation of explosives in contact with the material, or by pulsed laser or other source of radiation. The various techniques will be discussed in the sections that follow. First, however, the behavior of materials at high strain rates will be briefly discussed in Section 12.2.

12.2 HIGH-STRAIN-RATE MECHANICAL RESPONSE

As early as 1905, Bertram Hopkinson conducted a series of dynamic experiments on steel and concluded that the dynamic strength was at least twice as high as its low-strain-rate strength [2]. He did not have at his disposal any sophisticated measuring methods, but his results were quite correct. His tragic death as a pilot in WWI cut short a brilliant career. It is also known that steels undergo a ductile-to-brittle transition when the strain rate is increased. Thus it

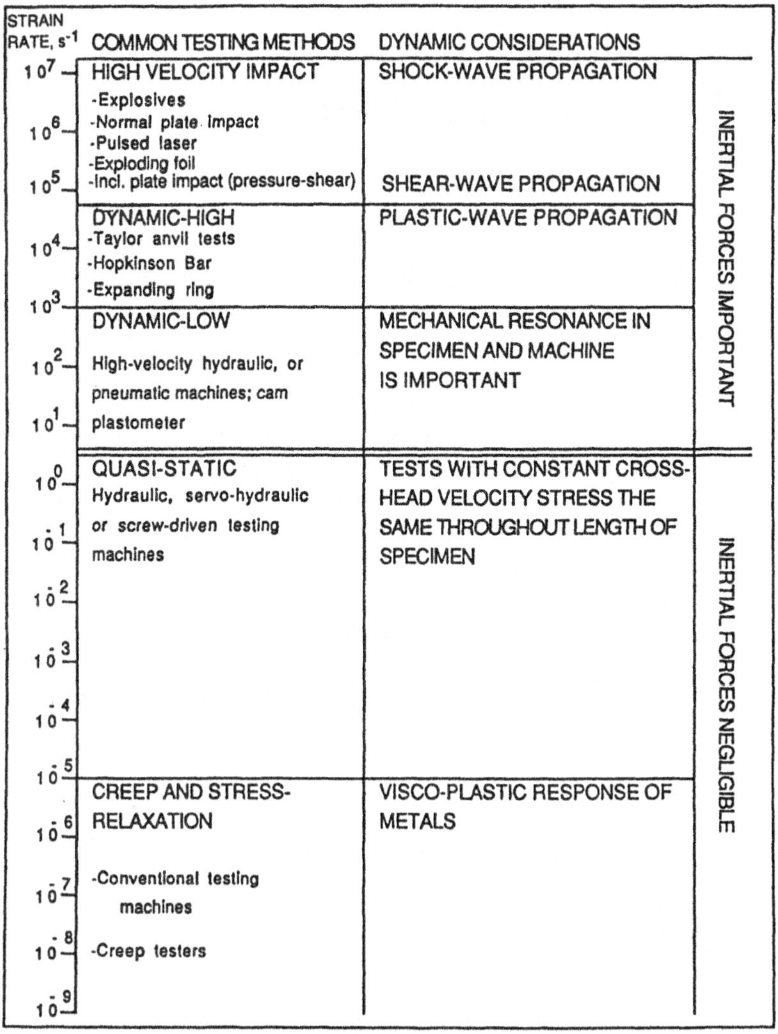

STRAIN RATE, s^{-1}	COMMON TESTING METHODS	DYNAMIC CONSIDERATIONS	
10^7 – 10^5	HIGH VELOCITY IMPACT -Explosives -Normal plate impact -Pulsed laser -Exploding foil -Incl. plate impact (pressure-shear)	SHOCK-WAVE PROPAGATION SHEAR-WAVE PROPAGATION	INERTIAL FORCES IMPORTANT
10^5 – 10^3	DYNAMIC-HIGH -Taylor anvil tests -Hopkinson Bar -Expanding ring	PLASTIC-WAVE PROPAGATION	
10^3 – 10^0	DYNAMIC-LOW High-velocity hydraulic, or pneumatic machines; cam plastometer	MECHANICAL RESONANCE IN SPECIMEN AND MACHINE IS IMPORTANT	
10^0 – 10^{-5}	QUASI-STATIC Hydraulic, servo-hydraulic or screw-driven testing machines	TESTS WITH CONSTANT CROSS-HEAD VELOCITY STRESS THE SAME THROUGHOUT LENGTH OF SPECIMEN	INERTIAL FORCES NEGLIGIBLE
10^{-5} – 10^{-9}	CREEP AND STRESS-RELAXATION -Conventional testing machines -Creep testers	VISCO-PLASTIC RESPONSE OF METALS	

FIGURE 12.2 Schematic classification of testing techniques according to strain rate.

is normal that scientists are curious about the effect of strain rate on the strength of materials. The responses vary considerably, and it is therefore necessary to test individual materials to obtain the specific information. The strain rate dependence of flow stress is necessary in computational codes if these are to be realistic. They are an important component in the constitutive models that are used in hydrodynamic codes when the pressures are sufficiently low so that material strength is an important parameter. As an example, we give a constitutive equation commonly used. It is known as the Johnson–Cook equation [3] and has the form

$$\sigma_{\text{eff}} = [\sigma_0 + B\varepsilon_{\text{eff}}^n][1 + C \ln \dot{\varepsilon}^*][1 - T^{*m}] \tag{12.2}$$

where σ_0, B, C, n, and m are parameters that allow us to establish the stress as a function of strain, strain rate, and temperature:

$$\sigma = f(\varepsilon, \dot{\varepsilon}, T) \tag{12.3}$$

The stresses σ_{eff} and strains ε_{eff} are effective (or equivalent) values and T^* is a normalized temperature. In Chapter 13 (Section 13.2) this equation will be presented in greater detail.

It should be noticed that this is only one form of the constitutive equation. There have been many attempts, but one does not have the ability to predict the parameters from first principles. Thus, we are forced to study different materials experimentally to determine σ_0, B, C, m, and n (or the equivalent). As an example of the variety of responses, Figure 12.3 shows compressive stress–strain curves (cylindrical specimens) for 7075-T6 aluminum and for the alloy titanium 6% Al–4% V [4]. Tests were conducted at strain rates between 3×10^{-2} and $5 \times 10^2\ s^{-1}$ for the aluminum alloy without an apparent effect on the stress–strain curve. For the titanium alloy, a marked effect is observed. The strain rates were varied between 4×10^{-3} and $2 \times 10\ s^{-1}$ and a considerable strain rate strengthening is observed.

Figure 12.4 shows the stress–strain response for commercially pure titanium and tantalum. The high-strain rate curve ($\varepsilon = 3.5 \times 10^3\ s^{-1}$) for tantalum is isothermal and was obtained by sequentially compressing Ta. The procedure described by S. Nemat-Nasser, Y. F. Li, and J. B. Isaacs, Mech. Matls. **17**

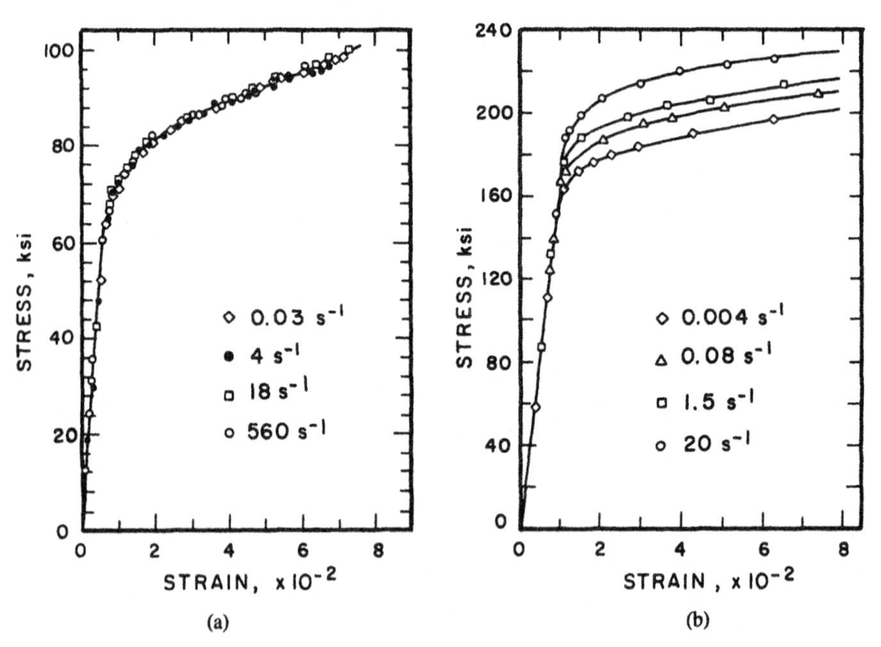

FIGURE 12.3 Effect of strain rate on the stress–strain response (in compression) of (a) 7075-T6 aluminum and (b) Ti–6% Al–4% V alloy.

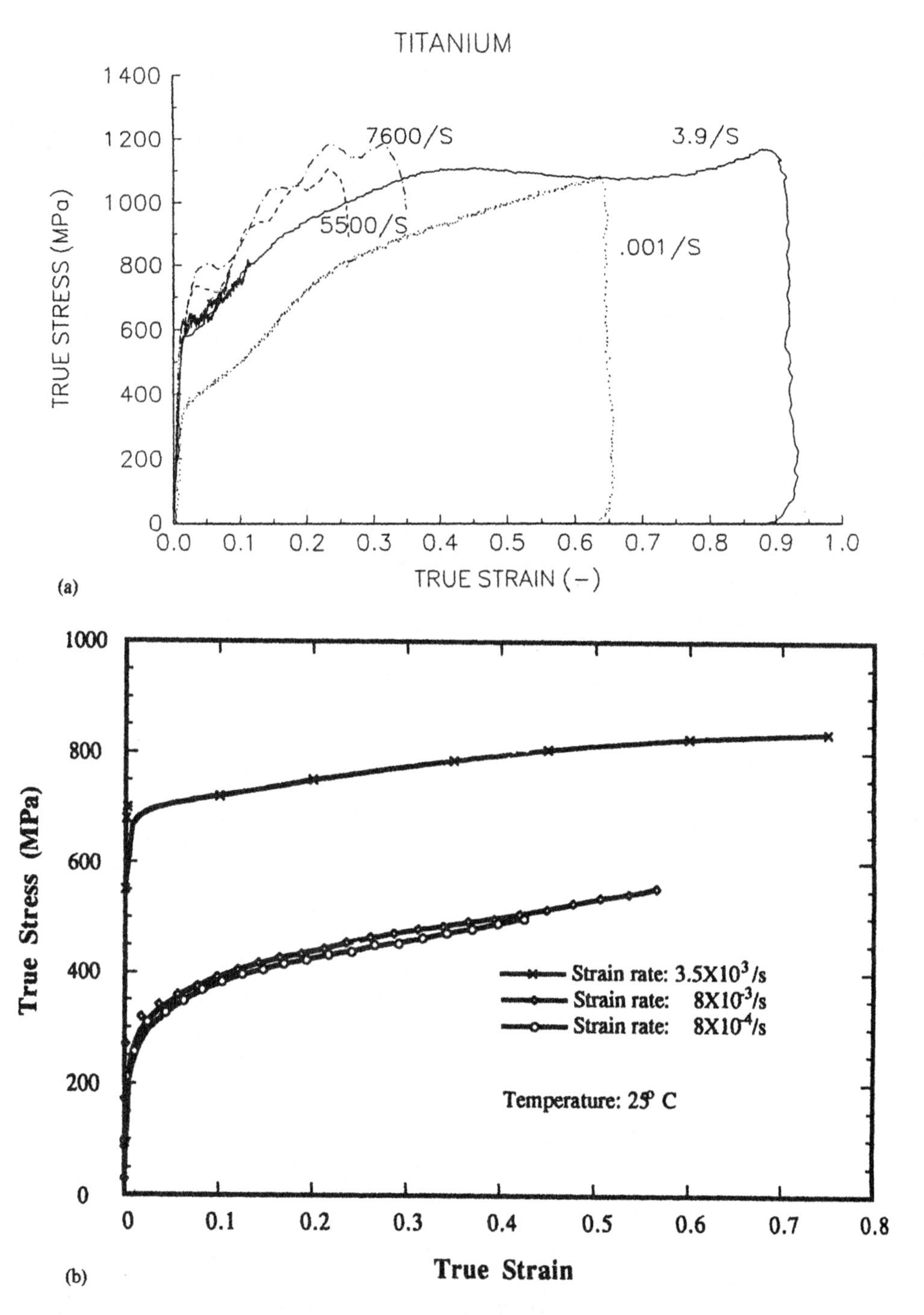

FIGURE 12.4 (a) Effect of strain rate on the stress–strain response (in compression) of commercially pure titanium. (From M. A. Meyers, G. Subhash, B. Khad, and L. Prasad, *Mechanics of Materials*, **17** (1994) 175.) (b) Effect on tantalum (from Y. G. Chen, M. A. Meyers, F. Marquis, and J. Isaacs, UCSD).

(1994) 111. The strain rates were in the range 10^{-3}–10^{3} s^{-1}. The gradual increase in flow stress as the strain rate is increased is obvious. This effect is known as *strain rate sensitivity.* Figures 12.3 and 12.4 show the response of FCC (aluminum), HCP (Ti), and BCC (Ta) metals. It is impossible to predict, from first principles, the response of the different classes of materials.

Clifton [14] reported dramatic increases in flow stress as strain rates of the order of 10^5 s^{-1} are attained. These results are very important, and they can have a profound effect on the computational predictions of shaped charges and other events. This dramatic "hardening" is clearly visible in Figure 12.5. This rise in strength is not accommodated by the conventional constitutive equations and requires special modification of the equation. This rise leads some scientists to believe that there is indeed a limiting strain rate at which the strength of the material becomes infinity.

One should take note of the following:

1. When the strain rate increases, the deformation process changes gradually from fully isothermal to fully adiabatic, because there is not enough time for the heat generated in the deformation to escape out of the body. This gives rise, in some cases, to adiabatic shear instabilities that have a profound effect on the mechanical response of the material. These adiabatic shear bands will not be treated here. Interested readers are referred to Rogers [5] or other sources in the literature [6, 7]. Chapter 15 is entirely devoted to shear instabilities.
2. Johnson's constitutive model is only one possible model. Many other equations have been and, undoubtedly, will be proposed [8–14]. Chapter 13 deals with these equations in greater detail and a specific example is worked out.

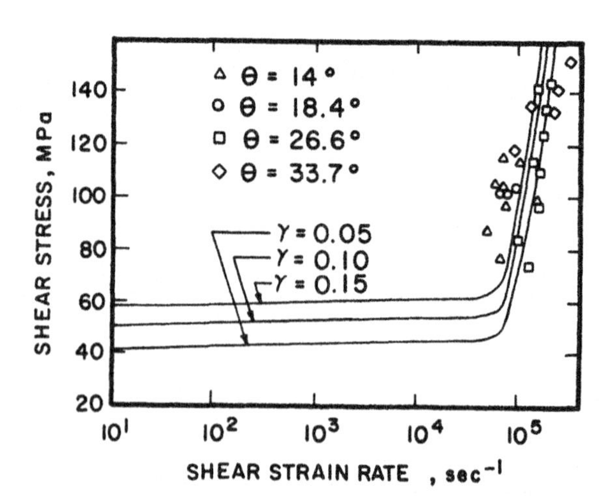

FIGURE 12.5 Dependence of flow stress (at different plactic strains) on the shear strain rate for 1100-0 aluminum. (From Clifton [14], Fig. 7, p. 948. Reprinted with permission of the publisher.)

12.3 HIGH-STRAIN-RATE MECHANICAL TESTING

This section briefly describes the techniques used to produce high strain rates. We will proceed in order of ascending strain rates. We will describe the methods in which impact is used (explosively accelerated flyer plates, gas guns, electromagnetic guns, and other means of rapid energy deposition) in Section 12.4. A more comprehensive presentation is given by Lindholm [15]. Figure 12.2 shows the common testing techniques.*

12.3.1 Intermediate-Strain-Rate Machines

We will start with a compressed gas machine, shown in Figure 12.6. The specimen (compression) is placed on the anvil of the machine. A movable piston passes through a reservoir. Support jacks keep the piston "cocked" [Fig. 12.6(a)]. When the auxiliary reservoir is pressurized after the support jacks are lowered [Fig. 12.6(b)], the piston is accelerated downward, impacting the anvil. This results in the compression of the specimen [Fig. 12.6(c)]. The machine shown in Figure 12.6, DYNAPAK is capable of velocities up to 15 m/s. Not shown in Figure 12.6 are the diagnostics required to record stress and strain. A simpler version of this machine is the "drop hammer," in which a large mass drops freely.

*For additional details, the reader is referred to the articles in the *ASM Handbook*, vol. 8, Mechanical Testing, ASM Intl., Metals Park, Ohio, 1985, pp. 187–297.

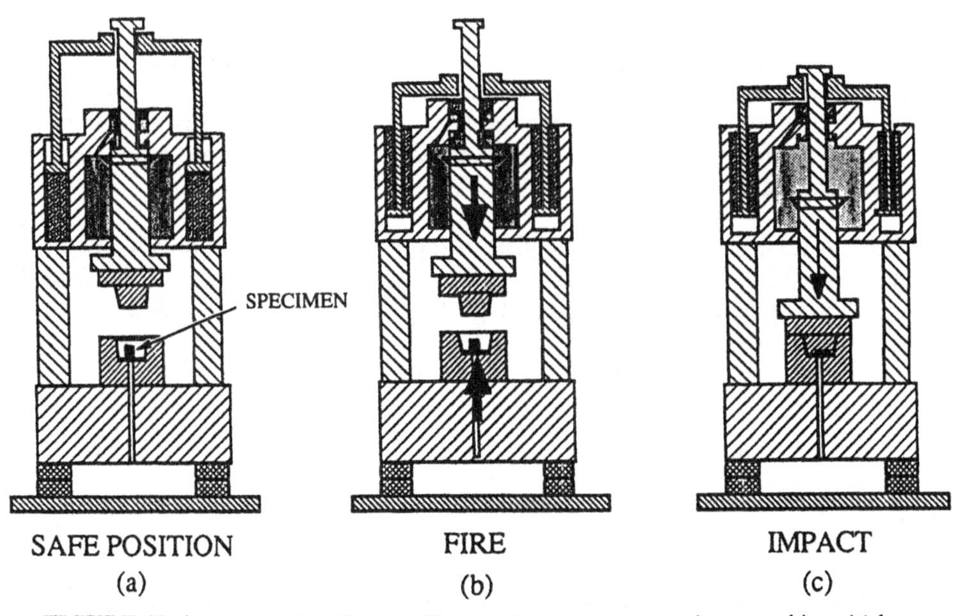

FIGURE 12.6 Schematic of intermediate-strain-rate compressed-gas machine. (a) hydraulic jacks keep piston in up position; (b) jacks lowered and reservoir activated, "firing" piston; (c) impact (DYNAPAK machine).

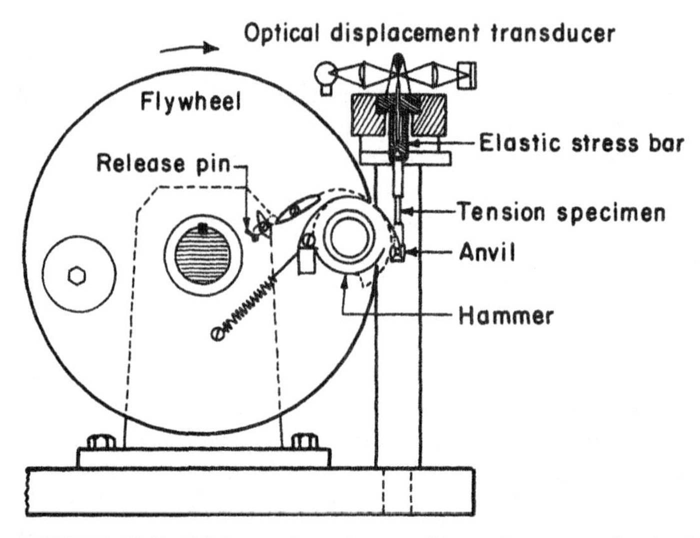

FIGURE 12.7 High-speed tension machine using rotary flywheel.

A rotary flywheel machine is shown in Figure 12.7 (Lindholm [15]). An electric motor drives a large flywheel in the clockwise direction. When the desired flywheel velocity is reached, a release pin trips the hammer (shown in Figure 12.7 in the two positions), which then impacts the bottom of a tensile specimen. The mass of the flywheel is sufficiently large to ensure a constant velocity. The release of the hammer is synchronized to the position of the specimen so that it occurs prior to impact with the specimen. The displacements of an elastic stress bar and the bottom part of the specimen are monitored by optical methods. The displacement of the stress bar provides the stress, whereas that of the bottom of the specimen provides the strain. Thus, it is possible to obtain a continuous stress–strain curve. Usually specimens with a gage length of 2.5 cm are used.

Example 12.2. Calculate the number of reverberations and strain rate assuming that a 2.5-cm specimen is deformed to a total strain of 0.2 by a flywheel with 1 m radius rotating at 1000 rpm. If the flywheel rotates at 1000 rpm and has a radius of 1 m, the velocity is

$$v = 2\pi \times \frac{1000}{60} \times 1 = 100 \text{ m/s}$$

If the plastic deformation undergone by the specimen is approximately 0.5 cm, one has a total duration of deformation of

$$t = \frac{0.5 \times 10^{-2}}{10^2} = 0.5 \times 10^{-4} \text{ s}$$

$$= 50 \times 10^{-6} \text{ s}$$

In 50 μs, we have a number of elastic oscillations on the specimen equal to (assuming an elastic wave velocity of 3000 m/s)

$$\# = \frac{t}{t_1} = \frac{50 \times 10^{-6}}{\frac{2.5 \times 10^{-2}}{3 \times 10^3}} = \frac{50 \times 10^{-6} \times 3 \times 10^3}{2.5 \times 10^{-2}}$$

$$= 60 \times 10^{-1} = 6$$

The strain rate for the above example is $(0.5/2.5)/(50 \times 10^{-6}) = 4 \times 10^3$ s^{-1}. Thus, approximately six elastic oscillations would traverse the specimen while it is being plastically deformed.

Another type of machine, providing a more efficient coupling between specimen and driver, is the cam plastometer (Fig. 12.8). A cam is rotated at a specific velocity. The compression specimen is placed on an elastic bar. At a certain moment, the cam follower is engaged (shown by the dashed line in Fig. 12.8). Thus, within one cycle the specimen is deformed. Strain rates between 0.1 and 100 s^{-1} have been achieved by this method.

12.3.2 Hopkinson (or Kolsky) Bar

The Hopkinson pressure bar has found a wide acceptance as the instrument for intermediate strain rate testing (10^2–10^4 s^{-1}). The Hopkinson bar has been described in Chapter 3. Figure 12.9 shows a compressional split Hopkinson bar. A projectile (striker bar) impacts the incident bar, producing in it a pulse with a length that is large with respect to the specimen (l_0). This elastic wave travels through the incident bar and then reaches the specimen, which is sandwiched between the incident and the transmitted bar. The amplitude of the wave is such that plastic deformation is imparted to the specimen. In Figure 12.9 one sees strain gages attached to both incident and transmitter bars. We measure the direct incident pulse, a reflected pulse, and a transmitted pulse,

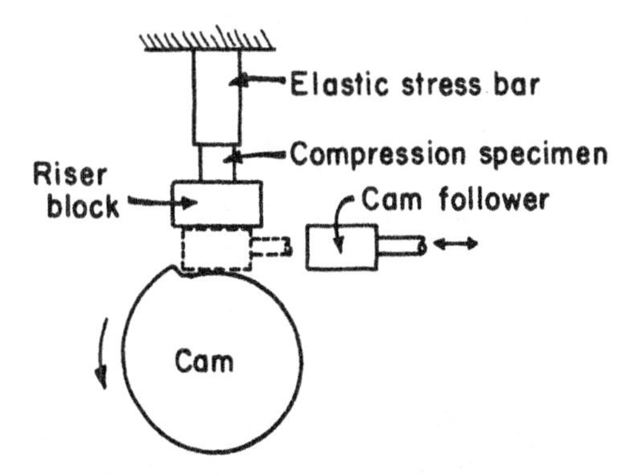

FIGURE 12.8 Schematic representation of cam plastometer. Strain rates between 0.1 and 100 s^{-1} can be achieved by this method.

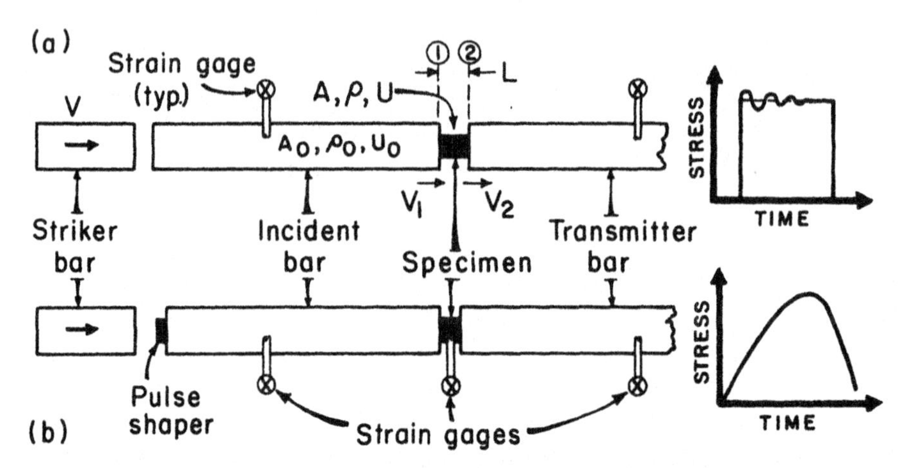

FIGURE 12.9 Split Hopkinson pressure bar for (a) ductile and (b) brittle materials (with pulse shaper); stress pulse shapes shown on the right.

which have amplitudes ε_I, ε_R, and ε_T, respectively. Figure 3.8 shows the three pulses. From these three pulses one can derive a stress–strain relationship for the specimen. Strain rates in the range 10^2–10^4 s^{-1} are achieved by means of this technique.

We will develop the basic equations that provide the stress, strain, and strain rate in a specimen when we know the strain in the bar (from the strain gage recordings). We have a metallic specimen with impedance lower than the bar: $\rho_0 A_0 C_0 > \rho AC$. The terms V_1 and V_2 are the interface velocities at 1 and 2. Since $V_1(t) > V_2(t)$, the distance L decreases with time and the specimen undergoes plastic deformation. The strain rate $\dot{\varepsilon}$ (also a function of time) is calculated as

$$\dot{\varepsilon} = \frac{d\varepsilon}{dt} = \frac{V_1(t) - V_2(t)}{L} \tag{12.4}$$

We will express the velocities as a function of strains in the strain gages. From Eqn. (2.45),

$$\sigma = \rho U_p C \tag{12.5}$$

$$\frac{\sigma}{E} = \varepsilon$$

$$C\varepsilon = U_p$$

Hence, we have, at the interfaces,

$$\begin{aligned} 1 &\rightarrow V_1 = C_0 \varepsilon_I \quad (\text{at } t = 0) \\ 2 &\rightarrow V_2 = C_0 \varepsilon_T \end{aligned} \tag{12.6}$$

At $t > 0$, V_1 is decreased because of the reflected wave. Thus

$$V_1 = C_0(\varepsilon_I - \varepsilon_R) \tag{12.7}$$

By substituting (12.6) and (12.7) into (12.4),

$$\frac{d\varepsilon}{dt} = \frac{C_0(\varepsilon_I - \varepsilon_R) - C_0\varepsilon_T}{L}$$

Thus

$$\dot{\varepsilon}(t) = \frac{C_0}{L}(\varepsilon_I - \varepsilon_R - \varepsilon_T) \tag{12.8}$$

The strain is found by integrating the strain rate from 0 to t:

$$\varepsilon(t) = \frac{C_0}{L}\int_0^t [\varepsilon_I(t) - \varepsilon_R(t) - \varepsilon_T(t)]\,dt \tag{12.9}$$

In order to obtain the stress in the specimen we assume equilibrium:

$$\sigma = \frac{P_1(t) + P_2(t)}{2A}$$

where $P_1(t)$ and $P_2(t)$ are the forces acting on the two interfaces 1 and 2. These forces are

$$P_1(t) = A_0E_0(\varepsilon_I + \varepsilon_R)$$

$$P_2(t) = A_0E_0\varepsilon_T$$

$$\sigma = \frac{A_0E_0}{2A}[\varepsilon_I(t) + \varepsilon_R(t) + \varepsilon_T(t)]$$

E_0 is the elastic modulus of the bars; A_0 is their cross-sectional area.

For equilibrium $P_1(t) = P_2(t)$ and $\varepsilon_I + \varepsilon_R = \varepsilon_T$:

$$\sigma(t) = E_0\frac{A_0}{A}\varepsilon_T(t) \tag{12.10}$$

Thus

$$\dot{\varepsilon}(t) = -\frac{2C_0}{L}\varepsilon_R \tag{12.11}$$

$$\varepsilon(t) = -\frac{2C_0}{L}\int_0^t \varepsilon_R\,dt \tag{12.12}$$

It is interesting to mention that we do not consider wave propagation in a specimen. Actually, when the wave initially enters the specimen there are reverberations. After approximately three reverberations equilibrium is reached in the specimen. It is seen in Chapter 2 (Fig. 2.15) that the wave front has a finite rise time. This rise time allows for the gradual increase in the stress in specimen if it is of the same order as the travel time of the wave in the specimen. The Pochhammer–Chree oscillations observed in real traces are also explained in Chapter 2 (Section 2.10). Ceramic (brittle) specimens cannot be tested in the same manner because failure occurs at very low plastic strain within the first reverberation of the wave. We have to increase the rise time of the wave, and this is shown in Figure 12.9; a copper "pad" (labeled "pulse shaper") is used. This copper pad deforms plastically upon impact and increases the wave rise time.

The Hopkinson bar is also used in tension, torsion, and shear. A number of experimental fixtures have been developed, and some of them are shown in Figure 12.10 (for tension). In Figure 12.10(b) the compressive pulse "by-

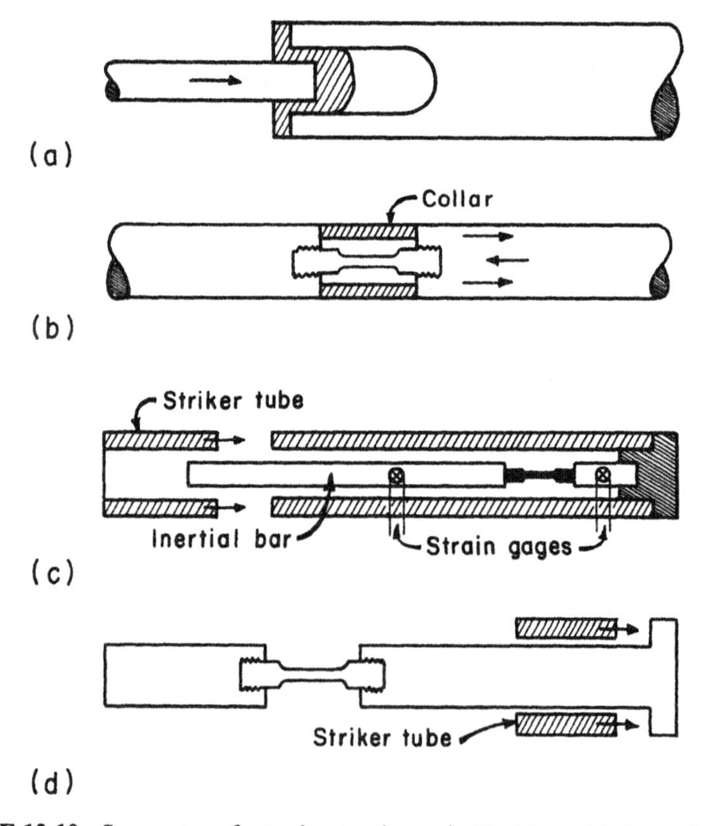

FIGURE 12.10 Some setups for testing specimens in Hopkinson bar in tension mode: (a) hat-shaped specimen; (b) collared specimen; (c) inertial-bar configuration; (d) collar impacting bolt head.

passes'' the specimen because of a collar. It is allowed to reflect at a free surface and returns as a tension pulse. Another arrangement, using an inertial bar, is shown in Figure 12.10(c). The striker tube (on the left) impacts the main tube, causing a compression pulse. This compression pulse acts on the specimen, pulling it to the right. The inertial bar counteracts this. The result is tension acting on the specimen.

For shear stress–shear strain data, the torsional bar [Fig. 12.11(a)] is the proper instrument. Torsional energy is released by braking a clamp. A tubular specimen is sheared by this wave. One can also create shear stresses in a compressional Hopkinson bar by using a hat-shaped specimen, shown in Figure 12.11(b). In recovery experiments the trapping of reflected waves is very important, especially in the case of brittle materials. Nemat-Nasser et al. [16] describe innovative testing and wave trapping procedures using the Hopkinson bar. The classic paper by Davies [17] is essential reading for Hopkinson bar users, as well as the *Metals Handbook* article by Follansbee [18].

Example 12.3. Calculate the stress, strain, and strain rate for three points using the following curve:

We apply Eqns. (12.10)–(12.12). Four points are read from the figure, and the Pochhammer–Chree oscillations are ignored, since they do not represent the material response. The velocity of the elastic wave, C_0, in the bar, is

$$C_0 = \sqrt{\frac{E}{\rho}} = \sqrt{\frac{210 \times 10^9}{7900}} = 5.16 \times 10^3 \text{ m/s}$$

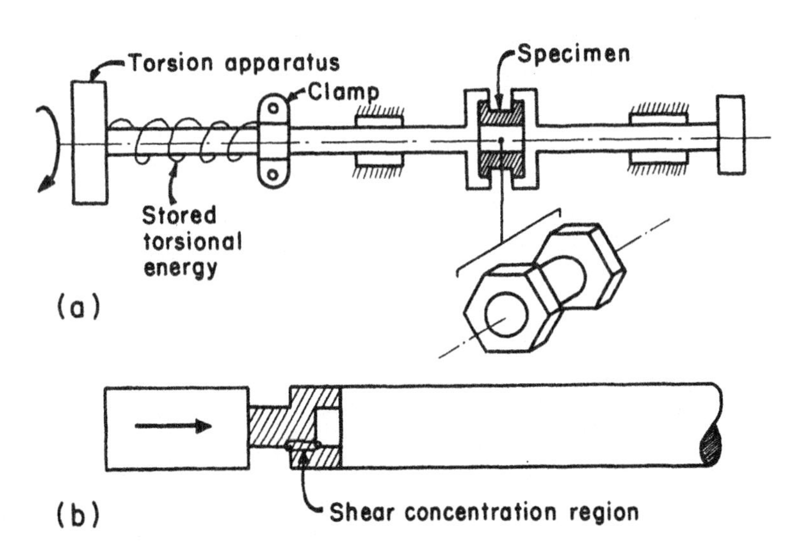

FIGURE 12.11 (a) Torsion Hopkinson bar for pure shear configuration; (b) hat-shaped specimen for shear stresses in compression Hopkinson bar. (This method was developed by L. W. Meyer and S. Manwaring, in "Metallurgical Applications of Shock-Wave and High-Strain-Rate Phenomena," M. Dekker, NY, 1986, p. 657.)

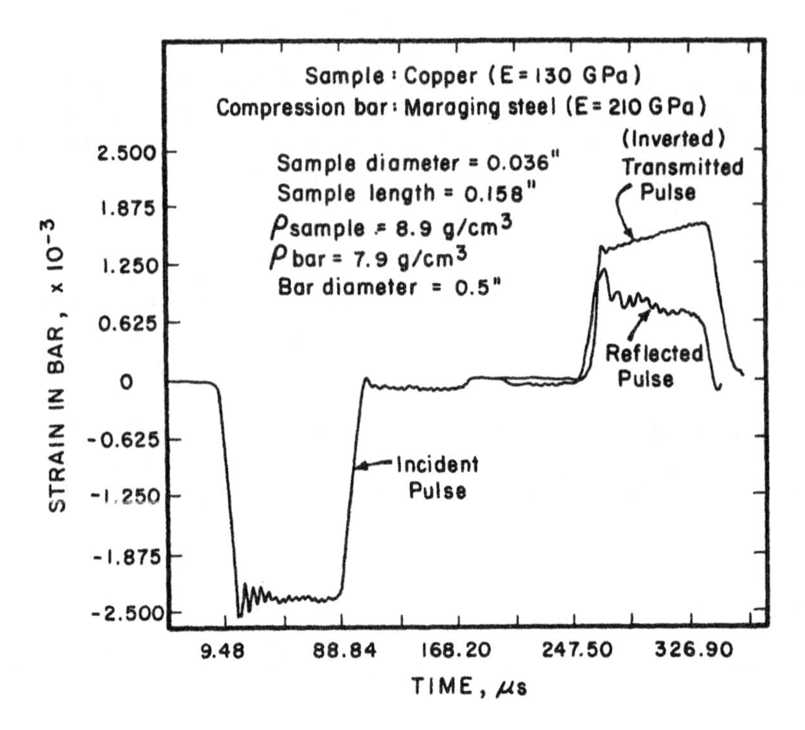

and

$$A_0 = 126.7 \times 10^{-6}\ \text{m}^2$$

$$A = 47.4 \times 10^{-6}\ \text{m}^2$$

Point	Time (μs)	ε_R ($\times 10^{-3}$)	ε_T ($\times 10^{-3}$)	$\dot{\varepsilon}(t) = -(2C_0/L)\varepsilon_R$
1	11.7	−0.88	−1.46	2.27×10^3
2	30.3	−0.79	−1.55	2.04×10^3
3	49.9	−0.69	−1.65	1.78×10^3
4	68.5	−0.65	−1.69	1.68×10^3

Point	$\sigma = E(A_0/A)\varepsilon_T(t)$ (MPa)	$\varepsilon(t) = -(2C_0/L)\int_0^t \varepsilon_R\, dt$
1	507	−0.028
2	538	−0.069
3	573	−0.108
4	586	−0.145

The integration to obtain the strain is done graphically (area under ε_R curve).

12.3.3 Expanding-Ring Technique

Another technique that has met with substantial success is the expanding ring, introduced by Johnson et al. [19]. Figure 12.12 shows a possible arrangement of the expanding ring. The steel cylinder has a core, in which we detonate an explosive. A shock wave travels outward and enters the metal ring, propelling it in a trajectory with an expanding radius [Fig. 12.12(c)]. By measuring the velocity history of the expanding ring by laser interferometry, one can determine the stress–strain curve of this ring at the imposed strain rate. The mathematical derivation of these equations is straightforward and is given below. Let us consider a small segment of a circular ring with radius r [Fig. 12.12(c)]. The ring is subjected, during flight, to Newton's second law, just like any other moving object. The stres acting on the ends of this segment is σ. If the instantaneous cross-sectional area is A, the force is σ_A. If ρ is the density of the material, we have

$$F = ma$$

$$\sum F_r = m\ddot{a}_r$$

$$2F \sin \frac{d\theta}{2} = m\ddot{a}_r$$

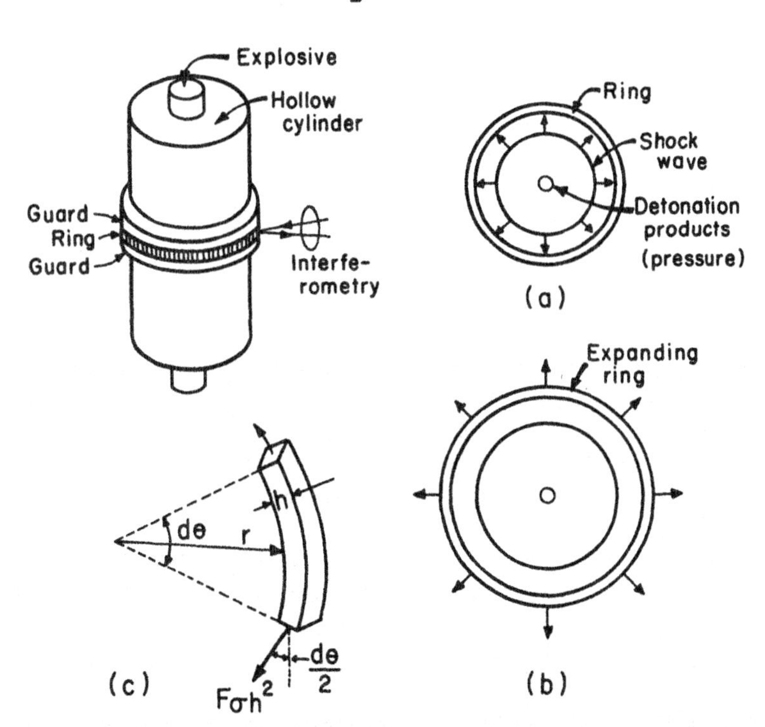

FIGURE 12.12 Expanding ring technique: (a) steel block with explosive in core; (b) sections of cylinder just after detonation and during flight of ring; (c) section of ring.

$$2\sigma h^2 \sin \frac{d\theta}{2} = \rho r \, d\theta \, h^2 \ddot{r}$$

For small $d\theta$,

$$\sin \frac{d\theta}{2} \approx \frac{d\theta}{2}$$

Thus,

$$2\sigma h^2 \frac{d\theta}{2} = \rho r \, d\theta \, h^2 \ddot{r}$$

$$\sigma = \rho r \ddot{r} \tag{12.13}$$

This simple equation tells us that if we know the deceleration history of the ring and its density, we can determine the stress in it. Velocity laser interferometry (see Chapter 11) gives us the velocity as a function of time.

For the true strain, we have (assuming a uniaxial strain)

$$\varepsilon = \ln \frac{r}{r_0} \tag{12.14}$$

where r_0 is the original ring radius.

From Eqn. (12.14) the strain rate is

$$d\varepsilon = \frac{dr}{r}$$

$$\frac{d\varepsilon}{dt} = \frac{1}{r} \frac{dr}{dt} = \frac{\dot{r}}{r} \tag{12.15}$$

It should be noticed that the velocity of the ring drops continuously from its initial value, given by the reflected stress pulse in the ring. Thus, the strain rate is continuously varying, and we have to conduct a number of tests with different explosive charges to develop stress–strain curves at constant strain rates. Figure 12.13 shows a characteristic velocity–time record obtained from laser interferometry measurements.

There are alternative ways of accelerating the ring. Investigators such as Fyfe and Rajendran [20] have used an exploding wire (by capacitor discharge) to produce the energy. A very strong electromagnetic pulse was used by Gourdin [21]; there is a heating induced by currents that should not be neglected.

Example 12.4. Determine the stress–strain curve for a copper specimen tested by the expanding-ring technique from the laser interferometry signal given in Figure 12.13. The initial ring diameter is 2 in.

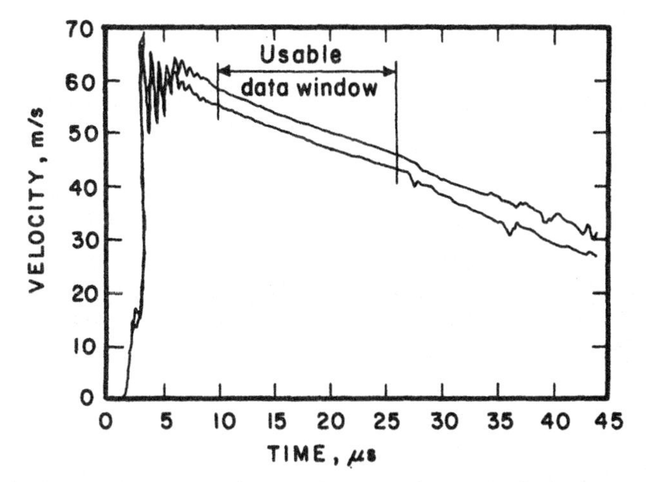

FIGURE 12.13 Velocity–time history of copper ring in explosively accelerated device.

One can see that the velocity rapidly rises to 70 m/s. Then, one observes early-time ringing of the stress waves, caused by wave reflections. After these wave interactions dampen, the ring slowly and steadily decelerates over the first 45 μs. We will have to reduce the data in an efficient manner, and a first stage is to tabulate it.

The position is found through the equation

$$r - r_0 = \int_{t0}^{t1} v\, dt$$

The integration is graphically conducted for the times of 5, 10, 15, . . . , 40 μs. The acceleration is taken as the slope of Figure 12.13. It is essentially constant and equal to 9.1×10^5 m/s^2. The results of the calculation are shown in Table 12.1.

TABLE 12.1 Tabulation of Ring Expansion Data: Experiment 2539

Time (μs)	Velocity (m/s)	Position, $r - r_0$ (mm)	r (mm)	Acceleration (m/s^2)	Stress (MPa)	Strain, $\ln(r/r_0)$
5	65	0.17	25.57	9.1×10^5	206.0	0.667×10^{-2}
10	58	0.47	25.87	9.1×10^5	208.5	1.83×10^{-2}
15	54	0.75	26.15	9.1×10^5	210.8	2.91×10^{-2}
20	50	1.01	26.41	9.1×10^5	212.9	3.89×10^{-2}
25	47	1.26	26.66	9.1×10^5	214.9	4.84×10^{-2}
30	42	1.48	26.88	9.1×10^5	216.6	5.66×10^{-2}
35	38	1.68	27.08	9.1×10^5	218.3	6.40×10^{-2}
40	34	1.86	27.26	9.1×10^5	219.7	7.07×10^{-2}

12.4 EXPLOSIVELY DRIVEN DEVICES

Explosively driven systems are the technique requiring least capital investment and, hence, are best suited for startup of a shock-loading program. It is assumed, of course, that the user is familiar with the safety requirements in the use of explosives. It is obvious that great care should be exercised in the handling of explosives; this is especially true when new systems are being tested. Different experimental arrangements have been developed throughout the years to transform a point detonation (produced by a detonator) into a plane detonation. Several of these will be reviewed here, and it will be shown how they can be designed. First, the point detonation is transformed in a line-wave generator. Then, the line detonation is transformed into a plane detonation by means of a plane-wave generator.

12.4.1 Line- and Plane-Wave Generators

Line-Wave Generator. Among the various systems designed, the perforated triangle is the most common. The perforated triangle is initiated at one of the vertices by a cap and a small booster charge. The detonation front has to pass between the holes, and the curved trajectory D_1 is the same as the trajectory at the edges. (Fig. 12.14). Hence, the wave front is linear. The diameter and spacing of the circles are chosen in such a way that the above conditions are met. Line-wave generators are commercially available from DuPont but can be easily fabricated by making steel dies with the appropriate hole pattern; the sheet explosive is placed between the dies and a punch is used to perforate the explosive. Typical dimensions are A = 0.25 in. (0.64 cm) and B = 0.3 in. (0.76 cm). These DuPont line-wave generators were tested, and it was found that there was a variation of about 0.5 μs in the arrival time of the wave on the side opposite to the detonation initiation apex; this corresponds to a deviation from a straight line of about 3.4 mm. There are other types of line-wave generators; one of them is discussed by Kestenbach and Meyers [22]; additional ones are presented by Benedick [23] and DeCarli and Meyers [24].

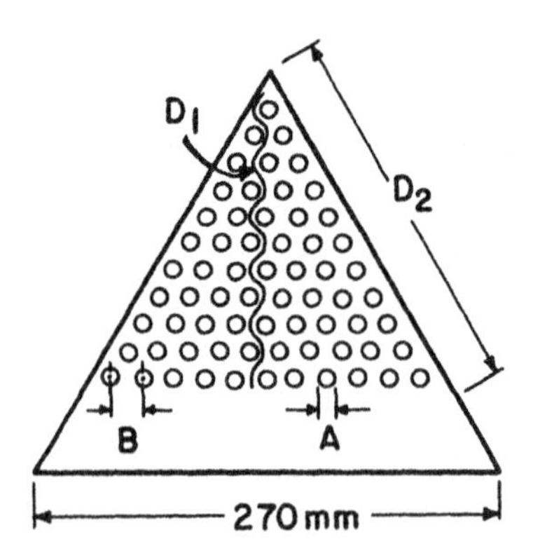

FIGURE 12.14 Triangular line-wave generator.

Plane Initiation. In order to transport a planar shock front to the flyer plate or to the system or to transform a point detonation into a desired surface configuration, special experimental configurations have to be used. These systems will now be discussed.

Explosive Lens. Figure 12.15 shows one of the possible designs for an explosive lens. The detonator transmits the front to two explosives that have different detonation velocities. The inside explosive has a detonation velocity V_{d2} lower than the outside one. The angle θ is such that

$$\sin \theta = \left(\frac{V_{d2}}{V_{d1}}\right) \tag{12.16}$$

The apex of the cone is not exactly straight; a certain curvature is introduced to compensate for initiation phenomena (the steady detonation velocity is not achieved instantaneously) and for the fact that the initiation source is not an infinitesimal point source. Because of these and other complications, such as the requirement for precision casting of explosives, fabrication of explosive lenses is best left to specialists.

Mousetrap Plane-Wave Generator. The mousetrap assembly is frequently used for metallurgical work. Figure 12.16 shows a common set up [25]. Two layers of Detasheet C-2 (2 mm thick each) are placed on top of the 3-mm-thick glass plate tilted an angle α to the main charge. The detonation of the explosive will propel the glass into the main charge; all glass fragments should simultaneously hit the top surface of the main charge, resulting in plane detonation. It has been experimentally found that one layer of Detasheet C-2 does not propel the fragments with a velocity high enough to initiate the main charge, if it is Detasheet. The angle α is calculated from the velocities of detonation and fragments, V_d and V_f, respectively. The velocity of fragments can be calculated

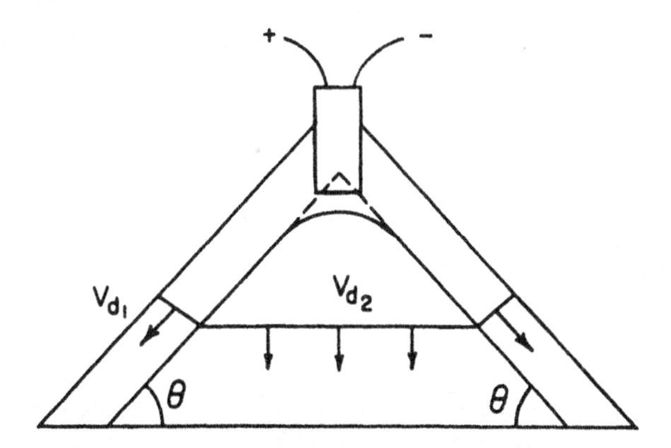

FIGURE 12.15 Conical explosive lens.

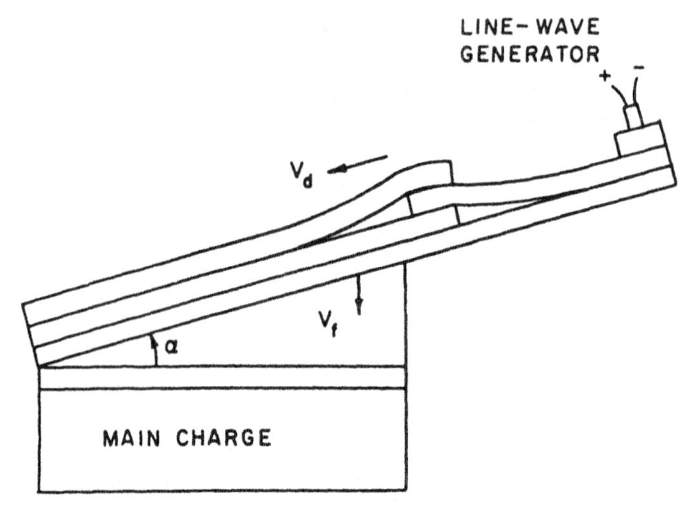

FIGURE 12.16 Mousetrap-type plane-wave generator.

using the Gurney equation, given in Chapter 8. For the system shown in Figure 12.16, it has been found that $\alpha = 11°$ provides a good planarity. The extremity of the line-wave generator should be inserted between the two plane-wave generator sheets. Usually the glass piece is cut so that the line-wave generator, booster, and plane-wave generator are glued to it. Metal plates can also be used instead of glass.

12.4.2 Flyer Plate Accelerating

The plane-wave generators are used to simultaneously initiate the entire surface of a main charge. The detonation of this main charge, in turn, produces the energy that propels a flyer plate at velocities that can be as high as 3 km/s for simple planar systems. Figure 12.17 shows two systems in which a plate is accelerated to impact a system in a parallel manner. In the first system, the flyer plate is placed at an angle to the target. The angle is such that the flyer plate velocity (V_{p}) creates a normal-parallel impact. This is achieved when

$$\sin \alpha = \frac{V_{\text{p}}}{V_{\text{d}}}$$

For larger charges one usually uses the mousetrap plane-wave generator shown in Figure 12.17(b).

Explosives are also used in different configurations. In Figure 12.18(a) the flyer plate is driven by the detonation of a main charge, initiated by a plane-wave lens. This system was developed by Sawaoka and co-workers (and is described by Graham and Sawaoka [26]) for the study of the shock response

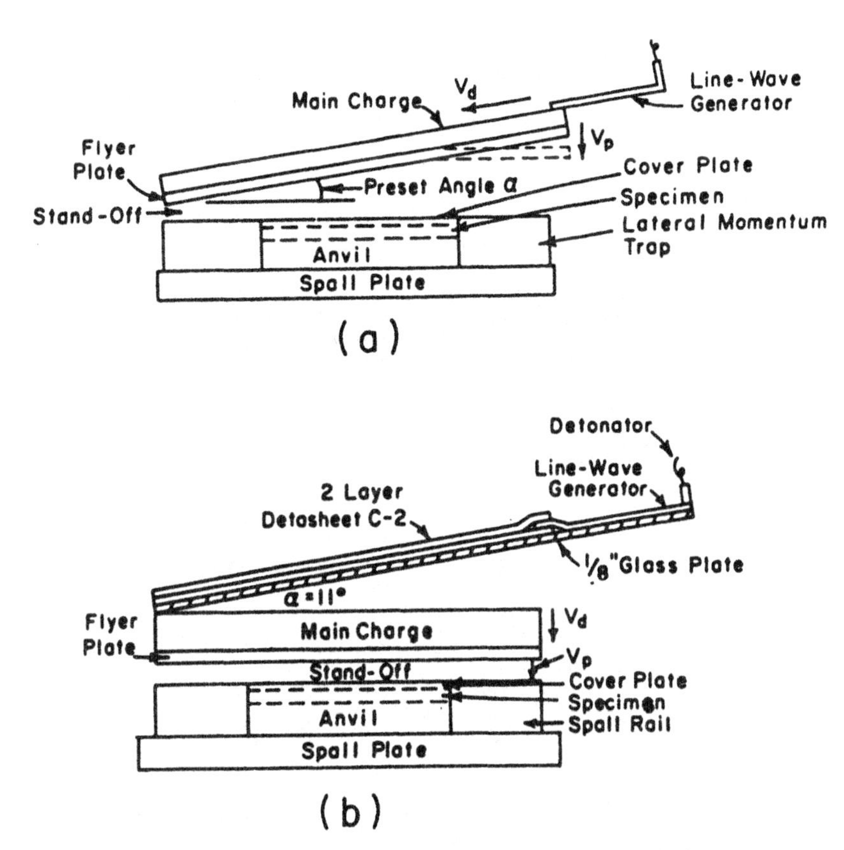

FIGURE 12.17 Shock-loading systems using HE: (a) inclined-plate geometry; (b) parallel-plate geometry with mousetrap plane-wave generator.

of powder. The target contains a number of powder capsules. Sawaoka obtained velocities of the order of 1–2 km/s. Yoshida [27] developed an explosively driven system consisting of two plane-wave lenses simultaneously initiated and driving two flyer plates (Fig. 12.18(b)). These flyer plates impact opposite sides of a main target simultaneously and generate two shock waves, which superimpose at the center, producing a corresponding pressure increase. The specimen to be studied is placed in the shock superposition region. Chapter 9 provides a more detailed discussion of this topic (shock wave superposition).

A system that is commonly used to generate the high pressures required in the shock consolidation of powders is the one shown in Chapter 17, (Fig. 17.36). An explosive charge is placed in a cylindrical container, coaxially with a cylindrical sample capsule. The sample capsule is surrounded by a flyer tube, which is accelerated inward by the detonation of the explosives. The detonation is initiated at the top and travels downward, propelling the flyer tube inward. The flyer tube impacts the sample capsule in an axisymmetric manner.

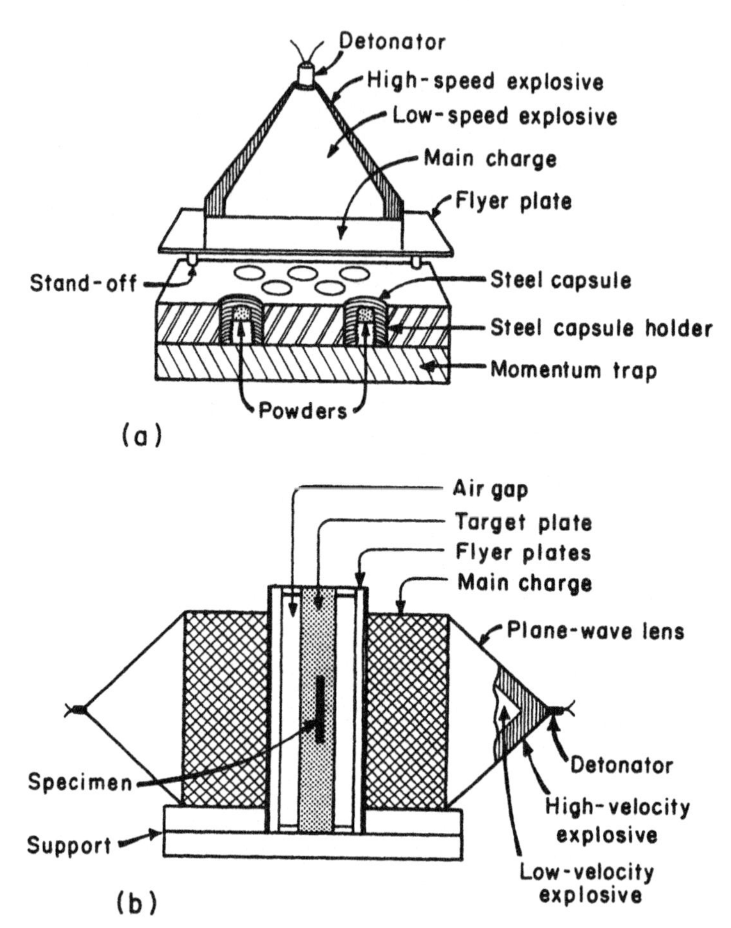

FIGURE 12.18 (a) Sawaoka system for producing shock waves traveling through capsules; (b) Yoshida system using shock superposition.

12.5 GUN SYSTEMS

Guns have been used for many years and will continue to be used as tools to generate impacts between 100 and 8000 m/s. The principal advantages of guns over other techniques are the reproducibility of results, the excellent planarity and parallelism at impact, and the relative ease to use sophisticated instrumentation and diagnostics. For lower velocity impacts one uses one-stage gas guns (maximum velocity 1100 m/s) or propellant guns (maximum velocity 2500 m/s). For velocities up to 8 km/s, one uses the two-stage gas gun. Even higher velocities can be reached by means of an electromagnetic gun known commonly as a "rail gun." Velocities of 15 km/s have been reported for this gun. The world record is currently held by Igenbergs of the Technical University of Munich [28], with claims of velocities of 25 km/s and higher using a plasma accelerator.

The one-stage gas gun is very simple in design. It consists of a breech, a barrel, and a recovery chamber (or target chamber). The high-pressure gas is loaded in the high-pressure chamber and the projectile, mounted in a sabot, is placed in the barrel. A valve is released (or a diaphragm is burst) and the high-pressure gas drives the projectile. One uses light gases for maximum velocity; hydrogen and helium are favorites, although air can also be used. The maximum velocity can be calculated from the maximum rate of expansion of the gas. Figure 12.19 shows a one-stage gas gun that can accelerate a projectile to a velocity of 1200 m/s using helium. It contains a fast-acting valve that is pulled back by a pressure differential and allows the high-pressure chamber to vent into the barrel. The projectile is accelerated in the barrel. Its velocity is measured by the interruption of two laser beams. The recovery chamber is used to capture the specimens with minimum postimpact damage.

Figure 12.20 shows a two-stage gas gun. The first two-stage gas gun was designed and built at the New Mexico School of Mines (now New Mexico Tech.) by Crozier and Hume [29]. It uses the first stage to drive a fairly massive piston. This piston is usually accelerated by means of the deflagration of a gunpowder load. In Figure 12.20 the four phases of operation of the gun are shown. The powder is placed in the breech. The first stage has a considerably larger diameter than the second stage. The first stage is filled (between projectile and piston) with a light gas (He or H). The deflagration of the gunpowder accelerates the piston, compressing the light gas. When the pressure in the first stage reaches a critical level, a diaphragm (rupture disk) opens and starts driving the projectile. The piston continues to move and to compress the gas that drives the projectile. The piston eventually partly penetrates into the launch tube. Very high pressures in the light gas are generated that can drive projectiles to velocities that can routinely reach 7 km/s. In some cases, higher velocities can be achieved.

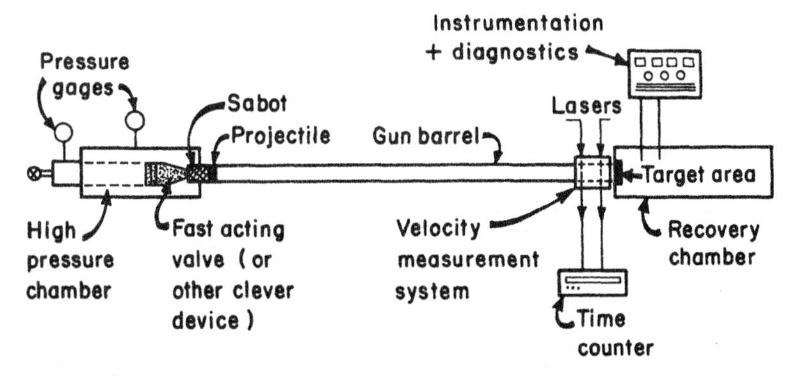

FIGURE 12.19 Schematic representation of one-stage light gas gun.

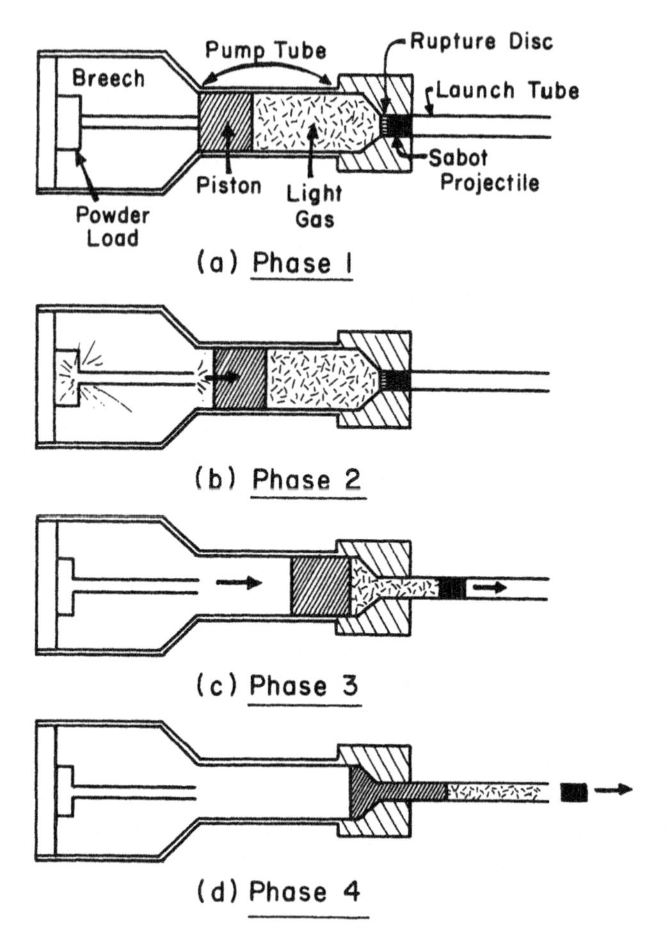

FIGURE 12.20 Two-stage gas gun. (Courtesy of A. Ferreira, Instituto Militar de Engenharia, Brazil.)

Even higher velocities can be reached with the electric rail gun [30, 31]. This gun uses EMFs that create a mechanical force (Lorentz force) to

$$F = \tfrac{1}{2}L'I^2$$

where L' is the inductance per unit length and I is the current passing through the circuit. Figure 12.21 shows, in schematic fashion, how a rail gun operates. Two parallel rails of a conductor (copper) bracket a projectile. This projectile is usually made of a very low density material (polycarbonate) with a metallic tail (armature). It can be seen that the velocity increases with the force F, which is proportional to the square of the current. High currents are achieved by capacitor discharge.

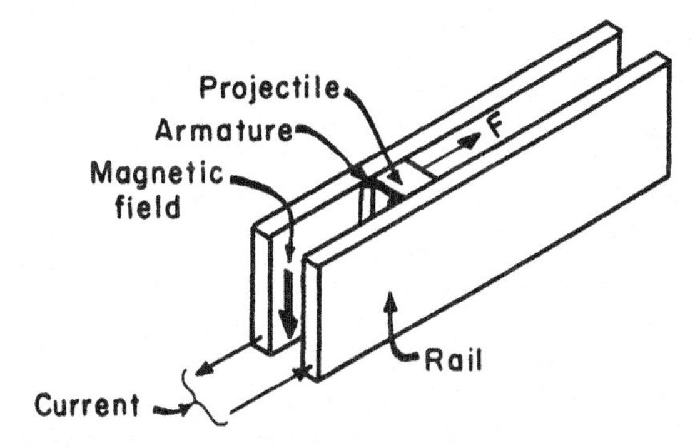

FIGURE 12.21 Basic operation of a DC electric-rail gun (From Swift, Fig. 9, p. 235.)

Other types of devices that have been developed to create high pressures are pulsed lasers and exploding wires and foils.

REFERENCES

1. G. E. Duvall, in *Metallurgical Applications of Shock-Wave and High-Strain-Rate Phenomena*, eds. L. E. Murr, K. P. Staudhammer, and M. A. Meyers, Dekker, New York, 1986, p. 3.
2. B. Hopkinson, *Proc. Roy. Soc. A*, **74** (1905), 498.
3. G. R. Johnson, *ASME J. Eng. Mater. Tech.*, **103** (1981), 201.
4. C. J. Maiden and S. V. Green, *ASME J. Appl. Mech.*, **33** (1966), 496.
5. H. C. Rogers, *Ann. Rev. Mater. Sci.*, **9** (1979), 283.
6. J. Mescall and V. Weiss, eds., *Material Behavior under High Stress and Ultrahigh Loading Rates*, Plenum, New York, 1983.
7. G. B. Olson, J. F. Mescall, and M. Azrin, in *Shock Waves and High-Strain-Rate Phenomena in Metals*, eds. M. A. Meyers and L. E. Murr, Plenum, New York, 1981, p. 221.
8. T. Vinh, M. Afzali, and A. Roche, in *Mechanical Behavior of Materials*, Intl. Conf. on Matls, 1981, Vol. 2, Pergamon, Elmsford, NY, p. 633.
9. J. D. Campbell, A. M. Eleiche, and M. C. C. Tao, in *Fundamental Aspects of Structural Alloy Design*, Plenum, New York, 1977, p. 545.
10. J. D. Campbell, *Dynamic Plasticity of Metals*, Springer, 1972.
11. J. Harding, in *Explosive Welding, Forming, and Compaction*, ed. T. Z. Blazynski, Applied Science Publishers, London, 1983, p. 123.
12. P. S. Follansbee, in *Metallurgical Applications of Shock-Wave and High-Strain-Rate Phenomena*, eds. L. E. Murr, K. P. Staudhammer, and M. A. Meyers, Dekker, New York, 1986, p. 451.
13. J. D. Campbell and W. G. Ferguson, *Phil. Mag.*, **21** (1970), 63.
14. R. J. Clifton, *J. Appl. Mech.*, **50** (1983), 941.

15. U. S. Lindholm, *Tech. Met. Res.*, **5** (1971), 199.
16. S. Nemat-Nasser, J. B. Isaacs, and J. E. Starrett, *Proc. Roy. Soc. Lond.*, **A435** (1991), 371.
17. R. M. Davies, *Phil. Trans. Roy. Soc. Lond.*, **A240** (1948), 375.
18. P. S. Follansbee, *Metals Handbook*, Vol. 8, American Society for Metals, Metals Park, OH, 1985, p. 198.
19. P. C. Johnson, B. A. Stern, and R. S. Davis, "Symposium on the Dynamic Behavior of Materials," Special Technical Publication No. 336, American Society for Testing and Materials, Philadelphia, PA, 1963, p. 195.
20. I. M. Fyfe and A. M. Rajendran, *J. Mech. Phys. Solids*, **28** (1980), 17.
21. W. H. Gourdin, in *Impact Loading and Dynamic Behavior of Materials*, eds. C. Y. Chiem, H.-D. Kunze, and L. W. Meyer, Informationsgesellshaft Verlag, Oberursel, Germany, 1988, p. 533.
22. H.-J. Kestenbach and M. A. Meyers, *Met. Trans. A*, **7** (1976), 1943.
23. W. Benedick, *Behavior and Utilization of Explosives in Engineering Design*, ASME, 1972.
24. P. S. DeCarli and M. A. Meyers, in *Shock Waves and High-Strain-Rate Phenomena in Metals*, eds. M. A. Meyers and L. E. Murr, Plenum, New York, 1981, p. 341.
25. R. N. Orava and R. H. Whittman, Proc. 5th Intl. Conf. High Energy Rate Fabrication, University of Denver, Colorado, 1975, p. 1.1.1.
26. R. Graham and A. Sawaoka, eds., *High Pressure Explosive Processing of Ceramics*, Trans. Tech. Publications, Basel, Switzerland, 1987.
27. M. Yoshida, CETR Report, New Mexico Institute of Mining and Technology, 1986.
28. E. B. Igenbergs, D. W. Jex, and E. L. Shriver, *AIAA J.*, **13** (1975), 1024.
29. W. D. Crozier and W. Hume, *J. Appl. Phys.*, **28** (1957), 292.
30. W. F. Feldon, *Intl. J. Impact Eng.*, **5** (1987), 671.
31. J. J. Zukas, T. Nicholas, H. F. Swift, L. B. Greszczak, and D. R. Curran, *Impact Dynamics*, Wiley, New York, 1982.

CHAPTER 13

Plastic Deformation at High Strain Rates

13.1 INTRODUCTION

High-strain-rate plastic deformation of materials is often described by constitutive equations that link stress with strain, strain rate, and often, temperature. We can express the stress schematically as

$$\sigma = f(\varepsilon, \dot{\varepsilon}, T)$$

where ε is the strain, $\dot{\varepsilon}$ is the strain rate, and T is the temperature. Since plastic deformation is an irreversible and thus path-dependent process, the response of the material at a certain (σ, ε) point is dependent on the deformation substructure created. Since there is a variety of deformation substructures, dependent on strain rate, temperature, and stress state, one has to add the general term called "deformation history" to the equation above:

$$\sigma = f(\varepsilon, \dot{\varepsilon}, T, \text{deformation history})$$

We will think here of scalar stress, although we know that stress (and strain) is a second-order tensor. This can be done by reducing the stress and strain to "effective" stress and strain according to the equations

$$\sigma_{\text{eff}} = \frac{\sqrt{2}}{2}\left[(\sigma_1 - \sigma_2)^2 + (\sigma_2 - \sigma_3)^2 + (\sigma_1 - \sigma_3)^2\right]^{1/2}$$

$$\varepsilon_{\text{eff}} = \frac{\sqrt{2}}{3}\left[(\varepsilon_1 - \varepsilon_2)^2 + (\varepsilon_1 - \varepsilon_3)^2 + (\varepsilon_2 - \varepsilon_3)^2\right]^{1/2}$$

Rather than dealing with tensors, we deal only with the scalar quantities σ_{eff} and ε_{eff}. Alternatively, we can deal with τ and γ, shear stresses and strains. We know that shear stresses are important components in plastic deformation. Metals and polymers deform plastically by shear and the shear stresses "drive" this deformation. More advanced treatments and computational schemes use

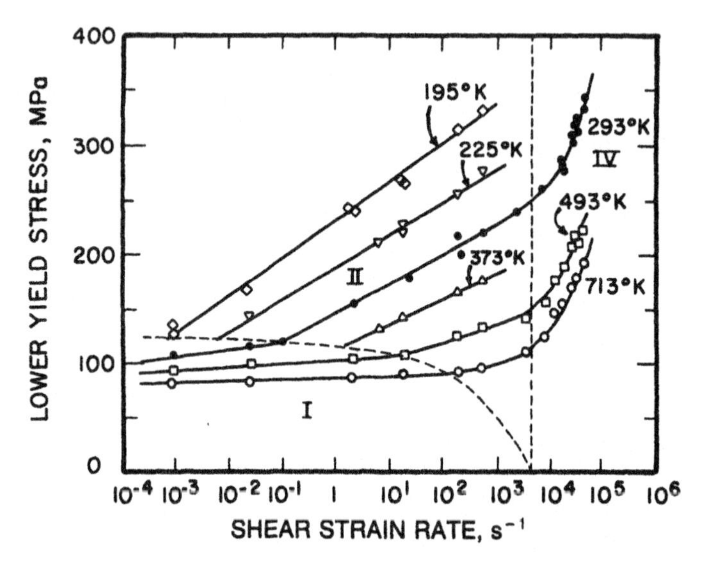

FIGURE 13.1 Lower yield stress vs. strain rate for mild steel. (From Campbell and Ferguson [2], Fig. 5, p. 68. Reprinted with permission of the publisher.)

the full tensorial approach [e.g., 1]. A solid foundation in mechanics is required from the student.

Figure 13.1 shows the classic plot by Campbell and Ferguson [2]. It represents the lower yield strength of mild steel plotted against the logarithm of the strain rate at different temperatures. Two observations can be made:

1. The yield stress increases with strain rate.
2. The increase of yield stress with strain rate is more marked at lower temperatures.

Observation 1 is almost universal and applies to most materials. In the literature, one often finds the yield stress plotted against the logarithm of the strain rate. A successful constitutive model will collapse the entire data of Figure 13.1 into a single equation and will have the capability of extrapolation and interpolation. It is customary to plot the logarithm of strain rates as the abscissa.

In this chapter, we first will restrict ourselves to the very simple constitutive equations, which will be described in Section 13.2. Then we will investigate the mechanisms by which materials deform plastically at high strain rates (dislocations, deformation twins, phase transformations, and their interactions) (Section 13.3). Following this we will present more recent constitutive equations based on the actual micromechanical deformation modes (Section 13.4). Note that these one-dimensional equations can be "tensorialized" to be incorporated in codes. This chapter will not discuss yield (or flow) criteria. These establish the stress level at which a material deforms plastically under a complex loading situation, such as von Mises or J_2, once we know the yield (or flow) stress in uniaxial tension or compression. An associated flow rule, such as Drucker's normality and convexity postulates, is also a necessary ingredient of full plasticity treatments. It dictates the plastic deformation path (the relative values by which the three principal stresses evolve) once the yield surface is penetrated. Work hardening of the material is a complex phenomenon, and two types of response are usually considered: isotropic hardening (the simple expansion of the yield surface) and kinematic hardening (the translation of the yield surface). These concepts comprise the intricacies of computational mechanics, a powerful tool in the solution of dynamic problems. We will leave this to the experts, or "jocks," as they are called. In Chapter 14, we will investigate the effects of shock waves on the microstructure of materials. Shock waves can be considered as a limiting case, in which the rate of increase of stresses at the front determines the strain rate. Estimated strain rates at the shock front are 10^7–10^9 s^{-1}.

13.2 EMPIRICAL CONSTITUTIVE EQUATIONS

There are a number of equations that have been proposed and successfully used to describe the plastic behavior of materials as a function of strain rate and temperature. Their objective is to collapse the data of Figure 13.1 into one single equation. At low (and constant) strain rates, metals are known to work harden along the well-known relationship (called parabolic hardening)

$$\sigma = \sigma_0 + \kappa \varepsilon^n$$

Here σ_0 is the yield stress, n is the work-hardening coefficient, and κ is the preexponential factor. The effects of temperature and strain rate are separately given, for mild steel, in Figure 13.2. The effect of temperature on the flow stress can be represented by

$$\sigma = \sigma_r \left[1 - \left(\frac{T - T_r}{T_m - T_r} \right)^m \right]$$

where T_m is the melting point, T_r is a reference temperature at which σ_r, a reference stress, is measured, and T is the temperature for which σ is calculated. This is simple curve fitting, and the equation above increases in "concavity" as m, an experimentally determined fitting parameter, is increased. The effect of strain rate, which will be discussed in much greater detail in Section 13.3, can be simply expressed by

$$\sigma \propto \ln \dot{\varepsilon}$$

This relationship is observed very often at strain rates that are not too high. (In Fig. 13.2, it breaks down at $\dot{\varepsilon} = 10^2$ s^{-1}.) Johnson and Cook [4] and

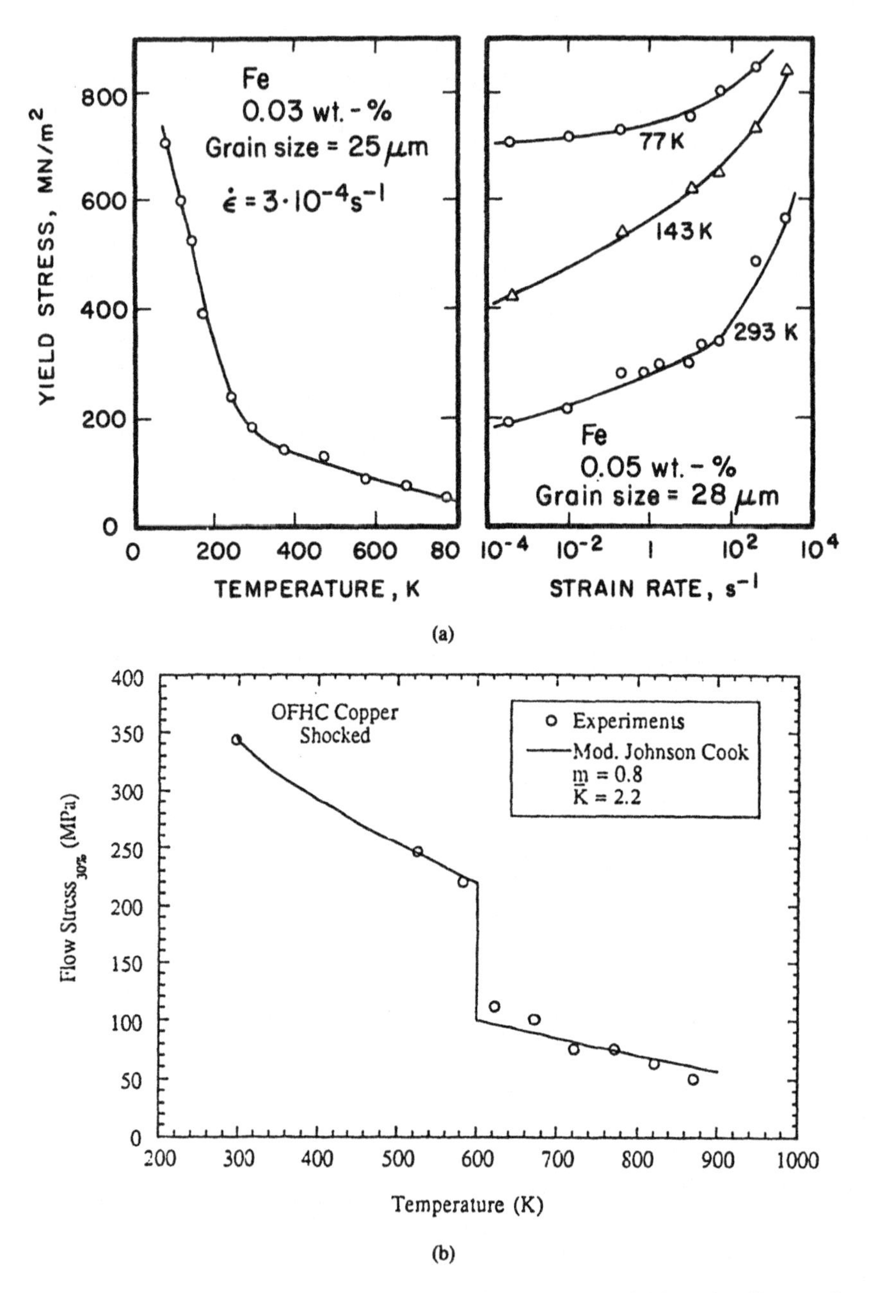

FIGURE 13.2 Effects of temperature and strain rate on the yield stress of iron. (From Vohringer [3], Fig. 31, p. 44); (b) experimentally determined and computed (using modified Johnson-Cook equation) flow stress vs. temperature for shock-hardened copper (from Andrade, Meyers, and Chokshi, *Scripta Met. et Mat.* **30** (1994) 933).

Johnson et al. [5] used these basic ingredients and proposed the following equation:

$$\sigma = (\sigma_0 + B\varepsilon^n)\left(1 + C \ln \frac{\dot{\varepsilon}}{\dot{\varepsilon}_0}\right)[1 - (T^*)^m] \tag{13.1}$$

This equation has five experimentally determined parameters (σ_0, B, C, n, m) that describe fairly well the response of a number of metals. The term T^* is calculated as:

$$T^* = \frac{T - T_r}{T_m - T_r}$$

where T_r is the reference temperature at which σ_0 is measured and $\dot{\varepsilon}_0$ is a reference strain rate (that can, for convenience, be made equal to 1). Johnson and Cook tested a number of materials and obtained these parameters, which are given in Table 13.1. The Johnson–Cook equation is a highly useful and successful constitutive model. One of the problems with this equation is that all parameters are coupled by being multiplied by each other. Nevertheless, the Johnson–Cook equation remains the "workhorse" of constitutive modeling. It has also been applied to ceramics (in modified form).

Additional empirical equations have been proposed by other investigators; they are reviewed by Meyer [6]. Klopp et al. [7] used the following equation:

$$\tau = \tau_0 \gamma^n T^{-\nu} \dot{\gamma}_p^m \tag{13.2}$$

where τ and γ are shear stress and strain, respectively, ν is a temperature-softening parameter, and n and m are work hardening and strain rate sensitivity, respectively.

Other investigators favor the equation

$$\tau = \tau_0 \left(1 + \frac{\gamma}{\gamma_0}\right)^N \left(\frac{\dot{\gamma}}{\dot{\gamma}_r}\right)^m e^{-\lambda \Delta T} \tag{13.3}$$

where τ_0 is the yield stress of the material ($\gamma = 0$) at the reference strain rate ($\dot{\gamma}_r$) and ΔT is the change in temperature from the reference value to $T - T_0$. This equation has an exponential thermal softening. This expression is often expressed explicitly by $\dot{\gamma}$. By making:

$$\sigma_r = \tau_0 e^{-\lambda \Delta T}\left(1 + \frac{\gamma}{\gamma_0}\right)^N$$

TABLE 13.1 Constitutive Constants for Various Materials ($\dot{\varepsilon}_0 = 1\ s^{-1}$)

Material	Description				Constitutive Constants for $\sigma = [\sigma_o + B\varepsilon^n][1 + C \ln \varepsilon^*][1 - T^{*m}]$				
	Hardness (Rockwell)	Density (kg/m^3)	Specific Heat (J/kg K)	Melting Temperature (K)	σ_o (MPa)	B (MPa)	n	C	m
OFHC copper	F-30	8960	383	1356	90	292	0.31	0.025	1.09
Cartridge brass	F-67	8520	385	1189	112	505	0.42	0.009	1.68
Nickel 200	F-79	8900	446	1726	163	648	0.33	0.006	1.44
Armco iron	F-72	7890	452	1811	175	380	0.32	0.060	0.55
Carpenter electrical iron	F-83	7890	452	1811	290	339	0.40	0.055	0.55
1006 steel	F-94	7890	452	1811	350	275	0.36	0.022	1.00
2024-T351 aluminum	B-75	2770	875	775	265	426	0.34	0.015	1.00
7039 aluminum	B-76	2770	875	877	337	343	0.41	0.010	1.00
4340 steel	C-30	7830	477	1793	792	510	0.26	0.014	1.03
S-7 tool steel	C-50	7750	477	1763	1539	477	0.18	0.012	1.00
Tungsten alloy (.07Ni, .03Fe)	C-47	17000	134	1723	1506	177	0.12	0.016	1.00
Depleted uranium-0.75% Ti	C-45	18600	117	1473	1079	1120	0.25	0.007	1.00

Source: From Johnson and Cook [4], p. 4.

one has the simplified form

$$\dot{\gamma} = \dot{\gamma}_r \left(\frac{\tau}{\sigma_r}\right)^{1/m}$$

This form is useful in large codes.

Two additional examples are given below:

$$\tau = \tau_0 \gamma^n \left(\frac{\dot{\gamma}}{\dot{\gamma}_0}\right)^m \exp\left(\frac{W}{T}\right) \tag{13.4}$$

where τ_0, W, n, and m are parameters that have to be determined experimentally and τ, γ, and $\dot{\gamma}$ are shear stress, strain, and strain rate, respectively. The equation above is due to Vinh et al. [8] and is essentially identical to Eqn. (13.3) Campbell and co-workers [9] used the following equation successfully for copper:

$$\tau = A\gamma^n[1 + m \ln(1 + \dot{\gamma}/B)] \tag{13.5}$$

For a more extensive discussion, the reader is referred to Harding [1], Clifton [11], and Meyer [6], who present descriptions of constitutive equations. It is important to recognize that the empirical constitutive equations above are basically a "curve-fitting" procedure and that each research group eventually develops its own formulation. The important part is to obtain the equation parameters from experiments. The Johnson–Cook equation has been most widely used, and the parameters are known for a large number of materials. Constitutive equations have also been developed for ceramic materials. Andrade et al. [12] modified the Johnson-Cook equation to incorporate dynamic recrystallization at higher temperatures.

They incorporated a reducer function $H(t)$ into the constitutive equation:

$$\sigma = (\sigma_0 + B\varepsilon^n)\left(1 + C \log \frac{\dot{\varepsilon}}{\dot{\varepsilon}_0}\right)\left[1 - \left(\frac{T - T_r}{T_m - T_r}\right)^m\right] H(T)$$

$$H(T) = \frac{1}{1 - \left[1 - \dfrac{(\sigma_f)_{rec}}{(\sigma_f)_{def}}\right] u(T)}$$

$u(T)$ is a step function of temperature defined as:

$$u(T) = \begin{cases} 0 & \text{for } T < T_c \\ 1 & \text{for } T > T_c \end{cases}$$

T_c is the temperature at which the critical phenomenon (dynamic recrystallization, phase transformation, etc.) occurs. $(\sigma_f)_{rec}$ and $(\sigma_f)_{def}$ are the flow stresses of the material just after and prior to recrystallization, respectively. Application of this equation to copper in the shock-hardened condition yielded the results shown in Figure 13.2(b). The drastic drop in flow stress at 600 K, corresponding to dynamic recrystallization, is well represented by the modified equation. This equation can also be applied to phase changes that are temperature activated.

13.3 RELATIONSHIP BETWEEN DISLOCATION VELOCITY AND APPLIED STRESS

13.3.1 Dislocation Dynamics

It is difficult to have broad generalizations and "unified" theories that describe the plastic behavior of all materials. Most rules have exceptions, and there are alternative deformation mechanisms for materials (dislocation glide, mechanical twinning, phase transformations). We will only consider dislocation movement as the agent for plastic deformation in this section. The reader not familiar with dislocations should read the appropriate section in the materials science textbook. It suffices here to recall that dislocations produce shear strain by movement and that they move under the action of shear stresses. Figure 13.3 shows an edge dislocation moving under the action of a shear stress τ. The force on the dislocation per unit length, under this ideal orientation arrangement, is given by

$$F = \tau b \tag{13.6}$$

where b is the dislocation Burgers vector (offset caused by dislocation). There are frictional forces resisting the movement of a dislocation; thus, a force is required to make it move. A more general formulation [11] is obtained considering vectors **F** and **b**:

$$\mathbf{F} = (\sigma \mathbf{b}) \times \mathbf{l} \tag{13.7}$$

where **l** is a vector parallel to the dislocation line. The movement of arrays of dislocations will produce a shear strain γ, as shown in Figure 13.3(b). It is assumed that the dislocations do not interact. This shear strain can be directly related to the number of dislocations, N, per unit area (or their density ρ):

$$\gamma = \tan\theta = \frac{Nb}{l} = \frac{Nbl}{l^2}$$

But $N/l^2 = \rho$, and we have

$$\gamma = \rho b l \tag{13.7a}$$

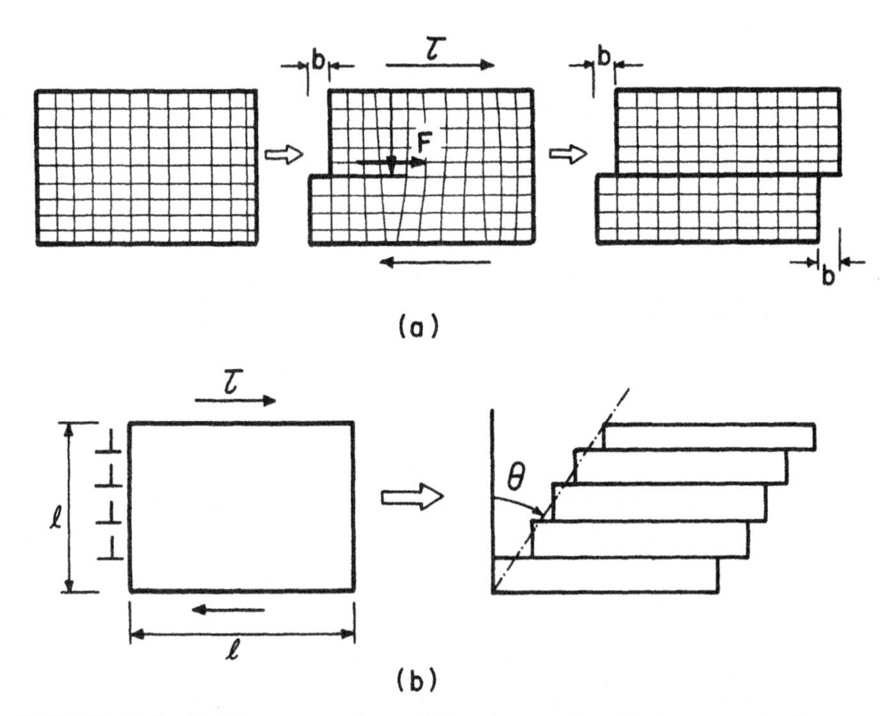

FIGURE 13.3 (a) Movement of one dislocation causing displacement b; (b) movement of array of dislocations causing shear strain γ (=tan θ).

Taking the time derivative,

$$\frac{dl}{d\gamma} = \rho b \frac{dl}{dt}$$

$$\dot{\gamma} = \rho b v \tag{13.8}$$

where v is the dislocation velocity. The shear strain can be converted into a longitudinal strain by adding an orientation factor M (see Meyers and Chawla [13]). This orientation factor is often taken as 3.1 (an average value that minimizes deformation energy for FCC crystals) and 2.75 (for BCC crystals):

$$\frac{d\varepsilon}{dt} = \dot{\varepsilon} = \frac{1}{M} \rho b v \tag{13.9}$$

The general tensorial formulation (since we have many slip systems in a crystal) is (see Clifton [11])

$$\dot{\varepsilon}^p = \sum_{k=1}^{n} \dot{\gamma}_k^p (\mathbf{m}^k \otimes \mathbf{n}^k)_s$$

By substituting Eqn. (13.8),

$$\dot{\varepsilon}^p = \sum_{k=1}^{n} b^k \rho^k v_d^k (\mathbf{m}^k \otimes \mathbf{n}^k)_s \tag{13.10}$$

Here, n is the number of slip systems, $\dot{\gamma}_k^p$ is the rate of plastic shearing in the kth slip system, which has slip direction ($\mathbf{m}^k$) and slip plane normal $\mathbf{n}^k$, ρ^k is the dislocation line length in slip system k per unit volume (density of dislocations k) and v^k is the dislocation velocity in system k.

The effect of applied stress on dislocation velocity has been established for a number of materials. Gilman and Johnston [14, 15], in their now classic experiments, measured the velocities of dislocations in LiF as a function of stress. They used an ingenious etch-pit technique, and their results are depicted schematically in Figure 13.4. This is, to the author's knowledge, the first study on dislocation dynamics. This pioneering work was followed by a succession of other papers. Some are Stein and Low [16] for Fe–Si, Ney et al. [17] for Cu and Cu–Al, Schadler [18] for tungsten, Chaudhuri et al. [19] for germanium and silicon, Erickson [20] for Fe–Si, Gutmanas et al. [21] for NaCl, Rohde and Pitt [22] for nickel, Pope et al. [23] for zinc, Greenman et al. [24] for copper, Suzuki and Ishii [25] for copper, Blish and Vreeland [26] for zinc, Gorman et al. [27] for aluminum, Parameswaran and Weertman [28] for lead, and Parameswaran et al. [29] for aluminum.

There is no experimental report, to the author's knowledge, of supersonic dislocation. The stress dependence of dislocation velocity has been fitted into different types of equations, some empirical, some with a theoretical backing. Johnston and Gilman [15] expressed the dislocation velocity as

$$v \propto \sigma^m e^{-E/kT} \tag{13.11}$$

where m varies between 15 and 25 when v is expressed in cm/s and σ in kgf/mm^2. At a constant temperature, we should have

$$v = K\sigma^m \tag{13.12}$$

Stein and Low [16], on the other hand, found their data to fit more closely the expression

$$v = A \exp\left(-\frac{A}{\tau}\right) \tag{13.13}$$

where τ is the resolved shear stress. Rohde and Pitt [22] found a good correlation with the expression

$$v = \frac{kT}{h} K \exp\left(-\frac{\Delta H}{kT}\right) \exp\left[\frac{B(\tau_a - \tau_l)}{kT}\right] \tag{13.14}$$

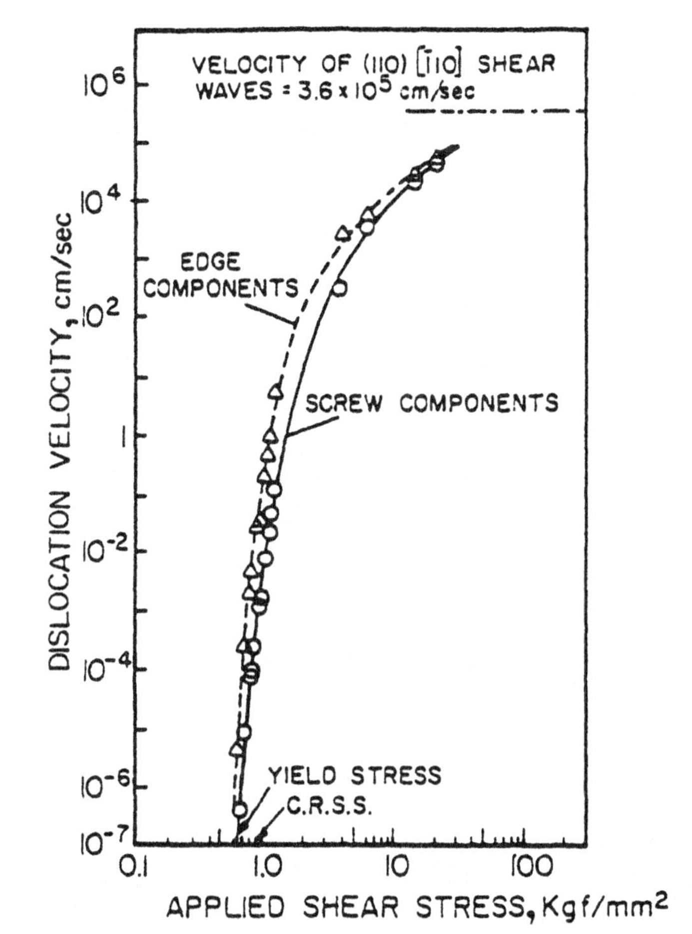

FIGURE 13.4 Johnston and Gilman's [15] classic experiments showing dislocation velocity vs. resolved shear stress for as-grown LiF single crystal. Units left unchanged (1 Pa = 9.8 × 10^6 kgf/mm^2). (From Johnston and Gilman [15], Fig. 5, p. 132. Reprinted with permission of the publisher.)

where h is Planck's constant, K a numerical constant, B an activation volume, ΔH the enthalpy of activation, τ_a the applied resolved shear stress, and τ_l the long-range internal stress. The term kT/h is the attempt frequency (see Section 13.3.2). This equation is based on the theory of reaction rates (Eyring theory) as presented by Glasstone et al. [30].

Greenman et al. [24] found the following relationship for copper:

$$\nu = \nu_0 \left(\frac{\tau}{\tau_0} \right)^m \tag{13.15}$$

where $m = 0.7$, ν_0 is unit velocity, and τ_0 is a material constant (= 0.25 × 10^3 N/m^2). But the data could also satisfactorily fit the equation with $m = 1$

and $\tau_0 = 2.7 \times 10^3\ \text{N/m}^2$:

$$\nu = \frac{\nu_0}{2.7 \times 10^3}\tau \quad \text{or} \quad \nu = K\tau \tag{13.16}$$

For the basal plane of zinc (and edge dislocations), Pope et al. [23] obtained a good fit by assuming that

$$\nu = 3.4 \times 10^{-5}\tau$$

where ν was expressed in cm/s and τ in dyn/cm^2. For the $\{\bar{1}2\bar{1}2\} > \langle\bar{1}\bar{2}13\rangle$ slip systems of zinc and varying the temperature, they were able to fit the data into an equation of the type

$$\nu = K\tau \exp\left(-\frac{E}{\kappa T}\right) \tag{13.17}$$

They found different K values for edge and screw dislocations. Notice that, at constant temperature, Eqns. (13.16) and (13.17) are special cases of Eqn. (13.15). For aluminum, Gorman et al. [27] were also able to obtain a linear relationship between the dislocation velocity and stress. Similarly, Weertman and co-workers [28, 29] were able to fit the data into Eqn. (13.16) for aluminum and lead.

As a conclusion it can be said that, although the correct form over a wide range of stresses is not a power function or a proportionality, these expressions can be used as an approximation. Gilman [31], in his book *Micromechanics of Flow in Solids*, presents the following equation to describe the behavior of almost any material:

$$\nu = \nu_s^*(1 - e^{-\tau/s}) + \nu_d^* e^{-D/\tau} \tag{13.18}$$

where D is the characteristic drag stress and ν_s^* and ν_d^* are limiting velocities. For $\tau \to \infty$,

$$\nu = \nu_s^* + \nu_d^*$$

So there is an upper limit for the velocity. On the other hand, the other models do not necessarily predict an upper limit for ν.

The approximation of $\nu \alpha \tau$ is valid at moderate stresses and is taken advantage of in the determination of the dislocation damping constant B. If we apply Newton's second law, we have

$$F = ma \tag{13.19}$$

Substituting Eqns. (13.6) into (13.19), we have

$$\tau b = ma$$

If we assume a certain dislocation mass, the dislocation would undergo acceleration under a constant velocity at a certain stress. Therefore, we would have to have something "holding back" the dislocations once a certain velocity is reached. There are two possibilities:

1. The mass m (or equivalent mass of a dislocation that will be defined later) increases with increasing velocity, causing the acceleration to decrease as the velocity increases (for the same applied stress).
2. In analogy with the movement of fluids, there is a drag stress, or damping of the dislocation motion.

These two effects are analyzed in Sections 13.3.3 and 13.3.4.

The various mathematical equations used to describe the dynamic behavior of dislocations can be confusing; for this reason, the log–log plot of Figure 13.5 was made. It is a compilation of data from several sources, and some trends are readily visible. The general conclusion at which one can arrive is that *the slope decreases as the velocity increases*. One can also see that the experimental results of Rohde and Pitt [22], despite the fact that they were adjusted to Eqn. (13.14), could very well be represented by Eqn. (13.12) (the log–log relationship of Gilman and Johnston [14]). If one wants to predict the response of nickel (or of any metal, for that matter) over the full velocity range, one should observe that in the medium-velocity range the velocity and stress have been found by several investigators to be linearly dependent. In the log–log plot of Figure 13.5, this corresponds to a unit slope found by Vreeland and co-workers [24, 26, 27], Weertman and co-workers [28, 29, 33], Preckel and Conrad [32], and Ney et al. [17]. This can be clearly seen from Eqn. (13.12), making $m = 1$. Accordingly, Figure 13.6 establishes a hypothetical stress dependence of dislocation velocities in nickel, in accord with the aforementioned observations. The regions of response are defined: I, II, and III. One has, for the coefficients of Eqn. (13.12), $m_{\mathrm{I}} > m_{\mathrm{II}} > m_{\mathrm{III}}$ and $m_{\mathrm{II}} = 1$. The limiting velocity of the dislocation is set as the velocity of propagation of elastic shear waves. This was established as the limiting velocity because both edge and screw dislocations generate shear stresses (and strains). These disturbances, in turn, cannot produce velocities higher than the velocity of elastic shear waves. Therefore, the dislocations, which are assemblies of stresses and strains, cannot travel faster than this.

These regimes of behavior, shown in Figure 13.6, are generally accepted as defining three different mechanisms governing plastic deformation. If one were to plot the logarithm of the dislocation velocity as the abscissa and the

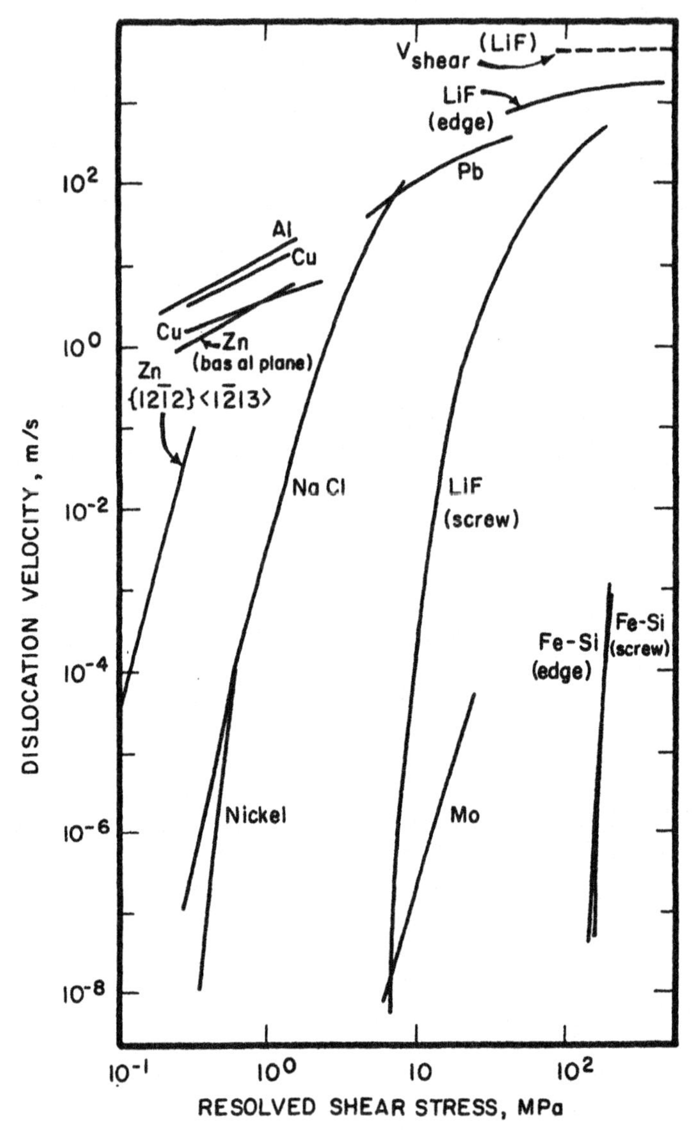

FIGURE 13.5 Compilation of results in the literature relating dislocation velocities and stress. (From M. A. Meyers and K. K. Chawla, *Mechanical Metallurgy, Principles and Applications*, © 1984, Fig. 8.9, p. 304. Reprinted by permission of Prentice-Hall, Englewood Cliffs, NJ.)

applied stress as the ordinate, one would obtain a plot of the same general shape as Figure 13.1. In the next three subsections, the three governing mechanisms will be delineated. These mechanisms are thermally activated dislocation motion (region I, Section 13.3.2), phonon drag (region II, Section 13.3.3), and relativistic effects (region III, Section 13.3.4).

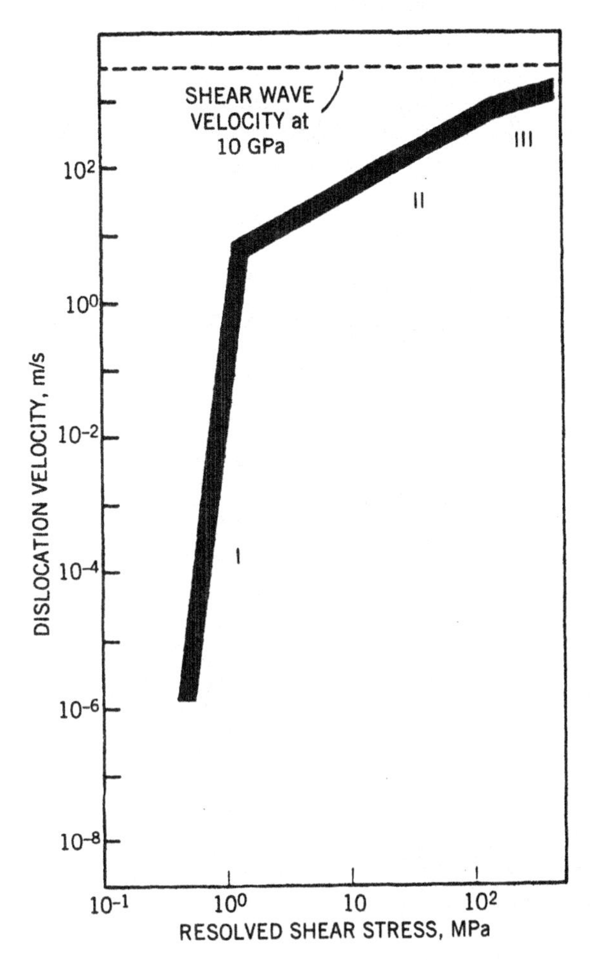

FIGURE 13.6 Schematic representation of the stress–velocity behavior of nickel. Three regions of response can be established: region I ($m_{\text{I}} > 1$), region II ($m_{\text{II}} = 1$), region III ($m_{\text{III}} < 1$). Region I was replotted from data by Rohde and Pitt [22].

13.3.2 Thermally Activated Dislocation Motion [34–41]

A dislocation continuously encounters obstacles as it moves through the lattice. Some of these obstacles are shown in Figure 13.7. These obstacles make the movement of dislocations more difficult. Figure 13.7 shows an array of obstacles: solute atoms (interstitial and substitutional), vacancies, small-angle grain boundaries, vacancy clusters, inclusions, precipitates, and so on. Dislocations themselves can oppose the movement of dislocations.

Not shown in Figure 13.7 are the Peierls–Nabarro forces, which are opposing the movement at the atomic level. When a dislocation moves from one equilibrium atomic position to the next, it has to overcome an energy barrier,

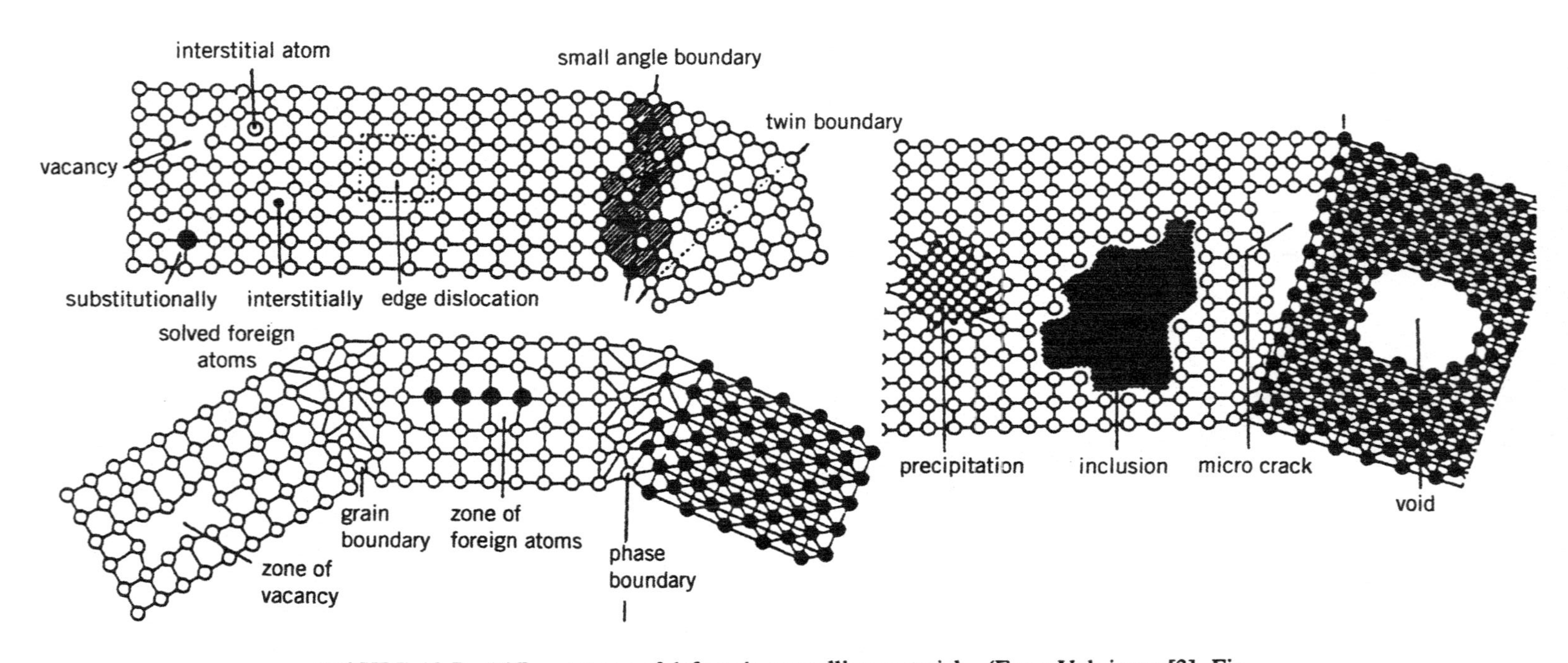

FIGURE 13.7 Different types of defects in crystalline materials. (From Vohringer [3], Fig. 12, p. 40.)

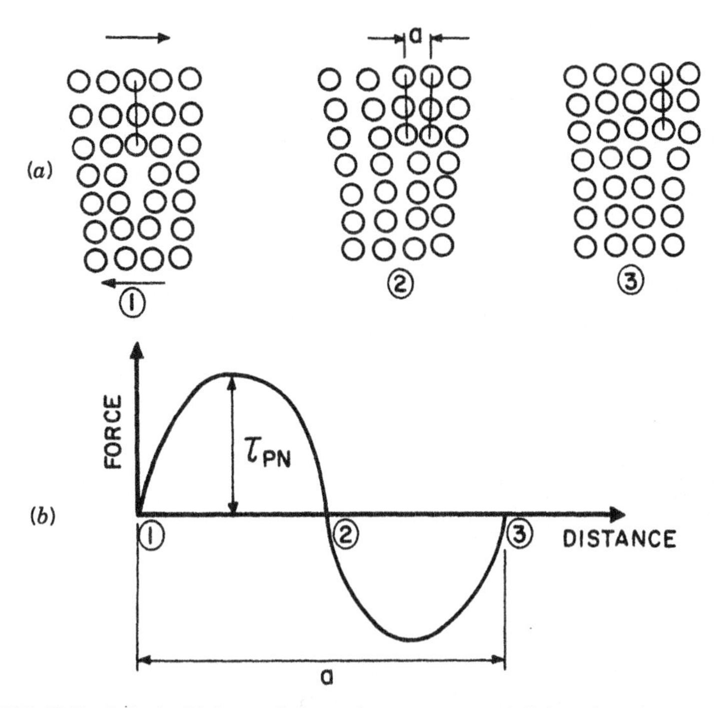

FIGURE 13.8 Peierls–Nabarro force: (a) movement of dislocation from one equilibrium position to next; (b) applied stress vs. distance.

that is, force has to be applied to it. Figure 13.8 shows the Peierls–Nabarro barrier. The stress required to move the dislocation without any other additional external help is the Peierls–Nabarro stress τ_{PN}. Position 2 is an unstable equilibrium position at the tip of the energy "hump." The wavelength of these barriers is equal to the periodicity of the lattice. Shown in Figure 13.9 is an array of dislocations intersecting a slip plane. The dislocations that "stand up" and through which the moving dislocation has to move are called forest dislocations.

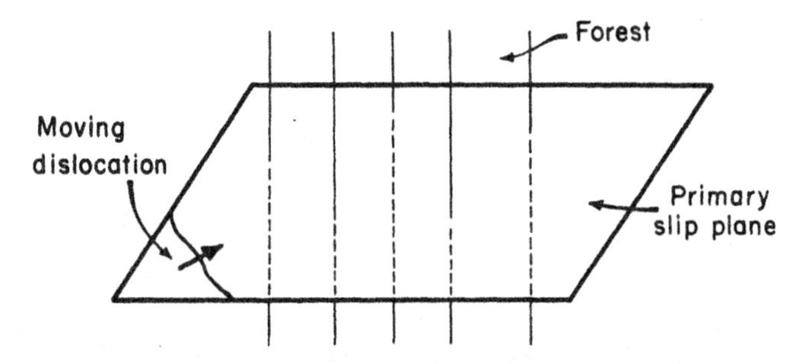

FIGURE 13.9 Dislocation cutting through a dislocation forest.

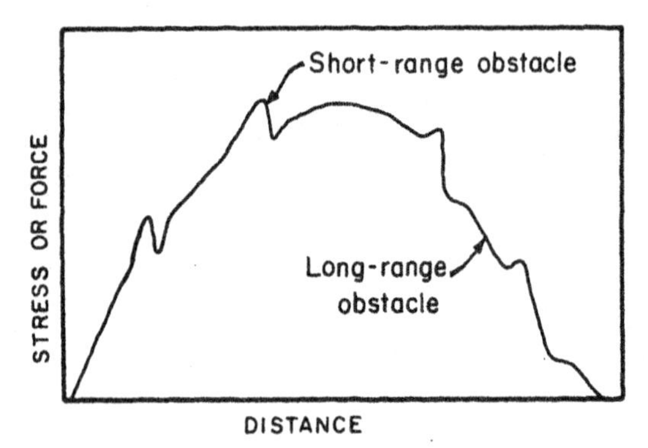

FIGURE 13.10 Barriers encountered by a dislocation on its course.

Thus, a moving dislocation encounters periodic barriers of different spacing and different lengths. Figure 13.10 shows schematically such a barrier field. We will now see how the length of these barriers and the thermal energy of the lattice influence the temperature and strain rate response of metals. The smaller, narrower barriers are called short-range obstacles, and the larger, wider barriers are called long-range obstacles. There are numerous treatments in the literature on how these obstacles are responsible for the temperature and strain rate sensitivity of crystalline materials. We will, in this book, trace a course that incorporates the principal ideas, avoiding controversial interpretations. We remind the reader that this concept is due to the physicist Becker, in the beginning of the century (1925) [34].

Thermal energy increases the amplitude of vibration of atoms. This energy can help the dislocation to overcome obstacles, as shown in Figure 13.11. The barrier is shown at four temperatures $T_0(= 0)$, T_1, T_2, T_3, $(T_0 < T_1 < T_2 < T_3)$. The thermal energies ΔG_1, ΔG_2, ΔG_3 have been shown by hatching. The reader is reminded that the area under a force–distance curve is an energy term. The effect of the thermal energy is to decrease the height of the barrier by successively increasing amounts as the temperature increases. Figure 13.11(b) shows the stress required to move the dislocation past that specific obstacle as a function of temperature. The effective height of the barrier decreases as the temperature rises. The effect of strain rate (or velocity) is similar: as the strain rate is increased, there is less time available to overcome the barrier and the thermal energy will be less effective. We will see later how this occurs. This stress eventually becomes zero. One should notice, however, that there are, in Figure 13.10, long-range obstacles that cannot be overcome by thermal energy; thus, one classifies the obstacles into short range (thermally activated) and long range (non-thermally activated). One therefore expresses the flow stress of a material as

$$\sigma = \sigma_G(\text{structure}) + \sigma^*(T, \dot{\varepsilon}, \text{structure}) \tag{13.20}$$

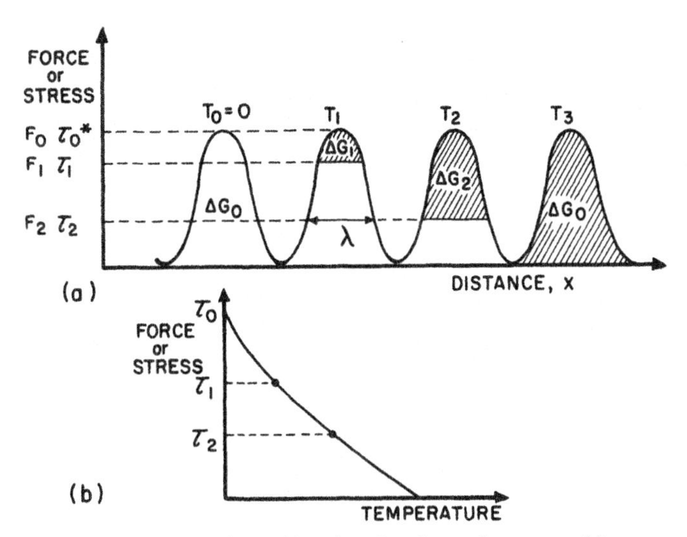

FIGURE 13.11 (a) Overcoming of barriers by thermal energy; (b) stress required to overcome obstacle as a function of temperature.

The term σ_G is due to the athermal barriers (long range) determined by the structure of the material. The term σ^* is due to the thermally activated barriers, that is, the barriers that can be overcome by thermal energy. The principal short-range barrier is the Peierls–Nabarro stress, which is very important for BCC metals. The Peierls–Nabarro stress is also the rate-controlling mechanism for ceramics. For FCC and HCP metals, dislocation forests are the primary short-range barriers at lower temperatures. The different nature of these barriers is *extremely important* and is responsible for the major differences in strain rate sensitivity between FCC and BCC metals. In ceramics, the Peierls–Nabarro stresses are so high that these barriers are not overcome at room temperature. The strong ionic and covalent bonds as well as their high directionality (bond angles fixed by electronic structure) are responsible for these extremely high Peierls–Nabarro stresses. Therefore, they fail by an alternative mechanism (crack nucleation and growth). But this is another can of worms, which will be opened in Chapter 16.

The probability of an equilibrium fluctuation in energy greater than a given value ΔG is given by statistical mechanics and is equal to [41] (the curious student should consult the appropriate texts; this is a seminal equation upon which the entire field of thermally-activated plastic deformation is based):

$$p_B = \exp\left(-\frac{\Delta G}{\kappa T}\right)$$

where κ is Boltzmann's constant. The probability that a dislocation will overcome an obstacle can be considered as the ratio of the number of successful jumps over the obstacle divided by the number of attempts. We assume that a dislocation will overcome the obstacle if it has an energy equal to or higher than the energy barrier ahead of it. Taking these values per unit time, we have frequencies. Thus, the frequency with which the dislocation overcomes the obstacle, ν_1, divided by the vibrational frequency of the dislocation, ν_0, is equal to p_B:

$$\nu_1 = \nu_0 \exp\left(-\frac{\Delta G}{\kappa T}\right) \tag{13.21}$$

Kocks et al. [41] discuss the vibrational frequency of dislocations in detail and conclude that it is $\sim 10^{11}\ \mathrm{s}^{-1}$. One should notice that this is 100 times less than the vibrational frequency of atoms, ν. This vibrational frequency, ν, is equal to kT/h, where T is the Debye temperature ($\nu \sim 10^{13}\ \mathrm{s}^{-1}$). If the spacing between obstacles is l, Kocks et al. [41] estimated the lower bound of ν_0 to be

$$\nu_0 = \nu \frac{b}{4l}$$

This is the ground frequency of a dislocation with wavelength $4l$.

Example 13.1. Find the vibrational frequency for dislocations if the density is approximately $10^7\ \mathrm{cm}^{-2}$:

$$\nu \sim 10^{13}\ \mathrm{s}^{-1}$$

$$b \sim 30\ \mathrm{nm} = 30 \times 10^{-9}\ \mathrm{m}$$

$$l = \rho^{-1/2} = (10^{11})^{-1/2} = 3 \times 10^{-6}\ \mathrm{m}$$

$$\nu_0 = 10^{13} \times \frac{3 \times 10^{-8}}{3 \times 10^{-6}} = 10^{11}\ \mathrm{s}^{-1}$$

The calculation was done for a dislocation density of $10^7\ \mathrm{cm}^{-2}$, typical of annealed copper. The obstacles were taken as the dislocations.

It is convenient to divide the time Δt taken by a dislocation to move a distance Δl between two obstacles into a waiting time in front of obstacles (t_w) and a running time between obstacles (t_r). Thus,

$$\Delta t = t_r + t_w \tag{13.22}$$

The average waiting time t_w is governed by the probability that an obstacle will be overcome by an adequately large thermal fluctuation of the free activation energy ΔG. From Eqn. (13.21) ($t_w = 1/\nu_1$)

$$t_w = \frac{1}{\nu_0} \exp \frac{\Delta G}{\kappa T} \tag{13.23}$$

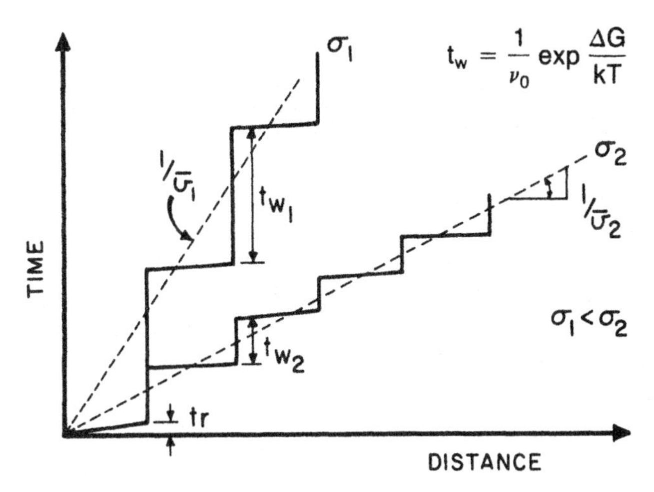

FIGURE 13.12 Distribution of running and waiting times for dislocation at two stress levels σ_1 and σ_2 ($\sigma_1 < \sigma_2$).

Figure 13.12 shows the distribution of running and waiting times for dislocations at two stress levels, σ_1 and σ_2; the waiting times are lower at the higher stress level, σ_2, and thus the mean dislocation velocity $\bar{v}$ is higher. In Figure 13.12, $t_w >> t_r$. This seems to be the case in actuality, and Eqn. (13.22) becomes

$$\Delta t \simeq t_w \tag{13.24}$$

Thus, using Eqn. (13.9),

$$\frac{d\varepsilon}{dt} = \frac{\Delta\varepsilon}{\Delta t} = \frac{1}{M}\rho b \frac{\Delta l}{\Delta t}$$

Substituting in Eqns. (13.23) and (13.24) yields

$$\dot{\varepsilon} = \frac{\nu_0 \rho b\, \Delta l}{M} \exp\left[-\frac{\Delta G}{\kappa T}\right]$$

where Δl is the distance between dislocation barriers and is assumed to be $\sim l$. The preexponential term can be conveniently called

$$\dot{\varepsilon}_0 = \frac{\nu_0 \rho b\, \Delta l}{M}$$

Thus

$$\dot{\varepsilon} = \dot{\varepsilon}_0 \exp\left[-\frac{\Delta G}{\kappa T}\right] \tag{13.25}$$

By expressing explicitly in terms of ΔG,

$$\Delta G = kT \ln \frac{\dot{\varepsilon}_0}{\dot{\varepsilon}} \tag{13.26}$$

This is a very important expression. We see that ΔG increases with T (as shown schematically in Fig. 13.11) and decreases with increasing strain rate. We schematically plot, in Figure 13.13, the combined effects of strain rate and temperature on the flow stress of a metal obeying a relationship of the type described above. Referring to Figure 13.11, one can see that, at a temperature T_3, the activation energy is sufficient to totally overcome the short-range barrier. At this temperature, the flow stress, given by Eqn. (13.20), becomes

$$\sigma = \sigma_G + \sigma^* = \sigma_G$$

Thus, the thermal component of stress becomes zero. At 0 K, on the other hand, the activation energy is zero, and the flow stress is

$$\sigma = \sigma_G + \sigma_0$$

where σ_0 is the thermal component of stress. In Figure 13.13, the thermal and athermal components of flow stress are marked. At 0 K, as well as above T_0 (which is strain rate dependent), the flow stress is strain rate independent. Between 0 K and T_0, the thermal component of flow stress is given by the formulation below. One has to calculate ΔG from the activation barrier. In Figure 13.11, one sees that ΔG is the hatched area. This area is equal to

$$\Delta G = \Delta G_0 - \int_0^{F^*} \lambda(F)\, dF \tag{13.27}$$

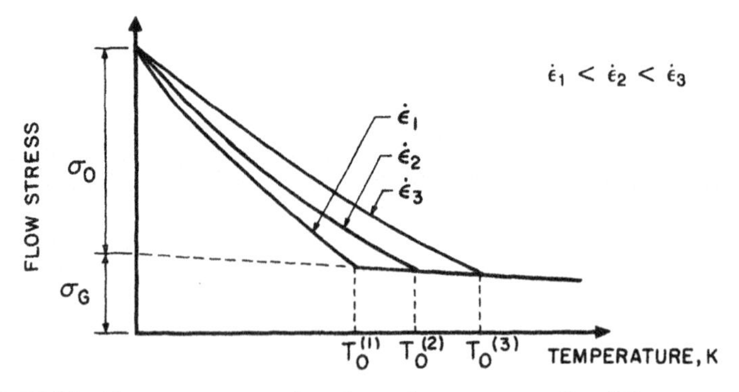

FIGURE 13.13 Flow stress as a function of temperature for different strain rates; thermal and athermal components of stress indicated.

where ΔG_0 is the activation energy at 0 K and the integral is the nonhatched area. Here, $\lambda(F)$ is the barrier width. The difference is the effective barrier. The shape of the activation barrier dictates the shape of the thermal portion of the curve. The athermal portion of the flow curve has a very low temperature dependence, equal to that of G (shear modulus). By appropriate manipulation of Eqn. (13.27), one obtains the relationship between stress and strain rate. Equations (13.26) and (13.27) yield

$$\kappa T \ln \frac{\dot{\varepsilon}_0}{\dot{\varepsilon}} = \Delta G_0 - \int_0^{F^*} \lambda(F)\, dF \tag{13.28}$$

Section 13.4 will delve into this in greater detail. This is the foundation for constitutive equations that is based on thermally assisted overcoming of obstacles. Figure 13.14 shows experimental data by Meyer [42] for a C–Cr–Si steel compared with Vohringer's [3] constitutive equation. It can be seen that the results are in good agreement with predictions in the 10^{-2}–10^{2} s^{-1} strain rate range.

13.3.3 Dislocation Drag Mechanisms

A detailed description of dislocation dynamics is given by Gilman [43] in his book *Micromechanics of Flow in Solids*. In region II of Figure 13.6 the velocity of dislocations is proportional to the applied stress, that is, the exponent in Eqn. (13.12) is equal to 1. It is well known that dislocation motion causes a temperature increase. The cold rolling of a piece of nickel increases its temperature substantially. It is also known that the energy stored in the material after deformation (as defects) is only a small portion of the energy spent to

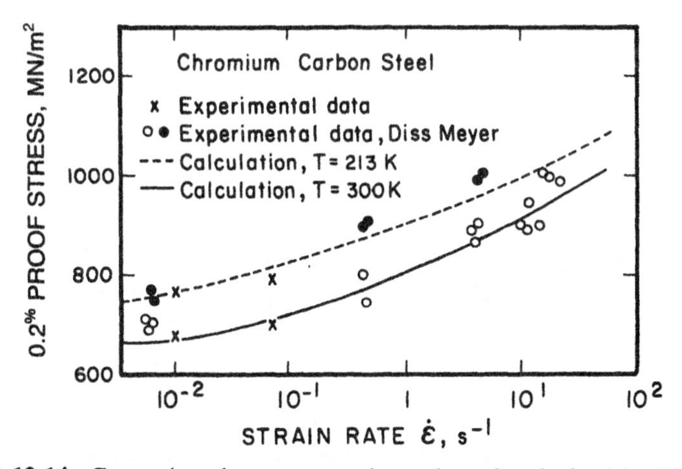

FIGURE 13.14 Comparison between experimental results obtained by Meyer [42] and constitutive equation developed by Vohringer [3] (Fig. 43, p. 47).

deform it. The energy of the substructure (residual) is usually only 5–20% of the total energy. So 90% of the energy is dissipated by forces opposing the applied stresses. They can be expressed as a viscous behavior of the solid. To a first approximation, the solid can be assumed to act as Newtonian viscous material with respect to the dislocation. Hence,

$$f_\nu = B\nu \tag{13.29a}$$

where B, the viscous damping coefficient, is independent of ν for a Newtonian fluid. Under the application of a certain external stress, the dislocation will accelerate until it reaches a steady-state velocity. Recalling Eqn. (13.6), equilibrium will be reached when

$$\tau b = B\nu \tag{13.29b}$$

Applying Eqn. (13.9), one obtains

$$\tau b = \frac{BM\dot{\varepsilon}}{\rho b}$$

Making $\tau = \frac{1}{2}\sigma$,

$$\sigma = \frac{2BM}{\rho b^2}\,\dot{\varepsilon} \tag{13.30}$$

Thus, the flow stress should be proportional to the strain rate if dislocation drag mechanisms are operative.

Some investigators (e.g., Gillis et al. [44]) consider B to be dependent on the velocity of the dislocation. This allows the use of B over a very wide range of velocities. The expression proposed by Gillis et al. [44] is

$$B = \frac{B_0}{1 - \nu^2/\nu_s^2} \tag{13.31}$$

where ν_s is the shear wave velocity and B_0 the viscosity at rest. It can be seen that viscous drag increases as the velocity increases.

The drag mechanisms that are not thermally activated are the interaction of the dislocation with thermal vibrations (phonon drag) and with electrons (electron viscosity); additionally, they also include relaxation effects in the dislocation core. A phonon is an elastic vibration propagating through the crystal. It is quantized because the lattice is discrete and not continuous. These mechanisms will be described briefly below. (The student should consult a good physics book for a more detailed explanation. They are discussed in detail by Gorman et al. [27], Dorn et al. [45], and especially Granato [46]).

1. *Phonon Viscosity.* A change in the compressive stress will trigger an increase in the modulus of rigidity. This increase in modulus relaxes with time, as the phonons approach equilibrium.
2. *Phonon Scattering.* There are several ways by which phonon-scattering mechanisms can operate. One of these is the scattering of phonons by the dislocation strain field in a manner similar to the refraction of light. Another is the absorption of energy from phonons by dislocations, with subsequent vibrations by dislocations.
3. *Thermoelastic Effect.* A moving dislocation alternatively strains adjacent regions in tension and compression; these regions show temperature decreases and increases, respectively. An irreversible process of heat flow ensues, increases the entropy, and depletes the dislocation of some of its energy.
4. *Electron Viscosity.* The free electrons in a metal will affect the dislocation motion in the same way as phonon viscosity.
5. *Anharmonic Radiation.* A dislocation, when undergoing positive or negative acceleration, emits elastic waves. This corresponds to an outflow of energy. Thus, the dislocation, although accelerating between two equilibrium positions, gives off energy.
6. *Glide Plane Viscosity.* There seems to be a certain consensus with respect to the relative importance of the various drag mechanisms under various conditions. Gilman's ideas are, in this respect, corroborated by Granato. They are given below (according to Gilman [43]):

   Covalent bonding: At low stress levels ($< G/100$) the motion is thermally activated (e.g., overcoming of Peierls stresses) and is stopped at low temperatures. However, stresses alone, when high enough ($> G/100$), cause motion at low temperatures.

   Ionic bonding: At low stress levels and high temperature, phonons are responsible for the drag.

   Metals: At high temperatures, phonons cause drag; at low temperatures, electrons cause drag. At high stresses, for both metals and salts, the viscosity increases because of relativistic effects.

Granato [46] adds that, for materials not containing an electronic cloud (covalent and ionic bonding), drag by radiation is the mechanism operating at low temperatures. They also comment on the interaction between dislocations and point defects if these are moving slowly enough. This belongs to the group of thermally activated mechanisms.

Hirth and Lothe [48] provide expressions for calculating the damping constant B from phonon viscosity. This estimated value is

$$B \cong \frac{bw}{10C} \tag{13.32}$$

where C is the elastic wave velocity (see Chapter 2) and w is the thermal energy density, equal approximately to (a^3 is the volume per atom)

$$w = \frac{3\kappa T}{a^3} \tag{13.33}$$

Kumar and Clifton [49] measured B for LiF at room temperature and obtained a value of approximately 3×10^{-5} N s/m^2 for an applied stress in the range 10–30 MPa, generating dislocation velocities of ~100 m/s. The inclined-plate experimental technique, described in Chapter 3, was used.

Example 13.2. Calculate B from the plot in Figure 13.4 and from the phonon viscosity equation by Hirth and Lothe [48] for LiF.

We first find the region in the log–log plot with slope equal to 1. This is the region in which viscous drag is operative. This is made clear by applying Eqn. (13.29b) twice:

$$\tau_1 b = B\nu_1 \qquad \tau_2 b = B\nu_2$$

The ratio is

$$\frac{\tau_1}{\tau_2} = \frac{\nu_1}{\nu_2}$$

By searching through the continuous change in slope of Figure 13.4, we find this condition satisfied for the following point:

$$\tau \approx 50 \text{ kgf/mm}^2 = 490 \text{ MPa}$$

$$\nu \approx 10^5 \text{ cm/s}$$

We can now establish the proportionality constant by applying Eqn. (13.29b):

$$\frac{B}{b} = \frac{\tau}{\nu} = 500 \times 10^3 \text{ N s/m}^3$$

But $b \approx 3 \times 10^{-10}$ m:

$$B = 1.6 \times 10^{13} \text{ N s/m}^2$$

This value is higher, by eighteen orders of magnitude, than the one measured by Kumar and Clifton [49]. This is due to the high stresses and velocities at which ν was established.

Using Hirth and Lothe's [48] equation,

$$B \approx \frac{bw}{10C_l} = \frac{3kbT}{10C_l a^3} \approx \frac{3kT}{10C_l a^2}$$

$$C_l = \sqrt{\frac{E}{\rho}} = 4.3 \times 10^3 \text{ m/s}$$

where $E \approx 50$ GPa, $\rho = 2635$ kg/m^3, and $a \approx 0.3$ nm. Thus $B \approx 2.9 \times 10^{-8}$ N s/m^2. This value is lower by five orders of magnitude, from the value measured by Kumar and Clifton [49].

Differences between the measured and calculated B values (assuming phonon viscosity) led Clifton [12] to suggest phonon scattering as a possible mode of damping dislocation motion in LiF. He suggested possible additional damping mechanisms, due to the inertial resistance of a dislocation to motion. From elastodynamics considerations, Eshelby [50] and Clifton and Markenskoff [51] have computed the drag coefficient B for edge and screw dislocations, respectively. The values of these drag coefficients are

$$B^{c}_{\text{screw}} \simeq \frac{Gb}{4\pi t C_s^2} \quad (\text{for } v < 0.4C_s)$$
$$B^{c}_{\text{edge}} \simeq \frac{Gb[1 + (C_s/C_l)^4]}{4\pi t C_s^2} \quad (\text{for } v < 0.4C_s) \tag{13.34}$$

where C_l and C_s are the elastic longitudinal and shear velocities, respectively. This drag is only important at very short times, on the order of 10 ns or lower. It can be seen above that $B^c \propto 1/t$.

Parameswaran and Weertman [28, 29] estimated B for aluminum and found the dependence on temperature shown in Figure 13.15. According to the theory of electron and phonon viscosity, the predictions are also shown in the figure. It can be seen that electron viscosity becomes negligible at temperatures higher than 100 K, whereas phonon viscosity becomes significant at this temperature. The results for aluminum indicate that electron viscosity is the damping mechanism at low temperatures, whereas phonon viscosity is the mechanism at ambient and higher temperatures.

13.3.4 Relativistic Effects on Dislocation Motion

The third stage of dislocation motion, stage III in Figure 13.6, is the one at which the velocity asymptotically approaches the shear wave velocity. Supersonic dislocations have not been observed yet, although they have been postulated by Eshelby [52].

In this section it will be shown analytically that this is not possible. The derivation will be conducted for a screw dislocation and was originated by Eshelby [50]. The treatment given here is based on Weertman [53]. For a basic treatment of stress fields around dislocations, the reader is referred to standard texts [54, 55]. Weertman and Weertman [56] have presented a more comprehensive treatment. The equations for the stresses and strains around a screw

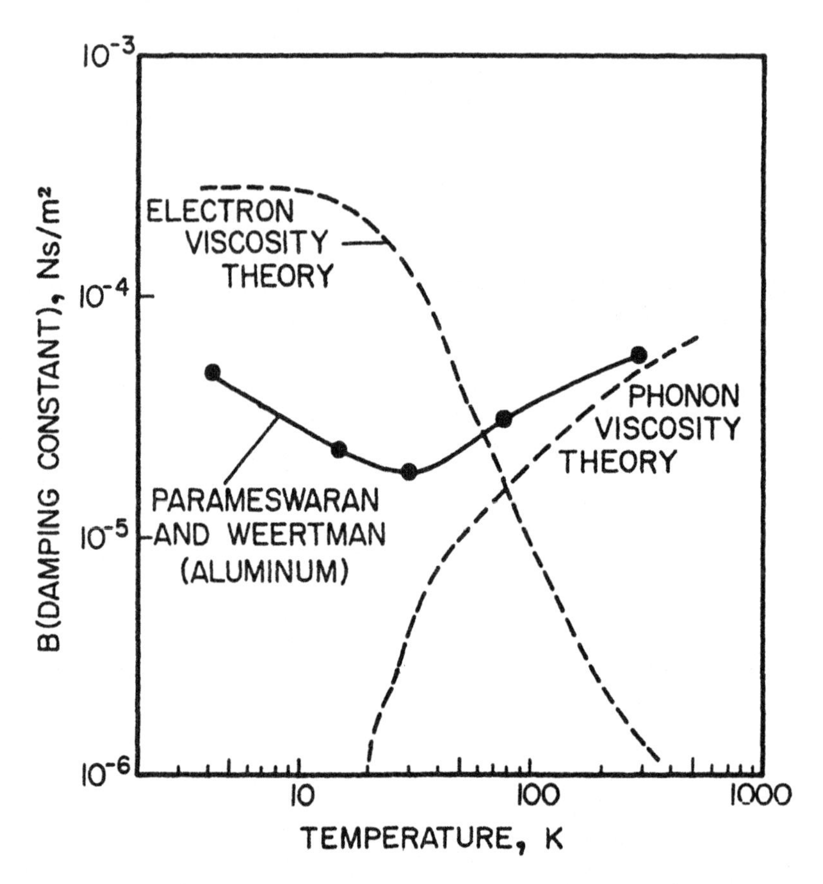

FIGURE 13.15 Dislocation damping constant B as a function of temperature for aluminum, compared with theoretical calculations from Mason's theories. (From Parameswaran et al. [29], p. 2985. Adapted with permission of the author.)

dislocation are used in the development. The equation of equilibrium expressed as $\Sigma F = 0$ is now replaced by Newton's second law, since one does have a dynamic situation (for more information see Meyers and Chawla [13]):

$$\frac{\partial \sigma_{13}}{\partial x_1} + \frac{\partial \sigma_{23}}{\partial x_2} + \frac{\partial \sigma_{33}}{\partial x_3} = \rho \frac{\partial^2 u_3}{\partial t^2} \tag{13.35}$$

From elasticity, one obtains the stresses in terms of the displacements:

$$\sigma_{13} = G \frac{\partial u_3}{\partial x_1} \qquad \sigma_{23} = G \frac{\partial u_3}{\partial x_2} \tag{13.36}$$

For a screw dislocation, $\sigma_{33} = 0$.

Substituting the displacements for the stresses [Eqns. (13.36) into (13.35)] yields

$$\frac{G}{\rho}\left(\frac{\partial^2 u_3}{\partial x_1^2} + \frac{\partial^2 u_3}{\partial x_2^2}\right) = \frac{\partial^2 u_3}{\partial t^2} \tag{13.37}$$

The equation for a wave disturbance ω traveling at a velocity ν along x, as given in Chapter 2, is expressed as

$$\nu^2 \frac{\partial^2 \omega}{\partial x^2} = \frac{\partial^2 \omega}{\partial t^2} \tag{13.38}$$

From the consideration of Eqn. (13.38) it can be seen, by analogy, that Eqn. (13.37) represents a disturbance u_3 traveling at a velocity

$$\nu = \left(\frac{G}{\rho}\right)^{1/2} \tag{13.39}$$

It has been shown in Chapter 2 that this is the velocity of an elastic shear wave. If the dislocation is at rest, Eqn. (13.37) reduces to

$$\frac{\partial^2 u_3}{\partial x_1^2} + \frac{\partial^2 u_3}{\partial x_2^2} = 0 \tag{13.40}$$

This is the Laplace equation, and the solution for a screw dislocation is known to be

$$u_3 = \frac{b}{2\pi} \tan^{-1} \frac{x_2}{x_1} \tag{13.41}$$

We will now determine the solution for Eqn. (13.37) when the dislocation is moving. The solution of this equation was found to be very similar to the treatment employed in the special relativity theory. By an appropriate transformation—similar to the one made in special relativity theory—Eqn. (13.37) is transformed to the Laplace equation. This transformation is made:

$$x_1^* = \frac{x_1 - \nu t}{(1 - \nu^2/C_s^2)^{1/2}} \qquad x_2^* = x_2 \qquad x_3^* = x_3 \tag{13.42}$$

This transformation is such that the system of reference moves with the dislocation; C_s is the velocity of the shear wave. This transformation is called a Lorentz transformation in special relativity theory. Calling $(1 - \nu^2/C_s^2)^{1/2} =$

β, we have

$$x_1^* = \frac{x_1 - vt}{\beta} \quad \text{and} \quad \partial x_1^* = \frac{\partial x_1}{\beta}$$

so

$$\frac{\partial}{\partial x_1^*} = \beta \frac{\partial}{\partial x_1} \quad \text{and} \quad \frac{\partial}{\partial x_1^*}\left(\frac{\partial}{\partial x_1^*}\right) = \beta \frac{\partial}{\partial x_1^*}\left(\frac{\partial}{\partial x_1}\right)$$

and

$$\frac{\partial^2}{\partial x_1^{*2}} = \beta^2 \frac{\partial^2}{\partial x_1^2} \tag{13.43}$$

For the derivation with respect to $\partial/\partial t^2$, we have

$$v = \frac{\partial x_1}{\partial t} \quad \text{and} \quad v \frac{\partial}{\partial x_1} = \frac{\partial}{\partial t}$$

at constant v; applying the operator $\partial/\partial t$, we have

$$\frac{\partial}{\partial t}\left(v \frac{\partial}{\partial x_1}\right) = \frac{\partial^2}{\partial t^2} \qquad v \frac{\partial}{\partial t}\left(\frac{\partial}{\partial x_1}\right) = \frac{\partial^2}{\partial t^2}$$

from

$$v^2 \frac{\partial^2}{\partial x_1^2} = \frac{\partial^2}{\partial t^2} \tag{13.44}$$

Substituting (13.37) for (13.43) and (13.44), we have

$$\frac{G}{\rho}\left(\frac{1}{\beta^2}\frac{\partial^2 u_3}{\partial x_1^{*2}} + \frac{\partial^2 u_3}{\partial x_2^2}\right) = v^2 \frac{\partial^2 u_3}{\partial x_1^2} = \frac{v^2}{\beta^2}\frac{\partial^2 u_3}{\partial x_1^{*2}}$$

Recall that

$$\beta = \left(1 - \frac{v^2}{C_s^2}\right)^{1/2}$$

Therefore

$$-(\beta^2 - 1) = \frac{v^2}{C_s^2}$$

$$\frac{v^2}{-(\beta^2 - 1)}\left(\frac{1}{\beta^2}\frac{\partial^2 u_3}{\partial x_1^{*2}} + \frac{\partial^2 u_3}{\partial x_2^{*2}}\right) = \frac{v^2}{\beta^2}\frac{\partial^2 u_3}{\partial x_1^{*2}}$$

Simplifications will lead to

$$\frac{\partial^2 u_3}{\partial x_1^{*2}} + \frac{\partial^2 u_3}{\partial x_2^{*2}} = 0 \tag{13.45}$$

This is the Laplace equation, whose solution is given by Eqn. (13.41) with the appropriate changes:

$$u_3 = \frac{b}{2\pi}\tan^{-1}\frac{x_2}{(x_1 - vt)/\beta} \tag{13.46}$$

Differentiation of Eqn. (13.46) yields

$$\begin{aligned}\frac{\partial u_3}{\partial x_1} &= \varepsilon_{13} = -\frac{b}{4\pi}\frac{x_2}{x_2^2 + [(x_1 - vt)/\beta]^2}\\ \frac{\partial u_3}{\partial x_2} &= \varepsilon_{23} = \frac{b}{4\pi}\frac{(x_1 - vt)/\beta}{x_2^2 + [(x_1 - vt)/\beta]^2}\end{aligned} \tag{13.47}$$

The corresponding stresses are

$$\begin{aligned}\sigma_{13} &= -\frac{Gb}{2\pi}\frac{x_2}{x_2^2 + [(x_1 - vt)/\beta]^2}\\ \sigma_{23} &= \frac{Gb}{2\pi}\frac{(x_1 - vt)/\beta}{x_2^2 + [(x_1 - vt)/\beta]^2}\end{aligned} \tag{13.48}$$

Figures 13.16 and 13.17 show the stress fields around a screw dislocation as a function of its velocity. These plots were generated for three velocities: 0, $0.5C_s$, and $0.995C_s$. The dislocation axis is assumed to be along $0x_3$. Figure 13.16 shows the shear stress σ_{13} as a function of dislocation velocity, and the same is shown for σ_{23} in Figure 13.17, where the dislocation is moving along the horizontal axis $0x_1$. The isostress lines clearly show that the stress fields (initially circular) are compressed along the $0x_1$ axis. At $0.5C_s$ the compression is barely visible. However, as C_s is approached, the compression becomes very pronounced and the stresses will eventually be reduced to the $0x_2x_3$ plane at v

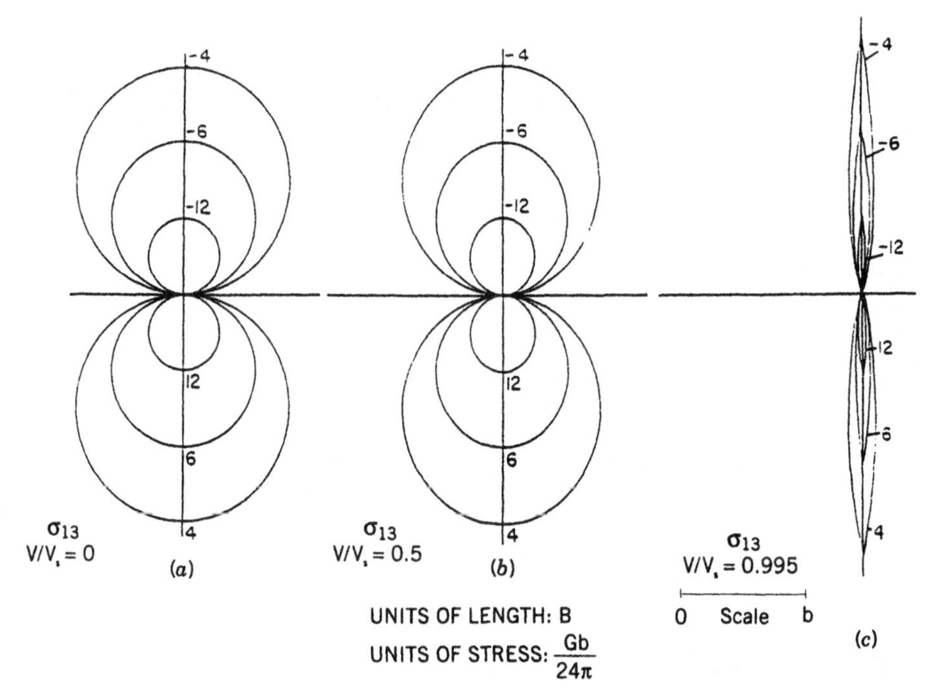

FIGURE 13.16 Variation of σ_{13} of a screw dislocation with velocity: (a) $v = 0$; (b) $v = 0.5v_s$; (c) $v = 0.995v_s$ (v_s is the velocity of elastic of shear waves). (Courtesy of A. R. Pelton and L. K. Rabenberg.)

$= C_s$. Concurrently with the deformation of the stress fields, there is an increase in the dislocation self-energy, as will be demonstrated below. So there seems to be a shear velocity wall, just as there is a sound wall for airplanes at Mach 1. For edge dislocations the situation is similar. The shear stresses and strains cannot propagate faster than C_s. However, the edge dislocations also generate normal stresses σ_{33} (and the associated longitudinal strains ϵ_{33}). These longitudinal strains can travel at velocities less than or equal to the velocity of longitudinal waves, C_l. To a first approximation, C_l is twice as large as C_s. So, edge dislocations are subjected to two sonic walls: at C_s and C_l. It can also be proven that their self-energy becomes infinite at $\nu = C_s$ and $\nu = C_l$. Weertman defines three ranges of velocity for dislocations:

$$\nu < C_s \rightarrow \text{subsonic dislocation}$$

$$C_s < \nu < C_l \rightarrow \text{transonic dislocation}$$

$$\nu > C_l \rightarrow \text{supersonic dislocation}$$

It should be noticed that a supersonic screw dislocation was proposed by Eshelby [57]; it is different from a normal dislocation and is known as an

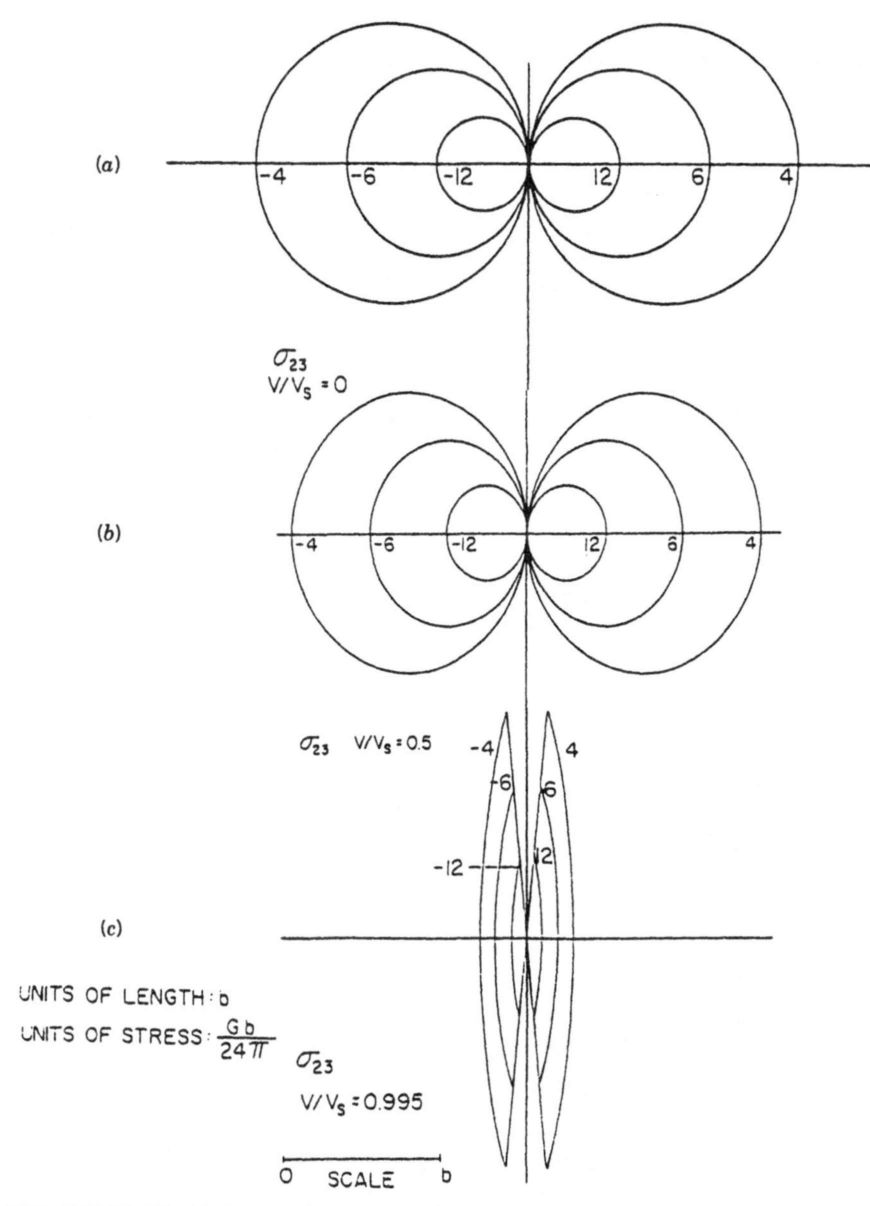

FIGURE 13.17 Variation of σ_{23} of a screw dislocation with velocity: (a) $v = 0$; (b) $v = 0.5v_s$; (c) $v = 0.995v_s$ (v_s is the velocity of elastic of shear waves). (Courtesy of A. R. Pelton and L. K. Rabenberg.)

Eshelby dislocation. According to Eshelby [52], it should propagate along planes that give up finite amounts of energy. Weertman [58] extended Eshelby's mathematical treatment to gliding and climbing edge dislocations (both in the transonic and supersonic range) and mixed dislocations. According to the mathematical treatment presented, these dislocations are possible. Weiner and Pear [59] and Earmme and Weiner [60], instead of treating the metal as a continuum, have used the concept of atoms vibrating. The simplest visualization is to consider the atoms connected by springs. Vibrations are transferred throughout the lattice as a propagating wave. Using atomistic models and lattice vibrations, they found, mathematically, that a dislocation breakdown was observed when $\nu = C_s$. These studies are, however, inconclusive because at present experimental support is lacking.

The total energy of a moving dislocation is calculated in a manner similar to the total energy of a stationary dislocation. Now, however, both kinetic and potential terms have to be added. The energy per unit volume of a screw dislocation is

$$U = \tfrac{1}{2}\sigma_{ij}\varepsilon_{ij} \tag{13.49}$$

Substituting Eqns. (13.47) and (13.48) into (13.49) and integrating with respect to r and θ (after appropriate change of coordinate system to cylindrical), we obtain

$$U_p = \frac{Gb^2}{4\pi} \ln\left(\frac{R}{r_0}\right) \frac{1+\beta^2}{2\beta} \tag{13.49a}$$

The kinetic energy, given by $\frac{1}{2}m\nu^2$, is expressed as (per unit volume)

$$U_k = \tfrac{1}{2}\rho\left(\frac{\partial u_3^*}{\partial t}\right)^2$$

Integration yields

$$U_k = \iint \frac{1}{2}\rho\left(\frac{\partial u_3^*}{\partial t}\right)^2 2\pi r\, dr\, d\theta$$

$$= \left(\frac{\nu}{C_s}\right)^2 \frac{1}{2\beta}\frac{Gb^2}{4\pi}\ln\frac{R}{r_0}$$

The total energy is

$$U_T = U_p + U_k = \frac{U_0}{\beta}$$

where U_0 is the energy of the stationary dislocation. It can be seen that as ν approaches C_s, U_T approaches infinity. So

$$\lim_{\nu \to C_s} U_T = \infty$$

The analogy to Einstein's relativity theory can be extended to the mass. Einstein's famous equation $E = mc^2$ stated that the energy depended on the mass. The mass at a velocity ν is related to the mass at rest by

$$m = \frac{m_0}{(1 - \nu^2/c^2)^{1/2}} \tag{13.50}$$

where c is the velocity of light. In our case, we can define an *equivalent (or effective) mass* for a dislocation.

Defining, for a stationary screw dislocation, $U_0 = m_0 C_s^2$, we have

$$\frac{Gb^2}{4\pi} \ln \frac{R}{r_0} = m_0 \left(\frac{G}{\rho}\right)$$

and

$$m_0 = \frac{\rho b^2}{4\pi} \ln \frac{R}{r_0}$$

Usually, $\ln (R/r_0)$ can be assumed to be $\sim 4\pi$, so that

$$m_0 \cong \rho b^2$$

Since $U_T \cong U_0/\beta$, we have

$$U_T = \frac{1}{\beta} m_0 C_s^2$$

If $U_T = mC_s^2$,

$$m = \frac{m_0}{(1 - \nu^2/C_s^2)^{1/2}} \tag{13.51}$$

This equation is in all respects analogous to Eqn. (13.50) from Einstein's relativity theory. It can be clearly seen that the acceleration caused by a certain stress would decrease at higher velocities.

Figure 13.18 presents, for nickel, a plot of the self-energy of a screw dislocation versus its velocity. In the calculations, the energy for the dislocation

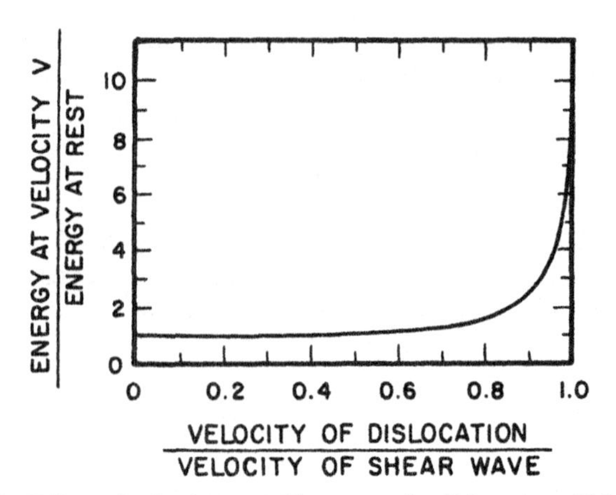

FIGURE 13.18 Effect of velocity on self-energy of a dislocation. This is a relativistic effect.

core was included, in addition to the energy for the surrounding strain field. The energy for the core was taken as $Gb^2/10$, and ln (R/r_0) was assumed to be equal to 4π. It can be seen that, beyond $0.5C_s$, the energy changes very rapidly. For an edge dislocation, the energy becomes infinite for $\nu = C_s$, and again for $\nu = C_l$. It is not known whether the edge dislocation traverses the first barrier and becomes transonic. This is treated in detail by Weertman [58].

Gillis and Kratochvil [61] treated the problem of dislocation acceleration and concluded that it is, under most conditions, extremely high. So the time interval and distance required to bring the dislocation to its steady-state velocity is very small and can, in many applications, be neglected. The acceleration of a dislocation can be calculated without too much difficulty. Figure 13.19 shows a dislocation segment; the applied force (F_{AP}) generated a certain displacement, which is opposed by the frictional force (F_{FR}) and by the back forces due to

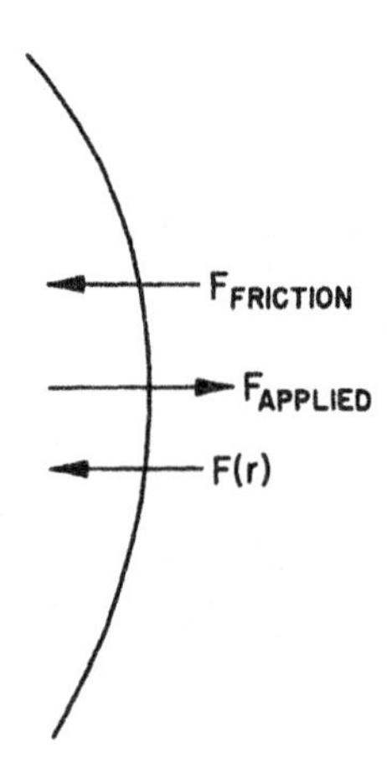

FIGURE 13.19 Forces on moving dislocation: applied force, frictional force, and force due to back stress produced by bowing.

the bowing $[F(r)]$. Applying Newton's second law, we have

$$F_{\mathrm{AP}} - F_{\mathrm{FR}} - F(r) = \frac{d}{dt}(mv) \tag{13.52}$$

Assuming a straight dislocation ($r = 0$), the back stresses vanish. Applying Peach–Koehler's equation [Eqn. (13.6)] yields

$$b\tau_{\mathrm{AP}} - b\tau_{\mathrm{FR}} = m\frac{dv}{dt} + v\frac{dm}{dt} \tag{13.53}$$

But the frictional stress is given by Eqn. (13.29) (Section 13.3.3):

$$b\tau_{\mathrm{FR}} = Bv \tag{13.54}$$

Equations (13.48) and (13.49) provide

$$m = \frac{m_0}{(1 - v^2/C_s^2)^{1/2}} = \rho b^2 \left(\frac{C_s^2}{C_s^2 - v^2}\right)^{1/2}$$

$$\frac{dm}{dt} = \rho b^2 C_s \frac{d}{dt}(C_s^2 - v^2)^{-1/2}$$

$$= \rho b^2 C_s v (C_s^2 - v^2)^{-3/2}\frac{dv}{dt} \tag{13.55}$$

Substituting Eqns. (13.54) and (13.55) into Eqn. (13.53), we have

$$\rho b^2 \left(\frac{C_s^2}{C_s^2 - v^2}\right)^{1/2} a + v^2 \rho b^2 C_s (C_s^2 - v^2)^{-3/2} a = b\tau_{\mathrm{AP}} - Bv \tag{13.56}$$

Substituting the value for B given in Section 13.3.3 into Eqn. (13.56), and remembering that $\beta^2 = 1 - v^2/C_s^2$, we obtain

$$a = \frac{C_s^2 \tau_{\mathrm{AP}} \beta^2 - BC_s^2 b^{-1}}{\rho b v_s^2 \beta + v^2 \rho b/\beta}$$

$$= \frac{v_s^2 \tau_{\mathrm{AP}} \beta^2 - BC_s^2 v b^{-1}}{\rho b(C_s^2 \beta + v^2/\beta)}$$

$$= \frac{\tau_{\mathrm{AP}} \beta^3}{\rho b} - \frac{Bv\beta}{\rho b^2} \tag{13.57}$$

Dimensional analysis provides a quick check of the correctness of this equation. It might be somewhat tricky, but the astute student will know how to untie the Gordian knot.

If only relativistic effects are considered, Eqn. (13.57) reduces to

$$a = \frac{\tau_{AP}}{\rho b}\left(1 - \frac{v^2}{C_s^2}\right)^{3/2} \tag{13.58}$$

Example 13.3. Calculate the acceleration as a function velocity for copper at an applied stress level of 10 MPa. The following values were used:

$$\rho = 8.92 \times 10^3 \text{ kg/m}^3$$

$$C_s = 2.92 \times 10^3 \text{ m/s} \quad \text{[from Chapter 2, Eqn. (2.1)]}$$

$$b = 2.55 \times 10^{-10} \text{ m}$$

$$B_0 = 7 \times 10^{-5} \text{ N s/m}^2$$

This is a reasonable level in conventional deformation processes; the yield stress of copper single crystal is a few megapascals. Since the value of B obtained by Greenman et al. [24] (7×10^{-5} N s/m^2) was obtained at low dislocation velocities, it was assumed equal to B_0. It does not have to be corrected because B, defined as the drag force per unit length of dislocation per unit velocity, was obtained per centimeter, and the unit velocity was taken as 1 cm/s; the two "cm" cancel each other. The data above are applied to Eqns. (13.57) and (13.58). Figure 13.20 shows that the initial acceleration of a dislocation is extremely high: 2.2×10^{19} m/s^2. If only relativistic effects

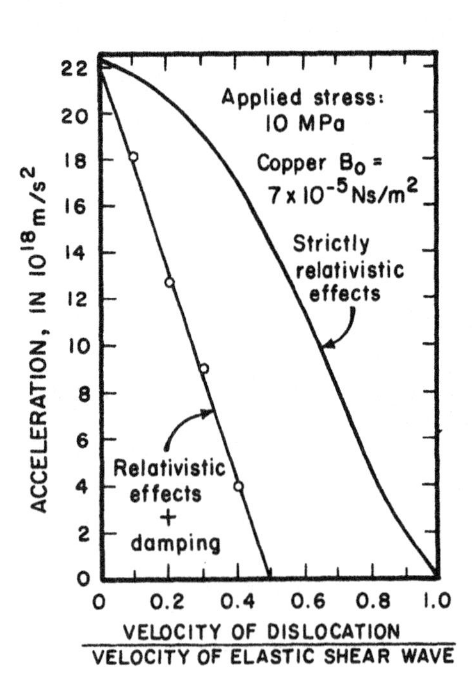

FIGURE 13.20 Acceleration of a dislocation in copper (with and without consideration of drag effects) as a function of velocity for an applied stress of 10 MPa.

were considered, the acceleration would steadily decrease until it becomes zero, when the dislocation velocity becomes equal to the shear wave velocity. On the other hand, the drag stress has an important effect: it establishes a steady-state velocity of about $0.5C_s$ at the applied stress level. The steady-state velocity is reached when the drag stress becomes equal to the applied stress. In any event, the acceleration to the steady-state velocity is extremely high. If one takes a mean acceleration of 10×10^{19} m/s^2, a dislocation will reach the steady-state velocity of $0.5C_s$ in approximately 1.5×10^{-16} s. This corresponds to a distance traveled of less than 1 Å. So for all intents it can be assumed that dislocations reach their steady-state velocity instantaneously.

13.3.5 Synopsis

The three regimes delineated in the preceding sections—thermally activated glide, drag-controlled dislocation motion, and relativistic motion—seem to realistically describe the motion of dislocations from $\nu = 0$ to $\nu = C_s$. One can represent these three regimes by a schematic plot based on Regazzoni et al. [62]. When the applied stress is lower than the threshold stress σ_0 (height of activation barrier), thermal activation plays a role in controlling the velocity of propagation. When the applied stress is higher than the short-range barrier height, drag (viscous and scattering) controls the resistance to dislocation motion. In Figure 13.21, the region in which the curve tangentiates the dashed line corresponds to the drag-controlled regime; this is a consequence of the $F \propto \nu$ or $\tau \propto \nu$ relationship. At even higher velocities, in the range of $0.8C_s$, relativistic effects start becoming important. We will, in a rapid exercise, calculate the strain rates required for these velocities.

Example 13.4 Copper ($\rho = 10^7$ cm^{-2}) is being deformed. Calculate the strain rate required to produce a dislocation velocity of $0.8C_s$, where relativistic effects start gaining importance; assume that all dislocations are mobile.

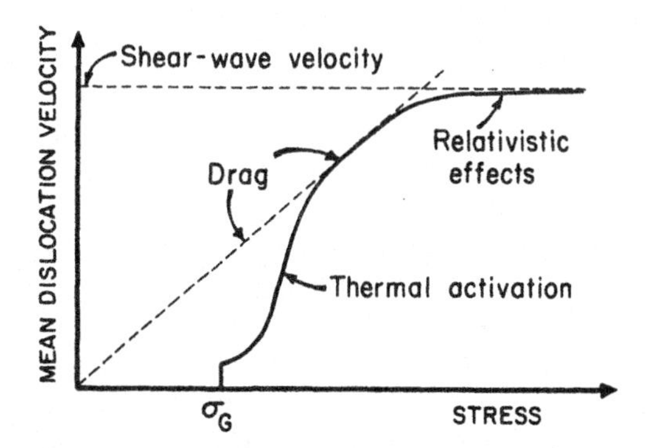

FIGURE 13.21 Three regimes of dislocation response. (Adapted from *Acta Met.*, vol. 35, Regazzoni et al., Fig. 4, p. 2868, Copyright 1987, with permission from Pergamon Press Ltd.)

We apply Eqn. (13.9): $\dot{\varepsilon} = \rho b v / M$

$$G \sim 48.3 \text{ GPa}$$

$$b \sim 3 \text{ Å}$$

$$\rho_0 = 8.9 \text{ g/cm}^3$$

$$C_s = \sqrt{G/\rho_0} \cong 2{,}000 \text{ m/s}$$

$$M = 3.1 \quad \text{(Taylor factor for FCC materials)}$$

$$\dot{\varepsilon} = \frac{10^7 \times 10^4 \times 3 \times 10^{-10} \times 2 \times 10^3 \times 0.8}{3.1}$$

$$= 2 \times 10^4 \text{ s}^{-1}$$

13.4 PHYSICALLY BASED CONSTITUTIVE EQUATIONS

The response of materials at high strain rates is intimately connected with the evolution of the microstructure. Defects, cracks, phase transformations, and their mutual interplay establish the mechanical performance. At high strains, lattice rotations and texturing start to play an important role. Section 13.3 reviews the three principal regimes of dislocation response: thermally activated overcoming of obstacles, drag effects, and relativistic regime. An additional plastic deformation mechanism is mechanical twinning, and the propensity of BCC and HCP metals and alloys to deform by twinning increases with strain rate. Thus, iron, tungsten, molybdenum, titanium, zirconium, and chromium twin profusely at high strain rates, whereas low-strain-rate plastic deformation is accomplished primarily by dislocation motion. Some materials undergo martensitic phase transformations induced by high stresses and strains; these transformations have a deviatoric (shear) component that results in plastic deformation.

A graphic representation of how the deformation mechanisms in a metal are dependent on temperature and strain rate has been developed by Sargeant and Ashby [63] and Frost and Ashby [64]. One of these plots is shown in Figure 13.22. The plot covers the strain rate range from 10^{-10} to 10^6 s^{-1} and temperature range from -200 to 1600°C. Normalized stress levels, indicated as σ/G (stress divided by shear modulus), are marked as families of curves in the plot; they vary from 10^{-6} to 10^{-2} (the ideal strength marked on top of the plot is 10^{-1}). The boundaries between the various regions are determined from the constitutive equations for the different mechanisms. According to Sargeant and Ashby [63] and Frost and Ashby [64], at ambient temperature one progresses from power law creep to obstacle-controlled plasticity (thermally activated motion of dislocations), to adiabatic shear, and finally, to phonon drag, as the strain rate is increased. There are also qualitatively different mechanisms as the scale of the deformation is changed from the micro to the macro level.

Harding [65, 66] reviews the constitutive equations proposed to explain in

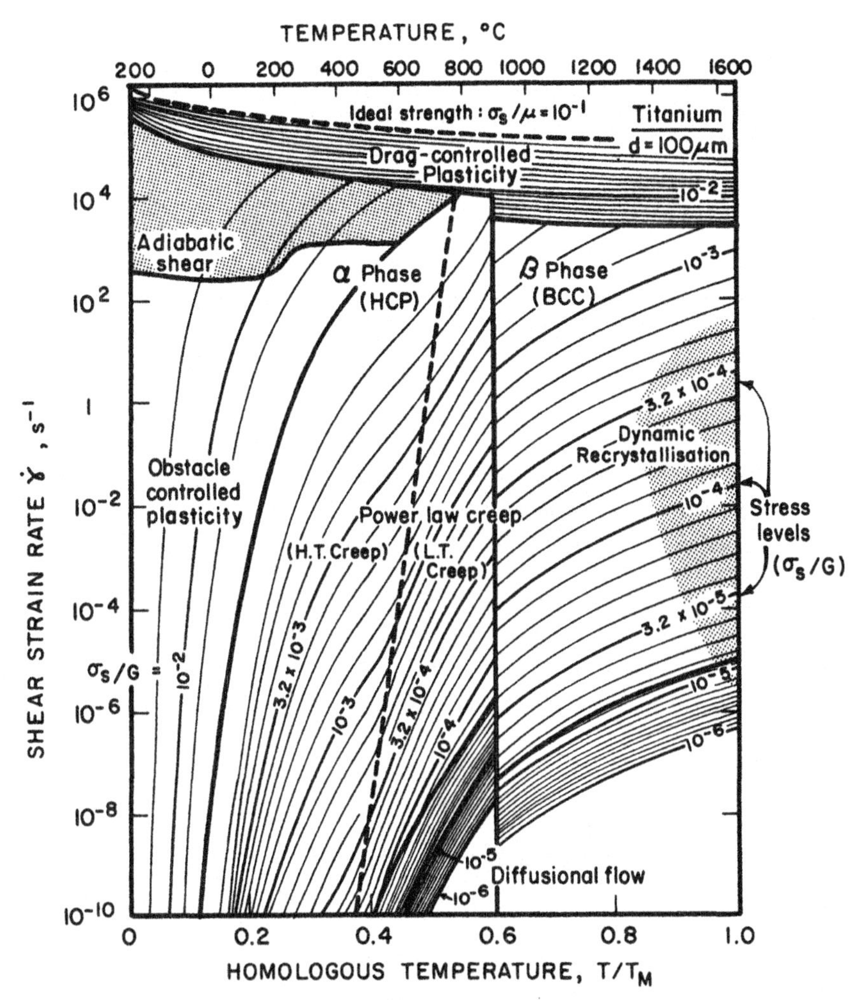

FIGURE 13.22 Strain rate/homologous temperature deformation map for titanium. (Reprinted from H. J. Frost and M. F. Ashby, *Deformation Mechanism Maps*, Copyright 1982, Fig. 17.4, with permission from Pergamon Press Ltd.)

detail the high-strain-rate response of metals. Physically based constitutive equations date from the 1960s, when Campbell [67], Lindholm [68], and Campbell and Harding [69] investigated the high-strain-rate response of steels. Using the thermal activation equation [Eqn. (13.28)], assuming a simple shape for the activation barrier (rectangular), and changing the integration limits from forces to stresses, we arrive at a constitutive equation expressed in terms of the activation volume V. Figure 13.23 shows three shapes of barriers. Notice barrier spacing l^*. Here V is the activation volume:

$$V = l^* \lambda b \tag{13.59}$$

where λ is barrier width and l^* barrier spacing.

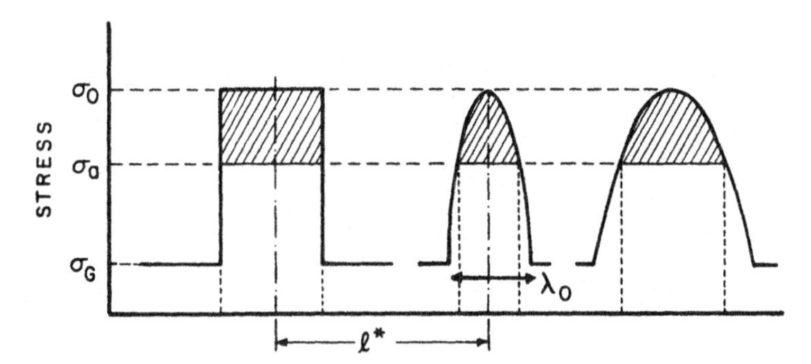

FIGURE 13.23 Different shapes of barriers: rectangular, hyperbolic, and sinusoidal.

The conversion of force into stress is made according to the procedure below. We assume that the spacing among obstacles is l^*. Figure 13.24 shows three barriers with spacing l^* and the position of dislocation before it overcomes the central barrier. The barrier width λ is marked in the figure:

$$F = \tau b \quad \text{(per unit length)}$$

$$\text{Force} = \tau b l^* \quad \text{(per barrier)}$$

The second term in Eqn. (13.28) becomes

$$\int_0^{F^*} \lambda(F)\, dF = \int_0^{\tau^*} \lambda(\tau) b l^*\, d\tau$$

$$= b l^* \int_0^{\tau^*} \lambda(\tau)\, d\tau$$

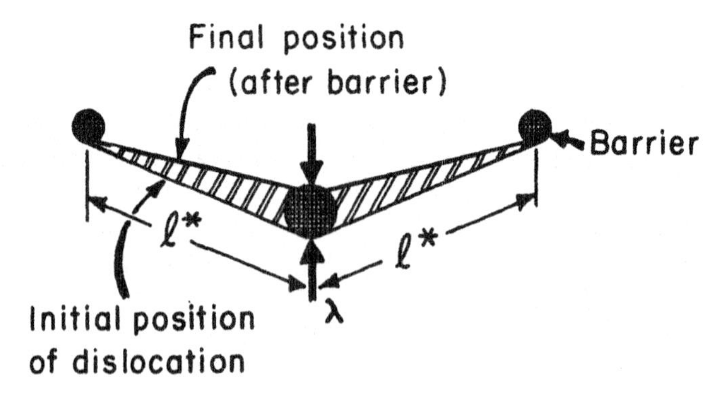

FIGURE 13.24 Dislocation overcoming a barrier; initial and final positions marked.

If the barrier is rectangular, λ = const (see Fig. 13.23):

$$bl^*\lambda \int_0^{\tau^*} d\tau = bl^*\lambda(\tau^* - 0)$$

Using Eqn. (13.59),

$$\int_0^{F^*} \lambda(F)\, dF = V\tau^* \tag{13.60}$$

But Figure 13.23 shows that the base level for the stresses is σ_G. The stress τ^* is therefore expressed as $\sigma - \sigma_G$. Substituting Eqn. (13.60) into Eqn. (13.28) yields

$$kT \ln \frac{\dot{\varepsilon}_0}{\dot{\varepsilon}} = \Delta G_0 - V(\sigma - \sigma_G) \tag{13.61}$$

and

$$\sigma = \sigma_G + \frac{\Delta G_0}{V} + \frac{kT}{V} \ln \frac{\dot{\varepsilon}}{\dot{\varepsilon}_0} \tag{13.62}$$

This form of the constitutive equation (based exclusively on thermal activation) was successfully used by Lindholm and Yeakley [68] for aluminum. It should be noticed that the shape of the activation barrier determines the form of the equation. For instance, if one assumes a hyperbolic barrier, shown in the center of Figure 13.23, one has the expression

$$\sigma = \frac{\sigma_0}{[1 + \lambda/\lambda_0]^2}$$

leading to

$$\Delta G = \Delta G_0 \left[1 - \left(\frac{\sigma}{\sigma_0}\right)^{1/2}\right]^2 \tag{13.63}$$

Kocks et al. [41] proposed a general expression for the activation energy dependence on σ of the form

$$\Delta G = \Delta G_0 \left[1 - \left(\frac{\sigma}{\sigma_0}\right)^{p}\right]^q \tag{13.64}$$

The parameters p and q determine the shape of the activation barrier. By using Eqn. (13.26), one arrives at the general constitutive equation

$$kT \ln \frac{\dot{\varepsilon}_0}{\dot{\varepsilon}} = \Delta G_0 \left[1 - \left(\frac{\sigma}{\sigma_0} \right)^p \right]^q \tag{13.65}$$

In 1977, Hoge and Mukherjee [70] developed a constitutive equation of the form

$$\dot{\varepsilon} = \left\{ C_0 \exp \left[\frac{\Delta G_0}{kT} \left(1 - \frac{\sigma}{\sigma_0} \right)^2 \right] + \frac{D}{\sigma} \right\}^{-1} \tag{13.66}$$

where the activation barrier was assumed to have a sinusoidal shape, leading to $p = 1$ and $q = 2$ [in Eqn. (13.64)]:

$$\Delta G = \Delta G_0 \left(1 - \frac{\sigma}{\sigma_0} \right)^2$$

The above equations do not incorporate work hardening. It is fairly universal to express hardening as γ^n, where $n < 1$. Thus, the addition of this term takes this into account: Campbell et al. [9] proposed Eqn. (13.5), which incorporates this work-hardening term.

It is important to recognize that the strain rate is not the only parameter determining the stress at a fixed strain and temperature. Figure 13.25 shows two different responses: (1) that of a material that does not exhibit strain-rate-sensitivity of structure development and (2) that of a material that exhibits it. The strain-rate change test is a good technique to determine this sensitivity. Figure 13.25 shows two stress–strain curves, obtained at strain rates of $\dot{\gamma}_1$ and $\dot{\gamma}_2 > \dot{\gamma}_1$. If the strain rate is changed from $\dot{\gamma}_1$ to $\dot{\gamma}_2$ at a strain of γ_T, the stress will jump. If the two structures are identical at $\dot{\gamma}_1$ and $\dot{\gamma}_2$, the curve will merge with the upper curve along path ABC, shown in Figure 13.23(a). On the other hand, if the structure develops in a different mode (i.e., strain rate sensitive), one ends up with different structures produced at $\dot{\gamma}_1$ and $\dot{\gamma}_2$. Figure 13.25(b) exemplifies this behavior. A strain rate change from $\dot{\gamma}_1$ to $\dot{\gamma}_2$ at γ_T will produce path $AB'C'$ and not ABC. The structure generated by plastic deformation at $\dot{\gamma}_2$ from 0 to γ_T has a higher threshold stress. This is an important point recognized by many investigators. Klepaczko [71] proposed the parameter λ_h, called the *rate sensitivity of strain hardening*, defined as

$$\lambda_h = \frac{\Delta \tau_h}{\log \left. \dfrac{\dot{\gamma}_2}{\dot{\gamma}_1} \right|_{T, \gamma_T}} \tag{13.66a}$$

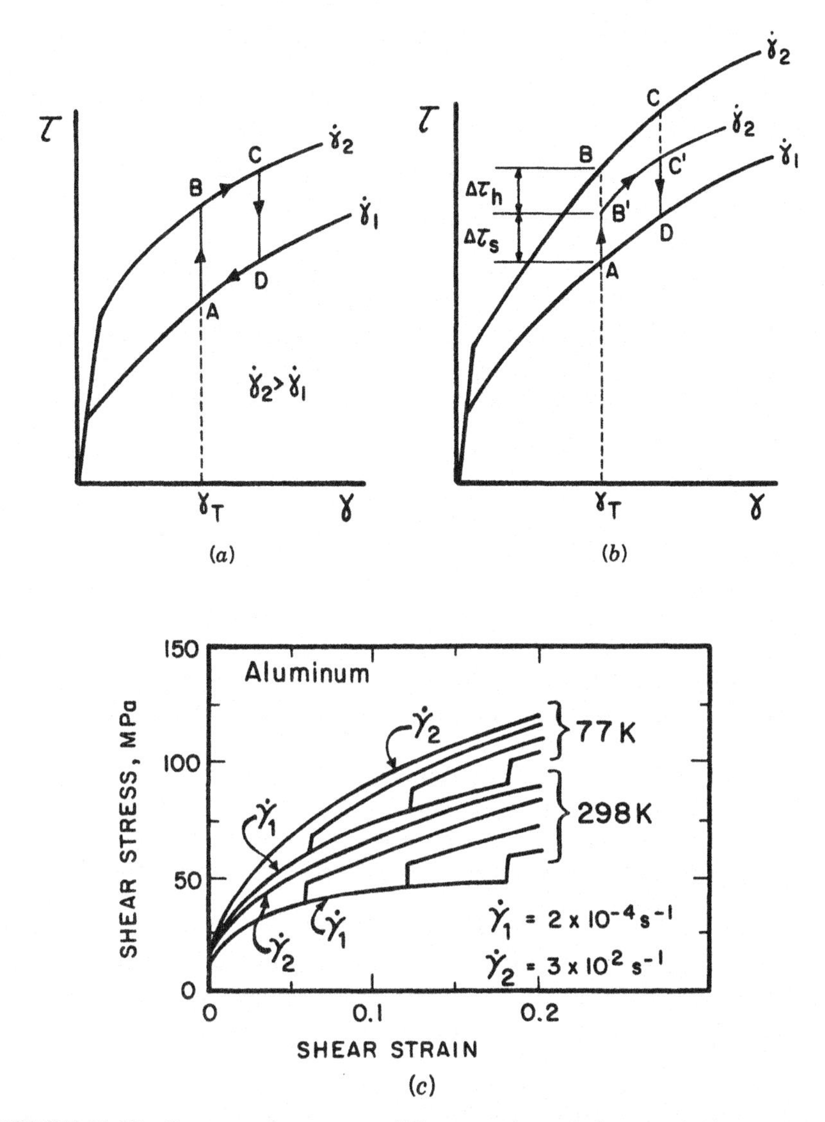

FIGURE 13.25 Stress–strain curves at different strain rates for material in which work hardening rate is (a) insensitive and (b) sensitive to strain rate. (c) Stress–strain curves (77 and 298 K) obtained for aluminum using Klepaczko's constitutive equation and predicted effects in strain rate change regime. (From Klepaczko [74], Fig. 6, p. 158. Reprinted by courtesy of Marcel Dekker, Inc.)

$\Delta\tau_h$ is the difference BB' in Figure 13.25(b). More recently, Klepaczko and Chiem [72] reanalyzed the subject in more detail. Klepaczko [73, 74] developed a constitutive method based on the kinetics of dislocation multiplication, glide and annihilation. He divided the total dislocation density ρ into a mobile dislocation density ρ_m and an immobile dislocation density ρ_i:

$$\rho = \rho_i + \rho_m \tag{13.67}$$

Then, he expressed the total shear stress τ as the sum of the athermal and thermal components:

$$\tau(\dot{\gamma}, T, S_j)_\gamma = \tau_G[S_j(\dot{\gamma}, T)]_\gamma + \tau^*[\dot{\gamma}, T, S_j(\dot{\gamma}, T)]_\gamma \tag{13.68}$$

where the S_j are the internal state variables, which are a function of strain rate and temperature. The principal internal state variables are the mobile and immobile dislocation densities. The first (athermal) component of stress is simply given by the Taylor equation:

$$\tau_G = \alpha_1 Gb(\rho_i + \rho_m)^{1/2} \tag{13.69}$$

For the thermal component of stress, on the other hand, the rates of dislocation generation and annihilation are strain rate and temperature dependent. Klepaczko developed analytical expressions for the change in mobile and immobile dislocation densities with strain, $\partial\rho_i/\partial\gamma$, $\partial\rho_m/\partial\gamma$, and incorporated these effects into the thermal component of the stress. This leads directly to predicted responses in line with the results of Figure 13.25(b). Figure 13.25(c) shows the computed effects of strain rate jump tests on the stress–strain response of aluminum. The computations are conducted for 77 and 298 K and the strain rates were increased from 2×10^{-4} to 3×10^2 s^{-1}. For each temperature, the bottom and top curves represent the computations for the constant strain rate tests (2×10^{-4} and 3×10^2 s^{-1}).

The models proposed by Armstrong and Zerilli [75–78] and Follansbee [79] and Follansbee and Kocks [80] for the mechanical response of metals at high strain rates are comprehensive treatments on the subject and will be reviewed here in some detail (Sections 13.4.1 and 13.4.2).

13.4.1 The Zerilli–Armstrong Model

Zerilli and Armstrong [75–78] proposed two microstructurally based constitutive equations that show an excellent match with experimental results; a simple way to calibrate constitutive equations is to compare the predicted and observed deformation pattern in a Taylor specimen (test described in Chapter 3). Section 13.5 provides a simple description. Zerilli and Armstrong [75–78] based their model on the framework of thermally activated dislocation motion described in Section 13.3.2. They analyzed the temperature and strain rate response of typical FCC and BCC metals and noticed a significant difference between these materials. The BCC metals exhibit a much higher temperature and strain rate sensitivity than the FCC metals. They observed that the activation area A was dependent on strain for FCC metals and independent of strain for BCC metals.

This activation area A is obtained from the activation volume V [Eqn. (13.59)]:

$$V = Ab = l^*\lambda b \tag{13.70}$$

It represents the area swept by the dislocation in overcoming an obstacle. Zerilli and Armstrong [75–78] concluded that overcoming Peierls–Nabarro barriers was the principal thermal activation mechanism for BCC metals. The spacing of these obstacles is equal to the lattice spacing and is, obviously, not affected by plastic strain. On the other hand, the activation area A decreased with increasing strain for FCC metals. In Figure 13.9 one sees that the spacing between the forest dislocations will decrease as this forest dislocation density increases. Thus, l^* (Fig. 13.24) decreases with plastic strain for FCC metals. This size dependence on plastic strain was obtained from the well-known relationship between dislocation density and spacing (see, e.g., Meyers and Chawla [13]):

$$\rho \cong \frac{1}{l^{*2}} \tag{13.70a}$$

But the dislocation density increases with the plastic strain as [Eqns. (13.7a) and (13.9)]

$$\gamma = \rho b l$$

If l is assumed constant,

$$\varepsilon = k\rho \quad \text{where } k = \frac{bl}{M} \tag{13.70b}$$

Thus, from Eqns. (13.7a) and (13.70b)

$$A = \lambda l^* = \lambda \left(\frac{bl}{M}\right)^{1/2} \varepsilon^{-1/2}$$

Zerilli and Armstrong used this expression to determine the activation area at 0 K. They applied this to their expression for the thermal portion of stress:

$$\sigma^* = \frac{M\,\Delta G_0}{Ab}\, e^{-\beta T}$$

where

$$\beta = -C_3 + C_4 \ln \dot{\varepsilon}$$

and M is an orientation (Schmid) factor, ΔG_0 is the height of free energy barrier at 0 K, A_0 is the activation area at 0 K, b is Burgers vector, and β is a parameter dependent on strain and strain rate.

Since A = const for BCC metals and is proportional to $\bar{\varepsilon}^{1/2}$ for FCC metals,

$$\sigma^* = C_1 \exp(-C_3 T + C_4 T \ln \dot{\varepsilon}) \quad \text{(for BCC)} \tag{13.71}$$

$$\sigma^* = C_2 \varepsilon^{1/2} \exp(-C_3 T + C_4 T \ln \dot{\varepsilon}) \quad \text{(for FCC)} \tag{13.72}$$

They added to these expressions a term that gives the athermal component of flow stress [σ_G in Eqn. (13.20)] and a term that describes the flow stress dependent on grain size. It is known that the yield stress increases as the grain

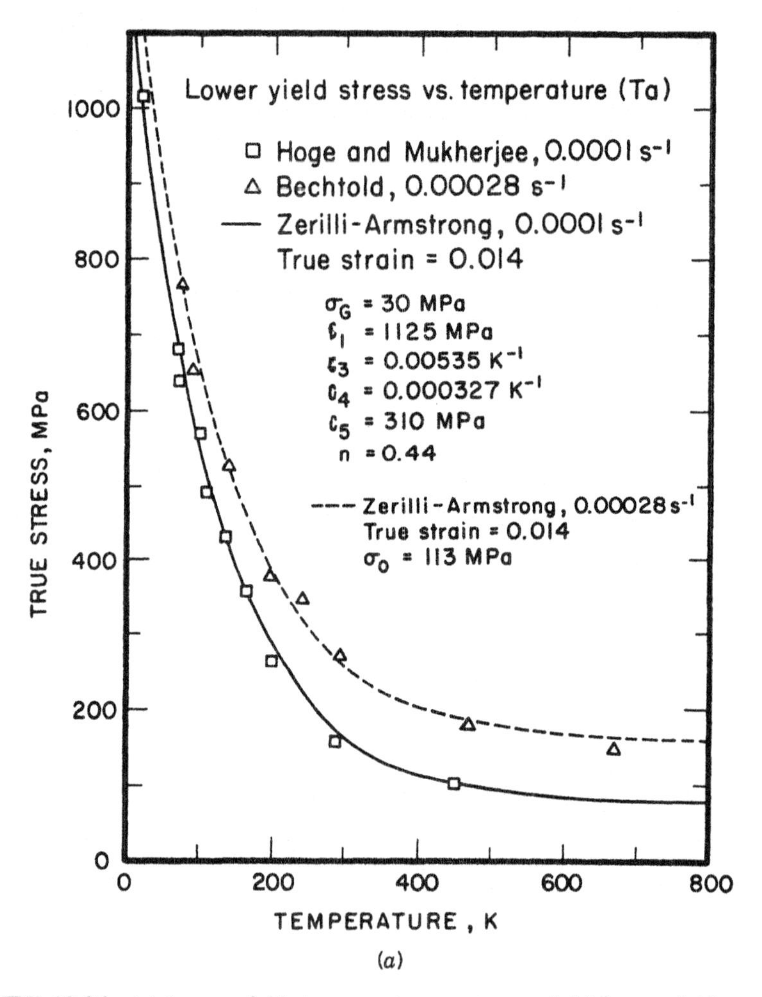

(a)

FIGURE 13.26 (a) Lower yield stress vs. temperature and (b) lower yield stress vs. strain rate for tantalum; experimental data by Hoge and Mukherjee, Bechtold, and Mitchell and Spitzig are compared with the Zerilli–Armstrong constitutive equation (continuous lines). (From Zerilli and Armstrong [77], Figs. 8 and 9, p. 1586. Reprinted with permission of the publisher.)

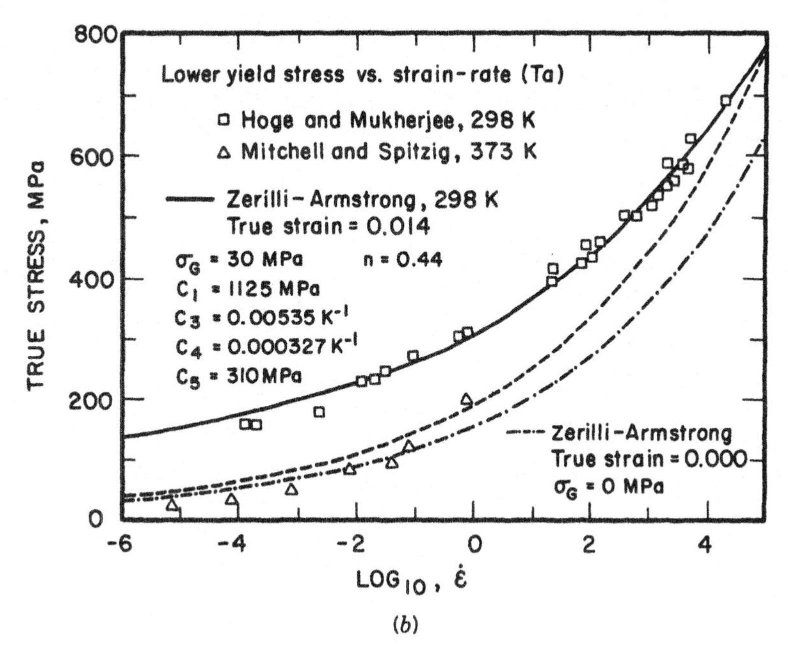

(*b*)

FIGURE 13.26 (*Continued*)

size decreases, and this dependence is usually described by the Hall–Petch equation ($d^{-1/2}$, where d is the grain diameter). Thus

$$\sigma = \sigma_G + \sigma^* + kd^{-1/2}$$

where σ^* is given by either Eqn. (13.71) or (13.72):

$$\text{FCC: } \sigma = \sigma_G + C_2 \varepsilon^{1/2} \exp(-C_3 T + C_4 T \ln \dot{\varepsilon}) + kd^{-1/2} \quad (13.73)$$

$$\text{BCC: } \sigma = \sigma_G + C_1 \exp(-C_3 T + C_4 T \ln \dot{\varepsilon}) + C_5 \varepsilon^n + kd^{-1/2} \quad (13.74)$$

The principal difference between the two equations resides in the fact that the plastic strain is uncoupled from the strain rate and temperature for the BCC metals. Figure 13.26 shows the application of the ZA equation to experimental results for tantalum (BCC) [70, 79]. Ta has a high temperature and strain-rate sensitivity, and the ZA equation represents the behavior well. Since the work-hardening component ($C_5 \varepsilon^n$) is added, the stress-strain curve is simply translated up and down with $\dot{\varepsilon}$ and T.

13.4.2 Mechanical Threshold Stress Constitutive Model

The mechanical threshold stress (MTS) model uses the same basic ingredients of thermally activated dislocation motion described in this chapter and focuses on the determination of the threshold stress (height of barrier in Fig. 13.11: F_0, τ_0, or σ_0). This threshold stress is the flow stress of a certain structure at 0 K. Follansbee [80] and Follansbee and Kocks [81] defined a normalized total activation energy g_0 in the following manner:

$$\Delta G_0 = G(T)b^3 g_0 \tag{13.75}$$

where $G(T)$ is the temperature-dependent shear modulus. By considering barriers with $p = \frac{1}{2}$ and $q = \frac{3}{2}$ [Eqn. (13.64)], they obtained

$$\Delta G = \Delta G_0 \left[1 - \left(\frac{\sigma}{\sigma_0}\right)^{1/2}\right]^{3/2} \tag{13.76}$$

Substituting Eqn. (13.75) into (13.76) yields

$$\Delta G = G(T)b^3 g_0 \left[1 - \left(\frac{\sigma}{\sigma_0}\right)^{1/2}\right]^{3/2}$$

substituting this into Eqn. (13.26) leads to (the student should work through this)

$$\left(\frac{\sigma}{G(T)}\right)^{1/2} = \left(\frac{\sigma_0}{G(T)}\right)^{1/2} \left[1 - \left(\frac{kT}{Gb^3 g_0} \ln \frac{\dot{\varepsilon}_0}{\dot{\varepsilon}}\right)^{2/3}\right] \tag{13.77}$$

By applying Eqn. (13.77) and conducting tests at strain rates ranging from 10^{-4} to 10^4 s^{-1} and final strains of 0.05, 0.1, 0.15, 0.2, and 0.25, they obtained the results plotted in Figure 13.27(a). It can be seen that the intercepts of these plots with the ordinate axis provide values of σ_0. The tests were conducted by reloading the prestrained specimens (to the different strains of 0.05, 0.1, . . . , 0.25) at varying temperatures T. When $T = 0$, in Eqn. (13.77),

$$\sigma = \sigma_0$$

Since it is not possible to perform 0 K tests, the intercepts are obtained by extrapolation. It can be seen that a straight-line behavior is obeyed, in accordance with Eqn. (13.77). Another important observation is that σ_0/G (or σ_0) is dependent on the strain rate. The linear behavior indicates that the correct barrier shape was assumed. The values of σ_0 increase with strain rate at a constant strain (of 0.10), as indicated in Figure 13.27(b). This lends support to one of the cardinal statements of the MTS model: *Structure evolution is rate sensitive*. Thus, the simple relationship of the type

$$\sigma = \sigma_0 + \kappa\varepsilon^{1/2}$$

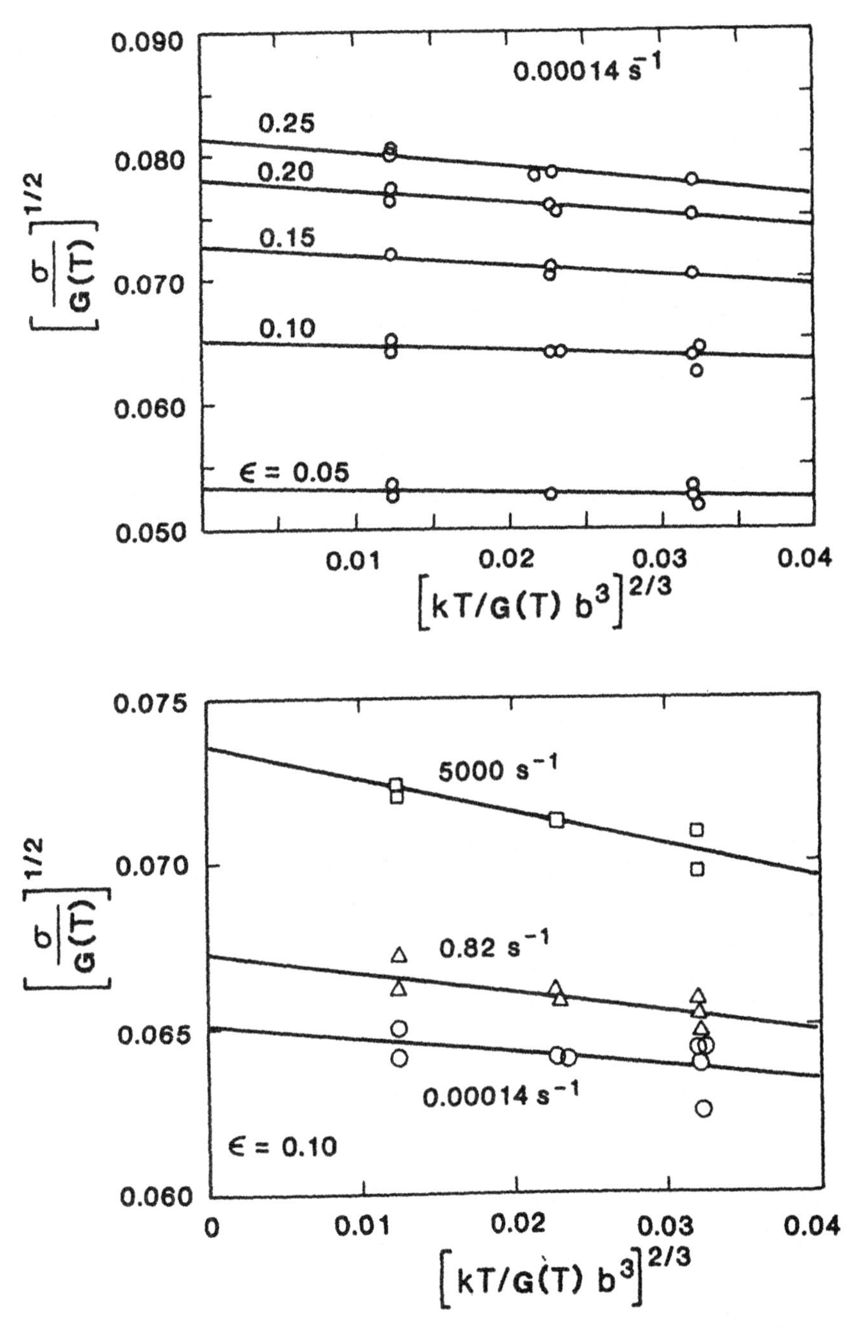

FIGURE 13.27 (a) Reload yield stress for copper deformed to different plastic strains (0.05, 0.1, . . . , 0.25) and constant strain rate. (b) Reload yield stress for copper at a plastic strain of $\varepsilon = 0.1$ at different strain rates as a function of temperature according to normalized coordinates provided by Eqn. 13.77. (From Follansbee [79], Figs. 11 and 13, pp. 467, p. 469. Reprinted by courtesy of Marcel Dekker, Inc.)

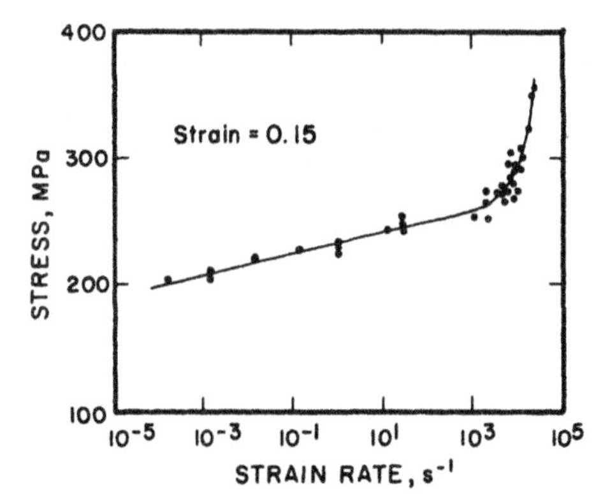

FIGURE 13.28 Flow stress for 99.99 Cu at plastic strain of $\varepsilon = 0.15$ as a function of strain rate. (Reprinted from *Acta Met.*, vol. 36, P. S. Follansbee and U. F. Kocks, Fig. 1, p. 82, with permission from Pergamon Press Ltd.)

which is obtained directly from Taylor's theory of work hardening, is only applicable at a constant strain rate (κ is strain rate dependent). The threshold stress is higher (at a constant strain ε) for higher strain rates. This is indicative of the stronger barriers created by high-strain-rate deformation. Based on this argument, Follansbee and Kocks [81] concluded that, for copper, dislocation drag mechanisms are not operative up to strain rates of 10^4 s^{-1}. The increased threshold stress σ_0 is responsible for the increase of strain rate sensitivity (as defined by $m = \partial\sigma/\partial \ln \dot{\varepsilon}$) at strain rates of 10^3 s^{-1} and higher. Figure 13.28 shows the experimental results obtained by Follansbee and Gray [82] for copper. It is indeed clear that at 10^3 s^{-1} it starts to rise.

The procedure used in the establishment of the threshold stress is to first perform the dynamic test to a fixed strain on a number of specimens and then to test these predeformed specimens at a very slow strain rate and decreasing temperatures. The flow stress is then the stress required to overcome the dynamically generated barriers minus the thermal activation. As the testing temperature is decreased, the thermal activation component decreases until it vanishes, at 0 K. Thus, one can find the elusive threshold stress by extrapolation if the correct barrier shape (p and q) is used.

13.5 EXPERIMENTAL VALIDATION OF CONSTITUTIVE EQUATIONS

The constitutive equations described in these chapters have a number of parameters:

Johnson–Cook	5
Klopp–Clifton–Shawki	4
Hoge–Mukherjee	4
Campbell	4
MTS	2–4
Zerilli–Armstrong	5

These parameters are experimentally determined in tests conducted over a range of temperatures and strain rates. Once the parameters are determined, it is useful to validate the constitutive equation by comparing its predictions with experimental results. The Taylor test is well suited for these experiments, since it provides a range of plastic strains and strain rates. Whereas the simple analysis by Taylor (Chapter 3, Section 3.5) predicted reasonably well the reduction of length of a projectile, it did not predict the correct shape of the specimen. The work-hardening and strain rate sensitivities of the material play an important role, as demonstrated by the comparison of two hydrocode calculations shown in Figure 13.29. The projectiles are copper, with initial length of 10 cm, diameter of 3 cm, and flight velocity of 150 m/s. The plastic deformation patterns are radically different. The plastic deformation is primarily localized at the specimen–target interface for the ideal elastoplastic specimen, whereas plastic deformation propagates through the specimen for the case of work hardening (assumed to be parabolic).

Zerilli–Armstrong [75] compare their equation to the Johnson–Cook equation by seeing how well their predictions match the profile of Taylor specimens. Figure 13.30 shows this comparison. The Zerilli–Armstrong model shows good correlation with the experimental results for a BCC metal (Fe) and for FCC metal (Cu). The advantage of the Zerilli–Armstrong over the Johnson–Cook model is that it is anchored in the physical processes taking place in the material. Its advantage over the MTS model is that the testing methodology is simpler (no low-temperature tests are required).

13.6 TEMPERATURE RISE DURING PLASTIC DEFORMATION

Plastic deformation at low strain rates can often be treated as an isothermal process. A tensile specimen pulled in tension at a strain rate of 10^{-4}–10^{-3} s^{-1} does not undergo a significant temperature rise. The distance that heat can "travel" during a certain time is often referred to as the thermal diffusion distance and is given by $2\sqrt{\alpha t}$, where α is the thermal diffusivity. In the case of the copper specimen (α = 1.14 cm^2 s^{-1}) being pulled at 10^{-4} s^{-1}, the diffusion distance is 150 cm assuming that the duration of the test is

$$\dot{\varepsilon} = \frac{\varepsilon}{t} \qquad \therefore\ t_1 = \frac{0.5}{10^{-4}} = 5 \times 10^3 \text{ s}$$

However, if the test were conducted at 10^3 s^{-1}, the same strain would be imparted in $t_2 = 5 \times 10^{-3}$ s, and the corresponding thermal diffusion distance would be 1.5 mm.

Thus, high-strain-rate deformation is often adiabatic, and the deformation work is transformed into heat with the attendant temperature increase of the specimen. This temperature rise has a profound effect on the constitutive behavior of the material because of the thermal softening. In Chapter 15 we deal with thermal softening, which may lead to shear instability. The temperature

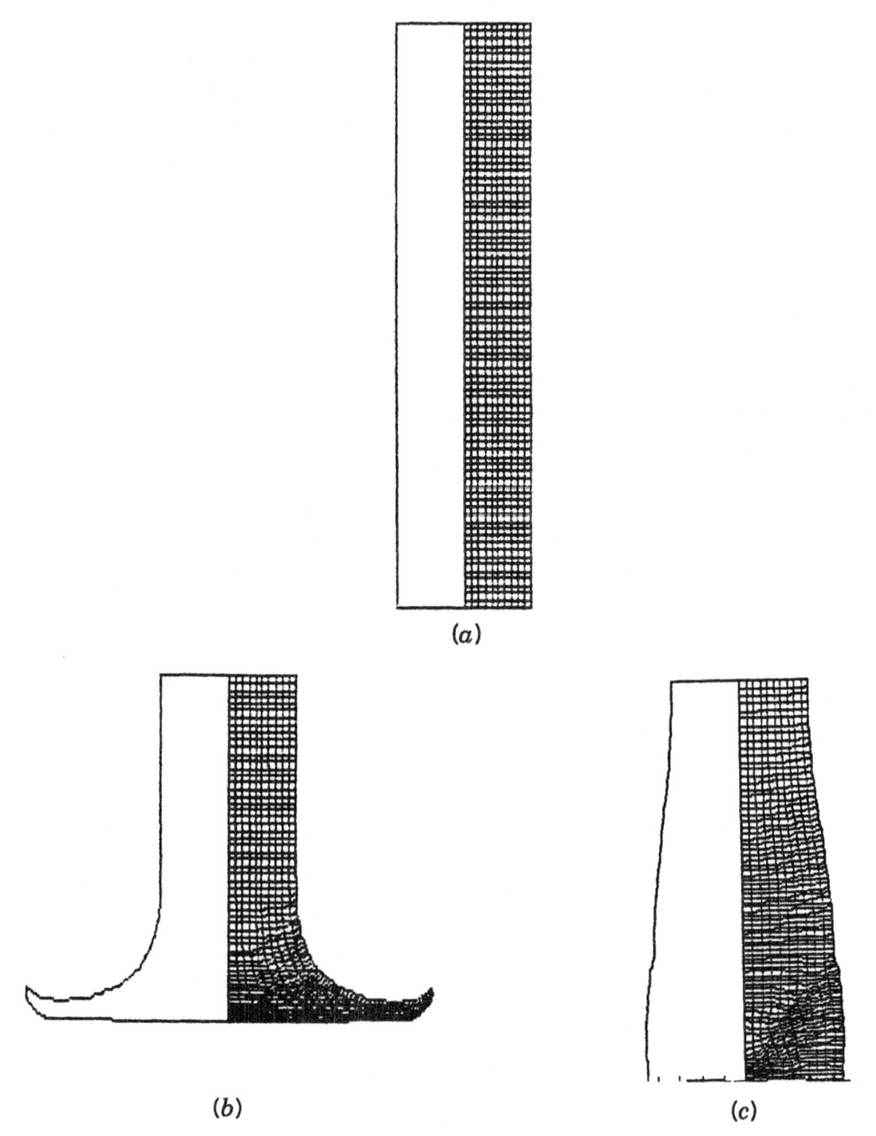

FIGURE 13.29 Simulation of Taylor impact test of copper specimen at a velocity of 150 m/s and 100 μs after impact: (a) initial specimen; (b) ideal plastic response (no work hardening; $\sigma_y = 150$ MPa); (c) parabolic hardening: $\sigma = (69 + 323.5\varepsilon^{0.54})$ MPa. (Courtesy of D. Chung, Taejon, S. Korea.)

rise associated with plastic deformation can be directly obtained from the constitutive equation by considering that a fraction of plastic deformation is converted into heat. These values have traditionally been obtained by measuring the stored energy of deformation in calorimetric measurements and subtracting it from the total work of deformation. Manson et al. [84] performed high-strain-rate experiments in a Hopkinson bar coupled with infrared detectors to

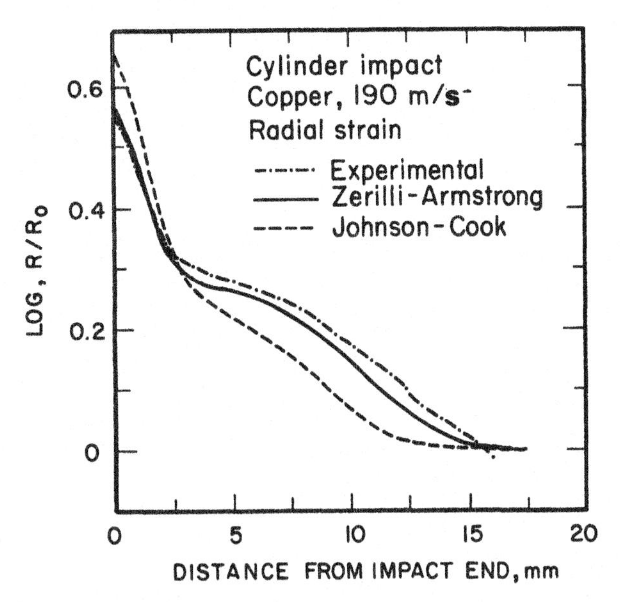

FIGURE 13.30 Profile of Taylor specimen (copper, impacted at 190 m/s) plotted as radial strain vs. distance; experimental result (dotted line) compared with Johnson-Cook (continuous line) and Zerilli–Armstrong constitutive equations. (From Zerilli and Armstrong [75], Fig. 5, p. 1822. Reprinted with permission of the publisher.)

calculate the work rate to heat rate conversion fraction, β. For 2024 Al and 4340 steel, they obtained values around 0.8–0.9, whereas for the Ti-6Al-4V alloy a drop from 1 to 0.6 was observed as the strain increased from 0.05 to 0.2. This author would like to suggest that this anomalous conversion fraction could be related to deformation mechanisms where dislocations are not the primary carriers of plastic strain. Twinning and martensitic phase transformations are probably contributory factors. For most metals, the conversion factor is usually taken as 0.9 (this implies that 10% of the work of deformation is stored in the material as defects). Thus, the temperature rise can be expressed as:

$$\Delta T = \frac{\beta}{\rho C_P} \int_0^{\varepsilon_f} \sigma \, d\varepsilon = \frac{0.9}{\rho C_P} \int_0^{\varepsilon_f} \sigma \, d\varepsilon$$

By substituting σ for its value in Eqn. (13.1) and solving, we arrive at

$$\int_{T_0^*}^{T_f^*} \frac{dT^*}{1 - T^{*m}} = \frac{0.9(1 + C \ln \dot{\varepsilon}/\dot{\varepsilon}_0)}{\rho C_P(T_m - T_r)} \int_0^{\varepsilon_f} (\sigma + B\varepsilon^n) \, d\varepsilon \qquad (13.78)$$

where the strain rate $\dot{\varepsilon}$ is assumed constant and the T^* are homologous temperatures. The integral on the left-hand side has to be numerically evaluated if

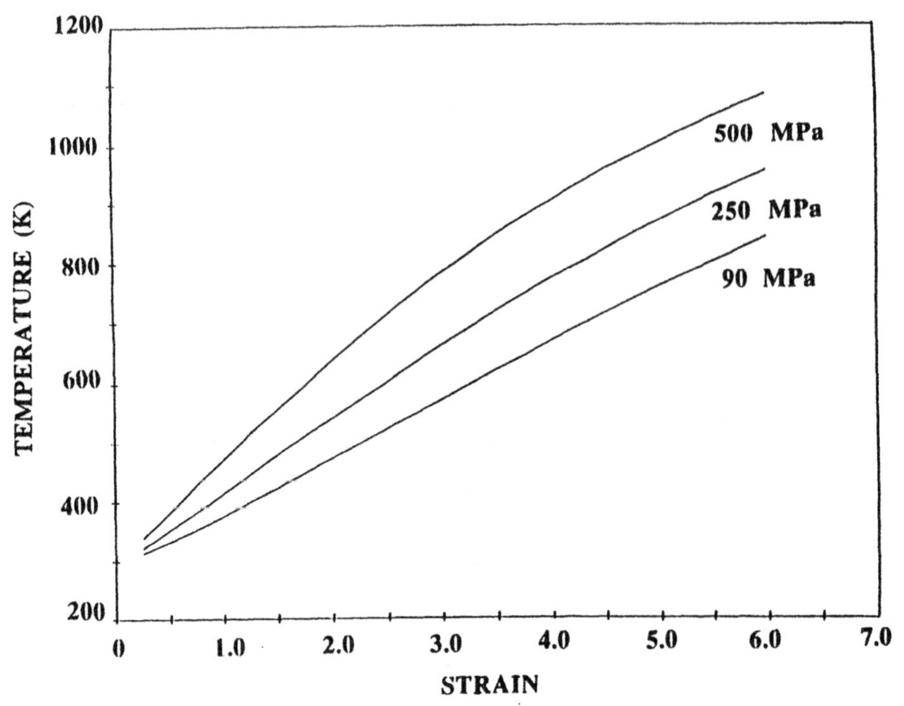

FIGURE 13.31 The variation in the temperature with plastic strain during the dynamic adiabatic deformation of copper at a strain rate of $10^4\ s^{-1}$; initial temperature 298 K. (From U. Andrade, et al. [83]).

$m \neq 1$. For $m \cong 1$, a simpler solution is found:

$$T^* = 1 - \exp\left[-\frac{0.9(1 + C \ln \dot{\varepsilon}/\dot{\varepsilon}_0)}{\rho C_P(T_m - T_r)}\left(\sigma_0 \varepsilon + \frac{B\varepsilon_f^{n+1}}{n+1}\right)\right] \quad (13.79)$$

Example 13.5 Estimate the temperature rise as a function of strain for copper deformed at $10^4\ s^{-1}$.

From Table 13.1, we obtain the values of the various Johnson–Cook parameters. By making the approximation $m = 1$, the integral on the left-hand side is simplified to

$$\ln \frac{1 - T_f/T_m}{1 - T_0/T_m} = \frac{0.9(1 + 0.025 \ln 10^4)}{383 \times 8960}\left(\sigma_0 \varepsilon_f + \frac{B\varepsilon_f^{n+1}}{n+1}\right)$$

By solving this equation for T_f, the values plotted in Figure 13.31 are obtained. The temperature rise is indeed very significant.

REFERENCES

1. S. Nemat-Nasser, *Appl. Mech. Rev.*, **45** (1992), 519.
2. J. D. Campbell and W. G. Ferguson, *Phil. Mag.*, **21** (1970), 63.
3. O. Vohringer, in *Deformation Behavior of Metallic Materials*, ed. C. Y. Chiem, International Summer School on Dynamic Behavior of Materials, ENSM, Nantes, September 11–15, 1989, p. 7.
4. G. R. Johnson and W. H. Cook, *Proc. 7th Intern. Symp. Ballistics*, Am. Def. Prep. Org. (ADPA), Netherlands, 1983.
5. G. R. Johnson, J. M. Hoegfeldt, U. S. Lindholm, and A. Nagy, *ASME J. Eng. Mater. Tech.*, **105** (1983), 42.
6. L. W. Meyer, in *Shock-Wave and High-Strain-Rate Phenomena in Materials*, eds. M. A. Meyers, L. E. Murr, and K. P. Staudhammer, Dekker, New York, 1992.
7. R. W. Klopp, R. J. Clifton, and T. G. Shawki, *Mech. Mater.*, **4** (1985), 375.
8. T. Vinh, M. Afzali, and A. Rocke, in *Mechanical Behavior of Materials*, Proc. ICM, eds. A. K. Miller and R. F. Smith, Pergamon, New York, 1979, p. 633.
9. J. D. Campbell, A. M. Eleiche, and M. C. C. Tsao, in *Fundamental Aspects of Structural Alloy Design*, Plenum, New York, 1977, p. 545.
10. J. Harding, in *Impact Loading and Dynamic Behavior of Materials*, eds. C. Y. Chiem, H. D. Kunze, and L. W. Meyer, DGM—Informationsgesellschaft, Oberursel, Germany, 1988, p. 21.
11. R. J. Clifton, *J. Appl. Mech.*, **50** (1983).
12. U. Andrade, M. A. Meyers, and A. H. Chokshi, Scripta Met. et Mat., in press (1993).
13. M. A. Meyers and K. K. Chawla, *Mechanical Metallurgy: Principles and Applications*, Prentice-Hall, Englewood Cliffs, NJ, 1984.
14. J. J. Gilman and W. G. Johnston, *Dislocations and Mechanical Properties of Crystals*, Wiley, New York, 1957, p. 116.
15. W. G. Johnston and J. J. Gilman, *J. Appl. Phys.*, **33** (1959), 129.
16. D. F. Stein and J. R. Low, Jr., *J. Appl. Phys.*, **32** (1960), 362.
17. H. Ney, R. Labusch, and P. Haasen, *Acta Met.*, **25** (1977), 1257.
18. H. W. Schadler, *Acta Met.*, **12** (1964), 861.
19. A. R. Chaudhuri, J. R. Patel, and L. G. Ribin, *J. Appl. Phys.*, **33** (1962), 2736.
20. J. S. Erickson, *J. Appl. Phys.*, **33** (1962), 2499.
21. E. Y. Gutmanas, E. M. Nadgornyi, and A. V. Stepanov, *Soviet Phys. Solid State*, **5** (1963), 743.
22. R. W. Rohde and C. H. Pitt, *J. Appl. Phys.*, **38** (1967), 876.
23. D. P. Pope, T. Vreeland, Jr., and D. S. Wood, *J. Appl. Phys.*, **38** (1967), 4011.
24. W. F. Greenman, T. Vreeland, Jr., and D. S. Wood, *J. Appl. Phys.*, **38** (1967), 3595.
25. T. Suzuki and T. Ishii, *Suppl. Trans. Jpn. Inst. Met.*, **9** (1968), 687.
26. R. C. Blish II and T. Vreeland, Jr., *J. Appl. Phys.*, **40** (1969), 884.
27. J. A. Gorman, D. S. Wood, and T. Vreeland, Jr., *J. Appl. Phys.*, **40** (1969), 833.
28. V. R. Parameswaran and J. Weertman, *Met. Trans.*, **2** (1971), 1233.

29. V. R. Parameswaran, N. Urabe, and J. Weertman, *J. Appl. Phys.*, **43** (1972), 2982.
30. S. Glasstone, K. J. Laidler, and H. Eyring, *The Theory of Rate Processes*, McGraw-Hill, New York, 1941.
31. J. J. Gilman, *Micromechanics of Flow in Solids*, McGraw-Hill, New York, 1969, p. 179.
32. H. L. Preckel and H. Conrad, *Acta Met.*, **15** (1967), 955.
33. J. Cotner and J. Weertman, *Disc. Faraday Soc.*, (38) (1964), 225.
34. R. Becker, *Z. Phys.*, **26** (1925), 919.
35. A. Seeger, *Z. Naturforsch*, **9A** (1954), 758, 819, 856.
36. A. Seeger, *Phil. Mag.*, **1** (1956), 651.
37. A. Seeger and P. Schiller, *Acta Met.*, **101** (1962), 348.
38. H. Conrad and H. Wiedersich, *Acta Met.*, **8** (1960), 128.
39. H. Conrad, *J. Metals*, **16** (1964), 582.
40. H. Conrad, *Mater. Sci. Eng.*, **6** (1970), 265.
41. U. F. Kocks, A. S. Argon, and M. F. Ashby, *Progr. Mater. Sci.*, **19** (1975), 1.
42. L. W. Meyer, Ph.D. Dissertation (82/106), University of Dortmund, Germany, 1982.
43. J. J. Gilman, *Micromechanics of Flow in Solids*, McGraw-Hill, New York, 1969, p. 195.
44. P. P. Gillis, J. J. Gilman, and J. W. Taylor, *Phil. Mag.*, **20** (1969), 279.
45. J. E. Dorn, J. Mitchell, and F. Hansen, *Exp. Mech.*, (5) (1965), 353.
46. A. V. Granato, in *Metallurgical Effects at High Strain Rates*, AIME Proc., eds. R. W. Rohde, B. M. Butcher, J. R. Holland, and C. H. Karnes, Plenum, New York, 1973, p. 255.
47. P. P. Gillis and J. Kratochvil, *Phil. Mag.*, **21** (1970), 425.
48. J. P. Hirth and J. Lothe, *Theory of Dislocations*, McGraw-Hill, New York, 1968, p. 193.
49. P. Kumar and R. J. Clifton, *J. Appl. Phys.*, **50** (1970), 4747.
50. J. D. Eshelby, *Phys. Rev.*, **90** (1953), 248.
51. R. J. Clifton and X. Markenscoff, *J. Appl. Phys. Solids*, **29** (1981), 227.
52. J. D. Eshelby, *Proc. Phys. Soc. (Lond.)*, **A12** (1949), 307.
53. J. Weertman, in *Response of Metals to High Velocity Deformation*, AIME Proc., P. G. Shewmon and V. F. Zackay, eds., Interscience, New York, 1961, p. 205.
54. J. Weertman and J. R. Weertman, *Elementary Dislocation Theory*, MacMillan, New York, 1964.
55. D. Kuhlmann-Wilsdorf, in *Physical Metallurgy*, 2nd ed., ed. R. W. Cahn, Elsevier/North-Holland, New York, 1970, p. 787.
56. J. Weertman and J. R. Weertman, in *Dislocations in Solids*, Vol. 3, ed. F. R. N. Nabarro, Elsevier/North-Holland, New York, 1980, p. 1.
57. J. D. Eshelby, *Proc. Phys. Soc. (Lond.)*, **B69** (1956), 1013.
58. J. Weertman, *J. Appl. Phys.*, **38** (1967), 5293.
59. J. H. Weiner and M. Pear, *Phil. Mag.*, **31** (1975), 679.

60. Y. Y. Earmme and J. H. Weiner, *J. Appl. Phys.*, **48** (1977), 3317.
61. P. P. Gillis and J. Kratochvil, *Phil. Mag.*, **21** (1970), 425.
62. G. Regazzoni, U. F. Kocks, and P. S. Follansbee, *Acta Met.*, **35** (1987), 2865.
63. P. M. Sargeant and M. F. Ashby, "The Presentation of High-Strain Rates on Deformation Mechanism Maps," Report No. CUED/C/MATS/TR.98, Cambridge U., Cambridge, U.K., March 1983.
64. H. J. Frost and M. F. Ashby, *Deformation Mechanism Maps*, Pergamon, Oxford, 1982.
65. J. Harding, "Constitutive Modelling of Material Mechanical Behavior at High Rates of Strain," International Summer School on Dynamics Behavior of Materials, Ecole Centrale de Nantes, Nantes, France, September 11–15, 1989.
66. J. Harding, in *Explosive Welding, Forming, and Compaction*, ed. T. Z. Blazinsky, Applied Science, London, 1983, p. 123.
67. J. D. Campbell, *Dynamic Plasticity*, Oxford University Press, New York.
68. U. S. Lindholm and L. M. Yeakley, *Exp. Mech.*, **7** (1967), 1.
69. J. D. Campbell and J. Harding, in *Response of Metals to High Velocity Deformation*, P. G. Shewmon and V. F. Zachay (eds), Interscience, New York, 1961, p. 51.
70. K. G. Hoge and A. K. Mukherjee, *J. Mater. Sci.*, **12** (1977), 1666.
71. J. R. Klepaczko, *Mat. Sci. Eng.*, **18** (1975), 121.
72. J. R. Klepaczko and C. Y. Chiem, *J. Mech. Phys. Solids*, **34** (1986), 29.
73. J. R. Klepaczko, in *Proc. Int. Conf. DYMAT*, Les Editions de Physique, Les Ulis, 1988, p. C3533.
74. J. R. Klepaczko, in *Shock Wave and High-Strain-Rate in Materials*, eds. M. A. Meyers, L. E. Murr, and K. P. Staudhammer, Dekker, New York, 1992, p. 145.
75. F. J. Zerilli and R. W. Armstrong, *J. Appl. Phys.*, **61** (1987), 1816.
76. F. J. Zerilli and R. W. Armstrong, in *Shock Compression of Condensed Matter—*1989, eds. S. C. Schmidt, J. N. Johnson, and L. W. Davison, Elsevier, Amsterdam, 1990, p. 357.
77. F. J. Zerilli and R. W. Armstrong, *J. Appl. Phys.*, **68** (1990), 1580.
78. F. J. Zerilli and R. W. Armstrong, *Acta Met. Mat.*, **40** (1992), 1803.
79. T. G. Mitchell and W. A. Spitzig, in *Refractory Metals and Alloys*, Vol. IV, Gordon and Breach, NY, 1967. p. 25.
80. P. S. Follansbee, in *Metallurgical Applications of Shock-Wave and High-Strain Rate Phenomena*, eds. L. E. Murr, K. P. Staudhammer, and M. A. Meyers, Dekker, New York, 1986, p. 451.
81. P. S. Follansbee and U. F. Kocks, *Acta Met.*, **36** (1988), 81.
82. P. S. Follansbee and G. T. Gray III, *Mater. Sci. Eng.*, **A138** (1991), 23.
83. U. R. de Andrade, M. A. Meyers, A. H. Chokshi, and K. S. Vecchio, Acta Met. et. Mat., **42** (1994), 3183.
84. J. J. Manson, A. J. Rosakis, and G. Ravichandran, *Mech. of Matls.*, **17** (1994) 135.

CHAPTER 14

Plastic Deformation in Shock Waves

14.1 STRENGTHENING DUE TO SHOCK WAVE PROPAGATION

The effects of shock wave passage through metals have been known since the 1950's and are even commercially exploited. Shock waves harden most metals, and this hardening is due to the generation of large densities of defects. Hadfield steel (an austenitic high-manganese steel), railfrogs (switching components in railways), and mining equipment are commercially hardened by shock waves. The shock waves are applied by the detonation of explosives in direct contact with the metal components. A plastic sheet explosive, such as Detasheet (DuPont), is used for that purpose. More details are given in Chapter 17. Murr [1] reviews a large volume of data generated by him and his colleagues. In research, shock hardening is usually imparted by planar parallel normal impact of a projectile (accelerated by either explosives or compressed gases). Figure 14.1 shows the hardening produced in a number of metals as a function of pressure. It should be noticed that this increase in hardness occurs with practically negligible strain, because shock wave propagation is a uniaxial strain process. On the other hand, rolling, forging, extrusion, and other operations producing work hardening require substantial plastic strains. This is illustrated in Figure 14.2, which shows the Vickers hardness (HV) of a nickel–manganese steel with ≈25% Ni precipitation hardened by the Ni_3 (Ti, Ni) intermetallic. A 20-GPa shock wave increases the yield stress by 85%. This was followed by aging at different temperatures, for 1 h, to optimize the overall mechanical properties. The hardness increased from HV 200 to HV 300 by the application of a 33-GPa shock wave. On the other hand, rolling to a 30% thickness reduction (obtained by interpolation in Fig. 14.2) is required for the same hardness. One can, by going through the literature, find numerous other alloys that have been shock hardened.

The shock-hardening effect has been found to be not only pressure dependent but also duration dependent. Figure 14.3 shows the effect of pulse duration on the hardness of a number of metals. It can be seen that, in the 1–10-μs range, there is no noticeable effect. At pulse durations lower than 1 μs, there is a significant effect for some materials. Figure 14.4 shows an anomalous variation in hardness observed in a Cu–8.7% Ge alloy by Wright and Mikkola [3]. At a pressure of 20 GPa, these fluctuations of hardness are due to time-dependent

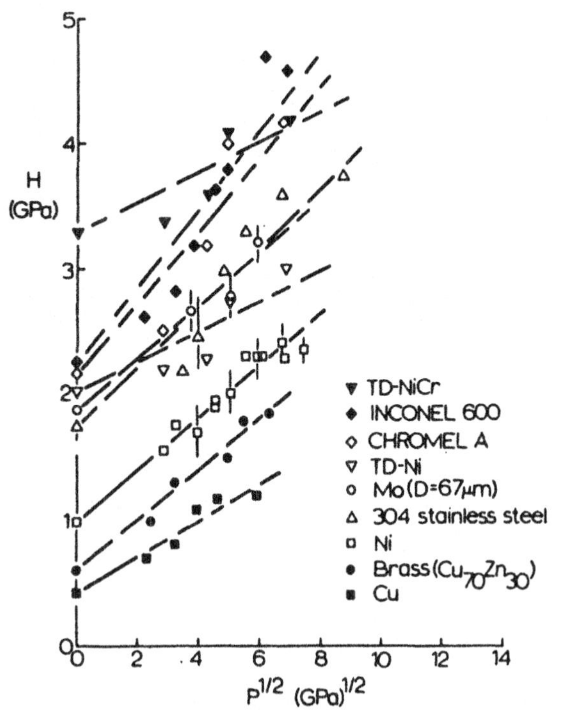

FIGURE 14.1 Effect of pressure on shock hardening for a number of metals and alloys. (From Murr [1], Fig. 8, p. 633. Reprinted with permission of the publisher.)

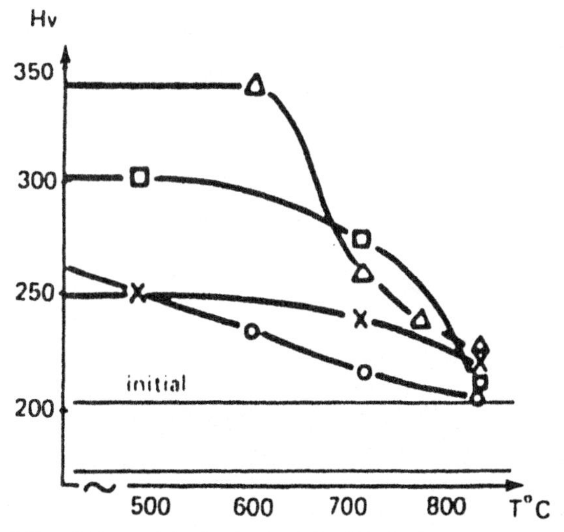

FIGURE 14.2 Microindentation hardness of nickel-manganese steel after thermal treatments; △ rolled by 40%; ○ rolled by 20%; □ shocked 33 GPa; × shocked 16 GPa. (From Deribas et al. [2], Fig. 2, p. 348. Reprinted by courtesy of Marcel Dekker, Inc.)

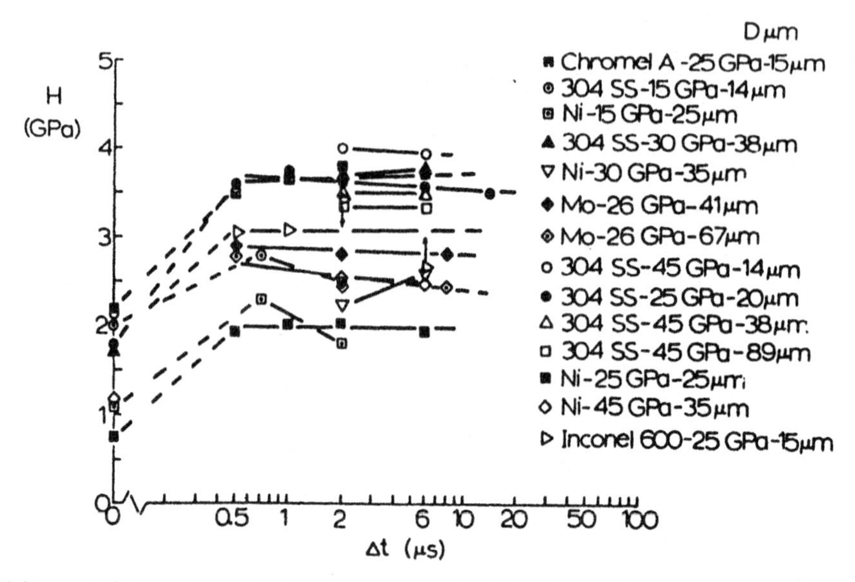

FIGURE 14.3 Effect of pulse duration on shock hardening for a number of metals and alloys. (From Murr [1], Fig. 13, p. 772.)

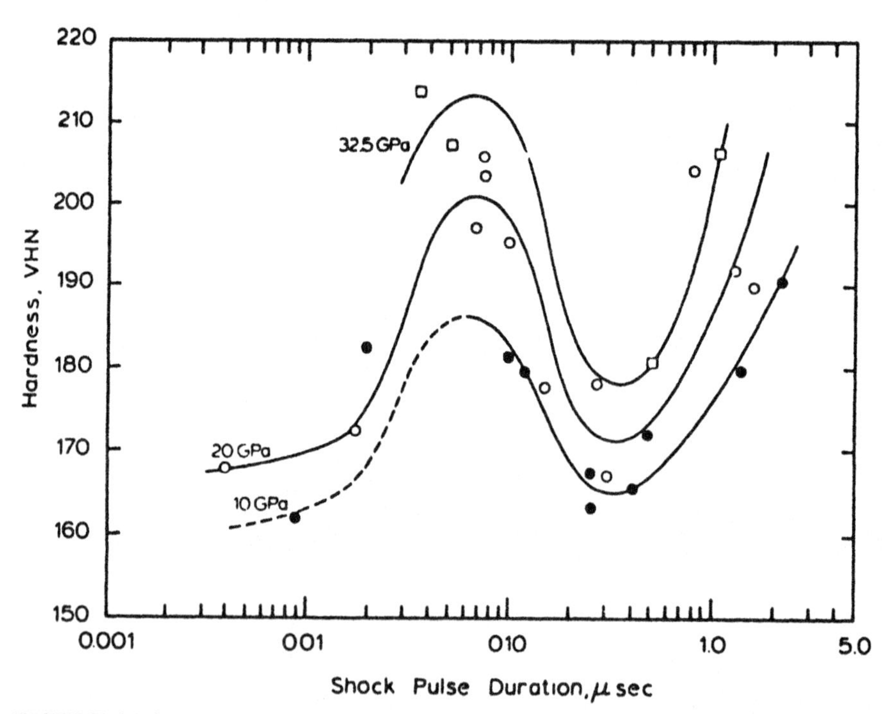

FIGURE 14.4 Effect of pulse duration on shock hardening of a Cu–8.7% Ge alloy. (From Wright and Mikkola [3], Fig. 1, p. 706. Reprinted with permission of the publisher.)

plastic deformation processes accompanying the shock process. The response of individual materials to shock waves has to be separately studied. The pulse duration effects are due to different mechanisms; whereas it is thought that dislocation generation is only pressure dependent, twinning is duration dependent.

The effect of shock hardening does not increase indefinitely. It reaches a saturation level and starts decreasing at increasing pressures. This is due to the shock-induced heating producing recovery and recrystallization in the structure, which eliminates the effects of shock-induced defects. Figure 14.5 shows this effect for iron. The hardness makes a significant jump at 13 GPa, which is the pressure at which the $\alpha \rightarrow \varepsilon$ transformation occurs (see Chapter 8). At a pressure of 50 GPa the hardness saturates, and at higher pressures it starts to drop. The pressure at which hardness saturation is observed varies from material to material and depends on the thermal stability of the shock-induced substructure.

The shock-induced strength enhancement measured in materials is due to the generation of defects, which will be described in the subsequent sections of this chapter. Point, line, and interfacial defects are routinely formed, and tridimensional defects (phase transformations) are often observed. The mechanisms responsible for the formation of these defects are not incontrovertibly established. Thus, a number of mechanisms will be discussed here, because, at the present stage, it is difficult to establish which mechanism (or mechanisms) are operational.

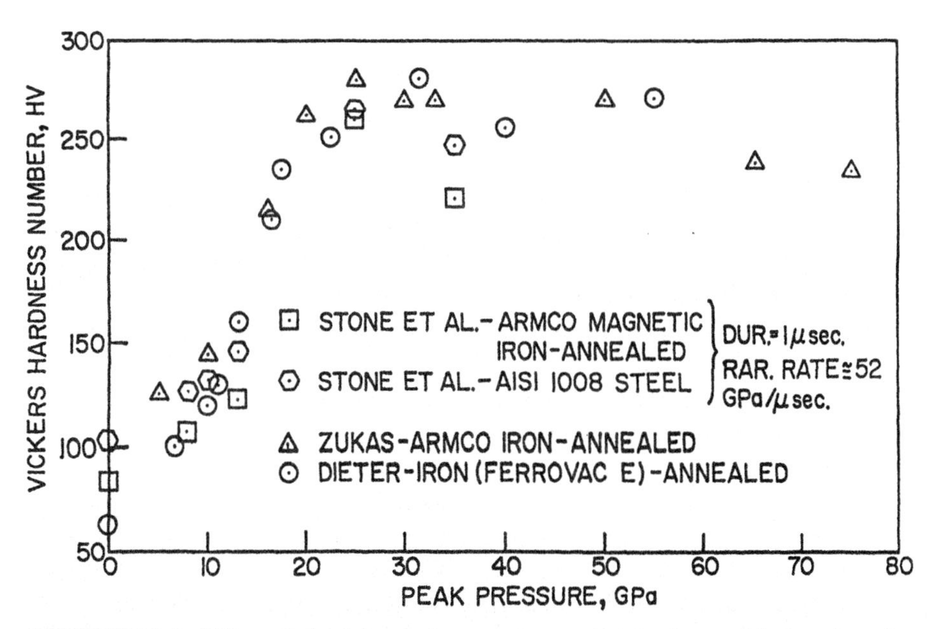

FIGURE 14.5 Effect of shock-hardening pressure on the hardness of iron and steel. (From Stone et al. [4]. Adapted with permission.)

The metallurgical effects associated with dynamic loading were first described in reasonable detail by Rinehart and Pearson [5]. The first systematic investigation of the substructural changed in terms of fundamental deformation modes induced by the passage of shock waves is described in the classic paper by Smith [6]; this work was conducted in cooperation with the Los Alamos Scientific Laboratory. During the past three decades, the activity in this specialized area of metallurgy has steadily continued, and a few hundred papers have resulted. Reviews by Dieter [7, 8], Zukas [9], Orava [10], Leslie [11], Davison and Graham [12], Murr [13, 14], Meyers and Murr [15, 16], Murr and Meyers [17], and Gray [18, 19] should guide the reader to a more in-depth knowledge.

It is important to recognize that the state of stress at the shock front is not hydrostatic. A uniaxial strain compression is applied to the material, as explained in Section 7.1. Thus, shear (or deviatoric) stresses and strains are produced, and these generate defects. In Figure 7.3, one has that $\varepsilon_x \neq 0$ and $\varepsilon_y = \varepsilon_z = 0$. The importance of separating deviatoric from hydrostatic stresses cannot be overemphasized. Different phenomena are controlled by different stresses. Hence, one has:

Dislocations. Generation and motion are controlled by deviatoric stresses; stacking-fault energy is affected by hydrostatic stresses.

Second-Phase Particles. These are a source of dislocations if they have different compressibilities than the matrix. If they are coherently bonded to matrix, coherency might be destroyed by pressure with an attendant loss in strength. These are effects of hydrostatic pressure.

Individual Grains. In materials that do not exhibit cubic symmetry, individual grains have anisotropic compressibilities and hydrostatic stresses will establish incompatibility stresses at their boundaries. This will lead to defect generation (in ductile materials) or cracking (in brittle materials).

Displacive Diffusionless Phase Transformation. A number of phase transformations are induced in materials by the hydrostatic component of stress [20]. Chapter 8 is fully devoted to this topic. Martensitic transformation can also be induced by shear stresses or strains.

Twinning. This is activated by shear stresses. The hydrostatic stresses might have an indirect effect.

Point Defects. Their generation is due to shear stress; their diffusion rate is affected by hydrostatic stresses.

Recovery and Melting Point. These are affected by hydrostatic stress.

Shock and Residual Temperatures. These are affected by both hydrostatic and deviatoric stresses, but for different reasons. The temperature rise due to the hydrostatic stresses is calculated in Section 5.5. It is strictly the conversion of PV into heat. The deviatoric stresses, on the other hand, generate dislocations that move. This internal friction (plastic work) is transformed into heat with an attendant temperature rise. As a result, the total temperature rise in a shock wave is the sum of these two components. In Chapter 15 the temperature rise due to localization of plastic deformation is calculated.

Researchers have attempted to compare the strength of shock-hardened materials with that of the same material processed conventionally (rolling, extrusion, etc.). There is always the very important question of whether this comparison has to be done at the same plastic strain. Meyers and Orava [21, 22] introduced the comparison based on the same effective strains. The equivalent (or effective) strain is defined as (Section 13.1)

$$\varepsilon_{\text{eff}} = \frac{\sqrt{2}}{3}\left[(\varepsilon_x - \varepsilon_y)^2 + (\varepsilon_x - \varepsilon_z)^2 + (\varepsilon_y - \varepsilon_z)^2\right]^{1/2}$$

This strain has three components, $(\varepsilon_x - \varepsilon_y)$, $(\varepsilon_x - \varepsilon_z)$, and $(\varepsilon_y - \varepsilon_z)$, which measure shear along three bisecting planes. Thus, if ε_x, ε_y, and ε_z are principal strains, ε_{eff} is a one-dimensional estimate of the total amount of shear strain undergone by a material. If we compare rolling to shock hardening, we have different stress (or strain) states. Rolling is a biaxial strain state: $\varepsilon_x \neq \varepsilon_y \neq 0$, $\varepsilon_z = 0$ and uniaxial tension is a state where $\varepsilon_x \neq \varepsilon_y = \varepsilon_z \neq 0$. A correct way of comparing the strengthening due to these different processing methods is to compare them at the same effective (or equivalent) strain. In shock hardening we have $\epsilon_x \neq \epsilon_y = \epsilon_z = 0$. Thus:

$$\varepsilon_{\text{eff}} = \frac{\sqrt{2}}{3}\left[2\varepsilon_x^2\right]^{1/2} = \frac{2}{3}\,\varepsilon_x = \frac{2}{3}\ln\frac{l}{l_0}$$

Since

$$d\varepsilon_x = \frac{dl}{l} = \frac{dV}{V}$$

then

$$\varepsilon_x = \ln\frac{V}{V_0}$$

This can be seen from Figure 7.3. However, the material goes through a full stress–strain cycle, that is, it is compressed at the front and expanded at the release (or rarefaction) part of the wave. If the residual strain is zero, then

$$(\varepsilon_x)_{\text{front}} = (\varepsilon_x)_{\text{release}}$$

$$(\varepsilon_{\text{eff}})_{\text{Total}} = \frac{4}{3}\ln\frac{V}{V_0} \tag{14.1}$$

For comparison purposes, the effective strain in a tensile test is, neglecting elastic strains (constant volume):

$$\varepsilon_x + \varepsilon_y + \varepsilon_z = 0;\ \varepsilon_y = \varepsilon_z$$

$$\varepsilon_x + 2\varepsilon_y = 0$$

$$\varepsilon_y + \varepsilon_z = -\varepsilon_x$$

$$\begin{aligned} \varepsilon_{\text{eff}} &= \frac{\sqrt{2}}{3}\left[\left(\varepsilon_x + \frac{\varepsilon_x}{2}\right)^2 + \left(\varepsilon_x + \frac{\varepsilon_x}{2}\right)^2\right]^{1/2} \\ &= \frac{2}{3}\left(\frac{\varepsilon_x^2}{4}\right)^{1/2} = \varepsilon_x \end{aligned} \tag{14.2}$$

It is possible to compare the relative strengthening effects of shock loading and conventional processing by using this effective strain. This is done by setting Eqn. (14.1) equal to Eqn. (14.2):

$$\varepsilon_x = \frac{4}{3} \ln \frac{V}{V_0} \tag{14.3}$$

Figures 14.5 and 14.6 show these effects. In Figure 14.6 Inconel 718, a nickel-based superalloy, was subjected to a shock pressure of 51 GPa by an impact with a copper flyer plate of 2.1 km/s [21, 22]. This resulted in an effective strain ($\frac{4}{3} \ln V/V_0$) of 0.25. By comparison, the same material was cold rolled to the same effective strain and was also pulled in tension until failure. The shocked and rolled materials were subsequently subjected to uniaxial stress tension testing. They are marked U (for undeformed), R (for rolled), and S (for shocked) in Figure 14.6; two solutionizing heat treatments (E → solution treated at 1950°F and D → solution treated at 1750°F) were used. The rolled material (R) is slightly weaker than the control tensile specimen at the same strain, whereas the shocked material (S) is slightly stronger. Figures 14.7(a) and (b) show the shock-strengthening effects for copper and an age-hardenable aluminum alloy. The results are again plotted with the origin for the shocked material at the abscissa corresponding to the appropriate effective stress. Shock hardening of copper results in a flow stress that considerably exceeds the one produced by work hardening in tension. For the aluminum alloy, the shock-strengthening effect depends on the initial microstructure; when the alloy is a solution, that is, there are no precipitates, shock hardening produces a flow stress that exceeds that of the material work hardened in tension. On the other hand, when the material is in the aged condition, shock hardening is less effective than work hardening in tension. This behavior can be explained by the loss of coherency of the precipitates induced by shock loading. The over-aged alloy is strengthened by precipitates that are coherent with the matrix.

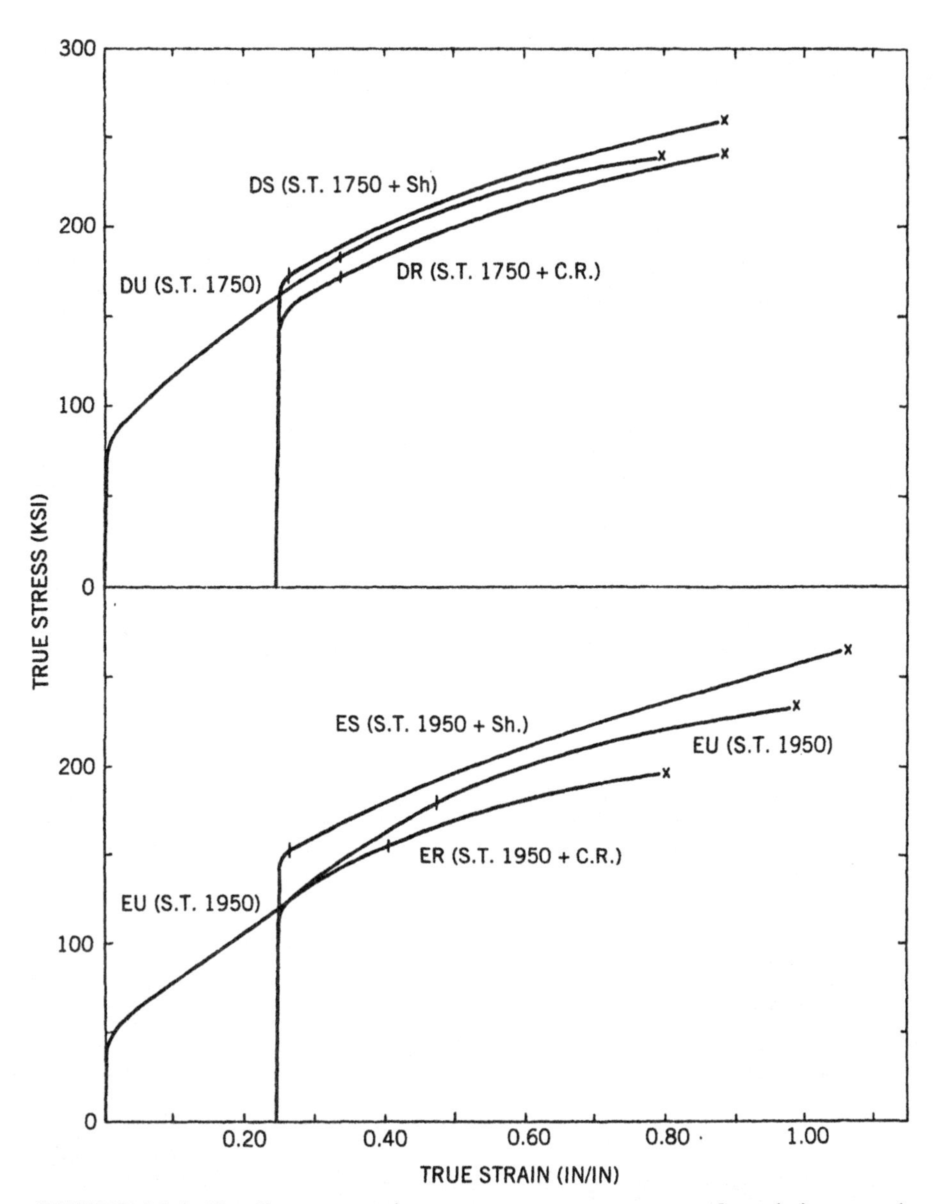

FIGURE 14.6 Tensile stress–strain curves at room temperature for solution-treated (U), cold-rolled (R) or shock-loaded (S) Inconel 718 metal-based superalloy; schedules D and E refer to solutionizing temperatures (1750 and 1950°F) prior to deformation. (From Meyers [21] and Meyers and Orava [22], Fig. 11, p. 187.)

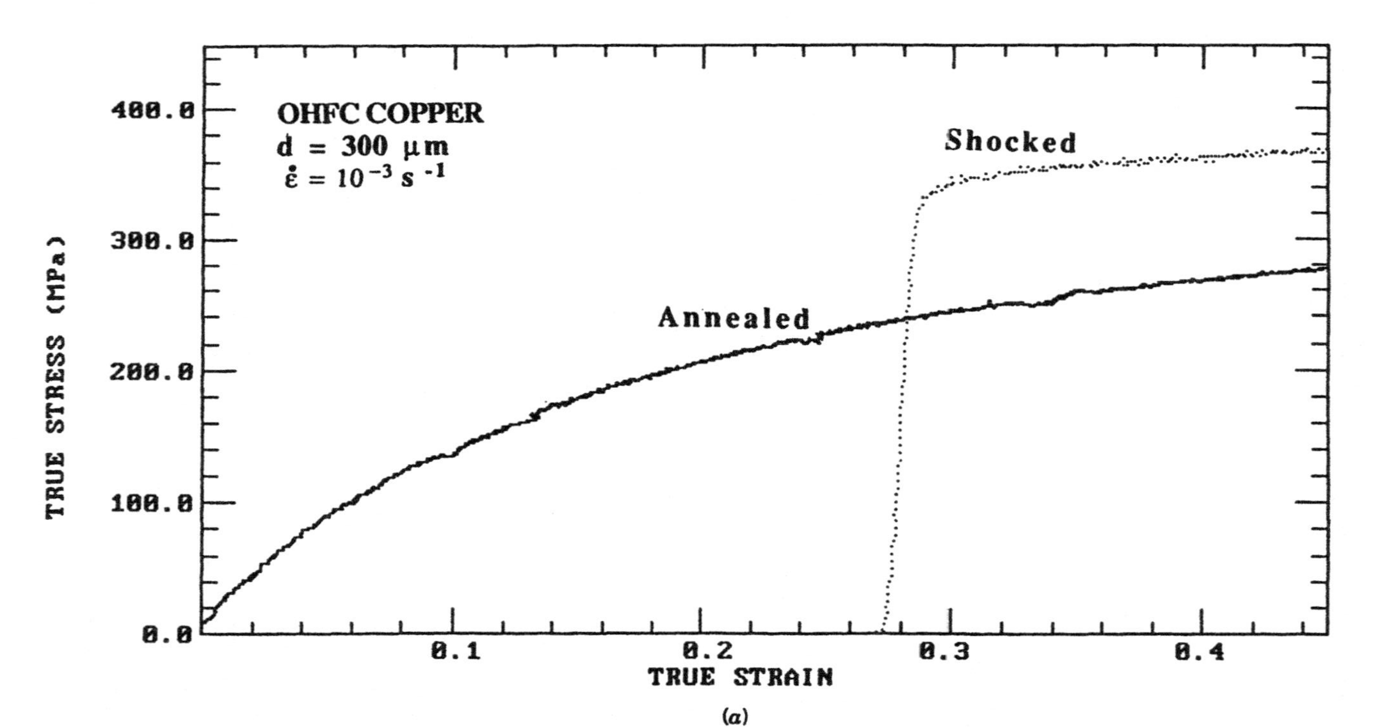

(*a*)

FIGURE 14.7 Comparison between strengthening produced by shock and uniaxial extension. (a) Copper. (From U. Andrade, A. H. Chokshi, and M. A. Meyers, unpublished results. (b) Al–4 wt % Cu alloy. (From Gray [18], Fig. 6, p. 907. Reprinted by courtesy of Marcel Dekker, Inc.) (c) Effect of shock pressure on strength of a titanium alloy RMI 38644. (From Rack [23], Fig. 1, p. 1571. Reprinted with permission of the publisher.)

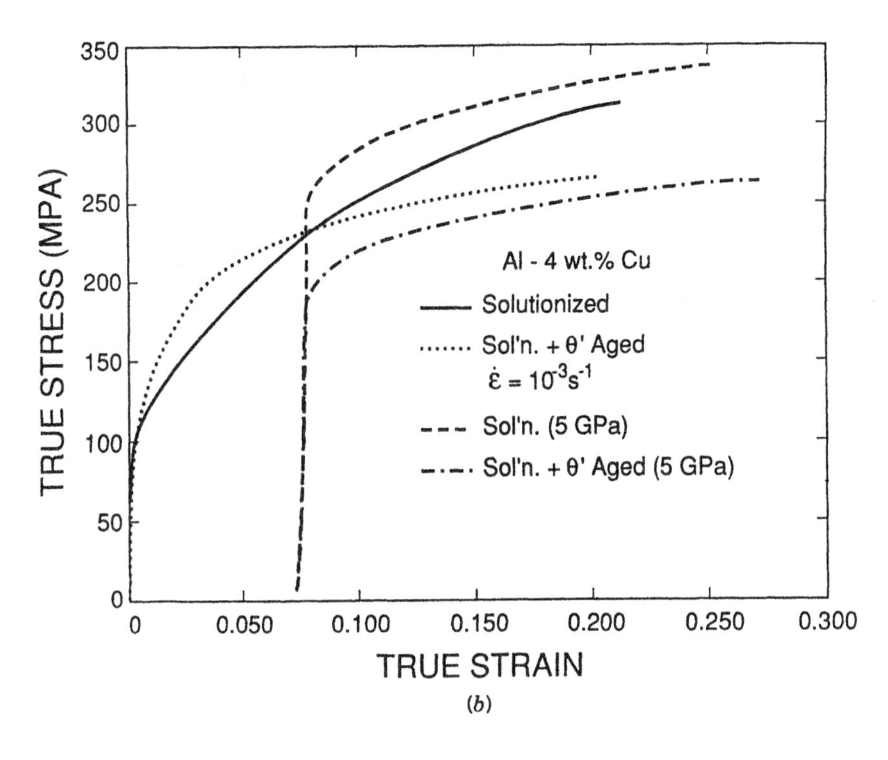

(b)

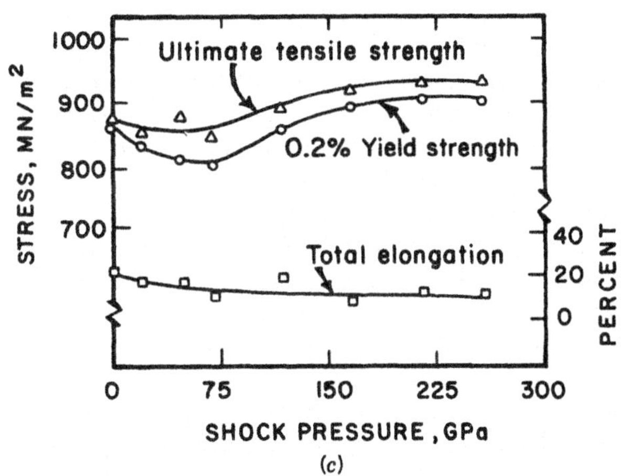

(c)

FIGURE 14.7 *(Continued)*

These coherency stresses are a powerful strengthening mechanism, and they are obstacles to the motion of dislocations. Since the precipitate and matrix have different compressibilities, the high pressures can destroy this coherency and the associated strengthening mechanism. Thus, the strengthening due to dislocations (plastic deformation) is decreased by loss of coherency. Another possible cause for the low strengthening by shock waves of alloys containing brittle phases is the breakup of these phases by the shock wave, creating microcracks that subsequently lead to failure.

There are other situations in which shock waves do not contribute to significant strengthening. High-strength martensitic and ferritic steels are not strengthened significantly, whereas austenitic steels are very sensitive to shock strengthening. In Section 14.6 we will see that martensitic transformation induced by shock loading can be a contributing factor. Rack [23] observed that a titanium alloy (RMI 38644) had a reduction in strength after being subjected to a shock wave of 7 GPa. This alloy, being initially martensitic (omega phase) underwent a reverse transformation induced by the high pressures, with a decrease in yield stress. At higher pressures the strengthening effect of dislocations overcomes the weakening effect due to the reverse transformation, with a net strengthening effect. This effect is shown in Figure 14.7(c).

If the shock-loading system is not properly designed, the state of strain can deviate significantly from a uniaxial one. Momentum trapping (discussed in Section 12.4) eliminates the unwanted reflected waves in the system. If proper care is not taken, the specimens are subjected to a complex loading history, where the shock pulse is only the first stage. Mogilevsky and Teplyakova [24] mention that subsequent plastic deformation (after the first shock pulse) can alter the flow stress of copper dramatically. Gubareva et al. [25] showed that the strength of pure copper at a pressure of 2.8 GPa was increased by a factor of 2.5 by the relaxation of the uniaxial strain condition. Gray [26] showed that the reload compressive strength of copper shocked to 10 GPa increased from 210 to 350 MPa when the residual strain increased from <2 to 26%. Figure 14.8 shows the electron micrographs of the shock substructure, which changes from equiaxial cells to elongated subgrains as the residual strain is increased. Thus, unwanted wave trapping is essential for controlled experiments. Another important effect in explosive experiments is the high pressure created by the detonation products. This pressure can generate stresses above the flow stress of the shocked material, resulting in additional plastic deformation, after the shock wave has traversed the specimen.

In the sections that follow we will look at the mechanisms by which dislocations are generated (Section 14.2), at point defects (Section 14.3), deformation twinning (Section 14.4), and displacive/diffusionless transformations (Section 14.5). We also look at work softening that may be induced by shock loading (Section 14.7). Several sections of this chapter were adapted from a review article by Meyers and Murr [16] on defect generation by shock waves. In Section 14.8, the effects of shock waves on ceramics will be briefly presented.

FIGURE 14.8 Transmission electron micrographs showing microstructure of copper subjected to a 10-GPa shock pulse with different residual strains: (a) $<2\%$; (b) 7%; (c) 26% ε_{res}. (From Gray [19], Fig. 9, p. 409. Reprinted with permission of the publisher.)

14.2 DISLOCATION GENERATION

The dislocation substructures generated by shock loading depend on a number of shock wave and material parameters. Among the shock wave parameters, the pressure is the most important one. As the pressure is increased, so is the dislocation density. This has been shown by a number of investigators. In high-stacking-fault-energy (SFE) metals the partial dislocations are very closely spaced, enabling a great extent of cross slip. The residual microstructures often consist of dislocation cells, that is, regions with high dislocation densities surrounding regions with low dislocation densities. Kuhlmann-Wilsdorf [27] has developed a work-hardening theory based on these cells. In shock-loaded high-SFE materials (copper, silver, nickel) these characteristic cells are also formed by shock loading. As the dislocation density increases for high-SFE FCC metals, the cell size decreases. Murr and Kuhlmann-Wilsdorf [28] found that the dislocation density varies as the square root of pressure ($\rho \alpha P^{1/2}$). This dependence breaks down at pressures close to 100 GPa due to shock-induced heating.

The effect of pulse duration is principally to allow more time for dislocation reorganization. The cell walls become better defined as the pulse duration increases, because there is more time for dislocation reorganization. Experimental observations at very low pulse durations do not seem to be in line with the above realization. Mikkola and co-workers [3, 29, 30] have observed dislocation density increases with increasing pulse duration in the submicrosecond range. These results are consistent with the concept of a threshold time for dislocation generation. The montage of Figure 14.9 illustrates well the combined effects of pressure and pulse duration on the residual microstructure of nickel. Murr [1] conducted systematic experiments establishing these effects for a number of materials and constructed these P–t_p maps (where t_p is the pulse duration). The decrease in the size of cells with increasing pressure is evident, as well as the onset of twinning at 30 GPa. As the pulse duration is increased, no significant changes in the dislocation structure are noticed. There are some systematic differences and similarities between shock-induced and conventionally induced dislocations; some of these will be briefly reviewed here. In FCC metals, the SFE determines the substructure to a large extent. This is the case for both quasi-static and shock deformation. In any case, however, the dislocations seem to be more uniformly distributed in shock than in conventional deformation. In high-SFE alloys, the cell walls tend to be less well developed after shock loading than after conventional deformation, especially by creep or fatigue, which allow time for dislocations to equilibrate into more stable configurations.

Figure 14.10 shows the substructure of (a) shocked and (b) cold-rolled aluminum. It can be seen that the distribution of dislocations is much more uniform in shocked aluminum, whereas the cold-rolled material exhibits subgrain boundaries. It should be noticed that aluminum has an extremely high SFE and that its shock substructure is not too typical of FCC metals. The more common feature observed for copper and nickel is a cell-like structure with poorly defined cell walls, shown in Figure 14.11. In addition, if the shock pulse duration is low, the substructures are more irregular because there is insufficient time for the dislocations generated by the peak pressure (in the shock front) to equilibrate. There is usually a preponderance of dislocation loops associated with the residual shock microstructures, and this is largely unique to shock

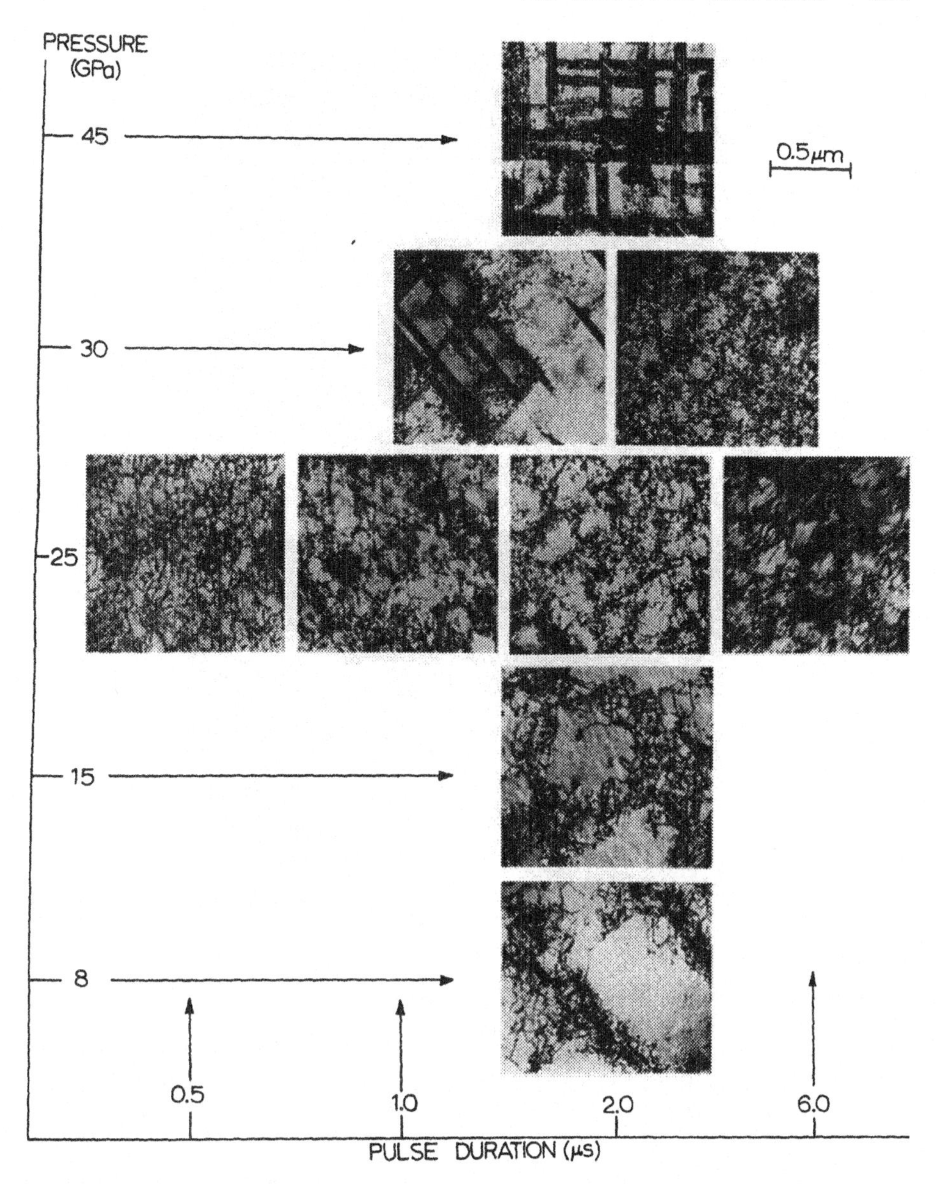

FIGURE 14.9 Effect of shock pressure and pulse duration on residual substructure of nickel. (Courtesy of L. E. Murr, University of Texas at El Paso.)

loading of high-stacking-fault-free-energy metals and alloys. At shock pressures above about 10 GPa, for most FCC metals having stacking-fault free energies above about 50 mJ/m^2, a transitional range gives rise to tangles of dislocations, poorly formed cells, and sometimes more planar arrays of dislocations (associated with the $\{111\}$ slip planes). Figure 14.12 (from Murr [13]) shows, in a schematic fashion, the defect structures found in FCC materials. As the SFE

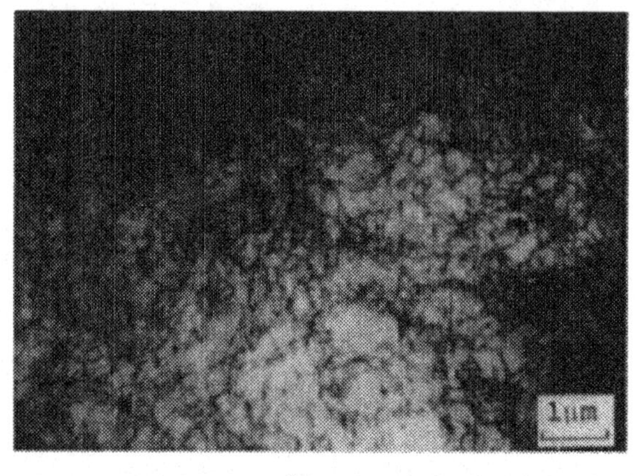

(a)

(b)

FIGURE 14.10 Comparison of dislocation substructures in shock-loaded and cold-rolled aluminum. (a) Aluminum shock loaded at 3 GPa peak pressure, 2 μs pulse duration. (b) Aluminum cold rolled reduced 60%. The microstructure is dominated by subgrain boundaries or low-angle (low-energy) dislocation interfaces. (From Dhere et al. [31], Figs. 2 and 3, pp. 117, 118. Reprinted with permission of the publisher.)

increases, so does the critical pressure for twinning (this will be discussed further in Section 14.4). Aluminum, which has a SFE of approximately 200 mJ/m^2, has not yielded deformation twins in shock-loading experiments yet. Figure 14.12 shows well how the shock-induced structure transitions from stacking faults to cells as the SFE is increased for a constant pressure.

For lower stacking-fault free-energy metals and alloys, there is a tendency toward planar dislocation arrays below about 40 mJ/m^2; with stacking faults and twin faults becoming prominent for stacking-fault free energies below about 25 mJ/m^2. This is illustrated in Figure 14.13 for 70/30 (cartridge) brass, 304

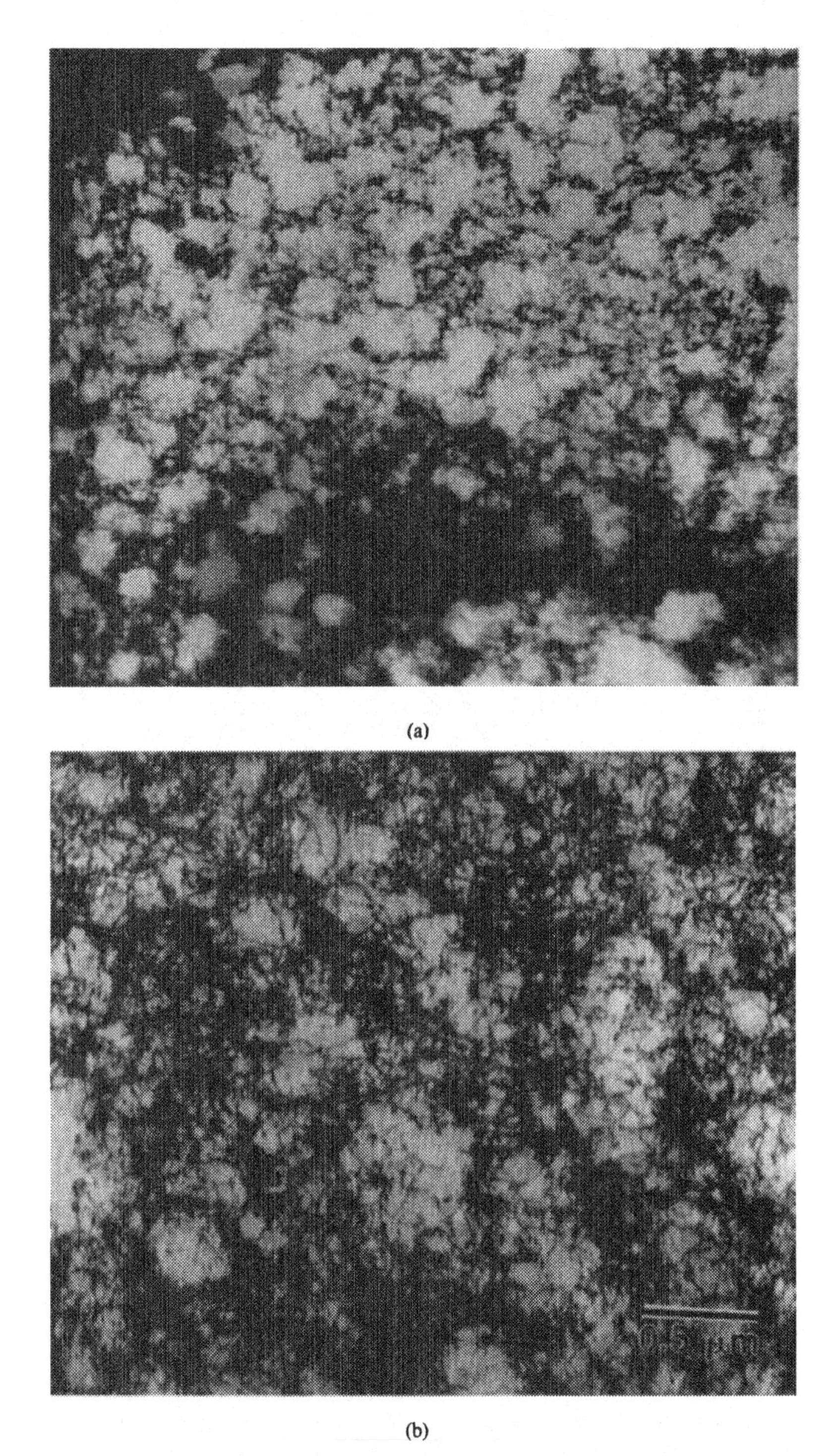

FIGURE 14.11 Dislocation substructures in high-stacking-fault free-energy FCC metals and alloys: (a) copper shock loaded at 5 GPa, 2 μs; (b) nickel shock loaded at 10 GPa, 2 μs.

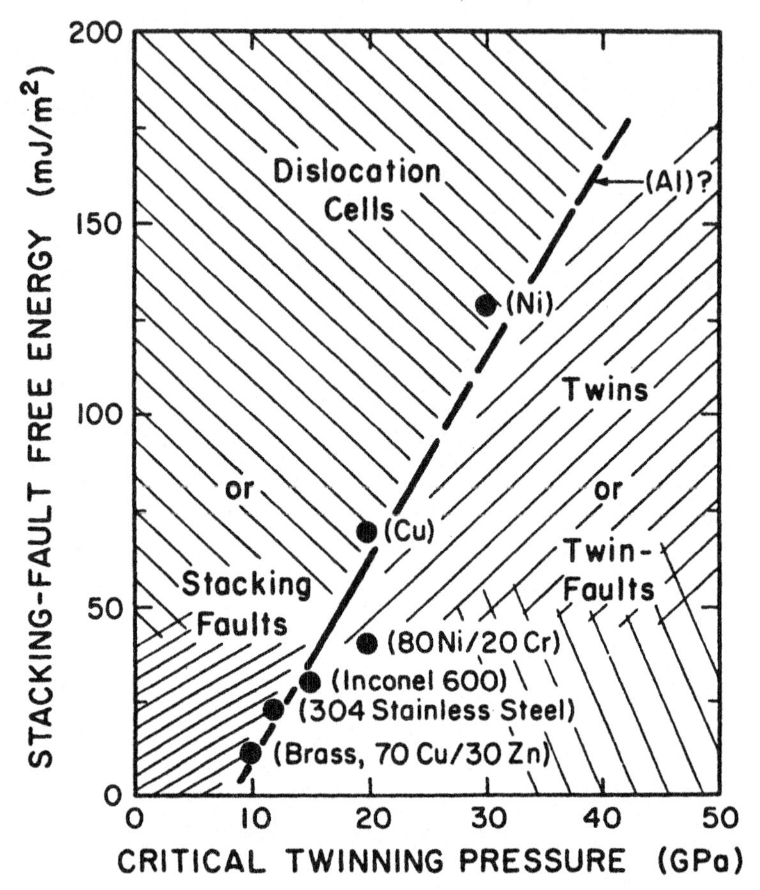

FIGURE 14.12 Schematic representation of shock-induced substructures in FCC metals as a function of shock pressure and SFE. (From Murr [13], Fig. 5, p. 319. Reprinted with permission of the publisher.)

stainless steel, and an Fe–Cr–Ni alloy, each shock loaded to roughly the same pressure at the same pulse duration.

The effect of shock loading on BCC metals is illustrated in Figure 14.14 for iron. Shock-loaded iron is characterized, at pressures below 13 GPa, by arrays of straight and parallel screw dislocations in the grains properly oriented. In molybdenum, the substructure is one of homogeneously distributed dislocations [1].

Shock-loaded HCP metals have not been as extensively studied by transmission electron microscopy. At 7 GPa, Koul and Breedis [33] found dislocation arrays that they described as being intermediate between those of FCC metals such as Cu and Ni and BCC metals such as iron and Fe_3Al. The substructure also exhibited twins and phase transformations at higher pressures. Galbraith and Murr [34] studied shock-loaded beryllium that exhibited dislocation substructures similar to those of BCC metals. No twins were observed

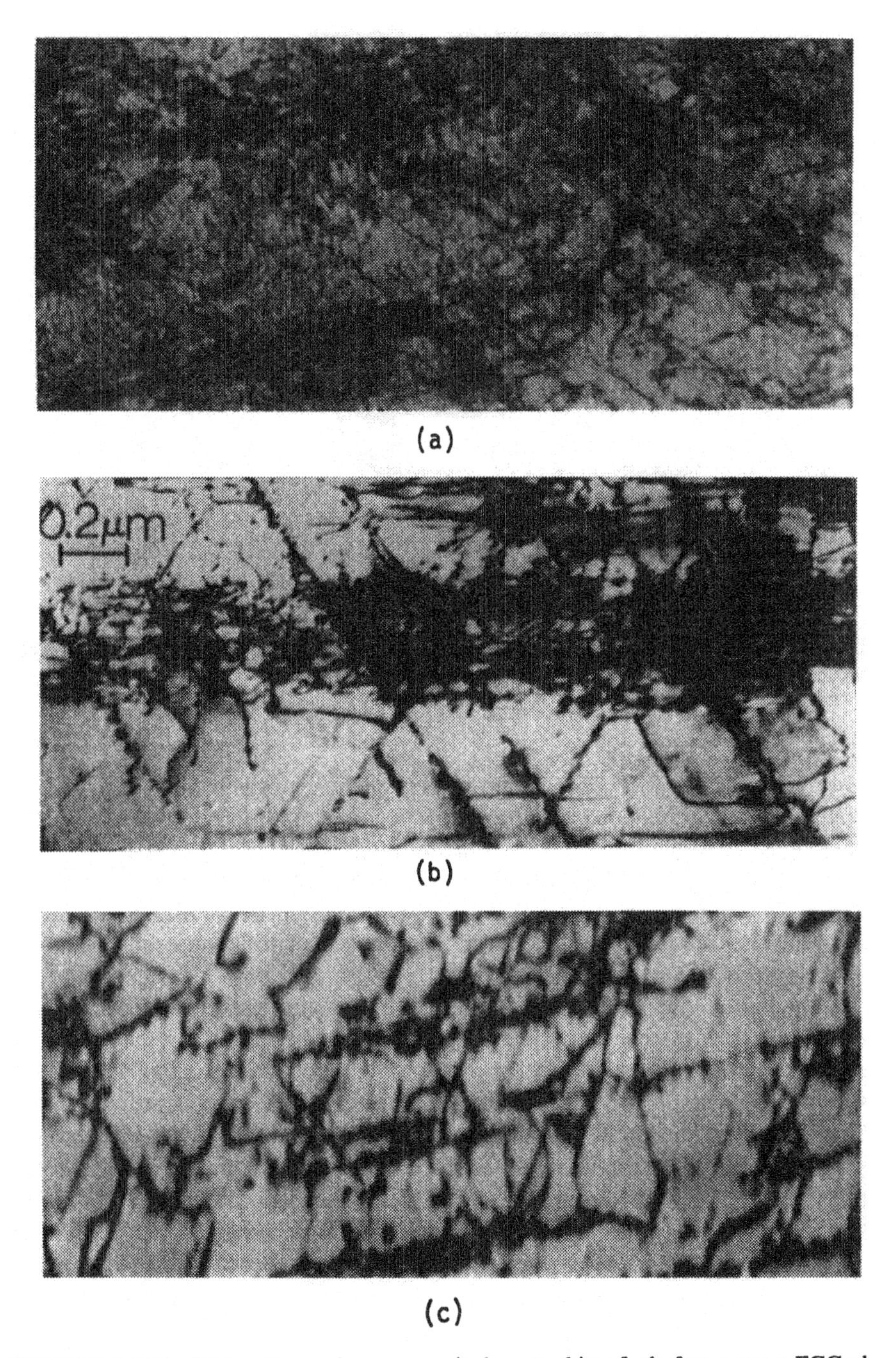

FIGURE 14.13 Dislocation substructures in low-stacking-fault free-energy FCC alloys; (a) 70/30 brass shock loaded at 5 GPa, 2 μs; (b) 304 stainless steel shock loaded at 15 GPa, 2 μs; (c) Fe–15% Cr–15% Ni alloy shock loaded at 10 GPa, 2 μs. (Courtesy of L. E. Murr, University of Texas at El Paso.)

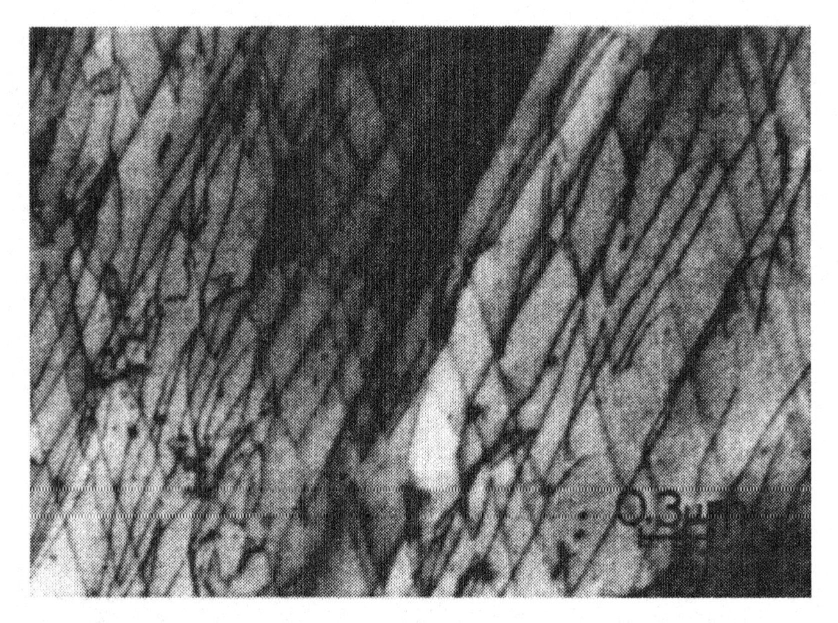

FIGURE 14.14 Dislocation substructure in BCC iron shock loaded at 7 GPa peak pressure. (From of Leslie et al. [32], Fig. 3, p. 122.)

up to a pressure of 0.9 GPa. A strong dependence of dislocation density was noted for grain boundary structure. This feature is consistent with the establishment of incompatibility stresses at the interfaces due to the anisotropy of compressibility, as noted above, giving rise to a preponderance of dislocations from grain boundary sources as well as alterations in the boundary structure. Beryllium has a high anisotropy ratio, yielding large differences in strains for adjacent grains at the same stress.

Intermetallic compounds (principally titanium and nickel aluminides) are materials that present a considerable potential for use in jet turbines. It is of interest to establish the impact response of these materials. Gray and others [18, 19, 35] have characterized their microstructure and mechanical response. They found that dislocations, stacking faults, and twins that increase in density with the peak shock amplitude are characteristic features. Figure 14.15 shows typical microstructures observed in Ni_3Al and NiAl subjected to shock loading. Gray and co-workers [18, 19, 35] report that shock hardening exceeds conventional hardening in tension of the same equivalent strain for these intermetallic compounds. This is similar to earlier results found for copper [35]. Figure 14.15(b) shows stacking faults and twins as prominent features; the yield stress is equal to 1250 MPa, as compared with the initial yield stress of 250 MPa for the same material.

A number of models have been proposed for the generation of dislocations in shock loading. They will be reviewed next. It is not known, at present, if

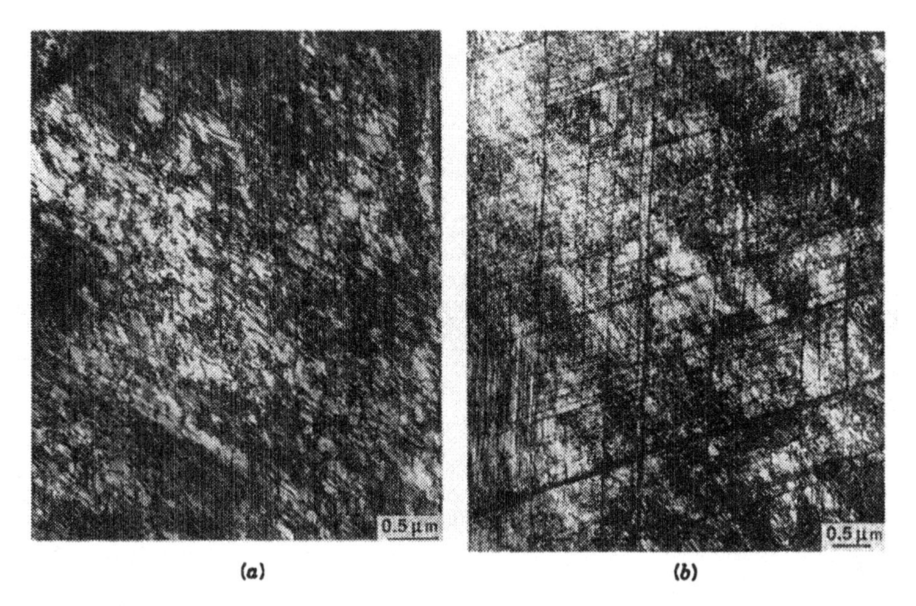

FIGURE 14.15 Substructures generated by shock loading intermetalic compounds: (a) Ni_3Al shocked to 14 GPa; (b) NiAl shocked to 23.5 GPa. (Courtesy of G. T. Gray III, Los Alamos National Laboratory.)

the dislocation generation mechanisms operating under shock loading are uniquely different from conventional mechanisms or whether they are the same.

14.2.1 Smith's Model

Smith [6] made the first attempt to interpret the metallurgical alterations produced by shock waves in terms of fundamental deformation modes. He depicted the interface as an array of dislocations that accommodates the difference in lattice parameter between the virgin and the compressed material. In this sense the Smith interface resembles an interface between two phases in a transformation; the interface is shown in Figure 14.16(b). Figure 14.16(a) shows the interface if no dislocations were present; the deviatoric stresses cannot be relieved. This interface of dislocations would, according to Smith, move with the shock front. Since the density of dislocations at the front is, according to Smith, 10^3–10^4 times higher than the residual density, sinks and sources, moving at the velocity of the shock, were postulated.

14.2.2 Hornbogen's Model

Hornbogen [36] modified Smith's model because it could not account for the dislocation substructures found by the former in shock-loaded iron. Figure 14.14 shows a typical microstructure, and one can see the screw dislocations lying on $\langle 111 \rangle$ directions. Hornbogen's explanation is shown schematically in

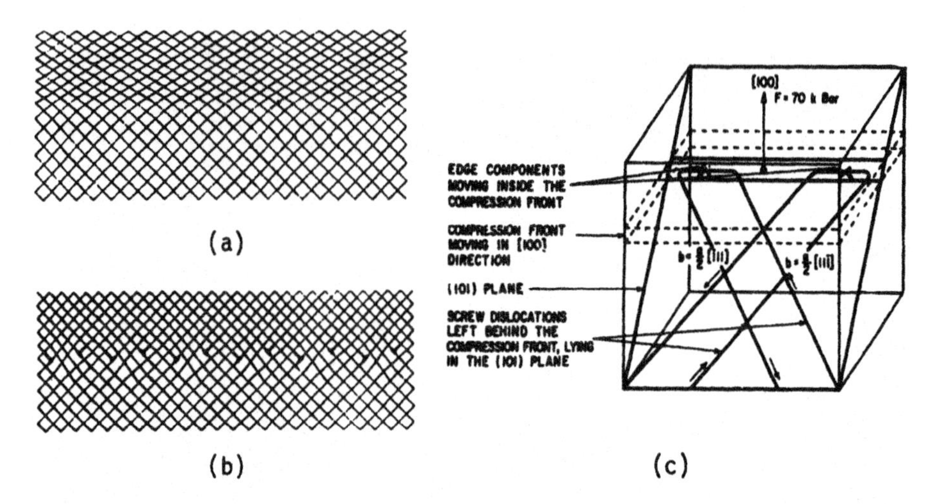

FIGURE 14.16 (a,b) Smith and (c) Hornbogen's models for dislocation generation on shock loading.

Figure 14.16(c): Dislocation loops are formed as the compression wave enters the crystal; the edge components move with the velocity of the shock front, so that their compression zone forms the wave; and screw components remain and extend in length as the edge components advance.

14.2.3 Criticism of the Smith and Hornbogen Models

Hornbogen based his model on the observation of only one metal, iron. Different metals and alloys exhibit widely different substructures, as shown in Figures 14.8–14.15, and Hornbogen's model would not be applicable to them. Additionally, the dislocation substructure depicted in Figure 14.14 is not unique; it is also typical of iron deformed at low temperature. Indeed, the substructures generated by shock loading tend to resemble the low-temperature substructures induced by conventional deformation. The large incidence of screw dislocations is indicative of their low mobility at low temperature; this topic is discussed in greater detail in reference [37]. This phenomenon is characteristic of BCC metals and is the reason for the large temperature dependence of their flow stress at low temperatures.

Smith's model requires dislocations moving with the shock front; it will be shown, for a specific example, that this requires supersonic dislocations. Figure 14.17 shows the velocity of a dislocation as a function of its orientation with respect to the shock front [38]. Calculations were conducted for nickel shocked at 9.4 GPa. The velocity of an elastic shear wave at the above pressure, V_s, was taken as the unit. The velocity of the dislocation (independent of orientation) in a Smith or Hornbogen interface substantially exceeds V_s.

Many attempts have been made to fit the velocity–stress relationships to

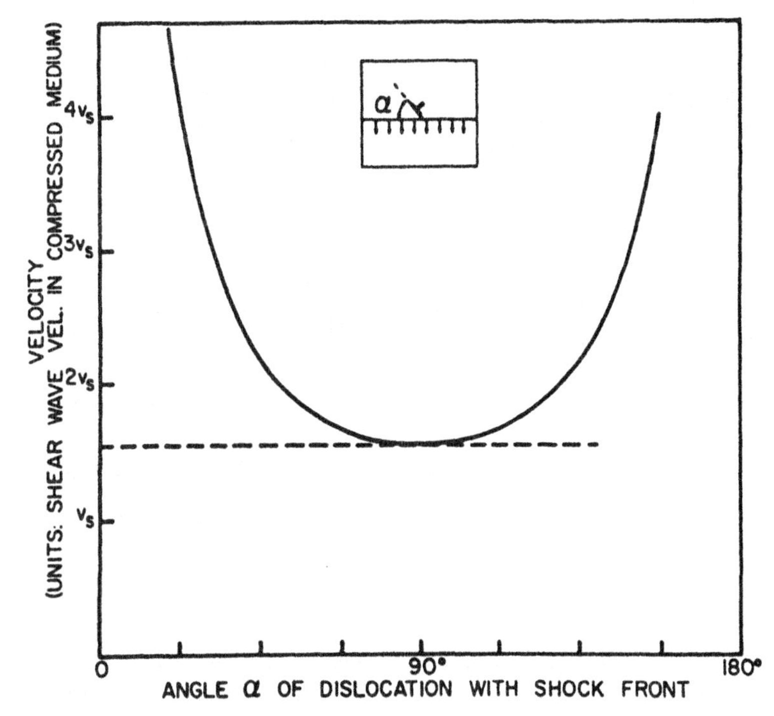

FIGURE 14.17 Velocity of dislocation in Smith interface as a function of orientation with shock front.

mathematical equations. Chapter 13 deals extensively with this topic, and a number of formulations are derived. Figure 13.6 shows a hypothetical prediction of the stress dependence of dislocation velocities in nickel, in accord with the aforementioned observations; it is based on measurements by Rohde and Pitt [39]. It can be seen that the stress required to move a dislocation at a velocity equal to the sound velocity is infinite. Another extremely important aspect of dislocation motion is that no supersonic dislocations have been observed to date. This is in line with the relativistic effects (infinite dislocation self-energy when velocity becomes equal to the shear velocity of sound) and the phonon scattering ideas. Friedel [40] stated that a dislocation moving under a constant applied external stress cannot exceed the speed of sound because it is a signal that can be Fourier analyzed into plane waves of strain (sound waves, or phonons), and one knows that a signal cannot travel faster than the waves that carry it. The possible existence of supersonic dislocations is discussed in Chapter 13. However, there is no experimental evidence for their existence.

A few simple calculations will show that the energy released by a moving Smith–Hornbogen interface would be of such magnitude that exceedingly high temperatures would be generated. Let us assume that nickel is being shock loaded at 10 GPa. The ratio between compressed and uncompressed specific volumes is 0.953; the density of dislocations geometrically required along the

Smith interface is determined from the misfit between the two lattices (the virgin and compressed ones). Following the calculations delineated by Meyers [41], one finds that there exists one dislocation doublet for every 21.3-unit cells, or one dislocation for every 53 Å. The work involved in moving them is given by the Peach–Koehler equation:

$$W = \tau b l \tag{14.4}$$

where τ is the applied shear stress, b the Burgers vector (3.5×10^{-10} m), and l the distance moved by each dislocation. A pressure of 10 GPa generates a shear stress of 9.25 GPa; this is assumed to be the shear stress driving the interfacial dislocations. It is assumed that the dislocations are at an angle of 45° with the shock front; they consequently move 1.41 cm when the front advances 1 cm. The work per dislocation and per centimeter of front advance is 5.8×10^{-6} J. The work per cubic centimeter is obtained by multiplying W by the dislocation density along the front. In order to compute it, one has to consider the two-dimensional grid of dislocations at the front. So, the number of dislocations per square centimeter of front is 3.77×10^6, or a total length of 3.77×10^6 cm. The total work required to move them is 21.9 J. Assuming that all this energy is dissipated as heat, one obtains (density is 8.87 g/cm^3 and heat capacity is 0.44 J/Kg) a temperature increase of 441 K. For pressures of 15 and 20 GPa one obtains, repeating the calculations residual temperature rises of 717 and 1045 K, respectively. However, experimental measurements by McQueen et al. [43] and Taylor [42] show that the residual temperatures in iron and copper, respectively, do not differ too much from thermodynamic predictions; this is shown in Figures 14.18 and 14.19. The full lines indicate the predictions from hydrodynamic theory. The measured residual temperatures tend to be equal to or higher than the calculated values (assuming that there are no internal, dissipative processes in the material), but the differences are very small and could not account for the energy dissipation of a Smith interface. Raikes and Ahrens [44] conducted measurements for shocked aluminum and stainless steel which are in line with the results of

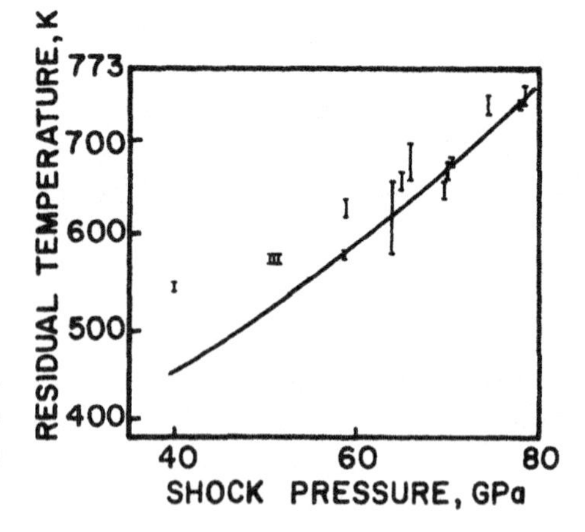

FIGURE 14.18 Experimentally determined and calculated (continuous line) residual temperatures in shock-loaded iron [43].

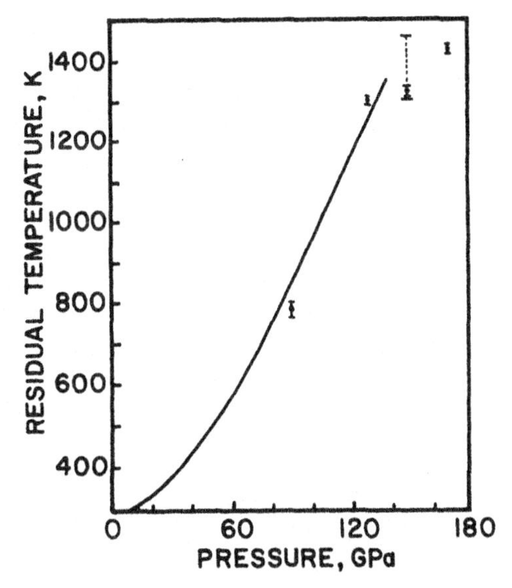

FIGURE 14.19 Residual temperatures in shocked copper. (From Taylor [42], Fig. 4, p. 2730. Reprinted with permission of the publisher.)

minum and stainless steel which are in line with the results of McQueen et al. [43] and Taylor [42]. An additional limitation is that the Smith interface would require the periodic intersection of dislocations that generate jogs and kinks. Thus, it is physically impossible.

14.2.4 Homogeneous Dislocation Nucleation

The limitations of Smith and Hornbogen's proposals led Meyers [38, 41] to propose a model whose essential features are as follows:

1. Dislocations are homogeneously nucleated at (or close to) the shock front by the deviatoric stresses set up by the state of uniaxial strain; the generation of these dislocations relaxes the deviatoric stresses.
2. These dislocations move short distances at *subsonic speeds*.
3. New dislocation interfaces are generated as the shock wave propagates through the material.

This model presents, with respect to its predecessors, the following advantages:

1. No supersonic dislocations are needed.
2. It is possible to estimate the residual density of dislocations.

Figure 14.20 shows the progress of a shock wave throughout the material in a highly simplified manner. As the shock wave penetrates into the material, high

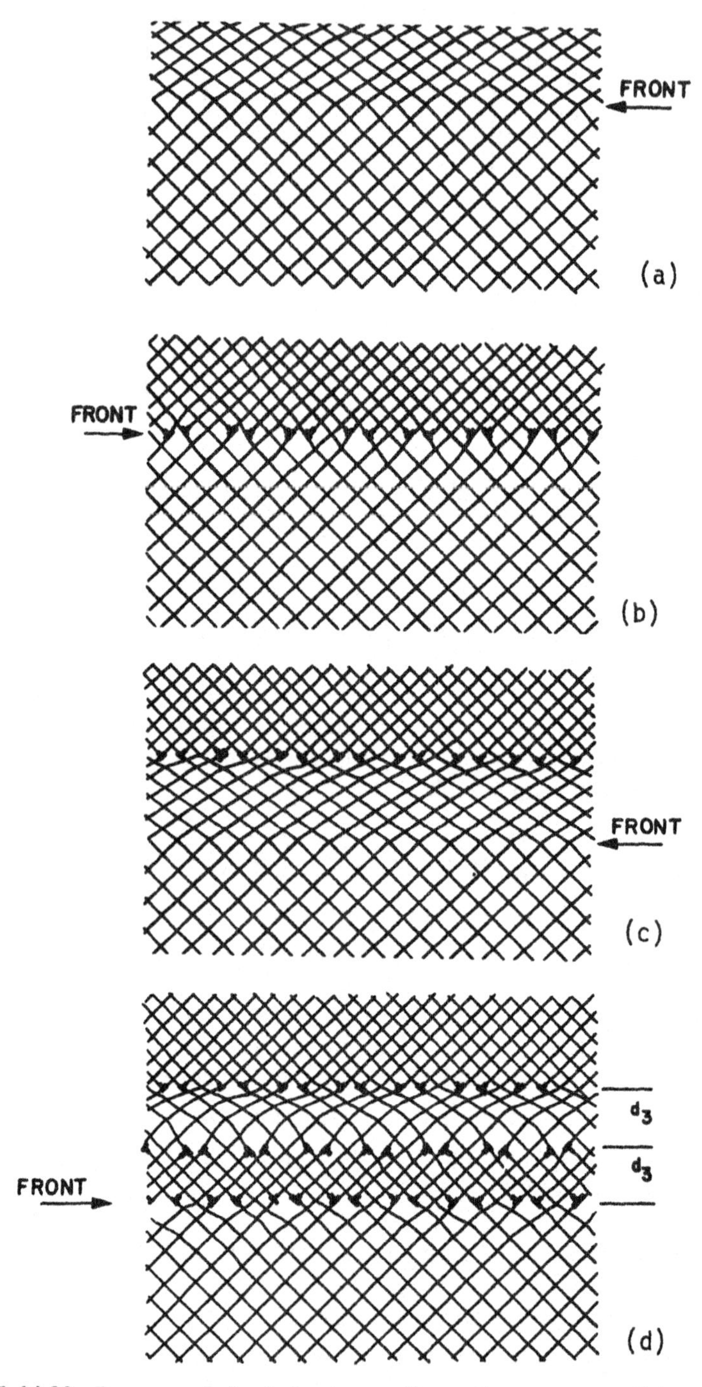

FIGURE 14.20 Progress of shock front according to homogeneous dislocation nucleation model. (Reprinted from *Scripta Met*. vol. 12, M. A. Meyers, Fig. 2, p. 24, Copyright 1978, with permission from Pergamon Press, Ltd.)

deviatoric stresses effectively distort the initially cubic lattice into a monoclinic lattice. When these stresses reach a certain threshold level, homogeneous dislocation nucleation can take place. Hirth and Lothe [45] estimate the stress required for homogeneous dislocation nucleation. The nucleation mechanism at the shock front is unique and is different from homogeneous nucleation in conventional deformation. In shock loading, the dislocation interface separates two lattices with different parameters. However, it will be assumed that the stress required is the same (as a first approximation). From Hirth and Lothe [45] one has

$$\frac{\tau_h}{G} = 0.054 \tag{14.5}$$

where τ_h is the shear stress required and G the shear modulus, assumed to be pressure independent as a first approximation. When the maximum shear stress becomes equal to τ_h (and is acting in the correct orientation), homogeneous dislocation nucleation takes place. The maximum shear stress is approximately $2GP/K$, where $K = E/3(1 - 2\nu)$; thus the pressure (threshold) at which homogeneous dislocation generation occurs is

$$P = 0.027K \tag{14.6}$$

For Ni ($E = 200$ GPa; $\nu = 0.3$) it corresponds to a pressure of approximately 6 GPa. Figure 14.20(b) shows the wave as the front coincides with the first dislocation interface. The density of dislocations at the interface depends on the difference in specific volume between the two lattices and can be calculated therefrom. In Figure 14.20(c) the front has moved ahead of the interface and the deviatoric stresses build up again; other layers are formed in Figure 14.20(d). It should be noted that since the macroscopic strain is ideally zero after the passage of the wave, the sum of the Burgers vectors of all dislocations has to be zero. This is accomplished, in the simplified model presented here, by assuming that adjacent dislocation layers are made of dislocations with opposite Burgers vectors. Figure 14.21 shows two adjacent layers under the effect of shear stresses still existing in the lattice after the dislocations were nucleated; a group of dislocations move away from it. It is possible to estimate the velocity at which these dislocations move if one knows τ_{res}. As these dislocations move, they locally accommodate and decrease τ_{res}. The total amount of internal friction and heat generation due to dislocation motion can be calculated by knowing the difference between the measured and the thermodynamically calculated residual temperature. A simplified calculation is presented by Hsu et al. [46].

It is believed that the rarefaction part of the wave plays only a minor role in dislocation generation. The main reason for this is that the rarefaction part of the wave enters into a material that is already highly dislocated. It was found by Kazmi and Murr [47] that when nickel is shock loaded repeatedly, the increase in dislocation density is much less pronounced for the succeeding

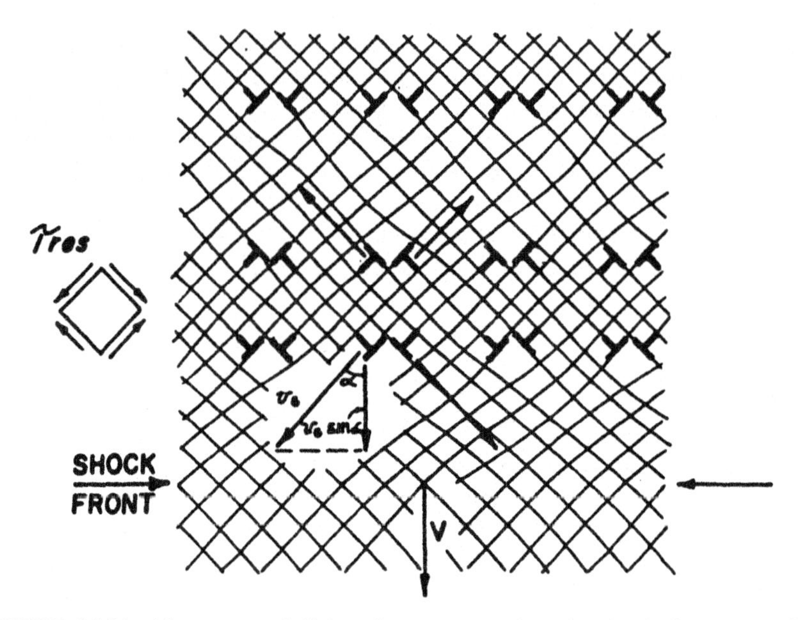

FIGURE 14.21 Movement of dislocations generated at the shock front. (Reprinted from M. A. Meyers, *Strength of Metals and Alloys*, Copyright 1979, Fig. 9, p. 552, with permission from Pergamon Press, Ltd.)

events. The shock wave passing through a highly dislocated material is not such an effective dislocation generator. Thus when a prestrained material is shock loaded, part of the deviatoric stresses at the shock front could be accommodated by existing dislocations; in this case the number of dislocations that would be generated at the front would be reduced. The same argument can be extended to the rarefaction portion of the wave; it can accommodate the deviatoric stresses by the movement of the existing dislocations.

14.2.5 Mogilevsky's Model

Mogilevsky* [48–51] performed molecular dynamics computations by subjecting a lattice of 500 atoms to stress levels several times higher than the theoretical shear strength of the material (copper). This resulted in two-dimensional strains of up to 18.6% in the ⟨211⟩ direction. The shear stresses as a function of time were monitored and are shown in Figure 14.22. Figure 14.22(a) shows the loading–unloading path (2 ps loading; 2 ps peak pressure duration; 4 ps unloading). Figure 14.22(b) shows the resultant shear stress. The shear stress drops to zero during the holding time at the peak pressure, showing that

*There have been more recent and complex computations using molecular dynamics for shock-wave propagation: B. L. Holian (Phys. Rev. A37 (1988) 2562); P. A. Taylor and B. W. Dodson (in "Shock Compression of Condensed Matter-1989," Elsevier, Netherlands, 1990, p. 165); and F. A. Bandak, R. W. Armstrong, A. S. Douglas, and D. H. Tsai (in "Shock Compression of Condensed Matter-1991," Elsevier, Netherlands, 1992, p. 559).

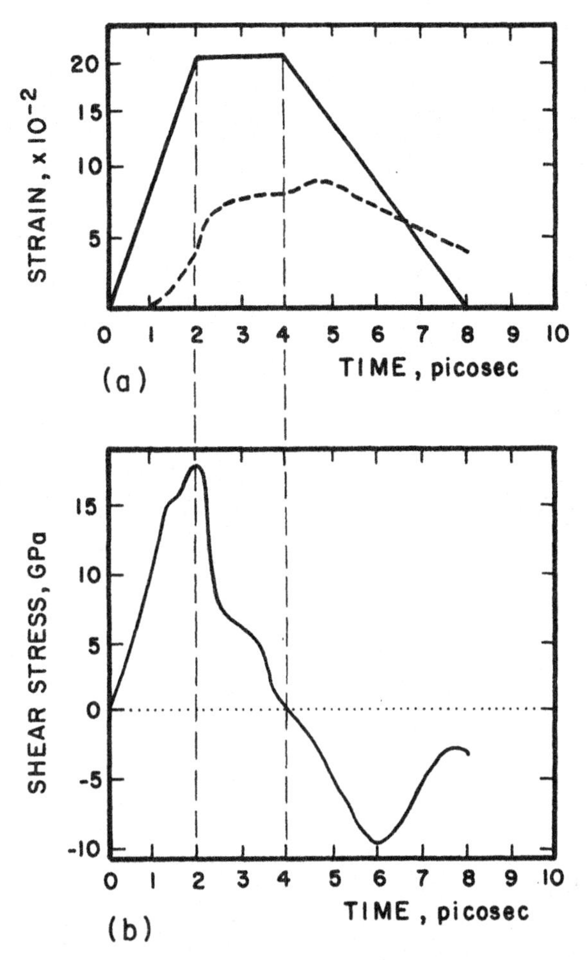

FIGURE 14.22 (a) Loading and unloading paths followed in molecular dynamics computations of Mogilevsky; (b) associated shear stress fluctuation. (From Mogilevsky [50], Figs. 1 and 2, p. 959. Reprinted with permission of the publisher.)

dislocations are generated, decreasing the deviatoric components of stress. Upon unloading, the reverse occurs, with dislocations reorganizing themselves. No new dislocation generation was observed during the unloading stage. The positions of the atoms during loading are shown in Figure 14.23. Imperfections in the form of vacancies, interstitials, and substitutional impurities were added to the structure, and there is a clear preference for nucleation of dislocations at these points. In Figure 14.23 only substitutional impurities are shown (single, pair, and triplet); they are indicated as hollow atoms. At 1.32 ps, a dislocation nucleates at the substitutional pair; it is decomposed into partials. At 1.48 ps, dislocations nucleate in the absence of defects, as can be seen at 1.84 ps. These dislocations can also be seen to move out of the atomic array.

Mogilevsky's [48–51] computations are in complete agreement with the model proposed by Meyers [38, 41]. Mogilevsky [48–51] also found that the

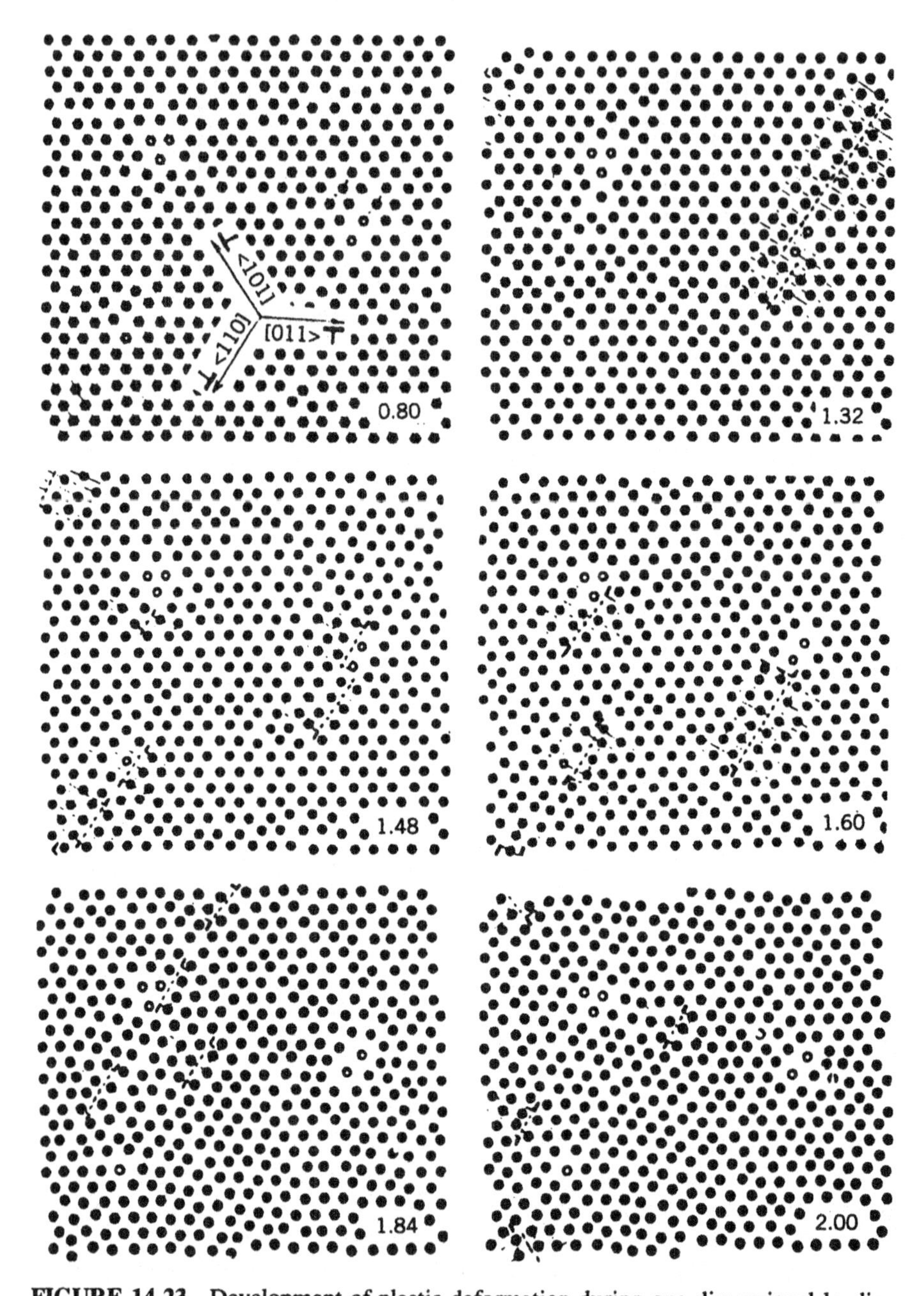

FIGURE 14.23 Development of plastic deformation during one-dimensional loading of copper by continuous compression in vertical ⟨211⟩ direction. Notice hollow atoms, which are substitutional impurities. Times: 0.80, 1.32, 1.48, 1.60, 1.84, 2.00 ps. (From Mogilevsky [50], Fig. 3, p. 960. Reprinted with permission of the publisher.)

rate of release of the shock wave had no effect on the hardness of copper single crystals; this was done by experimentation and is in accord with the molecular dynamics predictions that no additional dislocations are generated during the release portion of the shock wave. Mogilevsky's model has, however, severe limitations: (a) the total time is a few picoseconds, while actual shock waves have durations of microseconds (10^6 ps); (b) the cell, with ~ 500 atoms, is uniformly compressed and there is no shock front running through it.

14.2.6 Weertman–Follansbee Model

Weertman [52] and Weertman and Follansbee [53] investigated the formation and movement of dislocations in shock wave propagation by applying the concepts of drag-related effects and relativistic effects discussed in Chapter 13. They divided shock waves into two classes: weak shocks (when driving stress is small with respect to the elastic modulus) and strong shocks. Two different analytical treatments are applied for these two cases. For the weak or moderate shock the loading and unloading waves are decomposed into a series of small-amplitude plastic waves that are superimposed. These loading and unloading waves are considered to be "spread out" over a certain distance x. This is shown schematically in Figure 14.24. They developed a continuum mechanics formulation for these superimposed plastic waves and applied the dislocation dynamics equations to these plastic waves. From Chapter 13 [Eqn. (13.9)]

$$\dot{\varepsilon}_x^{\mathrm{p}} = \frac{\dot{\gamma}_{\mathrm{p}}}{M} = k\rho b v \tag{14.7}$$

They used a modified form of Eqn. (13.29b), which assumes that dislocation drag is operative, by subtracting from the applied stress τ, the frictional stress

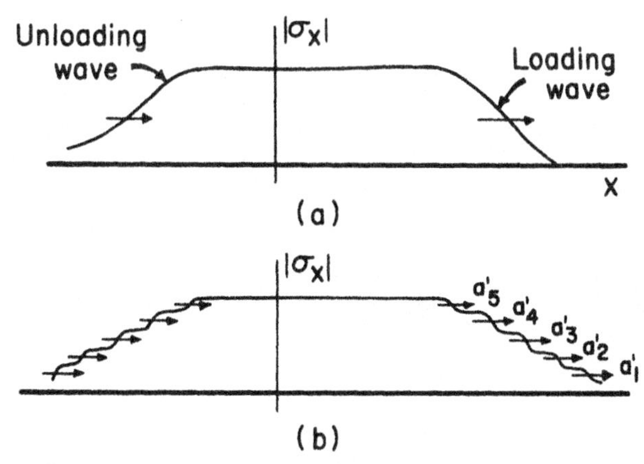

FIGURE 14.24 (a) Loading and unloading wave with finite rise and release distances; (b) decomposition of wave into a series of superimposed small-amplitude waves, each with a different propagation velocity a'. (From Weertman and Follansbee [53], Fig. 2, p. 266. Reprinted with permission of the publisher.)

τ_i, and the internal back stress produced by all other dislocations acting on one dislocation, τ_b:

$$v = \frac{b}{B}(\tau - \xi\tau_i - \tau_b) = \frac{b}{B}(\tau - \xi\tau_i - \xi\zeta\alpha Gb\sqrt{\rho}) \qquad (14.8)$$

where ζ, ξ, and α are parameters and the back stress τ_b is related to the dislocation density ρ by

$$\tau_b \sim \xi\zeta\alpha Gb\sqrt{\rho} \quad \text{(Taylor equation)}$$

They performed the melding of continuum mechanics and dislocation dynamics by applying Eqns. (14.7) and (14.8) to the equation

$$\dot{\sigma}_x = \frac{3K(1-\nu)}{1+\nu}\left(\dot{\varepsilon}_x - \frac{1-2\nu}{1-\nu}\dot{\varepsilon}_x^{p}\right) \qquad (14.9)$$

where $\dot{\varepsilon}_x$ is the total strain rate ($= \dot{\varepsilon}_x^{e} + \dot{\varepsilon}_x^{p}$) and K is the bulk modulus. The velocity of the plastic wave, V_p, is given by

$$V_p = a + U_{pe}\int_{\sigma_{xe}}^{\sigma_x} \frac{V}{a}\, d\sigma_x \qquad (14.10)$$

where

$$a = \sqrt{\left(\frac{KV}{1+\nu}\right)[3(1-\nu) - 2\eta(1-2\nu)]}$$

$$\eta = \frac{3}{2}\frac{\dot{\varepsilon}_{px}}{\dot{\varepsilon}_x}$$

V is the specific volume of the material, and U_{pe} is the particle velocity in the elastic precursor. By assuming a steady-state plastic wave, they arrived at the prediction given in Figure 14.25. The elastic stress σ_{xe} is subtracted from the total stress in the ordinate. In the abscissa, the normalization faction x^* is defined as

$$x^* = \frac{4^3 a_p' B\alpha^2 K^{*2}}{3^{10}\alpha(1-2\nu)^2 f\,|\sigma_{max} - \sigma_{xe}|^5} \qquad (14.11)$$

and

$$K^* = \frac{2(1-2\nu)K}{1(1+\nu) + (q-1)[3(1-\nu) - 2\eta_e(1-2\nu)]}$$

where $q \cong 4, \ldots, 6$ and f is the fraction of mobile dislocations. η_e is the value of η at the start of the steady-state part of the plastic wave. Weertman

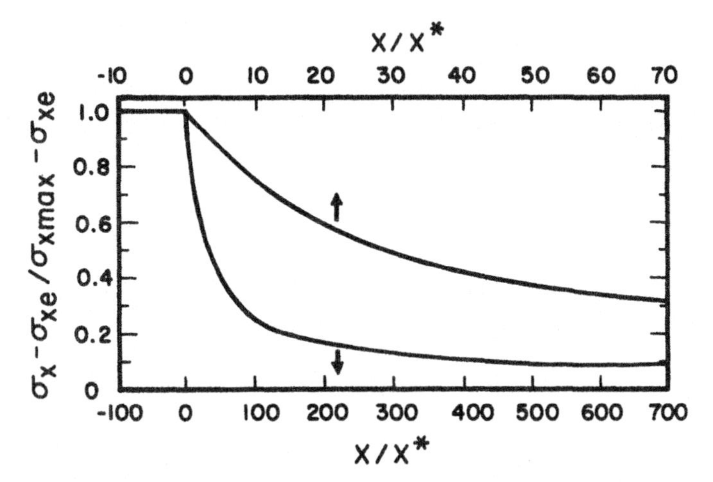

FIGURE 14.25 Normalized stress vs. distance in steady-state plastic wave. (From Weertman and Follansbee [53], Fig. 4, p. 281. Reprinted with permission of the publisher.)

and Follansbee [53] were able to calculate shock front widths for metals. For aluminum, at a pressure of 1 GPa, they used (or found)

$$x^* = \frac{5.6 \times 10^{-5}}{f}$$

$$K = 7.5 \times 10^{10} \text{ Pa}$$

$$V_p = 5.5 \times 10^3 \text{ m/s}$$

$$q = 4.4$$

$$B = 5 \times 10^{-5} \text{ Pa} \cdot \text{s}$$

$$\alpha = 0.3$$

Taking $f = 3 \times 10^{-4}$, they obtained $x^* = 0.22$ mm. This leads to a front width of approximately 60 mm, in good agreement with experimental results. One extremely important feature of the formulation above is that

$$x^* \alpha |\sigma_{\max} - \sigma_{xe}|^5 \qquad (14.12)$$

that is, the shock front thickness decreases with the fifth power of the imposed stress, $\sigma_{\max}$ (σ_{xe} is small at large stresses). Thus, the front steepens as pressure is increased, and this is exactly what is experimentally observed. Chapter 4 discusses this aspect in more detail. Another conclusion reached by Weertman and Follansbee [53] is that the fraction of total dislocations that is mobile at these high strain rates is very low ($\sim 10^{-4}$). This was predicted previously by Follansbee and Weertman [54].

The treatment by Weertman and Follansbee [53] also allows the prediction of dislocation density (ρ/ρ_{max}) as a function of normalized distance. This is shown in Figure 14.26. This plot also gives v/v_{max}, the normalized dislocation velocity. Using $|\sigma_{x,\max} - \sigma_{xe}| = 1.0$ GPa, $b = 2.87 \times 10^{-10}$ m, and the x^* = 0.22 mm, one obtains

$$\rho_{max} = 6.3 \times 10^{14}\ \text{m}^{-2}$$
$$= 6.3 \times 10^{10}\ \text{cm}^{-2}$$
$$v_{max} = 335\ \text{m/s}$$

These are very reasonable values. The maximum dislocation density is given by

$$\rho_{max} = \left[\frac{3(1+\nu)(1-2\nu)}{2[3(1-\nu) - 2\eta_e(1-2\nu)]^2}\right]^2 \left[\frac{(\sigma_{x,\max} - \sigma_{xe})^2}{\alpha G b K^*}\right]^2 \tag{14.13}$$

The maximum dislocation velocity is given by

$$v_{max} = \left[\frac{3b(1+\nu)(1-2\nu)}{3B[3(1-\nu) - 2\eta_e(1-2\nu)]^2}\right] \frac{(\sigma_{max} - \sigma_{xe})^2}{K^*} \tag{14.14}$$

It can be seen that it increases with the square of the applied stress.

Thus, a stress of 3 GPa would generate dislocation velocities of 3000 m/s. This high value, considered unreasonable, led Weertman [55] to the develop-

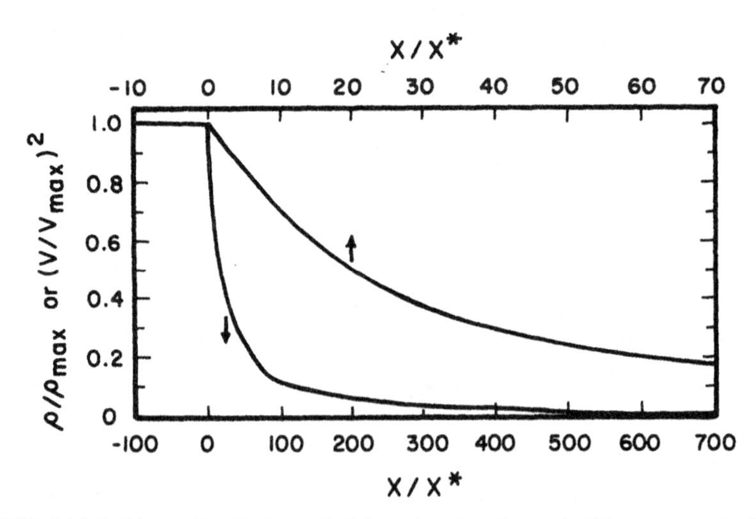

FIGURE 14.26 Normalized plots of dislocation density and dislocation velocity vs. distance for a steady-state plastic wave. (From Weertman and Follansbee [53]. Reprinted with permission of the publisher.)

ment of a theory for dislocation behavior in strong shocks. He basically divided this behavior into two ranges:

1. the shock front, where he assumed the existence of a Smith interface, and
2. the region immediately behind the shock front, where he assumed that dislocation generation, motion, and multiplication occur by conventional processes.

The inclusion of the Smith interface, discarded by Meyers [38, 41] and Mogilevsky [48–51], comes as a result of the computations for weak and intermediate shocks, where the shock front thickness ($\propto \sigma^{-5}$), was reduced to subatomic dimensions when the stress (pressure) is on the order of the bulk modulus, ≈75 GPa. So, the Smith interface had to be introduced by Weertman [55] because conventional dislocation multiplication mechanisms would not take place.

14.3 POINT-DEFECT GENERATION

The experimental investigations conducted to date are unanimous in indicating that shock loading produces a high density of point defects. In the first systematic comparison between point defects generated by shock loading and cold rolling, Kressel and Brown [56] reported vacancy and interstitial concentrations three to four times higher after shock loading than after cold rolling. Figure 14.27 reproduces these results. Direct quantitative evidence of vacancies and vacancy-type defects was obtained by Murr et al. [57]; vacancy-type dislocation loops were identified by transmission electron microscopy. Field-ion microscopy showed that these loops accounted for only a small portion of the shock-induced vacancies; the majority exist as single vacancies or small clusters difficult to resolve by conventional electron microscopy.

Graham [58] reviews the research done on shock-induced point defects and emphasizes this unique aspect of shock loading. Mogilevsky [59] performed shock experiments on copper at liquid nitrogen temperature and stored the recovered specimens in liquid nitrogen. He measured the electrical resistance and obtained the following estimates for the concentration of point defects:

Pressure	Vacancy Concentration	Interstitial Concentration
7	7×10^{-5}	5×10^{-5}
22	7×10^{-5}	5×10^{-5}
38	2×10^{-5}	6×10^{-5}

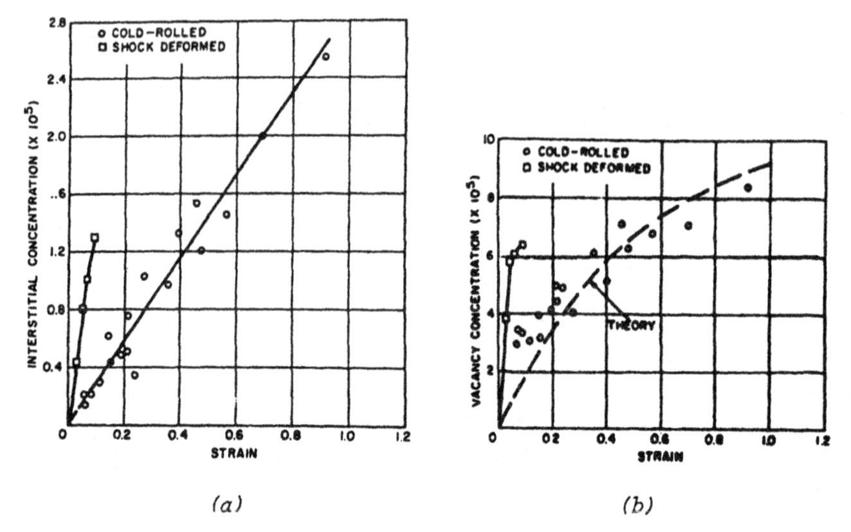

FIGURE 14.27 Comparison between estimated (a) interstitial and (b) vacancy concentrations in nickel deformed by shock loading and conventional rolling to different strains. (From Kressel and Brown [56], Figs. 6 and 7, p. 1622. Reprinted with permission of the publisher.)

These are considered reliable measurements. Vacancies and interstitials are mobile at ambient temperature, and every care should be taken to preserve the specimens at low temperatures.

Figure 14.28 shows vacancy-type dislocation loops in shock-loaded aluminum; the specimen was shocked to a pressure of 13 GPa at liquid nitrogen temperature (77 K). These loops are assemblages of vacancies with a disk shape. The edges of the disk (one atomic layer thick) have a contrast that is analogous to the dislocation image in the transmission electron microscope.

It is relatively simple to understand why shock loading induces a large concentration of point defects. The primary source of point defects is the nonconservative motion of jogs. These jogs are generated by the intersection of screw or mixed dislocations. A few simple calculations will be conducted below in order to estimate the concentration of point defects. These calculations are based on the model described in Section 14.2.4. Figure 14.21 shows the directions of motion of dislocations under the effect of the residual shear stresses. As they move, τ_{res} decreases, but in the process the dislocations intersect each other, generating jogs. The nonconservative motion of these jogs produces strings of either vacancies or interstitials. These can reorganize themselves as dislocation loops, as shown in Figure 14.28. The subject is treated in detail by Hirth and Lothe [45]. We reproduce here the essential elements of a model proposed by Meyers and Murr [16]. Figure 14.29 shows dislocations containing jogs. Assuming equilibrium of forces, the force due to the applied stress τ_{AP} is balanced by the friction force F_{f}, the force due to the jogs, F_{p}, and the force

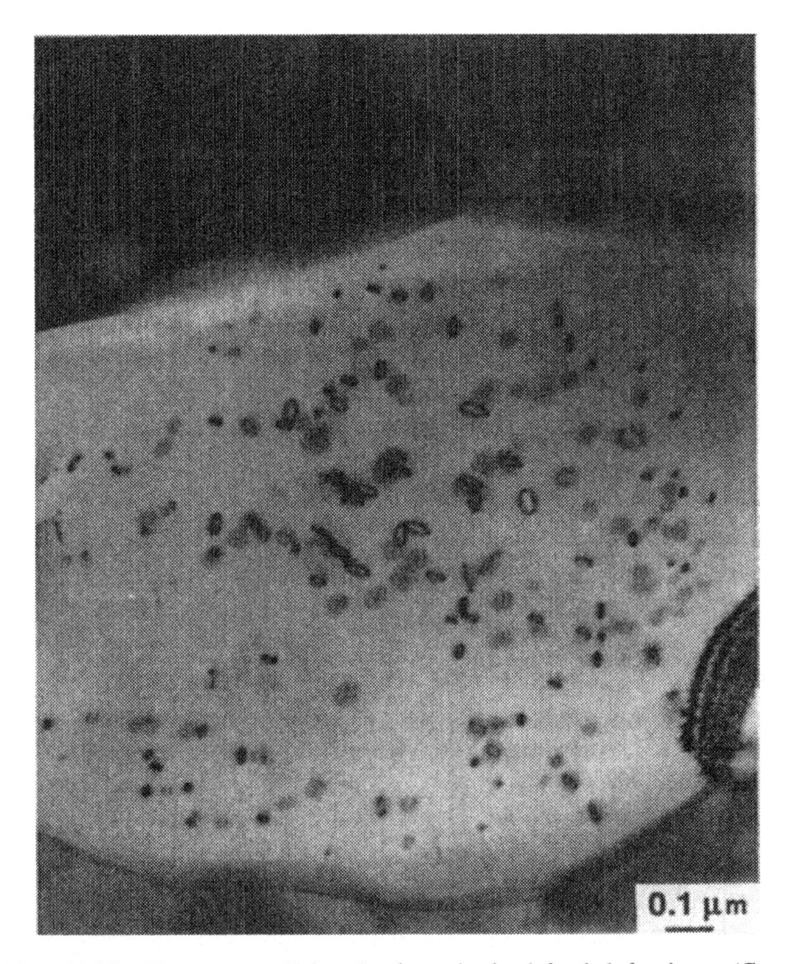

FIGURE 14.28 Vacancy-type dislocation loops in shock-loaded aluminum. (Courtesy of G. T. Gray III, Los Alamos National Laboratory.)

due to the binding of the dislocation, $F(r)$, one has (l is the spacing between obstacles and $1/l$ is their density along dislocation line)

$$F_{AP} - F_f - \frac{F_p}{l} - F(r) = 0 \tag{14.15}$$

It is assumed, sensibly, that acceleration is very high and that the equilibrium state is reached instantaneously. The frictional forces (mainly phonon and electron drag) are dependent on velocity. One can assume a constant, viscous drag acting on the dislocation (this is explained in detail in Chapter 13):

$$F_f = Bv \tag{14.16}$$

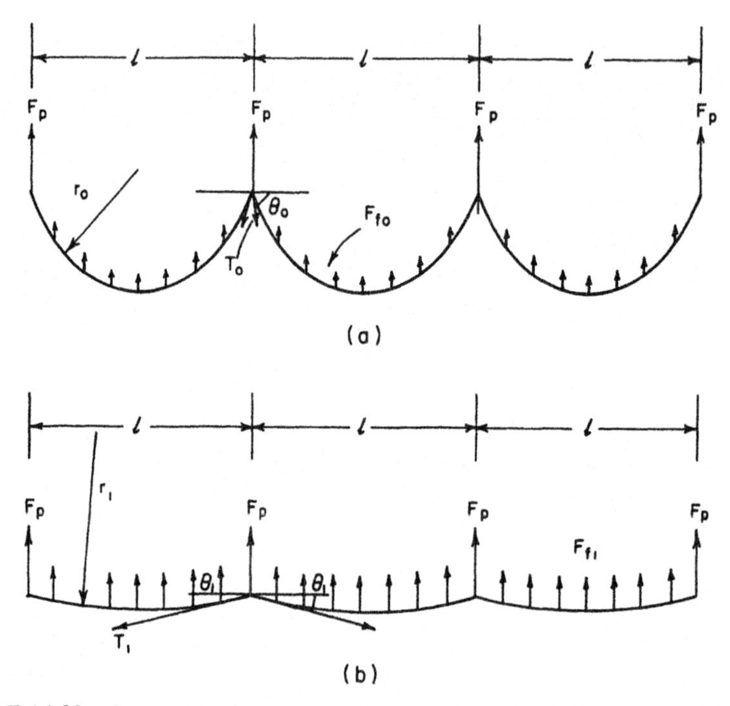

FIGURE 14.29 Comparison between (a) slow-moving and (b) fast-moving dislocation in periodic field of obstacles. (From Meyers and Murr [16], Fig. 18, p. 510. Reprinted with permission of the publisher.)

where B is the viscous damping coefficient and v the velocity of the dislocation. Peach–Koehler's equation states that if the applied stress is ideally oriented (Burgers vector and shear stress parallel):

$$F_{AP} = \tau_{AP} b \tag{14.17}$$

F_p is independent of velocity at high enough velocities (no diffusion needs to be considered) and can be computed from the energy of the vacancies (or interstitials) generated. If E_p is the energy of the point defect, and one defect is generated if the jog advances by b, one has

$$W = E_p = F_p b \tag{14.18}$$

Hence

$$F_p = \frac{E_p}{b} \tag{14.19}$$

The back force due to the circular shape of the dislocation, $F(r)$, can be expressed as

$$F(r) = \frac{E_d}{r} \tag{14.20}$$

where E_d is the energy of the dislocation line. The energy of the dislocation line is in turn dependent on its velocity. Weertman has treated this subject in detail [60]:

$$E_d = E_{d0}\left[1 - \frac{V^2}{C^2}\right]^{-1/2} \tag{14.21}$$

where E_d and E_{d0} are the self-energies of the dislocation at rest and at velocity V, respectively, and C is the velocity of elastic shear waves in the metal. Substituting Eqns. (14.16)–(14.21) into Eqn. (14.15) yields

$$\tau_{AP} b - Bv - \frac{E_p}{bl} - \frac{E_{d0}}{r}\left[1 - \frac{V^2}{C^2}\right]^{-1/2} = 0$$

$$r = \frac{E_{d0}\left[1 - \dfrac{v^2}{c^2}\right]^{-1/2}}{\tau_{AP} b - Bv - \dfrac{E_p}{bl}} \tag{14.22}$$

and

$$\lim_{v \to c}(r) = \infty$$

This leads to the following significant conclusion: As the velocity increases, the effectiveness of jogs as barriers to dislocation motion decreases. They are effectively "dragged along" by the dislocation, generating in the process a large number of point defects. It is even possible to estimate the concentration of point defects generated if one knows the density of dislocations generated and the amount that each dislocation will move. A simplified calculation is conducted below. The number of jogs per unit length of dislocations is estimated from the number of intersections with other dislocations. If the distance between adjacent layers is d (equal to the distance between dislocation duplets at the front), and if each dislocation moves by a distance l, it will have l/d^2 jogs per unit length at the end of its trajectory. The average distance by which a jog was dragged along is $l/2$. If one point defect is generated for each displacement equal to b, the total number N of point defects generated per dislocation is

$$N = K\frac{l}{d^2}\frac{l}{2}\frac{1}{b} = \frac{Kl^2}{2bd^2} \tag{14.23}$$

The factor K was introduced to account for the fact that only a certain percentage of the intersections generate jogs that cannot move conservatively. To a first approximation, one can take K as 0.25. The total dislocation density being ρ, the concentration of point defects generated by shock loading is

$$C = \frac{K\rho l^2}{2bd^2n} \tag{14.24}$$

where n is the number of atoms per unit volume. But $1/d^2$ is equal to the dislocation density and

$$C = \frac{K\rho^2 l^2}{2nb} \tag{14.25}$$

Some of these parameters are not known, at present, and some very rough approximations will be made. Making the calculation for nickel shock loaded at 20 GPa, one finds [1] that $\rho \simeq 5 \times 10^{10}$ cm^{-2}, $b = 3.5$ A, $K = 0.25$, and $n \simeq 2.5 \times 10^{23}$ cm^{-3}; Hsu et al. [46] observed that $l = 0.7$ μm. The input of these parameters provides

$$C = 7 \times 10^{-5}$$

Kressel and Brown [56] estimate the sum of vacancy and interstitial concentration to be approximately equal to this value. The closeness between calculated and measured point-defect densities seems to indicate that the critical assumptions in the calculation and the distance $l = 0.7$ μm are reasonable. This result is also consistent with measurements by Mogilevsky [59].

14.4 DEFORMATION TWINNING

14.4.1 Effect of Material and Shock Wave Parameters

The most important and self-consistent comment that can be made about deformation twins is that twinning is a highly favored deformation mechanism under shock loading. Metals that do not twin by conventional deformation at ambient temperatures can be made to twin by shock loading. In this respect, as in the morphology of dislocation substructures, shock deformation resembles conventional deformation at low temperature: loose cell walls and a greater tendency toward twinning. The occurrence of twinning depends on several factors:

1. *Pressure.* Nolder and Thomas [61, 62] found that twinning occurred in nickel above 35 GPa pressure. This was generally confirmed by Greulich and Murr [63]. DeAngelis and Cohen [64] found the same effect (a threshold pressure) in copper. Figure 14.30 shows how the density of twins increases with pressure for nickel.

2. *Crystallographic Orientation.* It is the deviatoric component of stress that induces twinning. Hence, when the resolved shear stress in the twinning plane and along the twinning direction reaches a critical level, twinning should occur. DeAngelis and Cohen [64] found an orientation dependence for the threshold stress; copper single crystals twinned at 14 GPa when the shock wave traveled along [100] and at 20 GPa when it traveled along [111]. Greulich and Murr [63] found that for nickel at and above 35 GPa, twinning occurred preferentially for [100] grains. As the pressure was increased, the preponderance of twins increased along orientations other than [100].

3. *Stacking-Fault Energy.* As the SFE of FCC metals is decreased, the incidence of twinning increases. As a corollary the threshold stress for twinning should decrease. Pure aluminum has a very high SFE, and no deformation twins have been reported in shock recovery experiments. The effect of alloying elements generally is to decrease the SFE and to increase the susceptibility to twin; Figure 14.12 shows this effect very clearly.

4. *Pulse Duration.* The effect of pulse duration, first explored by Appleton and Waddington [65], was systematically investigated by Champion and Rhode [66] for an austenitic (Hadfield) steel. They found striking differences in twin densities for different pulse durations, at 10 GPa. Numerous twins were observed at 2 μs, whereas no twinning was present at 0.065 μs. They concluded that there must be a threshold time for twinning. Staudhammer and Murr [67, 68] investigated the effect of pulse duration (0.5, 1, 2, 6, 14 μs) on the substructure of AISI 304 stainless steel. They found an increase in twin density up to about 2 μs; beyond that the twin density seemed to be essentially constant. Stone et al. [69] found, systematically, an increase in twin density as the pulse duration was increased from 0.5 to 1.0 μs in both AISI 1008 steel and Armco magnetic ingot iron. The twins generated by the shock pulse should not be confused with the ones formed by the elastic precursor wave in iron. The latter were investigated by Rohde and co-workers [70, 71]. Although twins are generated by the elastic precursor waves, the volume percent of twins generated by the shock wave is an order of magnitude higher. Whereas the elastic precursor may produce a twin density of 3 vol %, a shock wave of 30 GPa peak pressure and 1 μs pulse duration has been shown to generate about 50 vol % of twins.

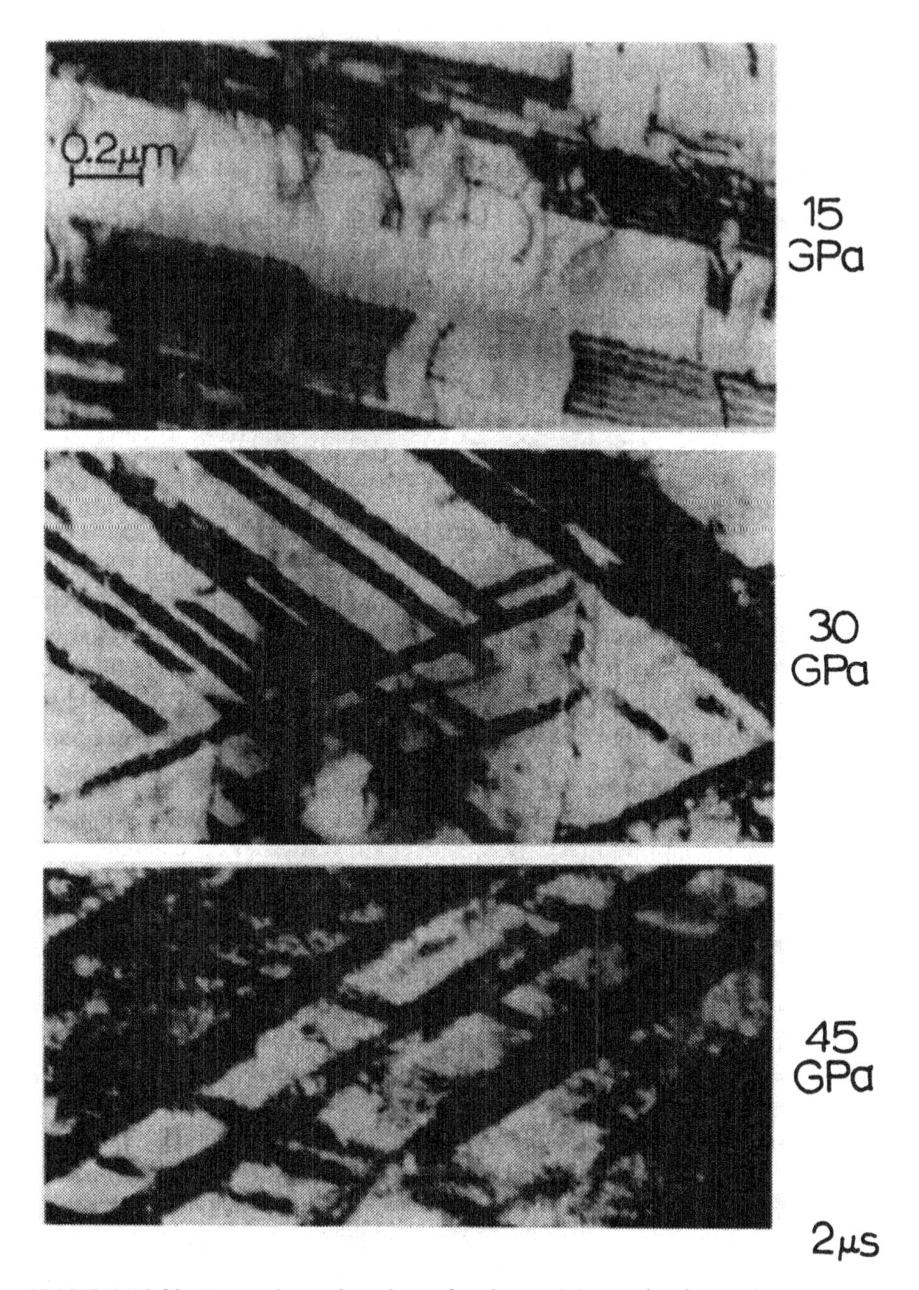

FIGURE 14.30 Increasing twin volume fraction and decreasing intertwin spacing, Δ, with increasing pressure in (001) grains in shock-loaded nickel sheet. The peak pressures are noted in each corresponding transmission electron micrograph. (From Murr [1], Fig. 23, p. 652. Reprinted with permission of the publisher.)

5. *Existing Substructure.* Rohde et al. [72] found profuse twinning upon shock loading titanium–gettered iron in the annealed condition. However, predeformed samples exhibiting a reasonable density of dislocations did not twin. The same results were obtained by Mahajan [73] for iron. Hence, if one looks at dislocation generation and motion and twinning as competing mechanisms, one can rationalize this response. The deviatoric stresses generated by a shock wave are accommodated by twinning when no dislocations are available and by motion of the already existing dislocations if iron is predeformed.
6. *Grain Size.* Wongwiwat and Murr [74] explained the incidence of twinning in molybdenum by showing that, at a certain pressure, large-grain-size specimens twinned more readily than small-grain-size ones. However, it should be emphasized that this response is not unique to shock loading; indeed, iron–3% silicon [75] and chromium [76] have been shown to exhibit a strong grain size dependence of the twinning stress (in conventional deformation). Figure 14.31 shows microstructures of copper with two widely ranging grain sizes subjected to shock loading at a pressure of 50 GPa. Twinning is generalized for the large grain size (200 μm) whereas it is absent for the small grain size (15 μm). Thus, both FCC and BCC materials exhibit a grain size dependence of twinning.

14.4.2 Mechanisms

The crystallographic and morphological features of shock-induced twins do not differ substantially from the conventionally formed ones. The rationale of considering twinning and slip by dislocations as competitive processes is a fair and helpful one. It is well established that either a decrease in temperature or an increase in strain rate tend to favor twinning over slip by dislocation motion (e.g., Peckner [77]). In this context, the graphical scheme proposed by Thomas [78] for martensite can be generalized [79]. This generalization is extended here to high strain rates, showing that they favor twinning. It is shown in Figure 14.32. The low-temperature dependence of the stress required for twin initiation is a strong indication that it is not a thermally activated mechanism. Hence, τ/G for twinning is not temperature dependent. On the other hand, the thermally activated dislocation motion becomes very difficult at low temperatures; T_t is the temperature below which the material will yield by twinning in conventional deformation. However, at high strain rates and in shock loading, dislocation generation and dynamics are such that the whole curve is translated upward, because of thermal activation; this is explained in detail in Chapter 13. As a consequence, the intersection of the two curves takes place at a higher temperature. Figure 14.33 shows deformation twins in nickel shocked to a pressure of 45 GPa at a pulse duration of 2 μs. As the shock pressure is

FIGURE 14.31 Effect of grain size on the residual microstructure of copper (P = 50 GPa). Notice profuse twinning for large-grain specimen and its virtual absence for small-grain specimen. (From U. R. Andrade, M. A. Meyers, and A. H. Chokshi, *Met. and Mat. Transactions*, to be published, 1994).

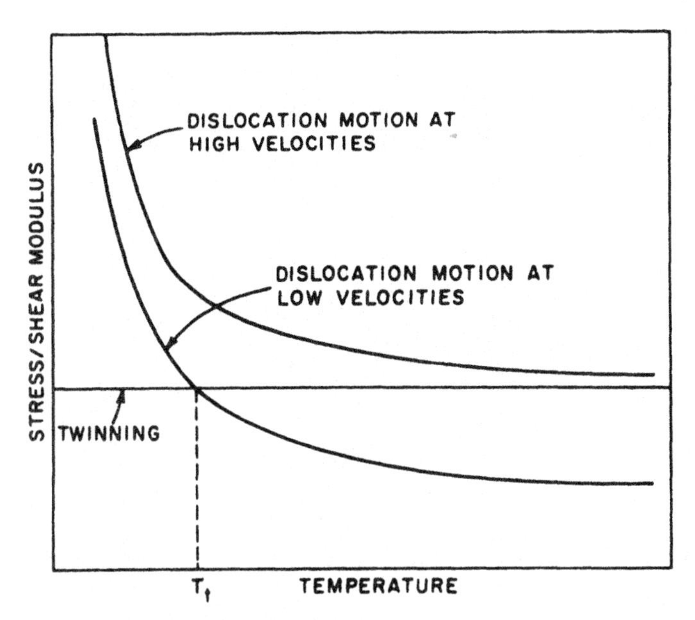

FIGURE 14.32 Effect of temperature on the stress required for twinning and slip (at low and high strain rates). (From Meyers and Murr [16], Fig. 20, p. 516. Reprinted with permission of the publisher.)

increased, the twin volume fraction increases, with a decrease in intertwin spacing.

The fact that deformation twins exhibit similar crystallographic and morphological features in both conventional and shock deformation leads to the conclusion that the mechanisms responsible for nucleation and growth should be similar. There are essentially two schools of thought regarding twin formation. The first is the pole mechanism proposed by Cottrell and Bilby [80] for BCC metals and extended by Venables [81] to FCC metals. The principal deterrent to the pole mechanism, especially in shock loading, is the maximum rate of growth predicted by this model; it is of the order of 1 ms^{-1}. This is about three orders of magnitude below values reported by Bunshah [82], Reid et al. [83], and Takeuchi [84]. Takeuchi found that a twin propagated at 2500 ms^{-1} in iron, and that this velocity was virtually independent of temperature in the interval -196 to $+ 126°C$. This latter observation is very important and is indicative of the fact that growth is not a thermally activated process. Hornbogen [36] has suggested that the propagation of twins in Fe–Si alloys occurs at such a rate as to generate shock waves. Twins are generated in shock loading at pulse durations much lower than the minimum pulse duration at which they could grow if a pole mechanism were operative. Indeed, Wright et al. [85] observed twinning at pulse durations as low as 0.07 μs.

The velocity limitation led Cohen and Weertman [86] to propose a much simpler model for FCC metals, involving the production of Shockley partials

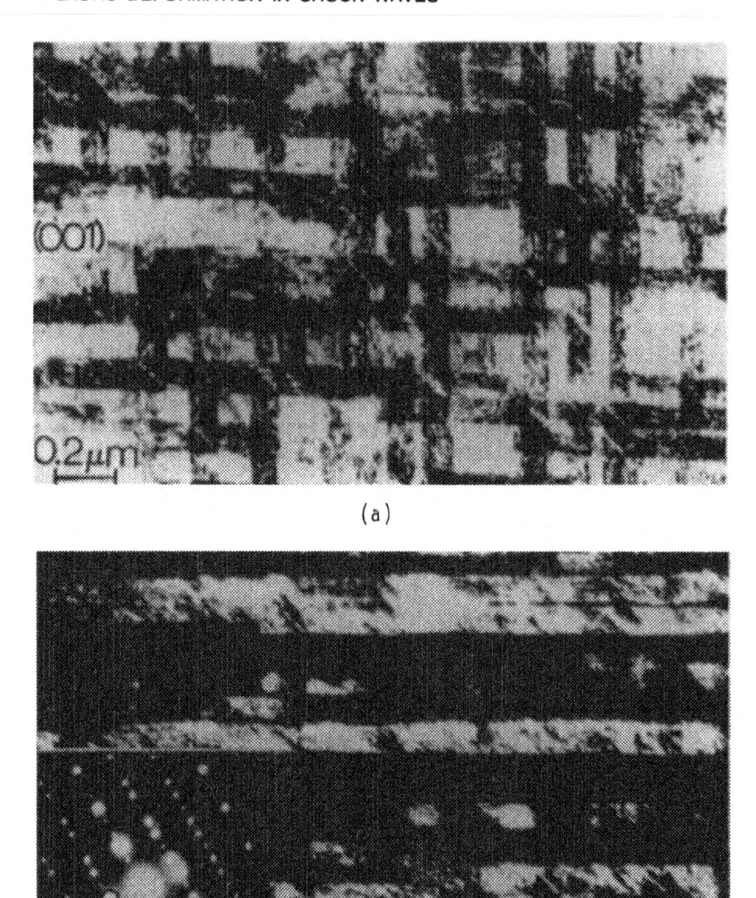

FIGURE 14.33 Deformation twins in shock-loaded nickel (45 GPa peak pressure, 2 μs pulse duration). (a) Bright-field electron transmission image. (b) Dark-field image using the twin spot shown in the selected-area electron diffraction pattern insert. Note the contrast irregularities along individual twins. (Courtesy of L. E. Murr, University of Texas at El Paso.)

at Cottrell–Lomer locks and their motion through the material. The velocity of propagation of a twin is in this case simply established by the velocity of motion of the Shockley partials. Hirth and Lothe [87] proposed a yet simpler model in which the dislocations are simply homogeneously nucleated; whereas stress required to homogeneously nucleate the first dislocation is of the order of 10% of the shear modulus, the subsequent loops would require stresses that are much lower (1% of G). This "homogeneous nucleation" concept was forwarded first by Orowan [88].

14.5 DISPLACIVE/DIFFUSIONLESS TRANSFORMATIONS

There are numerous instances in which a shock wave induces a phase transformation. A very detailed review is presented by Duvall and Graham [89]. Chapter 8 is totally dedicated to shock-induced phase transformations. The discussion here supplements Chapter 8. In most cases, there is not sufficient time for diffusional transitions; thus, martensitic transformations are the most common under shock conditions.

The effect of a shock pulse on a displacive/diffusionless transformation has to be analyzed from three points of view: (1) pressure, (2) shear stresses, and (3) temperature. The changes in these parameters are not independent; there are specific temperature rises and deviatoric stresses associated with a certain pressure level. Nevertheless, they have different effects on the thermodynamics of phase transformations. Patel and Cohen [90] have established a rationale for the effect of stresses on the M_s temperature in martensitic transformations. They found that, in Fe–30% alloy, the hydrostatic pressure decreased the M_s temperature. In these alloys, there is a dilatation of 5% associated with the martensitic phase. Hence, a pressure pulse should not favor the transformation, and this is reflected in the decrease in M_s. On the other hand, an alloy in the martensitic form should revert to austenite if a pressure pulse is applied because this results in a contraction of the lattice. Indeed, Rhode et al. [91] were able to confirm this effect. A negative pressure pulse inducing negative hydrostatic pressures would be the converse situation, and the γ(FCC) $\rightarrow$ α(BCC or BCT) transformation would be favored, with an increase in the transition temperature. Meyers and Guimaraes [92] were able to produce a tensile pulse and generate martensite in an Fe–31% Ni–0.1% alloy. Figure 14.34 shows the martensite tube generated by tensile waves; this tensile pulse was produced by a compressive shock wave, as it reflected a free surface. The region of the material traversed by the compressive wave exhibited only a dense array of dislocations organized in cells and occasional twins. This phenomenon was used to calculate a nucleation time for the martensitic transformation. The pulse duration was decreased, whether by impacting a target with thinner and thinner flyer plates or by selecting regions close to the free surface for observation. Meyers [93] and later Sano et al. [94] used shock experiments and found that virtually no martensitic transformation occurred below a tensile pulse duration of 50 ns. Thus, it was concluded that this is the nucleation time for martensite. These studies illustrate how dynamic shock experiments can be used to obtain fundamental information on the nature of phase transformations. Similar experiments were used to establish the kinetics of stress-induced martensitic transformations in Fe–Ni by Thadhani and Meyers [95] and in Fe–Ni–Mn alloys by Chang and Meyers [96]. The results are also presented in refs. [97, 98]. A tensile pulse of varying duration was applied to the alloys at temperatures higher than the M_s temperature (M_s is the martensite start temperature). This tensile pulse triggered the transformation; as the duration of the tensile pulse was increased, the fraction transformed increased. Figure 14.35 shows the experimentally measured results at −20, −30, −40, and −50°C for an Fe–32

FIGURE 14.34 Martensite generated by tensile hydrostatic stresses produced by a reflected pressure pulse in iron–nickel alloy. (From Meyers and Guimaraes [92], Fig. 3, p. 290. Reprinted with permission of the publisher.)

wt % Ni–0.035 wt % C alloy with M_s of $-61\,°C$. The transformation saturates at $\sim 1.8\ \mu s$ but is clearly time dependent for lower times. Prior to this research, it was thought that the phase transformation reached the saturation level "instantaneously." From these experimental results, it was possible to obtain fundamental parameters of the martensitic transformation in this alloy, such as the activation energy. The fraction transformed, f, is expressed as

$$\dot{f} = \frac{df}{dt} = \left[n_i + f\left(p + \frac{1}{\bar{v}}\right)\right] \nu \exp\left(-\frac{Q_a}{RT}\right)$$

where p is an autocatalytic factor, $\bar{v}$ is the mean volume of each martensite lens, n_i is the initial number of nucleation sites, ν is an attempt frequency, Q_a is the activation energy for the transformation, and T is the temperature. Shock experiments are needed to probe into the material at these short times. By

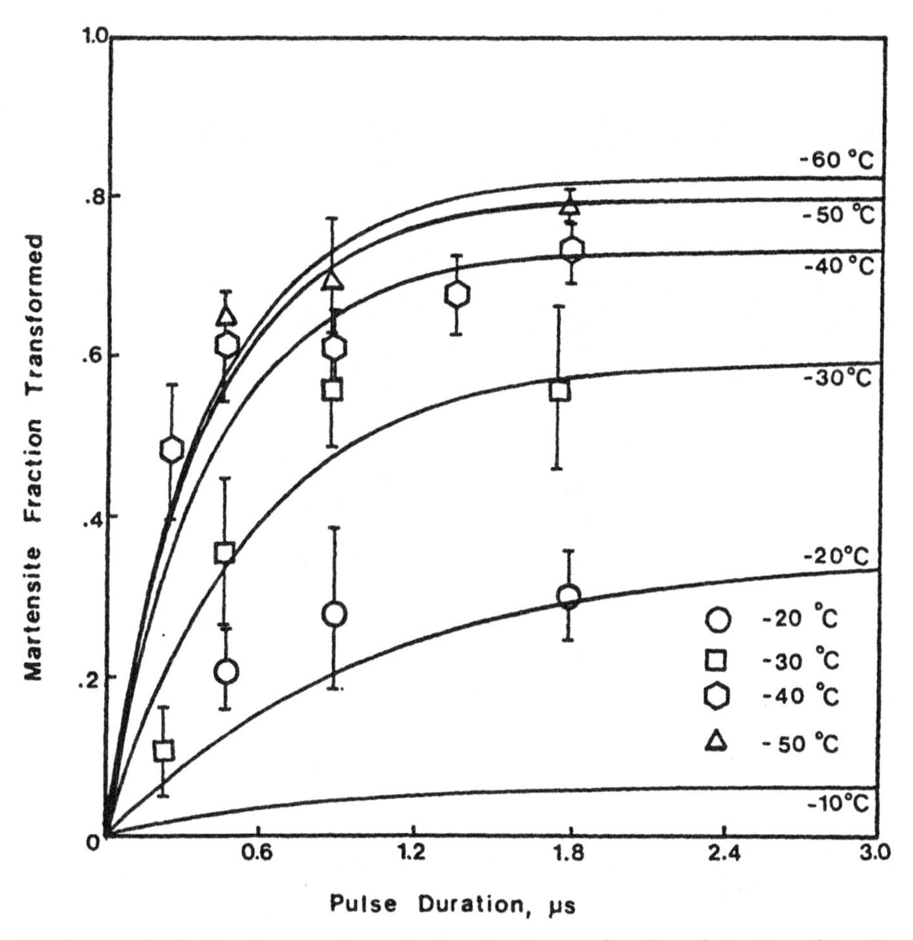

FIGURE 14.35 Fraction transformed of martensite as a function of duration of tensile pulse in Fe–32% Ni–0.035% C alloy impacted at temperatures above M_s. (Reprinted from *Acta Met.*, vol. 34, N. N. Thadhani and M. A. Meyers, Fig. 14, p. 1636, Copyright 1986, with permission from Pergamon Press Ltd.)

comparing experimental results with an analysis based on kinetic equations, it was possible to obtain the predicted values of Figure 14.35, represented by continuous lines. The martensitic transformations also acted as markers for regions in which plastic deformation by the shock wave preceding the tensile pulse was localized, preferentially forming in these regions. This is shown in Figure 14.36. These bands (shown in the bottom part of the figure) are evidence that the process of plastic deformation under shock is not homogeneous, but rather irregular. The orientation pattern for these lines is not fully understood yet.

In another study, Guimarães et al. [100] investigated the effect of preshocking an Fe–Ni–C alloy on the subsequent response to transformation of the lattice. The effect of shock loading was to introduce a high dislocation density

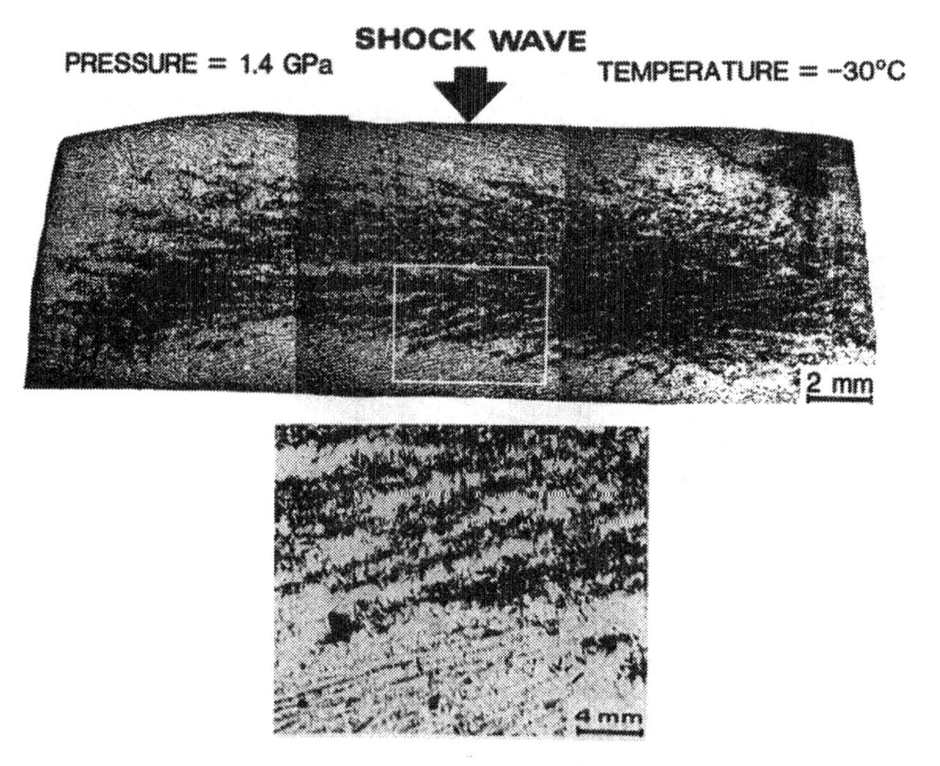

FIGURE 14.36 Cross section of specimen (Fe–32% Ni–0.035% C) subjected to tensile pulse after reflection of shock wave at free surface, showing irregular transformation markings indicative of stress wave inhomogeneities. (From Thadhani et al. [99], Fig. 2, p. 2792. Reprinted with permission of the publisher.)

in a cellular arrangement (as shown in Fig. 14.11) that sensitized the structure for martensitic transformation, increasing M_s.

14.6 OTHER EFFECTS

There are a number of additional effects due to the fact that the shock wave does not traverse an idealized monocrystalline metal but a medium in which there may be grain boundaries, twin boundaries, precipitates, and other microstructural features. They all affect the propagation of the shock wave and the generation of defects to a smaller or larger extent. The realization that the anisotropy of elastic and plastic properties of the individual grains in a polycrystal led to a "wavy-wave" model, which attempted to represent the irregularities in both the front and peak pressures of the elastic precursor and of a low-amplitude stress shock wave as they propagate throughout the metal [101–103]. Figure 14.37 represents a wavy-wave front after traversing polycrystalline nickel. The model used to represent the polycrystalline aggregate is shown in Figure 14.37(a); it was assumed to consist of cubes with three crystallographic orientations: [100], [110], and [111]. Since the elastic wave travels at a different characteristic velocity for each crystallographic orientation above, after it travels a certain length, the front becomes irregular. Computations, shown in Figure 14.37(b) and (c) for two different grain sizes, indicate that the irregularities of the front increase with grain size. The wavy-wave model is thought to represent well the behavior of the elastic precursor; however, the shock wave does not seem to be significantly affected by the polycrystallinity of the material.

Elastic precursor waves and low-amplitude shock waves are also reflected and refracted at grain boundaries. Indeed, Hornbogen [104] has shown that stress fields ahead of twins are reflected and diffracted at grain boundaries. Hence, there must be regions where the pressures are below the average value. During the passage of the pulse there must also be pulse irregularities propagating along different directions. These effects are, however, of secondary importance for very high amplitude pressure pulses.

Second-phase particles can play a significant role in the generation of defects:

1. If the particle has a different elastic modulus than the matrix, the hydrostatic component of the stress will produce interface stresses that will lead to the "punching out" of dislocations. Indeed, Das and Radcliffe [105] have observed this effect when applying hydrostatic pressure by conventional means to metals containing particles. The same process is certainly operative during shock loading.
2. The shock front is reflected, refracted, and disturbed when passing through the particle. This is eloquently illustrated in Figure 14.38 from Leslie et al. [106]. The inclusion has a clear effect on the distribution of deformation twins. The region "downstream" of the inclusion exhibits a higher DPH hardness (417) than the average value ($\approx$280). It is clear that the inclusion–matrix interface or the dislocations at the adjoining region acted as nucleation sites for additional twinning, which is responsible for the greater hardness.
3. When coherent precipitates are present in the lattice, the shock pressure, after equilibrating, produces different compressions of precipitate and matrix if they have different bulk moduli. This may result in the destruction of coherency with the creation of interfacial dislocations as well as plastic deformation in the (usually) softer matrix.*

There are additional effects of pressure that have not been considered in this review but that are worthy of analysis, since they have a bearing on the residual substructure of shock-loaded metals. Noteworthy among these are the effects

*L. E. Murr, H. R. Vydyanath, and J. V. Foltz (*Met. Trans.* A1, (1970) 3215) shocked Inconel 600, that contained originally coherent precipitates. Coherency was lost by the effect of the shock wave, with the creation of high densities of dislocations.

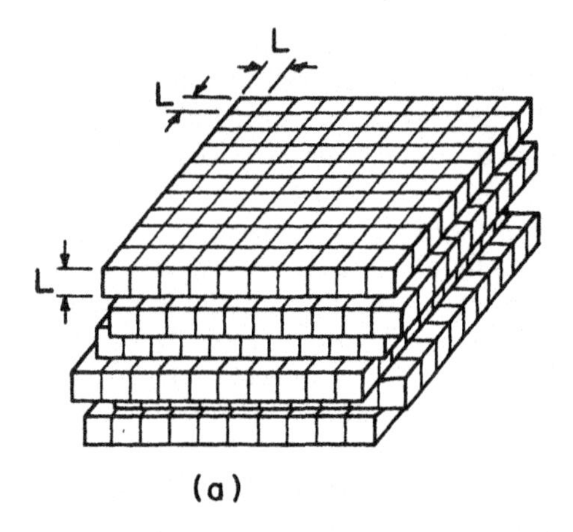

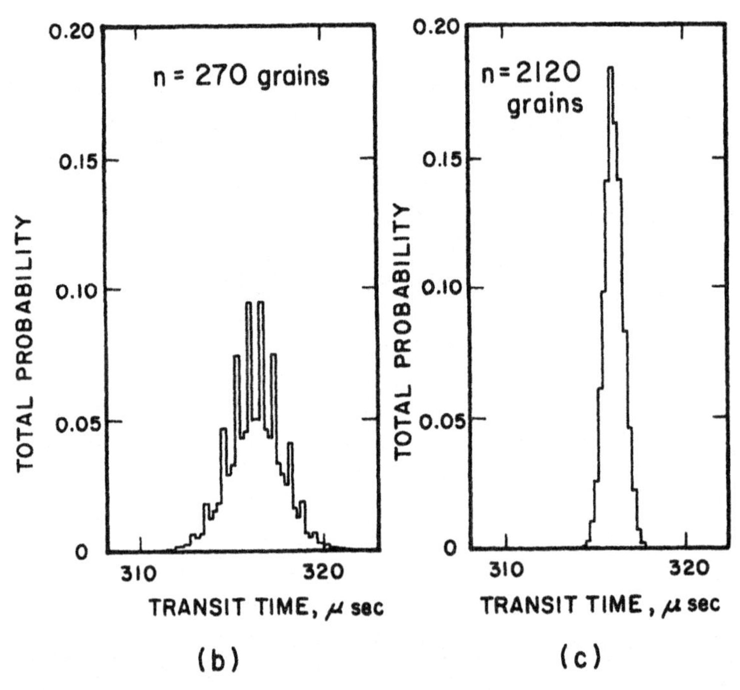

FIGURE 14.37 Proposed model for elastic stress wave (front at about 19 mm below surface). Propagation of stress wave in polycrystalline material. (a) Idealized grain arrangement. (b) Elastic precursor front for grain size of 70 μm. (c) Elastic precursor front for grain size of 9 μm. (From Meyers [103], Figs. 3, 5, and 9, pp. 104, 105. Reprinted with permission of the publisher.)

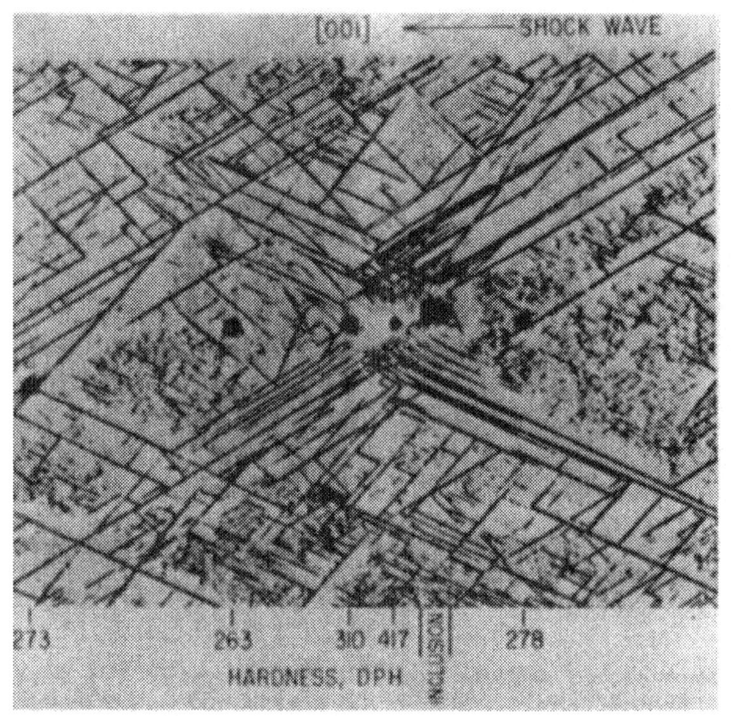

FIGURE 14.38 Hardness readings near an inclusion in a shock-loaded Fe–3% Si single crystal. Nital etch ×140. (From Leslie et al. [106], Fig. 10.9, p. 396.)

of pressure and shear stress on the SFE, on the diffusion coefficient, and on the mobility of dislocations.

The effect of polycrystallinity (or grain size) on the shock response of metals has been investigated and no broad generalizations can be made. We saw that the propensity for twinning is increased as the grain size is increased. The dislocation density is not significantly altered by the grain size. Braga et al. [107] looked at the effect of grain boundaries on shock-induced microstructures and found that the effects are second order. Figure 14.39 shows the residual microstructures of a copper bicrystal (boundary parallel to direction of propagation of shock wave) and a polycrystalline sample shocked to the same pressure. There is no significant difference in the residual dislocation density. Dhere et al. [31] shock loaded aluminum with different grain sizes and tried to determine whether grain rotation was produced by shock loading the smaller grain sizes. No grain rotation was observed, in contrast with an earlier report by DeAngelis and Cohen [108] that claimed that shock loading of copper to 43.5 GPa produced grain rotation with texture changes. Thus, it can be concluded that no grain rotation takes place.

Meyers et al. [109] investigated the effect of polycrystallinity on the shock

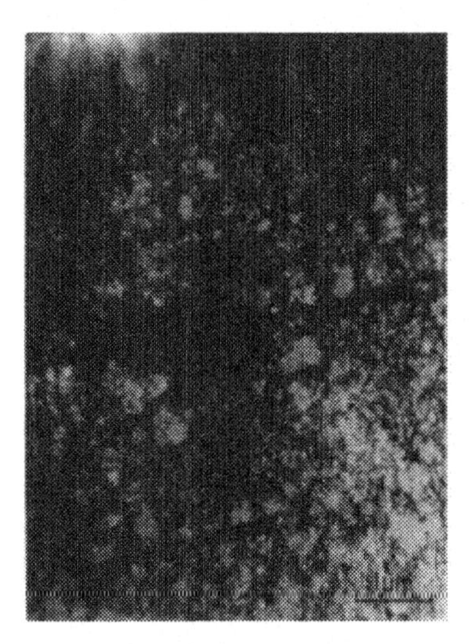

FIGURE 14.39 Effect of polycrystallinity on dislocation substructure generated in copper. (From Braga et al. [107].)

wave attenuation and strengthening of Fe–34% Ni and Fe–15% Ni–15% Cr alloys by producing polycrystals from deformed and recrystallized single crystals (P = 7.5 GPa; tp = 1.2 μs). They could not find any dramatic differences in the attenuation or in the dislocation densities. Figure 14.40 shows electron micrographs taken from the top and bottom surfaces of 20-mm-thick specimens. There seems to be a greater tendency toward forming planar arrays in the monocrystalline specimen. These are precursor to mechanical twinning, in accordance with the relatively low SFE exhibited by this alloy (30 mJ m^{-2}). Hsu et al. [46] and Meyers et al. [110] subjected nickel with two different grain sizes (32 and 150 μm) to shock loading and observed essentially no difference in the cell size and dislocation density for the two grain sizes. The attenuation of the shock wave was also analogous for the two cases.

Meyers [111] investigated the effect of shock front irregularities on hardening by using a plate on which grooves with a depth of 0.5 mm were introduced. This irregular plate was impacted by a flat flyer plate. These experiments conducted on AISI 304 stainless steel indicated that the irregularities at the shock front led to an increased hardening. Similar results were obtained by Gray and Morris [112] using a much more sophisticated technique. A "pillow" flyer plate yielding a ramped pressure front with a duration of 1 μs (time for pressure to reach top value) yielded a significantly higher yield stress for copper than a "stiff" flyer plate that produces a sharp shock front. The grooves made by Meyers [111] correspond to a spread of the rise time of 0.5/5, or 0.1, μs. This explains why the increase in strength obtained by Gray and Morris [112] was much more significant.

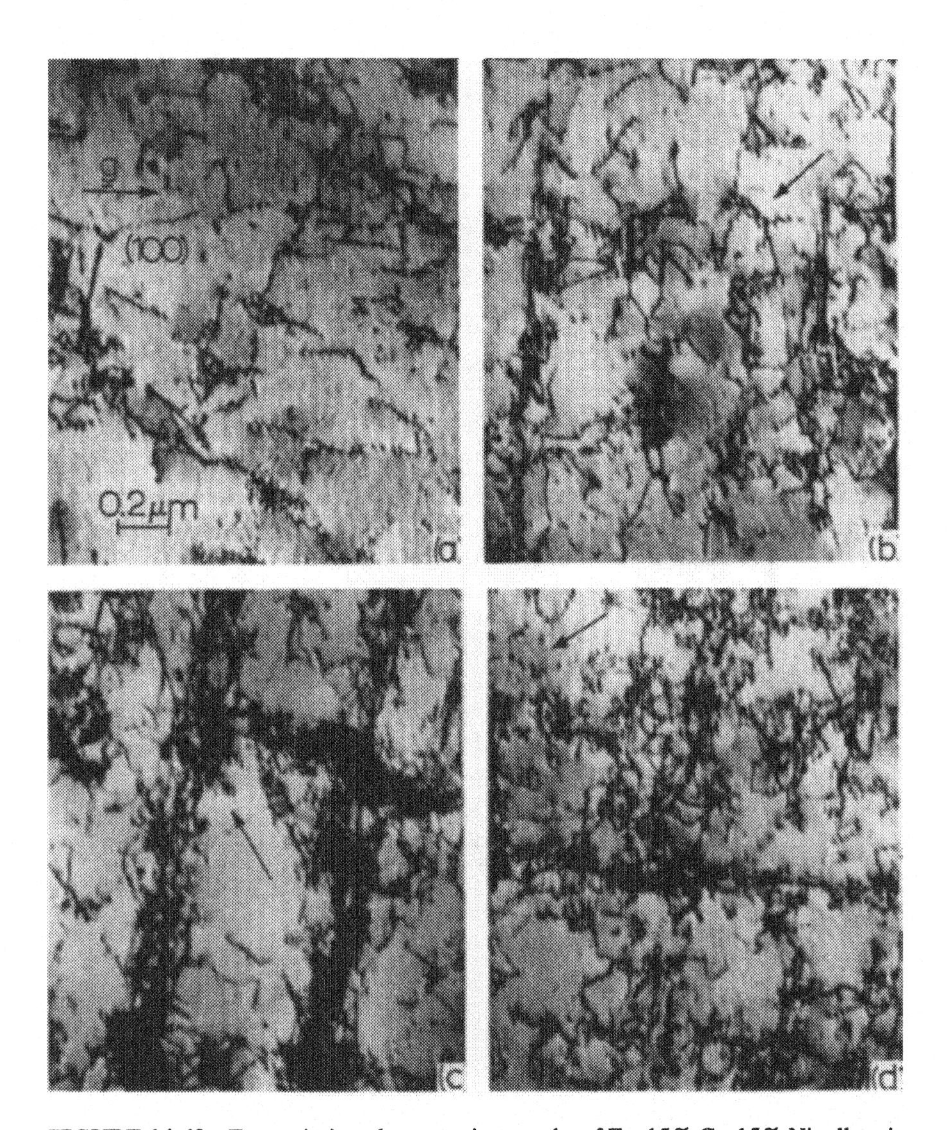

FIGURE 14.40 Transmission electron micrographs of Fe–15% Cr–15% Ni; all grains are (100) and the g-[002] is indicated by arrows in micrographs: (a) polycrystal, top; (b) polycrystal, bottom; (c) monocrystal, top; (d) monocrystal, bottom. All magnifications are the same as shown in (a). (From Meyers et al. [109], Fig. 13, p. 124. Reprinted with permission of the publisher.)

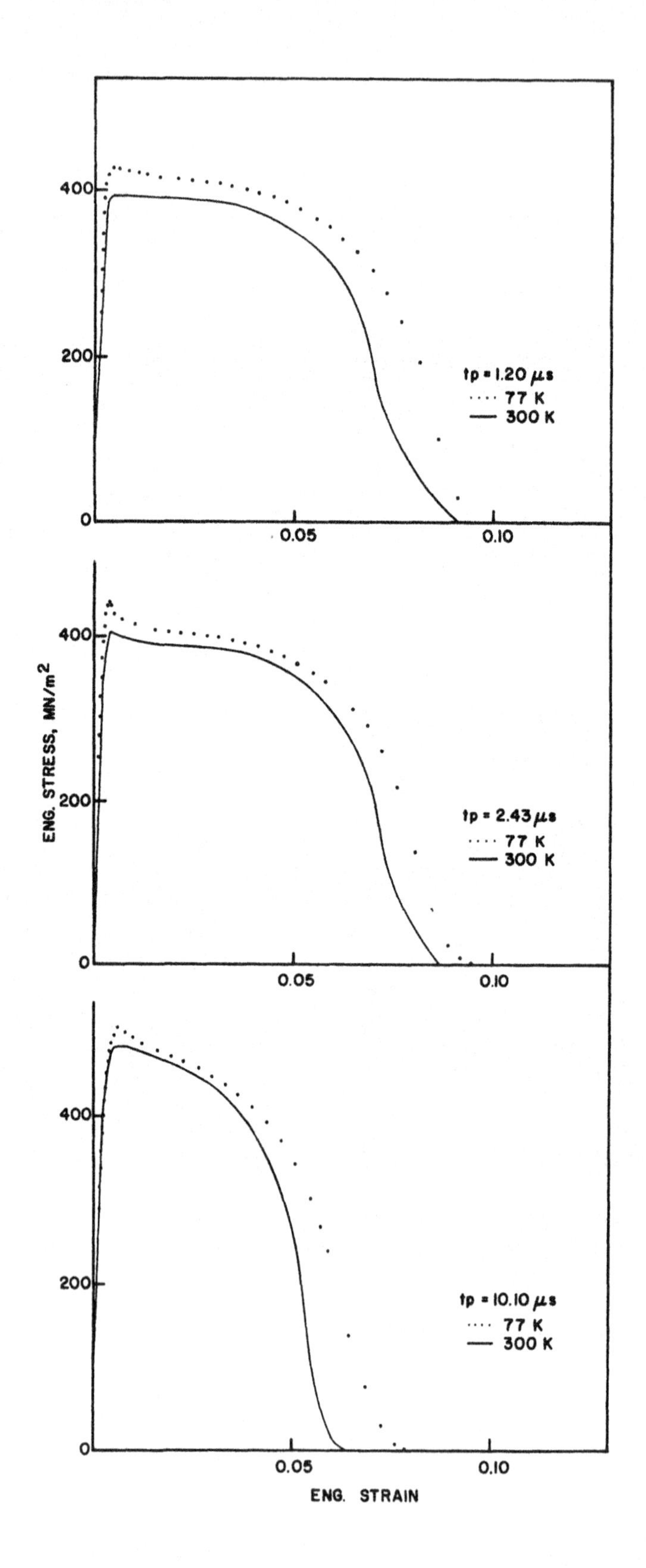

400
200
0
0.05
0.10
tp = 1.20 μs
77 K
300 K
ENG. STRESS, MN/m²
400
200
0
0.05
0.10
tp = 2.43 μs
77 K
300 K
400
200
0
0.05
0.10
tp = 10.10 μs
77 K
300 K
ENG. STRAIN

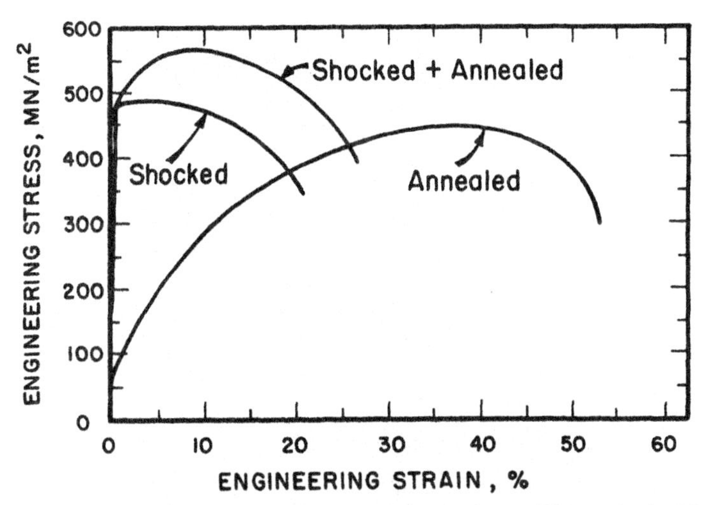

FIGURE 14.42 Effect of an annealing treatment of 1 hr at 573 K (which did not alter the hardness) on the stress–strain response of shocked nickel (double shocked to 10 and 25 GPa). (From Murr and Meyers [17], Fig. 26. Reprinted with permission of the publisher.)

14.7 MECHANICAL STABILITY OF SUBSTRUCTURE

The mechanical response of shock-hardened metals can become quite unstable under certain conditions. Meyers [113] found that nickel shock loaded to a pressure of 20 GPa exhibited a stress–strain curve (in tension) marked by the absence of work hardening. Figure 14.41 shows the stress–strain curves of material shocked at different pulse durations. Material was shocked both at room temperature and 77 K [113]. By taking the shocked specimens and heating them to a temperature characteristic of recovery, the work hardening characteristic of nickel was reintroduced. This is clearly evident in Figure 14.42, in which a heat treatment of 573 K for 1 h, which did not alter the hardness, completely changes the plastic response of the material, changing it from ''work softening'' to ''work hardening.'' At the microstructural level, this work softening can be understood if one looks at the dislocation configuration introduced by shock loading. The total Burgers vector of shock-deformed materials is zero or close to zero, since there is little or no net plastic deformation. Therefore, the cells that compose the deformation substructures can be considered to be assemblies of dislocations with roughly equal frequency of positive and negative Burgers' vector direction. Fourie et al. [115] provide a classification for these arrays of dislocations. Charsley and Kuhlmann-Wilsdorf [116] have investigated these dipolar walls and concluded that they are an unstable config-

FIGURE 14.41 Stress–strain plots of nickel shock loaded to 25 GPa at ambient temperature and 77 K at different pulse durations: (a) 1.2 μs, (b) 2.4 μs; (c) 10.16 μs. (From Meyers et al. [114], Fig. 8, p. 148. Reprinted with permission of the publisher.)

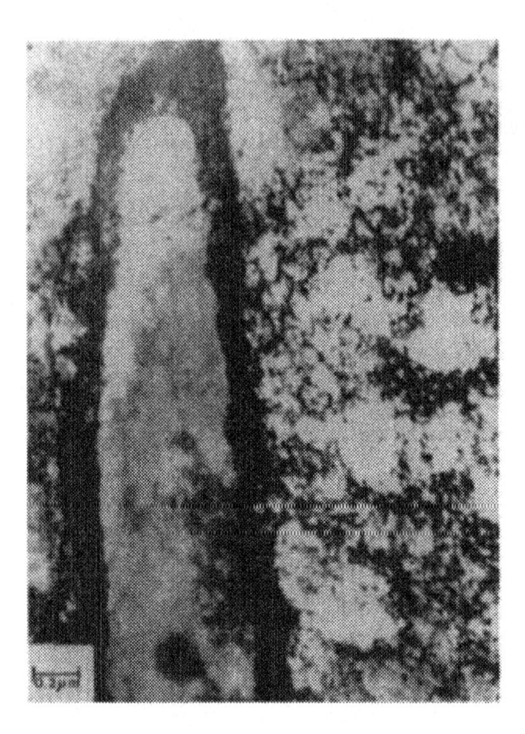

FIGURE 14.43 Transmission electron micrograph showing the breakdown of shock-wave-induced substructure occurring at the neck region of a tensile specimen (the effect of work softening). (From Meyers et al. [110], Fig. 13, p. 247. Reprinted with permission of the publisher.)

uration and that plastic deformation would tend to destroy this configuration. This is borne out by transmission electron microscopy observations on the neck region of specimens tested in tension, quasi-statically. The characteristic cell morphology is replaced by large elongated cells whose interiors are virtually dislocation free. Figure 14.43 shows the drastic alteration in the cell structure caused by the plastic deformation of the shock-hardened nickel specimen. The small cells are replaced by large, elongated cells; the process of reorganization of dislocations occurs without necessity for the generation of new dislocation. Thus, it occurs without work hardening, which takes place primarily by dislocation generation/multiplication.

14.8 SHOCK WAVE EFFECTS IN CERAMICS

Ceramics are brittle, and their shock response is clearly the result of a high hardness and brittle response. Chapter 15 discusses fracture of brittle materials in considerable detail, and we will limit ourselves here to defects generated by shock waves that do not involve extensive cracking. Ceramics have a high

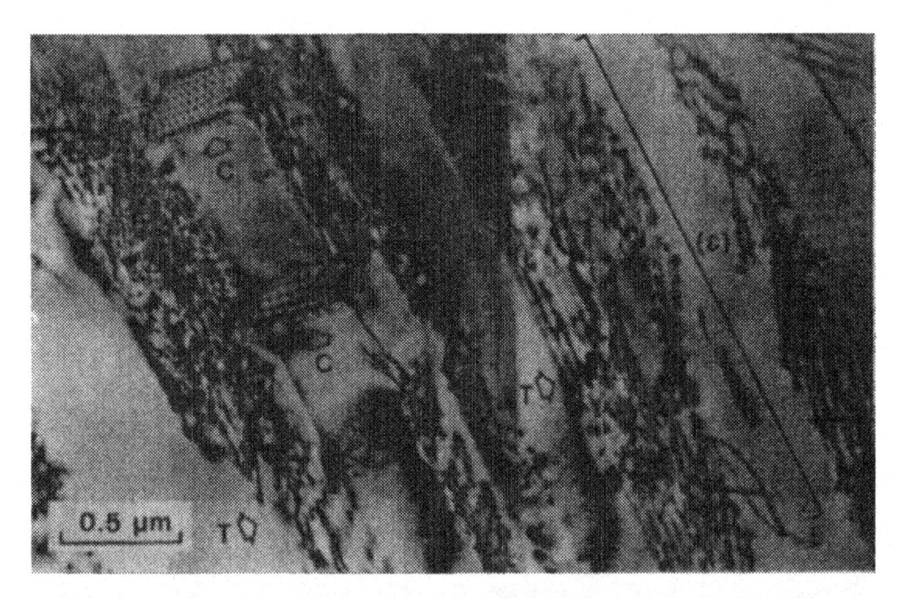

FIGURE 14.44 Alumina subjected to shock pulse ($P = 20$ GPa). The differently misoriented lattice bands are separated by parallel dislocation arrays lying on the basal plane. Basal plane twins and microcracks are also formed. Twins are indicated by T, microcracks or crack precursors by C, and a dislocation group by the bracket marked ε. The basal plane is tilted 40° to the plane of the photo. (From Yust and Harris [117], Fig. 3, p. 885. Reprinted with permission of the publisher.)

HEL. For example, the HEL of sapphire (monocrystalline Al_2O_3) is close to 25 GPa, and the passage of shock waves with a lower amplitude than 25 GPa does not produce significant damage. One of the first reports of shock-induced defects in ceramics is the work of Yust and Harris [117]. They performed transmission electron microscopy on alumina that was shock loaded to an unknown pressure (probably around 20 GPa). They observed profuse dislocations arranged in bands, twins, and microcracks in the structure. Figure 14.44 shows the microstructure. These bands are composed of dislocations on the hexagonal-based plane. Their Burgers vectors were $\{1120\}$ and $\{1010\}$. Basal plane twinning was also observed. Louro and Meyers [119] performed shock recovery experiments at much lower pressures on alumina. The shock amplitudes were actually below the HEL for polycrystalline alumina. They were able to identify dislocations and microcracks on isolated grains, although the majority of the grains did not exhibit any effect of the shock pulse. Figures 14.45 and 14.46 show dislocation arrays in the alumina, and Figure 14.47 shows microcracks that originate at a void within an alumina grain. The dislocation arrays in Figure 14.46 probably come from a Frank–Reed source; tilting of specimen produces their extinction. These specimens were subjected to a 7.5-GPa pressure pulse followed by a tensile release wave.

Wang and Mikkola [120] subjected mono- and polycrystalline alpha-alumina to shock compression (P = 5–23 GPa). Alumina is hexagonal and dislocation slip on basal ({0001}) and non-basal ({1010}) and ({1123}) planes was observed. Extensive twinning was observed at the higher pressures. In the polycrystals, cracks were often generated at the intersection of twins with grain boundaries, and propagated along the latter. In monocrystals, cracks often occurred along twin/dislocation planes.

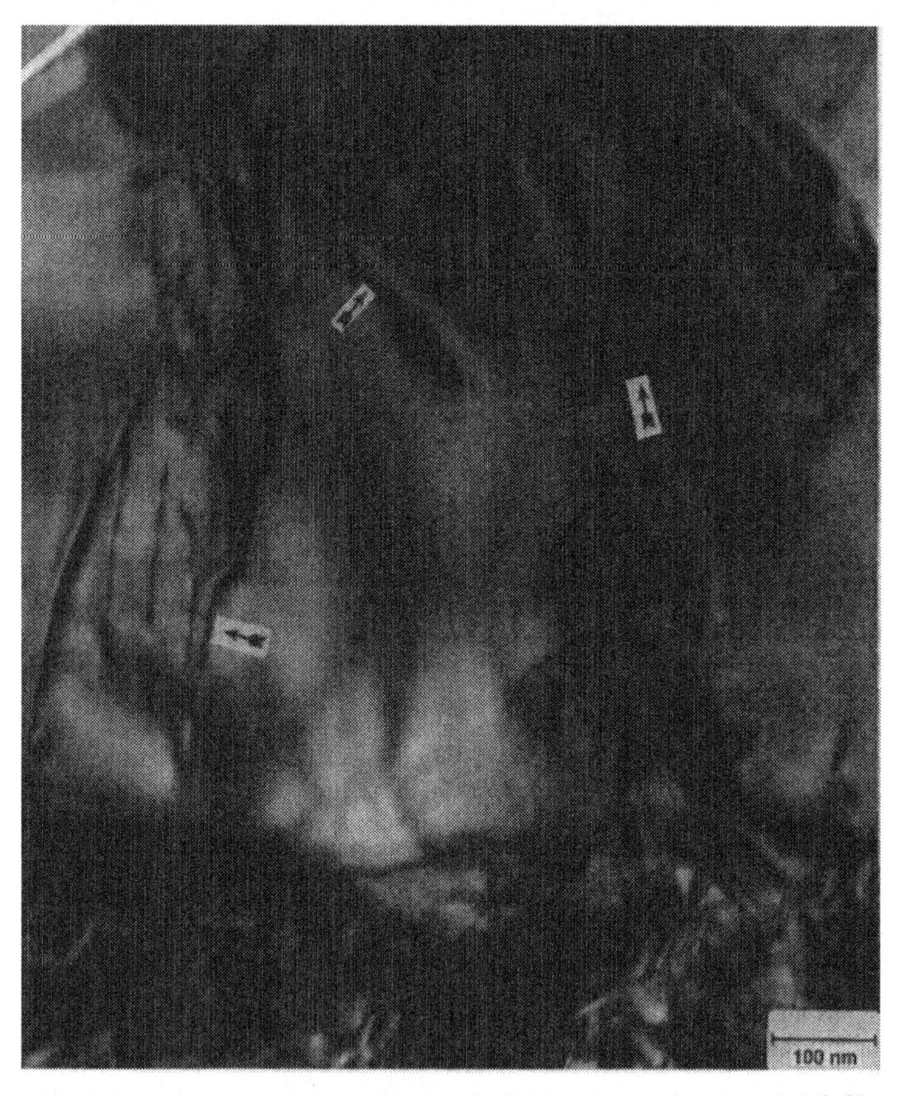

FIGURE 14.45 Dislocations in alumina shock loaded to peak stress of 4.6 GPa. (From Louro and Meyers [118], Fig. 17, p. 2528. Reprinted with permission of the publisher.)

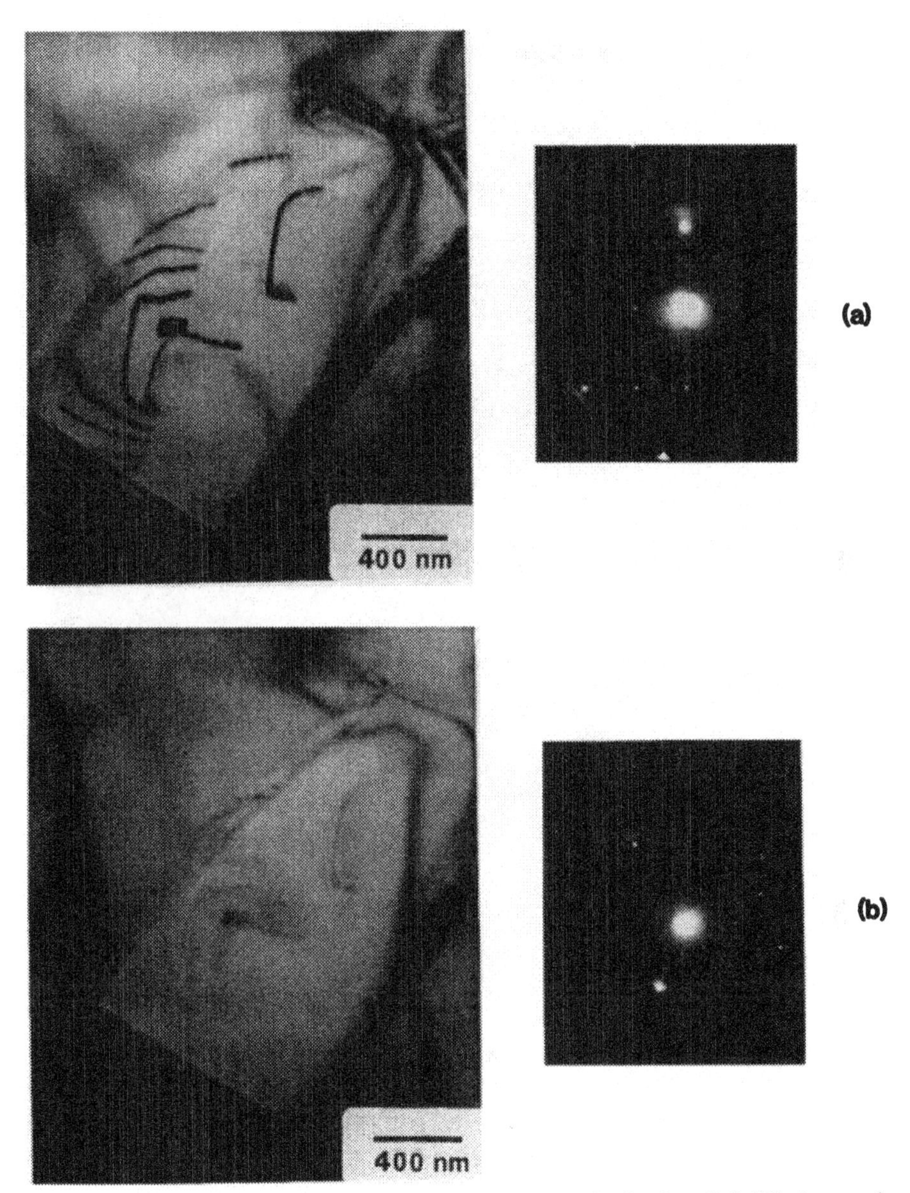

FIGURE 14.46 Dislocations in a grain of alumina submitted to 7.5 GPa impact in aluminum capsule. Subsequently generating tension: (a) dislocations; (b) extinction after tilt of specimen. (From Louro and Meyers [119].)

It can be concluded that in ceramics intense and generalized dislocation activity only occurs at very high stress levels and that it is associated with microcracking, since plastic deformation generates residual stresses that give rise to large tensile stresses once the shock pressure is removed.

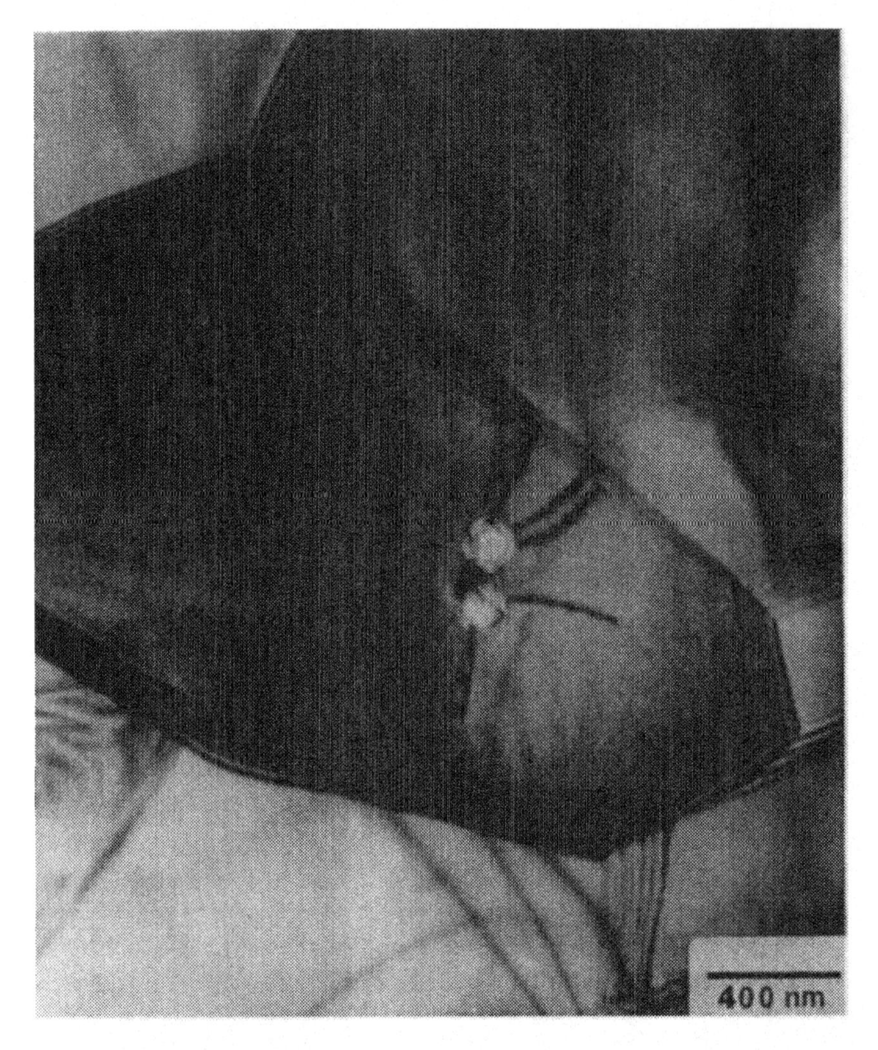

FIGURE 14.47 Cracks (or dislocations) associated with voids in alumina subjected to 7.5 GPa stress pulse. (From Louro and Meyers [118], Fig. 19, p. 2529. Reprinted with permission of the publisher.)

REFERENCES

1. L. E. Murr, in *Shock Waves and High-Strain-Rate Phenomena in Metals*, eds. M. A. Meyers and L. E. Murr, Plenum, New York, 1981, p. 607.
2. A. A. Deribas, I. N. Gavriliev, T. M. Sobolenko, and T. S. Teslenko, in *Metallurgical Applications of Shock-Wave and High-Strain-Rate Phenomena*, eds. L. E. Murr, K. P. Staudhammer, and M. A. Meyers, Dekker, New York, 1986, p. 345.

3. R. N. Wright and D. E. Mikkola, in *Shock Waves and High-Strain-Rate Phenomena in Metals*, eds. M. A. Meyers and L. E. Murr, Plenum, New York, 1981, p. 703.
4. G. A. Stone, R. N. Orava, G. T. Gray, and A. R. Pelton, "An Investigation of the Influence of Shock-Wave Profile on the Mechanical and Thermal Responses of Polycrystalline Iron," Report No. SMT-1-78, U.S. Army Research Office Final Report, Grant No. DAA G29-76-G-0180, September 1978.
5. J. S. Rinehart and J. Pearson, *Behavior of Metals under Impulsive Loads*, American Society for Metals, Cleveland, OH, 1954.
6. C. S. Smith, *Trans. AIME*, **212** (1958) 574.
7. G. E. Dieter, in *Strengthening Mechanisms in Solids*, American Society for Metals, Metals Park, OH, 1962, p. 279.
8. G. E. Dieter, in *Response of Metals to High-Velocity Deformation*, eds. P. W. Shewmon and V. F. Zackay, Interscience, New York, 1961, p. 409.
9. E. G. Zukas, *Metals Eng. Quart.*, **6** (1966), 1.
10. R. N. Orava and H. E. Otto, in *Principles and Practice of Explosive Metalworking*, Industrial Newspapers, Ltd., ed. A. A. Ezra, Chapter 11, (1973), 229.
11. W. C. Leslie, in *Metallurgical Effects at High Strain Rates*, eds. R. W. Rohde, B. M. Butcher, J. R. Holland, and C. H. Karnes, Plenum, New York, 1973, p. 571.
12. L. Davison and R. A. Graham, *Phys. Rep.*, **55** (1979), 255.
13. L. E. Murr, in *Shock Waves in Condensed Matter*, eds. S. C. Schmidt and N. C. Holmes, Elsevier, Amsterdam, 1988, p. 315.
14. L. E. Murr, in *Shock Waves for Industrial Applications*, ed. L. E. Murr, Noyes, Park Ridge, NJ, 1988, p. 60.
15. M. A. Meyers and L. E. Murr, in *Explosive Welding, Forming, and Compaction*, ed. T. Z. Blazynski, Applied Science, Elsevier, London, 1983, p. 17.
16. M. A. Meyers and L. E. Murr, in *Shock Waves and High-Strain-Rate Phenomena in Materials*, Plenum, New York, 1981, p. 487.
17. L. E. Murr and M. A. Meyers, in *Explosive Welding, Forming, and Compaction*, ed. T. Z. Blazynski, Applied Science, Elsevier, London, 1983, p. 83.
18. G. T. Gray, in *Shock-Wave and High-Strain-Rate Phenomena in Materials*, eds. M. A. Meyers, L. E. Murr, and K. P. Staudhammer, Dekker, New York, 1992, p. 899.
19. G. T. Gray III, in *High Pressure Shock Compression of Solids*, eds. J. R. Asay and M. Shahinpoor, Springer-Verlag, New York, 1993.
20. G. E. Duvall and R. A. Graham, *Rev. Modern Phys.*, **49** (1977), 523.
21. M. A. Meyers, "Thermomechanical Processing of a Nickel-Base Superalloy by Cold Rolling and Shock-Wave Deformation," Ph.D. Thesis, University of Denver, 1974.
22. M. A. Meyers and R. N. Orava, *Met. Trans.*, **8A** (1977), 1641.
23. H. J. Rack, *Met Trans.*, **7A** (1976), 1571.
24. M. A. Mogilevsky and L. A. Teplyakova, in *Metallurgical Applications of Shock-Wave and High-Strain-Rate Phenomena*, eds. L. E. Murr, K. P. Staudhammer, and M. A. Meyers, Dekker, New York, 1986, p. 419.

25. N. N. Gubareva, T. M. Sobolenko, and T. S. Teslenko, *Comb. Expl. Shock Waves*, **13** (1977), 543.
26. G. T. Gray III, in *Shock Compression of Condensed Matter*, eds. S. C. Schmidt, J. N. Johnson, and L. W. Davison, Elsevier, Amsterdam, 1989, p. 407.
27. D. Kuhlmann-Wilsdorf, *Trans. AIME*, **224** (1962), 1047.
28. L. E. Murr and D. Kuhlmann-Wilsdorf, *Acta Met.*, **26** (1978), 847.
29. E. T. Marsh and D. E. Mikkola, *Scripta Met.*, **10** (1976), 851.
30. J. A. Brusso, R. N. Wright, and D. E. Mikkola, in *Metallurgical Applications of Shock-Wave and High-Strain-Rate Phenomena*, eds. L. E. Murr, K. P. Staudhammer, and M. A. Meyers, Dekker, New York, 1986, p. 403.
31. A. G. Dhere, H. J. Kestenbach, and M. A. Meyers, *Mater. Sci. Eng.*, **54** (1982), 113.
32. W. C. Leslie, J. T. Michalak, and F. W. Aul, in *Iron and Its Dilute Solid Solutions*, Wiley, New York, 1963, p. 119.
33. M. K. Koul and J. F. Breedis, in *The Science, Technology, and Application of Titanium*, eds. R. I. Jaffee and N. E. Promisel, Pergamon, Oxford, 1978, p. 817.
34. J. Galbraith and L. E. Murr, *J. Mater. Sci.*, **10** (1975), 2025.
35. G. T. Gray III, R. S. Hixson, and C. E. Morris, in *Shock Compression of Condensed Matter 1991*, eds. S. C. Schmidt and R. D. Tasker, Elsevier, 1992, p. 427.
36. E. Hornbogen, *Acta Met.*, **10** (1962), 978.
37. E. Talia, L. Fernandez, V. K. Sethi, and R. Gibala, in *Strength of Metals and Alloys*, eds. P. Haasen, V. Gerals, and G. Kestorz, Pergamon, New York, 1979, p. 127.
38. M. A. Meyers, in *Strength of Metals and Alloys*, eds. P. Haasen, V. Gerals, and G. Kostorz, Pergamon, New York, 1979, p. 547.
39. R. W. Rohde and C. H. Pitt, *J. Appl. Phys.*, **38** (1967), 876.
40. J. Friedel, *Dislocations*, Addison-Wesley, New York, 1964, p. 63.
41. M. A. Meyers, *Scripta Met.*, **12** (1978), 21.
42. J. W. Taylor, *Appl. Phys.*, **34** (1963), 2727.
43. R. J. McQueen, E. G. Zukas, and S. P. Marsh, "Residual Temperature in Shock Loaded Iron," ASTM STP No. 336, American Society for Testing and Materials, Philadelphia, PA, 1962, p. 306.
44. S. Raikes and T. J. Ahrens, *Geophys. J.R. Astr. Soc.*, **58** (1979), 717.
45. J. P. Hirth and J. Lothe, *Theory of Dislocations*, McGraw-Hill, New York, 1968, p. 689.
46. K. C. Hsu, C. Y. Hsu, L. E. Murr, and M. A. Meyers, in *Shock Waves and High-Strain-Rate Phenomena in Metals*, eds. M. A. Meyers and L. E. Murr, Plenum, New York, 1981, p. 433.
47. B. Kazmi and L. E. Murr, in *Shock Waves and High-Strain-Rate Phenomena in Metals*, eds. M. A. Meyers and L. E. Murr, Plenum, New York, 1981, p. 733.
48. M. A. Mogilevsky, in *Shock Waves and High-Strain-Rate Phenomena in Metals*, eds. M. A. Meyers and L. E. Murr, Plenum, New York, 1981, p. 531.
49. M. A. Mogilevsky, *Phys. Rep.*, **97** (1983), 359.

50. M. A. Mogilevsky, in *Impact Loading and Dynamic Behavior of Materials*, eds. C. Y. Chiem, H. D. Kunze, and L. W. Meyer, DGM Informationsgesellschaft, Oberursel, Germany, 1988, p. 957.
51. M. A. Mogilevsky, in *Shock-Wave and High-Strain-Rate Phenomena in Materials*, eds. M. A. Meyers, L. E. Murr, and K. P. Staudhammer, Dekker, New York, 1992, p. 875.
52. J. Weertman, in *Shock Waves and High-Strain-Rate Phenomena in Metals*, eds. M. A. Meyers and L. E. Murr, Plenum, New York, 1981, p. 469.
53. J. Weertman and P. S. Follansbee, *Mech. Mater.*, **2** (1983), 265.
54. P. S. Follansbee and J. Weertman, *Mech. Mater.*, **1** (1982), 345.
55. J. Weertman, *Mech. Mater.*, **5** (1986), 13.
56. H. Kressel and N. Brown, *J. Appl. Phys.*, **38** (1967), 138.
57. L. E. Murr, O. T. Inal, and A. A. Morales, *Acta Met.*, **24** (1976), 261.
58. R. A. Graham, in *Shock Waves and High-Strain-Rate Phenomena in Metals*, eds. M. A. Meyers and L. E. Murr, Plenum, New York, 1981, p. 375.
59. M. A. Mogilevsky, *Comb. Expl. Shock Waves*, **6** (1970), 197.
60. J. Weertman, in *Response of Metals to High Velocity Deformation*, eds. P. G. Shewmon and V. F. Zackay, Interscience, New York, 1961, p. 205.
61. R. L. Nolder and G. Thomas, *Acta Met.*, **11** (1963), 994.
62. R. L. Nolder and G. Thomas, *Acta Met.*, **12** (1964), 227.
63. F. Greulich and L. E. Murr, *Mater. Sci. Eng.*, **39** (1979), 81.
64. R. J. DeAngelis and J. B. Cohen, *J. Metals*, **15** (1963), 681.
65. A. S. Appleton and J. S. Waddington, *Acta Met.*, **12** (1963), 681.
66. A. R. Champion and R. W. Rohde, *J. Appl. Phys.*, **41** (1970), 2213.
67. K. P. Staudhammer and L. E. Murr, Proc. 5th Intl. Conf. on High Energy Rate Fabr., University of Denver, Denver, CO, 1975, p. 1.7.1.
68. L. E. Murr and K. P. Staudhammer, *Mater. Sci. Eng.*, **20** (1975), 95.
69. G. A. Stone, R. N. Orava, G. T. Gray, and A. R. Pelton, "An Investigation of the Influence of Shock-Wave Profile on the Mechanical and Thermal Responses of Polycrystalline Iron," Final Technical Report, U.S. Army Research Office, Grant No. DAA629-76-0181, 1978, p. 30.
70. R. W. Rohde, *Acta Met.*, **17** (1969), 353.
71. J. N. Johnson and R. W. Rohde, *J. Appl. Phys.*, **42** (1971), 4171.
72. R. W. Rohde, W. C. Leslie, and R. C. Glenn, *Metall. Trans.*, **3A** (1972), 323.
73. S. Mahajan, *Phys. Stat. Sol.*, **33** (1969), 291.
74. K. Wongwiwat and L. E. Murr, *Mater. Sci. Eng.*, **35** (1978), 273.
75. M. J. Marcinkowski and H. A. Lipsitt, *Acta Met.*, **10** (1962), 951.
76. D. Hull, *Acta Met.*, **9** (1961), 191.
77. D. Peckner, *The Strengthening of Metals*, Reinhold, New York, 1964, p. 49.
78. G. Thomas, *Met. Trans.*, **2** (1971), 2373.
79. G. Thomas, University of California, Berkeley, private communications, 1978.
80. A. H. Cottrell and B. A. Bilby, *Phil. Mag.*, **42** (1951), 573.
81. J. A. Venables, *Phil. Mag.*, **6** (1961), 379.

82. R. F. Bunshah, in *Deformation Twinning*, eds. R. E. Reed Hill, J. P. Hirth and H. C. Rogers, Gordon and Breach, New York, 1964, p. 390.
83. C. N. Reid, G. T. Hahn, and A. Gilbert, R. W. Rohde et al., *Metall. Trans.*, **3A** (1972), 386.
84. T. Takeuchi, *J. Phys. Soc. Jpn.*, **21** (1966), 2616.
85. R. M. Wright, D. E. Mikkola, and S. LaRouche, in *Shock Waves and High-Strain-Rate Phenomena in Metals*, eds. M. A. Meyers, and L. E. Murr, Plenum, New York, 1981, p. 703.
86. J. B. Cohen and J. Weertman, *Acta Met.*, **11** (1963), 997, 1368.
87. J. P. Hirth and J. Lothe, *Theory of Dislocations*, McGraw-Hill, New York, 1968, p. 750.
88. E. Orowan, *Dislocations in Metals*, AIME, New York, 1954, p. 116.
89. G. E. Duvall and R. A. Graham, *Rev. Modern Phys.*, **49** (1977), 523.
90. J. R. Patel and M. Cohen, *Acta Met.*, **1** (1953), 531.
91. R. W. Rohde, J. R. Holland, and R. A. Graham, *Trans. Met. Soc. AIME*, **242** (1968), 2017.
92. M. A. Meyers and J. R. C. Guimarães, *Mater. Sci. Eng.*, **24** (1976), 289.
93. M. A. Meyers, *Met. Trans.*, **10A** (1979), 1723.
94. Y. Sano, S. N. Chang, M. A. Meyers, and S. Nemat-Nasser, *Acta Met. Mat.*, **40** (1992), 413.
95. N. N. Thadhani and M. A. Meyers, *Acta Met.*, **34** (1986), 1625.
96. S. N. Chang and M. A. Meyers, *Acta Met.*, **36** (1988), 1085.
97. M. A. Meyers, N. N. Thadhani, D. C. Erlich, and P. S. Decarli, in *Shock Waves in Condensed Matter—1983*, eds. J. R. Asay, R. A. Graham, and G. K. Stromb, North-Holland, Amsterdam, 1984, p. 411.
98. S. N. Chang, M. A. Meyers, N. N. Thadhani, and D. C. Erlich, in *Shock Waves in Condensed Matter 1987*, eds. S. C. Schmidt and N. C. Holmes, North-Holland, Amsterdam, 1988, p. 143.
99. N. N. Thadhani, M. A. Meyers, and D. C. Erlich, *J. Appl. Phys.*, **58** (1985), 2791.
100. J. R. C. Guimarães, J. C. Gomes, and M. A. Meyers, *Suppl. to Trans.*, J.I.M., (1946), 1741.
101. M. A. Meyers, Proc. Fifth Intl. Conf. on High Energy Rate Fabrication, University of Denver, Colorado, June 1975, p. 1.4.1.
102. M. A. Meyers and M. S. Carvalho, *Mater. Sci. Eng.*, **24** (1976), 131.
103. M. A. Meyers, *Mater. Sci. Eng.*, **30** (1977), 99.
104. E. Hornbogen, *Trans. AIME*, **221** (1961), 712.
105. G. Das and S. V. Radcliffe, *Phil. Mag.*, **20** (1969), 589.
106. W. C. Leslie, D. W. Stevens, and M. Cohen, in *High-Strength Materials*, ed. V. F. Zackay, Wiley, New York, 1965, p. 382.
107. F. Braga, H. J. Kestenbach, and M. A. Meyers, The Effect of Polycrystallinity on the Shock Response of Copper, Military Institute of Engineering, Rio de Janeiro, Brazil, unpublished results, 1978.
108. R. J. De Angelis and J. B. Cohen, *J. Metals*, **15** (1963), 681.

109. M. A. Meyers, L. E. Murr, C. Y. Hsu, and G. A. Stone, *Mater. Sci. Eng.*, **57** (1983), 113.
110. M. A. Meyers, K.-C. Hsu, and H. Couch-Robino, *Mater. Sci. Eng.*, **59** (1983), 235.
111. M. A. Meyers, *Scripta Met.*, **9** (1975), 667.
112. G. T. Gray and C. E. Morris, *J. Phys. IV, Colloque C3*, Suppl. to *J. Phys. III* (1991), C3-191.
113. M. A. Meyers, *Met. Trans.*, **8A** (1977), 1581.
114. M. A. Meyers, H.-J. Kestenbach, and C. A. Soares, *Mater. Sci. Eng.*, **45** (1980), 143.
115. J. T. Fourie, P. J. Jackson, D. Kuhlmann-Wilsdorf, D. A. Rigney, J. H. van der Merwe, and H. G. F. Wilsdorf, *Scr. Met.*, **16** (1982), 157.
116. P. Charsley and D. Kuhlmann-Wilsdorf, *Phil. Mag.*, **44** (1981), 1351.
117. C. S. Yust and L. A. Harris, in *Shock Waves and High-Strain-Rate Phenomena in Metals*, eds. M. A. Meyers and L. E. Murr, Plenum, New York, 1981, p. 881.
118. L. H. Leme and M. A. Meyers, *J. Mater. Sci.*, **24** (1989), 2516.
119. L. H. Leme Louro and M. A. Meyers, unpublished results.
120. Y. Wang and D. E. Mikkola, in "Shock-Wave and High-Strain-Rate Phenomena in Materials," eds. M. A. Meyers, L. E. Murr, and K. P. Standhammer, Dekker, New York, 1992, p. 1031.

CHAPTER 15

Shear Bands (Thermoplastic Shear Instabilities)

15.1 QUALITATIVE DESCRIPTION

Shear bands are regions where plastic deformation in a material is highly concentrated. The formation of these shear bands is extremely important in dynamic deformation of materials because they often are precursors to fracture. Thus, they have been the object of considerable study. There are comprehensive reviews [1–7] and in-depth studies, and this chapter will try to focus on the essential aspects that a student should understand. The more studious person should read the reviews by Rogers [1], Timothy [2], and Dormeval [4]. The book edited by Mescall and Weiss [3] contains several articles on adiabatic shear bands that can enhance the student's knowledge on the microstructural aspects of shear instabilities.*

Figure 15.1 illustrates the principal dynamic deformation events in which shear bands play a role. In the penetration of a target by a projectile, the formation of a shear concentration [as marked in Fig. 15.1(a)] alters the defeat mechanism. If no shear bands are formed, one has a more well distributed plastic deformation. The formation of the shear bands establishes the shear failure path and is responsible for the clean "plugging" of the target. In a similar manner, the projectile is fragmented by means of shear cracks preceded by shear bands [Fig. 15.1(b)]. In the expansion of hollow cylinders by explosives, the shear stresses can generate shear bands, which in turn determine the fragmentation pattern. Thus, the formation of the shear bands [Fig. 15.1(c)] can determine the size and distribution of fragments.

Figure 15.2 illustrates the importance of shear band formation in fabrication processes. They can form in forging operations [Fig. 15.2(a)] and introduce regions of weakness and internal flaws in forged products. Titanium forgings are susceptible to shear band formation. In rolling [Fig. 15.2(b)] regions of concentrated plastic deformation can be formed at high plastic strains, with effects on the final properties. The machining of metals is significantly affected

*The book by Y. Bai and B. Dodd (Adiabatic Shear Localization, Pergamon, 1992) and the special issue of *Mechanics of Materials* (Vol. 17, 1994, pp. 83–328) are very complete sources of information.

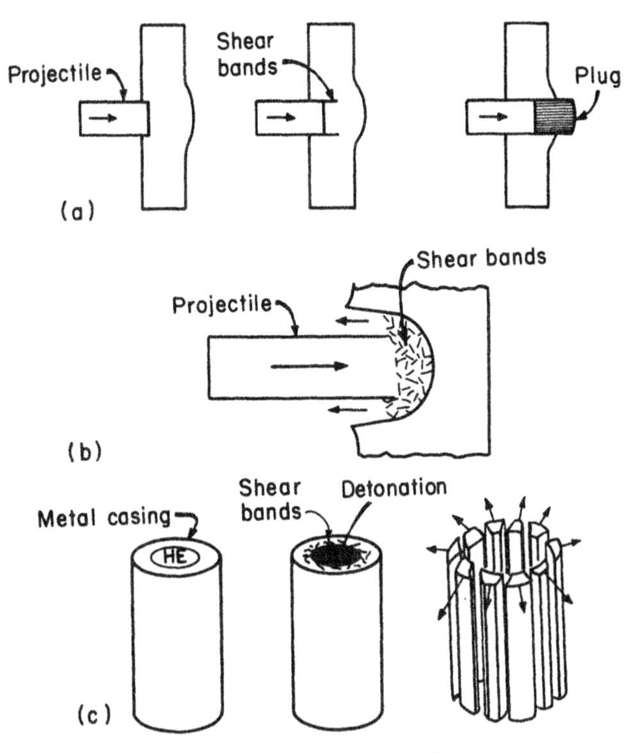

FIGURE 15.1 Formation of shear bands in dynamic deformation events (military applications): (a) defeat of armor by plugging; (b) shear bands breaking up projectile; (c) shear bands determining fracture in exploding cylinders.

by shear band formation; shear bands are responsible for the breakup of the chips formed by the cutting tool. This is a very positive aspect of shear band formation, because it enables higher machining rates. In punching and shearing operations in steels, shear bands are often formed and are responsible for smooth, clean cuts.

A significant number of metals, alloys, and polymers exhibit shear bands. The conditions under which they occur and their microstructures vary widely. Figure 15.3 illustrates shear localization obtained in steels, nickel, and titanium alloys by ballistic impact at velocities in the 200–800-m/s range. Upon microscopic observation these bands take on a variety of morphologies that are dependent upon the initial microstructure of the material and the thermomechanical excursion undergone by it during deformation. Photomicrographs (c) and (d) do not show clear shear bands. Only shear localization is seen, as a result of the imposed strain concentration at the projectile edges. Figure 15.4 shows two different shear bands, in a titanium alloy and in steel. The shear band in the titanium alloy is seen as a clear band, whereas the one in steel corresponds to a region of intense localized deformation. These microstructural differences will be explained in Section 15.4.

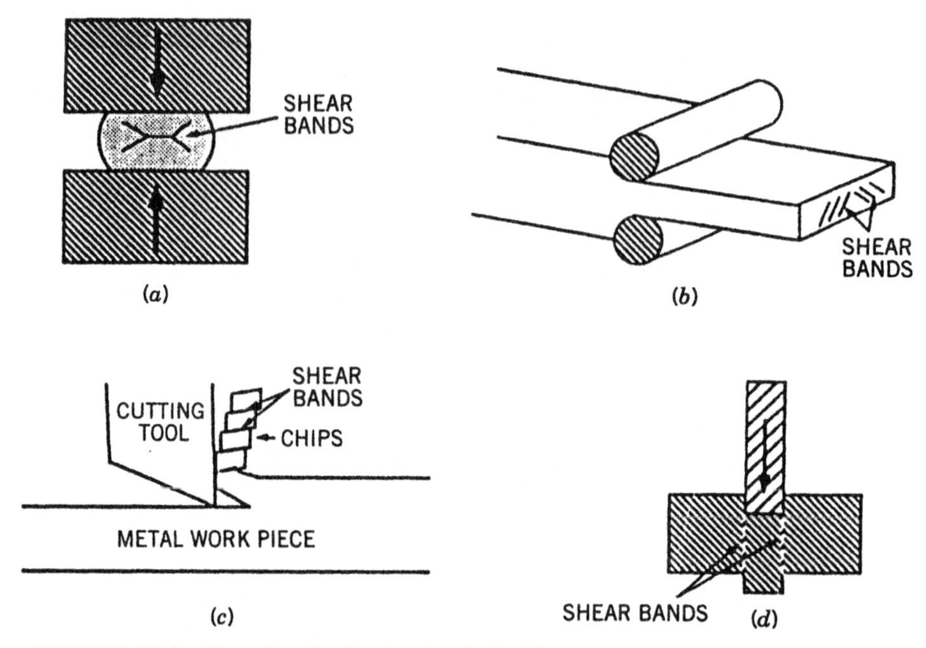

FIGURE 15.2 Shear localization in plastic-forming processes: (a) upset forging; (b) rolling; (c) machining; (d) punching and shearing.

A simple rationale for the formation of shear bands is presented below. As early as 1943, Zener and Hollomon [8] presented a simple explanation, that still holds: "When the material undergoes an element of strain, dE, adiabatically, the stress is raised by the strain hardening, and lowered by the associated rise in temperature." They attributed the formation of adiabatic shear bands to the thermal softening overcoming work (or strain) hardening. The various theories developed to explain and predict shear localization are based on the same precepts. Additional considerations that have been added are strain rate hardening (as seen in Chapter 13, the flow stress is stain rate dependent), heat conduction out of the band, and geometric softening. Thus, we have the following effects:

Opposing factors: strain hardening, strain rate hardening
Favoring factors: geometric softening, thermal softening

Figure 15.5 provides a simple explanation for the mechanics of shear band formation. A parallelepiped is being homogeneously sheared by τ [Fig. 15.5(a)]. At a certain strain γ_c, deformation localizes in a band, as shown in Figure 15.5(b). Figure 15.5(c) shows the progression of strain (or temperature) as a function of the increasing applied stress τ. The strain is initially homogeneous throughout the specimen: γ_0, γ_1, γ_2. At γ_c a small fluctuation appears that is

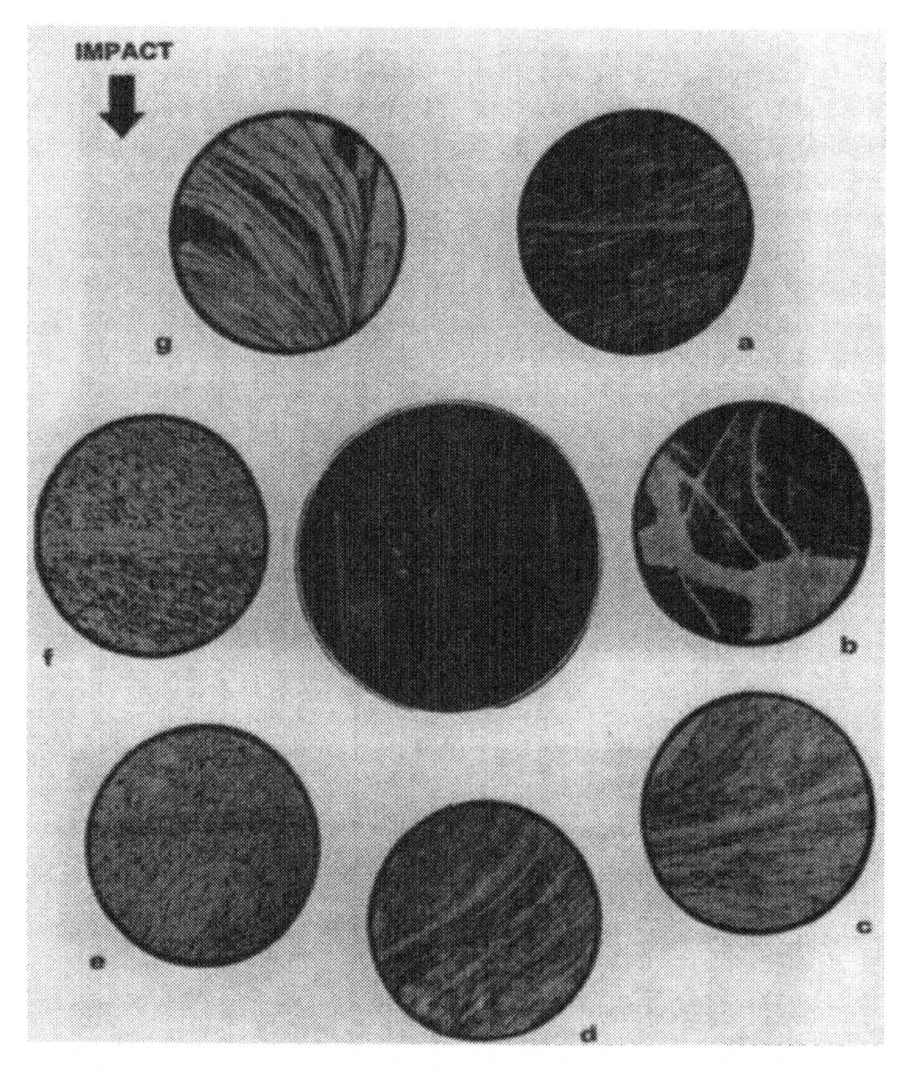

FIGURE 15.3 Shear localization in a number of materials produced by ballistic impact: (a) AISI 1040 steel; (b) 1090 steel; (c) Ni (annealed); (d) Ni (rolled); (e) Ti; (f) Ti_6Al_4 V; (g) AISI 1526 (From H. A. Grebe, M.Sc. Thesis, New Mexico Institute of Technology, Socorro, New Mexico 1984.)

accentuated as the applied stress is further increased: γ_4, γ_5, γ_6, γ_7. This localization results from softening dominating hardening. A generic stress–strain curve in Figure 15.5(d) shows that softening dominates beyond γ_c, the strain at which the stress is maximum. The results obtained by Marchand and Duffy [9] fully confirm the above schematic description. They subjected steel specimens to dynamic shear deformation at a strain rate of approximately 10^3 s^{-1} in a torsional Hopkinson bar. The specimens had a grid of parallel lines

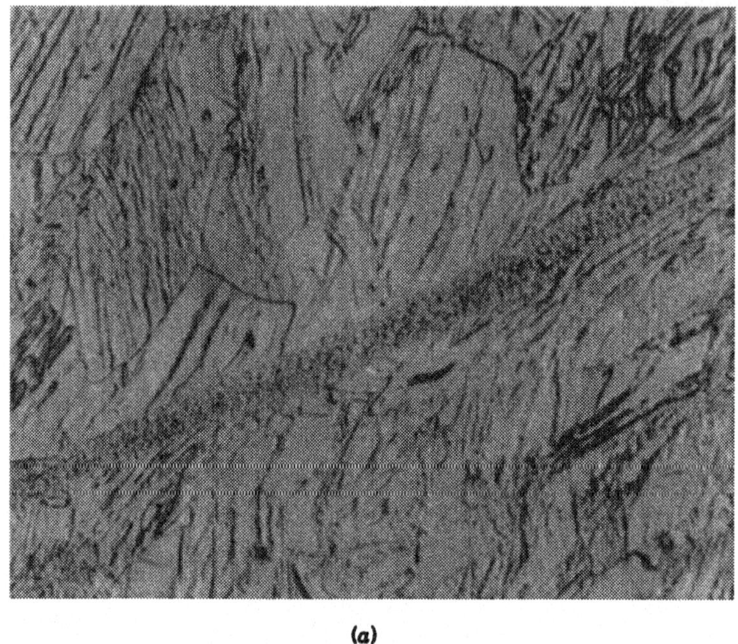

(a)

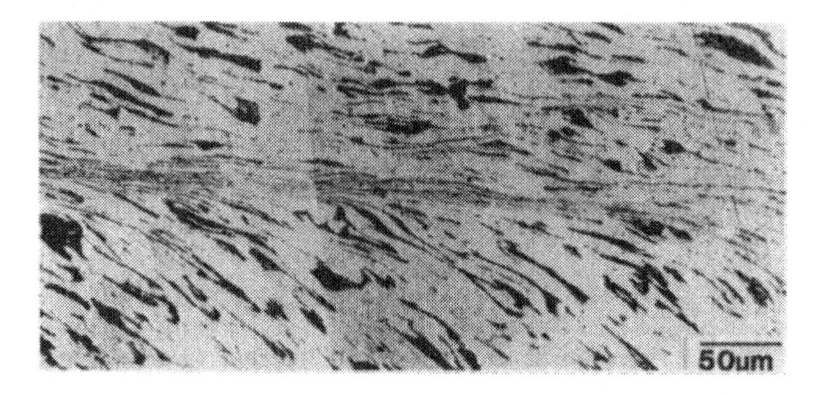

(b)

FIGURE 15.4 Shear band in (a) titanium alloy and (b) 1020 steel (notice deformation of pearlite).

etched onto them. High-speed photography conducted simultaneously with the shear deformation revealed that the onset of localization, clearly seen by the formation of an inflection in the parallel lines, is close to the maximum in the stress–strain curve. One can see this by observing the sequence of photographs in Figure 15.6. The localization rapidly increases in severity until total fracture occurs. At the same time, the strain within the shear band rapidly rises. This

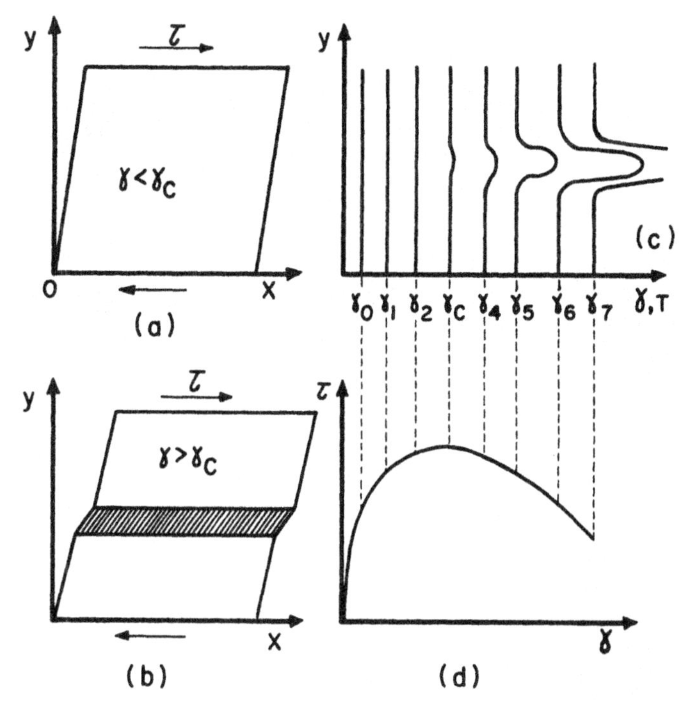

FIGURE 15.5 Formation of shear instability when a parallelepiped is sheared by stress τ in an adiabatic mode: (a) homogeneous shear of parallelepiped; (b) shear instability in deformed parallelepiped; (c) strain (or temperature) profiles (notice localization forming at γ_c; (d) adiabatic stress–strain curve with critical strain γ_c.

localization starts at a shear strain of approximately 0.4. The strain in the localized region is also plotted in the figure. It rapidly rises to 1400%, or 14, whereas specimen failure takes place at $\varepsilon \sim 0.5$.

The profound effects of adiabatic shear band formation in the armor–antiarmor situations alluded above will be illustrated now. The mechanism of armor perforation is altered by the localized shear, as shown in Figure 15.7. Three different conditions of the same AISI 8620 steel are shown. One can see the deformation of the armor, its spalling at the back surface, and the formation of the shear band in (a). These effects are marked in the photographs. Although the deformation is approximately the same for the three conditions, condition (a) is close to its ballistic limit. The shear bands have penetrated deeply into the target, and a modest increase in projectile velocity will result in perforation by a plugging mode. On the other hand, armor plates (b) and (c) will resist penetration by plastic deformation of a homogeneous nature. Thus, an increase in the hardness of the armor plate might result in the decrease of its ballistic limit (velocity at which perforation occurs).

The defeat of the projectile occurs by the same mechanism. Under the high shear stresses imparted by impact, it may undergo localized shear and break

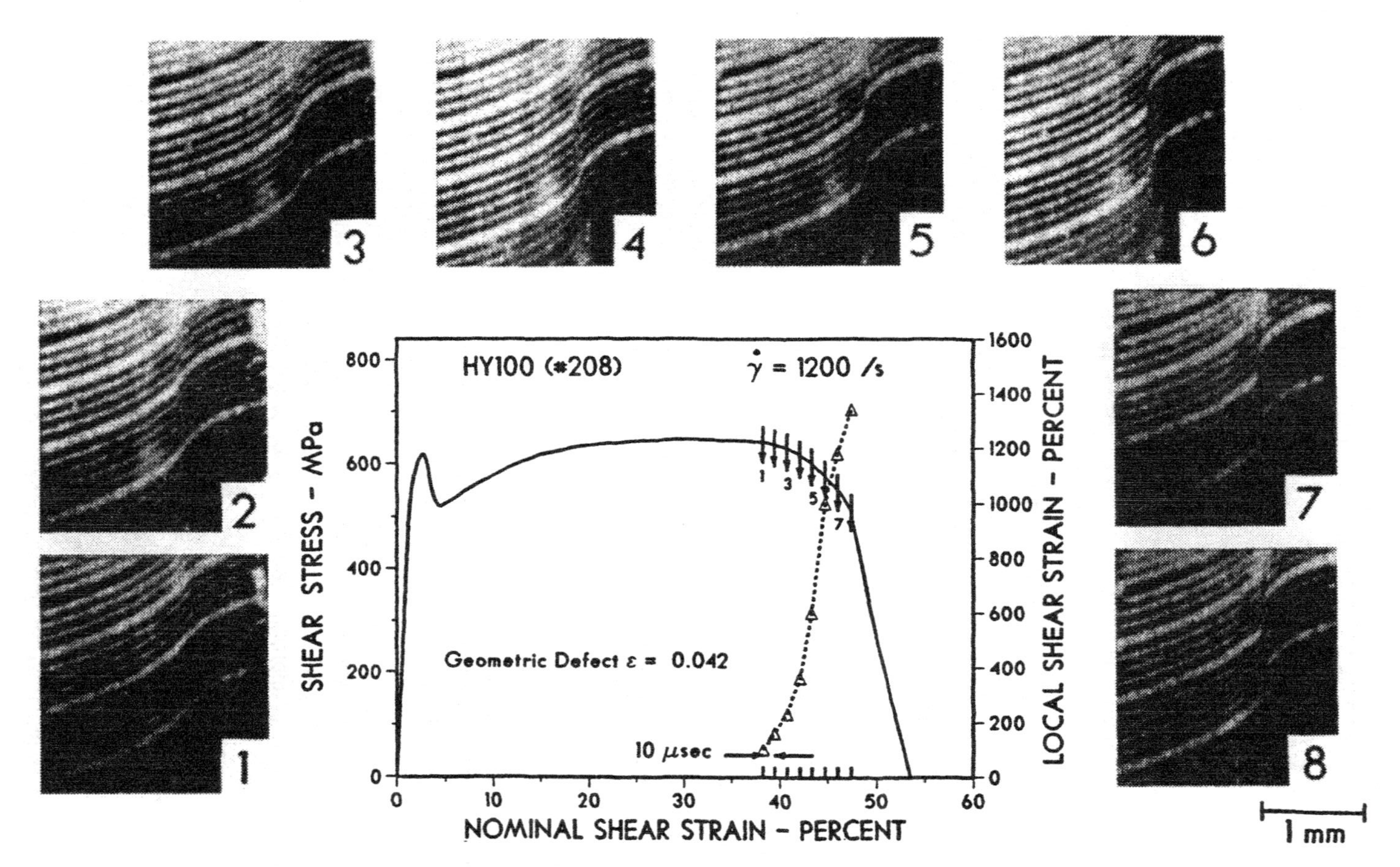

FIGURE 15.6 High-speed photographs of the grid pattern taken during the formation of a shear band in HY-100 steel. The stress–strain curve shows the strain at which the photographs

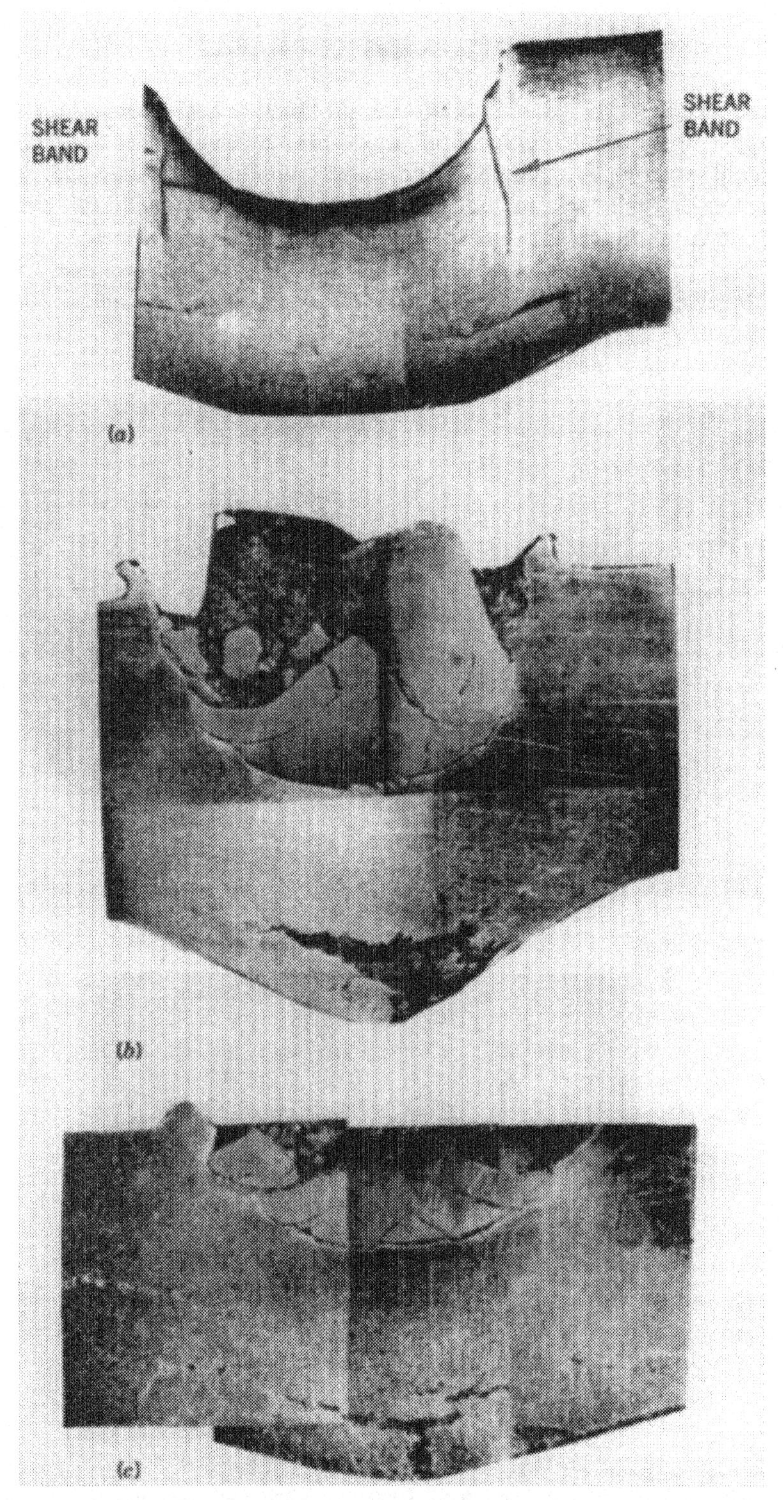

FIGURE 15.7 Effect of predisposition to shear band formation on the defect mechanism of AISI 8620 steel armor; (a) quenched and tempered; (b) normalized; (c) work-hardened condition. Impacts between 700 and 1000 ms. (From C. Wittman, M.Sc. Thesis, New Mexico Institute of Mining and Technology, 1986.)

apart gradually. This is particularly true in kinetic energy penetrators, and the example illustrated in Figure 15.8 shows this mechanism in operation. This projectile was initially cylindrical. In contact with the armor and under the high shear stresses with superimposed hydrostatic stresses, shear instabilities form. They break up projectiles into small fragments. One can also see a profusion of cracks produced most probably by tension when the tensile waves return to the projectile tip. This subject is treated in greater detail in Chapter 16. Figure

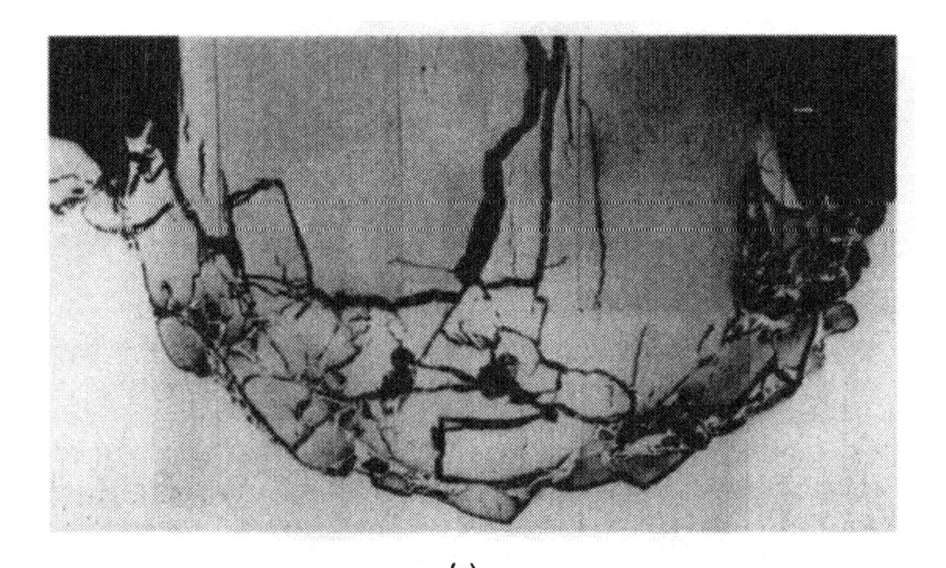

(a)

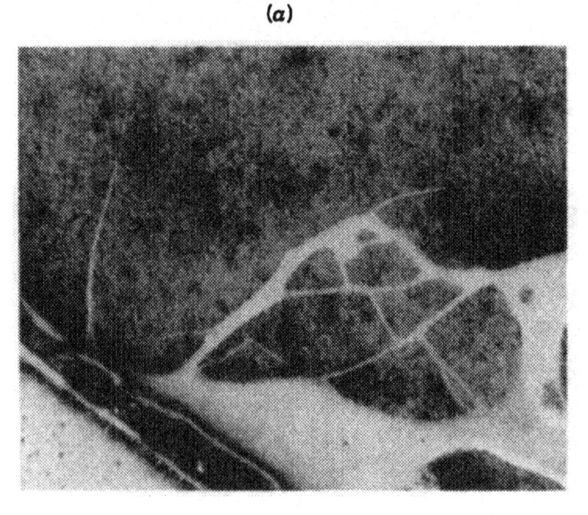

(b)

FIGURE 15.8 (a) Breakup of AISI 1090 quenched and tempered projectile as it penetrates targets; projectile was initially cylindrical; (b) detailed view of white shear bands.

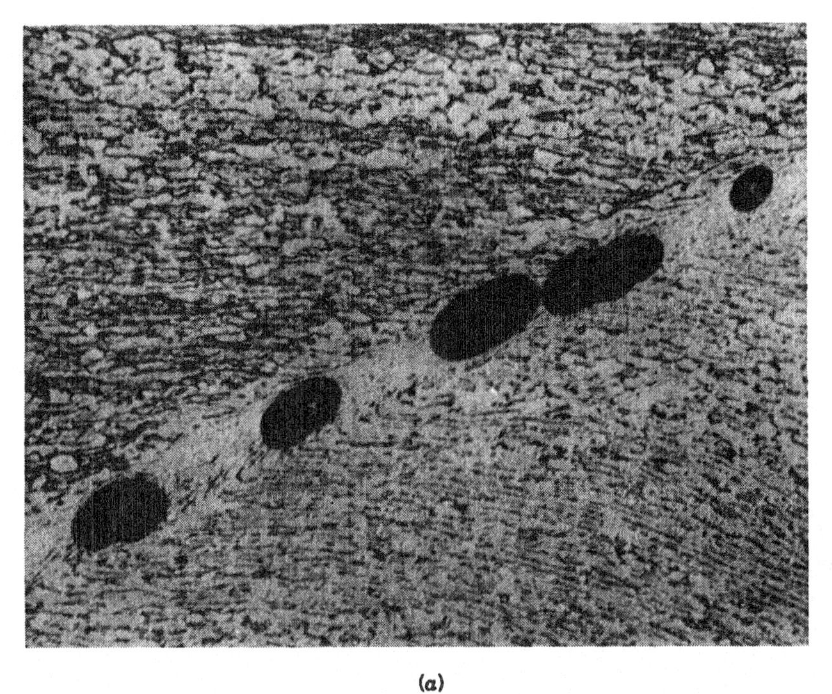

(a)

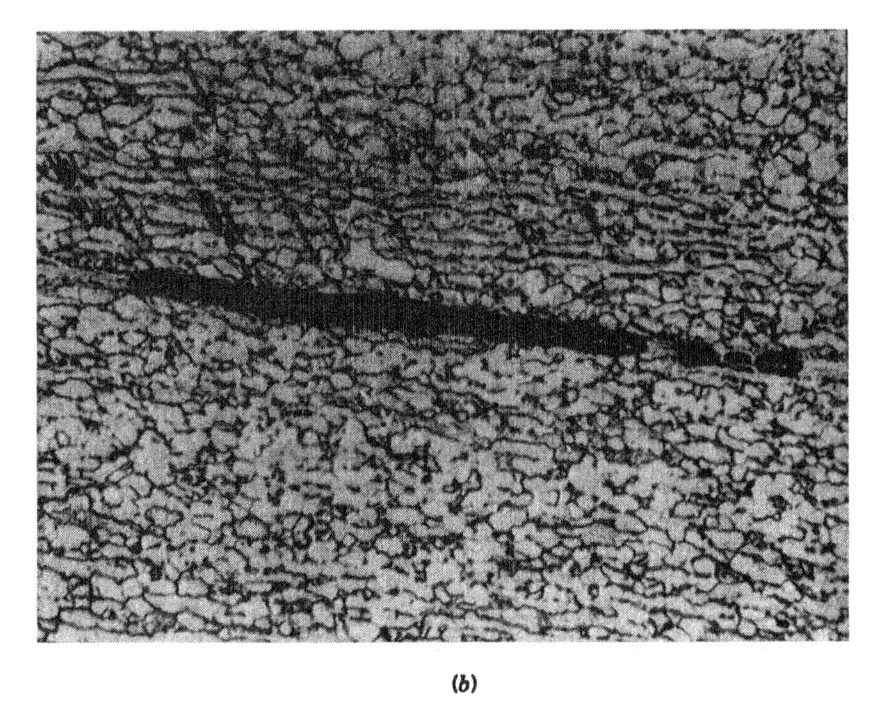

(b)

FIGURE 15.9 (a) Voids and (b) cracks forming along shear band due to tension in titanium.

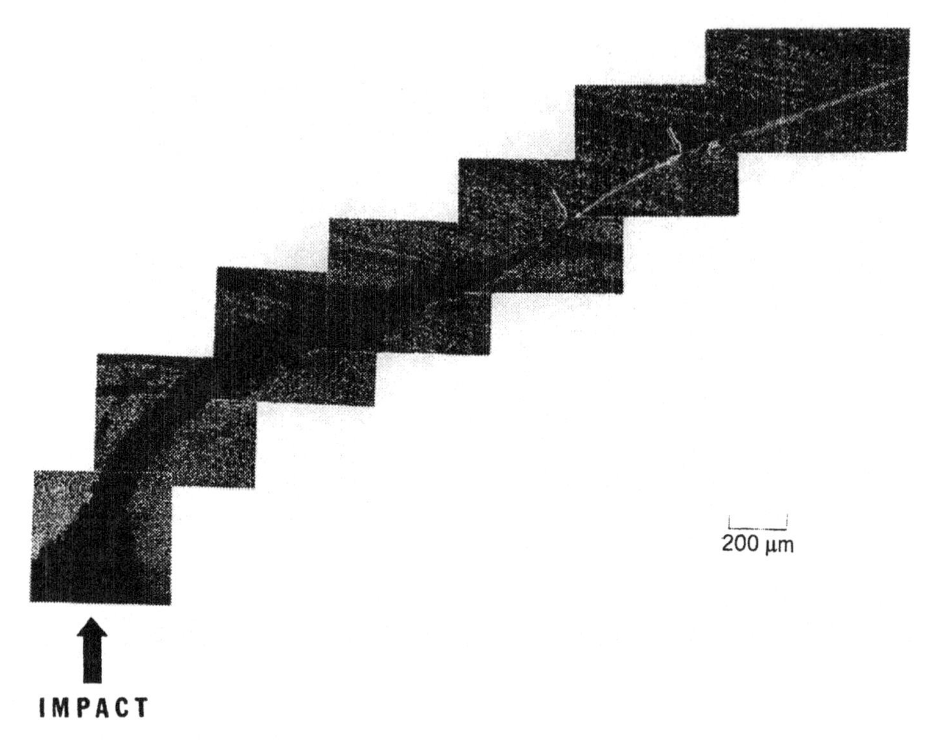

FIGURE 15.10 Void nucleation and coalescence at shear band leading to fracture in titanium.

15.8(b) shows a detail of the region close to the target. The white shear band regions form a pattern that eventually breaks up the projectile. One should realize that these are photographs of recovered specimens and that, during the penetration event, the white regions are much softer than the surrounding material.

The formation of voids and their coalescence into cracks along shear bands is illustrated in Figure 15.8. These voids are produced by tension due to residual stresses or tensile reflections. Since the flow stress of the material within the band is much lower than in the surrounding material, voids will form there; their diameter is equal to the thickness of the band [Fig. 15.9(a)]. In Figure 15.9(b) these voids have coalesced and are producing a crack. In Figure 15.10 the crack is fully opened. This example illustrates the importance of shear bands in fracture.

15.2 CONSTITUTIVE MODELS: ELEMENTARY

In 1943 Zener and Hollomon [8] correctly identified the reason for the formation of an adiabatic shear band: the competition between thermal softening and work hardening. Recht [10] expressed this quantitatively. He expressed the following relationship differentially:

$$\tau = f(T, \gamma)$$

$$d\tau = \left(\frac{\partial \tau}{\partial T}\right)_\gamma dT + \left(\frac{\partial \tau}{\partial \gamma}\right)_T d\gamma$$

$$\frac{d\tau}{d\gamma} = \left(\frac{\partial \tau}{\partial T}\right)_\gamma \left(\frac{dT}{d\gamma}\right) + \left(\frac{\partial \tau}{\partial \gamma}\right)_T$$

An adiabatic shear band can form when the material starts to "soften" (see Fig. 15.5d). This is mathematically expressed as:

$$\frac{d\tau}{d\gamma} \leq 0$$

$$\left(\frac{\partial \tau}{\partial \gamma}\right)_T = -\left(\frac{\partial \tau}{\partial T}\right)_\gamma \left(\frac{dT}{d\gamma}\right) \tag{15.1}$$

One can introduce the strain rate dependence and create a more general function:

$$\tau = f(\gamma, \dot{\gamma}, T)$$

$$d\tau = \left(\frac{\partial \tau}{\partial \gamma}\right)_{\dot{\gamma},T} d\gamma + \left(\frac{\partial \tau}{\partial \dot{\gamma}}\right)_{\gamma,T} d\dot{\gamma} + \left(\frac{\partial \tau}{\partial T}\right)_{\gamma,\dot{\gamma}} dT$$

$$\frac{d\tau}{d\gamma} = \left(\frac{\partial \tau}{\partial \gamma}\right)_{\dot{\gamma},T} + \left(\frac{\partial \tau}{\partial \dot{\gamma}}\right)_{\gamma,T} \frac{d\dot{\gamma}}{d\gamma} + \left(\frac{\partial \tau}{\partial T}\right)_{\gamma,\dot{\gamma}} \frac{dT}{d\gamma} \tag{15.2}$$

The condition for instability is $d\tau/d\gamma = 0$. If the experiment is conducted at a constant strain rate, $d\dot{\gamma} = 0$ and we are reduced to Eqn. (15.1).

Zener and Hollomon [8] assumed the process to be adiabatic and simply computed the adiabatic temperature rise in a material by converting the work of deformation into a temperature increase through the material heat capacity and density. Figure 15.11 illustrates the effect of incorporating the deformation energy into the temperature rise. Shear stress–shear strain curves for commercially pure titanium are shown up to 1000 K in 100-K intervals. These relationships are assumed to be linear as a first approximation. By starting at ambient temperature and converting 90% of the work of deformation into heat, one obtains the adiabatic curve. Whereas the isothermal curves show hardening, the adiabatic curve goes through a maximum (stress: 280 MPa, strain 1) and then decreases steadily. At this maximum, the temperature is approximately 400 K.

By having analytical expressions for $(\partial\tau/\partial\gamma)_T$ and $(\partial\tau/\partial T)_\gamma$ in Eqn. (15.1), one can obtain the instability strain, stress, strain, and temperature under adi-

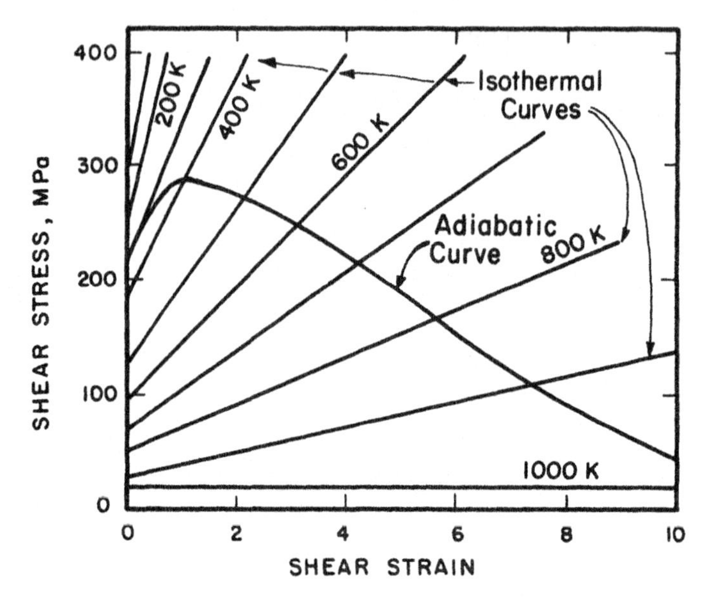

FIGURE 15.11 Isothermal (straight lines) shear stress–strain response of commercial purity titanium between 100 and 1000 K; adiabatic shear stress–shear strain curve showing maximum at $\gamma = 1.0$. (Reprinted from *Acta Metall.*, vol. 34, M. A. Meyers and H.-R. Pak, Fig. 6, p. 2496, Copyright 1986, with permission from Pergamon Press Ltd.)

abatic conditions. The power law is often used for isothermal work hardening:

$$\tau = A + B\gamma^n \tag{15.3}$$

where A, B, and n are temperature dependent; however, they will be assumed to be constant.

The increase in temperature generated by a shear strain increment $d\gamma$ is obtained by converting the deformation energy (per unit volume) into a temperature through the heat capacity and density:

$$dW = \tau \, d\gamma$$

$$dT = \frac{\beta}{\rho C_V} dW$$

$$dT = \frac{\beta}{\rho C_V} \tau \, d\gamma$$

$$\frac{dT}{d\gamma} = \frac{\beta}{\rho C_V} \tau = \frac{\beta}{\rho C_V} (A + B\gamma^n) \tag{15.4}$$

Integration yields

$$T = \frac{\beta}{\rho C_V} \int_0^\gamma \tau \, d\gamma \tag{15.5}$$

This assumes that C_V is independent of temperature in the regime investigated; β is the efficiency of the conversion of work into heat and is experimentally found to be 0.9–1 (see Section 13.6). The reader is referred to Chapter 13 (Section 13.2) for a more complete description of the constitutive behavior of metals.

The thermal softening component can be conveniently expressed by the linear relationship

$$\tau_T = \tau_{T0} \frac{T_m - T}{T_m - T_0} \tag{15.6}$$

The stress τ_T at temperatures T decreases linearly from τ_{T0}, at the initial temperature to the melting point T_m. Substituting Eqn. (15.3) into (15.6) yields

$$\tau_T = (A + B\gamma^n) \frac{T_m - T}{T_m - T_0}$$

$$\partial\tau = \frac{-(A + B\gamma^n)}{T_m - T_0} \partial T \quad \text{(constant } \gamma) \tag{15.7}$$

We can now substitute (15.4) and (15.7) into Eqn. (15.1):

$$\frac{d\tau}{d\gamma} = \left[-\frac{(A + B\gamma^n)}{T_m - T_0} \right] \left[\frac{\beta}{\rho C_V} (A + B\gamma^n) \right] \tag{15.8}$$

Equation (15.8) allows the determination of the critical strain γ_c at which instability sets in. The results shown in Figure 15.12 compare predictions with experimentally obtained values. One clearly sees that metals such as Cu, Ni, and Armco iron are very resistant to shear instabilities by virtue of their work hardening, whereas hard (quenched-and-tempered) steels are very susceptible to shear band formation. The correlation between experimental and calculated values is satisfactory. A weakness of the above analysis is that B is temperature dependent but was considered constant.

Similar analyses have been performed by Culver [13], Staker [14], and others.

15.3 CONSTITUTIVE MODELS: ADVANCED

The constitutive equations developed in Section 15.2 can predict the critical strain and temperature for the onset of shear localization. However, the complexity of the phenomenon is such that more comprehensive formulations are necessary. Figure 15.13 shows a stress–strain curve (idealized). At point I the curve reaches its maximum. This is the instability point. Homogeneous deformation will continue to occur in specimen beyond I if no perturbation is introduced. Mathematically, a perturbation in stress, strain, or temperature can be

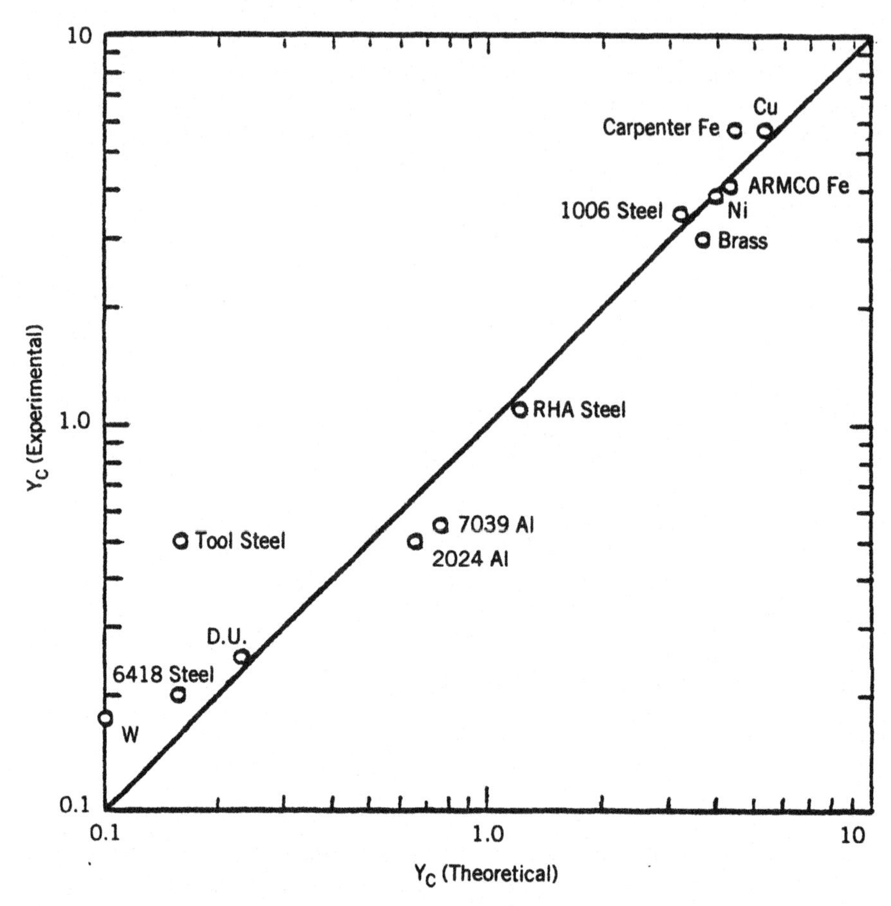

FIGURE 15.12 Correlation of experimental vs. theoretical value for critical strain. (From Lindholm and Johnson [12], Fig. 6, p. 72. Reprinted with permission of the publisher.)

incorporated in the calculation. The introduction of the perturbation at P leads to the dotted curve. At a certain strain, the stress starts to drop suddenly; this is the onset of localization. Thus, localization, not instability, marks the formation of a shear band. The temperature and strain rate are also plotted as a function of strain in Figure 15.13. The onset of localization corresponds to a drastic rise in both strain rate and temperature.

The rate at which the localization develops and its steady-state thickness are important characteristics. Heat conduction from the band to the surroundings plays an important role. In order to illustrate this, we reproduce results from an analysis by Wright [15], which predicts a decrease in the strain at the onset of severe localization as a function of strain rate (Fig. 15.14). This strain decreases from 0.8 to 0.2 as the strain rate increases from 10^0 to 10^4 s^{-1}. At higher strain rates, inertial factors dominate and the strain shows an increase with strain rate. This can be qualitatively understood if one considers that the

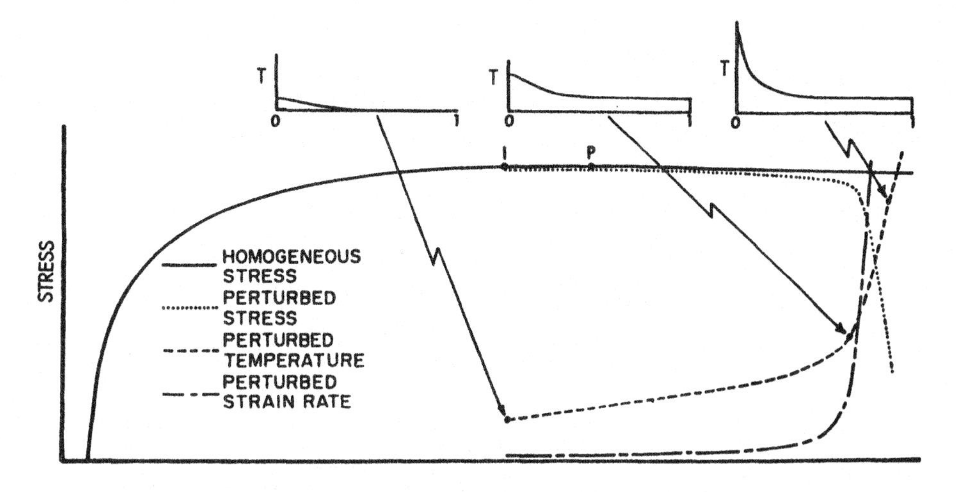

FIGURE 15.13 Sketch showing the evolution of stress, maximum temperature, and maximum strain rate; small temperature perturbation is introduced just before peak stress. (Reprinted from *J. Mech. Phys. Sol.*, vol. 35, T. W. Wright and J. W. Walter, Fig. 1, p. 702, Copyright 1987, with permission from Pergamon Press Ltd.)

time for heat conduction decreases as the strain rate increases, leading to higher temperatures (at the same strain) at higher strain rates. Another important prediction from a more advanced analysis is the thickness of the shear band. Bai et al. [16] obtained the following approximate equation for δ, the half-width of the shear band:

$$\delta \cong \left(\frac{\lambda T}{\tau \dot{\gamma}}\right)^{1/2} \tag{15.9}$$

where λ is the thermal conductivity and T, τ, and $\dot{\gamma}$ are the temperature, stress, and strain rate inside the shear band.

Wright and Walter [17] illustrate the evolution of temperature, strain rate, and stress at and after instability in Figure 15.13. If one homogeneously deforms the material, one proceeds along the full line for large strains beyond the instability ($d\tau/d\gamma = 0$) strain. The introduction of a temperature perturbation initiates a localization process, with the acceleration of the stress drop (dotted curve) and the associated temperature and strain rate increases. Thus, the perturbation *triggers* the formation of the localization after instability. The nature and extent of the temperature and strain rate increases can be mathematically calculated. We will present below the framework for this calculation, which uses the equations for the conservation of mass, momentum, and energy from mechanics and an appropriate constitutive equation for the material. This analysis has been presented, with minor differences, by Clifton [18], Bai [19],

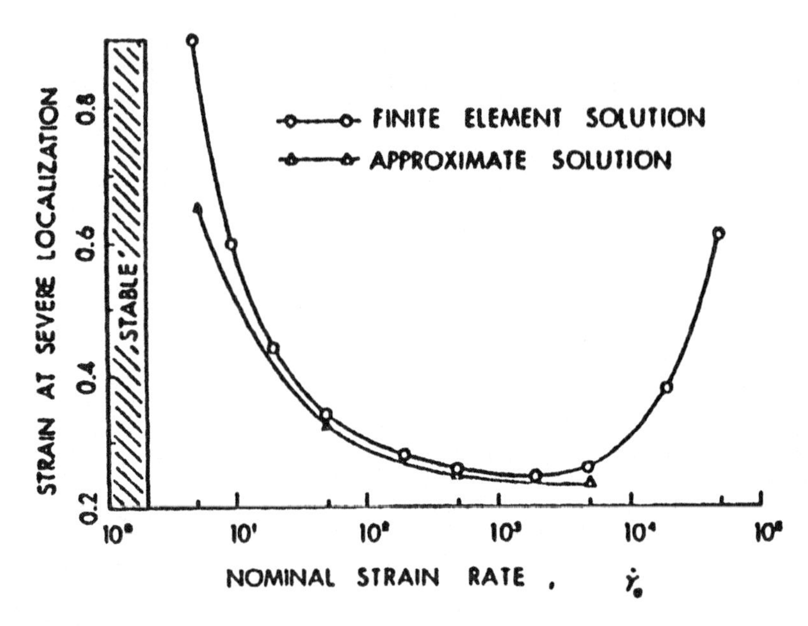

FIGURE 15.14 Strain at severe localization for the initial temperature defect $\vartheta_0 = 0.1(1 - y^2)^9 e^{-5y^2}$ as computed by finite elements (Wright and Walter [17]) and by the approximate method. (Reprinted from *J. Mech. Phys. Sol.*, vol. 35, T. W. Wright, Fig. 2, p. 269, Copyright 1987, with permission from Pergamon Press Ltd.)

Bai et al. [16], Grady and Kipp [20, 21], Burns and Trucano [22], Molinari and Clifton [23], and Wright [24]. Figure 15.15 shows, schematically, the geometry investigated. The material is considered to be under a homogeneous shear, the top surface being displaced at a velocity V with respect to the bottom surface. We will derive first the governing equations for this simplified configuration, with the following simplifying assumptions:

1. Elastic deformation is considered negligible.
2. Thermal expansion from heating is negligible.
3. Fourier's law of heat transfer holds.
4. The heat capacity is constant in the interval investigated.

The conservation equations are developed in Chapter 9 for a general state of deformation. We rederive them here. From the conservation of momentum we know that the change of momentum is equal to the impulse. Taking sections 1 and 2 in Figure 15.15(b), we have

$$\int_1^2 \rho A \Delta V \, dy = [F(y_2, t) - F(y_1, t)] \, dt \tag{15.10}$$

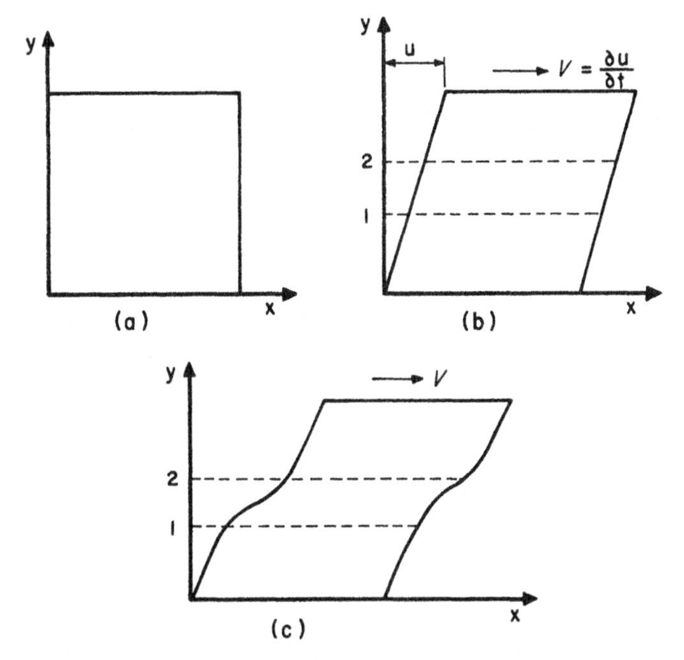

FIGURE 15.15 (a) Idealized deformation of a parallelepiped subjected to a displacement of upper surface of $V = \partial u/\partial t$; (b) homogeneous deformation; (c) localization.

Rearranging yields

$$\int_1^2 \rho \frac{\Delta V}{\Delta t}\, dy = \tau(y_1, t) - \tau(y_2, t) = \int_1^2 \frac{\partial \tau}{\partial y}\, dy$$

$$\rho \dot{V} = \frac{\partial \tau}{\partial y} \tag{15.11}$$

Bai et al. [16] express the conservation of momentum by using the strains:

$$\rho \frac{\partial \dot{V}}{\partial y} = \frac{\partial^2 \tau}{\partial y^2}$$

Since

$$\frac{\partial u}{\partial y} = \gamma \quad \text{and} \quad V = \frac{\partial u}{\partial t}$$

Thus

$$\frac{\partial V}{\partial y} = \frac{\partial^2 u}{\partial y\, \partial t} = \frac{\partial^2 u}{\partial t\, \partial y} = \frac{\partial \gamma}{\partial t} = \dot{\gamma}$$

And

$$\rho \frac{\partial \dot{\gamma}}{\partial t} = \frac{\partial^2 \tau}{\partial y^2} \tag{15.12}$$

Equations (15.11) and (15.12) are equivalent.

The conservation-of-energy equation is directly obtained from the first law of thermodynamics, stating that the change in internal energy is equal to work done on a system minus the heat extracted from the system:

$$dE = \delta W + \delta q \tag{15.12a}$$

Since the volume is constant,

$$\delta W = \tau\, d\gamma$$

The heat evolution obeys Fourier's law and is simply expressed as (heat conduction in y direction only):

$$\dot{q} = -\lambda \frac{\partial T}{\partial y}$$

where λ is the heat conductivity.

The heat capacity C_V* is defined as (ρ is the density)

$$C_V = \left(\frac{\partial E}{\partial T}\right) \rho^{-1}$$

Taking the time derivative of Eqn. (15.12a) yields

$$\frac{\partial E}{\partial t} = \frac{\partial W}{\partial t} + \frac{\partial q}{\partial t}$$

$$\rho C_V \frac{\partial T}{\partial t} = \tau \dot{\gamma} - \lambda \frac{\partial T}{\partial y} \tag{15.13}$$

This is the well-known equation for the conservation of energy. These equations [(15.12) and (15.13)] are complemented by a constitutive model that we can

*For solids, C_P and C_V are very close at low temperatures.

$$C_P - C_V = \alpha^2 VT/\beta$$

where α and β are the coefficients at thermal expansion and compressibility.

assume to have the general form

$$\tau = f(\gamma, \dot{\gamma}, T)$$

These differential equations have a homogeneous solution that Bai et al. [16] expressed as

$$\dot{\gamma}_h t = \text{const}$$

$$\gamma_h(t) = \dot{\gamma}_h t + \text{const}$$

$$T_h(t) = \frac{1}{\rho C_V} \int_0^t \tau_h(t)\dot{\gamma}_h \, dt + T_h(0) \quad \text{(assuming } \beta = 1\text{; see Section 13.6)}$$

$$\tau_h(t) = f[\gamma_h(t), \dot{\gamma}_h(t), T_h(t)]$$

The stability of the homogeneous solution can be checked by applying a perturbation in the temperature, stress, or strain. One then follows the progression of this perturbation. Different forms of the perturbation were used by Bai [19] and Clifton [18]. For a small perturbation $\delta T'$ with respect to the homogeneous temperature T_h, one has, for example [18],

$$T = T_h + \delta T'$$

This temperature inhomogeneity will disappear or be accentuated, depending on the other parameters. Clifton [18] derived the following criterion for stability:

$$\left[\frac{1}{\tau}\left(\frac{\partial \tau}{\partial \gamma}\right) + \frac{\alpha}{\rho C_V}\left(\frac{\partial \tau}{\partial T}\right)\right] m\dot{\gamma} + \frac{\lambda \xi^2}{\rho C_V} \gtreqless 0 \tag{15.14}$$

where m is the strain rate sensitivity ($\partial \ln \tau / \partial \ln \dot{\gamma}$) and ξ is the wave number for the initial perturbation (the wave number is the wavelength divided by 2π). In Eqn. (15.14) the effects of work hardening, strain rate hardening, and thermal softening can be seen. Bai et al. [16] arrived at a similar expression:

$$\frac{\alpha\tau(\partial \tau / \partial T)}{\rho C_V(\partial \tau / \partial \gamma)} - \left[\frac{4\alpha\lambda\dot{\gamma}_0(\partial \tau / \partial T)}{\rho C_V^2(\partial \tau / \partial \gamma)}\right]^{1/2} \gtreqless 1 \tag{15.15}$$

In the absence of heat transfer ($\lambda = 0$), the above expression reduces to

$$\frac{\alpha}{\rho C_V} \tau \left(\frac{\partial \tau}{\partial T}\right) \gtreqless \left(\frac{\partial \tau}{\partial \gamma}\right) \tag{15.16}$$

By inserting Eqn. (15.4) into Eqn. (15.2) one arrives at this equation. The criterion proposed by Bai [19] reduces itself to the simple adiabatic criterion derived in the elementary treatment of Section 15.2 in the absence of heat transfer. The same occurs with Clifton's [18] criterion. The second term in Eqn. (15.14) vanishes, and since $\dot{\gamma} \neq 0$, the sum within the first term is zero at the point of instability.

The foregoing analyses predict the effect of perturbations on the onset of shear band formation and provide a guideline to the prediction of the evolution of a shear band. The assumptions are rather severe and do not really depict the band with sufficient realism. A band propagates by extension of its tip, as shown in Figure 15.16(b), and not by simultaneous shearing over the full extent of the band, as shown in Figure 15.16(a). Thus, one should envisage the shear band as a shear crack (modes II and III; see Chapter 16, p. 491), in which the crack surfaces are replaced by a thermally softened layer; this was indeed emphasized by Clifton [18]. Figure 15.17 shows the extremities of two shear bands in an AISI 4340 steel. Figure 15.17(b) shows the bifurcation at the band tip. This is especially significant, since this mechanism of bifurcation is responsible for the spreading of two shear bands instead of one, contributing to the acceleration of the material failure.

Kuriyama and Meyers [25] modeled this process of shear band extension at the tip by the finite-element method. They assumed that the material within the existing shear band was softened to the point where its flow stress was zero. They compared the stresses and strains created at the tip of a wedge with thickness t equal to that of a shear band in HY-TUF steel (20 μm). They considered two responses (adiabatic and isothermal) and used the effective stress–effective strain curves obtained and calculated by Olson et al. [7]. Figure 15.18(a) shows the two curves. Whereas the isothermal curve shows monotonic

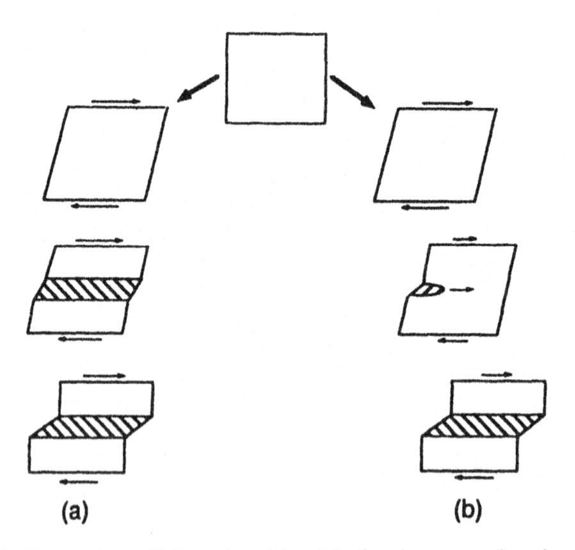

FIGURE 15.16 Formation of shear band by (a) simultaneous shearing over a planar region and (b) initiation and propagation of a shear instability.

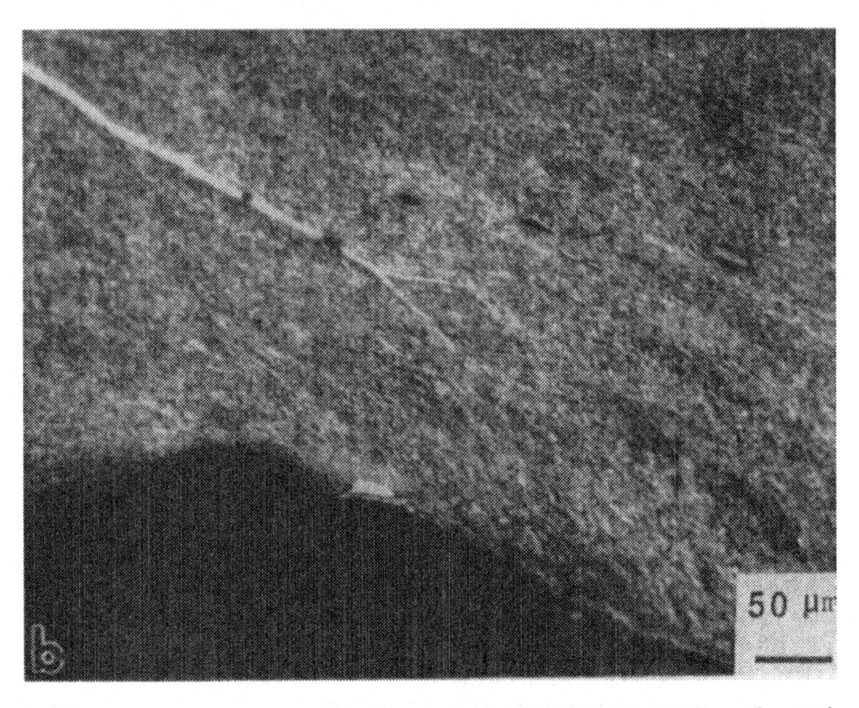

FIGURE 15.17 (a) Shear band extremities in AISI 4340 steel; (b) bifurcation at shear band tip resulting in formation of two bands. (From Wittman et al. [33], Fig. 5, p. 710. Reprinted with permission of the publisher.)

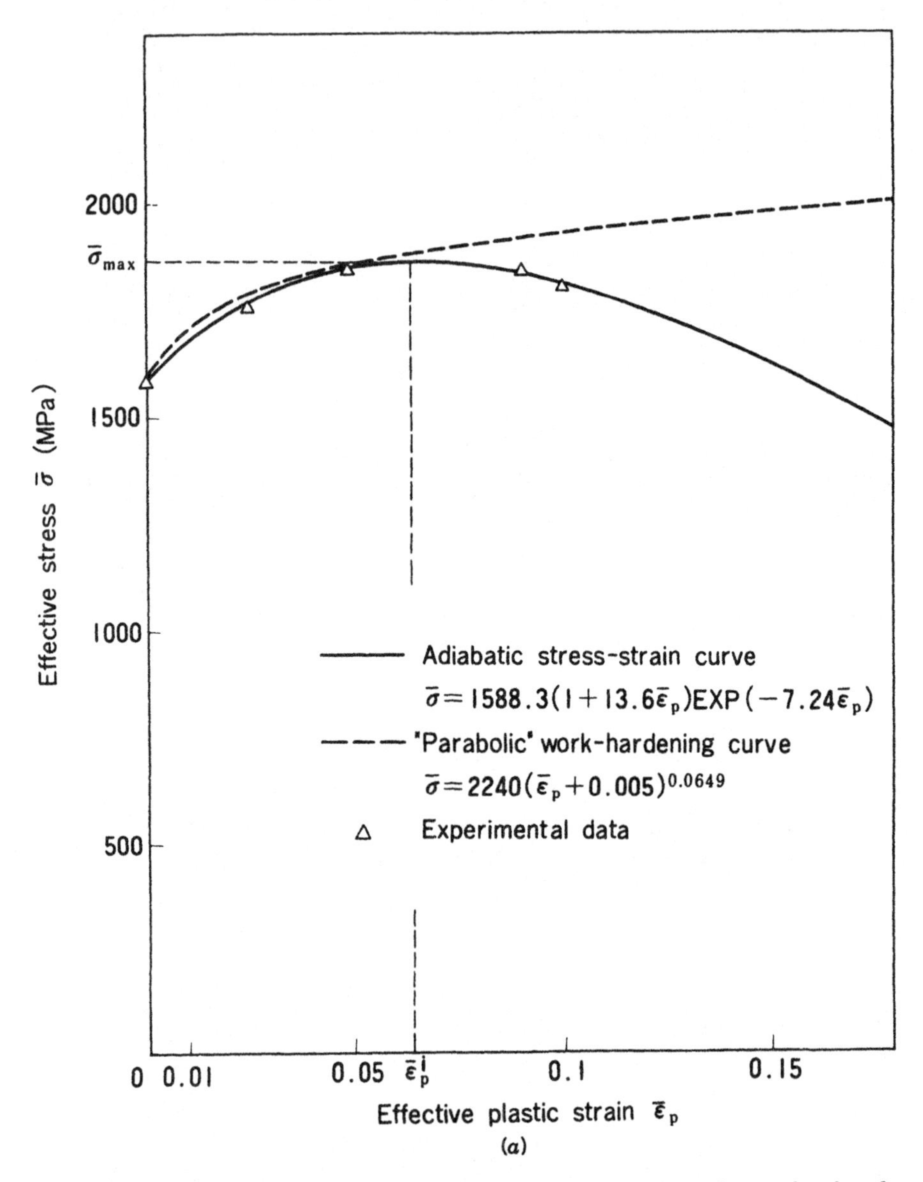

FIGURE 15.18 Finite-element computations of stresses and strains at the tip of a shear band. (a) Effective stress–strain curves for HY-TUF steel in quenched and tempered conditions. (From Olson et al. [7] Fig. 3, p. 298. Reprinted with permission of the publisher.) (b) Strain fields at tip of instability for adiabatic (i) and isothermal (ii) curves (notice more extended range for region with $\varepsilon_p = 0.1$ for adiabatic curve) and (c) length of region, S, with strain $\varepsilon_p > 0.0656$ [instability strain in (a) as a function of tangential displacement d]. (From Kuriyama and Meyers [25], Figs. 5 and 8, pp. 448, 449. Reprinted with permission of the publisher.)

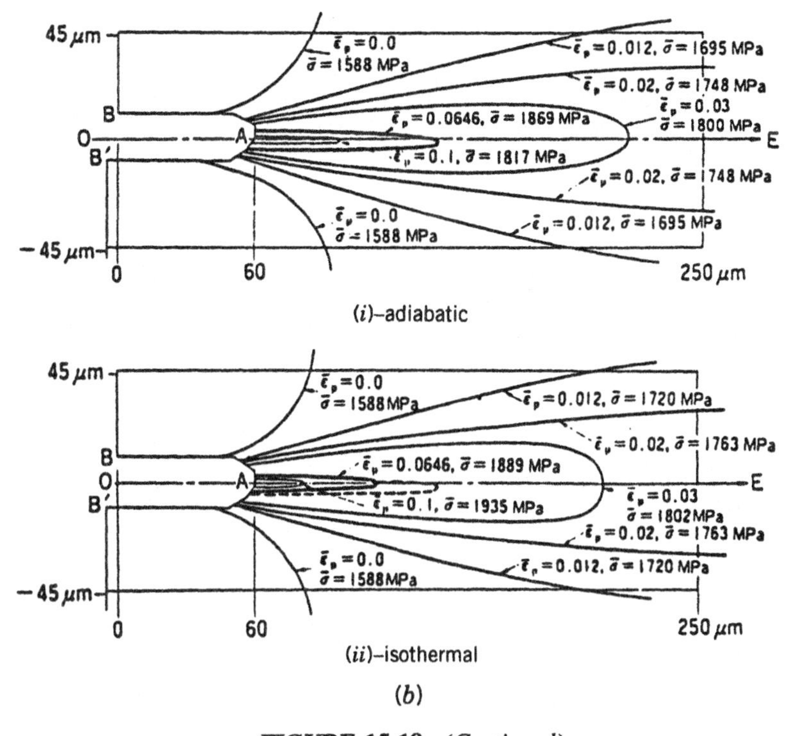

(*b*)

FIGURE 15.18 (*Continued*)

hardening, the adiabatic curve shows a maximum at $\varepsilon_p = 0.0656$. By applying an increasing displacement to the specimen (putting it under a shear parallel to the wedge), the stresses ahead of the wedge could be computed. The results are very interesting and are partially shown in Figure 15.18(b). The far-field stresses are almost identical for both cases. It is only for strains above $\varepsilon_p =$ 0.0656 that distinct differences are observed. The regions immediately ahead of the tip (isostrains of 0.0646 and 0.1) are expanded for the adiabatic case. Figure 15.18(c) shows the length of these regions, S, as a function of applied lateral displacement d for the two cases. Up to a displacement $d = 7$ μm, the two materials have the same behavior. However, when the plastic strain ε_p reaches the instability value (0.0656), there is an acceleration of plastic deformation with a stretching, ahead of the wedge, of the strain concentration region. This occurs for the adiabatic material, whereas the relationship between S and d is roughly linear for the isothermal curve. Thus, the process of shear band extension for adiabatic plastic deformation (in which an instability occurs) is clearly demonstrated. Notwithstanding the assumptions made in calculation (no heat conduction, mechanical equilibrium) the process of shear band propagation is one of progressive softening at the band tip.

Curran and Seaman [27] developed a model for shear band propagation based

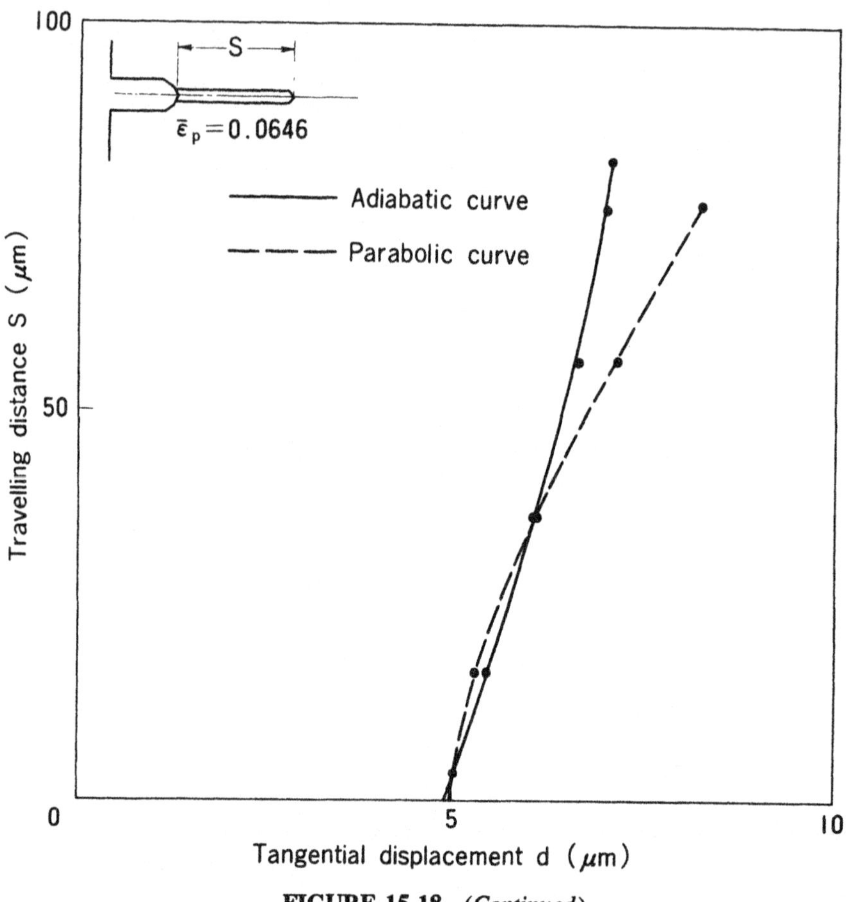

FIGURE 15.18 *(Continued)*

on the same assumptions.* They assumed a distribution of flaws within the material from which shear bands could nucleate and grow. Both the nucleation and growth rate were analytically expressed, based on a number of assumptions. This led to expressions predicting the loss of strength as the network of shear bands coalesced and fragmented the material. They incorporated this treatment of shear band–induced damage into large-scale computational codes [28]. This failure model is known as Shear 4. The basic equations are given below. The requirement for a flaw to have a size sufficient to generate an adiabatic shear band is (R is the "crack" radius)

$$R_1 \geq \left(\frac{10\lambda}{\dot{\varepsilon}_R}\right)$$

*D. Grady (*Mech. of Matls.*, **17** (1994) 289) developed a criterion for shear-band propagation by considering it as a Mode II crack, in analogy with Kuriyama and Meyers [25] and Curran and Seaman [27].

where λ is the thermal conductivity and $\dot{\varepsilon}_R$ is applied strain rate. For steel ($\lambda = 0.1$ cm^2/s) at an applied strain rate of 10^4 s^{-1}, the minimum size of R is 0.1 mm. A distribution of existing flaws was assumed as follows:

$$N_g = N_0 \exp\left(\frac{-R}{R_1}\right)$$

where N_0 is the total number per unit volume and N_g is the number of flaws per unit volume with a radius greater than R ($R > R_1$). From these expressions a nucleation rate was estimated for adiabatic shear bands, assuming an incubation time for nucleation. This incubation time τ is a function of the strain rate:

$$\tau \simeq \frac{(\varepsilon_m - \varepsilon_{cr})}{M\dot{\varepsilon}_R}$$

where $M\dot{\varepsilon}_R$ is the local strain rate ($M \sim 2$), ε_m is the strain required for complete softening of the band, and ε_{cr} is the strain of instability. Thus, $\varepsilon_m - \varepsilon_{cr}$ is the strain (local) for which total softening occurs and $M\dot{\varepsilon}_R$ is the strain rate (local) at the tip of the flaw. The nucleation rate is (for widely spaced flaws)

$$\dot{N} = \frac{N_0 - N}{\tau} = \frac{(N_0 - N)}{(\varepsilon_m - \varepsilon_{cr})} M\dot{\varepsilon}_R H\left(\varepsilon_R - \frac{\varepsilon_{cr}}{M}\right)$$

where H is a Heaviside function that establishes a lower limit for the applied strain, that is, $\dot{N} = 0$ if ε_R (the remote strain) is less then ε_{cr}/M (the local strain at which instability occurs). The growth rate was established as

$$\dot{R} = \frac{dr}{dt} \cong \frac{R}{\tau} = \frac{rM}{(\varepsilon_m - \varepsilon_{cr})}\dot{\varepsilon}_R$$

where r is the size of the softened region ahead of the flaw.

Since $r = \beta R$ (see Fig. 15.19),

$$\frac{\dot{R}}{R} = \frac{\beta M}{(\varepsilon_m - \varepsilon_{cr})}\dot{\varepsilon}_R$$

From a nucleation rate $\dot{N}$ and growth rate $\dot{R}$, Curran and Seaman [27] were able to predict shear band populations. When shear bands intersect, the material loses its strength. These elements are the basic components of their Shear 4 code, which predicts the response of a material with the formation of shear bands. An analogous approach was used to describe dynamic fracture (calculating nucleation, growth, and coalescence rates). This is described in Chapter 16.

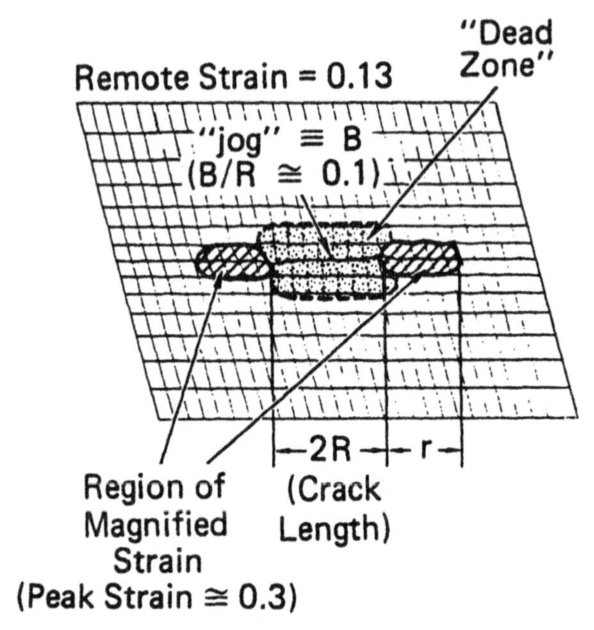

FIGURE 15.19 Computational simulation of plane strain crack (size $2R$) loaded in pure shear, generating shear band extensions at its tips of r due to localized softening. (From Curran and Seaman [27], Fig. 1, p. 317. Reprinted with permission of the publisher.)

15.4 METALLURGICAL ASPECTS

In Section 15.1 we presented the situations in which shear bands occur. Sections 15.2 and 15.3 dealt with the analysis of one individual shear band. In this section we will discuss the microstructural evolution within a shear band and the mechanical properties of shear bands. Very high shear strains have been observed across shear bands. Moss [29] reports a value of $\gamma = 572$ for a shear band. There is no limit to this shear strain, since the sliding across this region is determined by the overall geometry of deformation. Figure 15.20 illustrates the shear strain in a titanium alloy; two shear bands intersect. Shear band 1 formed prior to shear band 2, since the latter produces a step in the former. Shear band 2 produces the transverse displacement T in shear band 1. The thickness of shear band 2 is t. These quantities are marked in the figure. The engineering shear strain t/T is 15.

Shear bands have been observed in a variety of alloys; if one looks at Eqn. (15.2), which describes the formation of shear bands, a low work-hardening rate $(\partial\tau/\partial\gamma)$ and a large thermal softening are shear band predisposers. Thus, whereas pure annealed copper will deform homogeneously to very high strains, brass exhibits shear band behavior. The same applies to aluminum. Annealed

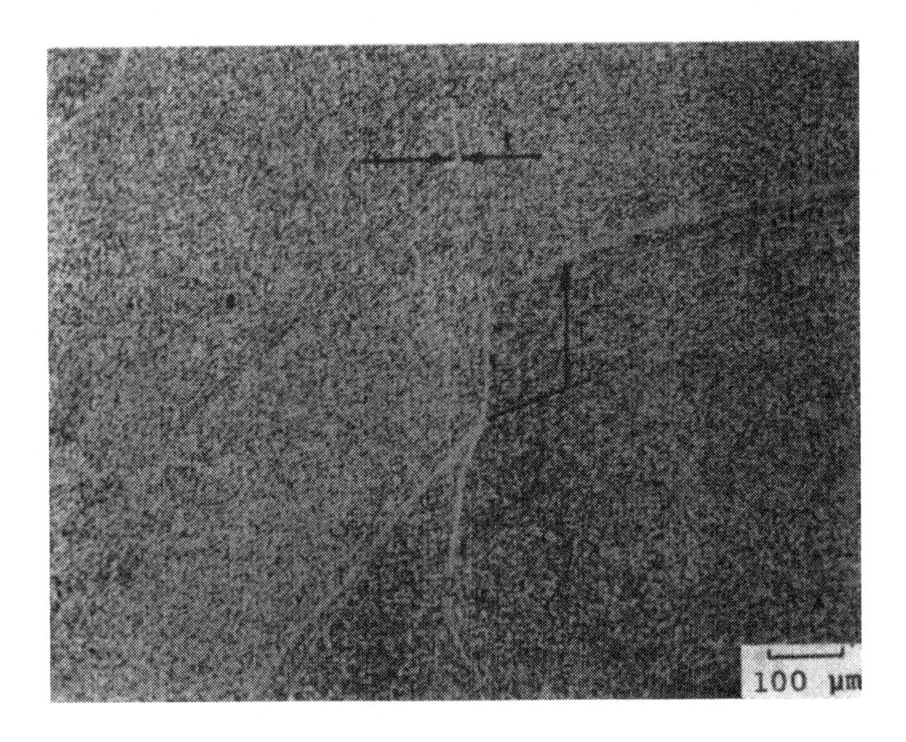

FIGURE 15.20 Shear band intersection in Ti_6Al_4V, t is the band thickness and T is the offset at shear band intersection. Engineering shear strain $\gamma = t/T = 15$. (From Grebe et al. [32], Fig. 7, p. 766. Reprinted with permission of the publisher.)

aluminum will deform homogeneously, whereas 2024-T4, 2014-T6, and 2014-overaged exhibit shear instabilities [1]. Additionally, Rogers [1] lists the following alloys as undergoing shear localization:

Steels: 1040, 1080, 4340, etc.
60-30 and 70-30 brass
Zirconium alloys
Titanium; Ti–6% Al–4% V
Tungsten heavy metal (W–Cu–Ni)
U–2% Mo
U–1% Mo–0.7% Nb–0.7% Z5Zr

Most quenched-and-tempered steels (martensitic) also exhibit shear bands. The number of alloys exhibiting shear localization under high-strain-rate deformation greatly exceeds the above list, and virtually any material can exhibit shear band formation if work hardening is appropriately depressed (by, e.g., plastic deformation).*

Shear bands have been classified into "deformed" and "transformed" based on optical observations. When the microstructure of the shear band appears to be very different from that of the surrounding region, the term *transformed* has been used. For instance, shear bands in quenched-and-tempered steels have a shiny appearance, quite different from the surrounding matrix (see Figs. 15.3, 15.8, and 15.17). On the other hand, shear bands in normalized or annealed steel have a dark appearance, exemplified by Figure 15.21. Similar differences in appearance are seen in titanium and its alloys: in some cases, the shear bands appear clear and shiny (Fig. 15.10), whereas in other cases they are dark. Metals and alloys have limits of phase stability, and temperature excursions can change the stable phase. Thus, AISI 1080 (eutectoid composition) undergoes a phase transition from α(BCC) to γ(FCC) at 723°C. If the temperature within the band reaches this value, then the phase transformation can occur, and this phase can be retained at room temperature, since the material within the band is rapidly quenched by the surrounding material once plastic deformation ceases. As an illustration, Figure 15.22 shows the cooling of shear bands in titanium and steel once deformation ceases. The cooling time is on the order of 1–10 ms. Since the temperature drops in Figure 15.22 are ap-

*Shear localization has also been observed in densified granular alumina by V. F. Nesterenko and M. A. Meyers (unpublished results, 1994).

FIGURE 15.21 Shear band in steel; dark appearance that has been attributed to "deformed" bands is clearly seen. (From Meyers and Wittman [34], Fig. 12, p. 3160. Reprinted with permission of the publisher.)

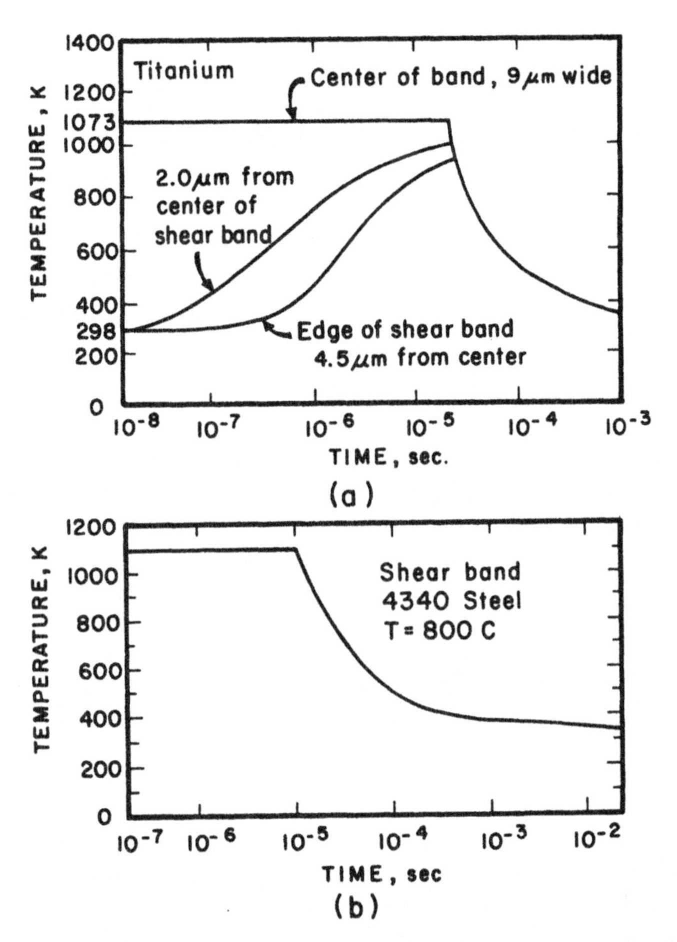

FIGURE 15.22 Cooling curves for shear bands. (a) In titanium. (Reprinted from *Acta Metall.*, vol. 34, M. A. Meyers and H.-r. Pak, Fig. 8, p. 2497, Copyright 1986, with permission from Pergamon Press Ltd.); (b) In 4340 steel (From Wittman et al. [33], Fig. 15, p. 715. Reprinted with permission of the publisher.)

proximately 800 K, an average cooling rate of 10^5–10^6 K/s is obtained. These rates are equal to the cooling rates obtained in rapid solidification processing, and one would expect a high degree of retention of the microstructures that exist during shear band formation. Transmission electron microscopy has shed some light on the mechanisms of microstructure evolution in shear bands. In 1971, Glenn and Leslie [30] studied shear bands on 0.6% carbon steel produced by ballistic impact. The structure of the white-etching shear bands was difficult to resolve by transmission electron microscopy, because the grain size was less than 0.1 μm. Glenn and Leslie [30] postulated that the white-etching region was very rapidly quenched martensite. Wingrove [31] also observed the white-etching bands by transmission electron microscopy but could not clearly identify the product phase. Wittman et al. [33] analyzed a band in AISI 4340 and could not find any evidence for a transformation to austenite. And Beatty et al. [36] analyzed a 4340 steel, finding a structure composed of grains with diameters ranging from 10 to 50 nm. The diffraction pattern formed an almost continuous ring, showing that a large number of these micrograins were simultaneously imaged. For titanium, Meyers and Pak [11] observed a structure composed of small, equiaxed grains (0.05–0.3 μm) with well-defined high-angle grain boundaries; a relatively low density of dislocations was observed in the shear bands. Similar results were independently obtained by Stelly and Dormeval [35] for a Ti–6% Al–4% V alloy: a microcrystalline structure with equiaxed grains having diameters less than 1 μm. Meyers et al. [37] have subjected shock-hardened copper to concentrated shear deformation under dynamic conditions, generating shear strains of 5 at strain rates of $\sim 10^4\ s^{-1}$. The microstructure of these highly deformed regions ($\sim$200 μm thick) consisted of fine grains with diameters of approximately 0.1 μm. Figure 15.23 shows the microstructures within the shear band for titanium, AISI 4340 steel, and shock-hardened (σ_y = 500 MPa) copper. The diffraction patterns of regions within the bands form almost continuous circles, indicating that numerous grains are imaged in each pattern. The dislocation density within the grains is rather low, and the grains are equiaxed. Figure 15.24 shows a dark-field transmission electron micrograph of the region within a shear band in commercially pure titanium. The bright micrograins are evident, and their morphology and size are more easily imaged by this method. The commonality among these microstructures in materials with diverse structures (Ti: HCP; AISI 4340: BCC; Cu: FCC) is strongly indicative of a mechanism of dynamic recrystallization taking place concurrently with plastic deformation. Thus, new recrystallization grains are nucleated within the deformed material once a critical strain is reached. This is made possible by the adiabatic heating generating temperature excursions within the bands that are equal or exceed $0.4T_m$, the melting point in Kelvin. At these temperatures diffusional processes responsible for recrystallization set in. Thus, the material is continuously being deformed while new grains nucleate and grow within it. Figure 15.25 shows, in a schematic fashion, the sequence of stages that occurs when a material recrystallizes dynamically. Once plastic deformation reaches a critical level (c), new, dislocation-free grains form in the deformed structure. They grow and are plastically deformed, and successive generations of dislocations are thus formed. As the strain rate increases, the size of the recrystallizing grains decreases. Sandstrom and Lagneborg [38] and Derby and Ashby [39] predict recrystallized grain sizes (d_{ss}) that vary as strain rates ($\dot{\gamma}$) as

$$d_{ss} \propto \dot{\gamma}^{-1/2}$$

More detailed mechanisms are discussed by Meyers et al. [40] and Andrade et al. [41]. As pointed out by Wright and Walter [17], instability (the onset of the decrease in the stress–strain curve) is a necessary but not sufficient condition for shear-band formation. Dynamic tests conducted on tantalum [42] and tung-

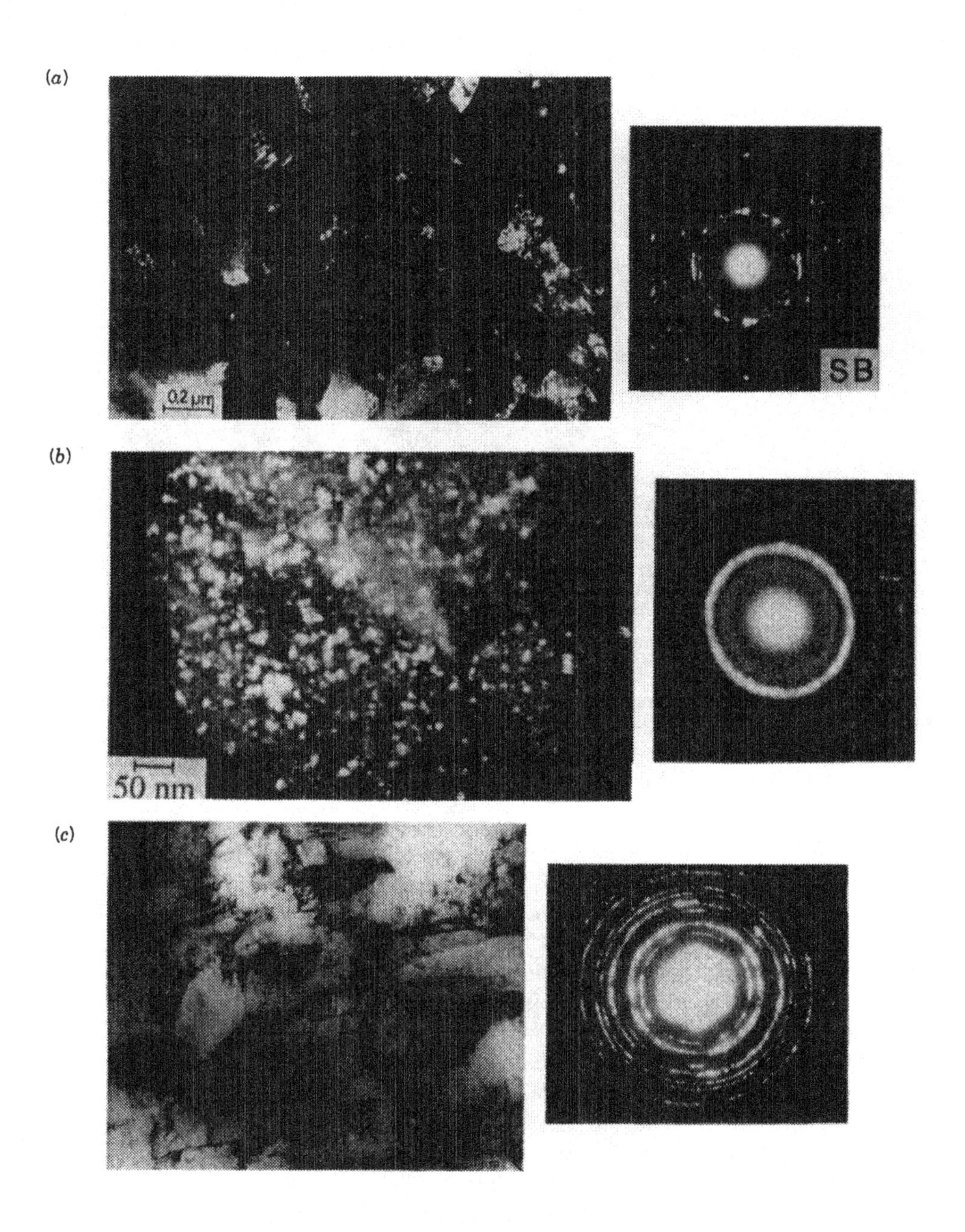

FIGURE 15.23 Transmission electron micrographs of shear bands showing a microcrystalline structure. (a) In titanium. (b) In quenched-and-tempered AISI 4340 steel. (From Beatty et al. [36], Fig. 10, p. 653. Reprinted by courtesy of Marcel Dekker, Inc.) (c) In copper, (Reprinted from *Acta Metall.*, vol. 34, M. A. Meyers and H.-R. Pak, Figs. 2 and 3, pp. 2494, 2495, Copyright 1986, with permission from Pergamon Press Ltd.) (From Meyers et al. [37], Fig. 5, p. C-3-15. Reprinted with permission of the publisher.)

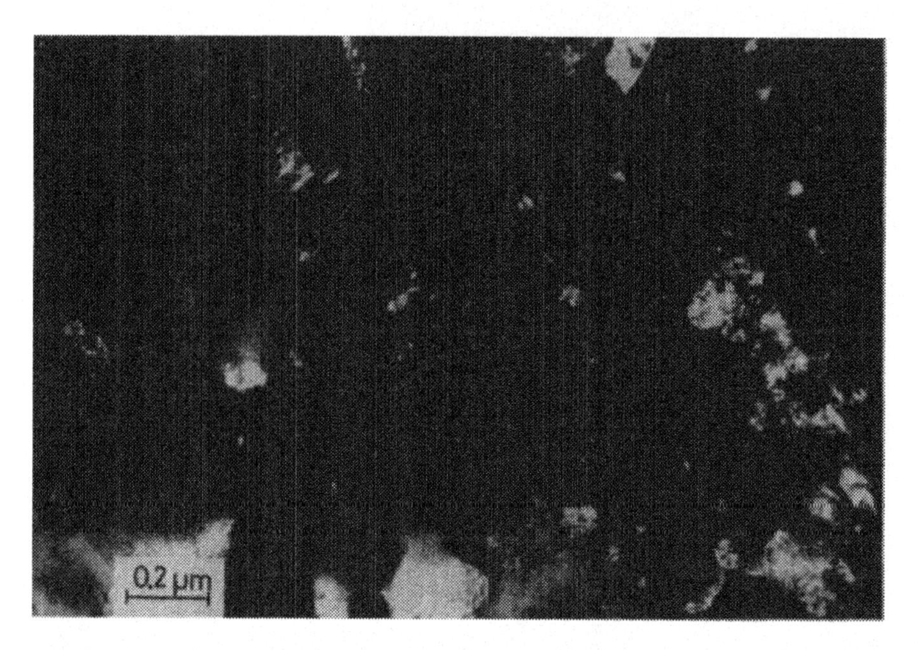

FIGURE 15.24 Dark-field transmission electron micrograph of region within a shear band in a commercially pure titanium alloy. (Reprinted from *Acta Metall.*, vol. 34, M. A. Meyers and H.-r. Pak, Fig. 4, p. 2495, Copyright 1986, with permission from Pergamon Press Ltd.)

sten [43, 44] show instability without shear-band formation. Similar results on titanium have led Meyers et al. [40] to suggest that, whereas instability is governed by the action of thermal energy on dislocations, decreasing the effective barrier height (see Chapter 13), localization is the result of a more drastic microstructural process involving dislocation annihilation by either dynamic recovery or recrystallization; this leads to a flow stress discontinuity that produces the thin and sharply deformed bands. For titanium, the onset of instability occurs at 350 K, while localization takes place at 776 K. Both Ta and W are high melting-point materials, and the temperatures for recrystallization (0.4Tm) are 1,300 K and 1,500 K, respectively. Therefore, the instability region $((d\tau/d\gamma) < 0)$ can be very extended and considerable plastic deformation takes place prior to localization.

The hardness of the material within the shear band varies according to the material. For steels, "transformed" shear bands are usually much harder than the surrounding material. Figure 15.26(a) shows the microhardness traverse for an AISI 4340 steel. The hardness inside of a shear band in titanium and a Ti–6% Al–4% V alloy did not exhibit the same high value. Microhardness values for these materials are shown in Figures 15.26(b) and (c), respectively; there is no difference between the band and bulk material hardness. The importance of the postdeformation shear band hardness stems from the fact that these bands can be embrittling factors in the microstructure, providing favored paths for fracture propagation (see Fig. 15.10). Titanium forgings possessing

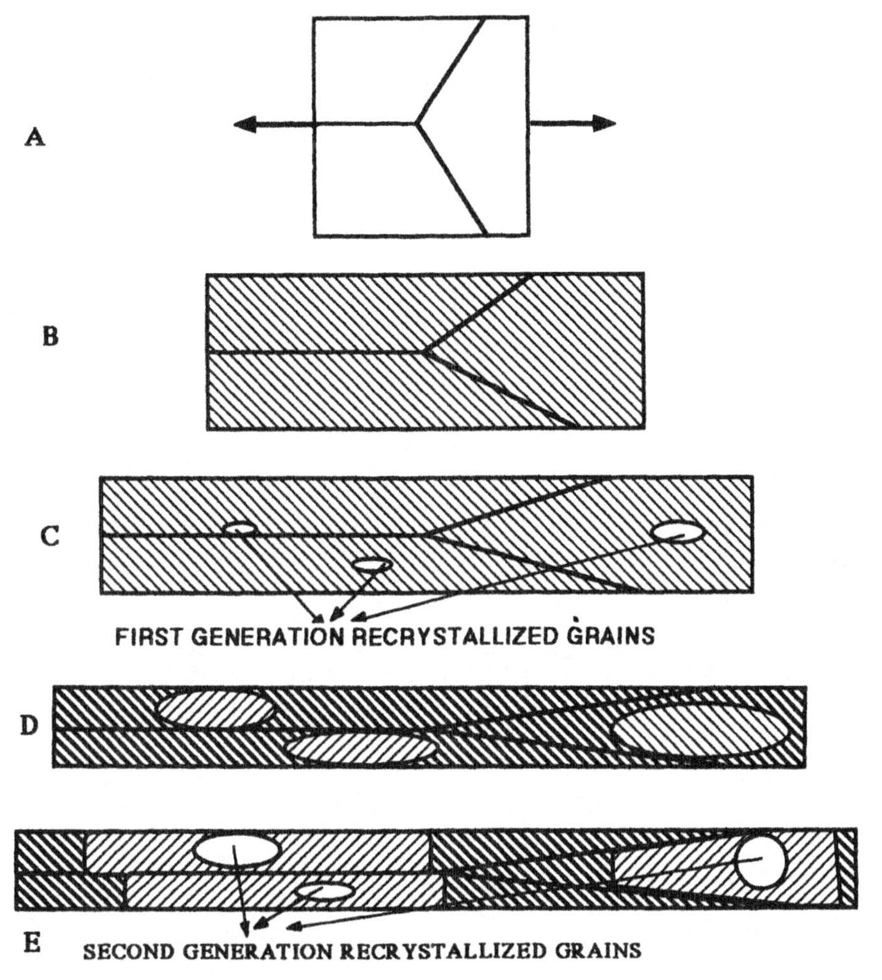

FIGURE 15.25 Schematic illustration showing sequence of deformation–recrystallization steps in dynamic deformation. (From Meyers et al. [37], Fig. 6, p. C-3-16. Reprinted with permission of the publisher.)

shear bands due to the forging process can be embrittled. The work conducted by Rogers and Shastry [45] provided results that are of considerable significance in understanding the hardness of the shear bands. They heat treated four steels with four different carbon contents (0.2, 0.4, 0.8, and 1%) to the same hardness and produced shear bands in them. The shear band hardness varied linearly with carbon concentration, as shown in Figure 15.27. Carbon is a powerful strengthener of BCC iron, and this increase in hardness with carbon content is consistent with solid-solution or precipitation strengthening of a common microstructure. The results shown in Figure 15.28 provide additional information on the nature of the adiabatic shear bands. These bands were produced by Rogers and Shastry [45] by impact with a cylindrical projectile on an AISI 1040 target. These targets were heat treated at different temperatures, yielding

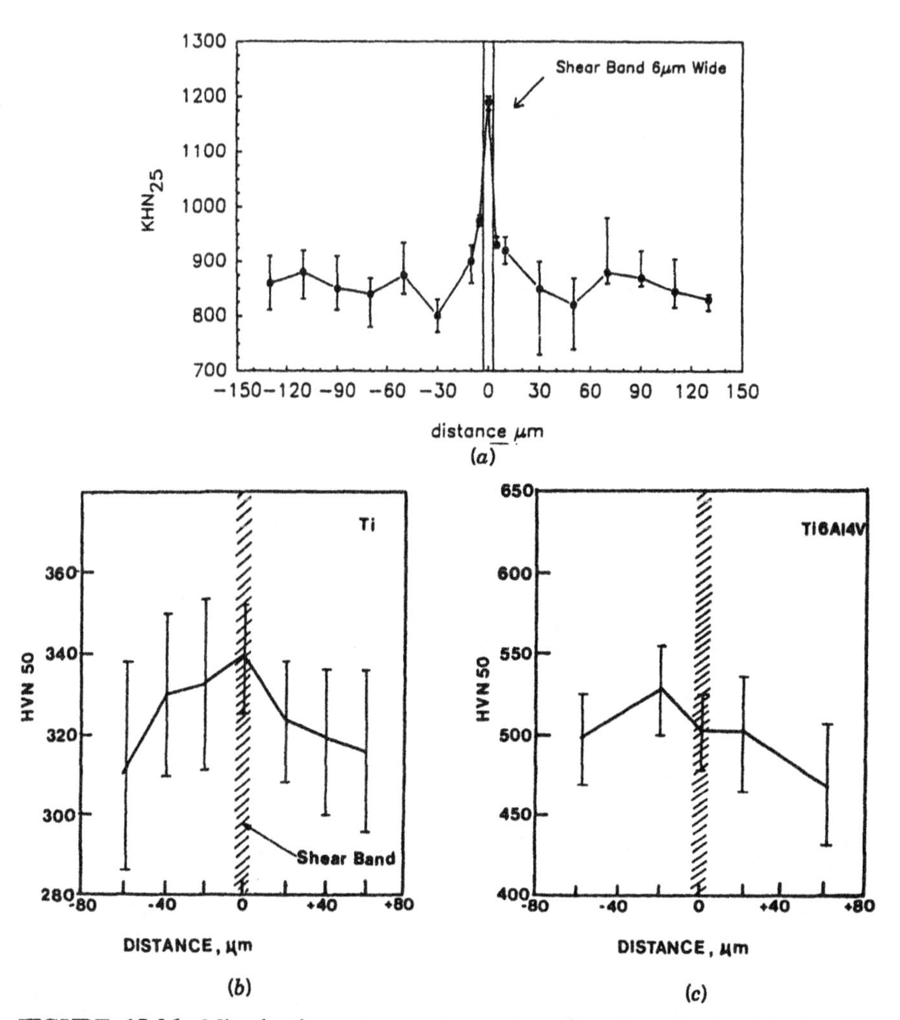

FIGURE 15.26 Microhardness traverses encompassing shear bands. (a) For AISI 4320 quenched-and-tempered steel. (From Meyers and Wittman [34], Fig. 16, p. 3162). (b) For titanium. (c) For a Ti–6% Al–4% alloy. (From Grebe et al. [32], Fig. 10, p. 769. Reprinted with permission of the publisher.)

hardnesses ranging from Rockwell C20 to C45. The hardness within the band is independent of both impact velocity and plate hardness. This indicates that this hardness is solely determined by the microstructure generated by the thermomechanical excursion inside the band.

Example 15.1. Estimate the thickness of shear bands for Ti, AISI 4340 steel, copper, and 2024 aluminum using Bai's equation. The parameters needed can be obtained from the literature. We take the temperature as $\frac{1}{2}T_m$, where T_m is the absolute melting point.

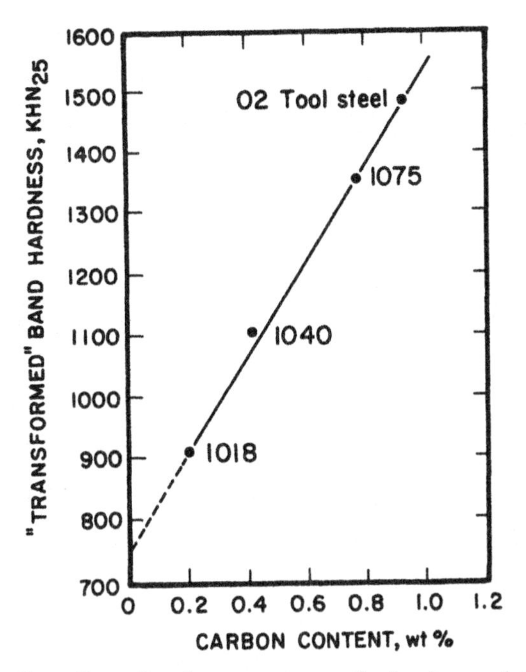

FIGURE 15.27 The effect of carbon content on the hardness of transformed shear bands in several steels. (From Rogers and Shastry [45], Fig. 4, p. 290. Reprinted with permission of the publisher.)

Material	τ(MPa)	T (K)	λ (W/m K)	2δ (μm)
Ti	450	1040	20	60
4340 Steel	450	860	80	110
Copper	250	670	390	290
2024 Al	100	460	250	304

Example 15.2. Determine the instability strain and temperature for a commercially pure titanium whose mechanical response is given in Figure 12.4. Use the Johnson–Cook equation.

1. We first obtain analytical expressions for T_{ins} and ε_{ins}. The Johnson–Cook equation is given as Eqn. (13.1) (We are using decimal logarithm here):

$$\sigma = (\sigma_0 + B\varepsilon^n)\left(1 + C\log\frac{\dot{\varepsilon}}{\dot{\varepsilon}_0}\right)[1 - (T^*)^m] \tag{1}$$

where

$$T^* = \frac{T - T_r}{T_m - T_r}$$

and T_m is the melting temperature and T_r is the test temperature (in this case room temperature). Assume $m = 1$. Let us take $\dot{\varepsilon} = 10^{-3}\ \text{s}^{-1}$ as the reference

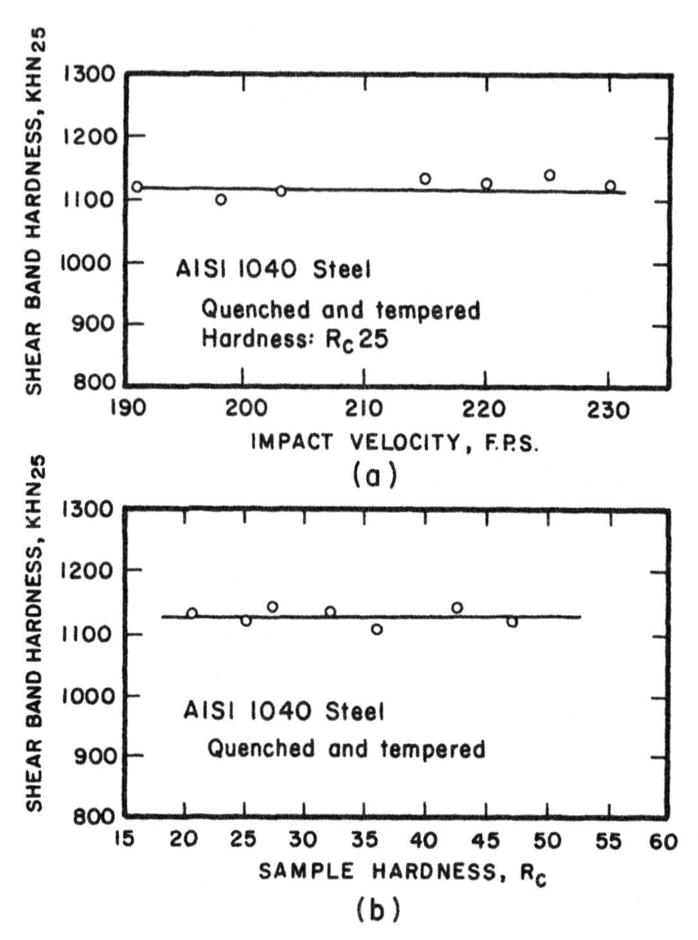

FIGURE 15.28 Hardness of shear band in AISI 1040 steel as a function of (a) impact velocity and (b) specimen hardness. (From Rogers and Shastry [40], Figs. 2 and 3, p. 289. Reprinted with permission of the publisher.)

strain rate. Let us assume that 90% of the "plastic work" is converted into heat (see Section 13.6 for more details).

$$\rho C_p \, dt = 0.9\sigma \, d\varepsilon \tag{2}$$

Section 13.6 provides the temperature as a function of plastic strain. Equation (13.79) is rewritten as

$$T^* = 1 - \exp\left[-\frac{0.9(1 + C \log \dot{\varepsilon}/\dot{\varepsilon}_0)}{\rho C_p (T_m - T_r)}\left(\sigma_0 \varepsilon + \frac{B\varepsilon^{n+1}}{n+1}\right)\right] \tag{3}$$

When the instability occurs,

$$\frac{d\sigma}{d\varepsilon} = 0$$

Substituting the value of T^* [from Eqn. (3)] in the Johnson–Cook [Eqn. (1)] gives

$$\sigma = (\sigma_0 + B\varepsilon^n)\left(1 + C\log\frac{\dot{\varepsilon}}{\dot{\varepsilon}_0}\right) \cdot \exp\left[-\frac{0.9(1 + C\log\dot{\varepsilon}/\dot{\varepsilon}_0)}{\rho C_p(T_m - T_r)}\left(\sigma_0\varepsilon + \frac{B\varepsilon^{n+1}}{n+1}\right)\right] \tag{4}$$

Differentiating Eqn. (4) with respect to ε and equating it to zero gives

$$\frac{(\sigma_0 + B\varepsilon^n)^2}{nB\varepsilon^{n-1}} = \frac{C_p(T_m - T_r)}{0.9(1 + C\log\dot{\varepsilon}/\dot{\varepsilon}_0)} \tag{5}$$

Equation (5) gives the instability strain. By using this value in Eqn. (3), the corresponding temperature is found.

2. We now determine various parameters, of the Johnson–Cook equation. Assume $\dot{\varepsilon}_0 = 10^{-2}\ \text{s}^{-1}$. The stress–strain curve for strain rate $10^{-2}\ \text{s}^{-1}$ does not show any softening (for which T^* can be assumed to be zero). The Johnson–Cook equation simplifies to

$$\sigma = \sigma_0 + B\varepsilon^n \quad (\text{for strain rate } 10^{-2}\ \text{s}^{-1})$$

where σ_0 is the yield stress ~320 MPa. Also,

$$\sigma - \sigma_0 = B\varepsilon^n$$

$$\ln(\sigma - \sigma_0) = \ln B + n\ln\varepsilon$$

If $\ln(\sigma - \sigma_0)$ is plotted with $\ln\varepsilon$, the y intercept yields the value of B and the slope gives n. For a particular strain $\sigma_0 + B\varepsilon^n$ is constant. The Johnson–Cook equation can be written as

$$\sigma = M\left(1 + C\log\frac{\dot{\varepsilon}}{\dot{\varepsilon}_0}\right)(1 - T^*)$$

$$M = \sigma_0 + B\varepsilon^n = \text{const}$$

For small strain, say ~1%, $T^* \approx 0$. The equation simplifies to

$$\sigma = M + MC\log\frac{\dot{\varepsilon}}{\dot{\varepsilon}_0}$$

If flow stresses corresponding to 1% strain for different strain rates are given, plot σ versus $\log(\dot{\varepsilon}/\dot{\varepsilon}_0)$. The slope gives the value of MC since M is known for 1% strain and C can be determined. The calculated values of the parameters are

$$\sigma_0 = 320 \text{ MPa} \qquad B = 940 \qquad M = 0.4 \qquad C = 0.1$$

Instability ε_{ins} can be determined from Eqn. (5) by the iterative method and the instability temperature from Eqn. (3). The instability strain is found to be

$$\varepsilon_{ins} = 0.45$$

The instability temperature is

$$T_{ins} = 630 \text{ K}$$

REFERENCES

1. H. C. Rogers, *Ann. Rev. Mat. Sci.*, **9** (1979), 283.
2. S. P. Timothy, *Acta Metall.*, **35** (1987), 301.
3. J. Mescall and V. Weiss, eds., *Material Behavior under High Stress and Ultrahigh Loading Rates*, Plenum, New York, 1983.
4. R. Dormeval, "Adiabatic Shear Phenomena," in *Impact Loading and Dynamic Behavior of Materials*, eds. C. Y. Chiem, H.-D. Kunze, and L. W. Meyer, DGM Informationsgesellschaft, Oberusel, Germany, 1988, p. 43.
5. A. J. Bedford, A. L. Wingrove, and K. R. L. Thompson, *J. Austr. Inst. Met.*, **19** (1974), 61.
6. R. J. Clifton, "Material Response to Ultra High Loading Rates," Report No. NMAB-356, National Materials Advisory Board Com., Washington, DC, 1980, p. 129.
7. G. B. Olson, J. F. Mescall, and M. Azrin, in *Shock Waves and High-Strain-Rate Phonomena in Metals*, eds. M. A. Meyers and L. E. Murr, Plenum, New York, 1981, p. 221.
8. C. Zener and J. H. Hollomon, *J. Appl. Phys.*, **15** (1944), 22.
9. A. Marchand and J. Duffy, *J. Mech. Phys. Solids*, **36** (1988), 251.
10. F. R. Recht, *J. Appl. Mech.*, **31** (1974), 189.
11. M. A. Meyers and H.-r. Pak, *Acta Metall.*, **34** (1986), 2493.
12. U. S. Lindholm and G. R. Johnson, in *Material Behavior under High Stress and Ultrahigh Loading Rates*, eds. J. Mescall and V. Weiss, Plenum, New York, 1983, p. 61.
13. R. S. Culver, in *Metallurgical Effects at High Strain Rates*, eds. R. W. Rohde, B. M. Butcher, J. R. Holland, and C. H. Karnes, Plenum, New York, 1973.
14. M. Staker, *Acta Metall.*, **29** (1981), 683.
15. T. W. Wright, *J. Mech. Phys. Sol.* (1989).
16. Y. Bai, C. Cheng, and S. Yu, *Acta Mech. Sinica*, **2** (1986), 1.
17. T. W. Wright and J. W. Walter, *J. Mech. Phys. Sol.*, **35** (1987), 701.
18. R. J. Clifton, "Material Response to Ultra High Loading Rates," Report No. NMAB-356, National Materials Advisory Board, NAS, Washington, DC, Chapter 8, 1979.

19. Y. Bai, in *Shock Waves and High-Strain-Rate Phenomena in Metals: Concepts and Applications*, eds. M. A. Meyers and L. E. Murr, Plenum, New York, 1981, p. 227.
20. D. E. Grady, *J. Geophys. Res.*, **82-B2** (1980), 913.
21. D. E. Grady and M. E. Kipp, *J. Mech. Phys. Sol.*, **35** (1987), 95.
22. T. J. Burns and K. R. L. Trucano, *Mech. Mat.* (1982), 313.
23. A. Molinari and R. J. Clifton, *C. R. Acad. Sci. Paris* 296(1983)1.
24. T. W. Wright, *J. Mech. Phys. Sol.*, **35** (1987), 269.
25. S. Kuriyama and M. A. Meyers, *Met. Trans. A*, **17A** (1986), 443.
26. M. A. Meyers and S. Kuriyama, in *Shock Waves in Condensed Matter—1985*, ed. Y. M. Gupta, Plenum, New York, 1986, p. 321.
27. D. R. Curran and L. Seaman, in *Shock Waves in Condensed Matter—1985*, ed. Y. M. Gupta, Plenum, New York, 1986, p. 315.
28. D. A. Shockey, D. R. Curran, and L. Seaman, "Development of Improved Dynamic Failure Models," Final Technical Report to U.S. Army Research Office, Contract DAA-29-81-0123, February 15, 1985.
29. G. L. Moss, in *Shock Waves and High-Strain-Rate Deformation of Metals: Concepts and Applications*, eds. M. A. Meyers and L. E. Murr, Plenum, New York, 1981, p. 299.
30. R. C. Glenn and W. C. Leslie, *Met. Trans.*, **2** (1971), 2945.
31. A. L. Wingrove, *J. Aust. Inst. Met.*, **16** (1971), 67.
32. H. A. Grebe, H.-r. Pak, and M. A. Meyers, *Met. Trans.*, **16A** (1985), 761.
33. C. L. Wittman, M. A. Meyers, and H.-R. Pak, *Met. Trans.*, **21A** (1990), 707.
34. M. A. Meyers and C. L. Wittman, *Met. Trans.*, **21A** (1990), 3153.
35. M. Stelly and R. Dormeval, in *Metallurgical Applications of Shock-Wave and High-Strain-Rate Phenomena*, eds. L. E. Murr, K. P. Staudhammer, and M. A. Meyers, Dekker, New York, 1986, p. 607.
36. J. Beatty, L. W. Meyer, M. A. Meyers, and S. Nemat-Nasser, in *Shock-Wave and High-Strain-Rate Phenomena in Materials*, eds. M. A. Meyers, L. E. Murr, and K. P. Staudhammer, Dekker, New York, 1992, p. 645.
37. M. A. Meyers, L. W. Meyer, K. S. Vecchio, and U. Andrade, *J. Phys. IV, Coll. C3* (1991), C-3-11.
38. R. Sandstrom and R. Lagneborg, *Acta Met.*, **23** (1975), 307.
39. B. Derby and M. F. Ashby, *Scripta Met.*, **21** (1987), 879.
40. M. A. Meyers, G. Subhash, B. K. Kad, and L. Prasad, *Mech. of Matls.*, **17** (1994), 175.
41. U. R. Andrade, M. A. Meyers, A. H. Chokshi, and K. S. Vecchio, *Acta Met. et Mat.*, in press (1994).
42. Y. J. Chen, M. A. Meyers, F. Marquis, and J. Isaacs, unpublished results, 1994.
43. G. Subhash, Y. J. Lee, and G. Ravichandran, *Acta Met. et Mat.*, **42** (1994), 319.
44. G. Subhash, Y. J. Lee, and G. Ravichandran, *Acta Met. et Mat.*, **42** (1994), 331.
45. H. C. Rogers and C. V. Shastry, in *Shock-Waves and High-Strain-Rate Phenomena in Metals*, M. A. Meyers and L. E. Murr, Plenum, New York, 1981, p. 285.

CHAPTER 16

Dynamic Fracture

16.1 INTRODUCTION: FRACTURE MECHANICS FUNDAMENTALS

In this chapter, we will study rapidly propagating cracks and their interactions. We will investigate in detail dynamic fracture occurring when a tensile pulse is created by the reflection of a stress wave at a free surface. This is called spalling and was discussed in Chapter 7. When many cracks are formed and grow simultaneously, the body is eventually divided into many parts and we have fragmentation. The feature that distinguishes dynamic fracture from quasi-static behavior is the presence of stress waves. These waves arise due to either the stresses released from the crack tip at fracture or externally applied loads. When stress waves reflected from the specimen boundaries return to the crack tip, they alter the crack tip stress state, and this can result in a change of crack speed or cause crack branching if the intensity of the stress waves is sufficiently high. The consideration of rapid fracture is of considerable importance in several practical applications, including crack arrest in engineering structures, fracture plane control in blasting, fragmentation in mining; military applications, such as the fragmentation of bombs and shells and the fracture of projectiles and armor; and space applications, including impact of structures by meteorites. The design of armor and armor-defeating projectiles requires an intimate knowledge of the dynamic fracture response of a material. Since the progression of fracture is driven by the stress field in the local neighborhood of the crack front, it is necessary to know the relationship between the crack speed and the local stress field for any type of dynamic analysis.

Before we delve in greater detail into fracture at high rates, we will briefly summarize some fundamental concepts with which the distinguished reader might not be familiar. Recommended background reading on fracture mechanics are the books by Broek [1] and Knott [2].

Fracture occurs principally in two modes: *brittle* and *ductile*. Brittle fracture takes place by the propagation of a crack with a sharp front. Usually, this crack has definite crystallographic orientation and is therefore called a "cleavage" crack [Fig. 16.1(a)]. It may also propagate between the individual grains that comprise the materials, and this mode is called intergranular. Figure 16.1(b) shows these cracks. When the material exhibits substantial ductility, i.e., plastic deformation before failure, the propagation of the crack requires more energy

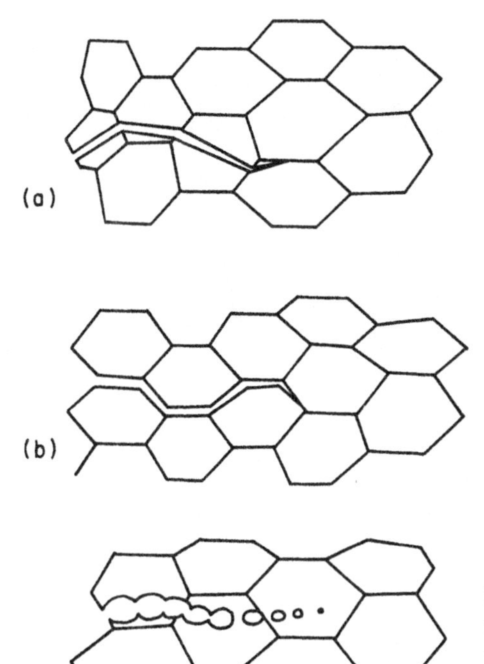

FIGURE 16.1 General classification of fracture: (a) brittle transgranular fracture; (b) brittle intergranular fracture; (c) ductile transgranular fracture.

and the tip of the crack becomes blunted. This is called ductile fracture. Ahead of the crack voids form and grow by plastic deformation. These voids eventually link with the crack tip, causing its extension. This process is responsible for the propagation of ductile failure. Figure 16.1(c) shows this ductile fracture propagation mode. Figure 16.2 shows optical micrographs of polished cross sections of an aluminum alloy and of Armco (pure) iron. This fracture is produced by spalling (reflection of a compressive pulse at a free surface, generating tension). The drastic difference between the ductile [Fig. 16.2(a)] and the brittle [Fig. 16.2(b)] modes is clearly evident. It is emphasized that the fracture patterns in Figure 16.2 are different from quasi-static fracture because fracture is independently nucleated at many sites. This is a unique aspect of dynamic fracture that can lead to marked differences in morphology.

The stresses at the tip of a crack are much higher than in the uncracked material. Figure 16.3(a) shows schematically the stress concentration at the tip of an elliptical hole, used to model a crack. The stress σ_{22} rises sharply at the front. It has been found that a critical stress intensity factor is needed for the propagation of a crack in a material. The stress intensity factor for the simple geometry shown in Figure 16.3 is given by

$$K_{\mathrm{I}} = \sigma\sqrt{\pi a} \tag{16.1}$$

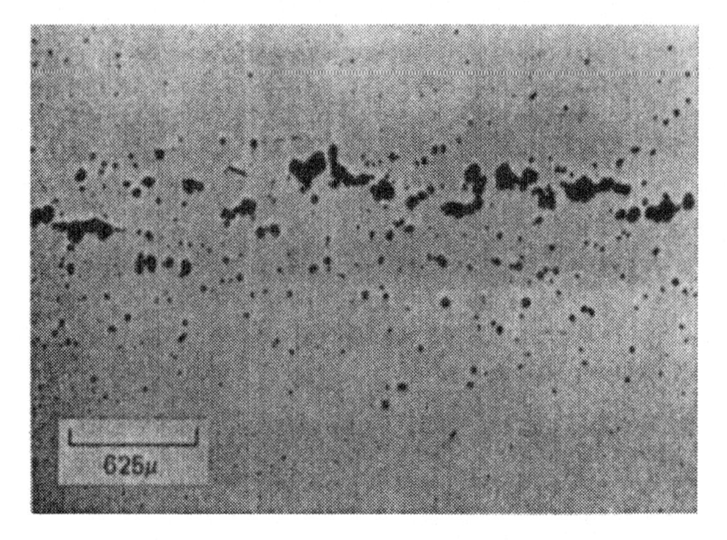

(a)

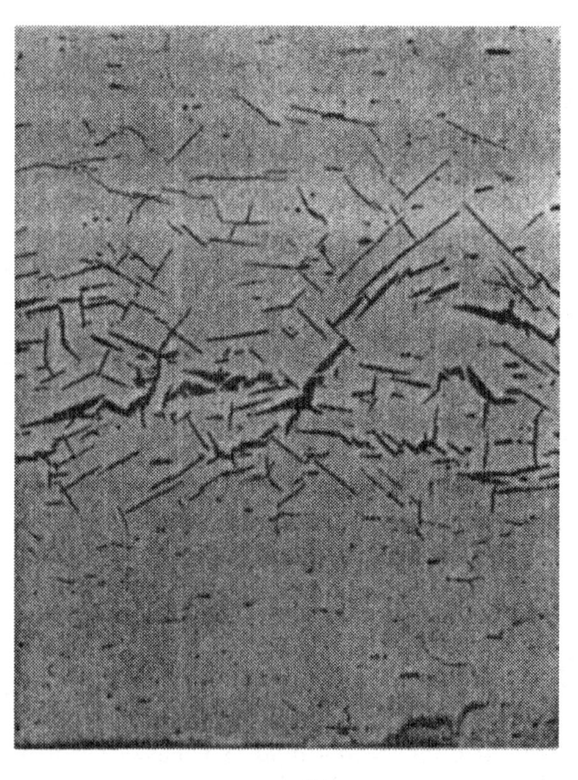

(b)

FIGURE 16.2 Polished sections of specimens subjected to dynamic tensile stress pulse experiments yielding incipient spalling (notice independent void and crack nucleation): (a) ductile fracture on aluminum alloy; (b) brittle fracture in iron. (Courtesy of D. A. Shockey, SRI International.)

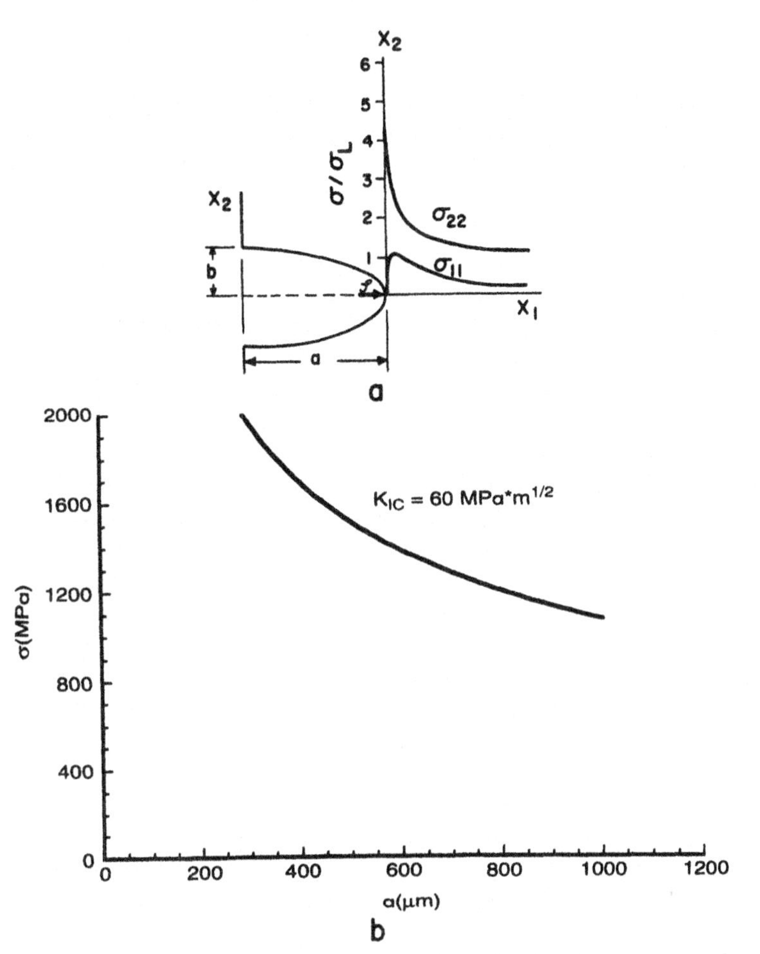

FIGURE 16.3 (a) Stress concentration at tip of crack; (b) stress required for crack propagation as a function of crack size.

When the stress intensity factor reaches a critical level, K_{IC}, the crack propagates:

$$K_{IC} = \sigma_{crit}\sqrt{\pi a} \tag{16.2}$$

For different boundary conditions Eqn. (16.2) takes the form

$$K_{IC} = k\sigma\sqrt{\pi a} \tag{16.2a}$$

where k is a parameter and K_{IC}* is a material property under quasi-static (low strain rate) conditions. Figure 16.3(b) shows how the stress required to activate

*Fracture is classified according to 3 modes of loading: I (tension); II (shear); and III (transverse shear). However, tension at the crack tip is needed for the crack to grow. Thus, we have: K_{IC}, K_{IIC}, and K_{IIIC}.

fracture would vary with crack size for an AISI 4340 steel with K_{IC} = 60 MPa $m^{1/2}$. For instance, Table 16.1 presents typical values for K_{IC}. Here K_{IC} is the fracture toughness of a material. It can be seen that metals can have quite large fracture toughness (~20–150 MPa $m^{1/2}$), whereas ceramics have low K_{IC} (~2–5 MPa $m^{1/2}$). This is due to the brittleness of ceramics. There is little or no plastic deformation at the tip of a propagating crack in a brittle material. Thus, little energy is consumed in the propagation of a crack; advanced ceramics are being developed that have much higher fracture toughnesses. Fracture toughness enhancement in ceramics is done by one of three ways, presented in a schematic fashion in Figure 16.4. The addition of fiber reinforcement is a method by which the fracture toughness can be increased by two- to three fold. The fibers act as crack arresters or, by detaching themselves from the matrix, as an additional fracture barrier. The sliding of the fiber inside the matrix jacket requires work. When a ductile fiber is added, its plastic deformation requires work and enhances the toughness of the ceramic. Table 16.1 shows a SiC/SiC fiber composite with K_{IC} = 25 MPa $m^{1/2}$. A second method, shown in Figure 16.4(b), consists of adding a second phase that undergoes a dilatational/shear (martensitic) phase transformation when subjected to the stresses existing at the crack tip. This phase transformation decreases the stresses at the crack tip and creates a "process" zone, increasing the toughness of the material. For instance, the fracture toughness of Al_2O_3 can be increased from 4 to 8 MPa $m^{1/2}$ by the addition of partially stabilized zirconia. This zirconia (PSZ) undergoes a stress-induced tetragonal-to-monoclinic transformation under the effect of the applied stresses at the crack tip and "dampens" these stresses. And a third method, microcracking, contributes to the creation of a process zone that absorbs energy and increases toughness.

In addition to the plane strain fracture toughness (K_{IC}), other parameters are used to describe the fracture response of a material. These parameters are briefly presented below.

1. The first parameter is the energy release rate, or crack extension force, G. When a cracked plate is under a load P [Fig. 16.5(a)], and external work ($F = Pv$) is done by this surface traction P, there will be an equivalent increase in internal strain energy U if there is no crack growth. This is a result of the energy principle (external work is identical to internal work). If the crack extends itself, the energy required for crack growth, W, has to be added to the equation. The condition for stable crack growth therefore is

$$\frac{d}{da}(U - F + W) = 0 \tag{16.3}$$

or

$$\frac{d}{da}(F - U) = \frac{dW}{da} \tag{16.4}$$

TABLE 16.1 Fracture Toughness for Some Common Materials

Alloy	σ_{ys} (MPa)	K_{IC} (MPa $m^{1/2}$)
Aluminum Alloys		
2020-T651	525–540	22–27
	530–540	19
2024-T351	370–385	31–44
	305–340	30–37
7075-T651	515–560	27–31
	510–530	25–28
Ferrous Alloys		
4330V (275°C temperature)	1400	86–94
4330V (425°C temperature)	1315	103–110
9-4-20 (550°C temperature)	1570	62
18 Ni(200) (480°C)	1280–1310	132–154
Titanium Alloys		
Ti6Al-4V	875	123
Ceramics		
Mortar	—	0.13–1.3
Concrete	—	2–2.3
Al_2O_3	—	3–5.3
SiC	—	3.4
SiN_4	—	4.2–5.2
Soda lime silicate glass	—	0.7–0.8
Electrical porcelain ceramics	—	1.03–1.25
WC(2.5–3 μm)-3 w/o Co	—	10.6
WC(2.5–3 μm)-9 w/o Co	—	12.8
WC(2.5–3.3 μm)-15 w/o Co	—	16.5–18
Indiana limestone	—	0.99
ZrO_2 (Ca stabilized)	—	7.6
ZrO_2	—	6.9
Al_2O_3/SiC whiskers	—	8.7
SiC/SiC fibers	—	25
Borosilicate glass/SiC fibers	—	18.9
Polymers		
PMMA	—	0.8–1.75
PS	—	0.8–1.1
Polycarbonate	—	2.75–3.3

Source: From Hertzberg [3], p. 410, Table 10.7a.

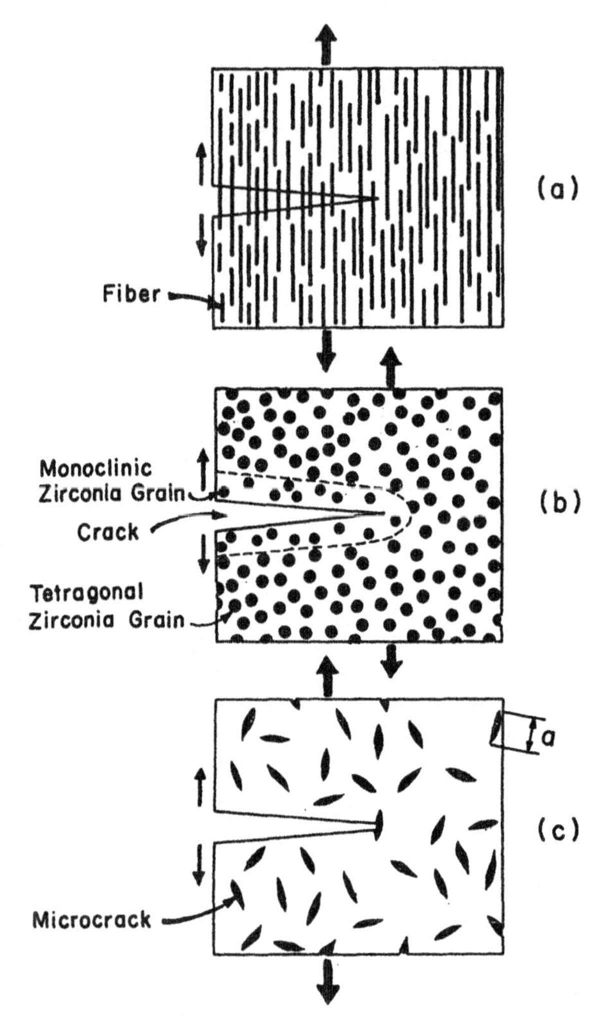

FIGURE 16.4 Toughening methods for ceramics: (a) fiber strengthening; (b) transformation toughening; (c) microcrack toughening.

where dW/da is also called R and is the crack resistance. The term $d(F - U)/da = G$ is called the energy release rate. We can assume, to a first approximation, that R is a constant for a crack of size a_1. The energy release rate G can be expressed as a function of the applied stress level σ for a crack size a as (for plane strain)

$$G = (1 - \nu^2) \frac{K_I^2}{E} = (1 - \nu^2) \frac{\pi \sigma^2 a}{E} \tag{16.5}$$

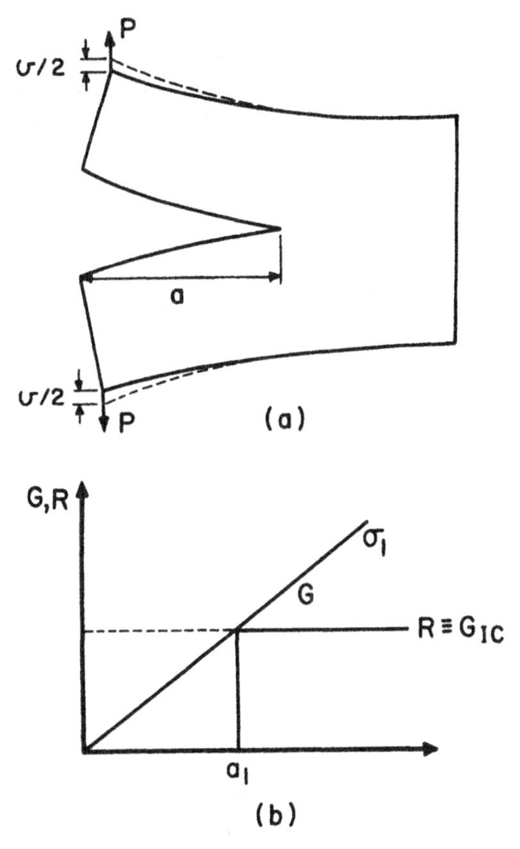

FIGURE 16.5 (a) Cracked plate (crack of size a) being pulled by force P; notice displacement v. (b) Energy release rate and crack resistance for a crack size a_1; when $G = G_{IC}$, crack grows.

2. The second equality in Eqn. (16.5) is due to Eqn. (16.1). Figure 16.5(b) shows the energy release rate G plotted for a crack size a at an applied stress σ_1. When $G = G_{IC}$, the crack starts to grow; since R is constant, the energy release rate G will exceed R for $a > a_1$, and the crack will continue to grow.

3. A third parameter that describes the crack propagation behavior that is particularly useful when the material exhibits considerable plasticity is the J integral. The J integral provides a means of determining the energy release rate when plasticity effects are not negligible. For the linear elastic case

$$J = G \tag{16.6}$$

When plastic deformation is significant, the above equality does not hold any longer. Cherepanov and Rice introduced the J integral to crack problems.

16.2 UNIQUE FEATURES OF DYNAMIC FRACTURE

Dynamic (high-velocity) crack propagation has unique aspects that differentiate it from quasi-static (low-velocity) crack propagation. In this section these differences will be briefly and qualitatively outlined. The peculiar behavior of dynamic fracture will be expanded upon in the subsequent sections. The main peculiarities are presented below:

1. There is a limiting velocity of propagating cracks, usually accepted as the Rayleigh wave velocity. Figure 16.6(a) shows a crack propagating at a velocity C driven by a load P. In order for the crack to grow, energy has to be "pumped" into its tip. This energy is provided by the external work performed by P. The energy arrives at the crack tip through the material in the form of stress (longitudinal and shear waves) and through surface waves resulting from the outward motion of the two halves. In Figure 16.6(a), the

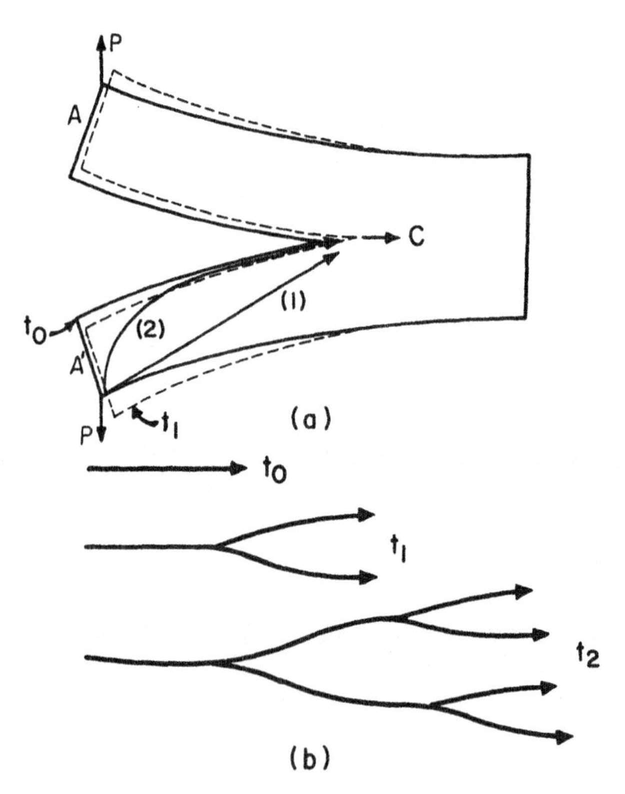

FIGURE 16.6 (a) Cracked plate being pulled by force P (positions marked at times t_0 and t_1) and propagation of crack. Notice two paths by which energy is being transferred from externally applied force to crack tip. (b) Branching of crack, as it propagates dynamically ($t_0 < t_1 < t_2$).

positions of the two sides and crack tips are shown at t_0 and t_1. By increasing the velocity of separation of AA', one increases the rate at which energy is pumped into the crack tip, contributing to its opening at an increased speed. Most of the energy arrives at the crack tip through path 2, that is, through the internal surfaces of the crack. Since the stresses can travel at a maximum velocity C_R at this surface, this establishes the upper bound for the velocity of propagation of a crack.

2. At a certain critical velocity, cracks tend to branch out (bifurcate) in order to decrease the overall energy of the system. This is shown in Figure 16.6(b). This continued branching is a primary cause of fragmentation. Thus, quasi-static failure will result in the propagation of a single crack, dividing a part into two, whereas dynamic crack propagation can produce fragmentation. An example is the shattering of glass under impact loading, whereas a single crack often propagates under quasi-static propagation conditions.

3. The fracture toughness of materials is often dependent on the rate of crack propagation. It is fairly complex to establish the stress intensity factor at the tip of a running crack, and therefore the fracture toughness determination is not as straightforward as under quasi-static conditions. In steel, the effect of crack propagation velocity on the fracture morphology is very clear. These important points will be analyzed in the following sections.

16.3 LIMITING CRACK SPEED

Dynamic (rapid) crack propagation is a nontrivial problem, and this has become a specialized research area since World War II. The first reports on rapid crack propagation are due to Hopkinson [4], who detonated an explosive charge in contact with a metal and produced spalling. Spalling is described in Chapter 7 and will be addressed in detail in Section 16.8. More recently, Rinehart and Pearson [5] and Kolsky [6] described spalling and dynamic fracture in their classic books. Since 1960, in parallel with the numerous studies on materials effects and spalling (which will be described in Section 16.8), the development of the highly mathematical field of elastodynamics has taken place, and the ability to rigorously describe the propagation of crack is ensuing. A solid foundation in solid mechanics is required of the student eager to penetrate into this somewhat exclusive domain. Nevertheless, the effort is definitely worthwhile and the derivation of the analytical expressions that describe rapid crack propagation is an essential part of the understanding of the dynamic response of materials. The mathematical approach requires familiarity with the indicial notation of stresses and strains, integral equations and their solution by the Wiener–Hopf technique, Laplace transforms, complex variables, and mixed boundary value problems. In this book an attempt has been made in previous chapters to derive all expressions presented. However, due to the highly specialized nature of the problem (dynamic crack propagation) only the most meaningful expressions will be presented here. The reader is referred to the two review articles by Achenbach [7] and Freund [8] as well as to the excellent specialized treatise by Freund [9].

An elementary treatment leading to a prediction of maximum crack velocity is given by Broek [1]. It is due to Mott and will be given here. An expression for the kinetic energy of a crack was developed based on dimensional analysis. Figure 16.7 shows a crack propagating at a velocity $v = da/dt = \dot{a}$. If we assume, at time t, that the crack has a length $2a$, and if the crack has grown to this size from a_c rapidly, the external boundary conditions can be assumed to be unchanged, and G = const. In Figure 16.7(b), G exceeds G_{IC}, as can be

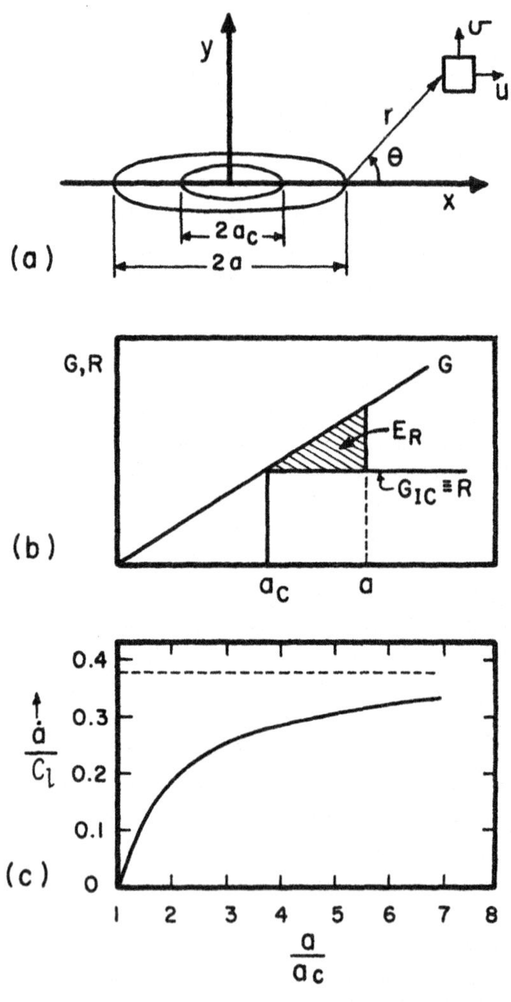

FIGURE 16.7 (a) Displacements u and v generated by propagating crack; (b) G, R, versus a plot showing energy excess as crack propagates dynamically; (c) normalized crack velocity versus length.

seen, at a. This excess energy can be expressed as

$$E_1 = \int_{a_c}^{a} (G - R)\, da$$

Assuming R to be constant and from Eqn. (16.5),

$$\begin{aligned} E_1 &= -R(a - a_c) + \int_{a_c}^{a} \frac{\pi\sigma^2 a}{E}\, da \\ &= \frac{-\pi\sigma^2 a_c}{E}(a - a_c) + \frac{\pi\sigma^2}{2E}(a^2 - a_c^2) \\ &= \frac{1}{2}\frac{\pi\sigma^2}{E}(a - a_c)^2 \end{aligned} \tag{16.7}$$

Another expression for the kinetic energy can be obtained from the displacement fields of u and v around a propagating crack [Fig. 16.7(a)]. The displacements are given by (see, e.g., Broek [1])

$$u = \frac{2\sigma}{E}\sqrt{ar} f_u(\theta) \qquad v = \frac{2\sigma}{E}\sqrt{ar} f_v(\theta)$$

where $f_u(\theta)$ and $f_v(\theta)$ are two functions of θ. As the crack moves away, the distance from a fixed element to its tip, r, will increase (assuming an initial small value of a):

$$r \propto a$$

Thus

$$u = C_1 \frac{\sigma a}{E} \qquad v = C_2 \frac{\sigma a}{E}$$

where C_1 and C_2 are proportionality constants. The total kinetic energy is obtained by integration of the square of the velocity fields:

$$\begin{aligned} E_k &= \frac{1}{2}\iint \rho(\dot{u}^2 + \dot{v}^2)\, dx\, dy \\ &= \frac{1}{2}\rho\dot{a}^2 \frac{\sigma^2}{E^2}\iint (C_1^2 + C_2^2)\, dx\, dy \end{aligned}$$

The proportionality constants C_1 and C_2 can be taken out of the integrand. The dimensions over which dx and dy are integrated are directly related to a, the crack size. Thus, the integration provides ka^2, where k is a proportionality constant:

$$E_k = \frac{1}{2}k(C_1^2 + C_2^2)\rho\dot{a}^2 \frac{\sigma^2}{E^2} a^2 \tag{16.8}$$

By equating (16.7) and (16.8),

$$E_1 = E_k$$

we obtain

$$\dot{a} = \sqrt{\frac{2\pi}{k(C_1^2 + C_2^2)}}\sqrt{\frac{E}{\rho}}\left(1 - \frac{a_c}{a}\right)$$

But $(E/\rho)^{1/2}$ is the longitudinal elastic wave velocity C_1:

$$\dot{a} = k'C_1\left(1 - \frac{a_c}{a}\right) \tag{16.9}$$

It is possible to estimate k' (=0.38). This yields

$$\dot{a} = 0.38C_1\left(1 - \frac{a_c}{a}\right)$$

The ratio $\dot{a}/C_1$ is plotted in Figure 16.7(c) as a function of the normalized crack size a/a_c. It can be seen that it asymptotically approaches 0.38 as a (and consequently K_I) increases. For a material with $\nu = 0.3$, the velocity of elastic shear waves is approximately equal to $0.38C_1$.

More rigorous solutions to this problem were found later, and the approaches by Broberg [10], Craggs [11], Baker [12], and Freund [13–16] should be pursued by the dedicated and intelligent student. Yoffé's treatment [17], from 1951, preceded the above ones, but she considered a crack of fixed length traveling through a body; however, this treatment had some limitations that will be discussed in the next section.

Broberg assumed a crack opening whose internal faces are loaded, as shown in Figure 16.8. The crack opens at a velocity v (each side) and the ratio between the stress intensity factor at a velocity $v[K_I(v)]$ and that of a stationary crack of the same size and under the same loading situation was calculated and is shown in Figure 16.8(b). Therefore, the crack exists in the region $-vt < x < vt$. The uniformly applied stress σ_{yy} is the only stress applied on the body. The ratio between the stress intensity factor at the velocity v and the stress intensity factor at $v = 0$, $K_1(0)$ was found to decrease almost monotonically from 1 to 0 when v increased from 0 to C_R, the Rayleigh wave speed. Thus, the driving force decreases to zero when the crack speed approaches C_R. This

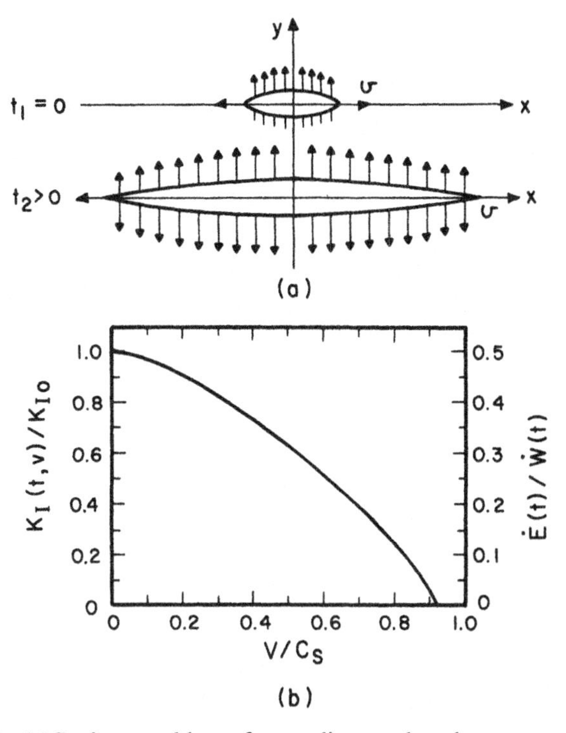

FIGURE 16.8 (a) Broberg problem of expanding crack under constant stress σ_{yy} acting on the crack faces; (b) stress intensity and energy deposited into crack tip as a function of velocity. (From Freund, [9], Fig. 6.4, p. 327. Reprinted with permission of the publisher.)

is equivalent to an upper bound for the velocity of mode I cracks. Broberg's early calculations have been confirmed by more sophisticated computations. The reader is referred to the reviews by Achenbach [18] and Freund [19] and the excellent treatise by Freund [9]. A small summary will be presented here of Freund's [13–16] development of the dynamic crack formulation. Freund considered a different problem: a semi-infinite crack grows at a constant velocity v under a time-independent (constant) loading σ_∞, shown in Figure 16.9(a). This external loading can be translated into a loading P on the crack faces. At time $t = 0$ the crack tip passes through the origin of the coordinate axes. The first step is to change the origin of the coordinate system to the tip of the advancing crack. This is shown in Figure 16.9(b). The original coordinates are (x, y), and the wave equations can be expressed in terms of the deformation potentials ϕ and ψ (see Section 2.7 for explanation) as

$$\begin{aligned} \frac{\partial^2 \phi}{\partial x^2} + \frac{\partial^2 \phi}{\partial y^2} &= \frac{1}{C_l^2} \frac{\partial^2 \phi}{\partial t^2} \\ \frac{\partial^2 \psi}{\partial x^2} + \frac{\partial^2 \psi}{\partial y^2} &= \frac{1}{C_s^2} \frac{\partial^2 \psi}{\partial t^2} \end{aligned} \tag{16.10}$$

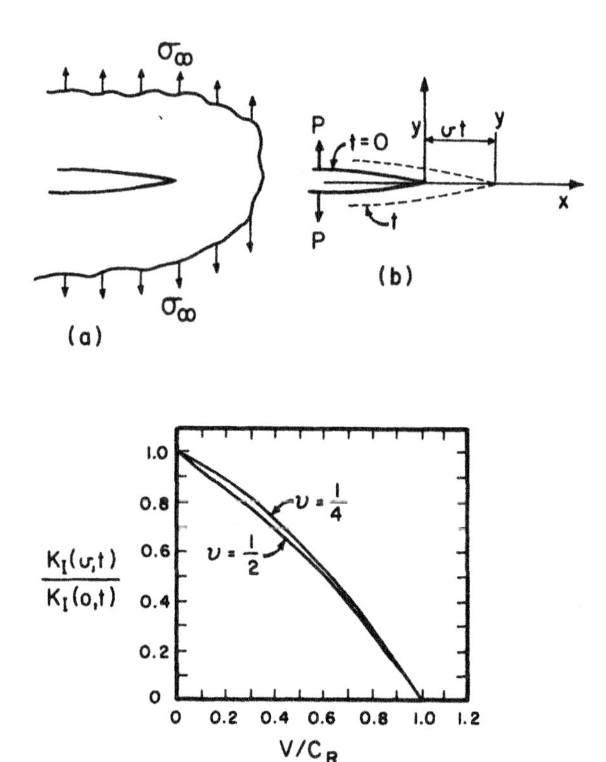

FIGURE 16.9 (a) Freund's analysis of a propagating crack. (b) Advancing crack and change of coordinate system. (c) Prediction of normalized stress intensity factor as a function of velocity. (From Freund [14], Fig. 3, p. 150. Reprinted with permission of the publisher.)

The boundary conditions are

$$\sigma_{yy}(x, 0, t) = -P\,\delta(x)H(t) \qquad -\infty < x < vt$$

$$\sigma_{xy}(x, 0, t) = 0 \qquad -\infty < x < \infty$$

$$u_y(x, 0, t) = 0 \qquad vt < x < \infty$$

where $\delta(x)$ is the Dirac delta function (equal to zero at all values of x except the one where the load P is acting) and $H(t)$ is the famous Heaviside function, a step function that is equal to zero for $t < 0$ and equal to 1 for $t > 0$. It simply states that the stresses are "turned on" at $t = 0$. The change in coordinate system is

$$\xi = x - vt \tag{16.12}$$

The wave equations (16.10) and (16.11) are expressed now as

$$\left(1 - \frac{v^2}{C_1^2}\right)\frac{\partial^2 \phi}{\partial \xi^2} + \frac{\partial^2 \phi}{\partial y^2} + \frac{2v}{C_1^2}\frac{\partial^2 \phi}{\partial \xi\, \partial t} - \frac{1}{C_1^2}\frac{\partial^2 \phi}{\partial t^2} = 0 \qquad (16.13)$$

and

$$\left(1 - \frac{v^2}{C_s^2}\right)\frac{\partial^2 \psi}{\partial \xi^2} + \frac{\partial^2 \psi}{\partial y^2} + \frac{2v}{C_s^2}\frac{\partial^2 \psi}{\partial \xi\, \partial t} - \frac{1}{C_s^2}\frac{\partial^2 \psi}{\partial t^2} = 0 \qquad (16.14)$$

The solution of this problem involves application of a one-sided Laplace transform to eliminate time t:

$$\hat{\phi}(\xi, y, s) = \int_0^\infty \phi(\xi, y, t)e^{-st}\, dt \qquad (16.15)$$

A two-sided Laplace transform is then applied to both the wave (governing) equations [Eqns. (16.13) and (16.14)] and boundary conditions:

$$\Phi(\zeta, y, s) = \int_{-\infty}^{\infty} \hat{\phi}(\xi, y, s)e^{-s\zeta\xi}\, d\xi \qquad (16.16)$$

$$\Psi(\zeta, y, s) = \int_{-\infty}^{\infty} \hat{\psi}(\xi, y, s)e^{-s\zeta\xi}\, d\xi \qquad (16.17)$$

The Wiener–Hopf technique, developed originally for the analysis of electromagnetic wave phenomena, is a method for the solution of integral equations. It is applied to the transformed governing equations. Once a solution is obtained in the transformed space, it has to be retransformed into the physical domain. These mathematical procedures are presented in detail by Freund [9] and the reader is referred to the source. For the sake of this book, it suffices to present the final transformed form of the stress intensity factor (Wiener–Hopf solution):

$$K_I(s, v) \cong P\sqrt{\frac{2}{vs}}\,\frac{(1 - v/C_R)}{\sqrt{1 - v/C_1}} \qquad (16.18)$$

Inversion of this stress intensity factor yields

$$K_I(v, t) = P\sqrt{\frac{2}{\pi v t}}\,\frac{(1 - v/C_R)}{\sqrt{1 - v/C_1}} \qquad (16.19)$$

The quasi-static value of $K_I(v, t)$ when $v = 0$ can be expressed by Eqn. (16.2a), which has an equivalent form for a concentrated load P (Broek [1], p. 79):

$$K_I = P\sqrt{\frac{2}{\pi a}} \tag{16.20}$$

Thus

$$\frac{K_I(v, t)}{K_I(0, t)} \cong \frac{1 - v/C_R}{\sqrt{1 - v/C_l}} \tag{16.21}$$

Since $C_R < \frac{1}{2}C_l$, the effect of the denominator is to lend a certain convexity to the plot of Figure 16.9(c). The relative values of the wave velocities are affected by the Poisson ratio as expressed in Chapter 2 [Eqns. (2.22)–(2.24)]. Thus, one has two curves that differ slightly, for $\nu = 0.25, 0.5$. An even simpler form of Eqn. (16.21) is

$$\frac{K_I(v, t)}{K_I(0, t)} \cong \frac{1 - v/C_R}{1 - 0.5v/C_R} \tag{16.22}$$

The derivations given above apply to *purely elastic* cracks. Real life is not as simple and elastoplastic crack propagation is more complex because of additional energy dissipation by plastic deformation at the crack tip. For ceramics, one also often has a process zone, composed of microcracks, close to the crack tip. These additional effects are compounded by the strain rate dependence of material flow stress (Chapter 14). As a result, the maximum observed crack propagation velocities for ductile materials are considerably lower, as will be seen later in this chapter.

16.4 CRACK BRANCHING (BIFURCATION)

It is observed that as cracks travel faster in glass, they tend to branch out. Yoffé [17]* analyzed a small crack of constant length $2a$, traveling at a velocity (constant) v in a body where the loading (Fig. 16.10) is uniform. Thus, the left edge of the crack heals itself as the right side opens. This is a fairly unrealistic situation; therefore, it did not predict an upper bound for the velocity. Yoffé calculated the stress $\sigma_{\theta\theta}$ as a function of crack velocity and orientation. Her predictions are shown in Figure 16.10(c). For low velocities, the maximum of $\sigma_{\theta\theta}$ occurs for $\theta = 0$. As the velocity v increases, the curve flattens out. Yoffé predicted that a maximum would occur at $\theta \neq 0$. She indicated that at this velocity the crack either becomes curved or branches into two, because

*G. Ravi-Chandar and W. Knauss (*Intl. J. Fract.*, **26** (1984) 141) make a critical analysis of branching mechanisms and conclude that branching occurs at velocities much lower than predicted by Yoffé [17]. They propose a mechanism based on the formation of microcracks ahead of main crack. The interactions of the microcracks among themselves and with the main crack dictate the branching response.

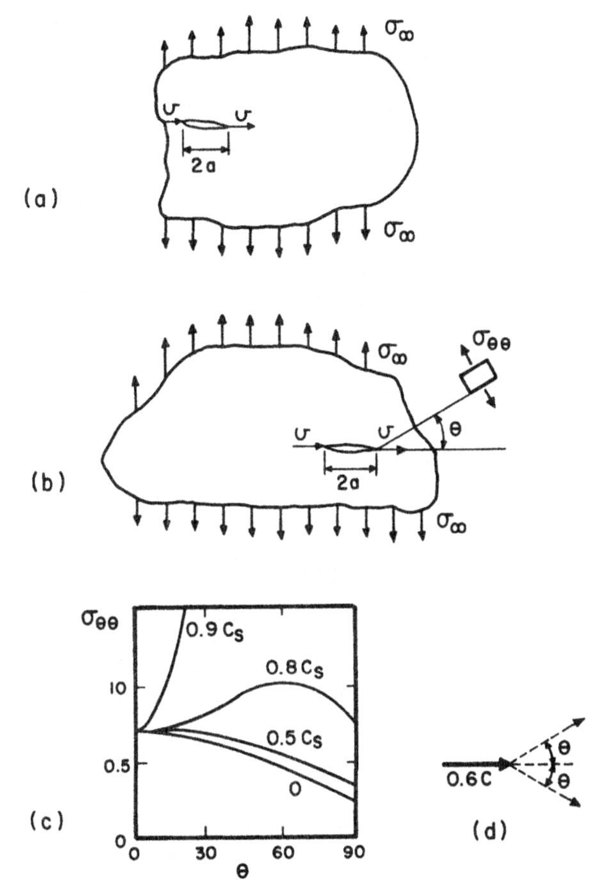

FIGURE 16.10 (a) Yoffé's problem, in which a crack of constant length $2a$ propagates at a steady-state velocity v; (b) prediction of stress; (c) crack tip stress $\sigma_{\theta\theta}$ as a function of velocity (0.0, $0.5C_s$, $0.8C_s$, $0.9C_s$); (d) bending and/or branching of crack at $v = 0.6C_s$. (From Yoffé [17], Fig. 1, p. 749. Reprinted with permission of the publisher.)

of the symmetry of the problem [Fig. 16.10(d)]. In Figure 16.10(c), $a/0.5C_s < V < 0.8C_s$, the maximum in the curve shifts from $\theta = 0$ to a positive value.

This change in the orientation of the maximum principal stress σ_{11} (or $\sigma_{\theta\theta}$) with velocity is accompanied by a net decrease in energy, as will be shown below in a schematic manner. The (G, R)-versus-Δa curve of Figure 16.11 shows that, for $\Delta a = a_c$, or $a = 2a_c$, there is sufficient energy to drive two cracks. Thus, the kinetic energy of the first crack provides the energy for driving a second crack. It is clear that the two cracks will propagate at velocities much lower than the first one. If there is sufficient energy stored in the material, the process can repeat itself indefinitely, leading to fragmentation. By applying Eqns. (16.7)–(16.9) one can find the velocity v at which the energetics would predict bifurcation. Since $a = 2a_c$, $v = 0.19C_s$. This velocity is quite smaller than the one predicted by Yoffé [17].

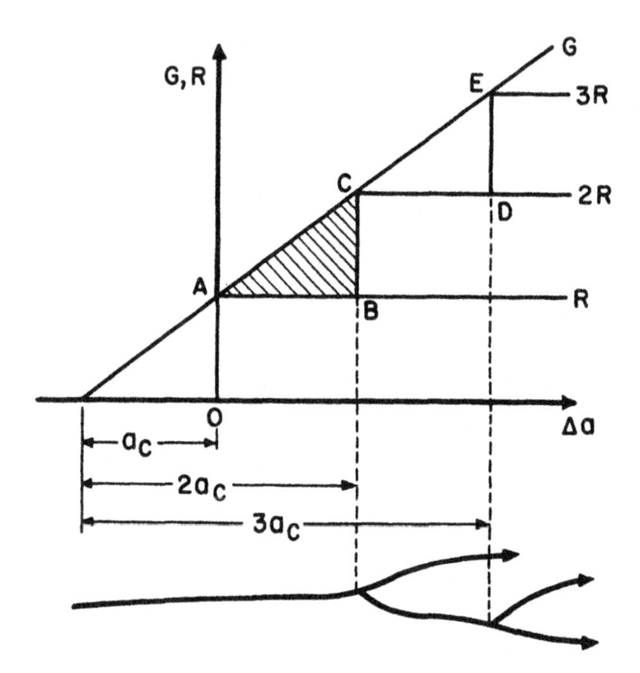

FIGURE 16.11 Crack bifurcation as energy release rate G exceeds crack resistance energy R because of excess energy available.

16.5 STRESS WAVE LOADING OF CRACKS

Up to now we assumed that the loading on the crack faces was constant as the cracks propagated. In dynamic loading situations one often has a tensile pulse propagating throughout the material and interacting with existing cracks. Figure 16.12 shows this situation schematically. A tensile stress wave (σ_0) travels at a velocity C_l toward the crack. Notice that, for simplicity, the front of the wave is parallel to the crack surface. The crack surface cannot transmit tension, and therefore the wave is reflected at the surface. At the edges of the crack, longitudinal and shear waves radiate (C_l and C_s). Eventually, they meet. During their expansion the loading of the crack increases. The problem for a general orientation relationship between the wave front and the crack was solved by Achenbach [7] and is presented in great detail by Achenbach [7] and Freund [19, 20]. The solution involves, as in the preceding section, Laplace transforms and the Wiener–Hopf method. The problem can equally well be solved by using Green's functions. The solution is

$$K_I(t) = 2\sigma_0 \frac{(1-\nu)^{1/2}}{1-\nu} \sqrt{\frac{C_l t}{\pi}} \tag{16.23}$$

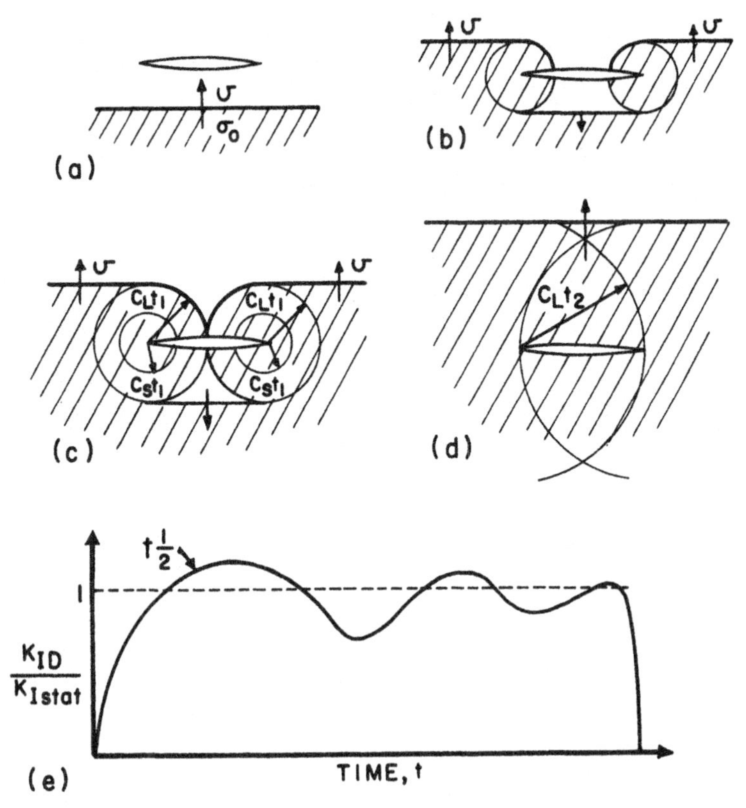

FIGURE 16.12 (a–d) Interaction of propagating tensile pulse and stationary crack; (e) resulting stress intensity. (Adapted from D. A. Shockey, SRI International, private communication.)

Thus, since all other parameters are constant, one has

$$K_I(t) \propto t^{1/2} \tag{16.24}$$

By means of a dimensional analysis applied to Figure 16.12, one can arrive, semiquantitatively, at a similar expression. Notice that the crack has a loaded region of $C_l t$. This region increases with t. Thus, the effective length of the crack is $C_l t$. By applying Eqn. (16.1), one has

$$K_I = \sigma\sqrt{\pi a} \cong \sigma\sqrt{\pi C_l t}$$

For $\nu = 0.3$, the two equations have predictions that are within a factor of 4:

$$K_I(t) \cong 6\sigma\sqrt{\pi C_l t} \quad \text{(Freund)}$$

$$K_I(t) \cong 1.7\sigma\sqrt{\pi C_l t} \quad \text{(preliminary analysis)} \tag{16.25}$$

After the maximum K_I is reached, fluctuations start taking place. These fluctuations are produced by secondary wave interactions (wave radiating from the right edge of the crack interacting with the left edge and vice-versa). These secondary interactions are shown in Figure 16.12(d). Figure 16.12(e) shows these fluctuations, which will eventually dampen around K_{stat}, the static stress intensity factor, assuming that the crack is stationary.

Ravichandran and Clifton [21] applied this concept to AISI 4340 steel and were able to develop an ultra-high-speed testing method providing rates of loading of $\dot{K}_I = 10^8$ MPa $m^{1/2}$ s^{-1}. Ravichandran and Clifton's [21] setup is schematically shown in Figure 16.13(a). It consists of a precracked target plate of thickness t that is impacted at a velocity v by a projectile of thickness $\frac{1}{2}t$. Thus, a tensile pulse is generated in the plane of the crack. The maximum length of this tensile pulse is equal to tC_l, where C_l is the velocity of propagation of the stress wave. The tensile pulses had less than 1 μs duration. The fracture

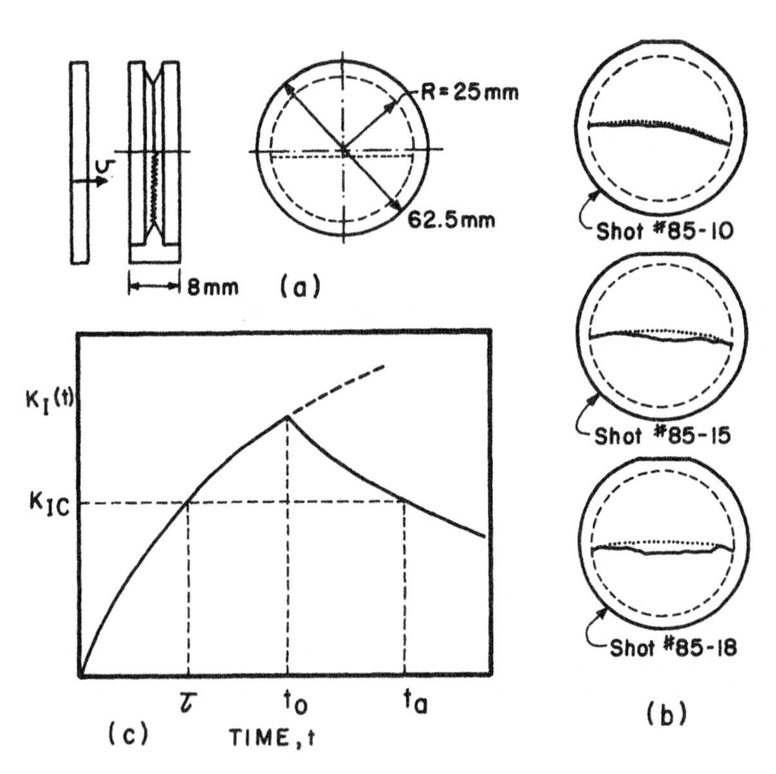

FIGURE 16.13 Ravichandran and Clifton's [21] dynamic loading method: (a) precracked specimen configuration (target) and flyer plate; (b) specimens subjected to tensile stress pulses of increasing amplitude (dotted line is initial crack position, solid line the final crack position); (c) schematic stress intensity. (From Ravichandran and Clifton [21], Figs. 3, 9, 15, pp. 165, 175, 182. Reprinted by permission of Kluwer Academic Publishers.)

toughness of the material was approximately 100 MPa $m^{1/2}$ at room temperature (see Table 16.1). This provided a rate of increase of the stress intensity factor of approximately

$$\dot{K}_I = \frac{K_{max}}{t} \cong \frac{100 \text{ MPa} \sqrt{\text{m}}}{10^{-6} \text{ s}^{-1}}$$
$$\cong 10^8 \text{ MPa m}^{1/2} \text{ s}^{-1}$$

Figure 16.13(b) shows the initial (dashed line) and final positions (solid line) of the crack fronts. It can be seen that the crack front advanced in the central region of the disk. By means of calculations it was possible to establish the curve of Figure 16.13(c); it shows the increase in $K(t)(\alpha t^{1/2})$ as well as its decay. Since we are dealing with a pulse whose length is smaller than the crack length, one has a simple exponential decay of the stress intensity factor. The student can establish this by reproducing the schematics of Figure 16.12 with a pulse width much smaller than the crack length. Ravichandran and Clifton [21] were able to establish the growth time $(t_a - \tau)$ based on a threshold stress intensity factor for growth, K_{IC}. For the three experiments shown in Figure 16.13(b), the following velocities were found:

Experiment	Impact velocity (mm/μs)	$\dot{a}/C_R$
85-10	0.040	0.786
85-15	0.044	0.826
85-18	0.060	0.905

These velocities indeed very high and approach the theoretical limit, C_R.

16.6 IS THE FRACTURE TOUGHNESS STRAIN RATE DEPENDENT?

This is an important question, and there is, unfortunately, no simple answer to it. Some materials show an increased K_{IC} (or J_{IC} or G_C) at high loading rates, whereas others exhibit a decrease in fracture toughness. One has, additionally, the following complication. Dynamically applied loading may interact with stationary and traveling cracks. A preexisting, stationary crack may show a different response than a crack that is already traveling at (or close to) its maximum velocity.

First, let us analyze the effect of a stress pulse on a stationary crack and let us assume that the duration of the pulse is comparable to or smaller than the time for a wave to run the length of the crack. This problem has been treated in detail by Shockey et al. [22–25]. The first consideration is that the dynamic loading produces an "overshoot" of approximately 25% over static loading. This overshoot is shown in Figure 16.12(e). Shockey et al. [24] subjected epoxy, with existing flaws (that were artificially introduced) to stress pulses of fixed duration ($\sim 2\ \mu s$) and increasing amplitude, by means of the flyer plate technique. They obtained the results plotted in Figure 16.14(a). Cracks with radius greater than 2 mm had a fixed threshold tensile stress for growth: 20 MPa. Shorter cracks, on the other hand showed a strong stress dependence of size. Figure 16.14(b) shows a replot of the same data with different mechanisms. The dashed curve shows the static fracture mechanics relationship ($K_{IC} = \text{const} = \sigma\sqrt{\pi a}$). The data seem to follow closer a minimum time criterion (solid curve). This criterion states that there is a threshold time required to activate cracks. When the duration of the applied pulse is less than the threshold time, the crack does not grow. This threshold time decreases as the stress is raised. Thus, the plot (generic) of Figure 16.15 shows the response of a material. The dynamic fracture toughness K_{Id} is a function of the time of application of the pulse, in contrast with the static stress intensity factor. These trends have been confirmed for Homalite-100 by Ravi-Chandar and Knauss [26, 27]; for times lower than 50 μs, the fracture toughness increases with a decrease in the time of application of the load; for times greater than 50 μs, the fracture toughness was independent of time. Shockey and co-workers [22–25] obtained similar results for AISI 4340 steel and 6061-651 aluminum alloy. Figure 16.16 shows the cracks that grew and did not grow when subjected to a 40-μs tensile pulse as well as the prediction from static fracture mechanisms. It can be seen that there is a significant difference between the observed behavior and the one predicted by the relationship $K_{IC} = \sigma\sqrt{\pi a}$.

In addition to the complicating factor brought about by the threshold time required for fracture initiation, there are changes in deformation mechanisms brought about by the increase in strain rate. These changes are briefly discussed below. In Section 16.1 two modes of fracture propagation are discussed, ductile and brittle fracture, which occur by cleavage and by void nucleation, growth, and coalescence, respectively. As discussed in Chapter 15, the flow stress of materials is strain rate dependent. In ductile fracture, the toughness of the material is determined, to a large extent, by the plastic deformation around the growing voids [Fig. 16.1(c)].

Freund [20] assumes, to a first approximation, that ductile fracture is strain governed. When plastic strain at the tip of the crack reaches critical level ε_c, fracture propagates [Fig. 16.17(a)]. From Chapter 15, we know that the flow stress increases with strain rate. Freund's [20] qualitative argument is reproduced in Figure 16.17(a). The deformation energy at the fixed critical strain ε_c is higher at higher strain rates. If we equate, to a first approximation, this deformation energy to the energy release rate G_{IC} of a material (or its fracture toughness K_{IC}), we see that, for $\dot{\varepsilon}_2 > \dot{\varepsilon}_1$,

$$G_{IC}(\dot{\varepsilon}_2) > G_{IC}(\dot{\varepsilon}_1)$$

On the other hand, one would not expect substantial plastic deformation in brittle fracture and, therefore, the strain rate dependence should be less prev-

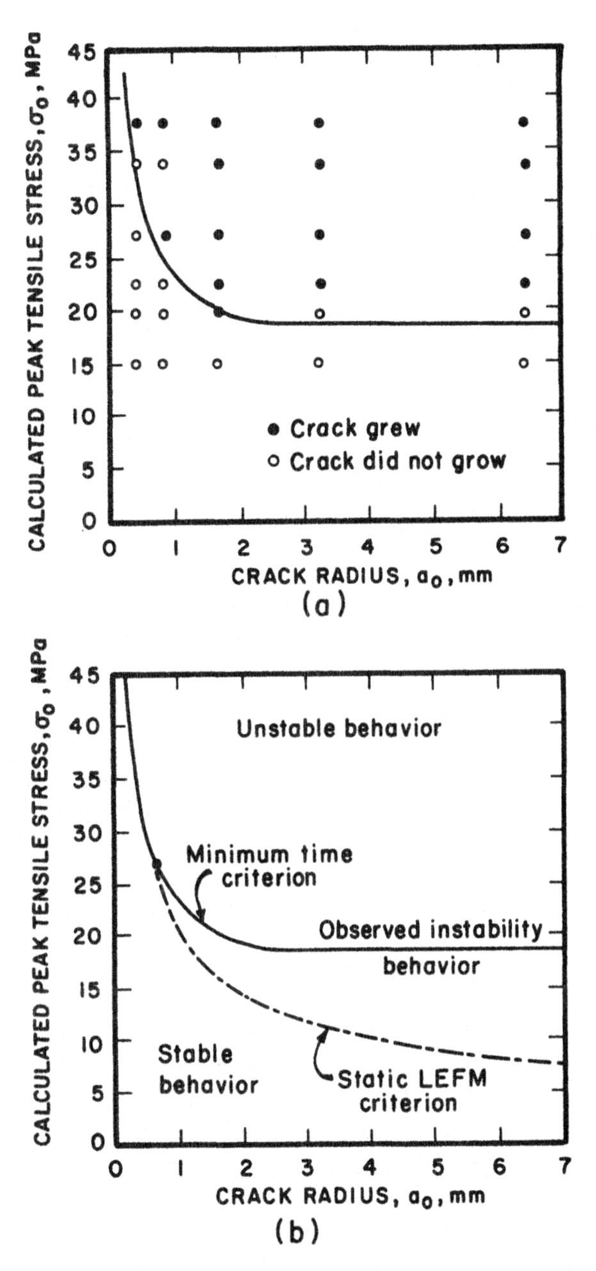

FIGURE 16.14 (a) Instability response of cracks of varying size in epoxy subjected to tensile stresses of 2.04 μs duration and varying amplitude. (b) Comparison of data with linear elastic fracture mechanics and minimum time criteria. (Courtesy of D. A. Shockey, SRI International.)

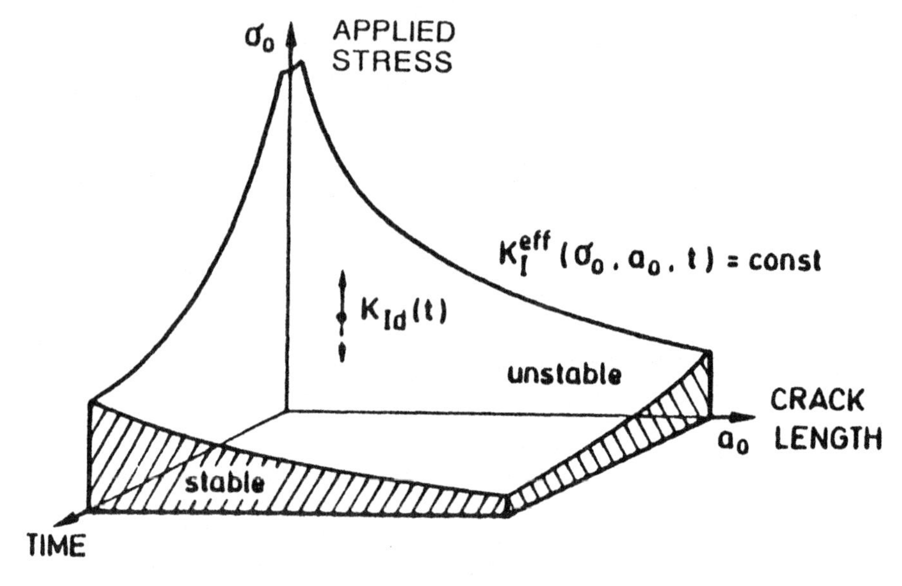

FIGURE 16.15 Relationship between applied stress, time, and crack length under dynamic conditions. (Courtesy of D. A. Shockey, SRI International, private communication, 1989.)

alent. Brittle fracture can be assumed to be produced, as a first approximation, when the stress at the crack tip reaches a critical level at which separation occurs, σ_c. If σ_c is in the elastic range, the areas in Figure 16.17(b) will be equal, and so will be the energy release rate G_{IC}. If plastic deformation occurs at the crack tip (and this is common even in brittle metals) as shown in Fig. 16.17(b), we will have a greater area at the lower strain rates, leading to

$$G_{IC}(\dot{\varepsilon}_1) > G_{IC}(\dot{\varepsilon}_2)$$

The effect of temperature on the flow stress of metals is opposite to the one of strain rate: as the temperature decreases, the flow stress increases. Figure 16.17(c) shows a schematic plot of G_{IC} as a function of both temperature and strain rate for both ductile and brittle fracture. For simplicity, G_{IC} for brittle fracture is assumed constant and is a plane perpendicular to the G_{IC} axis. If a material has two alternative fracture modes, it will choose the one that requires least energy. This is a direct extension of the minimum-energy principle. Body-centered-cubic metals and alloys tend to exhibit a ductile-to-brittle transition temperature. Steel, for example, fractures in a brittle mode at low temperatures and in a ductile mode at room temperature under quasi-static conditions. Under dynamic crack propagation conditions, on the other hand, this same steel may undergo brittle fracture. This behavior is qualitatively explained by the plot of Figure 16.17(c). The temperature at which the ductile-to-brittle transition occurs increases with increasing strain rate; it is the intersection of the G_{IC} sur-

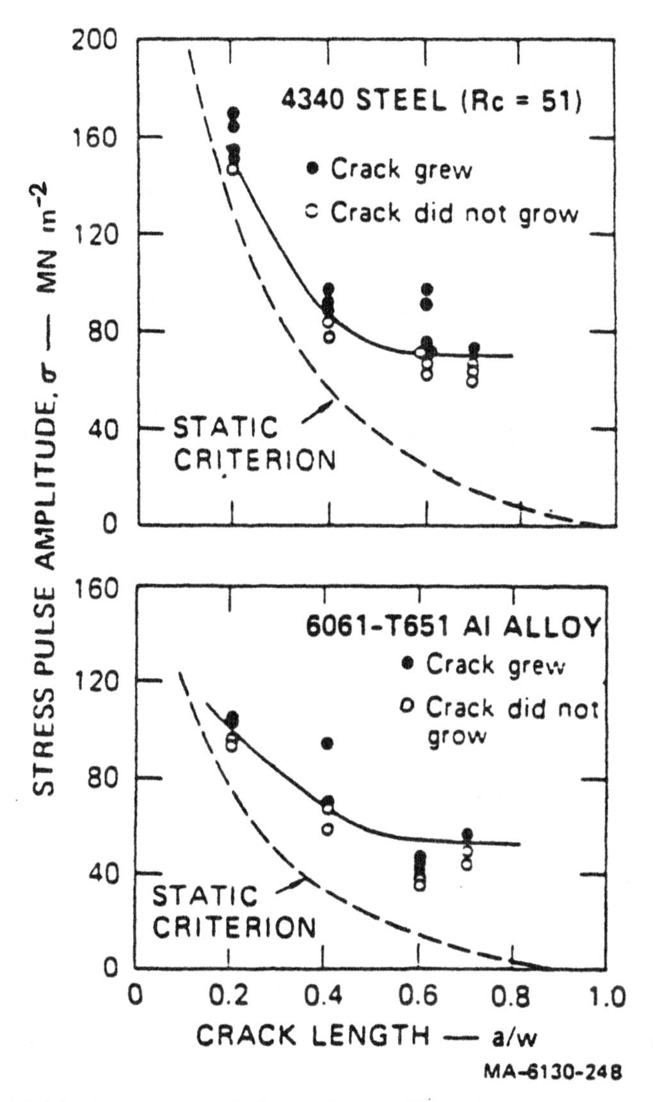

FIGURE 16.16 Comparison of observed instability behavior of cracks loaded by a 40-μs stress pulse with prediction from static linear elastic fracture mechanics. (Reprinted from *J. Mech. Phys. Sol.*, vol. 31, H. Homma et al., Fig. 3, p. 268, Copyright 1983, with permission from Pergamon Press Ltd.)

faces for ductile and brittle fracture. The above rationale, based on Freund's [20] explanation, explains the changes in fracture response of cold-rolled AISI 1018 steel. The results obtained by Wilson et al. [28] are in full agreement with the qualitative predictions of Figure 16.17(c). These results are shown in Figure 16.18(a). Fracture toughness was obtained as a function of temperature for two widely differing loading rates: 1 and 2×10^6 MPa $m^{1/2}$ s^{-1}. The

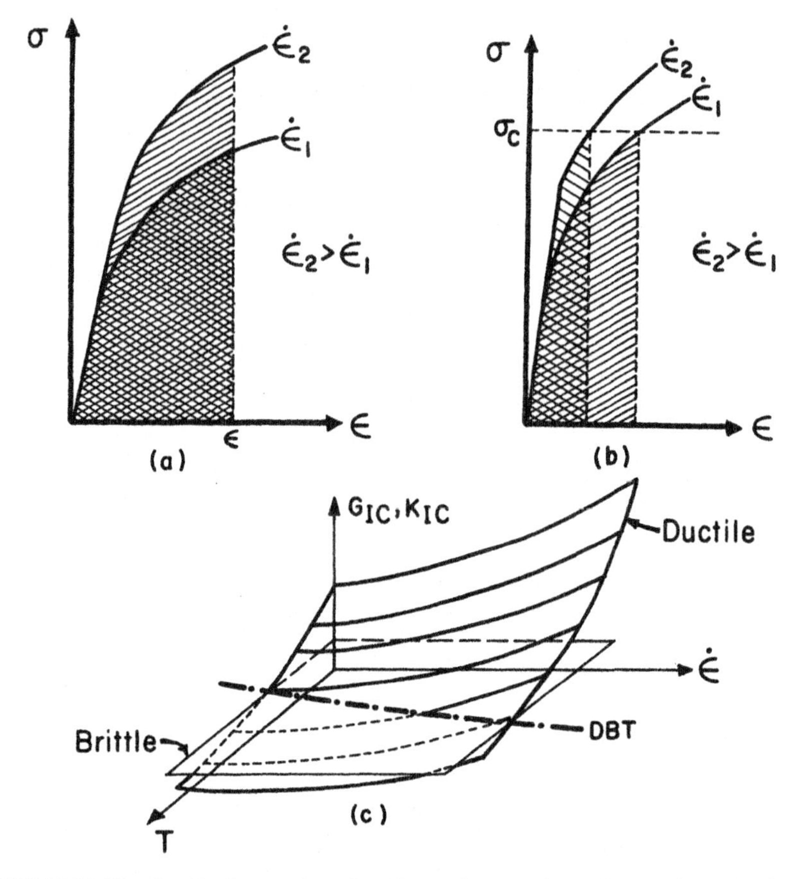

FIGURE 16.17 Qualitative explanation for strain rate dependence of (a) ductile and (b) brittle fracture; (c) G_{IC}–$\dot{\varepsilon}$–T plot showing how strain rate and temperature affect fracture mode. (Based on argument developed by Freund [20].)

following are important characteristics that determine the significant differences:

1. The ductile-to-brittle transition temperature is significantly higher (~150°C) for the dynamic tests.
2. For ductile fracture, the dynamic tests show higher toughness.
3. For cleavage fracture, the dynamic tests show lower toughness.

In the discussion above, an important aspect was not introduced. When plastic deformation occurs at the crack tip, additional kinetic energy is expended, at high crack propagation rates, to move the material that is plastically deformed. This kinetic energy term entails a higher energy expenditure to make the crack propagate. As an illustration of the effect of crack propagation velocity on the fracture toughness, the results obtained by Zehnder and Rosakis [29]

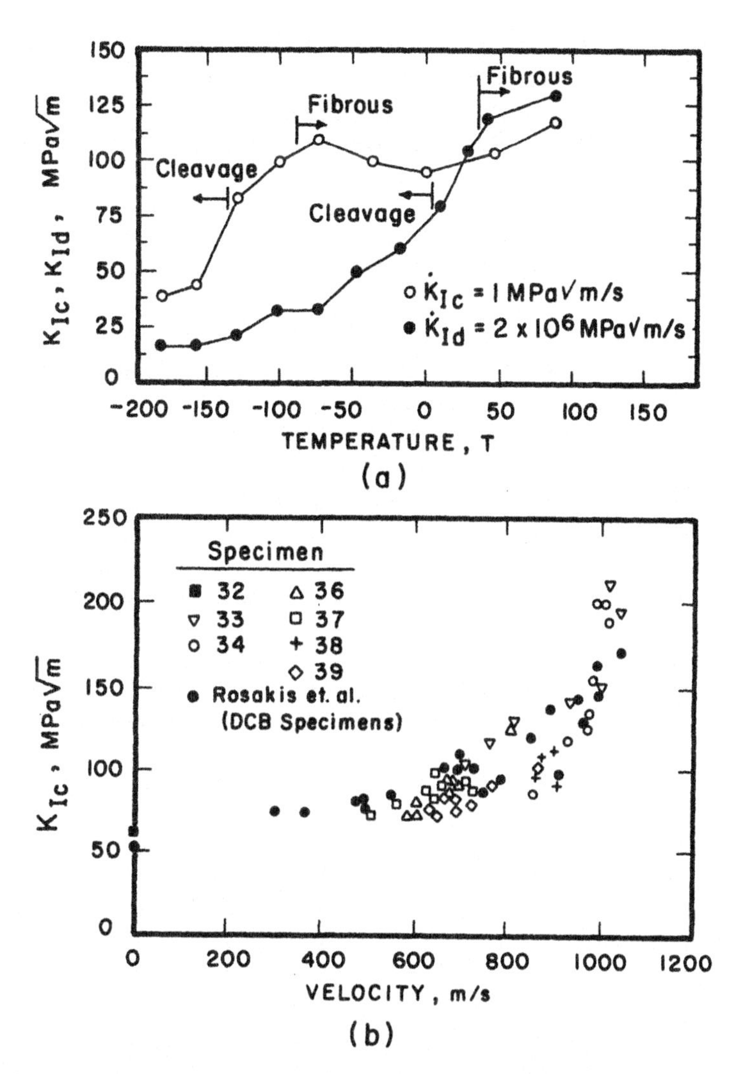

FIGURE 16.18 (a) Quasi-static and dynamic fracture toughness for a cold-rolled AISI 1018 steel as a function of temperature. (Reprinted from *Eng. Fract. Mech.*, vol. 13, M. L. Wilson et al., Fig. 13, p. 383, Copyright 1980, with permission from Pergamon Press Ltd.) (b) Dynamic fracture toughness versus crack velocity for AISI 4340 steel measured by the method of caustics. (From Zehnder and Rosakis [29], Fig. 11, p. 282. Reprinted by permission of Kluwer Academic Publishers.)

using the method of caustics are given in Figure 16.18(b). The steel investigated, AISI 4340, is a material of considerable technological importance. Here K_{IC} increases from ~60 MPa $m^{1/2}$ (quasi-static) to ~200 MPa $m^{1/2}$ at a propagation velocity of 1000 m/s. The technique used by Zehnder and Rosakis [29] is described in the next section.

As a concluding remark in this section, it should be emphasized that microstructural effects are of utmost importance in determining the resistance of a material to crack propagation. It is very difficult to develop a fully predictive capability of the response of materials due to the complexity of microstructural effects.*

16.7 DETERMINATION OF DYNAMIC FRACTURE TOUGHNESS

Under dynamic conditions, the stress field at the tip of the crack is not simply related to the applied stress (or force). Thus, one cannot obtain these stress fields from the measurement of the externally applied force if the loading rate exceeds a critical value; rather, the stresses at the tip of the propagating crack have to be directly measured. The determination of the dynamic stress intensity factor may be conveniently accomplished by two optical methods: (1) dynamic photoelasticity and (2) the dynamic caustic [30, 31]. The former extracts K_I from the isochromatic data obtained from the characteristic fringe loops at the tip of a crack propagating at high velocity and the latter from the shadow spots provided by the dynamic caustics.

The method of caustics was introduced by Schardin [32] and later developed by Manogg [33], Theocaris [34], and Kalthoff et al. [30]. It will be briefly described here. The basis of the method is that a light ray passing through a transparent stressed plate is deviated from its straight path partly due to thickness variation and partly due to the change in refractive index caused by the stress optic effect. If the plate contains a crack, the rays are deviated from the region around the crack tip and these form a singular curve called variously "stress corona," "shadow spot," or "caustic" on a reference plane placed some distance away from the specimen. The size of the caustic is related to the stress intensity factor. A curve called the initial curve is defined on the specimen plane.

Light rays from the outside of this initial curve fall outside the caustic, and rays from inside the initial curve fall on or outside the caustic curve. Hence, the caustic curve is a bright curve surrounding a dark region. The method generally followed in extracting K_I from the shadow spot is to measure its diameter D of the caustic and to relate it to K_I by the equation [35]

$$K_I = 0.56 \frac{F}{z_0 c d_{\text{eff}}} D^{5/2} \tag{16.26}$$

where F is a velocity correction factor that can be approximated by

*J. R. Brockenbough, S. Suresh, and J. Duffy (*Phil. Mag.* **58** (1988) 619) and S. Suresh, T. Nakamura, Y. Yeshrur, K.-H. Yang, and J. Duffy (*J. Am. Cer. Soc.* **73** (1990) 2457) studied dynamic fracture in brittle materials and concluded that the process zone ahead of the main crack, leading to an overall loss of stiffness, was affected by the velocity of the principal crack. Thus, the dynamic fracture toughness of brittle materials was higher than the static fracture toughness (K_{ID}/K_{IC} = 1.1–1.4).

$$F = 1 - 0.01\left(\frac{v}{C_R}\right) \quad \text{for } \frac{v}{C_R} < 0.4 \tag{16.27}$$

c is an optical constant for the material, z_0 is the distance of the shadow image plane from the object plane, and d_{eff} is the effective thickness of the plate. Figure 16.19 shows an optical configuration illustrating the method of caustics. The vector **OI** is deflected to **OR** by the stress and change in thickness of the cracked plate.

Ravi-Chandar and Knauss [26, 27] used the method of caustics for Homalite-100 and used a different factor F relating the dynamic stress intensity factor

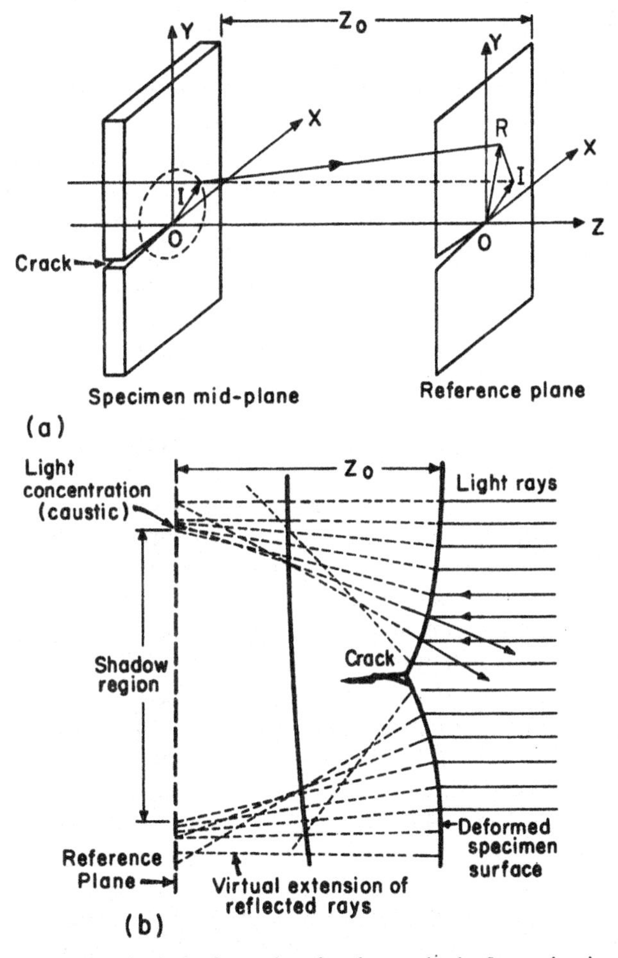

FIGURE 16.19 (a) Optical configuration for the method of caustics in the transmission mode. (From Ravi-Chandar and Knauss [26].) (b) Formation of caustic due to reflection of light from the polished, deformed specimen surface near the crack tip. (From Zehnder and Rosakis [29], Fig. 1, p. 273. Reprinted by permission of Kluwer Academic Publishers.)

the static one. Factor F is calculated as

$$F = \frac{4\alpha_1\alpha_2 - (1 + \alpha_2^2)^2}{(\alpha_1^2 - \alpha_2^2)(1 + \alpha_2^2)} \tag{16.28}$$

where

$$\alpha_1 = \left[1 - \left(\frac{v^2}{C_1^2}\right)\right]^{1/2} \qquad \alpha_2 = \left[1 - \left(\frac{v^2}{C_s^2}\right)\right]^{1/2}$$

where v is the crack velocity, C_l is the longitudinal wave speed, and C_s is the shear wave speed. Rosakis [36] used a similar but slightly modified expression for F. Figure 16.19(b) shows the method of caustics applied to a reflective area; this is from the work of Zehnder and Rosakis [29]. The crack deforms the initially flat surface and a parallel beam of light is reflected away from its source. The shadow is formed in the region marked in Figure 16.19(b). Around the shadow, an aura that is very bright due to the concentration of light is formed. Figure 16.20 shows a sequence obtained by combining the method of caustics with high-speed photography. A pulsed laser provided the light source, pulsed with exposure time of 15 ns. These sequential pictures were photographed in a rotating prism camera (of the Cordin type) operating in a streak mode; selected exposures are shown in Figure 16.20. The crack tip is at the bottom portion of the figures, and cracking was produced by the drop of a weight with a velocity of 10 m/s. The growth of the dark circle can be seen between 42 and 252 μs. From Eqn. (16.26), one sees that this corresponds to the increase in K_I at the crack tip $K_I \propto D^{5/2}$. The dark circle initiates motion at 259 μs. One sees concentric circles, which correspond to elastic waves emanating from the tip region. The lateral dimensions of the specimen are sufficiently large (much larger than in the photograph), ensuring that no reflected waves interact with crack propagation. Zehnder and Rosakis [29] were able to measure crack velocities of up to 1100 m/s in AISI 4340 steel and determined the corresponding stress intensity factors. This is possible because one can simultaneously determine the crack velocity (from the distance traveled between different frames) and the stress intensity factor (from the diameter of the caustic). The method of caustics has proved to provide satisfactory results under static loading [26, 27]. The reliability of the method of caustics for dynamic loading was investigated by Ravi-Chandar and Knauss [26] for Homalite in the transmission mode; they adopted two experimental configurations, shown in Figure 16.21: (1) a semi-infinite crack in an infinite medium being loaded by uniform pressure pulse P on the semi-infinite crack faces and (2) loading confined to a small length of the semi-infinite crack face at some distance a from the crack tip. They plotted the results in Figure 16.21. For the first configuration (loading with uniform pressure P), the crack position as a function of time is shown. For the second configuration (loading confined to a small region), the load was insufficient to initiate crack growth. From the first

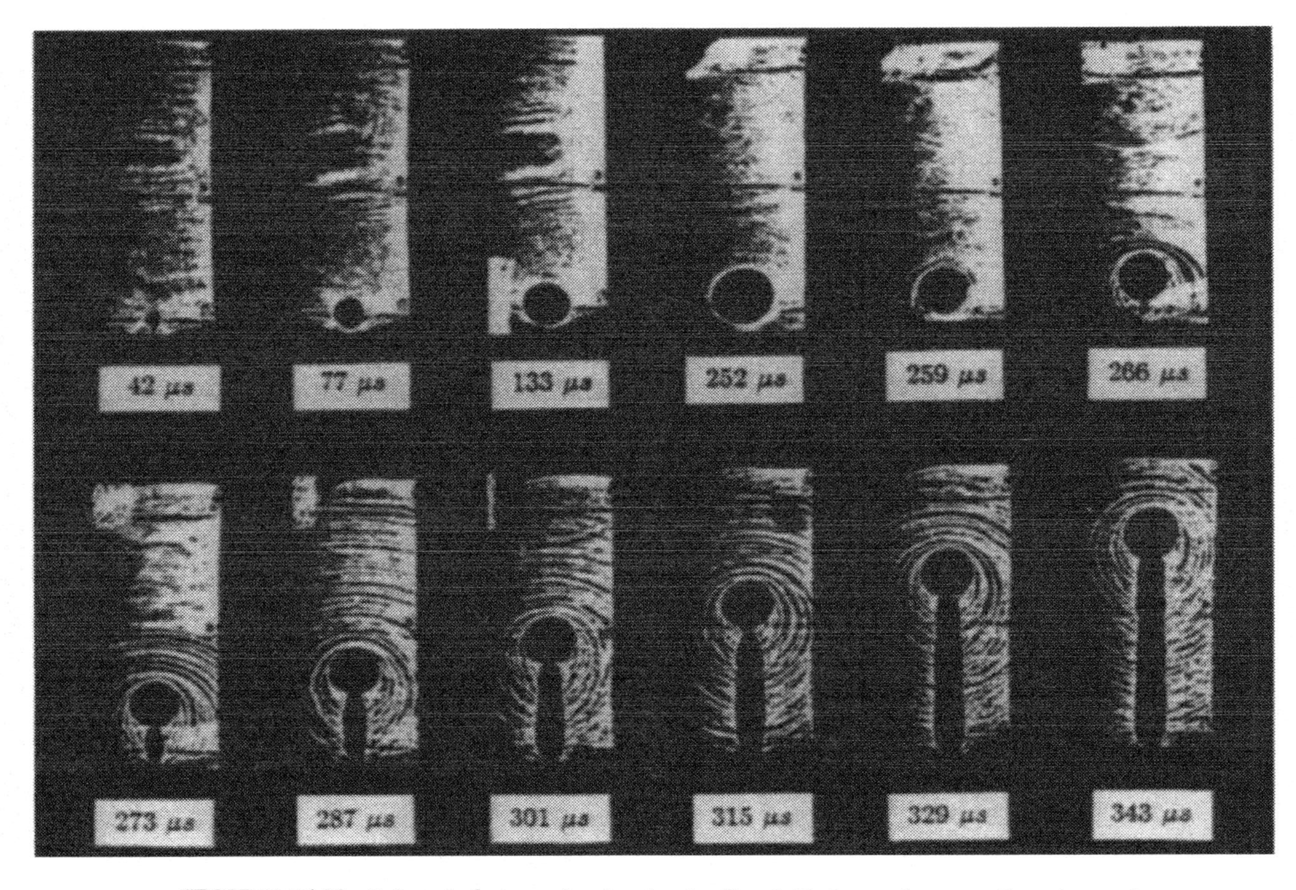

FIGURE 16.20 Selected photographs showing loading initiation and propagation stages of crack growth in a three-point bend specimen. (From Zehnder and Rosakis [29], Fig. 4, p. 276. Reprinted by permission of Kluwer Academic Publishers.)

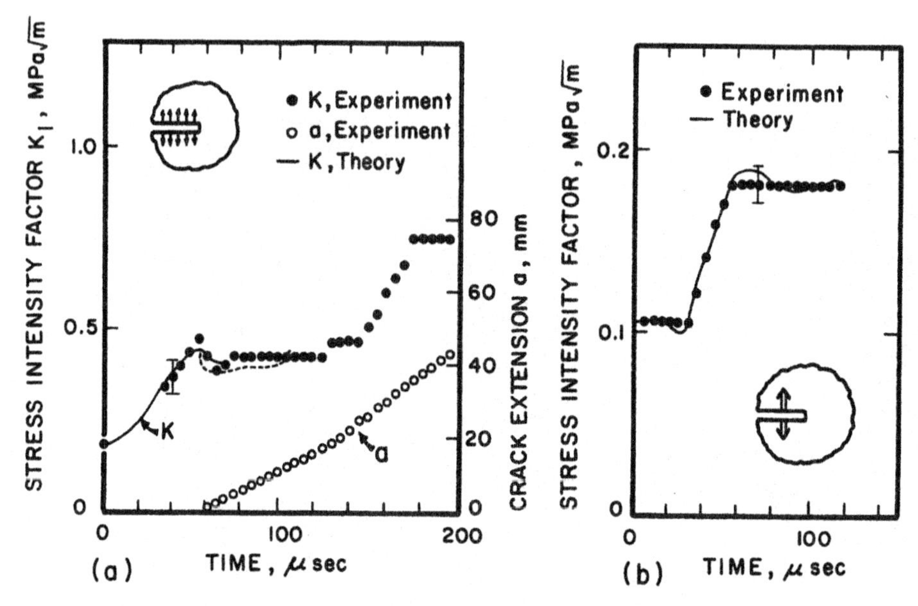

FIGURE 16.21 History of stress intensity factor for distributed and local pressure on crack surfaces; comparison of theory and experiments. (Reprinted from *Eng. Fract. Mech.*, vol. 23, W. G. Knauss and K. Ravi-Chandar, Fig. 1, p. 10, Copyright 1986, with permission from Pergamon Press Ltd.)

configuration, the crack attained a constant velocity of 240 m/s within the time resolution of the high-speed camera (5 μs). This velocity can be obtained from the slope of the crack length–time plot in Figure 16.21(a). These results showed that K_I increases with time of load application (both configurations) and with crack extension (first configuration). The idealized crack tip stress field and the assumption of a constant fracture energy establishes a relationship between the instantaneous stress intensity factor and the corresponding instantaneous crack speed. One form of this relationship is

$$\frac{E\gamma_f}{K^2} = \left(1 - \frac{v}{C_R}\right) \tag{16.29}$$

It is worth pointing out that many researchers believe that a unique relation exists between the instantaneous stress intensity factor and the instantaneous crack velocity. Often the effect of stress intensity factor on crack speed is presented in K–v curves. The results obtained by Zehnder and Rosakis [29] strongly support this unique relationship for the steel under study (AISI 4340). In experiments with Homalite-100, Ravi-Chandar and Knauss [26] found that a one-to-one correspondence does not appear to exist. Rather, in spite of sizable change in stress intensity, the crack is reluctant to change speed; the crack speed changed primarily when rapid changes in stress intensity occur due to waves impinging on the moving crack.

Figure 16.22 shows the stress intensity versus crack velocity obtained by many investigators working with Homalite-100. The upper portion of these curves (see, e.g., Dally's curve [38]) shows that a considerable range of velocities may be achieved depending on how rapidly the crack tip is loaded: the crack speed seems to be set by the crack tip conditions at initiation and to

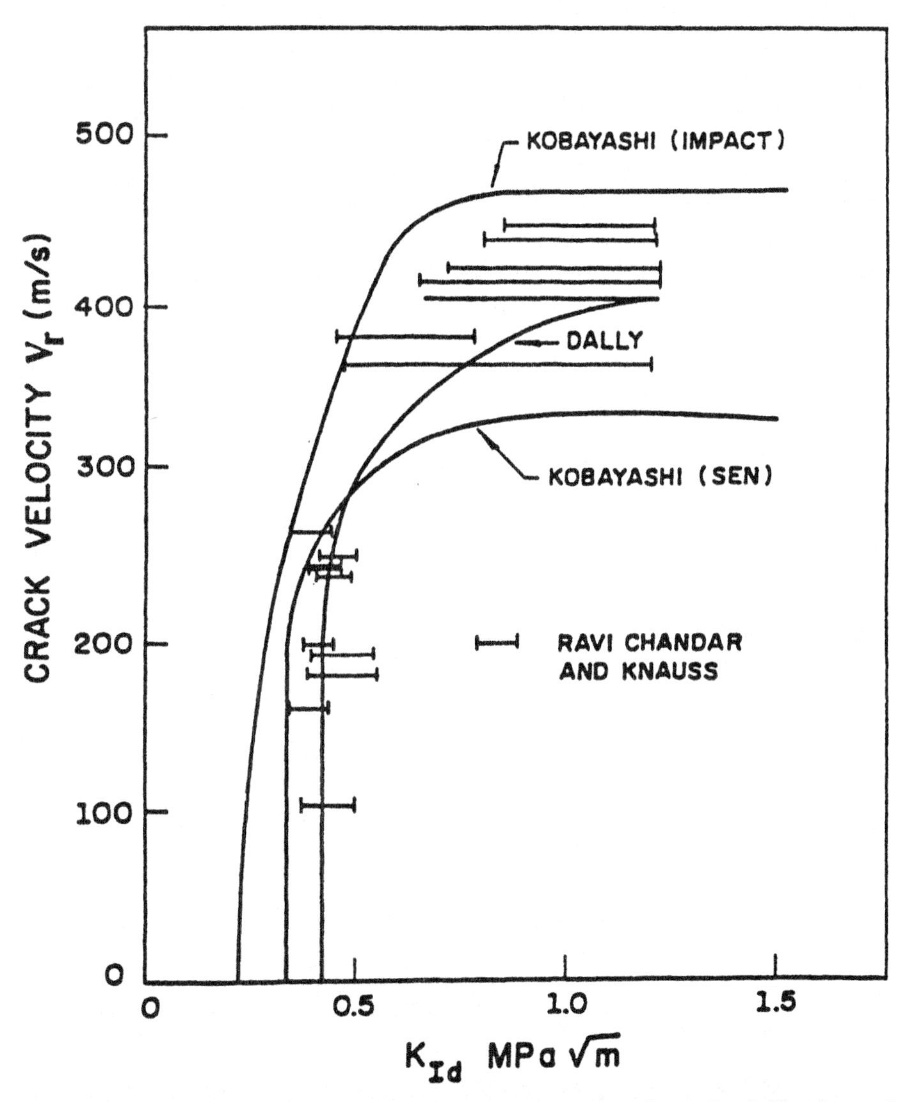

FIGURE 16.22 K_{Id}-vs.-v relations for Homalite-100. (From Ravi-Chandar and Knauss [26], Fig. 9, p. 136. Reprinted with permission of the publisher.)

contradict the existence of a clear upper velocity limit that is supposed to exist at less than the Rayleigh wave speed. While idealized fracture predicts the Rayleigh wave speed to be the limiting speed of crack growth, engineering materials exhibit much lower crack growth rates. The reason for the reduced rate depends on the material. Certainly the effects of plasticity, hole growth, and the viscoelastic behavior are important. In a polycrystalline aggregate, the crack follows an irregular path and cleavage planes change from grain to grain, with adjustments needed. However, regardless of these effects, the process of generating microfractures and their subsequent coalescence at the crack tip requires an interaction between these microfractures, which in turn requires time. It is thus readily understandable that cracks cannot propagate with near wave speed velocities as long as multiple microfractures play a role. One would expect that when fracture occurs by "fairly clean cleavage," as in a single crystal, crack speeds should be higher.

The experimentally observed crack velocities in all kinds of materials are always significantly lower than the Rayleigh surface wave velocity. A survey of the velocity of propagation of cracks in various media is given in Table 16.2. The formation of microcracks in the "process" zone of the major crack is, according to Ravi-Chandar and Knauss [26], the main reason for the limiting velocity $\sim 0.5C_R$.

TABLE 16.2 Survey of Brittle Crack Velocities

Material	v/C_0 [a]	v/C_R	v (m/s)
Glass [b]	0.29	—	1500
Steel [b]	0.20	—	1000
Steel [b]	0.28	—	1400
Cellulose acetate [b]	0.37	—	400
Glass [c]	0.29	0.51	
	0.28	0.47	
	0.30	0.52	
	0.39	0.66	
Plexiglas [c]	0.33	0.58	
	0.36	0.62	
	0.36	0.62	
Homalite-100 [c]	0.19	0.33	357
	0.22	0.38	411
	0.25	0.41	444
	0.27	0.45	487
AISI 4340 Steel [d]	0.21	0.30	1100

[a] C_0 = bulk sound velocity.
[b] From Roberts and Wells [20].
[c] Data collected by Knauss and Ravi-Chandar [26, 27].
[d] From Zehnder and Rosakis [29].

16.8 SPALLING*

16.8.1 Qualitative Description

Spalling is a dynamic material failure that occurs due to tensile stresses generated by the interaction (collision) of two release (or rarefaction) waves. Section 7.3 provides a brief description of the basic mechanism. The plot typified by Figure 16.23 is most helpful in understanding spalling. The left-hand side represents the projectile, and the right-hand side represents the target. At time $t = 0$, the projectile impacts the target. Elastic waves are emitted into the projectile and target; they are followed by plastic waves. In Figure 16.23, the inverse of the slope of the diagonal lines gives the velocities of the waves; it

*This section is based on an overview article coauthored by M. A. Meyers and C. T. Aimone, *Prog. Mater. Sci.*, 328 (1983), 1.

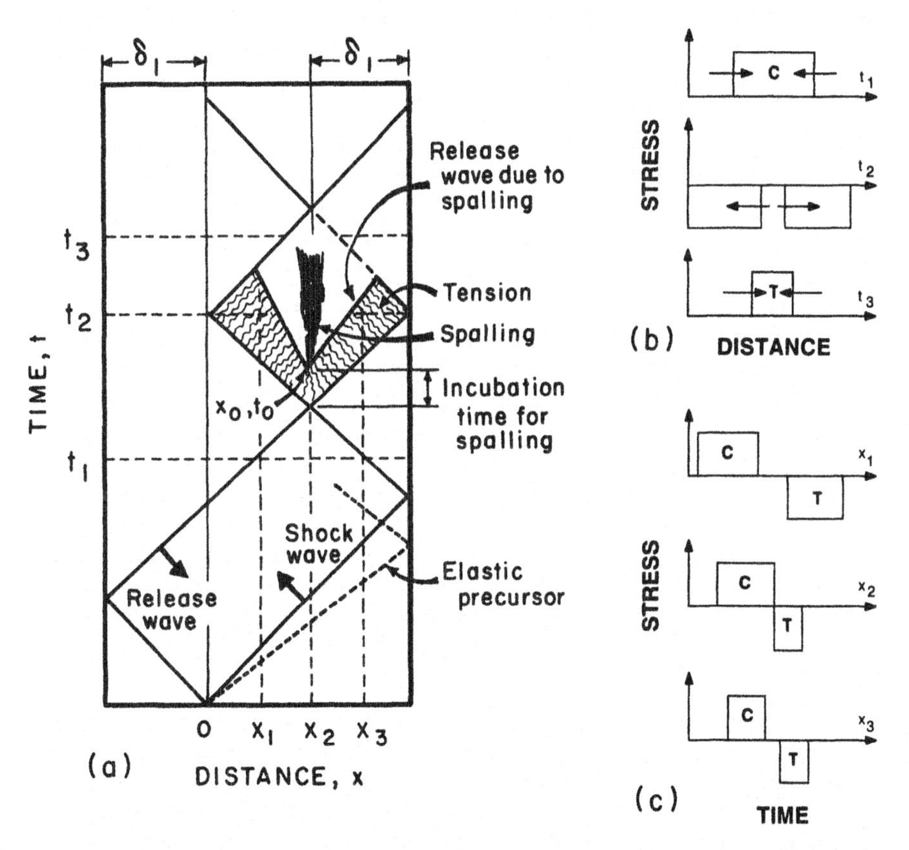

FIGURE 16.23 (a) Distance–time plot showing propagation of wave in target and projectile after impact and spalling. (b) Stress profiles at times t_1, t_2, and t_3. (c) Stress histories at positions x_1, x_2, and x_3.

can be readily seen that the elastic precursor has a higher velocity than the shock wave. As the elastic and plastic waves encounter the free surfaces of target and projectile, they reflect back. The x–t regions corresponding to tension and compression are shown in Figure 16.23(a). The elastic precursor is assumed, in this figure, to be small with respect to the total pulse amplitude. Figure 16.23(b) shows the stress pulses at three different times, whereas Figure 16.23(c) shows the stress histories at three different positions. These cross-plots show how the wave configuration changes. Figure 16.23(b) shows a compressive pulse t_1 at positions x_1, x_2, and x_3, the compressive and tensile pulses, which have a reduced duration, are again separated. Of importance is the fact that the first region to feel the tensile pulse is the one marked (x_0, t_0). It is presently known that spalling is dependent both on the amplitude and on the duration of the reflected tensile pulse. If this pulse is of sufficient magnitude to produce spalling, it should occur. Spalling will produce release waves at the newly created internal surfaces (spall surfaces) that alter the subsequent configuration of pulses. In Figure 16.23(a) an incubation time for the onset of spalling was assumed and is indicated. As spalling initiates, release pulses emanate from the newly created free surfaces and decrease the tensile stresses. If the projectile and target are of the same material, the distance of the spall from the free surface should be roughly equal to the thickness of the projectile. This can be seen by observing the interaction of the waves. In Figure 16.23(a), the thickness of the flyer plate is indicated by δ_1. The travel time of the wave in the flyer plate (the compressive wave and the tensile reflected wave) determines the duration of the compressive wave in the target; this latter one, in its turn, establishes the thickness of the spalled layer. Changes in the shape of the wave as it travels through the target as well as the elastic precursor and other factors complicate the picture in many cases.

Spalling was first studied by Hopkinson [39, 40], in the beginning of the century. The two excerpts below show that he correctly identified a spall:

> Iron or steel may become brittle under sufficiently great forces applied for very short times. No ordinary hammer blows will do, but the blow delivered by a high explosive is quick enough. If a slab of wet gun-cotton be detonated in contact with a mild steel plate, a piece will be blown out and the edges will show a sharp crystalline fracture with hardly any contraction of area. . . . Yet this same plate could be bent double in a hydraulic press.. . . (from "Scientific Papers" [40], p. 14).

> The fact that a blow involving only pressure may, by the effects of wave action and reflection, give rise to tension equal to or greater than the pressure applied, often produces curious effects. [As an illustration]. . . If such a cylinder of gun-cotton weighing one or two ounces be placed in contact with a mild steel plate, the effect, if the plate be half an inch thick or less, will be simply to punch out a hole of approximately the same diameter as the gun-cotton. . . But if the plate be three-quarters of an inch thick, the curious result. . . is obtained. . . Instead of a complete hole being made, a depression is formed on the gun-cotton side of the plate, while on the other side a scab of metal of corresponding diameter is

> torn off, and projected with a velocity sufficient to enable it to penetrate a thick wooden plank. . . The velocity in fact corresponds to a large fraction of the whole momentum of the blow. . . The separation of the metal implies, of course, a very large tension, which can only result from some kind of reflection of the original applied pressure, but the high velocity shows that this tension must have been preceded by pressure over the same surface, acting for a time sufficient to give its momentum to the scab. . . .
>
> I caused then a two-ounce cylinder of gun-cotton to be detonated in contact with a somewhat thicker plate. In this case no separation of metal was visible; the only apparent effects being a dint on one side and a corresponding bulge on the other. On sawing the plate in half, however, I was gratified to find an internal crack, obviously the beginning of that separation which in the thinner plate was completed. (from "Scientific Papers [40]," p. 423).

He also indicated the increased brittleness of steel under dynamic conditions; he described the brittle appearance (which he called "crystalline fracture") of the fractures produced dynamically and the very small amount of plastic deformation associated with them.

In the early 1950s, Rinehart [41, 42] reported the results of systematic experiments on steel, brass, copper, and an aluminum alloy using a modification of Hopkinson's [39, 40] technique. He found that there was a critical value of the normal tensile stress (σ_c) required to produce spalling and that this value was a characteristic of the material. He also observed and correctly explained the phenomenon of multiple spalling produced when a triangularly shaped pulse is reflected and has an amplitude substantially higher than σ_c. He proposed the following expression for the thickness of the spalled layers (direct contact explosives detonating on top of the metal were used):

$$\delta_1 = \tfrac{1}{2}x_1$$

where x_1 is such that $\sigma(x_1) = \sigma_0 - \sigma_c$. For the second spall, the thickness δ_2 is

$$\delta_2 = \tfrac{1}{2}(x_2 - x_1)$$

where $\sigma(x_1) - \sigma(x_2) = \sigma_c$. Figure 16.24 shows the sequence of spalling as well as the stress distribution function. The shape of the pulse determines the distances $\delta_1, \delta_2, \cdots$. The process repeats itself until the stress amplitude of the wave has been reduced to less than σ_c.

Figures 16.23 and 16.24 show the formation of spalling in simple unidimensional wave propagation. More complex loading situations can, obviously, generate internal tensile stresses in materials, leading to internal fracture, if the amplitude and duration are sufficient. An illustration of tensile stresses generated by impact is shown in Figure 16.25. This computer simulation shows how the reflected compressive wave and the lateral release waves interact in a cy-

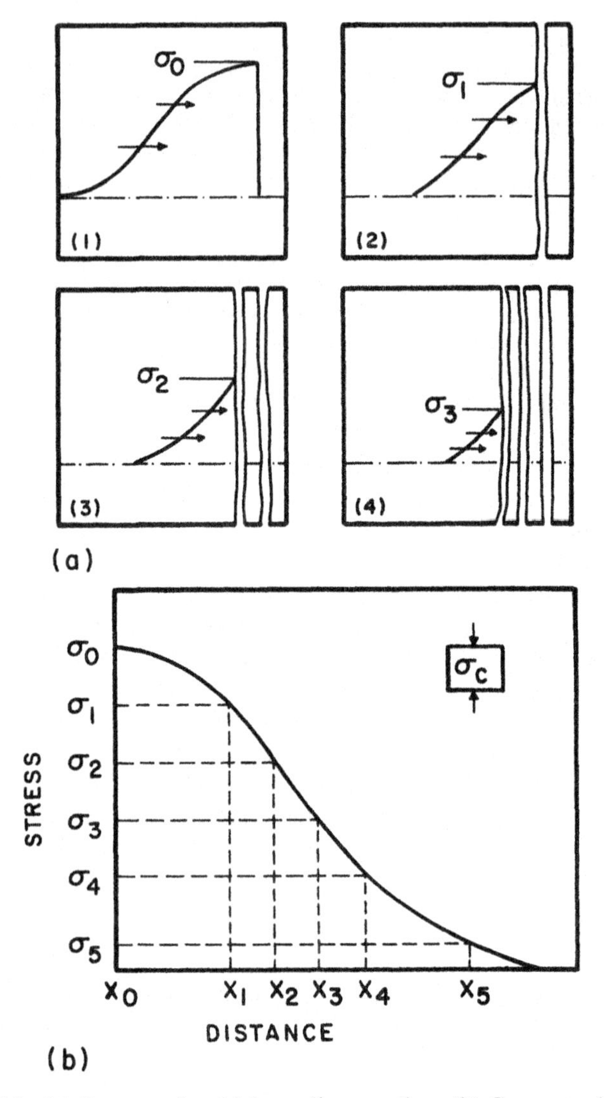

FIGURE 16.24 (a) Process of multiple spall generation. (b) Corresponding profile of stress with positions of various stress levels that are separated by c. (From Rinehart [42], Figs. 2 and 3, p. 1230. Reprinted with permission of the publisher.)

lindrical (Taylor) specimen subjected to normal impact, generating tensile stresses at $t = 0$.

16.8.2 Quantitative Spalling Models

In this section, the quantitative/predictive models developed by Sandia National Laboratories, Stanford Research Institute, Los Alamos, and Lawrence Liver-

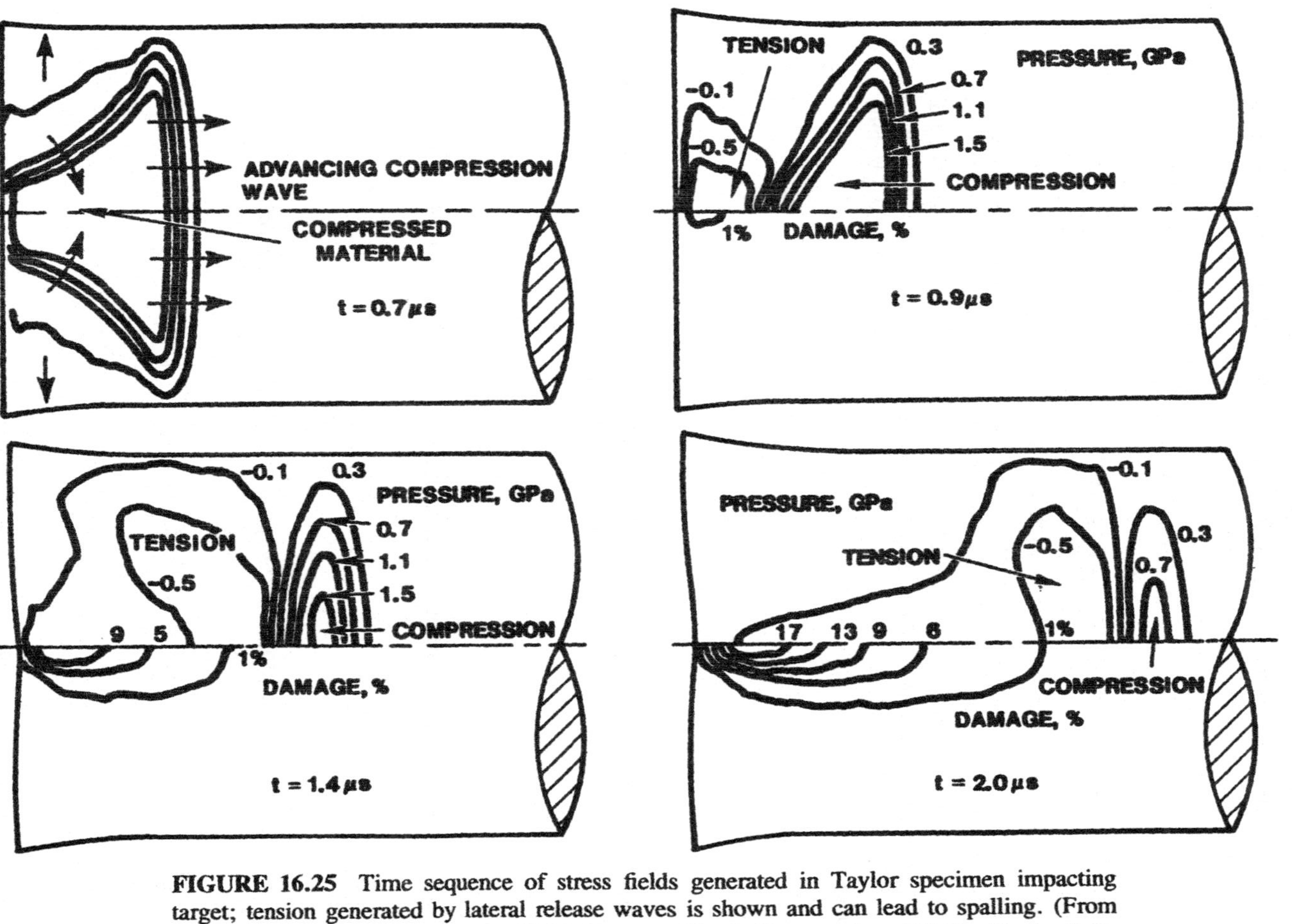

FIGURE 16.25 Time sequence of stress fields generated in Taylor specimen impacting target; tension generated by lateral release waves is shown and can lead to spalling. (From D. E. Grady and M. E. Kipp, in "High-Pressure Shock Compression of Solids," eds. J. R. Asay and M. Shahinpoor, Springer, NY, 1993, p. 265, Fig. 3.18, Fig. 8.36.)

more will be briefly reviewed. Davison and Stevens [43] introduced the concept of a continuum measure of spalling after reviewing the existing spall criteria and systematizing them. They classified them into instantaneous and cumulative, and local and nonlocal: instantaneous if spalling depends only on current values of the field variables and cumulative if it depends on the history of these variables; local if only values of field variables at a candidate spall plane enter into the determination of damage of this plane and nonlocal if field values at distant points also have a bearing on the determination. Further, the cumulative-damage criteria are classified into simple and compound: simple if the mechanism of damage accumulation does not depend on the amount of previously accumulated damage and compound if it does. Davison and Stevens [43] introduced a continuous measure of damage D, and proposed a theory of compound-damage accumulation. Their function D can be assumed to be the degree of separation along the spall interface and varies from 0 (no incipient spall) to 1 (complete spall). In *simple damage accumulation*, if a time τ_0 at the stress σ produces a damage D_0, then an increment in time of Δt will produce an increment in damage ΔD such that

$$\frac{\Delta D}{\Delta t} = \frac{D_0}{\tau_0} \tag{16.30}$$

In other words, damage increases monotonically with time. The stress dependence of damage accumulation can be introduced by the time τ_0, which is the time to produce a specific damage D_0; it decreases with increasing tensile stress σ:

$$\tau_0 = \hat{\tau}(\sigma) \tag{16.31}$$

For two or more load applications, one has, from Eqn. (16.30),

$$\Delta D_i = \Delta t_i \frac{D_0}{\hat{\tau}(\sigma)} \tag{16.32}$$

The sequence of load applications produces the following damage:

$$D = \sum_i \Delta D_i = \sum \left[\frac{\Delta t_i}{\hat{\tau}(\sigma)}\right] D_0 \tag{16.33}$$

or, integrating Eqn. (16.33),

$$D(x, t_f) = D_0 \int_{-\infty}^{t_f} \left[\frac{dt}{\hat{\tau}[\sigma(x, t)]}\right] \tag{16.34}$$

Tuler and Butcher [44] used an inverse relationship between σ and $\hat{\tau}(\sigma)$:

$$\hat{\tau}(\sigma) = \tau \left[\frac{(\sigma - \sigma_0) + |\sigma - \sigma_0|}{2\sigma_0} \right]^{-\lambda} \tag{16.35}$$

where σ_0 is a critical stress below which there is no damage. Thus, Eqn. (16.34) becomes, by using relationship (16.35),

$$D(x, t_f) = \frac{D_0}{\tau} \int_{-\infty}^{t_f} \left(\frac{\sigma(x, t) - \sigma_0 + |\sigma(x, t) - \sigma_0|}{2\sigma_0} \right)^{\lambda} dt \tag{16.36}$$

However, one may envisage that the presence of flaws (damage) accelerates the rate of damage accumulation, that is, a flawed microstructure fails more readily. This is the *compound-damage* accumulation hypothesis of Davison and Stevens [43]. Thus

$$\dot{D} = f(\sigma, D)$$

where f is called the damage rate function. It has been shown that the following form of the function f is most general (power expansion):

$$\dot{D} = f(\sigma, D) = \frac{D^*}{\tau_0} \left[f_0(\sigma) + f_1(\sigma) \frac{D}{D^*} + f_2(\sigma) \left(\frac{D}{D^*} \right)^2 + \cdots \right]$$

By eliminating second and higher order terms,

$$\dot{D} = f(\sigma, D) = \frac{D^*}{\tau_0} \left[f_0(\sigma) + f_1(\sigma) \frac{D}{D^*} \right] \tag{16.37}$$

where D^* is the damage at total separation. By solving Eqn. (16.37) (f_0 and f_1 are constants at the constant stress σ_1), one has

$$\frac{D}{D^*} = \frac{f_0}{f_1} \left[\exp\left(\frac{f_1 t}{\tau_0} \right) - 1 \right] \tag{16.38}$$

The following values are attributed for the functions f_0 and f_1:

$$f_0 = \tfrac{1}{2} \tau V_N B \sigma_G (\Sigma - \Sigma_N + |\Sigma - \Sigma_N|)$$

$$f_1 = 3\tau C \sigma_G \Sigma$$

where

$$\Sigma = \tfrac{1}{2}(\sigma - \sigma_G + |\sigma - \sigma_G|)$$

$$\Sigma_N = \tfrac{1}{2}(\sigma_N - \sigma_G + |\sigma_N - \sigma_G|)$$

The stresses were redefined as Σ so that $\Sigma = 0$ when $\sigma < \sigma_G$ and $\Sigma = \sigma - \sigma_G$ when $\sigma > \sigma_G$. Here σ_N and σ_G are the threshold stresses required for nucleation and growth, respectively, of damage ($\sigma_N > \sigma_G$) and τ, V_N, B, and C are material parameters. From Eqns. (16.37) and (16.38),

$$D = \frac{BV_N(\Sigma - \Sigma_N + |\Sigma - \Sigma_N|)}{6C\Sigma} [\exp(3C\sigma_G\Sigma t) - 1] \qquad (16.39)$$

If $\sigma_N > \sigma_G$, the term $0.5(\Sigma - \Sigma_N + |\Sigma - \Sigma_N|)$ is simply $(\sigma - \sigma_N)$ for $\sigma > \sigma_N$.

The predictions of Eqn. (16.39) with the values of the parameters $\sigma_N = 0.8$ GPa, $\sigma_G = 0.3$ GPa, $V_N B = 0.0116$ kbar^{-1} μs^{-1}, $C = 0.667$ kbar^{-1} μs^{-1}, typical for aluminum, are presented in Figure 16.26. One can see that, for a constant pressure, the damage increases with time. At 0.8 GPa, there seems to be a time delay of 0.3 μs before damage starts occurring. Davison and Stevens' [43] theory is phenomenological in the sense that no detailed mechanisms for the initiation and propagation of microfailures are incorporated. In

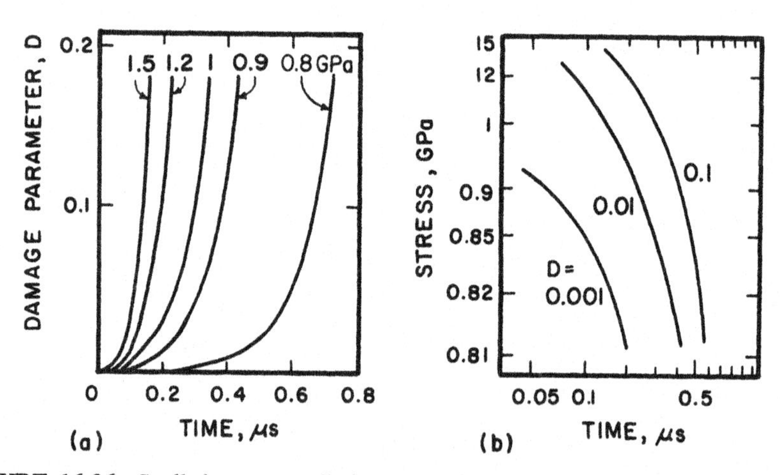

FIGURE 16.26 Spall damage prediction according to Davison and Stevens theory. (a) Time dependence of accumulated spall damage for five different pressures (0.8, 0.9, 1.0, 1.2, and 1.5 GPa). (b) Curves of constant damage (D) in time–stress plane. (From Davison and Stevens [43], Figs. 1 and 2. Reprinted with permission of the publisher.)

1973, Davison and Stevens [45] presented a more detailed theory for spall damage for the case where failure occurs by the initiation and propagation of cracks. The damage was represented by vector fields describing the size and orientation of the cracks. This continuous treatment eliminated the consideration of individual cracks, the stress concentration around each crack, its orientation, size, and location, and so on. The theory was applicable to alloys exhibiting brittle behavior under dynamic conditions, such as iron and beryllium. Later, in 1977, Davison et al. [46] published a detailed theory for the case where spalling occurs by void nucleation and growth (ductile spall damage). Davison et al. [46] assumed a viscoplastic behavior for the material in the establishment of the equations simulating spherical growth. They computed the damage D as the void volume percentage in the spall region. Incorporating their dynamic void nucleation-and-growth parameters into a WONDY code (a one-dimensional Lagrangian wave propagation code using the finite-difference method), they were able to compare experimental results with computations. The results obtained for a 3.5-mm-thick fused-quartz projectile impacting an 1100-0 aluminum alloy target (6.4 mm thick) are shown in Figure 16.27. The impact velocity of 142 m/s yielded a pressure in the target of 1.0 GPa. Figure 16.27(a) shows the calculated and observed velocity histories in the back surface for the cases of spalling and no spalling. The experimental measurements were made by laser interferometry. When no spall is observed, the free-surface velocity should return to zero after the passage of the shock pulse ($\sim$2.5 μs). The formation of the spall, on the other hand, generates a release wave that produces the hump behind the shock wave once it reaches the back surface ($\sim$3.0 μs). Davison et al. [46] call this hump the "spall cusp." One can see the good agreement between observations and calculations. It should be noticed that the WONDY code incorporates void growth dynamics. Figure 16.27(b) shows the calculated damage as a function of position after various times, and one can clearly see the peak at approximately 3.5 mm. This is, as discussed briefly in Sections 7.3 and 16.8.1, the thickness of the projectile. The final state, represented by the dashed line, corresponds to a time of approximately 3.5 μs. Figure 16.27(c) shows more clearly how the degree of damage drops slightly from its maximum value of 2.8 μs. In addition to the time dependence of damage, Fig. 16.27(c) shows more time dependence of stress and temperature for both the cases of spall and no spall. The stress required for nucleation of the voids is taken to be much larger than the one required for growth. There is also a considerable amount of heat generated in the growth of voids, as shown by the bottom plot.

In 1977, Cochran and Banner [47] (Lawrence Livermore National Laboratory) reported the results of an investigation into uranium. Using a gas gun, they systematically varied the stress and stress duration, keeping constant (at 0.5) the ratio between projectile and target thicknesses; this assured that the maximum tensile stress occurred in the middle of the target. Their results are shown in Figure 16.28. The flyer plate thicknesses are indicated in the plot. As the flyer plate thickness increases, the amount of damage increases at a

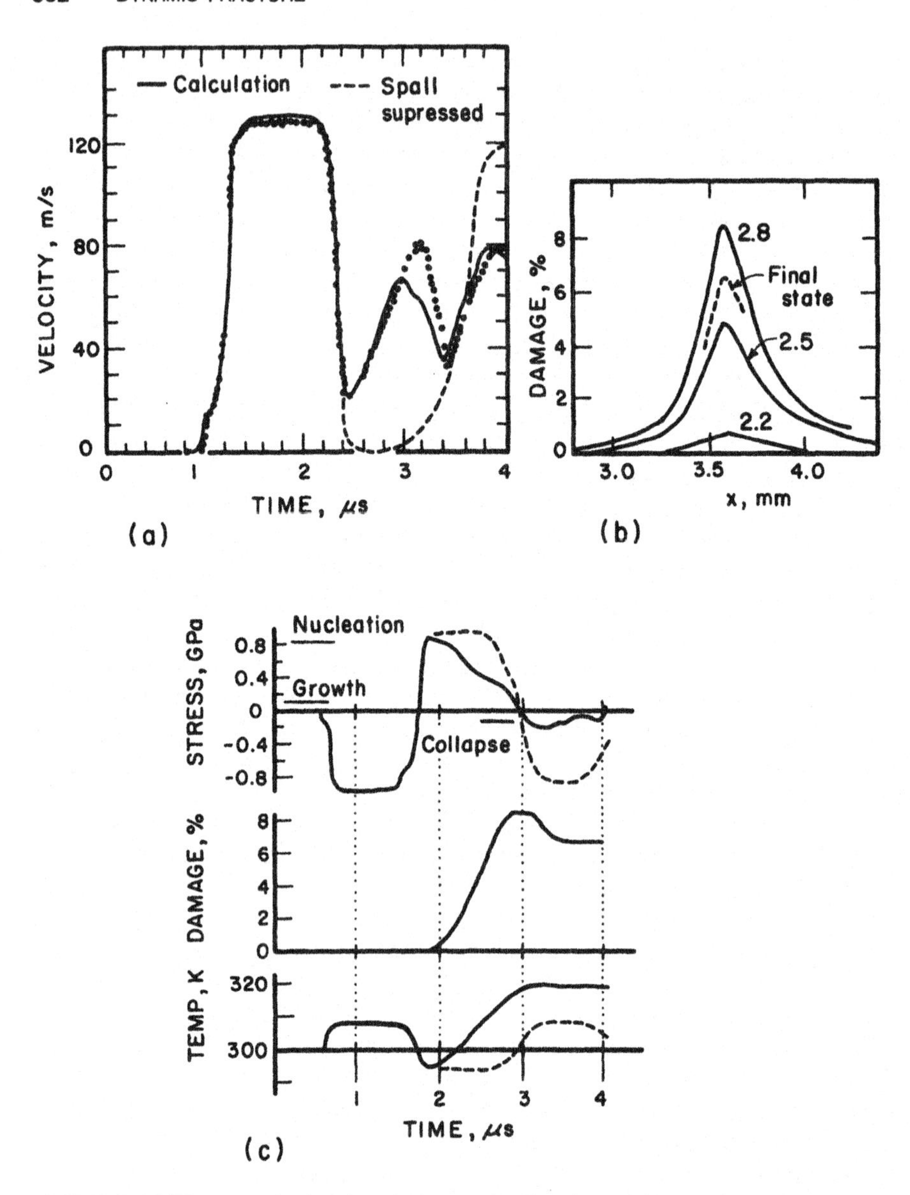

FIGURE 16.27 (a) Calculated and observed velocities of free surface of target. (b) Calculated void volume distributions in central region of 6.4-mm target. (c) Calculated histories of axial stress component, damage function D, and temperature (3.57 mm from impact surface). Dashed lines correspond to case where damage is suppressed. Stress thresholds for nucleation, growth, and collapse of voids marked in plot. (Reprinted from *J. Mech. Phys. Solids*, vol. 25, L. Davison et al., Figs. 3, 5, and 6, pp. 25, 26, Copyright 1977, with permission from Pergamon Press Ltd.)

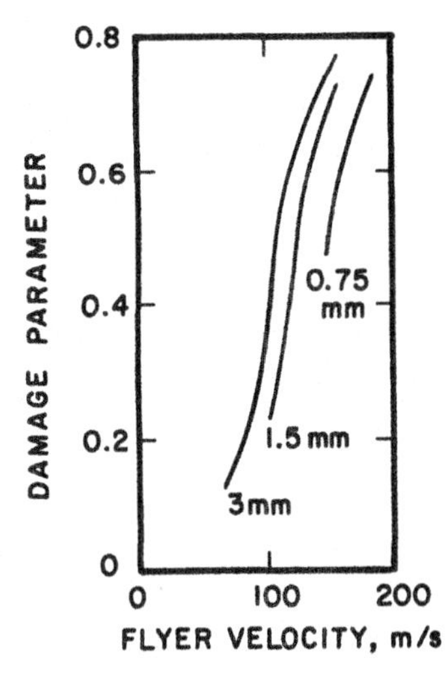

FIGURE 16.28 Damage parameter as a function of flyer plate velocity for three different flyer plate thicknesses in uranium. (From Cochran and Banner [47], Fig. 7, p. 2732. Reprinted with permission of the publisher.)

constant stress level (determined by the flyer plate velocity). Conversely, at a constant flyer thickness the damage increases, for increasing stresses. These results corroborate other investigations that established that both stress and stress duration are important. The damage parameter given in the ordinate axis was not obtained from direct measurement of void densities, but indirectly, from free-surface velocity measurements. Cochran and Banner [47] used, instead, the peak formed in the free surface of the target when a spall is formed. When a spall is formed, the velocity at the free surface does not return to zero. Cochran and Banner [47] found that the ratio between the free-surface velocities of the spall and compression pulse peaks provided a good correlation with void densities at the spall plane. Figure 16.29 shows a schematic diagram of the velocity trace; the damage parameter is V_c/V_a. This technique, if reliable, greatly simplifies the determination of spall damage, because tedious microflaw counting is eliminated.

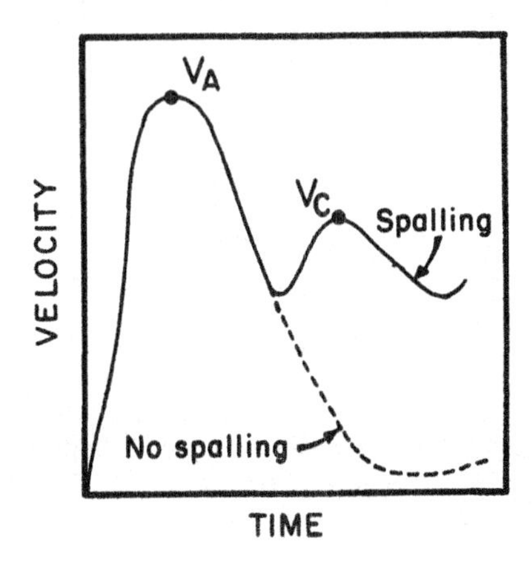

FIGURE 16.29 Free-surface velocity for spalling and no-spalling situations; V_C/V_A defines the damage parameter.

Cochran and Banner [47] developed a theoretical model and incorporated it into a one-dimensional finite-difference hydrodynamic code (KO) developed by Wilkins. They incorporated the Bauschinger effect into the model; the Bauschinger effect is the strength differential that a specimen exhibits when it is stressed plastically in tension and then in compression (or vice versa); the strength upon reversal of the stress may be lower. They also considered a threshold stress Σ below which no spalling occurs. Total spalling is obtained when D reaches a critical level D_0. By varying D_0 and Σ, Cochran and Banner try to match the experimentally observed free-surface velocity traces to the computed ones. Figure 16.30 shows a sequence of observed and calculated traces at increasing impact velocities for uranium (flyer plate thickness 3 mm; target thickness 6 mm). As the surface velocity increases, the spall peak becomes more and more evident; it occurs at $\sim 6\ \mu s$ and is marked by arrows.

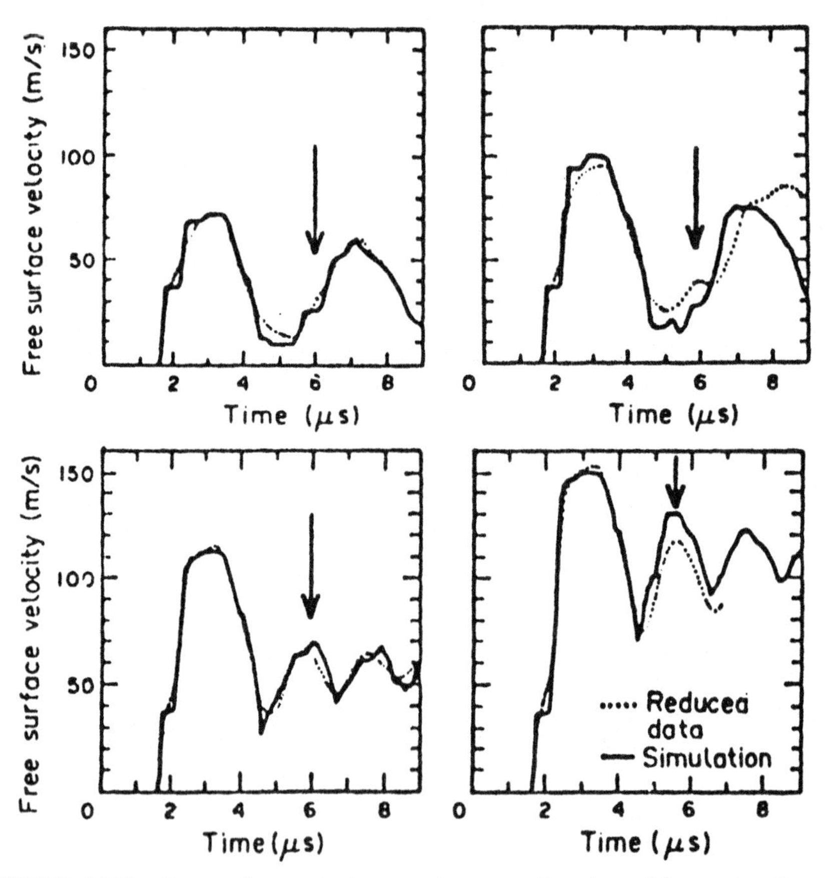

FIGURE 16.30 Free-surface velocity vs. time as a function of increasing impact velocity. Arrows indicate position of spall signal, which becomes stronger as impact (and free-surface) velocity increases. (From Cochran and Banner [47], Fig. 10, p. 2735. Reprinted with permission of the publisher.)

This fit was obtained with values of

$$\Sigma = 2.4 \text{ GPa} \qquad D_0 = 55 \ \mu\text{m}$$

This spall signal can also be seen in Figure 16.27(a) and results from the release waves produced in the spall plane [shown schematically in Fig. 16.23(a)].

The SRI (Curran-Seaman-Shockey) research effort in spalling characterizes itself by the definition and establishment of nucleation-and-growth parameters for describing the failure process. The resultant computer code, successfully used in a number of configurations, is called NAG. The systematic measurement of crack and void sizes and orientations after different degrees of spalling was used to establish the nucleation and growth equations. The concept of damage function, introduced by Tuler and Butcher [44] and Gilman and Tuler [48], was used by Barbee et al. [49] after careful determination of void and crack sizes. Figure 16.31(a) shows the cumulative number of voids with radius larger than R at various sections parallel to the free surface of the specimen (1145 aluminum alloy). Hence, $N(R) > R_1$ represents the number of voids (or cracks) per unit volume with radius larger than a certain value of R_1. The SRI group developed techniques for measuring the cracks and for transforming the apparent crack orientations, length, and numbers observed in a polished surface to true orientations, length, and numbers [50]. These metallographic calculations are of great importance in the development of quantitative parameters (and laws) describing the damage accumulation process. The techniques developed in quantitative metallography [51] are of great use in these transformations. Figure 16.31(b) shows the distribution of cracks in Armco iron at various distances (zones) from the back surface of the target.

In their work, Curran, Shockey, Seaman, and co-workers [52–62] developed the quantitative understanding of the rate of nucleation and the rate of growth of microcracks and voids from the systematic collection and processing of data for a number of materials. They found that the equations that describe the rate of nucleation $\dot{N}$ and the rate of growth $\dot{R}$ (time change of radius R) are

$$\dot{N} = \dot{N}_0 \exp\left[(\sigma - \sigma_{n0})/\sigma_1\right] \tag{16.40}$$

$$\dot{R} = \left(\frac{\sigma - \sigma_{g0}}{4\eta}\right) R \tag{16.41}$$

where $\dot{N}_0$ is the threshold nucleation rate, σ_{n0} is the tensile threshold stress for nucleation, σ_1 is the stress sensitivity for nucleation, σ_{g0} is the threshold stress for growth, and η is the viscosity of the medium. The equation for the nucleation rate is derived from earlier results by Zhurkov [63], who considered it as a statistical process. The viscosity term η represents the crack tip viscosity for cracks. It is generally accepted that the limiting growth rate for cracks is the Rayleigh wave velocity. Equation (16.40) expresses the fact that below σ_{n0}, no nucleation is observed; above the threshold stress the nucleation rate increases exponentially with the tensile stress σ. Equation (16.41) states that the growth

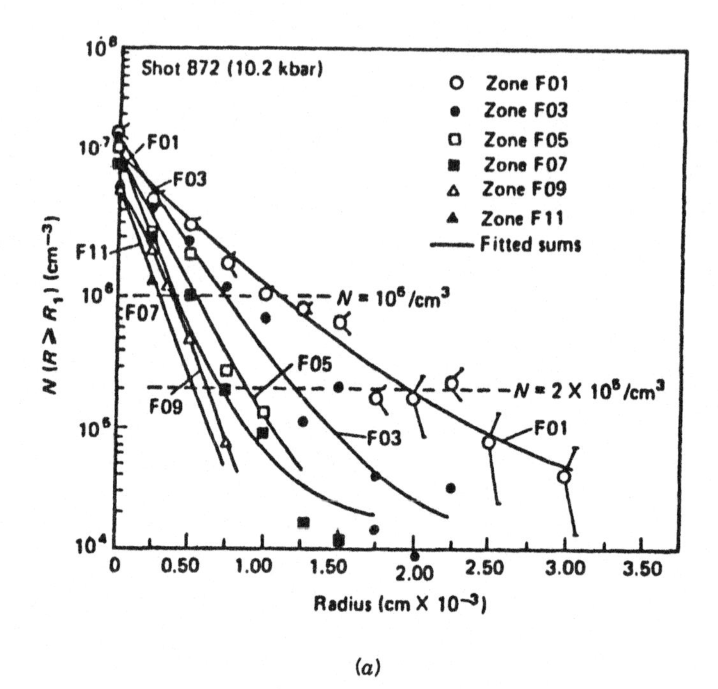

(a)

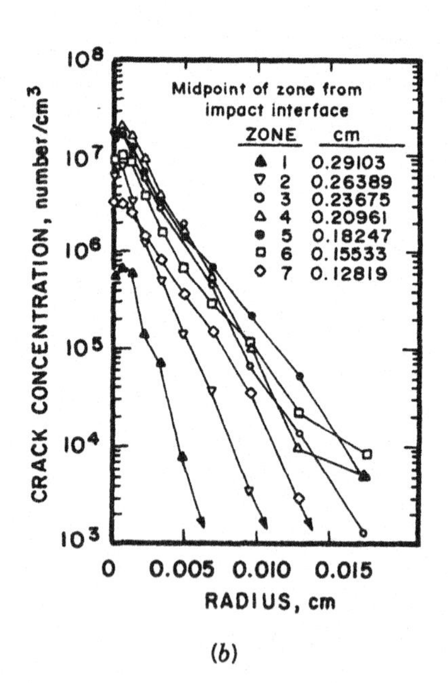

(b)

FIGURE 16.31 (a) Cumulative distribution of voids in 11145 aluminum. (b) Cumulative distribution of cracks in Armco iron. (From Seaman et al. [50], Figs. 8 and 9, pp. 396, 397. Copyright ASTM. Reprinted with permission.)

rate $\dot{R}$ is proportional to the radius R of a void (or crack); for cracks, this proportionality would break down as the Rayleigh wave speed is approached. The damage produced by spalling can be more simply described than in the plots of Figure 16.31, if one just considers the void density (number of voids per unit volume) or the relative void volume. In this case, the void size is neglected. It can be shown that for the case of spherical voids the relative void volume is given by

$$V_v = \frac{8\pi \dot{N} R_0^3}{[3(\sigma - \sigma_{g0})/4\eta]} \left[\exp\left(\frac{3(\sigma - \sigma_{g0})t}{4\eta} \right) - 1 \right] \tag{16.42}$$

where R_0 is the smallest visible void radius; Barbee et al. [53] assume it to be 1 mm. Once the parameters above are determined, it is possible, by inserting the nucleation and growth equations into Lagrangian (or Eulerian) codes, to obtain realistic predictions of the relative void volumes and void concentrations as a function of distance from the back surface. Seaman et al. [54] describe how these models were incorporated into the NAG-FRAG code. Figure 16.32 shows the distribution of voids (both density and relative volume) obtained by the use of the code compared with experimental results for S-200 beryllium. The spall plane corresponds, as expected, to the region of maximum damage, and the agreement between calculated and measured relative void volume and void concentration is excellent. The calculations take into account the change of bulk modulus and yield stress with the distension produced by the damage. There is also an important effect of stress relaxation as the cracks grow and coalesce. Figure 16.33 shows the variation of stress with nucleation, growth, and coalescence of cracks and final fragmentation as a specimen of Armco iron is extended at a constant strain rate of $1.5 \times 10^5\ \mathrm{s}^{-1}$. Once can see that the stress peaks at a value of 4.5 GPa and then starts to decay. In an analogous way, the stress in the spall regions peaks and then relaxes itself as the process of damage progresses.

Curran et al. [55] also studied dynamic failure by spalling in polycarbonate; this model material presents the advantage of being transparent, so that the crack morphology can be clearly observed. The cracks exhibited typically several morphologies in a ring pattern, which were interpreted in terms of the different regimes of propagation. They nucleated at flaws and first moved slowly, increasing in velocity as their size increased, up to the size determined by Eqn. (16.1) of fracture mechanics:

$$R \geq \frac{1}{\pi} \left(\frac{K_{IC}}{\sigma} \right)^2$$

This is the radius R of crack at which it propagates catastrophically; σ is the applied tensile stress. After the tensile pulse has passed, growth stops; however, the wave continues to reverberate throughout the specimen, and subsequent

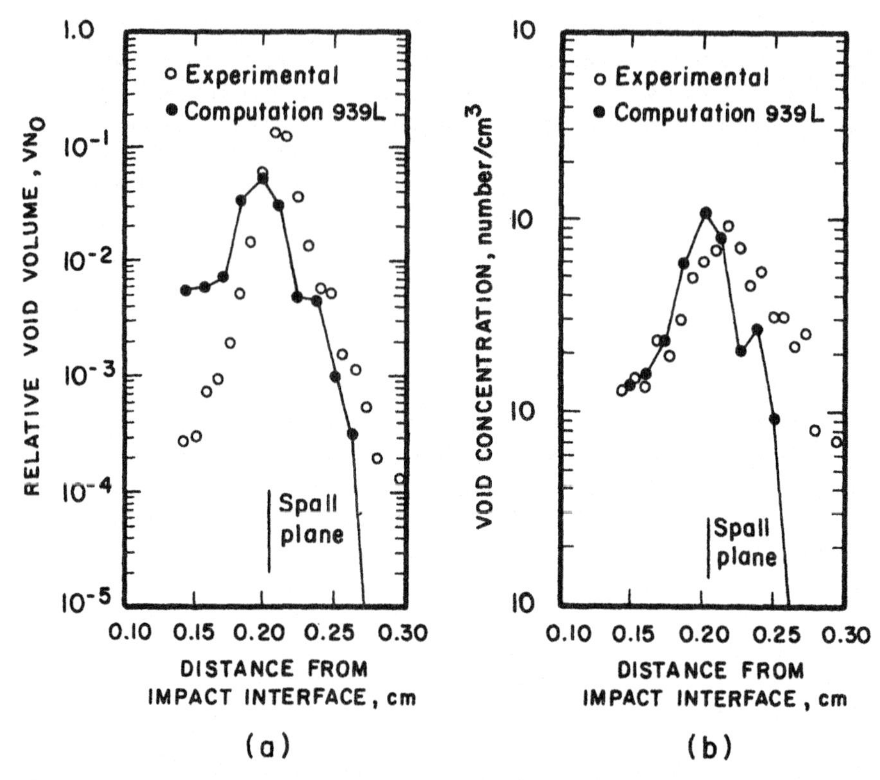

FIGURE 16.32 Calculated (NAG code) and measured fracture damage as a function of distance from free surface for S-200 beryllium impacted by electron beam irradiation: (a) Relative void volume; (b) void concentration. (From Seaman et al. [57], Fig. 19. Reprinted with permission of the publisher.)

tensile pulses promote further growth. The same type of behavior is expected to occur in brittle metals. A sequence is shown in Figure 16.34.

The third stage of fragmentation, which is coalescence of cracks and voids, is also discussed by Curran et al. [62]. They refer to the approach of McClintock [64] for ductile failure in which coalescence of elliptical voids is caused by their impingement. The impingement of voids takes place when the relative void volume is between 50 and 60%. For brittle crack propagation, impingement is the de facto criterion for coalescence. Although little is known at present of the exact mechanisms of coalescence, a model has been proposed by Shockey et al. [65] for the fragmentation of rocks.

Johnson [66], at Los Alamos, conducted a theoretical analysis of the development of voids produced by spalling and applied it to computer codes. His basic approach was to use the theoretical model of void collapse in a porous ductile material under pressure developed by Caroll and Holt (see Section 17.7.4) and invert the stress sign in such a way that void collapse is replaced by void growth. The changes in properties with distension are incorporated into the model as well as the material resistance to void growth. Figure 16.35

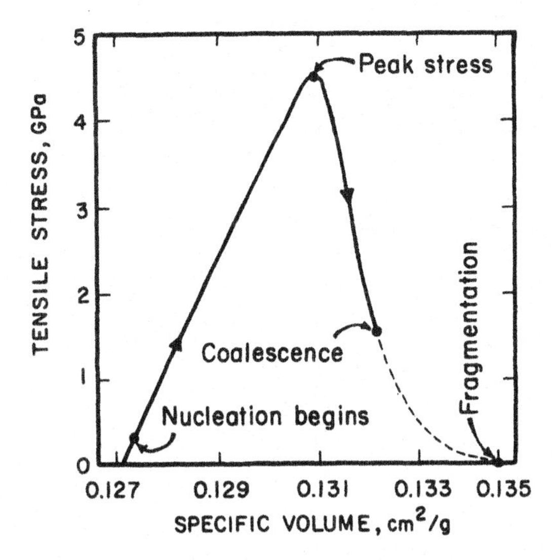

FIGURE 16.33 Stress–volume path of Armco iron loaded to fragmentation at constant strain rate. The dashed line marks the region in which fragmentation progresses by microcrack coalescence (From Curran et al. [58], Fig. 7, p. 54.)

shows the comparison of measured and predicted relative void volume (porosity) for a copper target (1.6 mm thick) impacted by a copper projectile (0.6 mm thick) at a velocity of 0.16 mm/μs, providing an initial pressure of 3 GPa. Up to 30% porosity is observed in the spall plane, and the calculated and observed results show good agreement. It should be noticed that there are a number of input parameters in the computer code; they have to be adjusted to provide a good fit and some of them (e.g., the viscosity η) cannot be independently obtained.

Davison and Graham [68], in a comprehensive review of the effects of shock waves and materials, summarized the information available on the spall strength. The highest reported value is for the AM 363 steel: 4.5 GPa. Davison and Graham [68] state that, although the most fundamental models are developed for spherical voids and cracks, technical alloys often exhibit "blunted cracks." The concept of spall damage is, in modern work, quantified as the number density and size distribution of the cracks and voids in the material. At low damage levels, they are typically $10^4/\text{mm}^3$ and 1–100 μm radius, respectively. Parallel Russian research* on spalling was carried out by Romanchenko and Stepanov [69]. The threshold spalling stress was determined by the observation of recovered specimens at a magnification of 100. They correlate the critical spalling stress with the tensile strains for aluminum, copper, and steel. Akhmadeev and Nigmatulin [70] developed numerical modeling approaches to spalling that incorporate shock-induced phase transformations (see Chap. 8) that produce more complex wave structures.

*Novikov [67] reviews the Russian research in spalling, that is extensive; he emphasizes the importance of microstructural aspects.

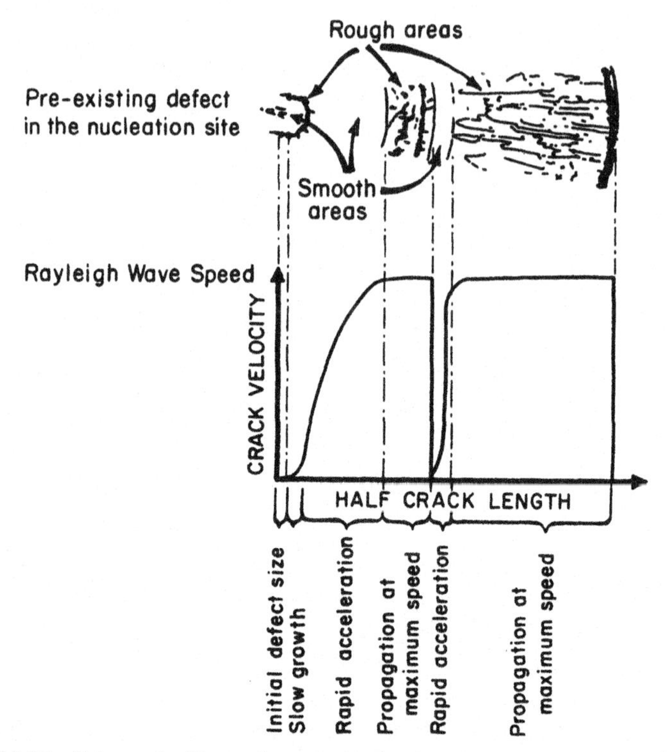

FIGURE 16.34 Schematic illustration of change in crack morphology with growth and interpretation in terms of stress history of crack. (From Curran et al. [55], Fig. 21, p. 4033. Reprinted with permission of the publisher.)

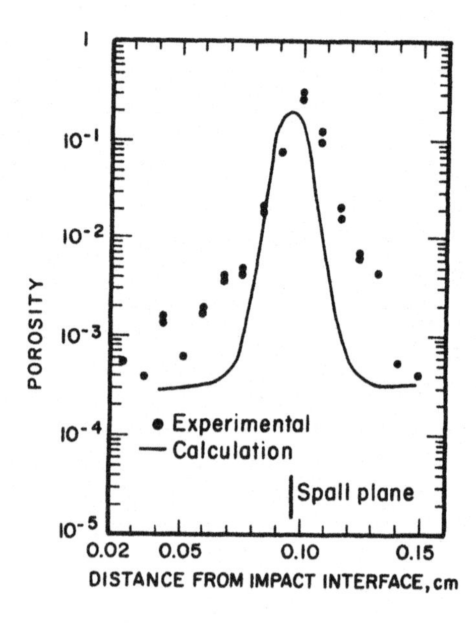

FIGURE 16.35 Comparison of calculated (solid line) and measured (data points, SRI) postimpact porosity for a copper sample. (From Johnson [66], Fig. 8, p. 2819. Reprinted with permission of the publisher.)

16.8.3 Microstructural Effects

Microstructural effects are of utmost importance in spalling. Some important aspects should be emphasized. Meyers and Aimone [71] have given a comprehensive description of microstructural effects in spalling, and the reader is referred to this article. Grain size, flow stress, the presence of second-phase particles, and inclusions and phase transformations alter the spall strength of a material. The study conducted by Christy et al. [72] illustrates the importance of microstructural parameters on the propensity to fracture by spalling. They subjected copper specimens with grain sizes of 20, 90, and 250 μm to dynamic tensile pulses produced by the reflection of a shock wave at a free surface. Additionally, they also had cold-rolled specimens and specimens with low purity (high concentration of second phases). Figure 16.36 shows the variation of damage across the cross sections of the different metallurgical conditions for a tensile stress of 3.8 GPa and pulse duration of ~2.8 μs. The specimen with the small (20-μm) grain size exhibited far lower damage than the large-grained specimens. The spall morphology was also markedly different, with the specimens with large grain size exhibiting an intergranular fracture. It is interesting that the differences in spall strength did not correlate directly with differences in hardness. The momentum trap consisted of copper of lower purity, in the rolled condition. The presence of impurities provide the initiation sites for spalling and are responsible for a lower spall strength.

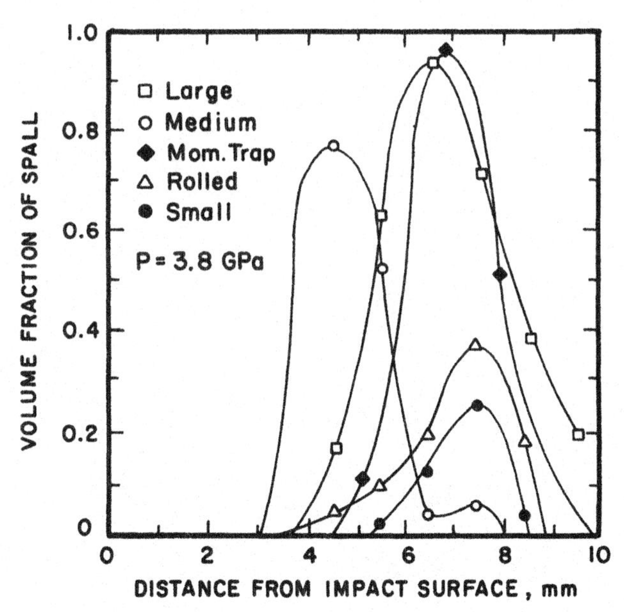

FIGURE 16.36 Volume fraction of voids formed in copper upon spalling at 3.8 GPa; large (250 μm), medium (90 μm), small (20 μm) grain sizes, cold-rolled specimen, and momentum trap (low-purity copper) (From Christy et al. [72], Fig. 15, p. 857. Reprinted by courtesy of Marcel Dekker, Inc.)

(*a*)

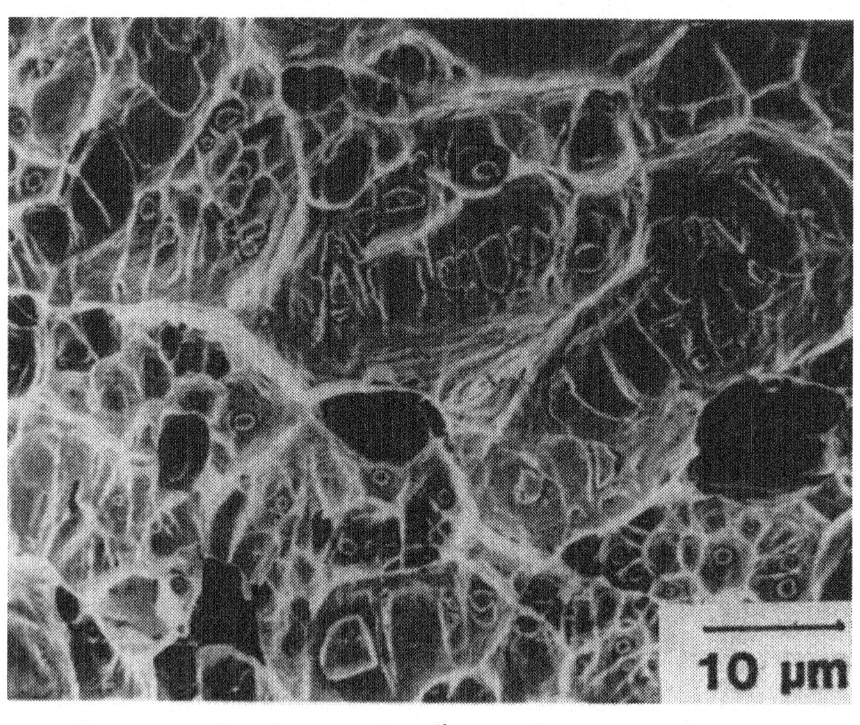

(*b*)

FIGURE 16.37 (a) Nucleation of voids at grain boundaries in copper subjected to a tensile stress pulse. (b) Scanning electron fractograph of spall in copper containing second-phase particles (copper oxide); (momentum trap in Figure 16.36). (From Christy et al. [72], Fig. 14, p. 854. Reprinted by courtesy of Marcel Dekker, Inc.)

The initiation of voids at grain boundaries is clearly evident in Figure 16.37(a). This mode of spall initiation for large-grain specimens was also observed by Brandon et al. [73] and Zurek et al. [74]. On the other hand, the presence of second-phase particles can strongly affect the spall strength if the particle–matrix interface strength is weak. Figure 16.37(b) shows a scanning electron micrograph of a spall surface of a copper specimen containing second-phase particles. It can be seen that most dimples contain a second-phase particle, indicating that these particles initiated spalling and, consequently, weakened the material. Similar observations have been made in steel, where the presence of MnS particles decreases their spall strength.

There is only one report, to the author's knowledge, of an incipient void in the spall region, by transmission electron microscopy. Christy et al. [72] used a 1-MeV electron microscope and were able to identify one void. A high dislocation density is indeed necessary to account for the plastic deformation involved in the expansion of a void. The heavily (plastically) deformed layer is predicted by theory.

Kanel et al. [75] confirmed this unique aspect of spalling in copper by conducting spalling experiments in mono- and polycrystals. Whereas polycrystals had a spall strength of 1.36 GPa, monocrystals had a spall strength between 3.3 and 4.6 GPa. This contradicts the quasi-static strength values,

FIGURE 16.38 High-voltage (1-MeV) transmission electron micrograph of peanut-shaped void in copper. (From Christy et al. [72], p. 860, Fig. 16.) This void is not of the classical spherical morphology, but is peanut-shaped. It is surrounded by a layer with very high dislocation density; this region is dark in TEM picture, because individual dislocations cannot be imaged due to their high density.

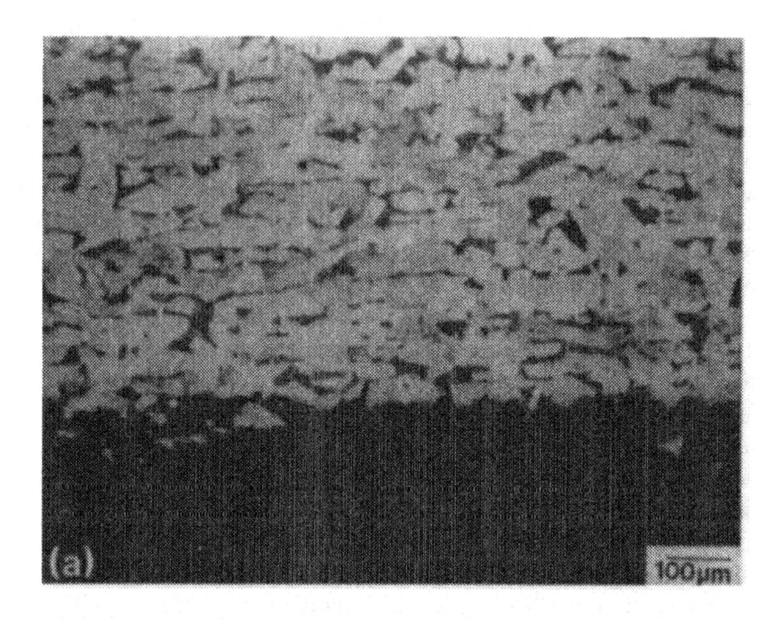

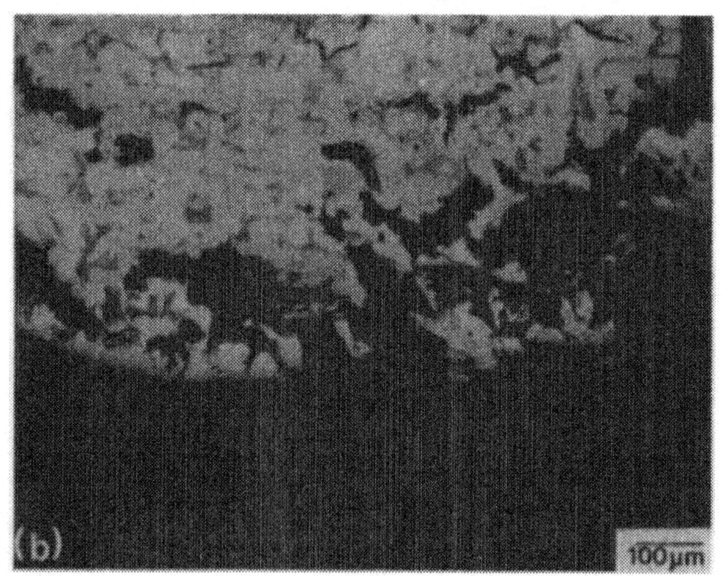

FIGURE 16.39 Spalling in steel: (a) smooth spall produced when pressure exceeds 13 GPa; (b) rough spall produced when pressure is below 13 GPa; (c) x–t diagram for two-wave structure and its interaction with free surface–smooth spall surface indicated. (From Erkman [76], Fig. 8, p. 943. Reprinted by permission of the publisher.)

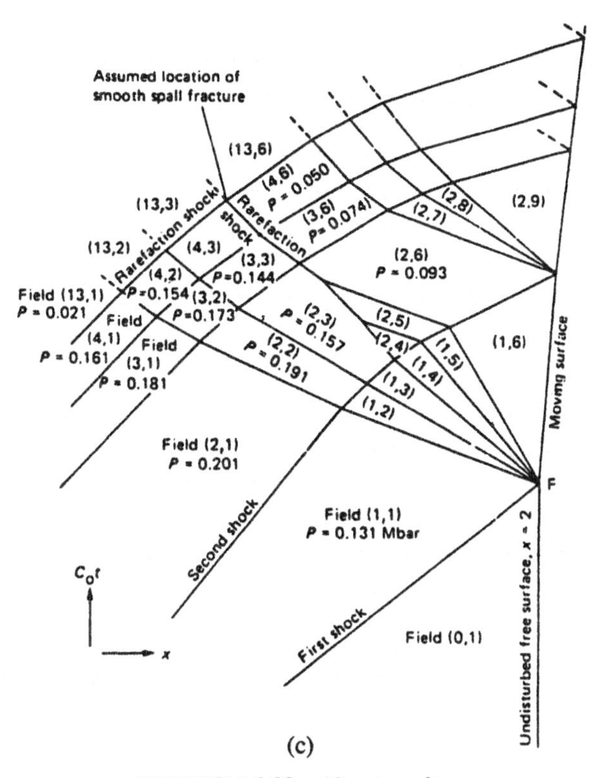

(c)

FIGURE 16.39 *(Continued)*

which are much higher for polycrystals than for monocrystals. The explanation provided for this anomaly is as follows: polycrystals contain grain boundaries, which are void initiation sites, whereas the stress required for nucleation of voids, σ_n, is much higher in monocrystals.

Another very interesting phenomenon is the existence of a smooth spall in iron.* Under certain conditions (above the 13-GPa threshold stress) the spall morphology in low-carbon steels changes drastically. No high-magnification observation is required, and the change in fracture morphology can be observed by the naked eye. The 13 GPa pressure marks the α(BCC) $\rightarrow$ ε(HCP) transformation in Fe (see Chapter 8 and Fig. 8.7). This leads to the splitting of the shock front into two waves, as explained in Chapter 8. The details of the fracture are shown in Figure 16.39 [71]. Figure 16.39(a) shows a section of the "smooth" spall, produced by a stress pulse exceeding 13 GPa; Figure 16.39(b) shows the "rough" spall, resulting from a lower pressure. In contrast, Figure 16.39(b) shows considerable damage inside the material and the damage region extending over a distance equal to several grain diameters. There is a

*This phenomenon was first observed by A. G. Ivanov and S. A. Novikov (*J. Exp. Theor. Phys.* (USSR) **45** (1961) 1880); they applied two explosive charges at opposite sides of cylinder and observed spalling in the region where the two waves superimposed, exceeding 13 GPa.

greater degree of residual deformation (twins and residues of the transformation) within the grains of Figure 16.39(a); this specimen was subjected to a pressure pulse of higher amplitude (> 13 GPa). Scanning electron microscopy of the smooth surface conducted by Meyers and Aimone [71] revealed significant differences. The fracture morphology of the smooth spall is definitely different from the one produced by the rough spall. One cannot see any clear cleavage planes at 1000×, whereas these are abundant in the rough spall. In some instances, one sees deep straight grooves tending to indicate intercrystalline fracture. One can also see ill-defined dimples as if the fracture occurred in a ductile manner.

The explanation provided by Erkman [76] for the formation of the smooth spall is based on the existence of a rarefaction shock. There is a sudden drop in pressure in the rarefaction part of the wave due to the allotropic $\alpha \rightarrow \varepsilon$ phase transformation. This transformation is described in Chapter 8. This sudden drop in pressure results (in the reflected portion of the sequence) in a sudden rise in the tensile pulse. This sudden rise in the tensile pulse occurs in a very narrow region; the resulting fracture region is highly localized and consequently the spall is smooth. On the other hand, the more gradual increase in the tensile stress pulse that occurs when the rarefaction portion of the pressure pulse is sloped allows the initiation and propagation of damage over a wider region, resulting in a more irregular pulse. Figure 16.39(c) reproduces a distance–time plot developed by Erkman [76] and used to calculate the distance of the smooth spall from the free surface of the target. The smooth spall should occur at the position where the first shock wave (the wave decomposes into two shock fronts at 13 GPa due to the phase transformation) intersects the rarefaction shock. This intersection should provide the tensile spike required for spalling. Erkman [76] compared the results of his calculations with experimental results and found reasonable agreement. However, one would expect spall failure at approximately that same position even if no rarefaction shock wave were involved, and the smooth and rough spalls occur in approximately the same plane, as shown by Figure 16.39. The smooth spall was also studied by Banks [77]. Phase transformations can also be induced by tensile pulses, as described in Chapter 8 (Section 8.6) and Chapter 14 (Section 14.5). Austenitic steels and Fe–Ni alloys can exhibit these transformations. Dremin et al. [78] found an anomalous increase in the spall strength when a martensitic transformation occurred in the tensile region. Thus, the dilatational component of the transformation inhibited void nucleation.

16.9 FRAGMENTATION

Sequential fracture leads to fragmentation. One of the most important aspects of dynamic fracture is that the body, at the end of the fracturing sequence, is divided into many parts. Quasi-static fracture often only breaks a loaded body into two parts. The reasons for fragmentation can be easily seen both from an energetic and a mechanistic point of view. Energetically dynamic loading provides the body with the kinetic energy not available quasi-statically. Mechanistically, independent crack nucleation occurs profusely because unloading due to the formation of new surfaces is hindered, and crack branching leads to smaller and smaller subdivision of the body. Both energetic and mechanistic arguments can be used to explain fragmentation. In this section, we will first present Mott's early (1946) [79] fragmentation theory (Section 16.9.1). Then, we will develop a generalization proposed by Louro and Meyers [80–82] and applicable to ceramics subjected to impact (Section 16.9.5). Grady and Kipp [83] formulated advanced fragmentation theories, which will be reviewed (Section 16.9.2). An important aspect of fragmentation is the fact that the fragments contain internal cracks that are incompletely grown. Thus, the measure of fragment size does not indicate all the damage to a body. Aimone et al. [86] and Jaeger et al. [88] deal with this. This will be treated in Section 16.9.4.

16.9.1 Mott's Fragmentation Theory

Mott [79] (a Nobel laureate) proposed,* in 1946, a theory to explain the formation of fragments in expanding rings. The problem addressed by Mott was the idealization of the fragmentation of a shell or bomb of cylindrical shape. An even simpler type of situation is the one where the cylinder is made up of stacked rings of equal diameter. Figure 16.40(a) shows the case of a shell before detonation and (b) at the moment of fracture. Mott stated that if the entire material exhibited the same fracture strain, fracture would occur simultaneously over the entire ring and the fragment size would be infinitely small. Thus, he assumed a distribution of fracture strains, expressed as the probability that an unfractured specimen of unit length fractures when the strain increases from ε to $\varepsilon + d\varepsilon$:

$$dp = Ce^{\gamma\varepsilon}\, d\varepsilon \tag{16.42a}$$

where C and γ are constants. As the strain ε increases, the probability that fracture occurs for an increment of strain $d\varepsilon$ increases exponentially. The probability that a specimen is unfractured at a strain ε is $1 - p$. Thus, the probability that a specimen will fracture when the strain is increased form ε to $\varepsilon + d\varepsilon$ is

$$dp = (1 - p)Ce^{\gamma\varepsilon}\, d\varepsilon$$

which leads to:

$$p = 1 - \exp\left(-\frac{C}{\gamma}e^{\gamma\varepsilon}\right) \tag{16.43}$$

*D. L. Wesenberg and M. J. Sagartz (*J. Appl. Mech.* **44** (1977) 643) and D. E. Grady (in "Shock Waves and High-Strain-Rate Phenomena in Metals" Plenum, NY, 1981, p. 181) present a detailed analysis of the Mott problem and comparison of its prediction with experimental results.

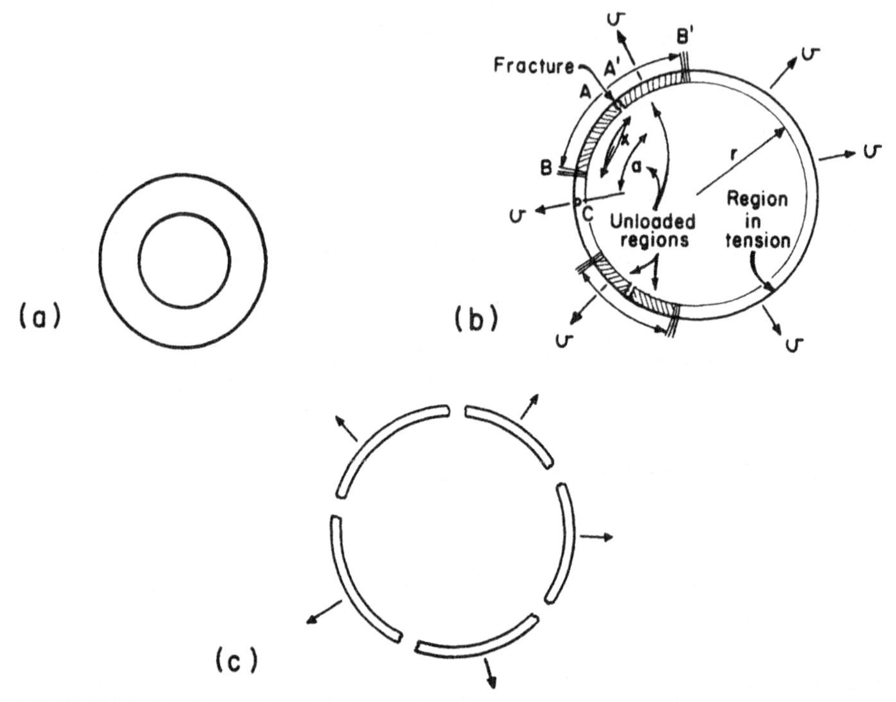

FIGURE 16.40 Expansion of circular ring leading to fragmentation; fractures produce unloading inhibiting further fragmentation in surrounding region (Mott problem).

Mott [79] considered the fracture strains, ε_f, to vary around an average

$$\varepsilon_f = \frac{1}{\gamma}\left[\log\frac{\gamma}{c} + 0.577\right]$$

Figure 16.40(b) shows two cracks forming and the unloading that resulted from it. The stress relief fronts advance and the unloaded material does not nucleate new cracks. Thus, it is the relationship between the rate at which the cracks are nucleated and the unloading of the material that determines the size of the fragments.

The radius of the ring at the instant of impact being r and the expansion velocity v, one has (see Section 12.3.3)

$$\dot{\varepsilon} = \frac{v}{r}$$

We now define a point C in the loaded region. This point is at a distance $a - x$ from the boundary with the unloaded region [Fig. 16.40(b)]. The velocity of the stress-free region AB with respect to C is:

$$\frac{v}{r}(a - x) = \dot{\varepsilon}(a - x)$$

Applying Newton's second law to the segment AB (conservation of movement) and assuming that the material density is ρ and fracture stress is σ_f yield

$$\sigma_f = -\rho x \frac{d}{dt}\left[\frac{v}{r}(a - x)\right] \quad (16.44)$$

Assuming v/r constant and differentiating yield

$$\frac{x^2}{t} = \frac{2r\sigma_f}{\rho v} \quad (16.45)$$

This is an approximate expression for the increase in the width of the unloaded region. No crack nucleation takes place in this region, and it is "shielded."

The velocity at which the interface between loaded and unloaded material moves is:

$$\frac{dx}{dt} = \left(\frac{r\sigma_f}{2\rho v}\right)^{1/2} t^{-1/2} \quad (16.45a)$$

This velocity decreases with $t^{-1/2}$. It is clear that the initial value is not realistic and that C_L is an upper bound for dx/dt. By using Eqns. (16.43) and (16.45a), Mott [79] arrived at an expression that gave the fragment length distribution as a function of the parameters C, γ, r, ρ, σ_f. He considered a distribution of fracture strains; the narrower the distribution, the greater the number of fragments.

16.9.2 Grady–Kipp's and Grady's Models

Grady and Kipp [83, 84] and Grady [85] developed models for fragmentation that attempt to relate the strain rate to the fragment size. Whereas the work of Grady and Kipp [83, 84] addresses the problem from a mechanistic angle, the work of Grady [85] uses a purely energetic approach.

Grady and Kipp [83, 84] consider that dynamic fracture is a complex interaction between the processes of fracture and fragmentation and the stress wave propagation properties of the material. They assumed that a fractured material subjected to a tensile stress will undergo a strain determined by a modulus K_f that is less than the intrinsic modulus K of the unfractured material. So,

$$K_f = K(1 - D) \quad (16.46)$$

where

$$D = NV$$

where N is the number of idealized penny-shaped cracks per unit volume, V is a spherical volume of material assumed to be affected by the crack, given by $\frac{4}{3}\pi r^3$ (where r is the average radius of the penny-shaped cracks), and D is defined as damage. They assumed that the N cracks, activated under the applied load, grow at a fixed velocity C_g. The damage is then, at time t:

$$D = N \tfrac{4}{3}\pi (C_g t)^3 \tag{16.47}$$

They considered that N, the number of activated cracks, is not fixed but depends on the applied stress conditions and geometry of the preactivated flaws. They assumed that crack activation was governed by the Weibull distribution, characterized by two parameters; so,

$$N = k\varepsilon^m \tag{16.48}$$

where N is the number of flaws that will activate at or below a tensile strain level ε. For a constant strain rate loading $\varepsilon = \dot{\varepsilon}\, t$, Grady and Kipp [84] developed an equation based on the following assumption: each crack generates an unloaded region around it equal to $V(t - \tau)$, where τ is the time at which it was activated. In this spherical unloaded region no subsequent activation takes place. Thus

$$\begin{aligned} D(t) &= \int_0^t \dot{N}(\tau) V(t - \tau)\, d\tau \\ &= \tfrac{4}{3}\pi k m C_g^3 \dot{\varepsilon}^m \int_0^t \tau^{m-1}(t - \tau)^3 (1 - D)\, d\tau \end{aligned} \tag{14.8a}$$

where

$$\dot{N} = N(\varepsilon)\dot{\varepsilon}(1 - D) \tag{14.8b}$$

The nucleation rate is a function $N(\varepsilon)$ of the strain (Eqn. 14.8) and only occurs in the material that is loaded $(1 - D)$.

Solving Eqn. (14.8a) by series expansion, Grady and Kipp [84] obtained:

$$D(t) = \alpha \dot{\varepsilon}^m t^{m+3} \tag{16.49}$$

where

$$\alpha = \frac{8\pi C_g^3 k}{(m + 1)(m + 2)(m + 3)}$$

Stress is related to strain through (see Eqn. 14.6)

$$\sigma(t) = K\varepsilon(t)(1 - D) \tag{16.50}$$

And then,

$$\sigma(t) = K\dot{\varepsilon}t(1 - \alpha\dot{\varepsilon}^m t^{m+3}) \tag{16.51}$$

Figure 16.41 [84] shows damage and stress histories defined by Eqns. (16.49) and (16.51), respectively. The dependence of the fracture stress on strain rate was determined by maximizing Eqn. (16.51) with respect to time $(d\sigma(t)/dt = 0)$. So,

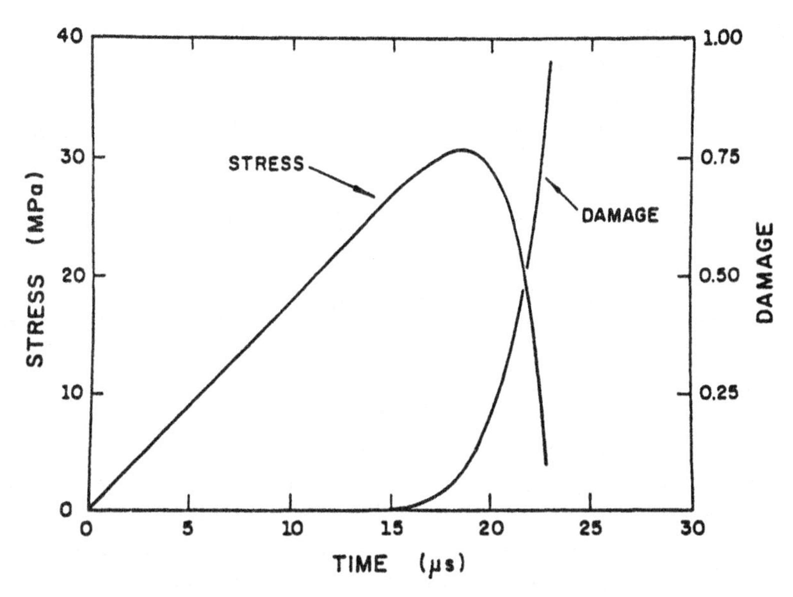

FIGURE 16.41 Tensile loading in oil shale at a constant strain rate of $10^2\ s^{-1}$. Both damage and stress history are plotted.

$$\sigma_M = K(m + 3)(m + 4)^{-(m-4)/(m-3)}\alpha^{-1/(m+3)}\dot{\varepsilon}^{3/(m+3)} \qquad (16.52)$$

Grady and Kipp [83] consider that fragmentation occurs at a time t_f, when growing cracks coalesce and the material loses its ability to support a tensile stress; namely,

$$D(t_f) = 1 \qquad (16.52a)$$

At fracture coalescence, that is, $t = t_f$, the mean fragment size was considered to be approximately $L_M = C_g t_f$, where t_f is obtained by applying the condition contained in Eqn. (16.52a) into Eqn. (16.49). An additional factor of $6/(m + 2)$ is taken into account, if more detailed calculation [84] is done. So,

$$L_M = \frac{6C_g}{m + 2}\alpha^{-1/(m+3)}\dot{\varepsilon}^{-m/(m+3)} \qquad (16.53)$$

The three fracture parameters k, m, and C_g in Eqns. (16.52) and (16.53) were obtained from experimental data through plots of fracture stress versus strain rate and fragment size versus strain rate by taking slopes and intercepts [84].

The important conclusion that can be drawn from Grady and Kipp's theory [84] is that both the maximum tensile stress σ_M and the fragment size-L_M can be simply expressed by

$$\sigma_M = f_1(k, m, C_g, K)\dot{\varepsilon}^{3/(m+3)} \qquad (16.54)$$

$$L_M = f_2(k, m, C_g, K)\dot{\varepsilon}^{-m/(m+3)} \qquad (16.55)$$

where m, the exponent in the Weibull flaw distribution, is a positive number that, for oil shale (for example), is equal to 8. Thus

$$\sigma_M \propto \dot{\varepsilon}^{0.27} \tag{16.56}$$

$$L_M \propto \dot{\varepsilon}^{-0.73} \tag{16.57}$$

At relatively low strain rates, the increase of tensile stress is low and the size of fragments is large (few flaws are activated). For rapid loading rates, the material fragments into many small pieces.

The model proposed by Grady and Kipp [83, 84] requires the independent experimental determination of a number of parameters that are very difficult to obtain: N, k, m. A second shortcoming of this model is that energetic considerations are totally ignored. Grady [85] developed a second approach in which the kinetic energy was introduced into the body under dynamic loading. Grady developed a general expression for the sizes of fragments based on an elegant treatment involving an energy balance. The basic precepts of this theory are that the interfacial energy generated by the fragmentation process is balanced by the local inertial or kinetic energy of the material. The method is first applied to the dynamic fragmentation in a fluid; it is then extended to brittle solids by incorporating fracture mechanics concepts. The general equation is developed from an expanding fluid which, at a time t, has density ρ, temperature T, and a rate of change of density $\dot{\rho}$. Surface tension alone resists the fracturing process. After fracturing, the fragments will fly apart at a certain velocity; this term also has a kinetic energy. Figure 16.42(a) shows (dashed circle) an element isolated for study. Since the fragments continue flying apart, the kinetic energy of the center of mass of each fragment is maintained and does not contribute to the generation of new surface. The component of the kinetic energy with respect to the center of mass is of importance. Figure 16.42(c) shows the two curves and the fragment size is determined by the minimum in the curve $(dU/dA = 0)$.

The kinetic energy available for fragmentation can be obtained by considering a fragment with a volume V. The kinetic energy of this fragment can be decomposed into a kinetic energy of the fragment with respect to its center of mass and a kinetic energy term due to the motion of the entire volume V. We will assume that the kinetic energy of the entire volume V is zero, due to the kinetic energy of the mass in the volume V with respect to its center of mass. In Figure 16.42(a) a volume is shown by a dashed line. This volume is assumed to expand with a rate of change of density $\dot{\rho}$. Figure 16.42(b) shows this spherical volume and an elemental volume embedded in its spherical coordinates. The kinetic energy of the elemental volume is

$$dE_k = \tfrac{1}{2}\rho \, dV \, v^2 = \tfrac{1}{2}r^2 \rho v^2 \, dr \, d\theta \, d\phi \tag{16.58}$$

From

$$m = \rho V$$

$$dm = 0 = \rho \, dV + V \, d\rho$$

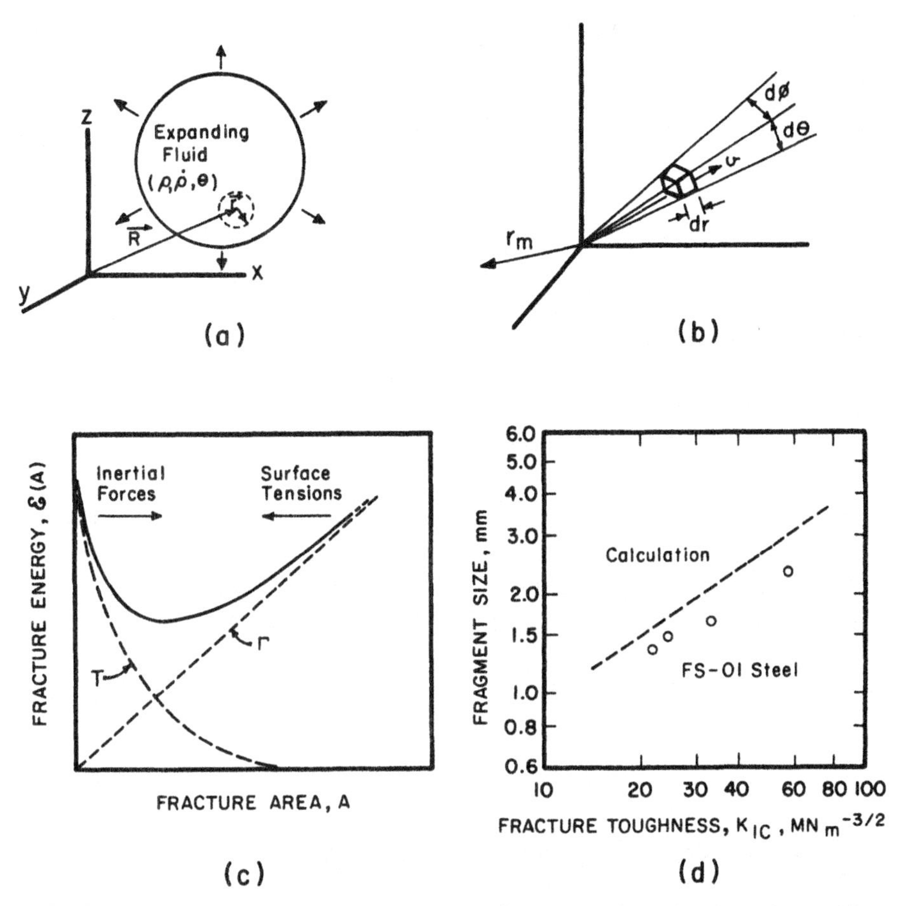

FIGURE 16.42 Grady's theory for fragment sizes. (a) Schematic view of expanding fluid that will, at time t, fragment. Dashed circle represents volume for which kinetic equations are applied. (b) Element isolated within sphere with coordinates. (c) Kinetic energy and surface energy terms; minimum establishes fracture area. (d) Application to fragment-size determination of steel. Fragment size increases with fracture toughness. (From Grady [85], Figs. 1, 2, and 4, pp. 322, 323, 324. Reprinted with permission of the publisher.)

$$\rho \frac{dV}{dt} = -V \frac{d\rho}{dt}$$

$$\dot{V} = -\frac{V}{\rho}\dot{\rho} \tag{16.59}$$

The volume is

$$V = \tfrac{4}{3}\pi r^3$$

Its change, with time, is

$$\frac{dV}{dt} = 4\pi r^2 \frac{dr}{dt} = 4\pi r^2 v$$

$$-\frac{V}{\rho}\dot{\rho} = 4\pi r^2 v \tag{16.60}$$

$$v = -\frac{\dot{\rho} r}{3\rho}$$

Hence, considering that ρ, $\dot{\rho}$ do not depend on the radius,

$$E_k = \frac{\dot{\rho}^2}{18\rho}\int_0^{r_m}\int_0^{2\pi}\int_0^{2\pi} r^4\, dr\, d\theta\, d\phi$$

$$= \frac{2\pi^2\dot{\rho}^2 r_m^5}{45\rho} \tag{16.61}$$

The surface area–volume ratio, assuming spherical fragments, is $A = 3/r$. Expressing the above kinetic energy in terms of A, one obtains

$$E_k' = \frac{E_k}{V} = \frac{\pi\dot{\rho}^2}{30\rho} r^2 \tag{16.62}$$

$$= \frac{3\pi\dot{\rho}^2}{10\rho A^2} \tag{16.63}$$

The surface energy of the new fragments generated is

$$\Gamma = \gamma A \tag{16.64}$$

where γ is the energy associated with the creation of fragment surface area (equal to or greater than the surface energy). The total energy is

$$U = \frac{3\pi\dot{\rho}^2}{10\rho A^2} + \gamma A \tag{16.65}$$

Grady and Kipp [84] assumed that during the fragmentation process the forces would seek to minimize the overall energy with respect to the fracture surface area density. The energy minimum provides ($dU/dA = 0$)

$$A = \left(\frac{3\pi\dot{\rho}^2}{5\rho\gamma}\right)^{1/3} \tag{16.66}$$

In terms of the fragment diameter d ($d = 6/A$), one has

$$d = 6\left(\frac{5\rho\gamma}{3\pi\dot{\rho}^2}\right)^{1/3} \tag{16.67}$$

Figure 16.42(c) shows the fracture energy-versus-A curve, composed of the surface energy and inertial components. The minimum of the curve leads to the fragment size $d(d = 6/A)$.

Grady and Kipp [84] compared the predictions of their theory with experimental results obtained in both solid and liquid materials. By using the expanding-cylinder technique, steels with different fracture toughnesses were tested and the average fragment size determined. In these tests, hollow cylinders were filled with explosives that were detonated. The predictions are compared with observed values for FS-01 steel in Figure 16.42(d). The agreement is satisfactory.

16.9.3 Internal Damage in Fragments

The only well-known type of energy expenditure that is innate in a fragmentation process is the energy of the new surfaces created during the process.

Aimone et al. [86] studied rock fragmentation by subjecting quartz monzonite to shock waves produced by plate impact. This study revealed that probably only a small fraction of the total energy consumed during fragmentation is used to create new surfaces and that no more than 15% of the total blast energy is utilized to do useful work, which is fragmenting and displacing the rock. They observed profuse microcracking within the fragments and established the total surface area of these internal cracks. It was found to greatly exceed the external surface area of the fragments. Figure 16.43 shows the principal results of this research. As the duration for the compressive pulse is increased, at constant pressure, fragmentation is increased (smaller fragment sizes) as shown in Figure 16.43(a). Very importantly, S_v, the crack density (surface area per unit volume), which includes the cracks within the fragments, also increases with increasing pulse duration. The external surfaces of the particles ($\sim 6/d$, where d is the crack diameter) were only a small fraction of the total surface area. This can be assessed by comparing Figure 16.43(a) and Figure 16.43(b). For instance, for $P = 2.7$ GPa and 1 μs, the internal S_v is 140 cm^2/cm^3 and the fragment S_v is 8×10^{-2} cm^2/cm^3. Thus, most of the damage is in the form of internal (or contained) cracks, and the fragment size is not really a good measure of the damage produced.

Jaeger et al. [88] also pointed out this additional deposition of the surface energy that is created during the fragmentation process, which is called internal and incompletely developed cracks in solids. They call these cracks internal damage and estimate their importance (in relation to the surface energy of the external boundaries) by using a simulation procedure to generate fragments. They examined the importance of internal damage in a two-dimensional model where fragments are formed by straight segments (representing crack in two dimensions) that were thrown on a plane randomly with respect to position and orientation. Figure 16.44 shows the resulting pattern obtained. Crack lengths were distributed in accordance with an empirical distribution for the number of cracks whose length exceeds l [$N_{\text{cum}}(l)$], so

$$N_{\text{cum}}(l) = N_c \exp\left(-\frac{l}{\lambda_c}\right) \tag{16.68}$$

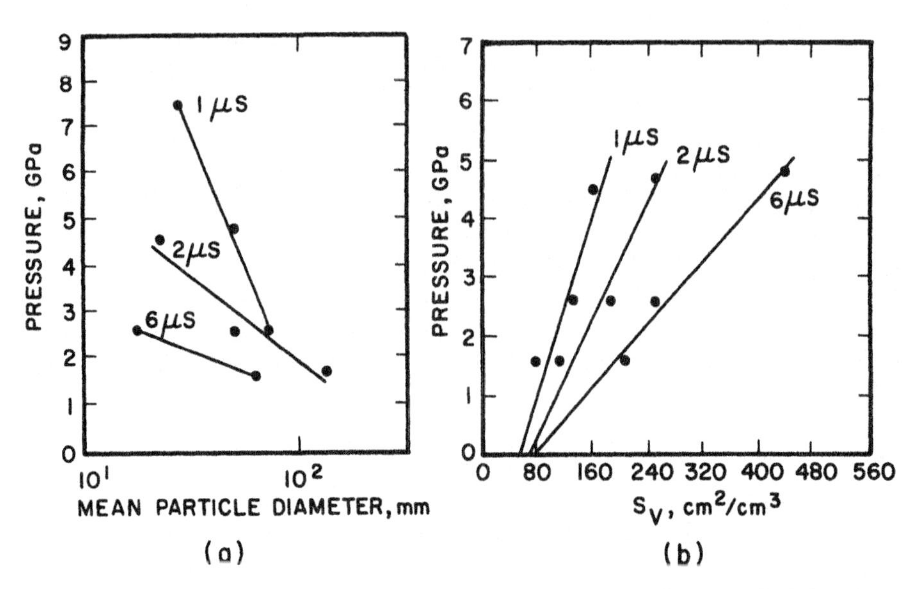

FIGURE 16.43 (a) Mean particle diameter vs. pressure for three nominal durations for quartz monzonite. (From Aimone et al. [86], Fig. 5, p. 985. Reprinted by permission of the publisher.) (b) Crack density (surface area per unit volume) vs. pressure for the same durations. (From Meyers [87], Fig. 6(b), p. 391. Reprinted by permission of the publisher.)

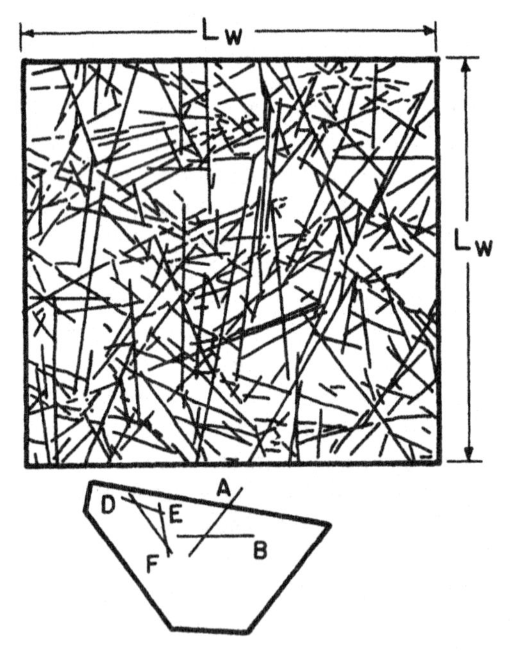

FIGURE 16.44 Fragmented medium resulting from about 400 randomly thrown linear cracks. In the bottom part the enlarged fragment illustrates the concept of internal damage. A fragment is shown within which five cracks exist (A, B, D, E, and F). These are "internal flaws" that do not directly contribute to the fragmentation. (From Jaeger et al. [88]. Reprinted by permission of the publisher.)

where N_c is the number of cracks formed in Arkansas–Novaculite rocks and λ_c is the average length of randomly thrown linear cracks.

They determined the number of fragments N_f from the expression

$$N_f = aN_c^{\beta}\lambda_c^{\alpha}L_w^{-\alpha} \tag{16.69}$$

where a, α, and β are constants (taken as π^{-1}, 2, 2, and 2 by [88]) and L_w is the window length, as shown in Figure 16.44, which also illustrates internal damage (lower part). From Fig. 16.44, the mean area, $\bar{s}$, of the fragments is the area L_w^2 divided by N_f

$$\bar{s} = \frac{L_w^2}{N_f} \tag{16.70}$$

The internal damage length D was quantified as being the sum of lengths of all cracks that are not external boundaries of fragments. This is shown in the r.h.s. part of Figure 16.44. The damage length D curve (Fig. 16.45) shows that D rises initially as more cracks appear, but, ultimately, at high crack

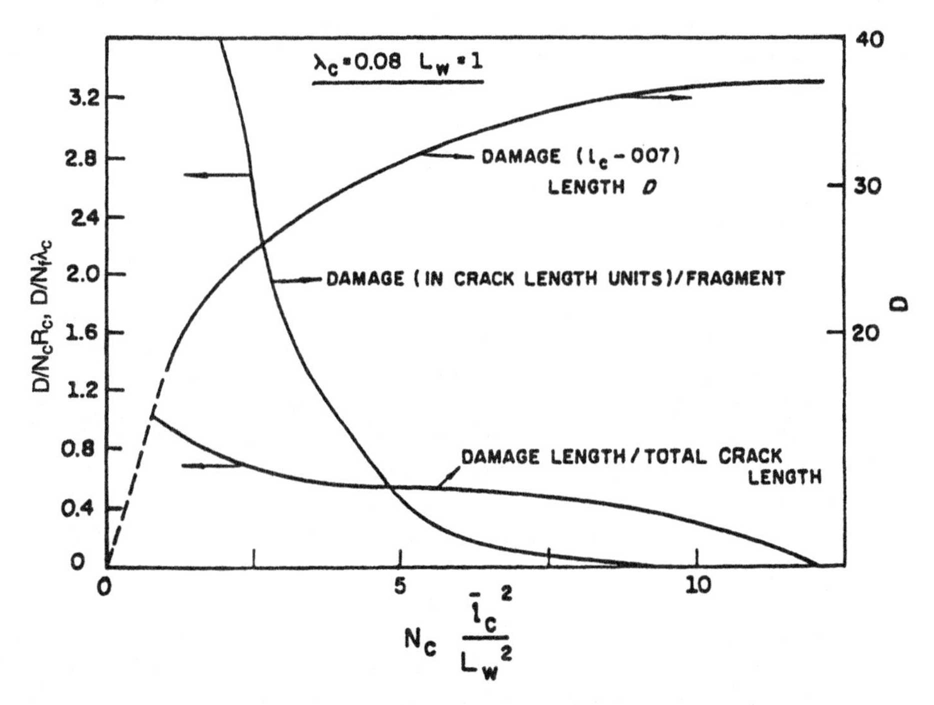

FIGURE 16.45 Internal damage length D (rhs scale) as function of the normalized crack density $N_c l_c^2/L_w^2$. For low crack densities, D is extrapolated linearly to the origin (broken line). Shown on the left are the damage in an average fragment ($D/N_f\lambda_c$) and the proportion of cracks contributing to internal damage ($D/N_c l_c$). (From Jaeger et al. [88]. Reprinted by permission of the publisher.)

densities, it approaches the value

$$D_\infty = 4.6 \frac{L_w^2}{\lambda_c} \qquad (16.71)$$

obtained for $L_w = 1$ and $\lambda_c = 0.08$ (average length). This is about 60. They interpreted this result as there being $60/\lambda_c \cong 800$ cracks that form dangling cracks or internal boundaries, whereas the rest of the cracks create fragment boundary. They found an equation for damage length as

$$D = 77[1 - 0.25 \exp (3.5/N_c^{0.2})] \qquad (16.72)$$

However, the (internal damage length)/(total crack length), $D/N_c l_c$, decreases from ~ 1.2 to zero as $N_c l^2/L_w^2$ is increased; the experimental values obtained by Aimone, Meyers, and Mojtabai [86] are much higher, indicating that damage has to be separately assessed at the micro, meso, and macrolevels.

16.9.4 Fragmentation of Ceramics Due to Impact

Louro and Meyers [80, 81] were able to develop a model that quantitatively predicts the fragment sizes in a ceramic subjected to uniaxial strain–stress wave loading. In the impact of a ceramic by a projectile, the three phenomena depicted in Figure 17.22(c) are important. First, a compressive stress wave travels through the ceramic. This wave, which we call a "shock wave," generates significant damage in the ceramic when the HEL is approached or exceeded. The compressive wave is not planar but has a front that resembles a spherical cap because the wave originates at the contact point. This wave, being a radially expanding wave, will generate tangential stresses that are tensile. This creates radial cracks emanating from the point of contact. When the compressive waves reach the limits of the ceramic, they are reflected as tensile waves, enacting additional cracking. A fourth damage mechanism in the ceramic is the formation of a thin layer of pulverized material at the ceramic–projectile interface and its expulsion from the ceramic. This is easily understood, since space (in the ceramic) has to be created in order for the projectile to penetrate into it.

The model developed by Louro and Meyers [81] only addresses the propagation of a simple stress pulse in a ceramic and of a subsequent reflected tensile pulse. Several potential microstructural processes of dynamic initiation were reviewed by Louro and Meyers [80]. They are shown in Figure 16.46. It is difficult to envisage fracture under compression because it is necessary to have tensile stresses to open up cracks. Nevertheless, *microstructural inhomogenities create conditions for tensile stresses under compressive loading*. The following are the most important mechanisms:

1. Spherical voids subjected under compression generate tensile stresses. In Figure 16.47 we have σ_r tensile. Therefore, cracks are generated. These cracks are parallel to the direction of load application. This problem was first solved by Goodier [89].

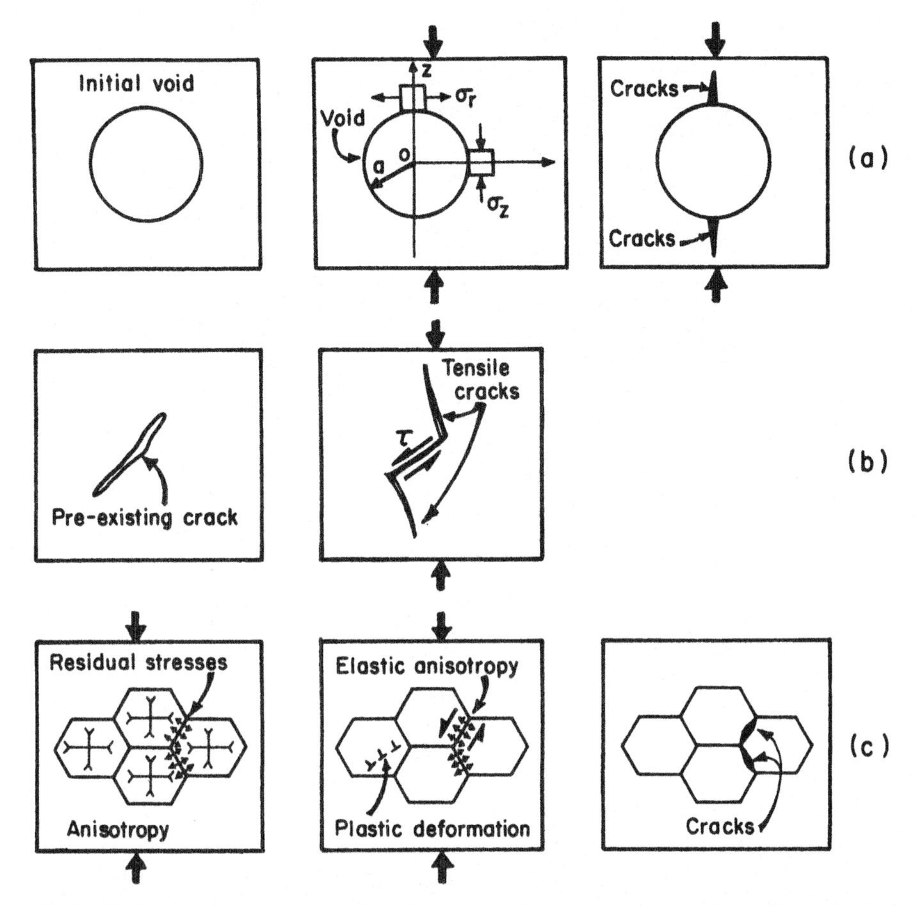

FIGURE 16.46 Mechanisms of crack propagation under compression: (a) crack generated by spherical flaw; (b) cracks generated by elliptical flaw inclined to compression axis; (c) cracks generated by dislocation formed on stress release by elastic anisotropy and stress concentration pile-ups or twinning.

2. For ellipsoidal flaws, the same phenomenon occurs. Indeed, the spherical case 1 is a special case of the elliptical flaw. Brace and Bombolakis [90] and Hori and Nemat-Nasser [91] studied the stresses generated by this configuration. The shear stress due to the applied compressive load τ generates tensile stresses at the extremities of the elliptical flaw, eventually cracking it [Fig. 16.47(b)].
3. Elastic anisotropy of a polycrystalline ceramic aggregate leads to incompatibility stresses at the boundaries. Figure 16.47(c) shows the "softer" and "stiffer" directions of grains marked by short and long lines, respectively. Plastic deformation can occur under compressive loading. When the stress is released, that is, after the passage of the stress pulse, localized tension regions are created by the residual stresses due to plastic deformation. This can open up cracks. These are not strictly compression

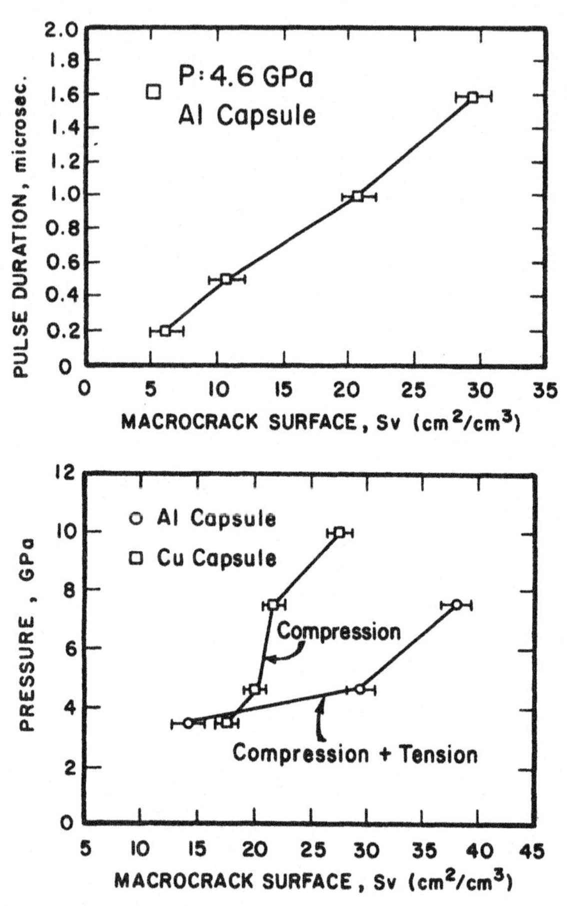

FIGURE 16.47 Macrocrack surface area as a function of (a) tensile pulse duration at a constant pressure and (b) pressure at a constant pulse duration. Notice increase in damage with tensile pulse. (From Louro and Meyers [82], Figs. 2, 3, p. 466. Reprinted with permission of the publisher.)

cracks but rather "postcompression cracks." Lankford [92–94] describes some of these concepts.

In plate impact experiments, Louro and Meyers [80] showed that cracking occurred in alumina at stress levels below the HEL. Figure 16.46(a) shows their measured macrocrack surface area (surface area of cracks observed at a low magnification) as a function of pressure, whereas Figure 16.46(b) shows the effect of pulse duration (of the tensile pulse) at a constant pressure. It is clear that both compressive and tensile stresses create cracks and that the crack density increases with the time of application of the pulse. The compression experiments in Figure 16.46(a) were done in copper capsules, which has the same shock impedance as alumina. The compression and tension experiments were conducted using aluminum capsules (impedance lower than alumina). The amplitude of the tensile pulses was approximately equal to one-third of the compressive stress amplitude. Louro and Meyers [80] also obtained another important information: the compressive pulse creates crack initiation sites. Thus, a tensile pulse traveling in a "virgin" material produces less damage than a tensile pulse traveling through a material that has seen a previous compressive pulse.

The above considerations served as a basis for the development of a model whose elements are depicted in Figure 16.48. The material is assumed to have preexisting flaws [Fig. 16.48(a)]. These flaws are activated by the compression pulse [Fig. 16.48(b)] and become larger. Upon tension, the flaws start to grow at a velocity dictated by fracture dynamics, that is, their maximum velocity is the Rayleigh speed. As these cracks grow, new cracks are nucleated [Fig. 16.48(c)]. Upon intersecting, the cracks define fragments whose size can be calculated [Fig. 16.48(d)].

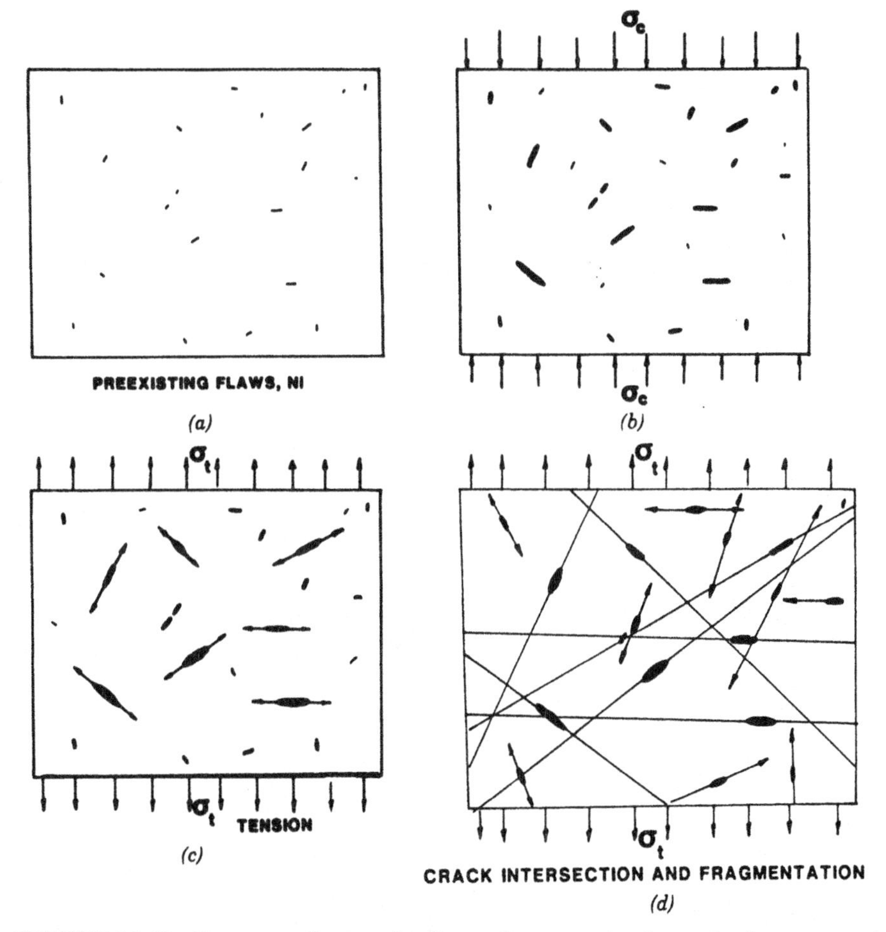

FIGURE 16.48 Sequence of events leading to fragmentation in mechanism proposed by Louro and Meyers: (a) preexisting flaws; (b) activation and stable growth of preexisting flaws under compression; (c) dynamic growth of activated flaws under tension as well a new flaw nucleation; (d) intersection of cracks and fragmentation.

Louro and Meyers [81] determined the surface area per unit volume of fragments, S_v, as

$$S_v = S_{v1} + S_{v2} + S_{v3} + S_{v4} \tag{16.73}$$

where

S_{v1} → cracks that are critical at the onset of tension
S_{v2} → increase in surface area of cracks that are critical at onset of tension and continue to grow
S_{v3} → cracks that become critical during application of tensile pulse
S_{v4} → increase in surface area due to cracks that become critical during the application of tensile pulse

From this total surface area per unit volume it is possible to estimate the mean fragment size through the expression (assuming spherical fragments).

$$D = \frac{6}{S_v} \tag{16.74}$$

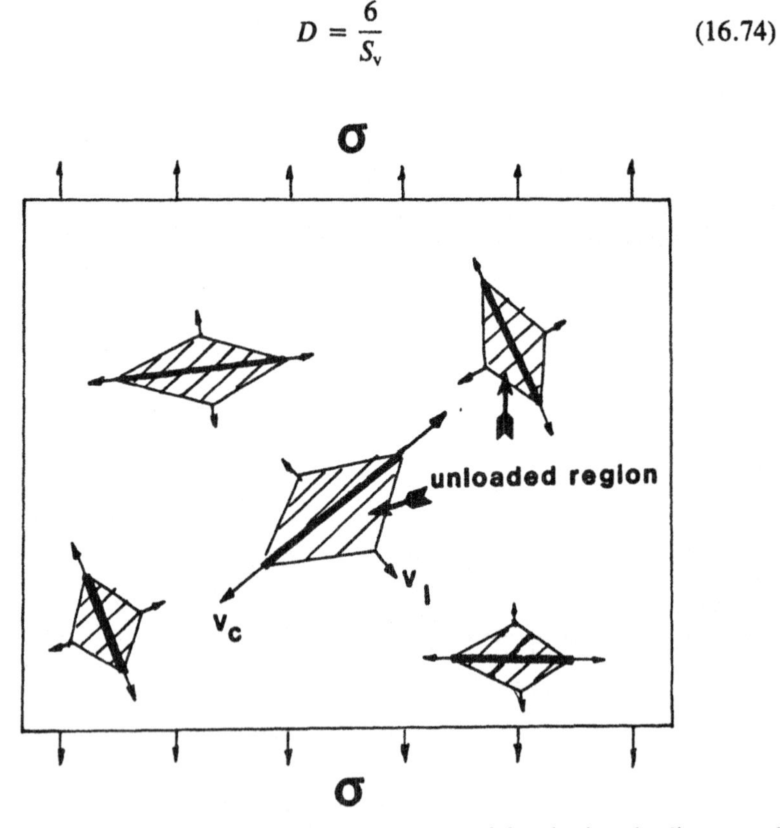

FIGURE 16.49 Shielding of crack nucleation and growth by elastic unloading around growing crack; hatched regions are "neutralized" and will not be fragmented.

It is possible, through analysis, to develop expressions for S_{v1}, S_{v2}, S_{v3}, and S_{v4}. Thus, the damage can be quantitatively estimated as a function of stress history (for a uniaxial strain state). The expressions developed by Louro and Meyers [82] are rather lengthy and contain a number of material properties. For this reason, they are not reproduced here. Nevertheless, these expressions explicitly show the effect of grain size, compressive pulse amplitude and duration, tensile pulse amplitude and duration, and maximum crack speed on the final fragment sizes. It is important to note that Louro and Meyers incorporated a "shielding factor" around every growing crack, shown in Figure 16.49. This shielded region is unloaded by the free surfaces created and does not provide any additional nucleation sites for new cracks. Therefore, fragmentation does not continue indefinitely and a final fragment size is reached when the material is unloaded. Thus, the density of active or activated crack nuclei (which is a function of the stress) and the growth kinetics of cracks determines the final fragment size.

REFERENCES

1. D. Broek, *Elementary Engineering Fracture Mechanics*, Martinus Nijhoff, The Hague, Netherlands, 1982.
2. J. F. Knott, *Fundamentals of Fracture Mechanics*, Butterworth, London, 1973.
3. R. W. Hertzberg, *Deformation and Fracture Mechanics of Engineering Materials*, 3rd ed., Wiley, New York, 1989.
4. B. Hopkinson, *The Effects of the Detonation of Guncotton*, Scientific Papers, Cambridge University Press, Cambridge, 1921.
5. J. S. Rinehart and J. Pearson, *Behavior of Metals under Impulsive Loads*, American Society of Metals, Metals Park, Ohio, 1954.
6. H. Kolsky, *Stress Waves in Solids*, Dover, NJ, 1963.
7. J. D. Achenbach, "Dynamic Fracture Effects in Brittle Materials," in *Mechanics Today*, Vol. 1, ed. S. Nemat-Nasser, Pergamon, New York, 1972, p. 1.
8. L. B. Freund, "The Analysis of Elastodynamic Crack Tip Stress Fields," in *Mechanics Today*, Vol. 3, ed. S. Nemat-Nasser, Pergamon, New York, 1976, p. 55.
9. L. B. Freund, *Dynamic Fracture Mechanics*, Cambridge University Press, Cambridge, 1990.
10. K. B. Broberg, *Arch. für Fysik*, **18** (1960), 159.
11. J. W. Craggs, *J. Mech. Phys. Sol.*, **8** (1960), 66.
12. B. R. Baker, *J. Appl. Mech.*, **29** (1962), 449.
13. L. B. Freund, *J. Mech. Phys. Sol.*, **20** (1972), 129.
14. L. B. Freund, *J. Mech. Phys. Sol.*, **20** (1972), 141.
15. L. B. Freund, *J. Mech. Phys. Sol.*, **21** (1973), 47.
16. L. B. Freund, *Int. J. Eng. Sci.*, **12** (1974), 179.

17. E. H. Yoffé, *Phil. Mag.*, **42** (1951), 739.
18. J. D. Achenbach, "Dynamic Effects in Brittle Fracture," in *Mechanics Today*, Vol. 1, ed. S. Nemat-Nasser, Pergamon, Elmsford, NY, 1974, p. 1.
19. L. B. Freund, "The Analysis of Elasto-dynamic Crack Tip Stress Fields," in *Mechanics Today*, Vol. 3, ed. S. Nemat-Nasser, Pergamon, Elmsford, NY, 1976, p. 55.
20. D. K. Roberts and A. A. Wells, *Engineering*, **171** (1954), 820.
21. G. Ravichandran and R. J. Clifton, *Intl. J. Fract.*, **40** (1989), 157.
22. J. F. Kalthoff and D. A. Shockey, *J. Appl. Phys.*, **48** (1977), 986.
23. H. Homma, D. A. Shockey, and Y. Murayama, *J. Mech. Phys. Sol.*, **31** (1983), 261.
24. D. A. Shockey, J. F. Kalthoff, and D. C. Erlich, *Intl. J. Fract.*, **22** (1983), 217.
25. D. A. Shockey, J. F. Kalthoff, H. Homma, and D. C. Erlich, *Eng. Fract. Mech.*, **23** (1986), 311.
26. K. Ravi-Chandar and W. G. Knauss, *Intl. J. Fract.*, **20** (1982), 209; **25** (1984) 247; **26** (1984), 65; **25** (1984), 141; **26** (1984), 189; **27** (1985), 127.
27. W. G. Knauss and K. Ravi-Chandar, *Eng. Fract. Mech.*, **23** (1986), 9.
28. M. L. Wilson, R. H. Hawley, and J. Duffy, *Eng. Fract. Mech.*, **13** (1980), 371.
29. A. T. Zehnder and A. J. Rosakis, *Intl. J. Fract.*, **43** (1990), 271.
30. J. F. Kalthoff, J. Beinert, and S. Winkler, *VDI Berichte*, **313** (1978), 791.
31. A. J. Rosakis, *Eng. Fract. Mech.*, **13** (1980), 331.
32. H. Schardin, in *Fracture*, ed. J. Averbach, MIT Press and Wiley, 1959, p. 297.
33. P. Manogg, in *Proceedings, International Conference on the Physics of Non-Crystalline Solids*, Delft, 1964, p. 81.
34. P. S. Theocaris, *Appl. Opt.*, **10** (1971), 2240.
35. J. W. Dally, W. L. Fourney, and A. R. Irwin, *Int. J. Fract.*, **27** (1985), 159.
36. A. J. Rosakis, *Eng. Fract. Mech.*, **13** (1980), 331.
37. M. Ramulu and A. S. Kobayashi, *Int. J. Fract.*, **27** (1985), 187.
38. J. W. Dally, *Exper. Mech.*, **19** (1979), 349.
39. B. Hopkinson, *Trans. Roy. Soc. (Lond.)*, **213A** (1914), 437.
40. B. Hopkinson, *Scientific Papers*, Cambridge University Press, London, 1910.
41. J. S. Rinehart, *J. Phys.*, **22** (1951), 131.
42. J. S. Rinehart, *J. Phys.*, **23** (1952), 1229.
43. L. Davison and A. L. Stevens, *J. Appl. Phys.*, **43** (1972), 988.
44. F. R. Tuler and B. M. Butcher, *Int. J. Fract. Mech.*, **4** (1968), 431.
45. L. Davison and A. L. Stevens, *J. Appl. Phys.*, **44** (1973), 688.
46. L. Davison, A. L. Stevens, and M. E. Kipp, *J. Mech. Phys. Solids*, **25** (1977), 11.
47. S. Cochran and D. Banner, *J. Appl. Phys.*, **48** (1977), 2729.
48. J. J. Gilman and F. R. Tuler, *Int. J. Fract. Mech.*, **6** (1970), 169.
49. T. W. Barbee, L. Seaman, R. Crewdson, and D. Curran, *J. Mater. JMLSA*, **7** (1972), 393.
50. L. Seaman, D. R. Curran, and R. C. Crewdson, *J. Appl. Phys.*, **49** (1978), 5221.

51. R. T. DeHoff and R. N. Rhines, eds., *Quantitative Microscopy*, McGraw-Hill, New York, 1968.
52. D. R. Curran, in *Shock Waves and the Mechanical Properties of Solids*, eds. J. J. Burke and V. Weiss, Syracuse University Press, New York, 1971, 121.
53. T. W. Barbee, L. Seaman, R. Crewdson, and D. Curran, *J. Mater. JMLSA*, **7** (1972), 393.
54. L. Seaman, D. A. Shockey, and D. R. Curran, in *Dynamic Crack Propagation*, ed. G. C. Sih, Noordhoff, Leyden, 1973, p. 629.
55. D. R. Curran, D. A. Shockey, and L. Seaman, *J. Appl. Phys.*, **44** (1973), 4025.
56. D. A. Shockey, L. Seaman, and D. R. Curran, in *Metallurgical Effects at High Strain Rates*, eds. R. W. Rohde, B. M. Butcher, J. R. Holland, and C. H. Karnes, Plenum, New York, 1973, p. 473.
57. L. Seaman, D. R. Curran, and D. A. Shockey, *J. Appl. Phys.*, **47** (1976), 4814.
58. D. R. Curran, L. Seaman, and D. A. Shockey, *Phys. Today*, January (1977), 46.
59. D. A. Shockey, K. C. Dao, and R. L. Jones, in *Mechanisms of Deformation and Fracture*, ed. K. E. Easterling, Pergamon, Oxford, 1977, p. 77.
60. D. A. Shockey, D. R. Curran, and L. Seaman, in *High Velocity Deformation of Solids*, eds. K. Kawata and J. Shiori, Springer-Verlag, Berlin, 1979, p. 149.
61. L. Seaman, D. R. Curran, and R. C. Crewdson, *J. Appl. Phys.*, **49** (1978), 5221.
62. D. R. Curran, L. Seaman, and D. A. Shockey, in *Shock Waves and High-Strain Rate Phenomena in Metals: Concepts and Applications*, eds. M. A. Meyers and L. E. Murr, Plenum, New York, 1981, p. 129.
63. S. N. Zhurkov, *Int. J. Fract. Mech.*, **1** (1965), 311.
64. F. A. McClintock, in *Fracture Mechanics of Ceramics*, Vol. 1, eds. R. C. Bradt, D. P. H. Hasselman, and F. F. Lange, Plenum, New York, 1973.
65. D. A. Shockey, D. R. Curran, L. Seaman, J. T. Rosenberg, and C. F. Peterson, *Int. J. Rock Mech. Min. Sci. Geomech. Abstr.*, **11** (1974), 303.
66. J. N. Johnson, *J. Appl. Phys.*, **52** (1981), 2812.
67. S. A. Novikov, *J. Appl. Mech. Tech. Phys.*, **22** (1981), 385.
68. L. Davison and R. A. Graham, *Phys. Rep.*, **55** (1979), 257.
69. V. I. Romanchenko and V. G. Stepanov, *J. Appl. Mech. Tech. Phys.* (translated from Russian), **21** (1981), 551.
70. N. K. Akhmadeev and R. I. Nigmatulin, *J. Appl. Mech. Tech. Phys.* (translated from Russian), **22** (1981), 394.
71. M. A. Meyers and C. T. Aimone, *Prog. Mater. Sci.*, **28** (1983), 1.
72. S. Christy, H.-r. Pak, and M. A. Meyers, in *Metallurgical Applications of Shock-Wave and High-Strain-Rate Phenomena*, New York, eds. L. E. Murr, K. P. Staudhammer, and M. A. Meyers, Dekker, 1986, p. 835.
73. D. G. Brandon, M. Boas, and Z. Rosenberg, *Mechanical Properties at High Rates of Strain*, Institute of Physics, London, 1984, p. 261.
74. A. K. Zurek and C. E. Frantz, in *Impact Loading and Dynamic Behavior of Materials*, eds. C. Y. Chiem, H.-D. Kunze, and L. W. Meyer, DGM Verlag, Oberursel, 1988, p. 785.
75. G. I. Kanel, S. V. Rasorenov, and V. E. Fortov, in *Shock Wave and High-Strain-Rate Phenomena in Materials*, eds. M. A. Meyers, L. E. Murr, and K. P. Staudhammer, Dekker, New York, 1992, p. 775.

76. J. O. Erkman, *J. Appl. Phys.*, **31** (1961), 939.
77. E. E. Banks, J. *Iron Steel Inst.*, **206** (1968), 1022.
78. A. N. Dremin, A. M. Molodets, A. I. Melkumov, and A. V. Kolesmikov, in *Shock-Wave and High-Strain-Rate Phenomena in Materials*, eds. M. A. Meyers, L. E. Murr, and K. P. Staudhammer, Dekker, New York, 1992, p. 751.
79. N. F. Mott, *Proc. Roy. Soc. Lond.*, **300** (1947), 300.
80. L. H. L. Louro and M. A. Meyers, "Stress Wave Induced Damage in Alumina," in *Proc. DYMAT 88 (Ajaccio, France), J. Phys.*, **49** (1988), C3-333 (Colloque-3).
81. L. H. L. Louro and M. A. Meyers, *J. Mater. Sci.*, **24** (1989), 2516.
82. L. H. L. Louro and M. A. Meyers, "Shock-induced Fracture and Fragmentation in Alumina," in *Shock Waves in Condensed Matter—1989*, eds. S. C. Schmidt, J. N. Johnson, and L. W. Davidson, North-Holland, Amsterdam, 1990, p. 465.
83. D. E. Grady and M. E. Kipp, *Proc. 20th Symposium on Rock Mechanics*, Austin, Texas, 1979, p. 403.
84. D. E. Grady and M. E. Kipp, *Int. J. Rock Mech. Min. Sci.*, **17** (1980), 147.
85. D. E. Grady, *J. Appl. Phys.*, **55** (1982), 322.
86. C. T. Aimone, M. A. Meyers, and N. Mojtabai, in *Rock Mechanics in Productivity Design*, eds. D. H. Dowding and M. M. Singh, AIME, Warrendale, PA, 1984, p. 979.
87. M. A. Meyers, *Mech. Mater.*, **4** (1985), 387.
88. Z. Jaeger, R. Englman, Y. Gur, and A. Sprecher, *J. Mater. Sci.*, **5** (1986), 577.
89. N. Goodier, *J. Appl. Mech.*, **1** (1933), 39.
90. W. F. Brace and E. G. Bombolakis, *J. Geophys. Res.*, **68** (1963), 3709.
91. H. Horii and S. Nemat-Nasser, *Trans. Roy. Soc. Lond.*, **319** (1986), 337.
92. J. Lankford, *J. Mater. Sci.*, **12** (1977), 791.
93. J. Lankford, *J. Mater. Sci.*, **16** (1981), 1517.
94. J. Lankford, *Mater. Sci. Eng.*, **A107** (1989), 261.
95. S. Nemat-Nasser and H. Deng, *Acta Met. et Mat.*, **42** (1994), 1013.

ADDENDUM

Nemat-Nasser and Deng [95] considered an array of wing cracks (shown in Fig. 16.46 (b)) in a ceramic subjected to compression. They obtained closed-form solutions for the ceramic, for dynamically growing and interacting cracks. They applied Eqn. 16.22 to these cracks, varying both the stress state (uniaxial stress and strain) and strain rate, $\dot{\varepsilon}$. Lateral confinement (represented by uniaxial strain) is very important and increases the compressive strength. The microstructural parameters were introduced through the crack length, $2a$, and spacing, $2w$. The failure stress, which was dependent on a, w, $\dot{\varepsilon}$, and stress state, was observed to increase significantly in the 10^4–10^6 s^{-1} range.

CHAPTER 17

Applications

17.1 INTRODUCTION

We briefly introduced the applications of dynamic processes in (or behavior of) materials in Chart 1.1 and Section 1.1 (Figs. 1.1 and 1.2). The reader is advised to look at these figures with attention. The multidisciplinary nature of the field is strikingly evident in Chart 1.1. Materials science, physics, chemistry, mechanics, combustion theory, applied mathematics, and large-scale computation are important contributory scientific disciplines. There are a wide range of phenomena and applications where these dynamic processes play an important role. Throughout this chapter, we will succinctly introduce each basic application and present, quantitatively, the principal governing equations, where applicable. We cannot cover the entire spectrum of aspects because for each application one monograph could be written. Indeed, these books are available, and the studious reader is referred to them. We present, in Table 17.1, the principal supplementary sources of information for these various applications.

We will now briefly introduce the topics that will be covered in this chapter. We will start with the military applications (Sections 17.2–17.4) and then cover the "civilian" applications (Sections 17.5–17.8). We will first discuss the different types of penetrators in this chapter (Sections 17.2 and 17.3). They are classified into:

1. *Kinetic Energy Penetrators.* The energy comes from their kinetic energy, which is imparted by compressed gases that are generated by the burning of propellant in a barrel. In this class are the armor-piercing projectiles and kinetic energy penetrators that will be described in Section 17.3.
2. *Chemical Energy Penetrators.* The energy is provided by the detonation of an explosive when the system approaches or is in direct contact with the target. Among these are the shaped charges, explosively forged projectiles (Section 17.2), plastic charges, and detonating rounds. Figure 17.1 shows the energy transfer process in a high-explosive (chemical energy) warhead. This charge is also called High Explosive Anti-Tank (HEAT). The portion of the energy that traverses the target can be considered to be of three classes: thermal, residual penetration energy, and debris energy. The term P_k is a euphemism for the gory fact that the intelligent reader will immediately see: the integrated residual energy.

TABLE 17.1 Sources of Information to the Principal Technological Applications of Dynamic Processes in Materials

Application	Source
Penetrators and penetration	1. W. P. Walters and J. A. Zukas, *Fundamentals of Shaped Charges*, Wiley, New York, 1989. 2. J. A. Zukas, ed., *High-Velocity Impact Dynamics*, Wiley, New York, 1990. 3. M. E. Backman and W. Goldsmith, "The Mechanics of Penetration of Projectiles into Targets," *Intl. J. Eng. Sci.* **16** (1978), 1–99. 4. T. W. Wright, "A Study of Penetration Mechanics for Long Rods," in *Computational Aspects of Penetration Mechanics*, eds. J. Chandra and J. E. Flaherty, Springer, 1982.
Explosive welding/cladding/ forming	1. B. Crossland, *Explosive Welding of Metals and its Applications*, Oxford University Press, Oxford, 1982. 2. A. A. Ezra, *Principles and Practice of Explosive Metalworking*, Ind. Newspapers Ltd., London, 1973. 3. A. A. Deribas, *Physics of Explosive Welding and Hardening* (in Russian), Nauka, Novosibirsk, 1972. 4. J. S. Rinehart and J. Pearson, *Explosive Working of Metals*, Pergamon, 1983. 5. T. Z. Blazynski, ed., *Explosive Welding, Forming, and Compaction*, Applied Sci. Publishers, 1983.
Shock consolidation	1. R. Prummer, *Explosivverdichtung Pulvriger Substanzen*, Springer, Berlin, 198. 2. V. F. Nesterenko, *High-Rate Deformation of Heterogeneous Materials*, *Nauka*, Sub. Div., Novosibirsk, Russia (in Russian), 1992. 3. L. E. Murr, ed., *Shock Waves for Industrial Applications*, Noyes, 1990. 4. W. H. Gourdin, *Prog. Matls. Sci.* **30** (1986), 39.
Shock synthesis	1. S. S. Batsanov, *Effects of Explosions on Materials*, Springer, NY, 1993. 2. R. A. Graham and A. B. Sawaoka, eds., *High Pressure Explosive Processing of Ceramics*, Trans. Tech., 1987. 3. N. N. Thadham, *Prog. Matls. Sci.* **37** (1993) 117.
Blasting	1. C. H. Dowding, *Blast Vibration Monitoring and Control*, Prentice-Hall, Englewood Cliffs, 1985. 2. *Blasters Handbook*, E. I. Dupont de Nemours. 3. P. A. Persson, R. Holmberg, and J. Lee, *Rock Blasting and Explosives Engineering*, CRC, Boca Raton, 1993.
Geological materials/space	1. H. J. Melosh, *Impact Cratering*, Oxford University Press, Oxford, 1989.

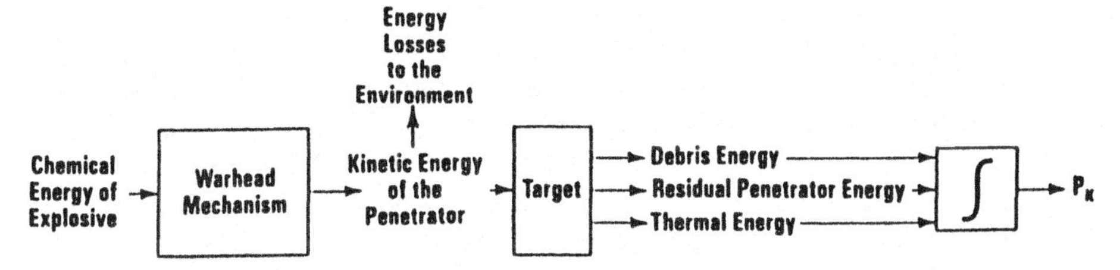

FIGURE 17.1 The energy transfer process that occurs in a chemical energy warhead. (Courtesy of J. Carleone, Aerojet.)

In Section 17.3 we will review the best known theories for penetration, starting with the de Marre equation developed by the French in the mid-1800s. The simple analysis presented by Birkhoff et al. [1] will be introduced, and the well-accepted theories developed by Alekseevskii and Tate will be covered. For more advanced work, the reader will be referred to the work of Bodner and to the modern marvels of large-scale computation. As a reminder to our most respected reader of what kinetic energy can do, the following are noted: (1) ceramics can be cut by high-speed water jets (indeed, this is a successful industrial process) and (2) I recall hearing the story of the tornado that sent a straw against a wooden post with such speed that it penetrated it.

Armor will be described, in very simple terms, in Section 17.4. The tide of battles and, as a consequence, the fate of nations, have often been determined by the armor/antiarmor balance. The long bow of the Saxon foot soldiers developed in the thirteenth century proved to be a tremendous weapon. It could pierce the armor of knights at 50 yards. The long bow (6 ft) delivered, at the hands of highly skilled archers, a long and heavy arrow. This was a formidable kinetic energy penetrator. The English trounced the French at Crécy and Poitiers. Shortly thereafter, the introduction of gunpowder helped to completely eliminate the heavily armored cavalry that was the principal battle force in the Middle Age. Thus, the requiem of feudalism and the ascension of a more centralized and modern form (albeit not less exploitative) of government was connected to the ability to dismount, by penetration, the medieval knight. The powerful tanks introduced into the battlefield at the end of World War I (1918) decided the fate of the battle and of the war. Hence, the discussion of Section 17.4 is very important. A large number of materials have been and are being used for armor. Japanese samurai used tortoise-shell shields! Modern vehicles used metals, ceramics, glass, and even explosive materials to render ineffective incoming projectiles. In nature, scarabees, turtles, shellfish, and crustaceans use the same stratagem. In Section 17.4, we will concentrate on the most significant aspects of metal, ceramic, composite, and reactive armor.

Sections 17.5–17.8 will concentrate on the "civilian" applications of the dynamic behavior of materials. Explosive welding and cladding is covered in Section 17.5. Explosive welding was discovered by accident, when it was observed that pieces of ordnance (shell fragments), after detonation, "stuck" to the target. Systematic experimentation led to the development of the technology of explosive welding. The first official account is by Carl, in 1944 [2]. Deribas [3] reports that, in the then USSR, explosive welding was independently discovered by Lavrentiev and co-workers, during World War II, and developed in the Institute of Hydrodynamics from the early 60s. Explosive welding (Section 17.5), forming (Section 17.6), hardening (Section 17.6), compaction, and synthesis (Section 17.7) are interconnected. Explosive welding (and cladding) has been very successful industrially, and there are companies in the United States, Japan, Russia, and the United Kingdom where products are routinely processed by this method. The market is limited, but there are some unique applications, such as the welding of titanium and aluminum plates to steel, where explosives are the only possible process. The world market for explosively welded products is approximately 100–200 million dollars/year. Explosive hardening has only been successful in a few applications. We saw, in Chapter 14, that shock waves generate a high concentration of defects in materials, increasing their flow stress. Hadfield steel rail frogs are routinely shock hardened in Russia, and this procedure increases their life by a factor of 3 or more. Explosive synthesis of diamond from graphite is successfully used in the United States (DuPont); in Japan, this process is being seriously considered; in Russia, there are a number of successful industrial operations. In spite of worldwide research, explosive compaction does not have a single successful product to date.* In Section 17.7 the principles of explosive compaction and main experimental methods will be reviewed. The impacts of meteorites on earth and other planets involve the classical phenomenology of dynamic processes such as shock waves, shock-induced phase transformations, melting, and crater formation. They are described in Section 17.8. An additional group of applications deals with the use of shock waves in the mining, construction, and prospection industries; a brief mention is made in Section 17.8.

17.2 SHAPED CHARGES AND EXPLOSIVELY FORGED PROJECTILES

A shaped (or hollow) charge consists of an explosive with a cavity lined with metal. The detonation of the explosive accelerates the metal, in a convergent trajectory, producing a jet that can travel at a velocity V_3 higher than the detonation velocity of the explosive (V_0). Figure 17.2 shows schematically the longitudinal section of the explosive charge with a hollow conical metal liner. Upon detonation, the metal liner is accelerated inward, toward the axis of the cone. Part of the liner is deflected forward, forming the jet, and part is deflected backward, forming the slug (that travels at a much lower velocity). The point that separates the forward from the backward flow is called the stagnation point.

*In Russia, it is used for the production of metal-ceramic-metal tubes for hinges in electric furnaces and in nickel plants (V. F. Nesterenko).

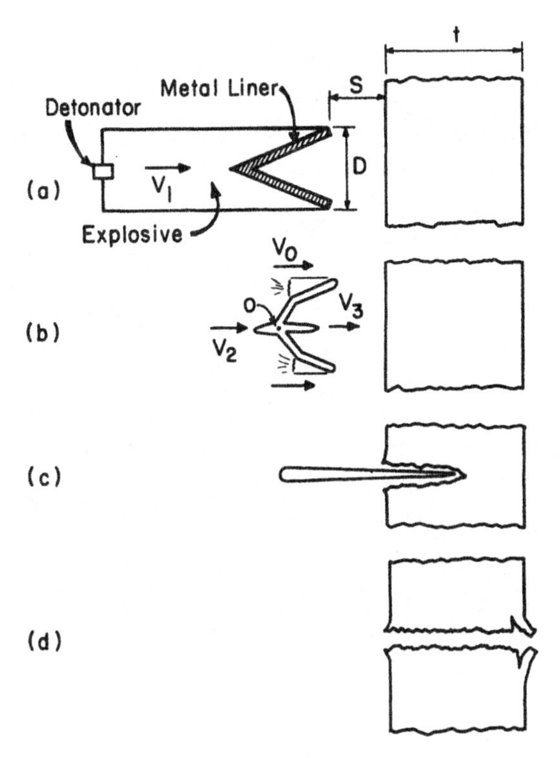

FIGURE 17.2 (a) Metal liner and explosive comprising shaped charge; (b) formation of jet and slug; (c) extension of jet in flight; (d) perforation of target.

Figure 1.3 (Chapter 1) shows the same sequence in greater detail. The penetration of a jet into a target depends on a number of factors. Some of these will be discussed in Section 17.3 (Armor penetration). Important parameters are the charge diameter D and the stand-off distance S. One tries to obtain as long a jet as possible, avoiding breakup. After the jet has stretched beyond a certain point, it fragments. This breakup is produced by the strain induced by the difference between the velocities of the jet tip, V_3, and the slug, V_2. Modern shaped charges use a number of shapes other than the conical one, for example, bell-shaped and spherical caps. Tandem shaped charges are also used to traverse targets that are designed to destroy the first incoming jet. Section 17.4 will briefly describe reactive armor, which uses this concept. The penetrations obtained by modern shaped charges are quite large. Penetration in rolled homogeneous armor (RHA) steel can exceed $8D$ (where D is the charge diameter).

Figure 17.3 shows a sequence of flash radiographs obtained by Raftenberg [7]. They show a shaped-charge jet prior to (a) and after (b and c) penetration of a 12.5-mm-thick rolled homogeneous armor steel. These flash radiographs were taken at times of 61, 82.2, and 137.4 μs after initiation of the explosive. The devoted student is asked to calculate the velocity of propagation of the jet forehead, assuming that it is not consumed by the target. In order to obtain

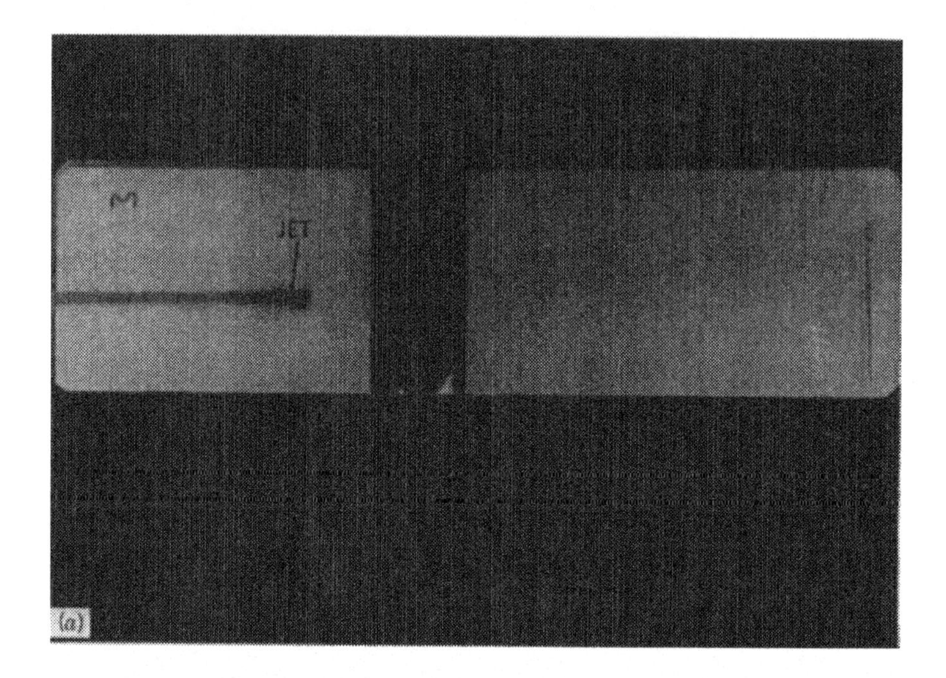

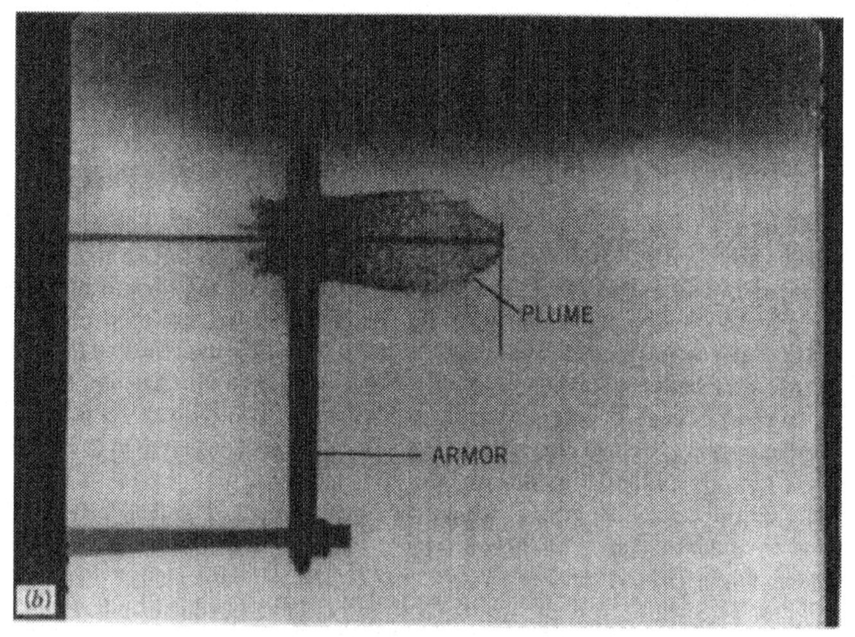

FIGURE 17.3 Flash radiographs from 81.3-mm-diameter shaped charge penetrating 12.5-mm-thick rolled homogeneous armor steel. Times after initiation of charge: (a) 61 μs (prior to impact with target); (b) 82.2 μs (penetration and plume formation); (c) 137.4 μs (fragments of target dispersed throughout radiograph). (Photographs courtesy of M. Raftenberg, U.S. Ballistic Research Laboratory, 1992.)

FIGURE 17.3 (*Continued*)

true distance, the target thickness should be taken as the scale. In Figure 17.3(b) one sees the "plume" of material that was formed by the penetrating jet. There is also material that is ejected backward. In Figure 17.3(c) the plume has broken up and the individual fragments can be clearly seen.

17.2.1 Theory of Jet Formation and Propagation

The analysis of shaped charges is currently conducted using hydrocodes. Nevertheless, a simple closed-form analysis published in 1948 by Birkhoff et al. [1] will be presented. It provides an excellent physical understanding of the principal phenomena involved. Apparently, a similar analysis has been developed by Lavrentiev in the former USSR [3]. This analysis is based on the geometrical elements of Figure 17.4(b) and on Bernoulli's equation. Figure 17.4(a) shows how the detonation changes the original cone apex angle from 2α to 2β. The material is accelerated to a velocity V_0, so that the stagnation point A moves to B. The flow of material into the stagnation points A and B to the right and to the left is shown in Figure 17.4(a). The flow of material to the right and left of the stagnation point is also indicated. The analysis by Birkhoff et al. [1] assumes that the pressure produced by the detonation of the explosive drops to zero shortly after the liner initiates its inward collapse and that the liner material behaves as an incompressible nonviscous fluid. It is assumed that V_0, the collapse velocity, is constant for the entire core. This velocity can be calculated, to a first approximation, using the Gurney equation (see Chapter 9). In Figure

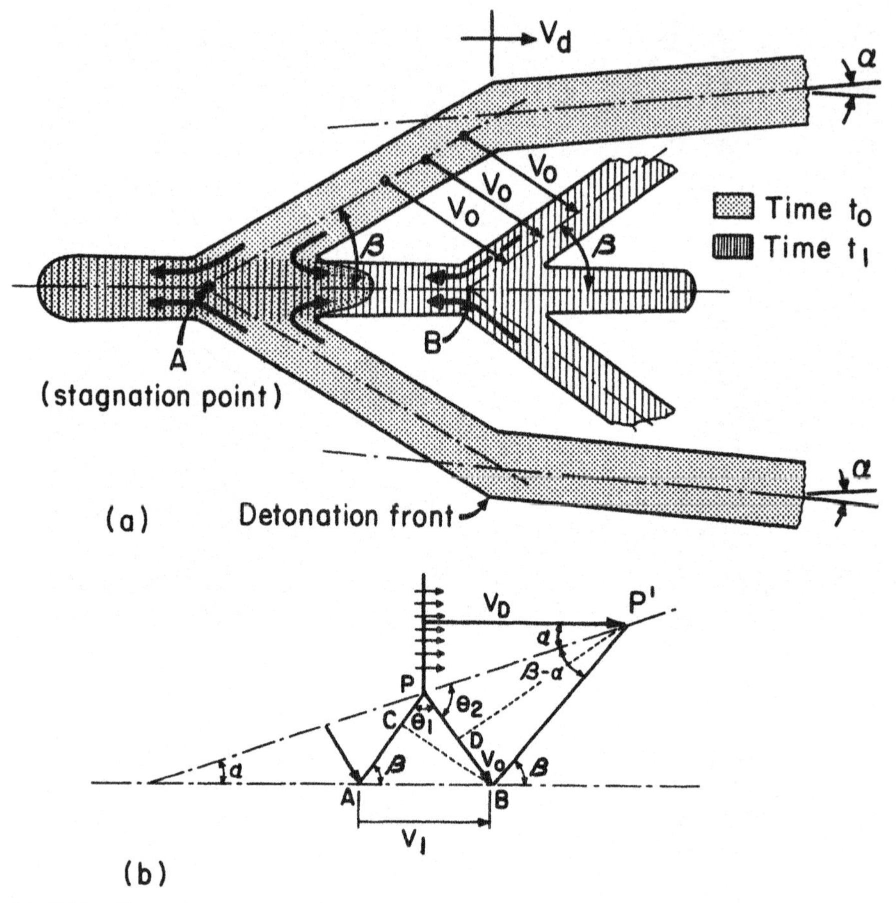

FIGURE 17.4 Geometrical aspects of cone collapse with formation of jet and slug.

17.4(b) the position of the liner at two time intervals separated by unit time is shown. This is a convenient interval, since the distance moved becomes numerically equal to the velocity. The velocity V_0 joins P, on the original cone surface, to B, the significant point at t_2. One important aspect of the derivation is the equality

$$P'P = P'B$$

which is based on the hypothesis that the explosive deflects the liner while not altering its thickness. We have, so to speak, a moving hinge traveling along PP'. Hence, we have the equality of angles:

$$\widehat{P'PB} = \widehat{PBP'} = \theta_2$$

Since $PA \parallel P'B$, we also have

$$\theta_1 = \widehat{APB} = \widehat{PBP'}$$

Hence, $\theta_1 = \theta_2$ and V_0 bisects the angle APP'. Here V_0 bisects the angle between the original liner and the new liner orientation. The velocity at which the stagnation point will move, V_1, can be found:

$$AB \sin \beta = PB \cos \widehat{CBP}$$

But

$$AB = V_1 \qquad PB = V_0$$

and

$$\widehat{CBP} = \tfrac{1}{2}\pi - \theta_1 = \tfrac{1}{2}\pi - \tfrac{1}{2}(\pi - \beta + \alpha)$$
$$= \frac{\beta - \alpha}{2}$$

Thus

$$V_1 = V_0 \frac{\cos [\frac{1}{2}(\beta - \alpha)]}{\sin \beta} \tag{17.1}$$

The velocity at which the material moves toward the stagnation point, V_2, in the stagnation point reference system is equal to AP:

$$AP = AC + PC$$
$$= V_1 \cos \beta + V_0 \sin [\tfrac{1}{2}(\beta - \alpha)] \tag{17.2}$$

Substituting (17.2) into (17.1) yields

$$V_2 = V_0 \left\{ \frac{\cos [\frac{1}{2}(\beta - \alpha)]}{\tan \beta} + \sin \left(\frac{\beta - \alpha}{2} \right) \right\} \tag{17.2a}$$

Birkhoff et al. [1] applied Bernoulli's equation to the stagnation point (streamline flow):

$$\int \frac{dP}{\rho(P)} + \tfrac{1}{2}u^2 = \text{const} \tag{17.3}$$

Equation (17.3) relates the flow velocity u to the pressure P. Assuming an incompressible flow and a constant pressure for incoming and outgoing flow, this expression reduces to

$$P + \tfrac{1}{2}\rho_0 u^2 = \text{const} \tag{17.4}$$

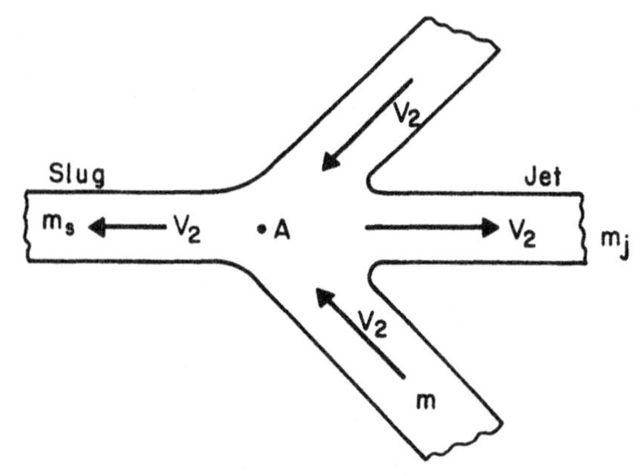

FIGURE 17.5 Streamline flow for constant-pressure assumption; jet and slug thickness are usually different. (From Birkoff et al. [1], Fig. 13, p. 571. Reprinted with permission of the publisher.)

The constant in Eqn. (17.4) can be found by taking into account that, far from the stagnation point, $P = 0$ and $u = V_2$. This enables the determination of P at stagnation point ($u = 0$).

The pressure at any point in the fluid determines its velocity. For this constant pressure, the boundary streamlines of constant pressure (and velocity) are shown in Figure 17.5. The entrance velocity V_2 and the exit velocities to the jet and slug are equal with respect to a referential A. These assumptions of constant pressure and steady state were modified by Pugh et al. [4], Godunov et al. [5], and Grace [6], providing more realistic results. With respect to a stationary frame of reference, the velocities of the jet and slug, V_j and V_s, respectively, are readily determined:

$$V_j = V_1 + V_2 \tag{17.4a}$$

$$V_s = V_1 - V_2 \tag{17.4b}$$

V_s and V_j can be obtained by substituting Eqns. (17.1) and (17.2a) into (17.4a) and (17.4b). The application of the conservation of mass and momentum to A enables the determination of the jet and slug masses, m_j and m_s, respectively. The mass per unit time entering A is m:

$$m = m_j + m_s$$

The horizontal components of the momentum equation give

$$-mV_2 \cos\beta + m_j V_2 - m_s V_2 = 0$$

This yields

$$m_j = \tfrac{1}{2}m(1 - \cos\beta) \qquad m_s = \tfrac{1}{2}m(1 + \cos\beta)$$

It is also helpful to relate the velocity V_0 to V_D, the detonation velocity. In triangle $PP'B$, the line $P'D$ is

$$P'D = PP' \cos [\tfrac{1}{2}(\beta - \alpha)]$$

But

$$PP' = \frac{V_D}{\cos \alpha}$$

$$P'D = \frac{V_D}{\cos \alpha} \cos [\tfrac{1}{2}(\beta - \alpha)]$$

This leads to

$$\frac{V_D}{\cos \alpha} = V_0 \frac{\cos [\frac{1}{2}(\beta - \alpha)]}{\sin (\beta - \alpha)} \qquad (17.5)$$

The simplified analysis by Birkhoff et al. [1] makes the following predictions:

1. The velocity of the jet increases with a decrease in α. In the limit, for $\alpha = 0$ and $\beta = 0$, $V_j = 2V_D$ [Eqns. (17.4a), (17.1), and (17.2a)].
2. The fraction of mass that goes to jet $[0.5(1 - \cos \beta)]$ increases with an increase in β.

The angle β is dependent on both the detonation velocity V_D and the velocity V_0 imparted to the liner and on α.

It is possible to obtain β, knowing V_D and V_0, through the modified expression below, obtained from Eqn. (17.5) by substitution of $\sin (\beta - \alpha)$ by $2 \sin [\frac{1}{2}(\beta - \alpha)] \cos [\frac{1}{2}(\beta - \alpha)]$:

$$\sin \left(\frac{\beta - \alpha}{2}\right) = \frac{V_0}{2V_D} \cos \alpha \qquad (17.6)$$

Example 17.1. For an explosive having detonation velocity $V_D = 7000$ m/s and producing a velocity $V_0 = 3000$ m/s in a shaped charge with $2\alpha = 42°$ (the most common apex angle, for U.S.-made shaped charges), find 2β.

One directly applies Eqn. (17.6) and obtains $2\beta = 87.2°$.

Example 17.2. For the BRL 81-mm precision charge shown in Figure 17.6, determine the jet and slug velocities and masses. Assume that the liner thickness $t = 2$ mm ($\rho = 8.9$). Given, for explosive RDX (Tables 9.1 and 10.1),

$$\rho_{\text{explosive}} = 1650 \text{ kg/m}^3$$

$$V_D = 8180 \text{ m/s}$$

$$\sqrt{2E} \text{ (Gurney energy)} = 2.93 \text{ mm/}\mu\text{s}$$

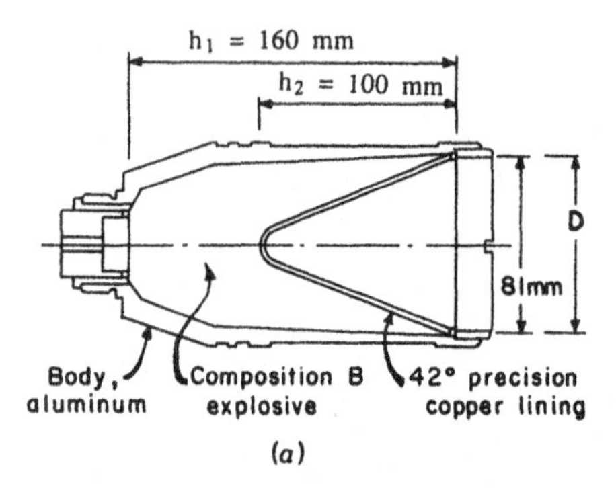

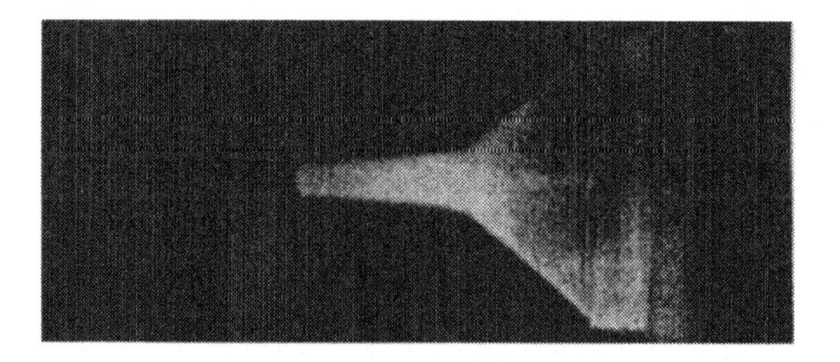

(b)

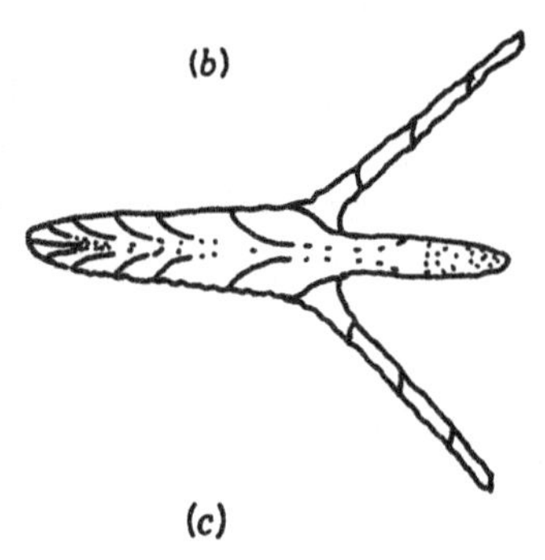

(c)

FIGURE 17.6 (a) Longitudinal section of shaped charge (81 mm diameter); (b) X-ray of liner during collapse; (c) computation of liner shape during collapse. (From Grace [8], Figs. 3, 4, and 5, pp. 497, 498. Reprinted by courtesy of Marcel Dekker, Inc.)

From Chapter 9, we have to select the proper equation for the calculation of V_0. As a first step, we compute the total volumes of explosive and copper. We will use (Fig. 9.8) the open-faced sandwich geometry as a first approximation:

$$\frac{V}{\sqrt{2E}} = \left[\frac{(1 + 2(M/C))^3 + 1}{6(1 + (M/C))} + \frac{M}{C}\right]^{-1/2}$$

We will use (Fig. 9.8) the open-faced sandwich geometry as a first approximation:

$$\frac{V}{\sqrt{2E}} = \left[\frac{(1 + 2(M/C))^3 + 1}{6(1 + (M/C))} + \frac{M}{C}\right]^{-1/2}$$

The volume of charge (V_c) and metal (V_m) are obtained from the equations below.

$$V_c = \frac{\pi D^2}{4} \times h_1 - \frac{1}{3}\frac{\pi D^2}{4} \times h_2 = 652{,}734 \text{ mm}^3$$

$$V_m = \frac{\pi D}{2}\left(\frac{D^2}{4} + h_2^2\right)^{1/2} t = 27{,}454 \text{ mm}^3$$

where t is the liner thickness.

The masses of copper and explosive are, respectively,

$$M = V_m \rho_m = 27{,}454 \times 8900$$

$$C = V_c \rho_c = 652{,}731 \times 1650$$

$$\frac{M}{C} = 0.226$$

Thus

$$V_0 = 2.93\left[\frac{(1 + 2 \times 0.226)^3 + 1}{6(1 + 0.226)} + 0.226\right]^{-1/2}$$
$$= 3.32 \text{ mm}/\mu\text{s}$$

The angle β is found through Eqn. (17.6)

$$\sin\left(\frac{\beta - 21}{2}\right) = \frac{3.32}{8.18}\cos 10.5$$
$$\beta = 68.9$$

From Eqn. (17.1), $V_1 = 3.63$ km/s
From Eqn. (17.2a), $V_2 = 2.52$ km/s
Thus,

$$V_j = V_1 + V_2 = 6.15 \text{ km/s}$$

$$V_s = V_1 - V_2 = 1.11 \text{ km/s}$$

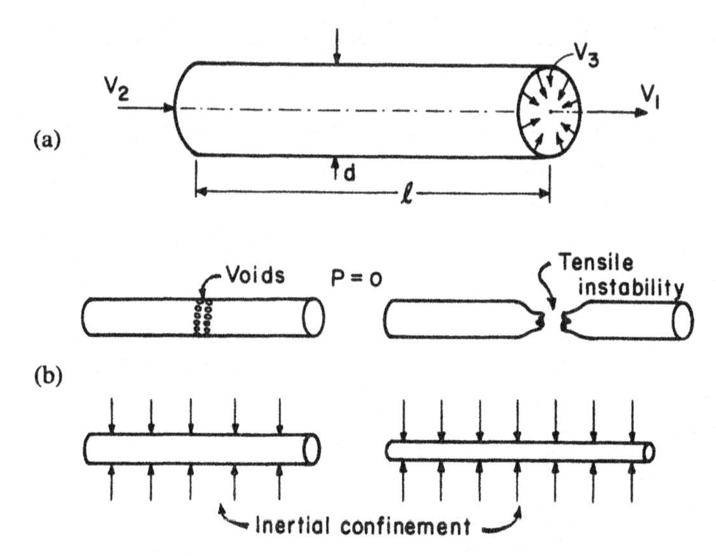

FIGURE 17.7 (a) Inertial confinement of jet creating an equivalent pressure that generates stable extension without necking; (b) effect on confining pressure on the tensile elongation of jet.

$$m_j = \frac{M}{2}(1 - \cos\beta) = 156 \text{ g}$$

$$m_s = M - m_j = 88 \text{ g}$$

After stretching to a certain point, shaped-charge jets break up. The time to breakup is an important jet property, and one tries to keep it as long as possible, because the individual fragments start to "tumble" and are less effective than the monolithic jet. Chou and Carleone [9] have developed detailed treatments on this. The ability of the jet to stretch without breaking up has been attributed to a number of causes. Chou and Flis [10], Fressengeas and Molinari [11], and Grady [12] have developed the concept of inertial stability, or a pressure induced by the radial flow of material during the stretching. They calculated this radial flow velocity and found that it can correspond to significant pressures. Figure 17.7(a) shows this effect. The gradient of velocity $\Delta V = V_1 - V_2$ creates a radial velocity V_3 that can be simply estimated. The example below shows this calculation.

Example 17.3. Calculate the confining pressure that would be created by the radial velocity gradient. For

$$V_1 = 9000 \text{ m/s} \qquad V_2 = 3000 \text{ m/s}$$

$$d = 1 \text{ cm} \qquad l = 10 \text{ cm}$$

we have, in plasticity, a constant volume. The volume is expressed as

$$V = \frac{\pi d^2}{4} l$$

$$\frac{dV}{dt} = \frac{1}{2} \pi \, dl \frac{dd}{dt} + \frac{\pi d^2}{4} \frac{dl}{dt} = 0$$

But

$$\frac{dd}{dt} = 2V_3 \quad \text{and} \quad \frac{dl}{dt} = \Delta V$$

Thus

$$V_3 = -\frac{d \, \Delta V}{4l} = 150 \text{ m/s}$$

Since the central axis has a velocity zero, one has a velocity gradient, radially, of 150 m/s. One can estimate the stress created by this velocity gradient by applying the conservation-of-momentum equation:

$$\sigma = \tfrac{1}{2} \rho_0 V_3^2$$

$$\sigma = \frac{8.9 \times 10^3}{2} \times 150^2$$

$$= 100 \text{ MPa}$$

This is of the same order of magnitude as the flow stress of copper at temperatures around 300°C (the adiabatic temperature). The creation of voids could be inhibited. It is well known that superimposed hydrostatic pressures increase the tensile ductility of materials and this is shown in Figure 17.7(b).

Material effects are extremely important in jet formation. Copper seems to be the most popular material, although higher density metals present distinct advantages. Tungsten (ρ = 19.2), tantalum (ρ = 16.7), and uranium (ρ = 18.5) are examples of materials used or considered for use in jets. By means of flash X-ray (diffraction), Jamet [13] has shown that the jet is fully solid in aluminum and at least partially solid in copper. Nevertheless, the temperature in the jet can reach considerable values, and Chokshi and Meyers [14] and Meyers et al. [15] have proposed that the copper undergoes dynamic recrystallization during jet formation. This dynamic recrystallization could lead to a "superplastic" response, which could explain the enhanced tensile ductility of the jet. Figure 17.8(a) shows the recrystallized structure found in jets and slugs, showing that the temperature reaches values of $0.4T_m$, where T_m is the melting

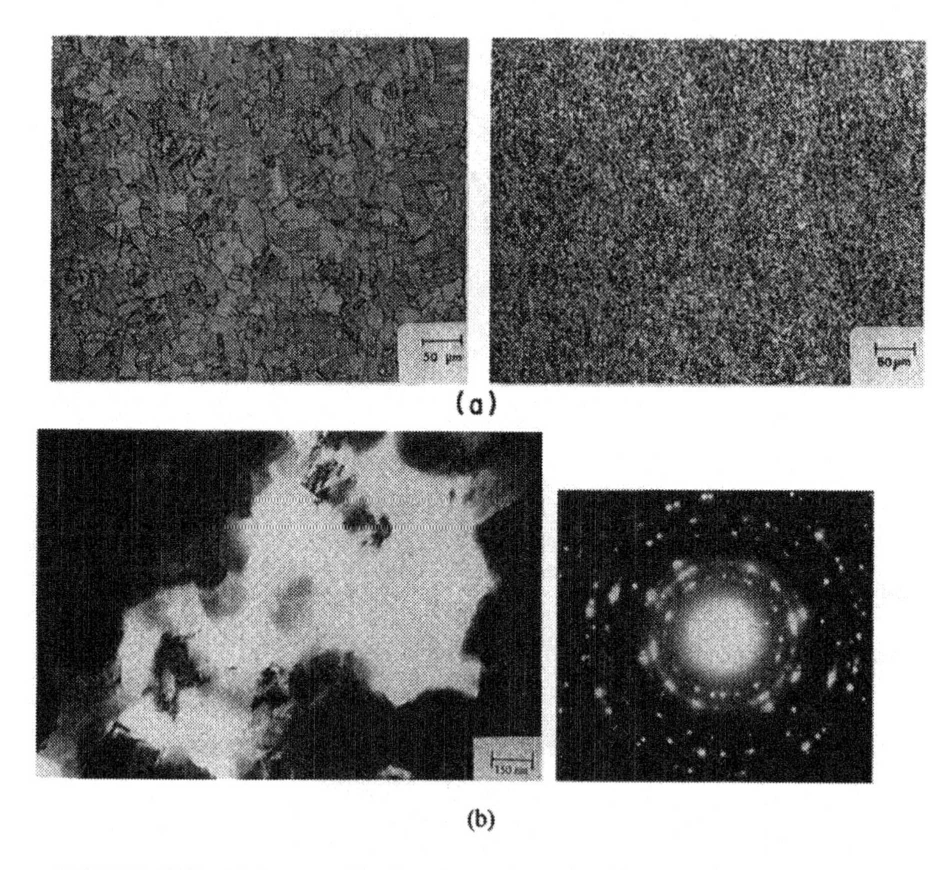

FIGURE 17.8 (a) Recrystallized grain structure from jet and slug.* (From Meyers et al. [15], Fig. 1, pp. 531. Reprinted by courtesy of Marcel Dekker, Inc.) (b) Transmission electron micrograph showing the presumed structure of jet during the plastic deformation process.

point in kelvin. One cannot see any grain elongation, although the material (Cu) underwent a considerable deformation. By transmission electron microscopy one can determine the fine details of the microstructure. Figure 17.8(b) shows a transmission electron micrograph from a region that underwent a high shear strain at a high strain rate in adiabatic conditions. It belongs to a copper shaped-charge slug; for greater details, see Chokshi and Meyers [14]. The structure is microcrystalline, as evidenced by the almost continuous rings in the diffraction pattern and by the grains (~ 0.1 μm diameter) seen in the micrograph. This is strongly suggestive of dynamic recrystallization.* The me-

*L. E. Murr and coworkers (*Scripta Met. et Mat.* **29** (1993) 567; *Matls. Charact.* **30** (1993) 201) have observed systematic grain reduction in copper shaped charges and attributed it to dynamic recrystallization, confirming results by Chokshi and Meyers [14] and Meyers et al. [15].

chanical response of this structure is very different from the initial structure of the material, which had a grain size of ~50 μm. This dynamic recrystallization is produced by the combined effects of plastic deformation and temperature rise.

The effectiveness of a shaped charge is highly dependent on the "stand-off distance," which is the distance from the cone to the target at the instant of detonation. If the stand-off is too high, the jet is particulated prior to impact; if it is too small, it is not stretched to its maximum length. Figure 17.9 shows the penetration of a jet into rolled homogeneous armor (RHA, a special armor steel) as a function of standoff. The maximum penetration, of 7 CD (charge diameters) occurs at a 7-CD standoff. Flash radiographs of jets from several materials, obtained by Held [17], are shown in Figure 17.10. These radiographs were taken at approximately the same time (130–160 μs) after initiation. The ability of the jet to stretch without breaking up is strongly dependent upon composition. If one compares nickel and copper, one sees that copper outperforms nickel. The Pb–Sb alloy seems to produce a diffuse cloud at the front. The copper–silver alloy shows considerable breakup. Within a fixed composition, the grain size, texture, purity levels, and prestrain affect the performance. Copper shows, in Figure 17.10, the jet with least fracture. Copper is known to be an excellent material for shaped charges.

Explosively forged projectiles or fragments (EFPs) are similar to shaped charges, but only a slug is formed because of differences in geometry. Figure 17.11(a) shows the evolution of the process. A metal dish (most probably, high-purity iron) is driven by an explosive charge. Upon detonation, the dish is "forged" into a rod. It is possible to create "fins" into the EFP. They are activated at a distance from the target that is much larger (10–100 CD) than

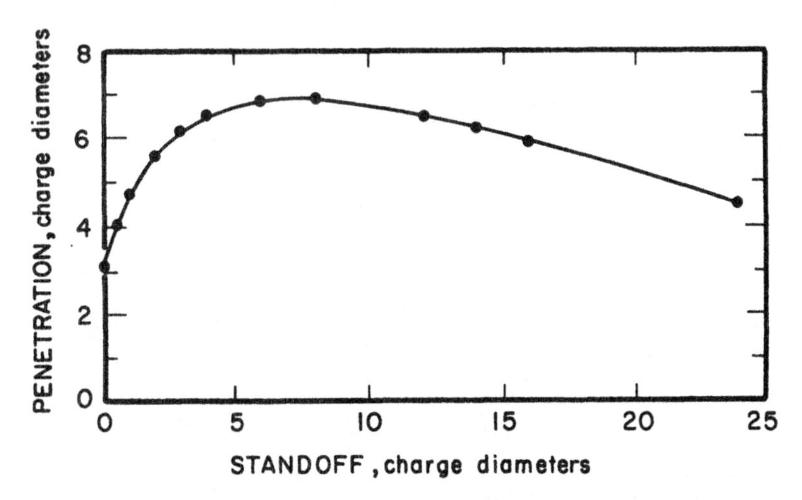

FIGURE 17.9 Calculated penetration–standoff curve in RHA of representative jet. (From Eichelberger [16], Fig. 4, p. 382. Reprinted with permission of the publisher.)

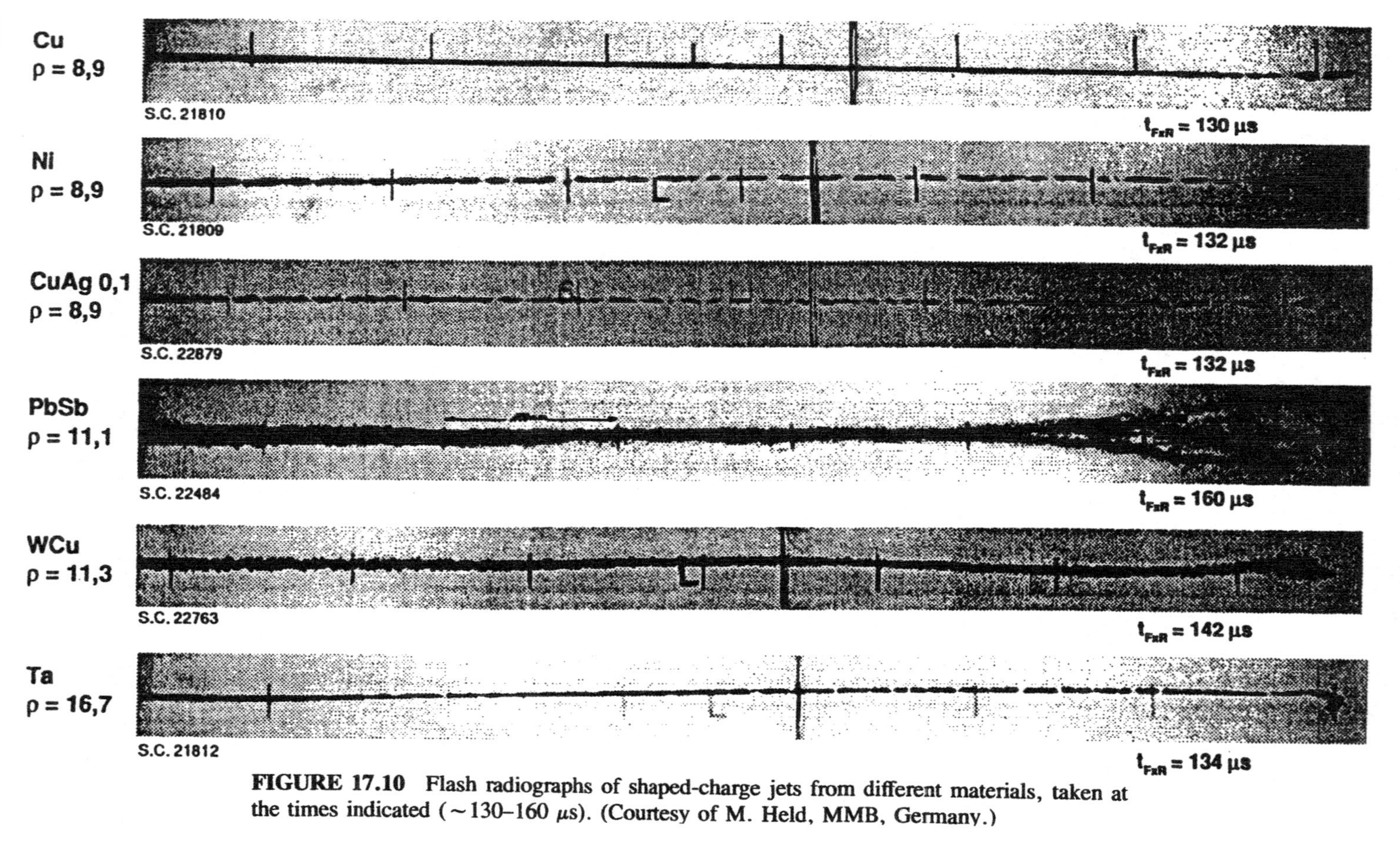

FIGURE 17.10 Flash radiographs of shaped-charge jets from different materials, taken at the times indicated (~130–160 μs). (Courtesy of M. Held, MMB, Germany.)

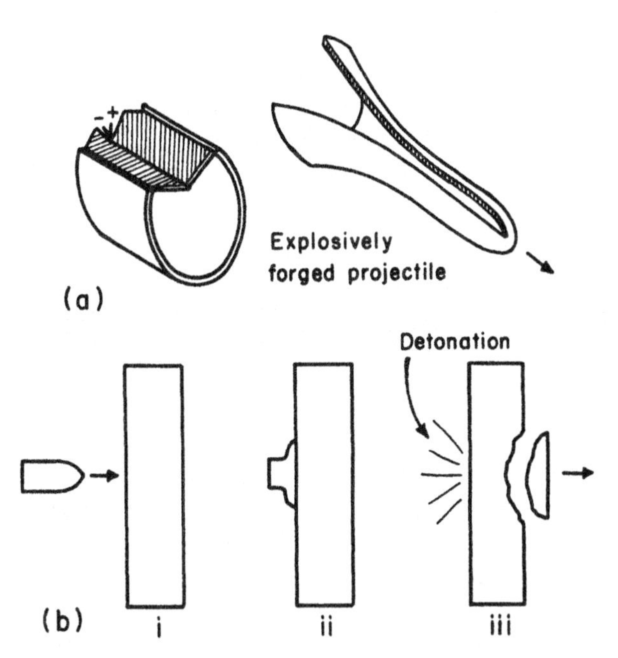

FIGURE 17.11 (a) Self-forging fragment; (b) high explosive squash head projectile (HESH): (i) explosive projectile in flight; (ii) explosive projectile plastically deformed (squashed); (iii) detonation of explosive and internal spalling.

the shaped charge (5–10 CD). The strains experienced by the material are much lower than for the shaped charge.

Figure 17.11(b) shows another type of chemical energy projectile that deforms plastically upon contact with the target and then detonates. The acronym HESH (high-explosive squash head) is used often. A shock wave is produced that traverses the target, reflecting at the internal surface and generating a tensile pulse. This tensile pulse can produce a spalling, therefore producing damage without perforation of the target.

17.3 PENETRATION

In Section 17.2 we saw two types of chemical energy projectiles: the shaped charge and the EFP. What makes them unique, from the tactical viewpoint, is that they can be launched through rockets/missiles and detonated in close proximity to the target, deriving their kinetic energy from a chemical (explosive) source. The other projectiles that will be described in this section derive their kinetic energy from the high-pressure gases within a confined volume: the barrel of a gun (*kinetic energy projectiles*). In contrast with the *chemical energy* projectiles, the kinetic energy projectiles need to be launched from a gun, and this entails a weight handicap.

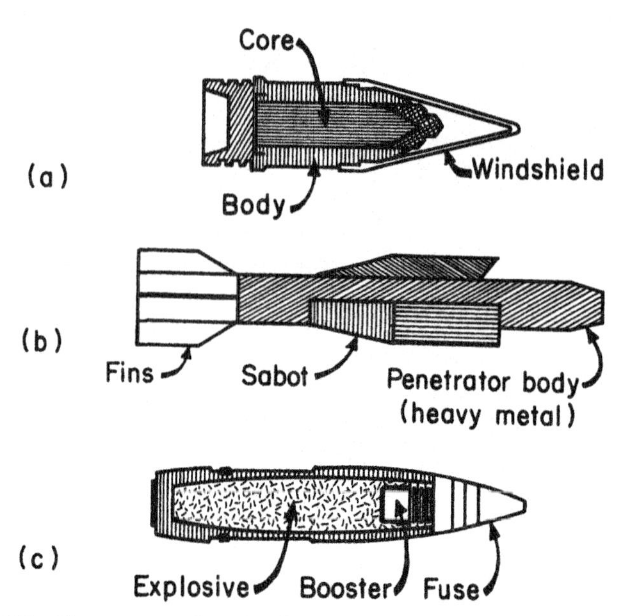

FIGURE 17.12 Three types of projectile: (a) 0.50-in. armor-piercing projectile; (b) kinetic energy penetrator; (c) projectile with explosive charge.

Figure 17.12 shows three common kinetic energy projectiles. The common 0.50 armor piercing projectile is a steel-jacketed lead projectile (density 14). The modern-day equivalent to the arrow is the kinetic energy penetrator, shown in Figure 17.12(b). It consists of a long (length–diameter ratio 10) cylinder of a high-density alloy (tungsten based or uranium based, with a density of 18) propelled from a smooth bore barrel. Flight stability is ensured by fins. A discardable sabot that ejects after the exit from the barrel projects the projectile during the internal ballistic period. The third projectile [Fig. 17.12(c)] is the armor-piercing charge-bursting type. The charge detonates after the projectile has penetrated the target.

The mechanisms by which kinetic or chemical energy penetrators defeat armor are varied and depend on a number of factors. Prominent among these is the velocity of the projectile. As we well know, the kinetic energy is proportional to the square of the velocity. Projectile–target interaction has captivated and stimulated the ingenuity of weapons designers, and numerous concepts and combinations have been developed. As the velocity of impact is increased, the region of damage decreases, and one progresses from *structural* effects to *local* effects. The stresses and strains produced by external tractions travel through a material at sonic speeds. Thus, quasi-static tractions create a situation of equilibrium. The externally applied loads are countered by internal stresses throughout the body. In high-velocity impact, the externally applied tractions can reach their maximum and decay long before the entire body has experienced the reactive stresses: thus, a hole is formed, as opposed to large-scale plastic deformation.

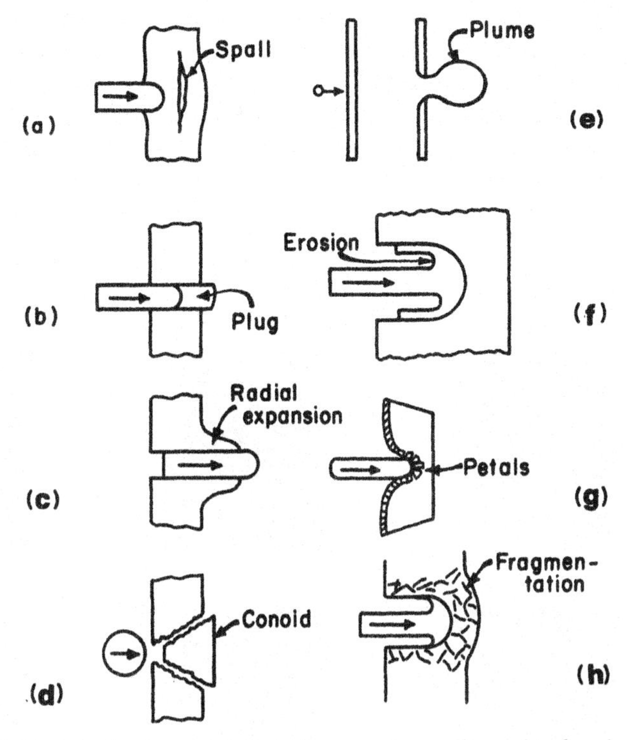

FIGURE 17.13 Mechanisms of defeat of armor: (a) spalling, (b) plugging, (c) ductile hole growth, (d) conoid formation, (e) melting/vaporization and plume formation, (f) concurrent erosion of projectile and target, (g) petaling, (h) comminution/cracking.

Depicted in Figure 17.13 are some modes of projectile–target interaction. Spalling (see Chapter 16) (a), plugging by shear band formation (b) (see Chapter 15), and ductile hole growth (c) occur in ductile materials. When the target is very thin, it usually fractures in a star pattern called "petaling" after it has been stretched [Fig. 17.13(g)]. When the target is brittle (i.e., glass), a projectile creates a conoid, which is ejected [Fig. 17.13(d)]. This can be readily verified by shooting a BB gun projectile into window glass. Impacts at hypervelocities (>3000 m/s) can melt and even vaporize the target or create a thin plume of fragments traveling at much lower velocities. Figure 17.13(e) shows this schematically. This concept is used as "bumper shields" in space vehicles that protect the vehicle from meteorite impacts, which can have velocities ranging from 10 to 20 km/s. Long-rod penetrators, traveling at velocities of 1-2 km/s, and shaped-charge jets, at velocities of 6–10 km/s, interact with targets in a mode in which there is concurrent deformation of both. When the pressure at the deformation interface exceeds significantly the dynamic flow stress of the material(s), deformation can be assumed to be guided by fluid dynamics, and one develops patterns similar to the ones of Figure 17.13(f). A ceramic target responds quite differently than a metallic one, and the mechanisms by which it is defeated are quite different. Figure 17.13(h) shows the ceramic having undergone damage by the stresses due to the projectile. The projectile tip is plastically deformed at the same time. The comminuted ceramic is ejected from the target through the orifice created by the projectile. This creates room for the continued penetration of the projectile.

Figure 17.14 shows the deformation pattern (predicted by large-scale computations using EPIC-2 code) resulting from the impact of cylinders (L/D =

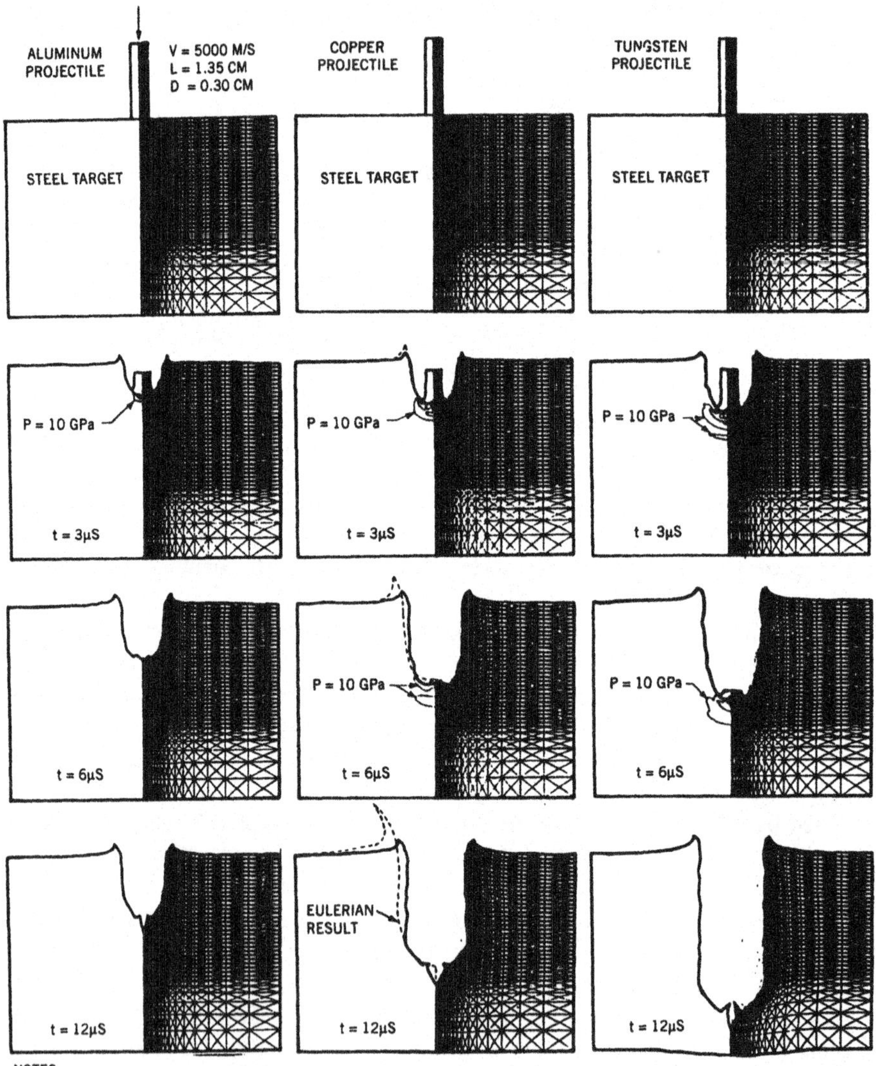

FIGURE 17.14 Two-dimensional computations of damage progression in steel target impacted by aluminum, copper, and tungsten projectiles at 5000 m/s (L/D = 4.5). (Reprinted from *J. Impact Eng.*, vol. 5, G. Johnson and R. A. Stryk, Fig. 3, p. 415, Copyright 1987, with permission from Pergamon Press Ltd.)

15) of three different materials: aluminum, copper, and tungsten [19]. The increase in density, at the same velocity of 5000 m/s used in computation, results in increase in the penetration. The initial pressure at the impact interface is of the order of 10 GPa. This pressure attenuates itself from its initial value because of lateral rarefaction waves coming from the free cylindrical surfaces. Figure 16.25 (Chapter 16) shows them more clearly.

We will present the analytical treatment for penetration in a brief manner. Wright [20] classifies the approaches into three levels:

1. data correlations,
2. engineering analyses, and
3. large-scale computational codes.

We have seen the computational approach in Chapter 6. Here, we will restrict ourselves to data correlations and simpler, closed-form analyses. From a historical viewpoint, de Marre's equation, dating from the 1800s and resulting from systematic experiments conducted by the French military in Metz, should be noticed. It correlates the kinetic energy of the projectile ($\frac{1}{2}MV^2$) to the maximum thickness of the target (T) that could be perforated:

$$MV^2 = CD^\beta T^\alpha \tag{17.7}$$

where C, α, and β are empirical parameters and D is the projectile diameter. The cannonball experiments by the Metz group are the first attempt to instill science into this area. More recently, Bruchey [21] proposed a more complex expression:

$$\frac{\rho_p V^2}{E_p} \frac{L_0}{D} = C \left(\frac{T \sec \theta}{D} \right)^\alpha \frac{E_p^\beta}{E_t} \left(\frac{\rho_p}{\rho_t} \right)^\gamma \tag{17.8}$$

The subscripts p and t refer to projectile and target, respectively; L_0 and D are the initial length of projectile and its diameter, respectively; ρ_p and ρ_t are the densities; T is the target thickness; θ is the angle at which the projectile strikes the target; and α, β, γ, and C are empirical parameters. Once these parameters are obtained, through experimentation, one can develop "master curves" from which the ballistic limit can be predicted.

The velocity beyond which a projectile perforates a target and below which it will not is called the ballistic limit. Different definitions for perforation are used: (1) the target must be such that light can pass through it; (2) a witness plate placed behind the target must be perforated by fragments; (3) one-half or more of the projectile must pass through the target. Two methods of determination of the ballistic limit are illustrated in Figure 17.15. This classification scheme is due to Zukas et al. [23]. In the first method, a large number of tests is done at various velocities. The probability of perforation is plotted as a function of velocity. The velocity at which this probability is equal to 0.5 is determined. This velocity is V_{50}. In the second method, called deterministic by Zukas et al. [23], the residual velocity, V_R, of the projectile is measured as a

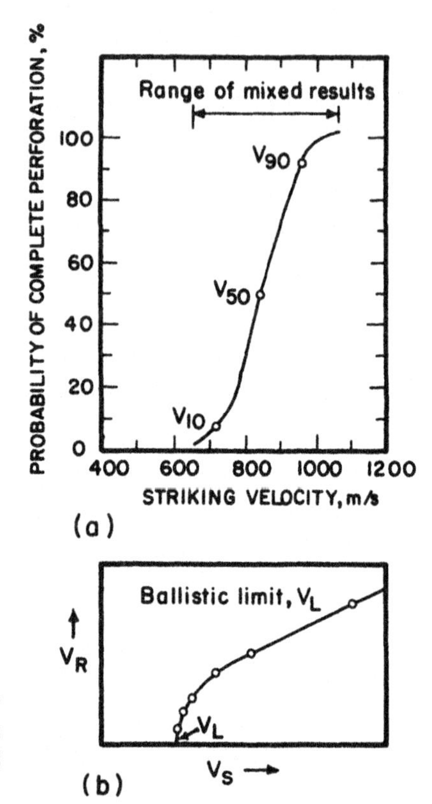

FIGURE 17.15 Ballistic limit determination by (a) statistical measurements and (b) residual velocity measurement. (From Zukas et at. [23], Fig. 12, pp. 172–173.)

function of impact velocity, V_S. The impact velocity at which the residual velocity becomes zero is called the ballistic limit V_L. This is shown in Figure 17.15(b).

Whereas the above equations consider only the kinetic energy of the projectile, studies have been performed in which the physical mechanisms of penetration are analyzed in greater detail. The comprehensive review by Backman and Goldsmith [22] as well as the book by Zukas [24] provide in-depth treatments. The physical processes depend on the impact velocity, shape of the penetrator, and materials of the penetrator and target.

The penetration of a long cylinder into a metallic target has been treated in a very preliminary manner by Bishop et al. [25] and by Birkhoff et al. [1] and in a more complete way by Alekseevskii [26] and Tate [27]. These analyses apply to impact velocities generating stresses much higher than the dynamic yield stress of either target or projectile. This treatment is applicable to shaped-charge jets and long-rod penetrators and is commonly known as the "eroding rod" model. Birkhoff et al. [1] considered, simply, a semi-infinite target impacted by a long rod of length L at a velocity V. The densities of the projectile and target are ρ_p and ρ_t, respectively. Figure 17.16 shows the advance of the tip of the projectile during penetration; its velocity is U. By changing the coordinate system so that the stagnation point (interface between target and

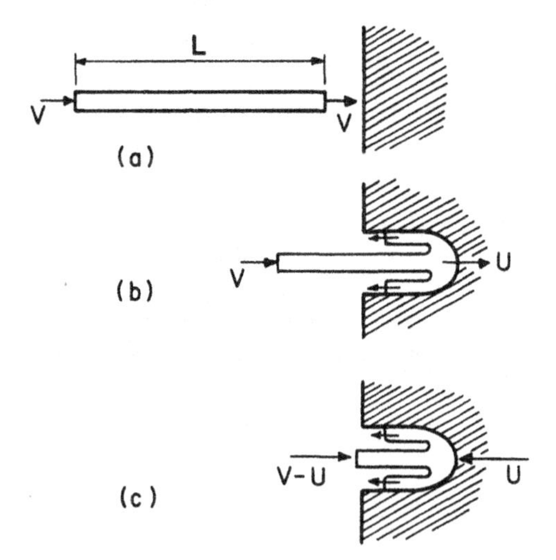

FIGURE 17.16 High-velocity penetration of projectile into target; projectile is consumed as it penetrates.

projectile) is at the origin, one has the target moving from right to left at a velocity U and the projectile moving from left to right at a velocity $V - U$. Birkhoff et al. [1] assumed that the pressure created at the interface was much higher than the strength of the materials, which could therefore be treated as fluids by applying Bernoulli's equation:

$$P = \tfrac{1}{2}\rho_p(V - U)^2 = \tfrac{1}{2}\rho_t U^2 \tag{17.9}$$

From this expression, we obtain

$$U = V\left[\frac{1 \pm (\rho_t/\rho_p)^{1/2}}{1 - (\rho_t/\rho_p)}\right] \tag{17.10}$$

The penetration depth is equal to the time that the projectile takes for penetration divided by the velocity at which it is penetrating, $V - U$:

$$p = \frac{LU}{V - U}$$

From Eqns. (17.9) and (17.10) one can eliminate U and obtain P:

$$p = \frac{LV\left[\dfrac{1 \pm (\rho_t/\rho_p)^{1/2}}{1 - (\rho_t/\rho_p)}\right]}{V - V\left[\dfrac{1 \pm (\rho_t/\rho_p)^{(1/2)}}{1 - (\rho_t/\rho_p)}\right]}$$

$$p = L\left(\frac{\rho_p}{\rho_t}\right)^{1/2} \tag{17.11}$$

Equation (17.11) is significant and important: it predicts that the depth of penetration is equal to the projectile length if the two materials have the same density. It also predicts that the penetration increases with the ratio ρ_p/ρ_t (ratio of projectile and target densities). The fact that the penetration depth is independent of impact velocity V is also significant. As expected, Eqn. (17.11) is an oversimplified representation of the penetration process.

Example 17.4 Calculate the penetration depth for a kinetic energy penetrator of tungsten ($\rho = 19.2$) into a steel target ($\rho = 7.9$) traveling at a velocity of 1.4 km/s:

$$(\sigma_y)_{\text{steel}} = 1.5 \text{ GPa} \qquad (\sigma_y)_p = 1.3 \text{ GPa}$$

$$L = 24 \text{ cm} \qquad D = 1.2 \text{ cm}$$

We first must establish whether the pressure created significantly exceeds the yield strengths of the target and projectile. The initial pressure, upon impact, is given by impedance matching. We will assume the EOS values for pure W and Fe. Following the procedure of Chapter 4 (see Table 4.1), for $P = 40$ GPa

$$(U_P)_{Fe} + (U_P)_W = 0.945 + 0.453 = 1.393 = 1.4 \text{ mm}/\mu\text{s} = 1.4 \text{ km/s}$$

Thus, the pressure generated is 40 GPa.

The steady-state pressure in the target is given by the Bernoulli equation:

$$P = \tfrac{1}{2}\rho_{Fe}U^2 = \tfrac{1}{2} \times 7.8 \times 1.4^2 \times 10^9 = 6.9 \text{ GPa}$$

Both values are much higher than the strengths of either target or projectile. Thus, the use of Eqn. (17.11) is justified. The penetration is calculated as

$$p = 24\sqrt{\frac{19.2}{7.9}} = 37.4 \text{ cm}$$

The predictions of Eqn. (17.11) are not entirely realistic and, therefore, the incorporation of material strength has been implemented. This has been done by Alekseevskii [26] in 1966 and by Tate [27] in 1967. Resistances to plastic flow of projectile and target are incorporated into the treatment. For the projectile, the material is unconstrained, and this resistance can be taken as the dynamic yield stress $(\sigma_{yd})_p$. For the target, the material is laterally constrained. Thus, the stress at which it flows is, to a first approximation, the stress to produce a hole through a material. There are elastoplastic solutions to this problem. Tate assumed this value to be R_t, where R_t was assumed by Bishop et al. [25] to be 4.5 $(\sigma_{yd})_t$, the target flow stress is calculated as

$$R_t = 4.5(\sigma_{yd})_t$$

Applying Bernoulli's equation [Eqn. (17.4)], which equates the pressures on the two sides of the stagnation point, and incorporating the strength of the material, yields

$$P = \tfrac{1}{2}\rho_t U^2 + R_t = \tfrac{1}{2}\rho_p (V - U)^2 + (\sigma_{yd})_p \qquad (17.12)$$

This is a modification of Bernoulli's equation. The difference in pressures is correlated with $(\sigma_{yd})_p$ and R_t. The solution of Eqn. (17.12), coupled with the equations for the conservation of mass and momentum, provides a realistic prediction of the penetration. Figure 17.17 shows Wright's [20] plot of the Alekseevskii–Tate equation for three different cases:

$$(\sigma_{yd})_p > (\sigma_{yd})_t$$

$$(\sigma_{yd})_p = (\sigma_{yd})_t$$

$$(\sigma_{yd})_p < (\sigma_{yd})_t$$

The predictions are in qualitative agreement with experimental results. The sigmoidal shape of the penetration curve predicts a saturation penetration at a maximum velocity. It also predicts that the curve is shifted to the right for increasing target strength and to the left for increasing penetrator strength, in agreement with observed results. If the mass of the projectile is constant, increasing the L/D ratio should increase penetration.

Tate [27] also compared the kinetic energy of the projectile with the crater volume, W. In support of earlier experiments by Christman and Gehring [28], he found the proportionality shown in Figure 17.18(a); the crater volume is divided by the projectile volume, $\pi L r_p^2$. This is indeed a sensible prediction, because the work required to create the crater is provided by the kinetic energy of the projectile. Figure 17.18(b) shows the proportionality between the two parameters.

Whereas long rods can penetrate targets by amounts that exceed their length (the maximum, in Fig. 17.18 is $P/L = 1.3$ at an equal density for penetrator and target), shorter rods, with an $L/D = 1$, can have considerably higher penetrations. Figure 17.18(b) shows the penetration of tungsten rods into steel. Two L/D ratios (1 and 10) are shown, at a constant diameter. The maximum penetration–length ratio, at a velocity of ~4000 m/s, reaches a value close to 3 for $L/D = 1$. This is due to the fact that, for a short rod, the initial impact effects are dominant. For the long rods, the pressure created upon impact decays to a steady-state value, as shown in Example 17.4 (it is given by the Bernoulli equation). There are two aspects to the results of Figure 17.18. First, for a

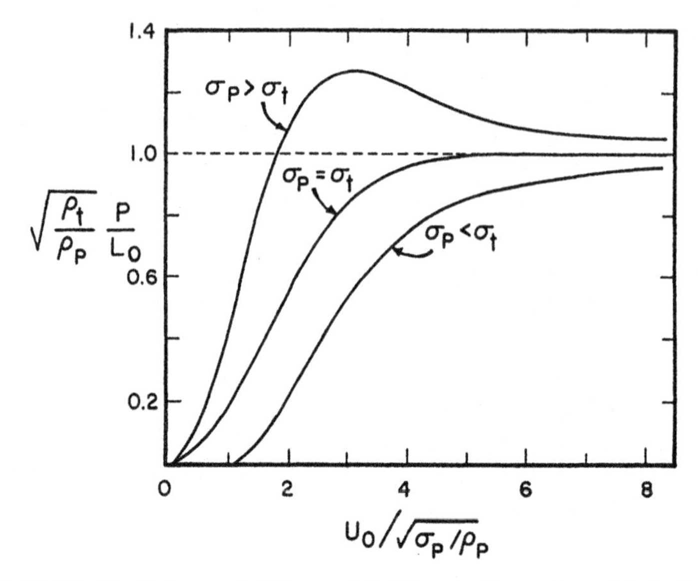

FIGURE 17.17 Predictions of Alekseevskii–Tate eroding rod model; both velocity U_0 and penetration P have been scaled to provide nondimensional axes. (From Wright [20], Fig. 4, p. 92. Reprinted with permission of the publisher.)

constant projectile mass and kinetic energy, the penetration increases with L/D ratio. This is shown by the simple exercise below.

Example 17.5. Two projectiles with the same mass but different L/D ratios (1 and 10) impact the target ($V = 4{,}000$ m/s) as shown in Figure 17.18. Calculate the relative penetrations. From Fig. 17.8

$$P_1/L_1 = 1.7 \quad \text{for } L/D = 10$$

$$P_2/L_2 = 2.7 \quad \text{for } L/D = 1$$

The mass is

$$L \times \tfrac{1}{4}\pi D^2 \times \rho$$

Thus

$$L_1 D_1^2 = L_2 D_2^2$$

$$L_1 \left(\frac{L_1}{10}\right)^2 = L_2 \left(\frac{L_2}{1}\right)^2$$

$$\frac{L_1^3}{100} = L_2^3 \qquad \therefore\ L_1 = L_2(100)^{1/3}$$

$$= 4.6 L_2$$

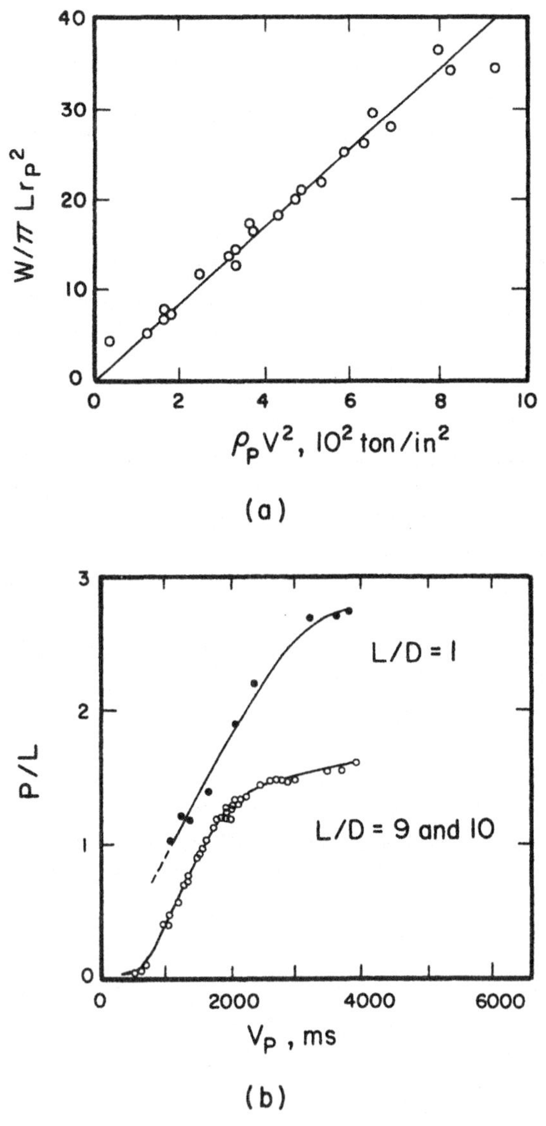

FIGURE 17.18 (a) Crater volume (W) normalized to projectile volume ($\pi L r_p^2$) as a function of kinetic energy of projectile; aluminum alloy projectile impacting lead target. (Reprinted from *J. Mech. Phys. Sol.*, vol. 17, A. Tate, Fig. 13, p. 146, Copyright 1969, with permission from Pergamon Press Ltd.) (b) Normalized, penetration depth P/L vs. projectile velocity V_p of tungsten alloy rods into steel targets $10 \geq L/D \geq 1$. (Reprinted from *Int. J. Impact Eng.*, vol. 9, A. Tate, Fig. 1, p. 328, Copyright 1990, with permission from Pergamon Press Ltd.)

$$P_1 = 1.7L_1 = 1.7 \times 4.6L_2 = 7.8L_2$$

$$P_2 = 2.72L_2$$

Thus, the penetration at $L/D = 10$ is 2.9 times higher than the one at $L/D = 1$.

There are practical limits to the L/D ratio because of the launch and flight stability of the projectile. The second aspect of Figure 17.18 is that one can increase the overall penetration by creating a segmented penetration, composed of a stacking of short ($L/D \sim 1$) segments separated by spaces, so that each impact event can be considered independently and their effects are additive. Figure 17.19 shows schematically such a penetration. During launching the segments are close to each other, and the problems associated with excessive L/D ratio are minimized. During the flight, the penetrator extends itself so that penetration occurs by a sequence of impacts. Tate [29] provides a detailed analytical treatment of segmented penetrators and discusses the increase in performance that one can expect. Using tungsten penetrators against a homogeneous steel target, increases of penetration between 20 and 30% can be expected at the higher impact velocities.

There are practical limitations on the length–diameter ratios of long rods. Above $L/D = 20$, buckling during launching and sabot discard asymmetries can produce significant flight perturbations [29]. The very long rods are also more susceptible to defeat by reactive armor (see Section 17.4).

There are more advanced analytical treatments of the penetration problem, and the reader is referred to the literature. In particular, Ravid et al. [30] and Wright and Frank [31] present these analyses. However, it is only through large-scale computations that one can arrive at a truly realistic predictive capability.

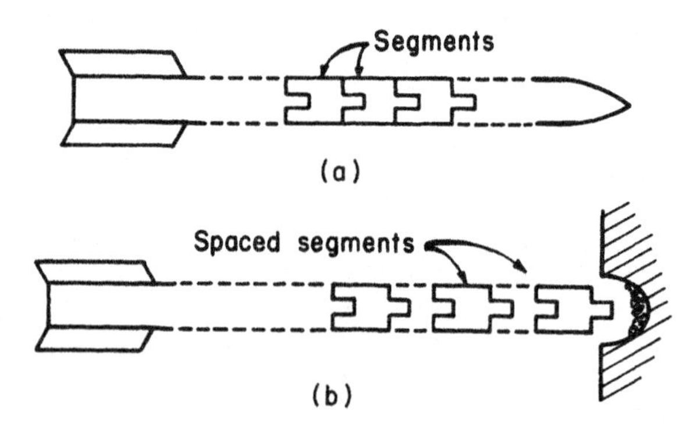

FIGURE 17.19 (a) Segmented rod with segments adjacent to each other during launching. (Reprinted from *Int. J. Impact Eng.*, vol. 9, A. Tate, Fig. 2, p. 330, Copyright 1990, with permission from Pergamon Press Ltd.) (b) Segmented rods during impact.

TABLE 17.2 Evolution of Armor and Armor-defeating Concepts

Armor	Projectile
↓	↓
Rolled homogeneous armor	Armor-piercing rounds
↓	↓
Spaced armor	Bursting head, armor-piercing rounds
↓	↓
Composite armor	Shaped charges
↓	↓
Reactive armor	Kinetic energy penetrators
↓	Explosively forged projectiles
Active ("smart") armor	↓
	"Smart" munitions
	Electromagnetic accelerators

17.4 ARMOR

The reader is referred to the reports by Viechnicki [32, 34]. The evolution of projectiles has been paralleled by the evolution of armor. Table 17.2 shows, in a schematic manner, the developments. As projectiles are developed that can perforate an armor, new armor concepts are introduced that can defeat the projectiles. Steel has traditionally been the major armor component. It is gradually being complemented or replaced by other materials: ceramics, composites, glass, and explosives. The armor concept to be used depends on the intended application, classified as

1. body armor (personnel armor),
2. light armor (vehicular and aircraft armor), and
3. heavy armor (tank armor).

Body armor is intended to protect individuals primarily against fragments from high-explosive artillery shells, grenades, fragmenting mines, as well as projectiles from small arms. In World War II, the percentage of casualties due to fragments was approximately 80%. In addition to the traditional steel helmet, nylon fabrics, metals, and ceramics have been used for the "flak jackets" that protect individuals. The Hadfield steel (an austenitic, high-manganese steel) of helmets has been replaced by Kevlar and Spectra fiber-reinforced composites. Spectra is a fiber that has a very high tensile strength (2.8 GPa) and a very low density (0.97 g/cm^3). Ceramic plates backed with aluminum or fiberglass-reinforced plastic have been used in body armor.

In the category of light armor we include applications such as seats in helicopters (protection against ground fire) and the protection of light vehicles and airplanes. A number of different systems have been developed using monolithic metals (steel, aluminum, and titanium alloys), ceramic composite armor, polymer composite armor, and laminated armor [32]. It is important to evaluate the armor performance in terms of its areal density, that is, the weight per unit area required to offer a specific ballistic protection. For vehicular and aircraft applications, the ceramic armor (alumina, boron carbide, silicon carbide, etc.) backed by a ductile organic composite such as graphite-reinforced plastic has been quite successful against small arms fire and shell fragments. The principal idea is to break the projectile with a very hard surface and then absorb the energy of projectile and/or armor fragments by using a soft, ductile backing material.

Heavy armor is primarily intended for tanks. Steel has been the principal armor material because of its low cost, ease of fabrication, and structural efficiency (it is at the same time armor and part of the structure). It is shown in Chapter 15 (Figure 15.6) that different plastic failure mechanisms can operate in steel: plastic deformation, shear band formation, and spalling. The predisposition to shear band formation depends on the thermomechanical response of the steel; this is explained in detail in Chapter 15. Figure 17.20 shows the effect of hardness on the ballistic performance of steel. Rolled homogeneous armor is the standard armor material. It is based on the AISI 4340 composition (quenched and tempered steel). There are limitations on the thicknesses that can be produced: cast steel can be produced with thicknesses in the 7–12-in. range; so can rolled steel. When one desires the full hardness throughout, the maximum section thickness is decreased, because the heat extraction rate (during quenching) is dependent on the size of the steel block. High hardness steel, electroslag-refined (ESR), and dual hardness are more advanced concepts.

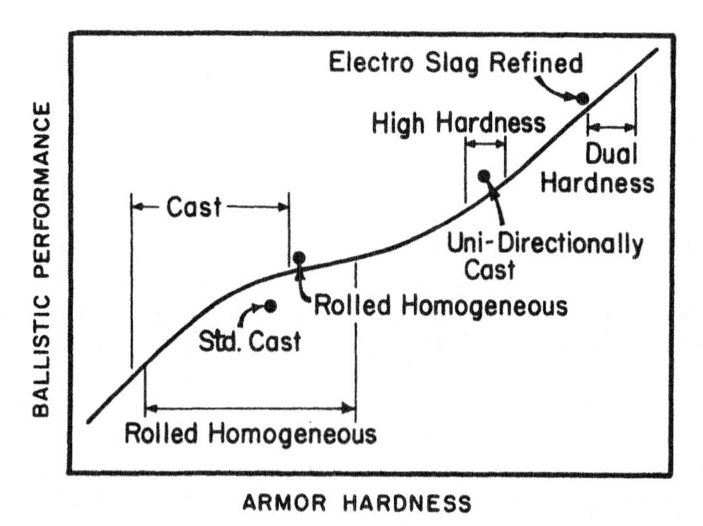

FIGURE 17.20 Hardness vs. ballistic performance for various armor steel. (From Viechnicki et al. [32]. Reprinted with permission.)

In Sections 17.2 and 17.3 we saw that shaped charges and kinetic energy penetrators can penetrate thicknesses of 25 cm or more. A 1-in.- (2.5-cm-) diameter copper-shaped charge can penetrate 20 cm of steel (Fig. 17.9); a 50-cm-long (5-cm-diameter) tungsten kinetic energy penetrator at a velocity of 1500 m/s can penetrate a 40-cm steel plate. Thus, the weight of a modern tank (40–50 tons) would have to be increased if protection against these threats is desired. The student should, as an exercise, calculate the weight required for a vehicle if frontal protection against these threats is required. The density of steel being 7.9 g/cm^3, a 1-m^3 (a plate of $2 \times 2 \times 0.25$ m) would weigh 8 tons. This is indeed a very high value. Ceramics possess two very important qualities that make them ideal candidates for armor materials: *high hardness* and *low density* [33]. Their low or negligible ductility should not be a factor if they are not used as structural components; thus, they are added to the structure. The French term *Appliqué* is used. Figure 17.21 shows the relative ballistic efficiencies of several ceramics against impact from a 0.3-in.-diameter projectile with a conical tip. The penetration of the projectile into the backup plate is measured. The mechanisms of damage are unique in ceramics and quite different from the ones in metals. One cannot assume that the ceramic has a low flow stress at high impact velocities. The penetration into ceramics is schematically shown in Figure 17.22. Viechnicki [32, 34] divided the mechanisms into three velocity regimes [Fig. 17.22(a)]: low, intermediate, and hypervelocity. At low velocity the ceramic deforms the projectile, while undergoing a conical fracture, which is described later in more detail. At intermediate velocities penetration into the ceramic is initiated. At high velocities, erosion of the ceramic occurs. This is shown in greater detail in Figure 17.22(c), in which a confinement (steel) is used. The confinement of the ceramic is very

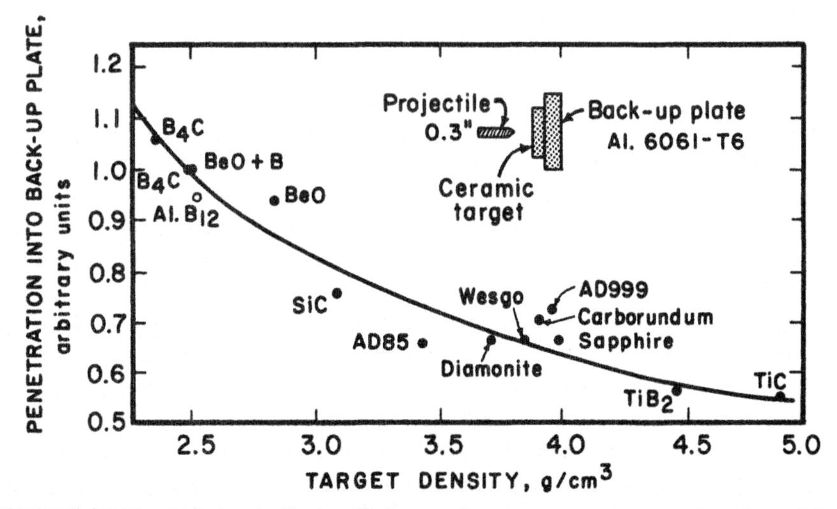

FIGURE 17.21 Relative ballistic efficiency of various ceramics as a function of their density. (Courtesy of D. Yaziv, Rafael, Israel.)

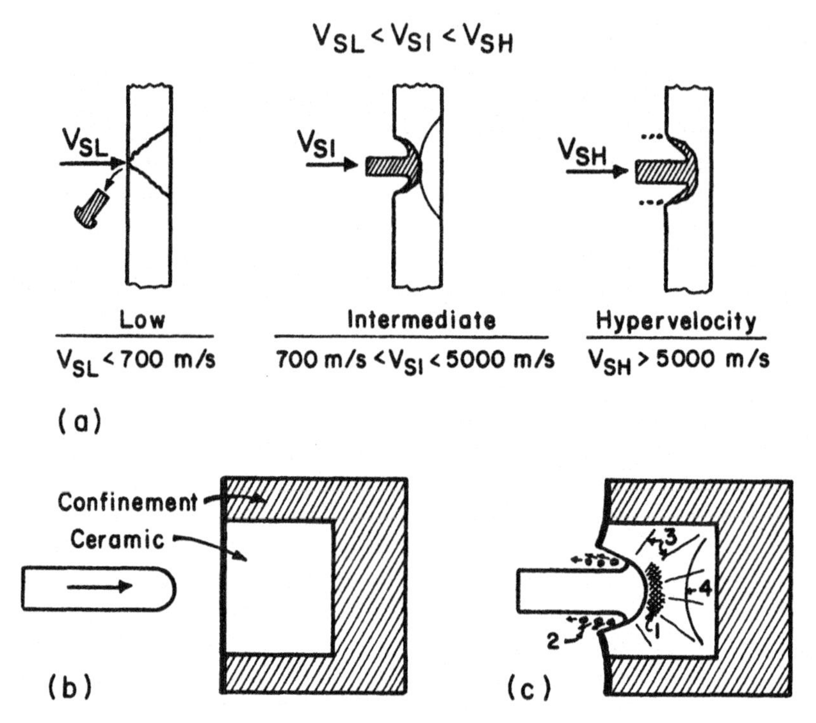

FIGURE 17.22 (a) Mechanisms of interaction between projectiles and ceramic target as a function of impact velocity. (From Viechnicki et al. [34], Fig. 2, p. 1037. Reprinted with permission of the publisher.) (b, c) Pattern of fracture of ceramic target impacted by projectile at high velocity: 1, comminuted (pulverized) region; 2, target and projectile ejecta; 3, conoidal cracks; 4, spall cracks from reflections.

important in heavy armor because it keeps the fragmented ceramic in place; it can therefore continue to be effective under the (primarily) compressive stresses produced by the projectile. Three types of failure are shown in Figure 17.22(c). A comminuted zone forms ahead of the projectile. This comminuted zone, also called "Mescall zone" [35], in honor of John Mescall [36], who predicted its formation in computer simulations, is a direct result of the high compressive stresses. This comminuted region can undergo shear localization during plastic deformation; indeed, this has been shown by V. F. Nesterenko, M. A. Meyers, and H. C. Chen (unpublished results, 1994) in controlled experiments. Radial cracks (3) emanate from the impact point. These radial cracks are produced by tensile stresses that result from the compression at the center (Hertzian stresses). When the stress wave produced by impact reflects at the back or free surface, reflection cracks may be produced (4). The comminuted material is ejected from the target, together with the highly deformed/fractured projectile (2).

Figure 17.23 illustrates the damage produced by low-velocity impact (250 m/s) on a soda–lime glass plate. These experiments were conducted by Field [37]. These pictures show the progress of failure. The dark region that forms

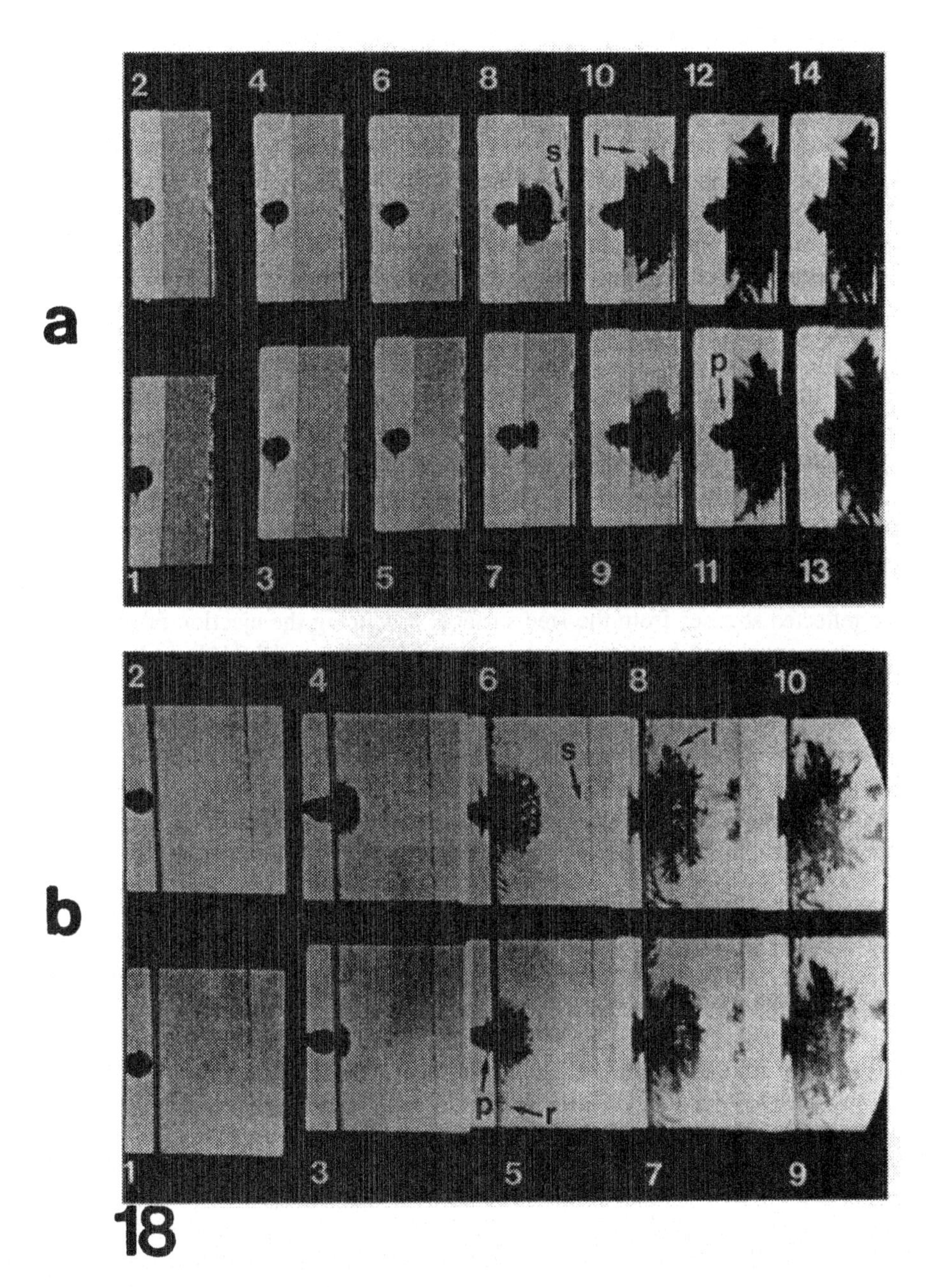

FIGURE 17.23 Sequence of high-speed photographs (IMACON camera) taken at 0.95-μs interval and showing the impact of a 3-mm steel sphere at 250 m/s on a soda-lime glass target. Prominent features are the formation of a failure front, damage from the reflected wave (s), lateral cracks (l) initiating at surface. (From Field [37], Fig. 18b. Reprinted with permission.)

is a fracture front. The images were at taken at time intervals of approximately 1 μs. Damage from the back surface (marked s) starts when the compressive wave reaches it. In Frame 8 for both Figure 17.23(a) and (b), damage is occurring from the back surface (s) and from the direct propagation of the compression from impact. The reader should compare Figure 17.23 with Figure 17.22. The fracture front propagates at a velocity that is slower than the stress wave front. At time 5, radial cracks form, away from the impact region. They are marked by r. It is possible to estimate the damage front velocity using the diameter of the sphere (3 mm) as a scale. It is approximately 1.5 mm/μs, or 1500 m/s. This is considerably lower than the velocity of longitudinal (5600 m/s) or shear (3300 m/s) waves. It represents, most probably, the velocity of the cracks emanating from the point of contact. The reader is referred to Chapter 16 for greater details.* Figure 17.24 shows the formation of a cone in a ceramic (alumina) by impact with a 5-mm-diameter steel sphere at 250 m/s. This cone is produced by tensile stresses generated by the high contact stresses. These stresses were studied by Hertz [38]. The same type of crack is observed under static indentation of glass; see Lawn and Wilshaw [39]. Tensile cracks emanate from the point of contact in a cone pattern. They interact with the reflected stresses from the free surface, producing the ejection of the cone. The cone produced by the impact is shown in Figures 17.24(d) and (f) (side and top views) while the orifice is shown in Figures 17.24(c) and (e). The apex and top portion of the cone consists of a very smooth surface, produced by the Hertzian stresses, while the base of the cone is formed by a different stress system; the reflected stresses play an important role in these. Figure 17.25 shows the same phenomenon in a steel specimen impacted at liquid nitrogen temperature (77 K) by Meyers and Wittman [40]. At these temperatures, AISI 4340 steel is very brittle and behaves like a ceramic, with the ejection of a cone. The student can produce these cone cracks in glass by using a compressed air (BB) gun and regular window glass (use double-thickness glass). The author remembers vividly the windows of his house, in Socorro, where the neighborhood kids would systematically shoot their BB guns. On the inside of the house the author would regularly find these cones. The bullets would not traverse the glass; they would bounce off, as shown schematically in Figure 17.22(a).

Figure 17.26 shows a cross section of a confined alumina disk that was impacted by a tungsten projectile at 560 m/s. The comminuted area ahead of the projectile (marked B) and the radial cracks (marked A) are easily visible. The complexity of the phenomena involved in penetration of ceramic armor by projectiles has deterred simple analytical models with a predictive capability. Large-scale computations are required to deal with this problem, and the comparison of the performance of different ceramics against impact is done primarily in an experimental fashion. Titanium diboride and titanium carbide seem

*Failure waves have been postulated and studied in glass by G. I. Kanel, S. V. Rasorenov, and V. E. Fortov (in "Shock Compression of Condensed Matter, eds. S. C. Schmidt et al., Elsevier, 1992, p. 451). Experiments by Raiser et al. (*J. Appl. Phys.* **75** (1994) 3862) provided a clear confirmation for the existence of failure waves. It is possible that failure waves are due to the time required to nucleate and grow flaws, behind the shock wave.

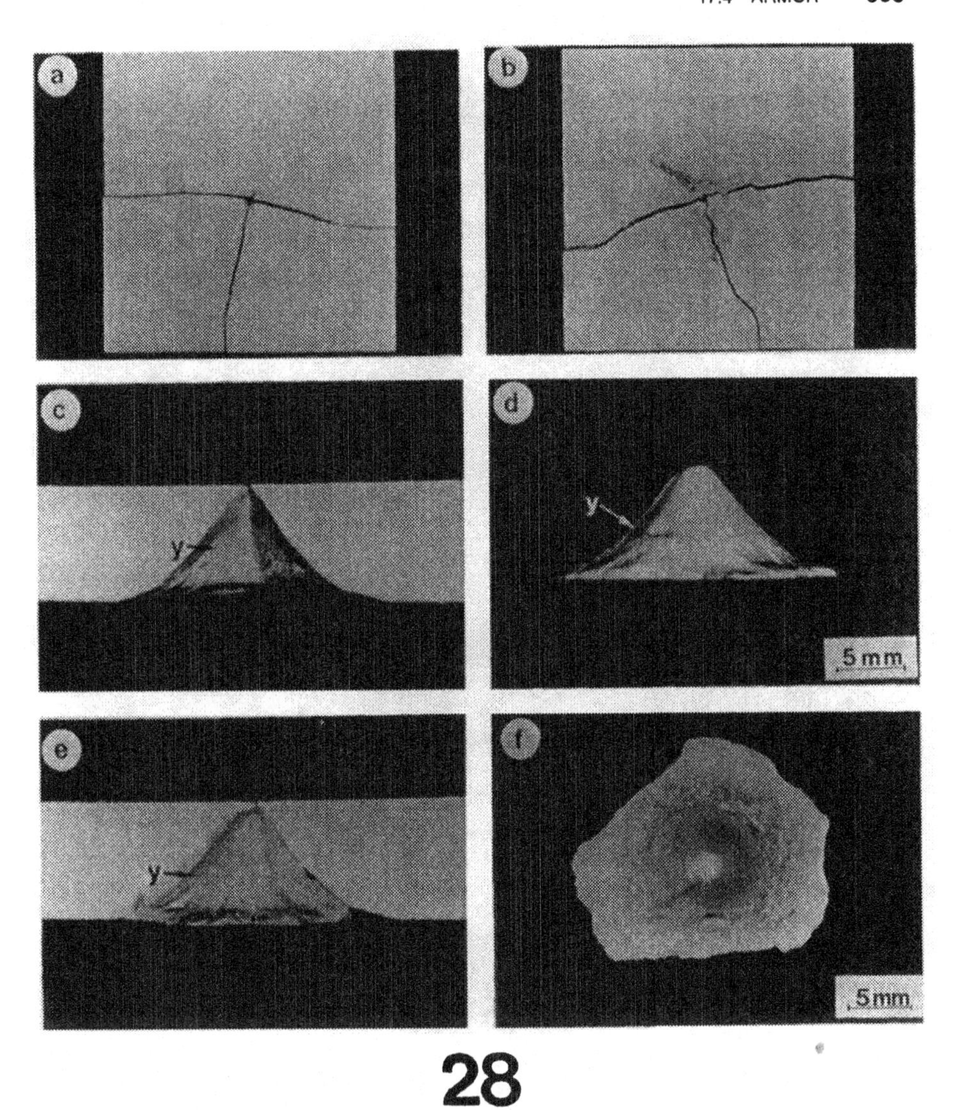

FIGURE 17.24 Formation of cone in alumina impacted by a 5-mm hardened-steel sphere at 225 m/s: (a) front target face; (b) back target face; (c) side view; (d, f) cone; (e) cone placed in target. (From Field [37], Fig. 28. Reprinted with permission.)

marily in an experimental fashion. Titanium diboride and titanium carbide seem to outperform other ceramics. Silicon carbide, boron carbide, and alumina also have a good performance. When designing ceramic armor, one has to consider the importance of the *areal density* required for protection against a certain threat. Due to the differences in density among ceramics, this will correspond to different thicknesses. Ceramics are used most often as *appliqué*. It is simple to obtain the efficiency of a ceramic against a certain threat by conducting a penetration test into a semi-infinite steel or aluminum target and obtaining the depth of penetration P_0 as a reference. One then applies the ceramic and repeats

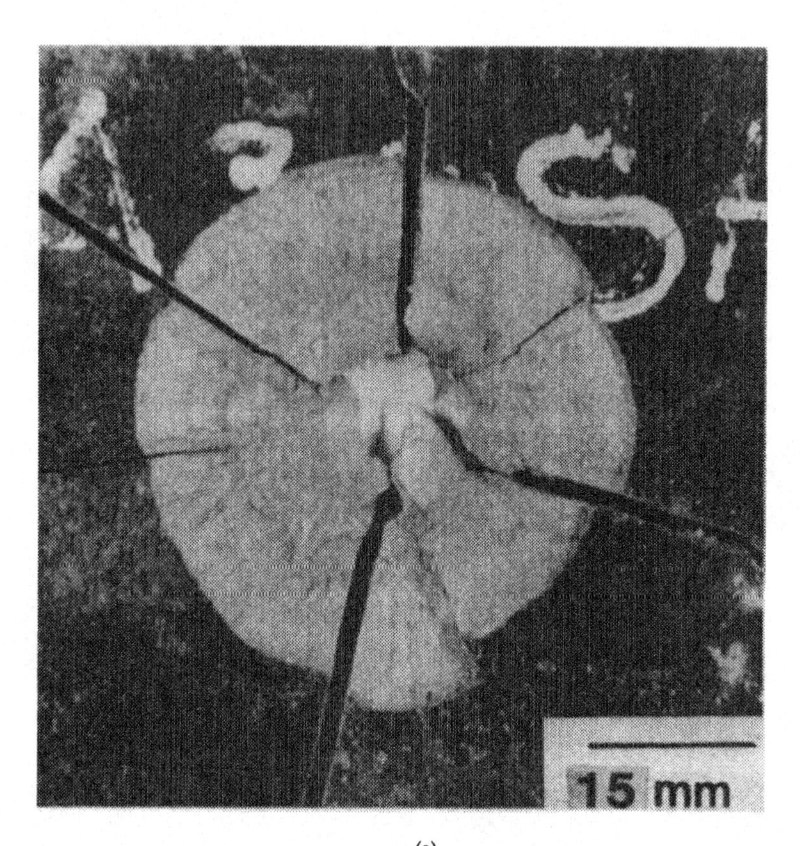

(a)

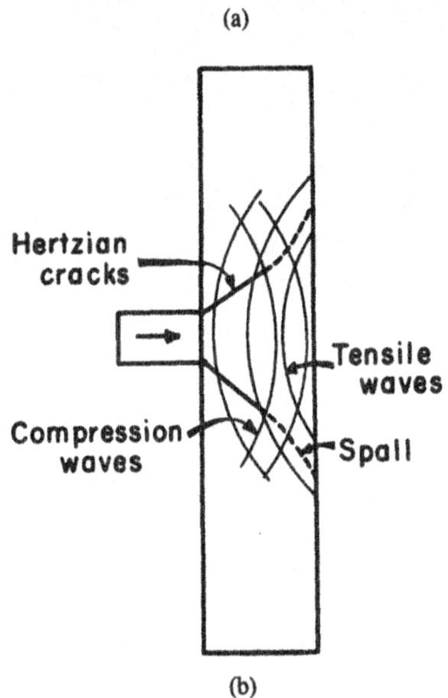

(b)

FIGURE 17.25 (a) Formation of a cone in AISI 8620 steel impacted at 77 K by a cylindrical steel projectile at 930 m/s. (From Meyers and Wittman [40].) (b) Schematic representation of Hertzian cracks and tensile reflection leading to formation and ejection of conoid.

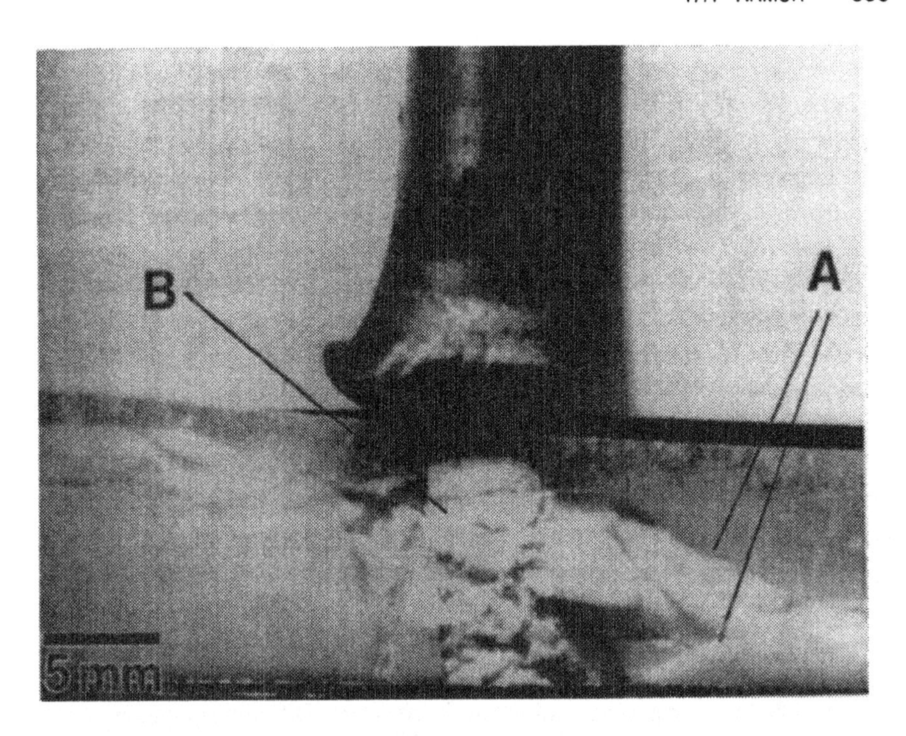

FIGURE 17.26 Cross section of alumina target impacted by tungsten penetrator at 560 m/s, showing radial cracks (A) and comminuted region (B). (Experiment conducted by J. Isaacs, CEAM, UCSD.)

the experiment, as shown in Figure 17.27. A reduction in the penetration of the steel (or aluminum) to P_x is obtained. By converting the dimensions into areal weights (using the densities), one obtains the total efficiency factor (TEF), as described by Yaziv [41], as

$$\text{TEF} = \frac{W_0}{W_1 + W_x} = \frac{P_1\rho_{\text{cer}} + P_x\rho_{\text{st}}}{P_0\rho_{\text{st}}} \tag{17.13}$$

where ρ_{st} and ρ_{cer} are the densities of the steel and ceramic, respectively, and P_1, P_x, and P_0 are marked in Figure 17.27. Alternatively, the differential efficiency factor (DEF) is defined as

$$\text{DEF} = \frac{W_0 - W_x}{W_1} \tag{17.14}$$

In the example shown in the right-hand side of Figure 17.27, an aluminum plate is used as a backup for the alumina ceramic. The resultant expression is

$$\text{DEF} = \frac{W_0 - W_x - W_{\text{Al}}}{W_{\text{Al}_2\text{O}_3}} = \frac{7.8(P_0 - P_x) - 2.7P_{\text{Al}}}{3.42P_{\text{Al}_2\text{O}_3}}$$

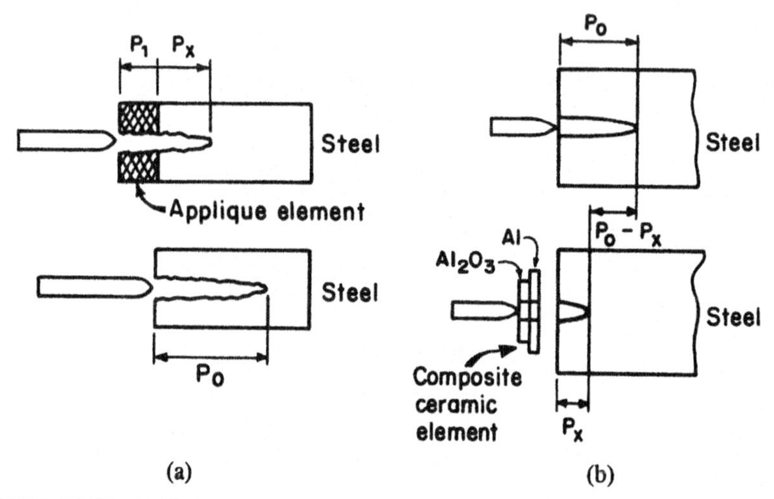

FIGURE 17.27 Ballistic test used to determine the efficiency of (a) ceramics and (b) ceramic-metal. (Courtesy of D. Yaziv, Rafael, Israel.)

where 7.8, 2.7, and 3.42 are the densities of steel, aluminum, and alumina, respectively. The test shown in Figure 17.27 can be used to assess the performance of different ceramics and to compare them. Different ceramics should be tested at the same areal weight. Lateral confinement should be added for the simulation of heavy armor conditions. An efficiency factor greater than 1 results in armor weight reduction; conversely, if TEF < 1, the ceramic is less effective than the same weight of steel. The right-hand side of Figure 17.27 shows the same test applied to a composite ceramic element (Al_2O_3 backed with Al).

Other novel armor concepts are also under development. Spaced armor, in which open spaces are left between armor elements, has the objective of deflecting the projectile as well as trapping the compressive waves. Armor sections are placed at different angles to the incoming projectile to deflect it. Laminated armor, composed of materials with highly different shock impedances, can also be used. Perforated armor is simply a perforated high-strength steel placed as an *appliqué* ahead of the armor. These perforations will rotate the projectile, so that its trajectory and longitudinal axis lose their parallelism. Upon tumbling, much of the penetration ability of the projectile is lost. One very imaginative development that has been successfully introduced is the reactive armor, invented by Held [42] and implemented by the Israelis. Its principle of operation is shown in Figure 17.28. A layer of plastic explosive such as Detasheet (DuPont) is sandwiched between two metal sheets. This system forms the frontal face of a box that is bolted to the conventional armor of the tank. The vulnerable surfaces are covered with these boxes, shown schematically in Figure 17.28. A projectile, upon traversing the explosive, initiates it. The explosive sensitivity is adjusted in such a manner that small arms cannot initiate it. The reader is referred to Chapter 10 for details on initiation and sensitivity of explosives. The expanding explosive will separate the two con-

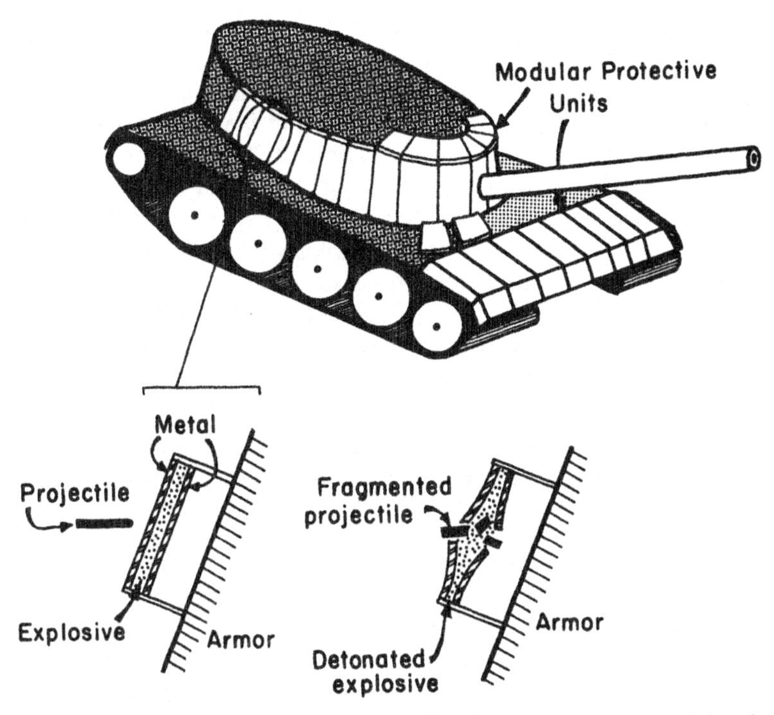

FIGURE 17.28 Reactive armor added to battle tank as *appliqué*; principle of operation of one box (modular protective element) shown on bottom.

fining metal plates, which will strike the incoming projectile, deflecting it and breaking it up. These modular protective elements are bolted to the existing armor, and therefore the detonation fragments do not harm the tank crew. They are, however, fairly noisy! Their detonation has been known to catastrophically affect the peristaltic movement of the intestinal tract of the crew.

17.5 EXPLOSIVE WELDING

Explosive welding is a fairly well developed technology and the most important and successful commercial application of explosives to materials. Small companies in the United States, Japan, FSU, Europe, and Eastern Europe produce a variety of welded and cladded parts. Explosive welding will never replace conventional welding entirely; rather, it has found a niche in the marketplace, primarily for welding metal combinations that cannot be welded conventionally. Examples are titanium/steel, copper/aluminum, and copper/steel. Hundreds of combinations have been successfully welded. For reviews on the subject, the reader should consult references 43–51. The process was "invented," or "discovered," by Carl [43] in 1944 and was developed by Rinehart and Pearson [44], in the 1950s. Russian contributions were very important and parallel U.S. developments [45]. The principle of explosive welding is the inclined impact between two metallic objects, creating a jet that cleanses the surfaces of all surface contaminants and forces them into direct contact. The pressure due to the explosives acts upon these two clean (and hot) surfaces for a time sufficient for bonding (as the result of shear localization at the surfaces) and cooling to occur. Thus, the materials are welded. Figure 17.29 shows the two most common geometries used in explosive welding: plate and tubular. In the plate geometry, two plates separated by a stand-off gap are covered by an explosive layer (usually, a low-detonation-velocity explosive that is subsonic with respect to the materials being welded). The explosive, upon detonation, accelerates the top plate against the bottom plate. In such a manner a welding front is established. For the tubular geometry, one has essentially the same situation. The explosive is placed in the center and detonation is initiated at the top. The inner tube is welded to the outer tube. The details of the welding process are shown in the circled region of Figure 17.29. The jet is shown; the welded interface often has a wavy configuration. In actual experiments a mandrel is needed,

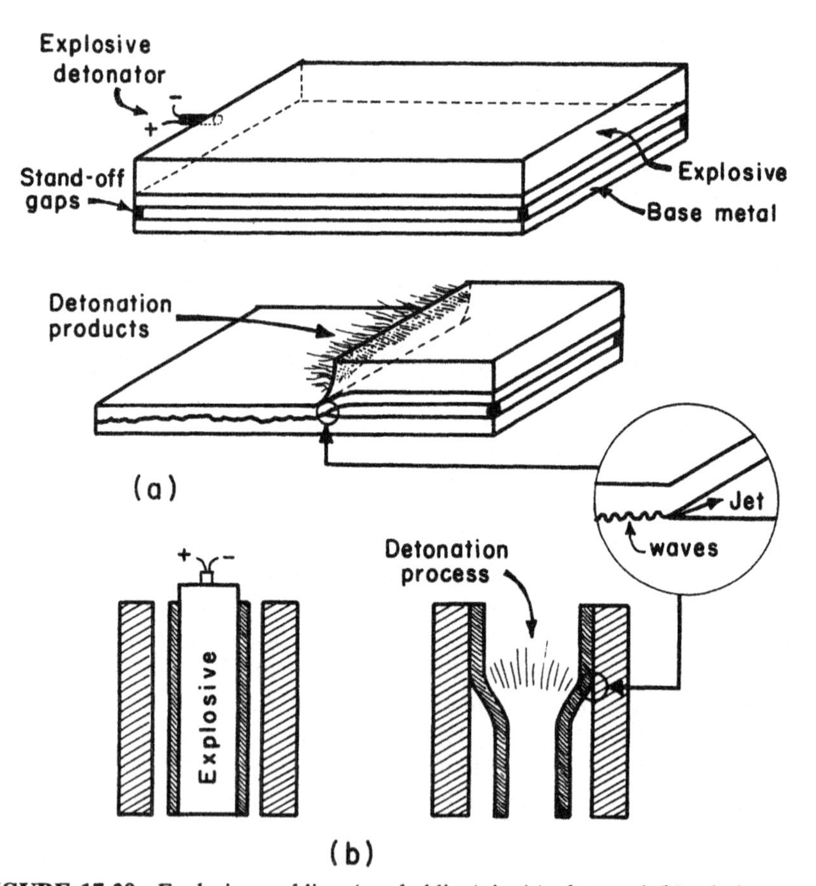

FIGURE 17.29 Explosive welding (or cladding) in (a) plate and (b) tubular geometries.

especially in the tube geometry, to inhibit the expansion of outer tube; it is now shown in Fig. 17.29.

Figure 17.30 shows the explosive welding configuration, in which the velocity of the collision front V_C is equal to the detonation velocity V_D. This is only true in the parallel-plate configuration. The dynamic bend angle β is obtained from the collision and plate velocities V_C and V_p, respectively, by simple geometric considerations. The construction in Figure 17.30(b) shows that V_p bisects the angle between the original plate orientation and the deformed plate. The construction is best understood if one considers unit time. Point *A* goes to *B*, along V_p when point *O* goes to *B*, along V_C. Proof for this is found in Section 17.2, where a similar situation occurs with shaped charges. Thus, the triangle *OAB* is isosceles, and for triangle *OBC*, one has

$$\tfrac{1}{2}V_p = V_D \sin \tfrac{1}{2}\beta \quad \text{or} \quad V_p = 2V_D \sin \tfrac{1}{2}\beta \tag{17.15}$$

For small dynamic bend angles, one has, approximately,

$$V_p = V_D \sin \beta \tag{17.16}$$

It should be noted that for the parallel geometry the collision point velocity V_C is equal to the detonation velocity V_D. The determination of the conditions for

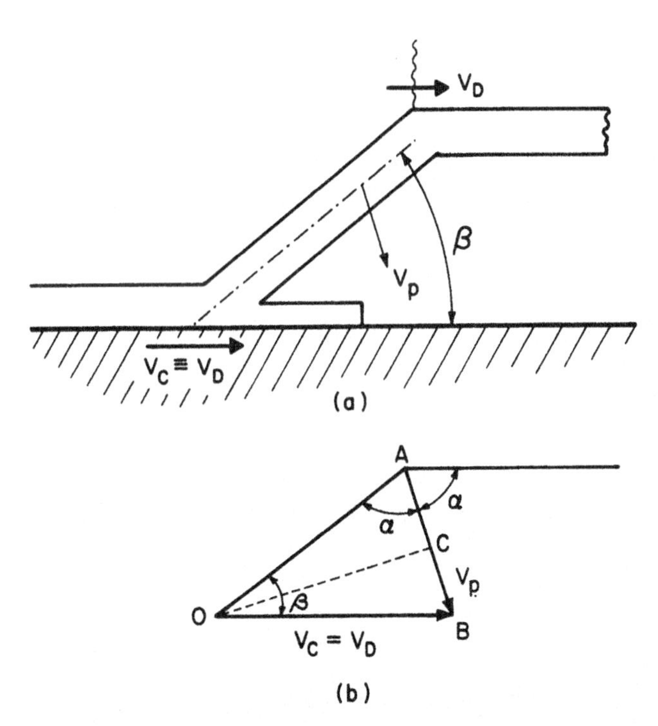

FIGURE 17.30 (a) Explosive welding in parallel plate geometry; (b) geometrical relationship between velocities.

explosive welding has been addressed by Wittman [52] and Deribas [53] in a systematic fashion. It has been possible to establish an "explosive welding window" from fundamental considerations. This window is a region in the β (collision angle)–V_C (collision point velocity) space. It is bounded by four lines.

1. *Jetting.* The formation of a jet is considered a prerequisite for explosive welding. When the detonation velocity is such that the collision velocity is supersonic with regard to the material, no jet can occur. A shock front is produced at the impact point that travels with this collision front. There is an angular dependence of this phenomenon. In Figure 17.31, this critical velocity increases with β; it forms the right-hand side of the window.

2. *Critical Impact Pressure.* If the pressure at the collision point is insufficient, no jetting occurs. This pressure depends on the strength of the material. Both Wittman [52] and Deribas [53] found that a pressure equal to five times the strength of the material was necessary for jet formation. They obtained the required pressure for hydrodynamic flow using Bernoulli's equation:

$$p = \frac{\rho U_p^2}{2}$$

Making U_p equal to V_p (the plate velocity) and the pressure P equal to $k\sigma_y (k \sim 5)$ yields

$$V_p = \left(\frac{k\sigma_y}{\rho}\right)^{1/2} \tag{17.17}$$

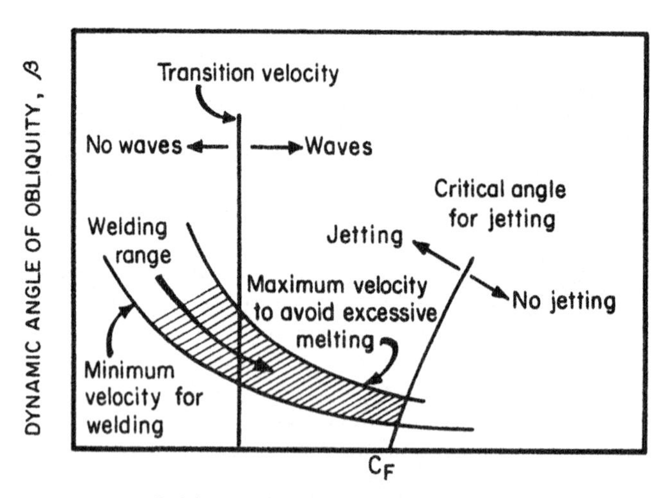

FIGURE 17.31 Welding window proposed by Wittman [52] and Deribas [53].

Inserting Eqn. (17.17) into Eqn. (17.16), we obtain (replacing k by $k_1 = k^{1/2}$)

$$V_C \sin \beta = k_1 \left(\frac{\sigma_y}{\rho}\right)^{1/2} \tag{17.18}$$

This equation is the lower boundary in the window of Figure 17.31.

3. *Maximum Impact Velocity.* The kinetic energy of the flyer plate is converted into thermal energy during impact, and consequently there is melting at the interface above a certain impact velocity. If the molten layer is too thick, the interface will have no strength. Thus, the deposition of energy at the interface and its cooling must be computed; a molten interface will have no strength and will debond. Thus, the upper boundary is the formation of a continuous interfacial molten layer. One of the expressions is [47, 52]

$$V_p = k_2 \frac{(T_{mp} C_0 k C_p)^{1/2}}{V_C t^{1/4}} \tag{17.19}$$

where k_2 is a constant, C_0 is the bulk sound velocity, t is the flyer plate thickness, T_{mp} the melting point, k the thermal conductivity, and C_p the heat capacity. This equation represents the upper boundary in Figure 17.31, by proper substitution of Eqn. (17.16).

4. *Critical Flow Transition Velocity.* The left-hand boundary is a vertical line separating two regimes of welding: a wavy from a straight interface. There is a critical collision point velocity above which the flow becomes turbulent, resulting in a wavy interface. This wavy interface is considered by many to have properties superior to the straight interface. Mechanisms for the formation of the wavy interface have been proposed by several investigators. These interfacial waves have been observed between fluids traveling at different speeds, at the surface of the sea, and due to erosive action of high-speed liquids. There is more than one theory as to the origin of the waves:

(a) Kelvin-Helmholtz instability due to a velocity discontinuity at the interface between the two materials: this is a classical mechanism for instabilities forming at the interface of two fluids moving at different velocities. The reader should consult a fluid dynamics book for detailed treatment.

(b) Vortex shedding mechanism: analogous to von Karman vortex arrays, initiated at the stagnation point and propagating behind. The stagnation point can be imagined as an obstacle in a moving fluid. It disrupts the flow, so that vortices are created behind. The student can verify this in a river, observing the flow pattern downstream from an obstacle.

(c) Jet indentation mechanism: the jet indents the base plate; it is rotated upward. This periodic rotation of the jet leaves behind a periodic array of indentations, which comprise the wave.

(a)

(b)

(c)

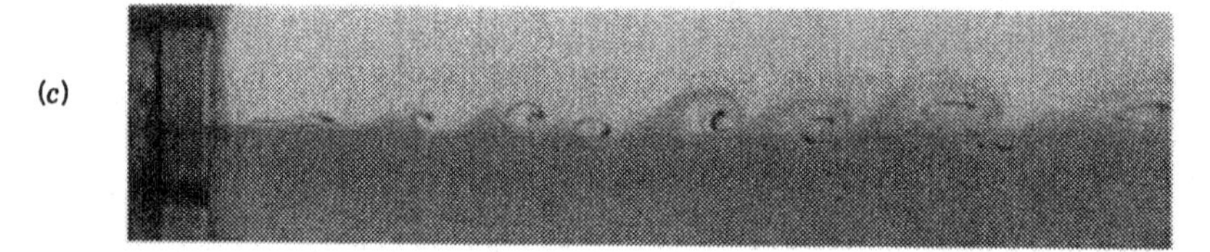

FIGURE 17.32 (a) Wave pattern produced at interface between copper and nickel in explosive welding; ignore the surfer, a graduate student that accidentally entered this picture by piercing "hyperspace." (Courtesy of R. Prummer, Ernst Mach Institute, Germany.) (b) Wave pattern produced between steel and aluminum in explosive welding. (From Wang et al. [54], Fig. 17, p. 1802. Reprinted with permission of the publisher.) (c) Wavy interface produced by the flow of two layers of liquid at velocities of 3.4 and 1.5 m/s; Kelvin–Hemholtz instability generated by velocity discontinuity at interface. (From Lasheras and Choi [55], Fig. 4(a), plate 1, p. 59. Reprinted with permission of the publisher.)

(d) Stress wave pulsing due to reflections at the free surface and interaction between waves.

Figure 17.32(a) shows the interface between copper and nickel welded by explosives. The clear wave pattern is obvious. The amplitude and wavelength are dependent on experimental conditions. Such a wavy interface can very effectively weld dissimilar materials. The wavy welds produce a mechanical gripping action between the two metals. Experiments conducted by Wang et al. [54] on steel–aluminum explosive welds show the same characteristic wavy pattern, seen in Figure 17.32(b). Figure 17.32(c) shows a wave pattern produced when two layers of liquid (the dark and light ones) move at different velocities. This creates the famous Kelvin–Hemholtz shear instabilities. These experiments, conducted by Lasheras and Choi [55], show very nicely the pattern; the velocities of the liquids were 3.4 and 1.5 cm/s in the top and bottom layers, respectively. It is difficult to negate the effect of Kelvin–Hemholtz instabilities in explosive welding if one looks at the almost identical wave pattern produced in experiments that are radically different.

The material at the interface between the two metals is subjected to high-strain-rate, high-strain deformation, and the resultant temperature increase is significant. There have been only a few reports of the structure of the interface, and they indicate that a microcrystalline structure is produced. Figure 17.33 shows schematically how this interfacial structure is generated in an Al–Al_3Cu

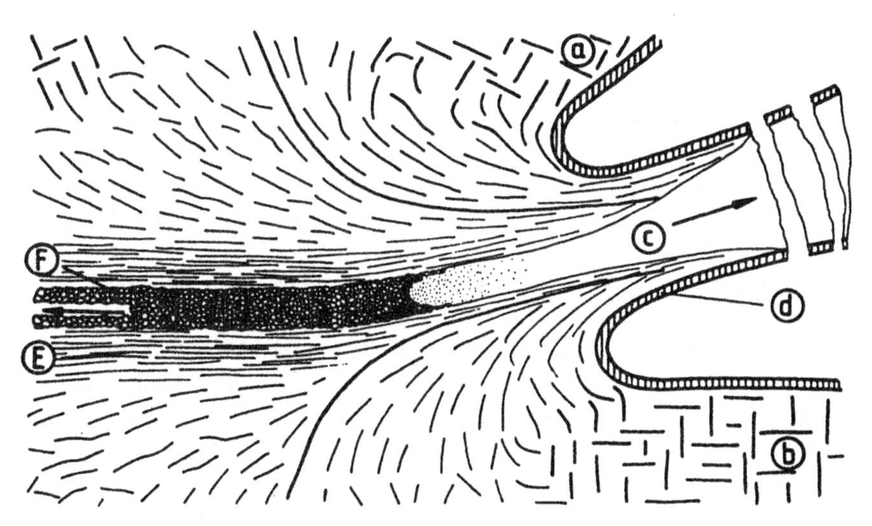

FIGURE 17.33 Schematic representation of plastic flow and microstructure in explosive welding of an Al–Cu alloy containing disk-shaped precipitates that serve as markers for plastic flow: (a) top plate; (b) bottom plate; (c) jet region; (d) surface layer of oxides; (e) highly deformed region; (f) ultrafine grain region possibly due to dynamic recrystallization. (From Hammerschmidt and Kreye [56], Fig. 6, p. 970. Reprinted with permission of the publisher.)

alloy. These experiments, conducted by Hammerschmidt and Kreye [56], show that intense plastic deformation has occurred at the regions adjoining the interface. The small lines represent the disk-shaped Al_3Cu precipitates, and their change in length and reorientation indicates the plastic strain. In the interface a layer of microcrystalline material is produced, most probably by a dynamic recrystallization process. These dynamically recrystallized, microcrystalline structures are a characteristic of high-strain, high-strain-rate deformation and are observed in shear bands, shaped charges, and EFPs and at interfaces between particles in shock consolidation. The reader is referred to Chapter 16, where dynamic recrystallization in shear bands is discussed.

17.6 EXPLOSIVE FORMING AND HARDENING

Explosive forming uses the high gas pressures created by the detonation of explosives to form metals. It has definite advantages over conventional forming methods when small batches of large pieces have to be produced, because investment in dies is minimized, or in instances where the metal exhibits an enhanced formability at high strain rates. Explosive forming can be accomplished in the air or in water. Figure 17.34 shows two methods. For forming of thin sheets to close tolerances, the system shown in Figure 17.34(a) is used. The detonation of the explosives causes high-pressure gases to expand, forming a component of the tube. For large parts, when close tolerances are not so important, water can be used as a transmitting medium. Figure 17.34(b) shows such a system. The detonation of the explosive produces a shock wave propagating through the water and a gas bubble that expands. Both the shock wave and the pressure in the water due to the gas bubble impart energy to the blank that is formed in this manner. Large diameters can be produced with little capital investment. The lower die can be made of a low-melting-point material, such as Kirksite, a zinc-based alloy with a melting point of 380°C. Reinforced concrete and plaster have also been used for one-shot dies.

High-amplitude shock waves produce plastic deformation in metals without altering their shape, because a state of uniaxial strain is established. Chapter 14 reviews the effects of shock waves in metals; Figure 14.1 shows the increase in hardness with shock pressure. Thus, shock hardening is well suited for applications where an increase in hardness is desired without shape change. The most successful industrial application of shock hardening has been for Hadfield steel. This is a high-manganese austenitic steel with high strength, toughness, and wear resistance. By shock hardening the surfaces, the wear resistance is tripled. The major applications in which shock hardening has been used are in railway connections ("frogs") and excavator teeth. Plastic laminated explosive (such as DuPont Detasheet) is laid on the surface of the steel, forming to its shape. It is detonated, producing an increase in hardness that has a considerable depth.

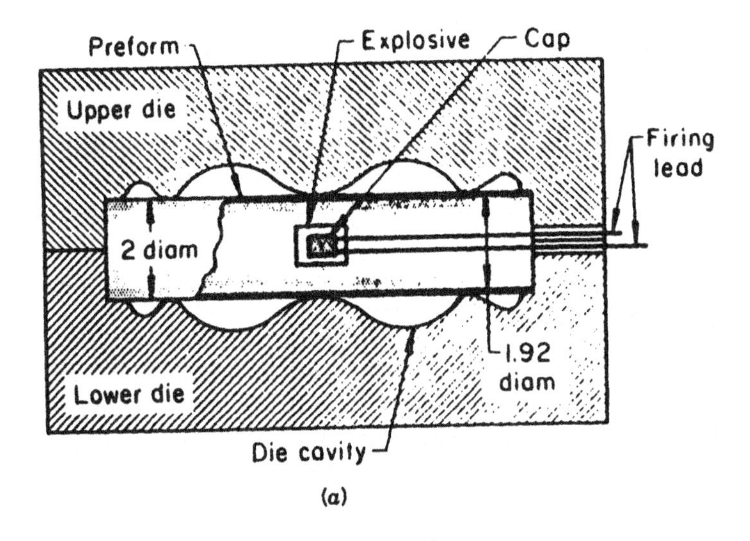

(a)

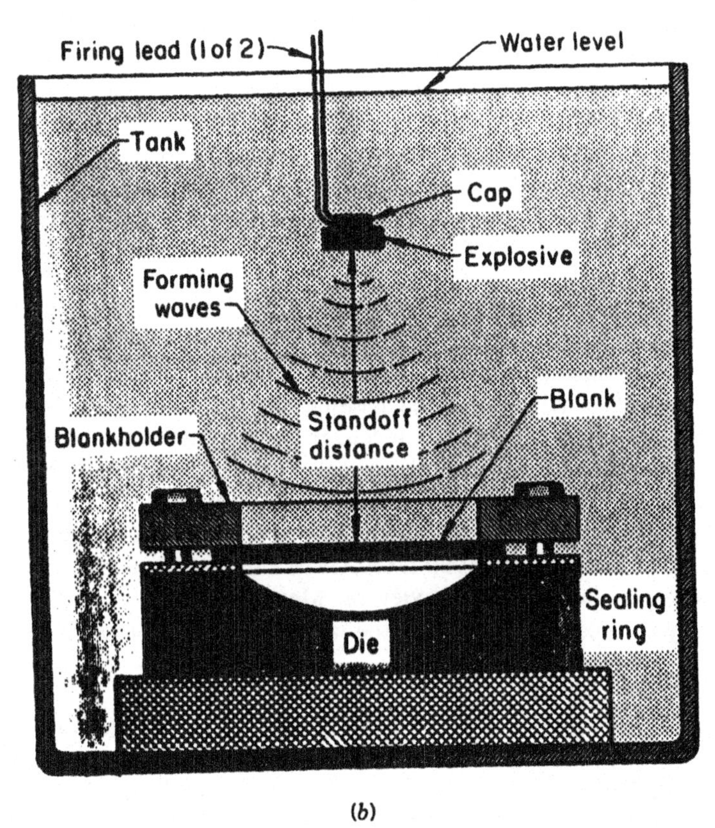

(b)

FIGURE 17.34 Explosive forming systems: (a) direct/air detonation in confined enclosure; (b) detonation in water. (From *Metals Handbook*, Desk Edition, ASM International, Metals Park, OH, 1985, Fig. 57, p. 26.23. Reprinted with permission.)

17.7 SHOCK PROCESSING OF POWDERS

17.7.1 Introduction

A variety of advanced materials are produced in the form of powders. Rapidly solidified metals possessing a microcrystalline, microdendritic, or glassy structure are synthesized in the form of fine powders. Similarly, many hard ceramics (diamond, boron nitride, silicon carbide, etc.) are produced in powder form. A method that can produce monoliths from powders without changing their unique properties would be very desirable.

Shock consolidation of powders is a one-stage densification/bonding process that presents potential for rapidly solidified powders and very hard and difficult-to-consolidate ceramics. This dynamic form of powder consolidation, which involves a very rapid and intense deposition of shock energy at powder particle surfaces resulting in interparticle bonding, has been known for over 20 years. LaRocca and Pearson were apparently the first to use this technique. Nevertheless, there is no commercial application of this process in the United States, except in the synthesis of diamond [57]. The production of the wurtzite form of boron nitride in Japan by shock energy is a modest industrial effort [58]. There are reports of extensive industrial applications of shock processing of powders in the FSU [59].

Shock wave consolidation of powders is a technique that avoids prolonged heating and can be used for producing bulk solids in the form of disks, plates, cylinders, tubes, and cones. The basic mechanism by which consolidation is achieved as a shock wave passes through a porous medium is either by one or both of the following: (1) deformation and high-velocity impact of particles, filling the interstices, breaking down surface oxides, leading to heating and melting of particle near-surface material and subsequent welding; (2) fracture of particles, filling interstices, cleansing of surfaces, preferential heating of particles surfaces leading to partial melting and welding or solid-state-diffusional bonding. Whereas the first mechanism seems to apply to metals, the second is more prevalent in ceramics. Shock consolidation of powders has been recently reviewed by Gourdin [60], Thadhani [61], Prümmer [62], Meyers [63], and an NMAB study [64]. A number of papers can also be found in the Proceedings of the 1985 and 1990 EXPLOMET International Conferences [65, 66]. This section is based on a review by Meyers [63].

A variety of processes can utilize the rapid energy deposition at the powder surface:

1. Shock compaction, in which consolidation of the powder occurs due to shock energy being preferentially deposited at particle surfaces during the passage of shock waves;
2. shock-enhanced sintering, in which the shock consolidated compact is statically compressed and heated in a normal sintering procedure to produce the final product;

3. shock conditioning, in which the powder is shocked in any convenient geometry, remilled, and conventionally sintered;
4. shock-induced chemical synthesis, in which a compound is formed from a powder mixture during the passage of the shock wave and is at the same time consolidated;
5. shock-induced phase transformations, in which novel structures with desirable properties can be formed under the high-pressure regime; and
6. chemically assisted shock consolidation, which is a combination of shock-induced chemical synthesis and shock consolidation.

In the last case, inert powders are mixed together with an exothermically reacting elemental mixture, and the passage of a shock wave induces a reaction between the elemental powders, promoting at the same time bonding between the initially inert and difficult-to-consolidate materials.

17.7.2 Experimental Techniques

There are several means of depositing the energy required for shock consolidation on the powder surface.

Detonation of explosive in direct contact with the powder or powder container

Rapid deposition of energy at the powder surface by high-amplitude pulsed laser

Impact of projectile against powder or container. This projectile can be accelerated by several means and velocities in the range of 200–3000 m/s are required to produce the pressures listed in Figure 17.35. These projectiles can be accelerated electromagnetically, by compressed gases, by deflagration of gun powder, or by the detonation of explosives.

Of the different concepts introduced above, explosive compaction is the technique lending itself best to industrial production [67]. Since the costs are low, the process lends itself to scaleup without major problems. The effort by Raybould at CERAC, a subsidiary of Atlas Copco, is noteworthy [68]. Raybould developed a semiautomated machine capable of accelerating a piston at velocities up to 700 m/s. Parts could be repeatedly made with this assemblage with a diameter up to 7 cm. Net shape or near net shape was accomplished by the development of appropriate dies and momentum traps. Gas gun experiments yield excellent control of both impact velocity and impact planarity. This method has been extensively used by Ahrens and co-workers [69], Nellis and co-workers [70], Sawaoka and Horie [72], and Mutz and Vreeland [71]. Different impact and wave trapping geometries have been developed by these investigators that address the specific requirements of their investigations. In the following sections, explosive compaction systems used by the author and his

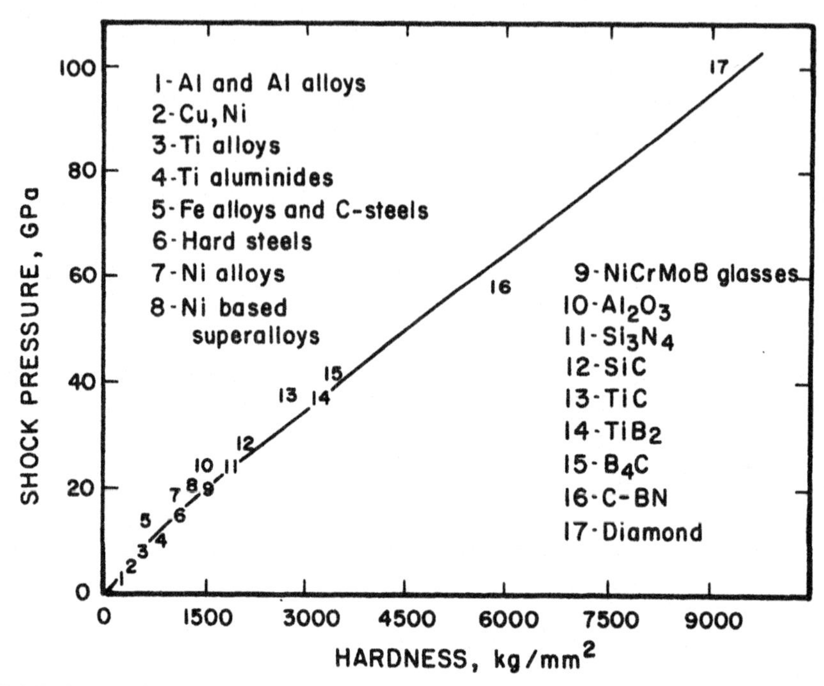

FIGURE 17.35 "Engineering" correlation between powder hardness and pressure required for shock consolidation. (From Meyers et al. [63], Fig. 8.1, p. 269. Reprinted with permission of the publisher.)

co-workers at New Mexico Institute of Mining and Technology will be described.

Cylindrical Configuration/Single-Tube SetUp. This is the most common configuration used for shock compaction experiments. It was extensively used by Prümmer [62]. It is a very simple design and very little tooling is required around the material. Figure 17.36 shows the basic components of this system. The pipe is filled with the chosen explosive and detonation is initiated at the top. The shock pressure on the powder can be varied by varying the amount and type of explosive. By increasing the detonation velocity from 3500 m/s [the lowest range of ammonium nitrate and fuel oil (ANFO)] to 7000 m/s (C-4 plastic explosive), the explosive pressure is quadrupled since the pressure varies with the square of the detonation velocity. The pressure pulse converges toward the central axis of the cylinder and, if excessive, a hole is generated along the cylinder axis. This is called a Mach stem and can be eliminated by using a solid metal rod along the axis or by adjusting the shock and detonation conditions. This system lends itself well to the consolidation of powders that are fairly soft and ductile (hardness below 500 kg/mm^2).

Cylindrical Configuration/Double-Tube Setup. This technique which was developed for the shock synthesis of diamonds, was used for the first time for shock consolidation by Meyers and Wang [67]. The powder is contained in

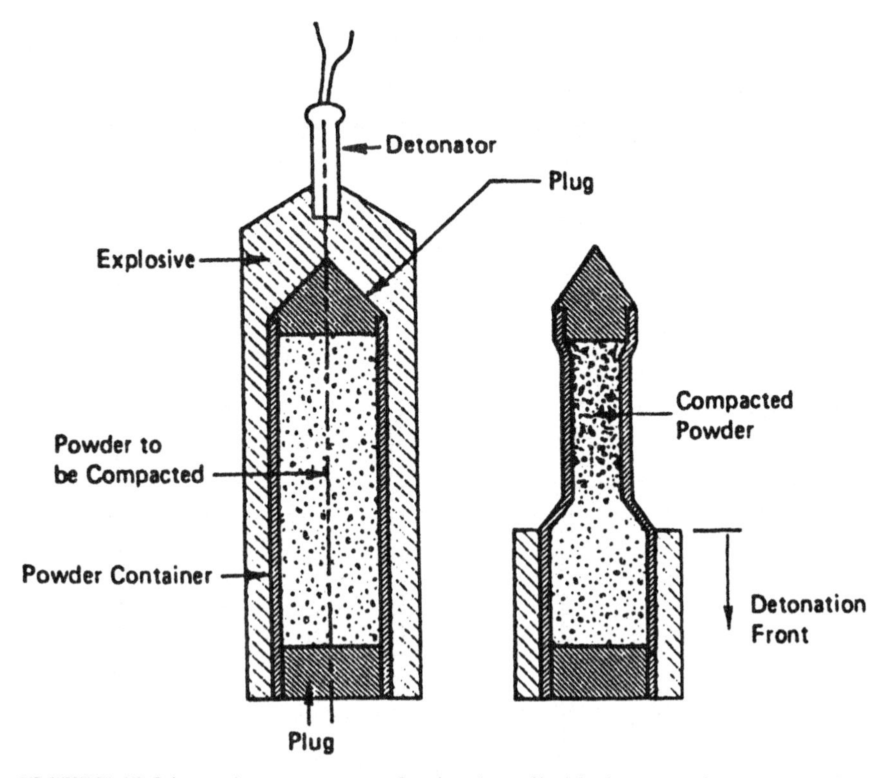

FIGURE 17.36 Basic components of a simple, cylindrical compaction system. (From Prümmer [62] Fig. 5.1, p. 34. Reprinted with permission of the publisher.)

the internal tube. The external tube is surrounded by the explosive charge, which is detonated at one end; this external tube acts as the flyer tube, impacting the internal tube. This technique generates pressures in the powder that can be several times higher than the ones generated by the single-tube technique. The main advantage of this technique is that it allows the use of low detonation-velocity explosives for consolidating hard powders. The lower detonation velocity explosives minimize cracking of compacts. Significant improvements in compact quality have been obtained in nickel-based superalloys, titanium alloys, and aluminum–lithium alloys.

The basic experimental setup is shown in Figure 17.37(a). It is similar to the single-tube setup. The explosive charge is detonated at the top; a Detasheet booster is used to create a more uniform detonation front. The explosive is placed in the cylinder, at the center of which is the assembly containing the powder. The central axis of the container may have a solid rod to eliminate Mach stem formation. The difference between the system shown in Figure 17.37, described above, and the conventional explosive consolidation system of Figure 17.36 is that a flyer tube is placed coaxially surrounding the container tube.

The radial cross-sectional view of the system employing the double-tube

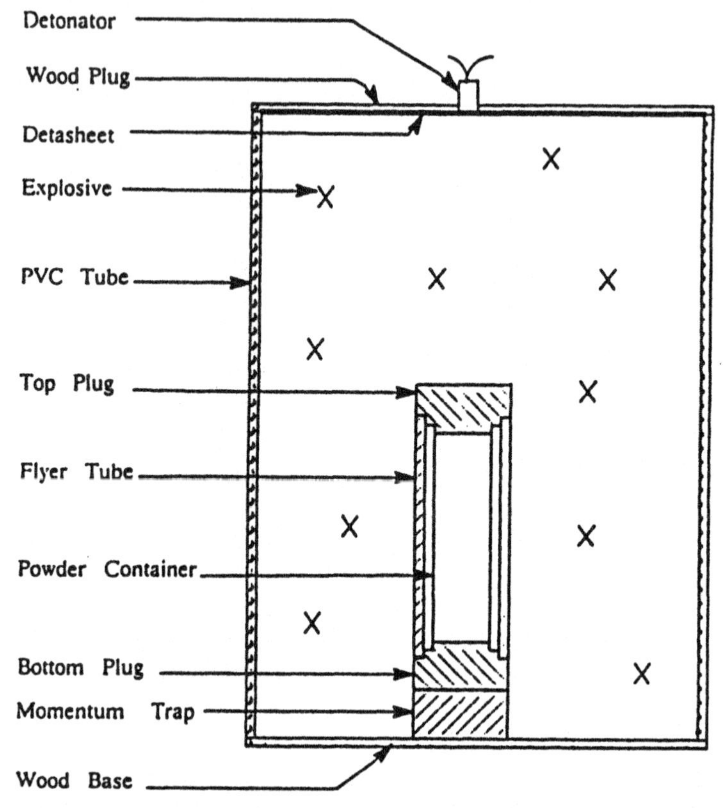

FIGURE 17.37 (a) Basic experimental setup for the cylindrical axisymmetric double-tube system. (From Ferreira et al. [75], Fig. 1, p. 686. Reprinted with permission of the publisher.) (b) Radial cross-sectional view of system employing the double-tube configuration.

configuration is shown in Figure 17.37(b). By applying an analysis similar to that developed by Gurney [73] (see Chapter 9) for the velocity of fragments, one can estimate the velocity at which the flyer tube is accelerated inward by the explosive. The chemical energy of the explosive is equated to the sum of the kinetic energies of the gases and of the tube. One obtains the following equation:

$$V_P = \sqrt{2E}\left(3\Bigg/\left[5(m/c) + 2(m/c)\frac{2R + r_0}{r_0} + \frac{2r_0}{R + r_0}\right]\right)^{1/2} \qquad (17.20)$$

where V_P is the velocity of the flyer tube and m/c is the ratio between the mass of the flyer tube and the mass of the explosive charge; R and r_0 are shown in Figure 17.37(b). The importance of this equation is that it allows the selection of a preestablished pressure in the powder. The shock pressure is directly related to the impact velocity. A more detailed description can be found in Meyers and Wang [67].

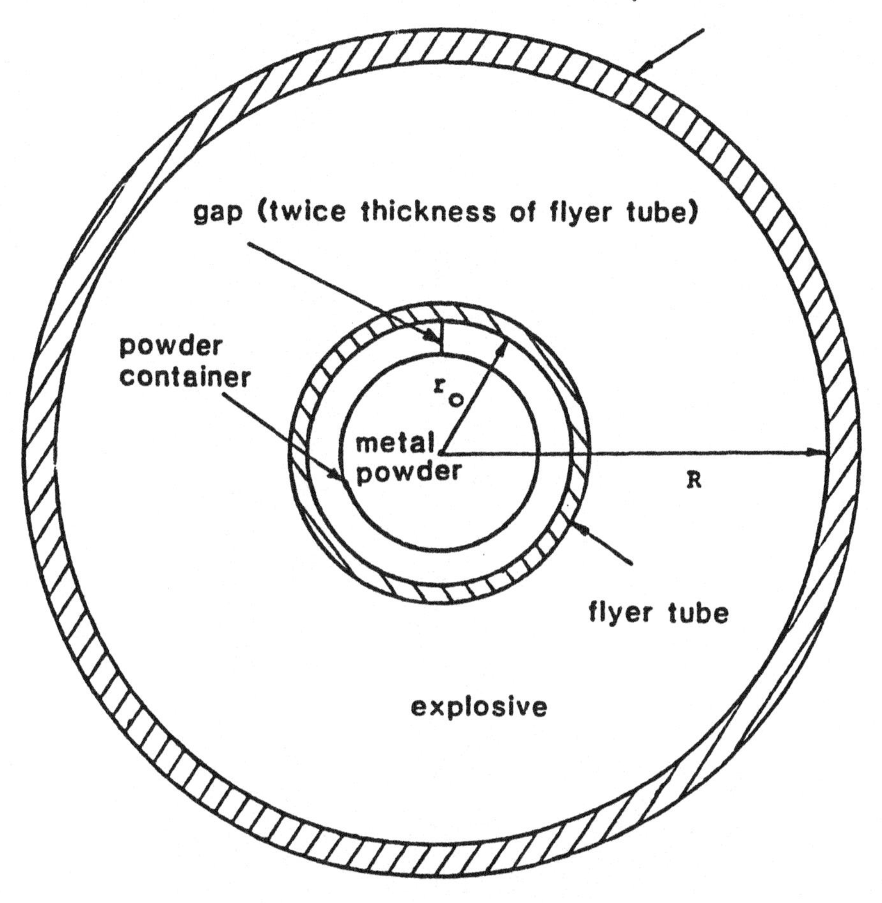

FIGURE 17.37 (*Continued*)

Figure 17.38 shows 2.5-, 5-, 7.5-, and 10-cm diameter cylinders of a rapidly solidified titanium alloy powder that were shock consolidated by the double-tube method. The microstructures of the scaled-up compacts were virtually identical to the smaller diameter compacts, showing that scaleup is feasible.

High-Temperature Setups. When excessive cracking is present after room temperature shock consolidation, often improved results are obtained by preheating the powders. The high temperature can induce additional ductility to the powders and reduce their strength and hardness. Another advantage of high temperatures is that the shock energy required to melt the powder surfaces is decreased. Two systems were developed by Wang et al. [54] and Ferreira et al. [75] for high-temperature shock consolidation experiments.

FIGURE 17.38 Shock consolidated titanium alloy powders; scaleup in size was successfully obtained. (From Coker et al. [76], Fig. 16, p. 1285.)

Sawaoka Shock Compaction Fixtures. Very high pressure impact experiments can be conducted for consolidation of certain very hard and difficult-to-bond materials like ceramics. The Sawaoka fixtures utilize explosively driven plates for impact to generate very high shock pressures (in the range of about 20–100 GPa [77]).

The setup employs flyer plates impacting stainless steel capsules at velocities of 1.5–3.0 km/s. The fixture configuration and the explosive assembly are shown in Figure 12.18 (Chapter 12). The flyer plate is accelerated downward by the detonation of an explosive charge resting on its top. The explosive charge is initiated simultaneously over the top surface by an explosive plane-wave generator. This can be either a mousetrap assembly or an explosive lens (see Chapter 12). Up to 8 or 12 individual capsules containing the powder can be utilized and are held in cavities in the recovery fixture.

TABLE 17.3 Basic Characteristics of Sandia Shock Recovery Fixtures

Fixture	Explosive	Pressure Range (GPa)	Average GPa	Duration (μs)
Bertha	B	3.0–4.5	4.	11.2
Papa Bear	B	3.6–6.0	5.	6.4
Big Bertha A	B	4.6–7.5	6.	15.0
Bertha A	B	5.0–10.0	7.5	11.2
Momma Bear	B	5.0–10.0	7.5	4.5
Papa Bear	C	6.5–9.5	8.0	6.4
Big Bertha A	C	10.0–16.0	13.0	15.0
Momma Bear A	B	14.0–20.0	16.0	4.5
Bertha A	C	14.0–26.0	17.0	9.8
Baby Bear	B	13.0–26.0	20.0	2.5
Momma Bear A	C	19.0–26.0	20.0	4.7
Baby Bear	C	22.0–32.0	27.0	2.6

Sandia Calibrated Shock Recovery Fixture. The Sandia recovery fixtures for controlled shock loading and preservation of sample were developed with the intent of generating controlled and reproducible high-pressure and pulse durations. They were developed by Graham and others [78–81] at Sandia National Laboratories. Table 17.3 shows the shock pressure and duration ranges achievable with these fixtures. The Baby Bear capsule contains 1 cm^3 of powder; the Momma Bear capsule contains 5 cm^3 of powder; the Pappa Bear capsule contains 10 cm^3 of powder; the Bertha capsule contains 60 cm^3 of powder; and the Big Bertha contains 100 cm^3 of powder. The copper capsule that contains the powder is surrounded by a 4340 steel block that provides containment support. The general arrangement of the explosive fixture is shown in Figure 17.39. Copper is used for the capsule because it permits sealing of the sample cavity by plastic flow. Furthermore, optically flat surfaces required

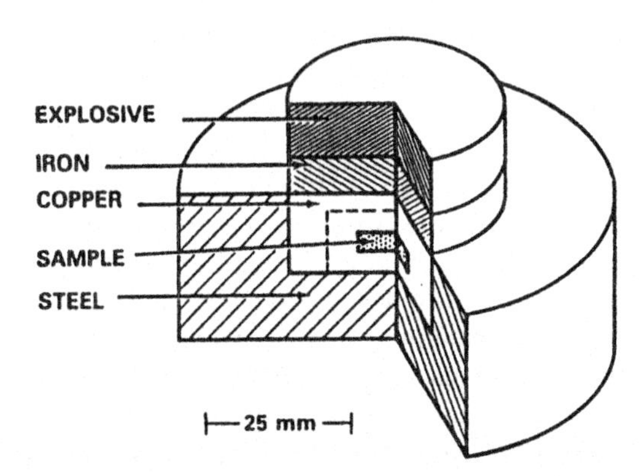

FIGURE 17.39 Section view of Sandia explosive compaction fixtures. (From Graham and Webb [79], Fig. 1, p. 212. Reprinted with permission of the publisher.)

for intimate contact between the surfaces can be achieved with copper. Plane-wave generators are used to initiate the explosive pads shown in Figure 17.39. Here, unlike the Sawaoka compaction system, the explosive is in contact with the copper capsule and is not used to accelerate a flyer plate.

Two-dimensional computer simulations of the shock propagation process have also been conducted on these fixtures, and pressure and temperature histories are available. Thus, one can establish, after recovery and analysis of the shock-modified material, the effects of pressure and temperature in a quantitative manner.

While the Graham fixtures are an excellent research tool, they cannot be used to produce compacts because of their exceedingly high cost and because the compacts undergo considerable fragmentation due to tensile release waves. These fixtures should be used to establish the shock pressure and temperature necessary for good consolidation. With this knowledge in hand, different fixtures (described in the preceding sections) should be used. Wang et al. [82] used the Bear series to independently vary the shock pressure and pulse duration and were able to determine the optimum consolidation conditions for the superalloy IN-100.

Consolidation fixtures utilizing explosively driven flyer plates were also developed by Korth et al. [83] and Yoshida [84]. Korth et al. [83] were able to trap reflected waves successfully by using rectangular specimens. Yoshida [84] used a special fixture for ultrahigh pressures. He used the simultaneous detonation of two explosive lenses, producing a symmetric impact of two flyer plates on the target. The powder (diamond) was subjected to the superimposed shock pulse produced by the two shock waves intersecting in the center. Yoshida's fixture is shown in Figure 12.18(b).

17.7.3 Materials Effects

In a successfully consolidated product, one usually finds interparticle regions that are molten and rapidly resolidified. This is much more obvious for metals than for ceramics. Figure 17.40(a) shows an optical micrograph of a well-consolidated superalloy powder. The black region is a perforation for transmission electron microscopy. A clear melt pool (resolidified) is seen and indicated by an arrow. This molten material bonds the powders. Similar results were obtained by the author and co-workers on aluminum alloys [85], titanium alloys [76], titanium aluminide [86, 87], and other superalloys [74, 88]. This rapidly resolidified microstructure is often microcrystalline or amorphous. Figure 17.40(b) shows the microcrystalline structure. The bottom left corner is microcrystalline, while the rest of the specimen exhibits typical shock substructure (deformation bands). The almost continuous diffraction rings, as compared with regular diffraction patterns in the particle interiors, are clear evidence of rapid (10^5–10^8 s^{-1}) resolidification. Figure 17.41(a) shows an optical micrograph of another nickel–base superalloy (Mar M 200) powder explosively consolidated. The white-etching regions, marked by arrows, indicate melting and resolidification. In ceramics, melting and resolidification are not so ob-

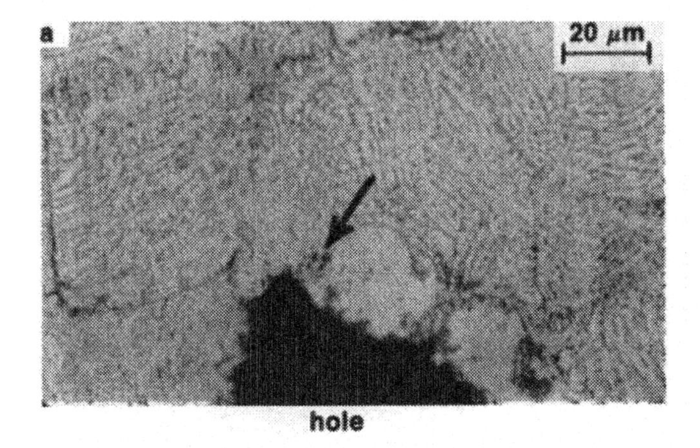

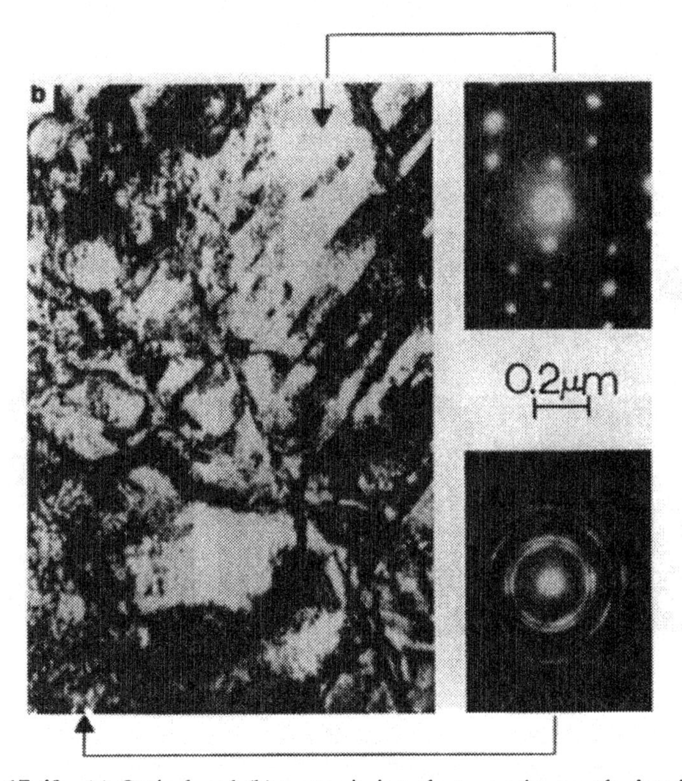

FIGURE 17.40 (a) Optical and (b) transmission electron micrograph showing interparticle regions that are molten and rapidly resolidified in nickel base superalloy. (From Meyers et al. [88], Fig. 8a, p. 25. Reprinted with permission from *JOM* (formerly *Journal of Metals*), a publication of The Minerals, Metals & Materials Society.)

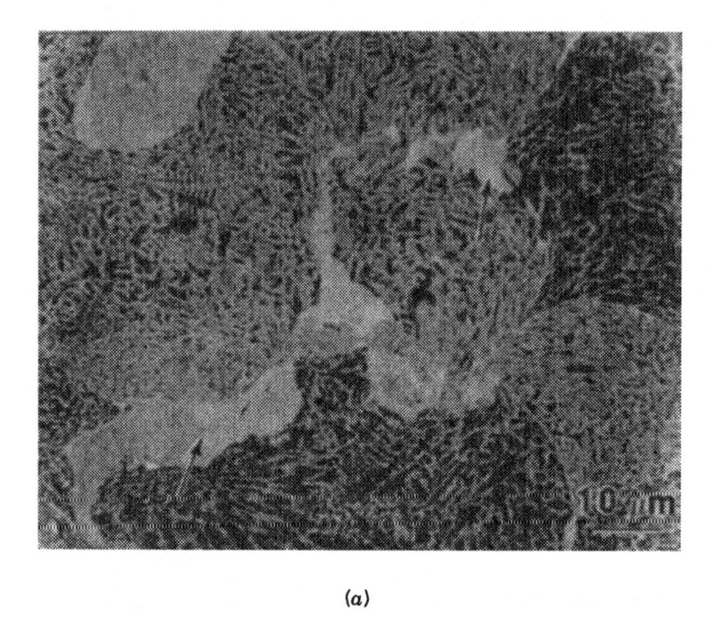

(a)

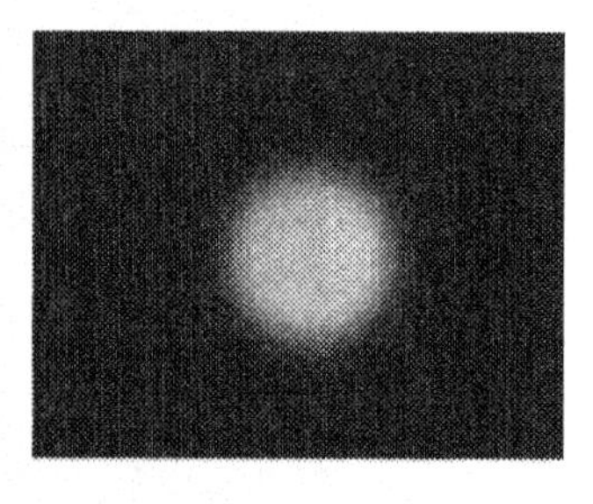

(b)

FIGURE 17.41 (a) Optical micrograph of shock-consolidated nickel-based superalloy. (b) Amorphous region between two particles in shock-consolidated silicon carbide. (Courtesy of S. S. Shang.)

viously necessary. However, molten regions have been identified in SiC, and one such region (between two particles) is shown in Figure 17.41(b). The region shown in the transmission electron micrograph is primarily amorphous whereas the material was initially polycrystalline. The thickness and orientation of the boundary are shown by the two arrows. The amorphous region has a thickness of approximately 0.5 μm, shown by the transverse arrow; within this amorphous region small crystals nucleated, probably during cooling.

The primary objective of the jet is to cleanse the surfaces of their oxides, and metals and nonoxide ceramics need to undergo this process. Nevertheless, oxide ceramics and diamond (which possesses a gaseous oxide) do not necessarily require melting for good bonding.

The most insidious problem in shock consolidation is cracking of the compacts. This problem is more severe for hard materials; these are the materials for which shock consolidation has the greatest potential. In order to minimize these problems, different approaches have been implemented. Preheating the compacts so that they are more ductile and soft is one approach (Wang et al. [54], Ferreira et al. [75, 91], Shang et al. [92]). Akashi and Sawaoka [77] introduced a novel method for shock consolidating ceramics. It uses shock-induced reactions to produce localized heating (at particle surfaces) and thereby enhance bonding. Figure 17.42 shows the sequence of events in reaction-assisted shock consolidation. A ceramic or intermetallic (A_xB) is mixed with elemental C and D. The shock wave triggers the reaction (exothermic) between C and D, which helps to bond the A_xB particles. This method has been successfully applied to aluminides and silicides by Yu and Meyers [86] and Yu et al. [87].

17.7.4 Theory

The theoretical understanding of shock processing of powders is still incomplete. Graham [89] has proposed the name CONMAH, which represents the various contributing factors in shock compression processing of porous materials. The acronym CONMAH (see Chapter 8, Section 8.1) means

CONfiguration
↓
Mass mixing
↓
Activation
↓
Heating

There are changes in configuration taking place at the shock front; intense plastic deformation and localized kinetic energy can produce mixing of the powders (if they have different compositions). These effects produce high defect concentrations that activate the powders and can render them more reactive.

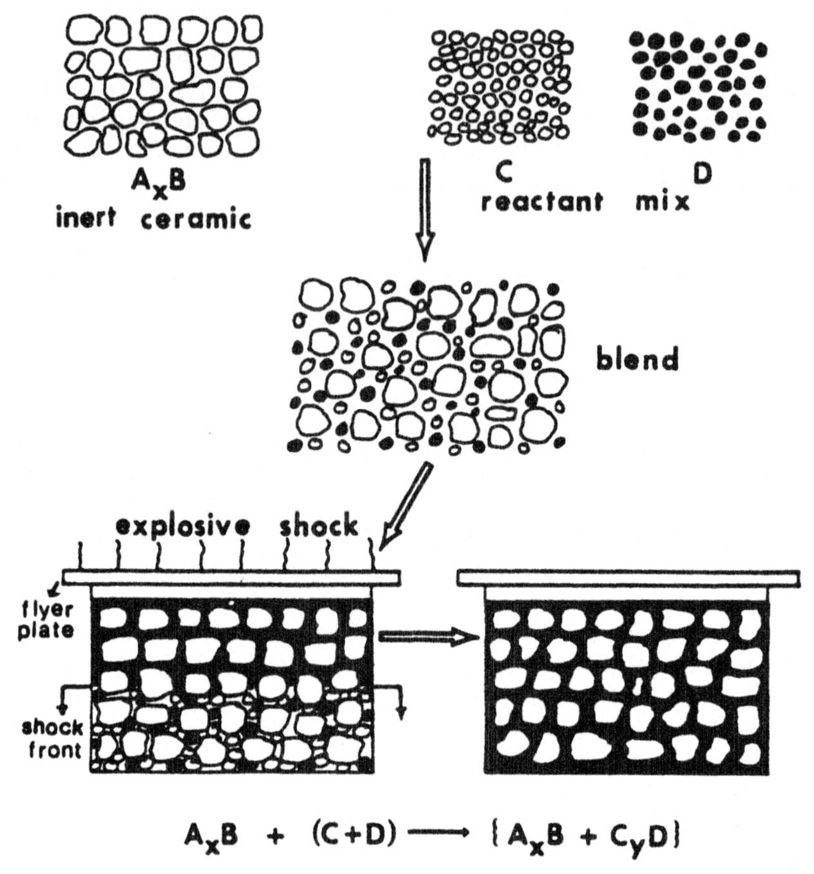

FIGURE 17.42 Sequence of events in chemically assisted shock consolidation of powders. From Yu et al. [87] Fig. 1, p. 303. Reprinted with permission of the publisher.)

Intense localized heating accompanies these processes. As a result, one can expect reaction kinetics that are higher than the ones observed in conventional processes.

In Chapter 5 we provide a thermodynamic treatment of shock wave passage through porous inert and reactive materials (powders). We will look in more detail, at the particle level, at what happens during the passage of the shock wave. Figure 17.43 shows the various modes of energy dissipation of a shock wave in a powder. The central portion of the figure shows the initial, uncompacted powders, whereas the different mechanisms leading to the final consolidated material are displayed in the periphery. The powder surfaces are accelerated into the pores at high velocities, impacting each other, with frictional energy release. This leads to melting of the surface regions with the associated bonding once this material is resolidified. If brittle materials are consolidated, particle fracture also occurs, leading to the filling of the gaps. As explained in

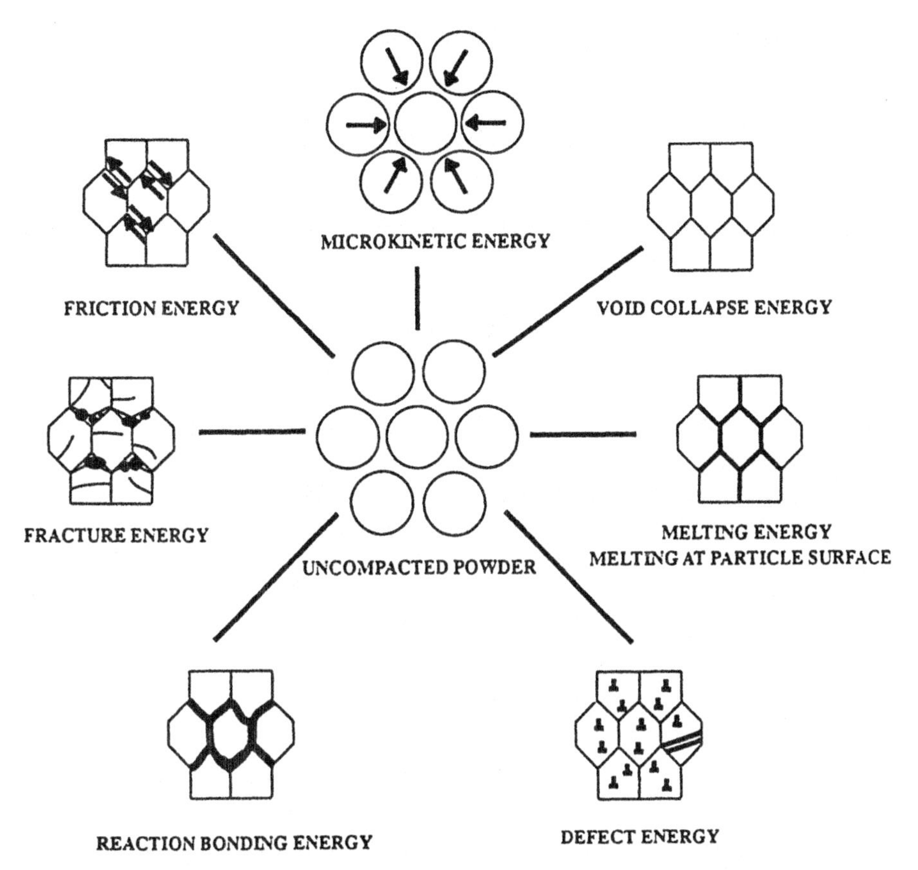

FIGURE 17.43 Various modes of energy dissipation in shock compression of powders. (From Meyers et al. [74], Fig. 2. Reprinted with permission of the publisher.)

Figure 17.42, reactive elements can also be added to help the bonding process. Schwarz et al. [97] developed a model in which the energy to melt a certain fraction of the particles is computed. From this energy, the pressure can be calculated at a certain porosity. They also calculated a shock pulse duration required for the interparticle melt region to resolidify. They considered that this solidification time has to be lower than the required pulse duration. The melting energy is assumed to be provided by total conversion of the shock energy into thermal energy at the particle surfaces. The shock energy is (see Chapter 5)

$$E_s = \tfrac{1}{2}P(V_{00} - V_0) = \tfrac{1}{2}PV_0(1 - \alpha) \tag{17.21}$$

where V_{00} is the initial specific volume of the powder equal to the specific volume at zero pressure of the solid material, V_0, multiplied by $1 - \alpha$, where

α is the distension (V_{00}/V_0). The energy required to heat the material and melt it is

$$E_m = C_p(T_m - T_0) + H_m \tag{17.22}$$

where C_p, the heat capacity, is assumed to be constant; H_m is the heat of melting; and T_m is the melting temperature. The melting fraction L is the ratio between E_s and E_m:

$$L = \frac{PV_0(\alpha - 1)}{2[C_p(T_m - T_0) + H_m]} \tag{17.23}$$

Schwarz et al. [93] were able to calculate the fraction of interparticle melting for a material as a function of distension α and pressure P. They also obtained the minimum pulse duration by computing a solidification time for the interparticle melt layer, t_s, and a cooling time, t_c, required for the interparticle layer to acquire sufficient mechanical strength. Thus, the pulse duration t_d must be

$$t_d \geq t_s + t_c$$

where

$$t_s = \frac{\pi\rho}{16K_m}\left[\frac{LdH_m}{(T_m - T')}\right]^2 \tag{17.24}$$

where d is the particle diameter (Ld is the thickness of the molten layer), K_m the thermal conductivity, and T' the temperature as $x \to \infty$. The above equation is for a semi-infinite slab having a layer of thickness $\frac{1}{2}L$ and temperature T_m.

The predictions of the model are compared with experimental results in Figure 17.44. The three boundaries that define the shock consolidation window are given by the above-discussed equations. The full points (circles and squares) denote good consolidation, whereas the hollow circles denote poor consolidation. Thus, the model of Schwarz et al. [93] predicts optimum consolidation conditions.

An important aspect is not considered in Schwarz et al.'s [93] model. A considerable amount of energy is consumed in plastically deforming the powders if their strength is high. This geometrically necessary plastic deformation is required to collapse the voids. The energy required to effect this plastic deformation does not contribute integrally to melt the interparticle regions, since it is more homogeneously distributed throughout the particles.

Carroll and Holt [94] developed a constitutive equation to describe the densification of porous materials. The Carroll-Holt model [94] is based on the collapse of a hollow sphere loaded, on the external surface, by a hydrostatic pressure P. The porosity of the actual material is represented by a distention, α, that can be calculated from the dimensions of Figure 17.45(a) (the densities

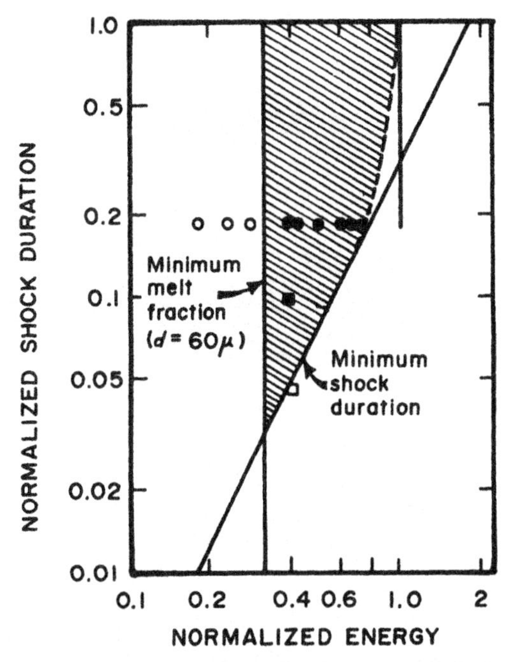

FIGURE 17.44 Map of dynamic consolidation for an AISI 9310 iron-based powder. The theory predicts that compacts with good mechanical properties are obtained when operating within the shaded area. (Reprinted from *Acta Met.*, vol. 32, A. B. Schwarz et al., Fig. 7, p. 1249, Copyright 1984, with permission of Pergamon Press Ltd.)

for solid and porous specimens are ρ_s and ρ_p respectively):

$$\alpha = \frac{\rho_s}{\rho_p} = \frac{b^3}{b^3 - a^3}$$

The viscoplastic Carroll-Holt model was expanded by Carroll et al. [95] to incorporate thermal effects. Thermal softening can be, as shown in Chapter 15, very important. Nesterenko [96] developed a model with greater physical basis (Fig. 12.45(b)). It consists of a solid sphere surrounded by (and concentric with) a hollow sphere. The hollow sphere collapses on the solid sphere; the size of the hollow sphere provides the scale for the events (in powders, it is equivalent to the diameter of the particles). This model correctly predicts that only on portion of the powder undergoes plastic deformation (the external shell). The internal solid sphere simulates the interior of the particles, that undergo negligible plastic deformation during the process of collapse. The dimensions of the spheres are defined as a function of α_0, the distention, in Nesterenko's [96] model; the pore volume should be equal to the volume of plastically deformed material in the collapse stage.

$$b_0 = a_0\alpha_0^{1/3}, \quad c = a_0(2 - \alpha_0)^{1/3}$$

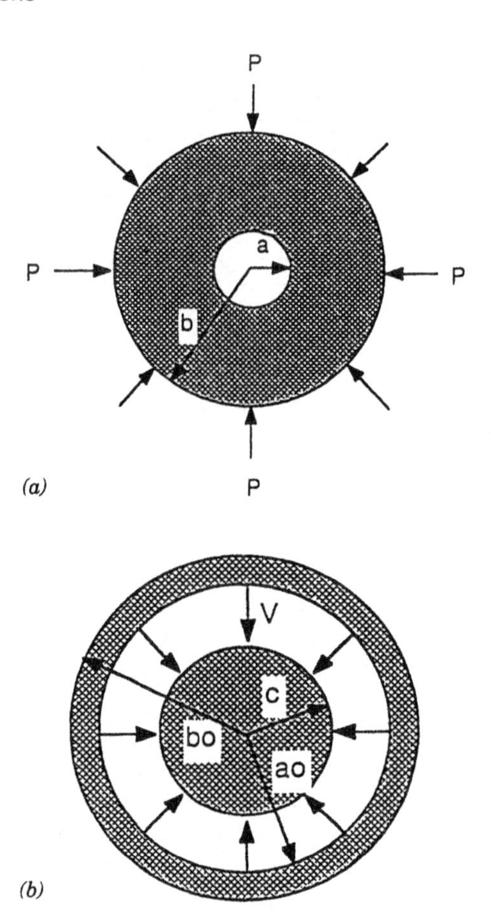

FIGURE 17.45 (a) Carroll-Holt [94] model of hollow sphere used in analytical modeling of densification of porous material; (b) Nesterenko's [96] modification incorporating a solid core onto which the hollow sphere is accelerated.

a_0 is the characteristic particle dimension. The Carroll-Holt formulation has the following form, for an ideal plastic material:

$$E_{vc} = \tfrac{2}{3}\sigma_y V_0\{[\alpha_0 \ln \alpha_0 - (\alpha_0 - 1) \ln (\alpha_0 - 1)] - [\alpha \ln \alpha - (\alpha - 1) \ln (\alpha - 1)]\} \qquad (17.25)$$

Where E_{vc} is the energy required to collapse a hollow sphere from an initial distention α_0 to a distention α. An analogous expression can be obtained from the Nesterenko model using $\alpha_0/\alpha_0 - 1$ instead of α_0 and $\alpha/\alpha_0 - 1$ instead of α. For total collapse of voids, from Eqn. 17.25:

$$E_{vc} = \tfrac{2}{3}\,\sigma_y V_0[\alpha_0 \ln \alpha_0 - (\alpha_0 - 1) \ln (\alpha_0 - 1)] \qquad (17.26)$$

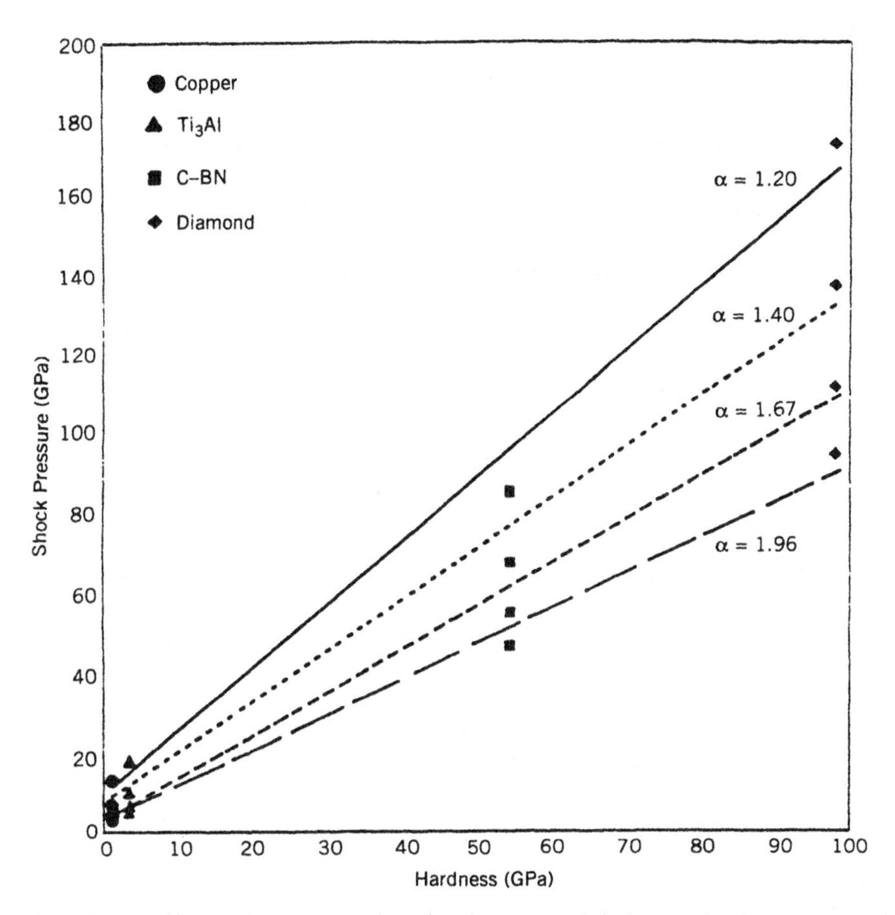

FIGURE 17.45 (c) Pressure required for shock consolidation vs. hardness at several distentions. (From Ferreira and Meyers [93], Fig. 5, p. 369. Reprinted by courtesy of Marcel Dekker, Inc.]

Tong and Ravichandran [97] applied the Carroll-Holt analysis to a material whose rate dependence is described by the Klopp-Clifton-Shawki constitutive equation (Eqn. 13.2, p. 327) and obtained predictions for pore collapse dynamics which differed significantly from the original Carroll-Holt predictions that considered only inertia. This shows that the constitutive behavior of material (strain-rate sensitivity, thermal softening, and work hardening) are important in the consolidation.

There are numerous other constitutive equations available, and the model developed by Helle et al. [98] is an example of a simple, yet physically realistic model:

$$E_{vc} = \frac{2.97\sigma_y}{1-\rho_0}\left[\frac{(-\rho_f^2 - \rho_0^2)}{2} + \rho_f\rho_0\right] \tag{17.27}$$

ρ_0 and ρ_f are the initial and current density of the material. The model by Helle et al. [98] is based on the plastic deformation of a surface by a cylindrical punch and contains, as do the Carroll-Holt and Nesterenko constitutive equations, some important physical assumptions. It is only through large-scale computational programs, such as the ones used by Williamson [99], and Benson and coworkers [100, 101] that the physical details of the deformation are captured. Ferreira et al. [91] calculated the energy required to to collapse the voids from Carroll and Holt's equation. They equated the shock energy [Eqn. (17.21)] to

$$E_{vc} = \tfrac{2}{3}\,\sigma_y V_0[\alpha_0 \ln \alpha_0 - (\alpha_0 - 1)\ln(\alpha_0 - 1)] \qquad (17.28)$$

They were able to predict shock pressures that are required for the consolidation of powders of different hardnesses. The results are shown in Figure 17.45(c). For the ceramic powders with the highest hardness (diamond and boron nitride), they assumed no interparticle melting. The results obtained by Ferreira and Meyers [103] are in good agreement with experimentally obtained pressures and provide a good groundline for the prediction of shock energies to consolidate powders. This can be seen by comparing Figures 17.35 and 17.44; to convert the flow stress into hardness, one multiplies it by 3 (this is an approximate conversion). Meyers, Shang, and Hokamoto [102] improved the model of Ferreira and Meyers [103] by considering melting as a result of the frictional, plastic deformation, and microkinetic energy processes. Nesterenko [96] was the first to propose the microkinetic energy as a new fundamental mesoscopic parameter for shock consolidation, which determines the process of plastic deformation *localization* on the particle surfaces. He also postulated the scaling of shock energy according to E_{vc} (instead of E_m in [93]) and introduced the determination of quasistatic and dynamic regimes of particle deformation in shock waves. For $\alpha_0 \sim 2$, it is possible to obtain an engineering criterion for consolidation: pore collapse requires $P > HV$ and good bonding requires (HV is the Vickers hardness of the material) $P > 2H$.

There are considerable problems with shock consolidation that have impeded it from finding a technological application. The complete bonding of the powder surfaces is difficult to accomplish, and there is usually a variation in the plastic deformation pattern around a particle leading to temperature fluctuations and inhomogeneous melting. Thus, often the compacts are only partially bonded, and unbonded regions are flaws which weaken the compact. This incomplete bonding problem is compounded by cracking resulting from tensile stresses generated during the shock consolidation process. The activation of existing flaws is dictated by the well known fracture mechanics equation (described in detail in Chapter 16):

$$K_{IC} = \sigma\sqrt{\pi a}$$

K_{IC}, the fracture toughness, is a material parameter. Figure 17.46(a) shows the flaw size, $2a$, plotted against the maximum tensile stress, σ, for materials with

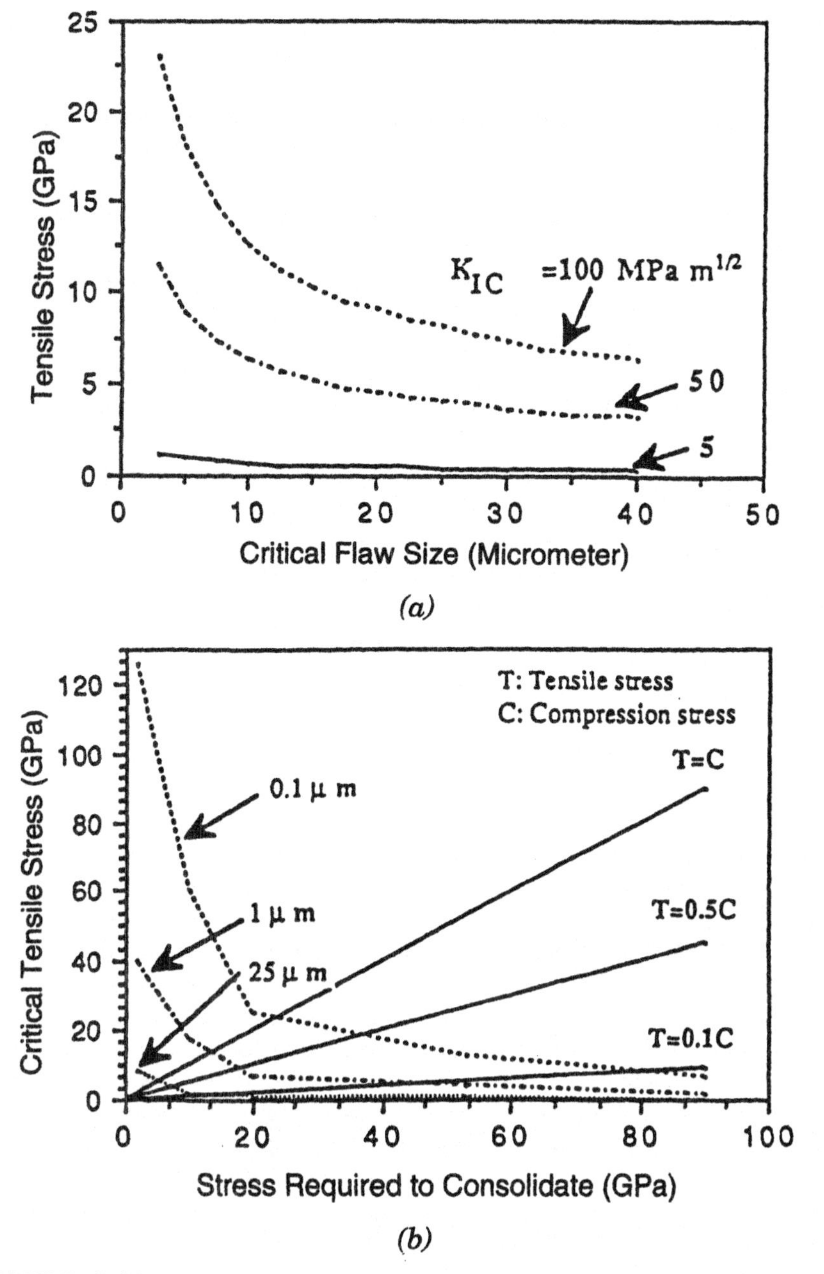

FIGURE 17.46 (a) Critical tensile stress vs. crack size for materials exhibiting different fracture toughnesses
(K_{IC} = 5 MN $m^{3/2}$-ceramic;
K_{IC} = 50 MN $m^{3/2}$-metal;
K_{IC} = 100 MN $m^{3/2}$-ductile metal)
(b) Critical tensile stress as a function of pressure required for shock consolidation for powders with different size.

three fracture toughnesses: 5, 50, and 100 MPa $m^{1/2}$. These values are characteristic of ceramics (5 MPa $m^{1/2}$), tough metals (steel, titanium alloys: 50 MPa $m^{1/2}$) and ductile metals (copper, nickel). Figure 17.46(b) shows the stresses required to consolidate these powders, in the abscissa. The values are calculated for a distention of 65% by the procedure described by Ferreira and Meyers [102]. For a known flaw size taken to be equal to the particle size (since we assume unbonded regions having the size of a particle) and fracture toughness it is possible to calculate the maximum tensile stress that the specimen will withstand without fracture. This is shown in the ordinate of Figure 17.46(b). Three curves are shown for different flaw sizes ($2a$): 25, 11, and 0.1 μm. For instance, SiC requires a pressure of $\sim$20 GPa for shock consolidation. If the powder size is 25 μm, a tensile stress of 0.8 GPa will generate failure. On the other hand, if the particle size is reduced to 0.1 μm, a tensile stress of 20 GPa will be needed for failure. The tensile stresses generated by the reflecting of the compressive pulses in the fixtures are often present, and it is very difficult to eliminate them. The various straight lines drawn in Figure 17.46(b) correspond to different levels of tensile stresses: 10%, 50%, 100% of the compressive stresses. If a point for a material falls below the chosen line, the compact should fail. If it falls above, the compact should retain its integrity. Therefore, the following are important considerations in shock consolidation:

1. Minimization of reflected tensile pulses.
2. Reduction in particle size. In the limit, nanocrystalline particles ($\sim$0.05 μm) would be more effective because of the smaller flaw size.
3. Reduction in shock energy.

Reduction of energy can be accomplished by shock consolidating at elevated temperatures [e.g., 54 and 75] and by adding chemically reactive mixtures to the powders [e.g., 77 and 87]. Hokamoto et al. [104] used hot shock consolidation to considerably improve the quality of diamond and boron nitride compounds. Ahrens et al. [105] used graphite in addition to diamond to aid in the consolidation of the latter and Kunishige et al. [106] used a reactive mixture to bond diamond. Kondo and Sawai [107–109] were successful in consolidating nonocrystalline diamond. The recovery fixtures developed by Mutz and Vreeland and coworkers [110] minimize reflections and result in improved compact quality. The diameter of the flyer plate (in the gas-gun setup) is equal to the powder container and this eliminates lateral waves. A similar setup was developed by Korth et al. [111]. They used a "piston" concept, that pushes directly into the powder container. The above examples address the three considerations enumerated above with different degrees of success.

17.7.5 Shock-Induced Phase Transformations and Synthesis

The industrial production of diamonds by Dupont using shock-induced transformation from grpahite is a unique process [57, 112]. This breakthrough has inspired a great deal of additional research. The Dupont process of producing diamonds consists of mixing graphite powder with copper powder in large cylindrical containers (several meters high). These containers are imploded with a flyer-tube arrangement (see in Figure 17.37). The high pressure generated by the shock wave (10–30 GPa) and temperature generated by the collapse of the voids creates a regime in which diamond is stable (see Figure 8.9, p. 216). The transformation from graphite to diamond is martensitic in nature. The diamonds produced by this method are not cubic, but hexagonal. They are polycrystalline, with a grain size of 100 A, forming particles with 60–90 μm. Each particle contains from 10 to 10,000 grains. This polycrystalline diamond has found applications as a polishing compound for specialized applications, such as sapphire, ceramics, and ferrites. The individual particles, classified into size groups, consist of polycrystalline aggregates, and the fracturing process during polishing is affected by the polycrystallinity. In Japan, Yoshida and coworkers [113–115] have investigated the transformation of graphite to diamond and have established that, in the absence of copper, most of the diamond reverts to graphite upon the release of the shock wave. The function of the copper in the mixture is to provide a local quenching medium. The diamond in contact with copper, at high pressure, is cooled more rapidly upon release and is therefore retained at ambient pressure. The yield of diamond has been increased to 70% by using a mixture of 10% graphite and 90%wt copper, spherical particles with ~100 μm diameter were used. Yoshida and Fujiwara [115] also report the formation of C—B—N compuonds by shock. The formation of wurtzitic BN from graphitic BN was accomplished by Adadurov, Dremin, Breusov and coworkers [116] and this has led to the industrialization of this process in Japan, by Nippon Oil and Fats. The wurtzitic BN is subsequently transformed into cubic BN by static pressure, and is at the same time consolidated. The synthesis of ultra-fine diamonds from the direct detonation of carbon-rich explosives has been carried out in Russia by several groups [117–119]. This process was initiated by Adadurov in 1972, and Adadurov, Breusov, and Dremin, in 1973. These dimaonds have a cubic structure and are not the result of a martensitic transformation, but of a nucleation-and-growth process from the vapor phase. It is also possible to form ultra-fine diamonds by mixing the explosive with graphite. It the latter case, the diamonds are somewhat larger and the diamonds could be produced from the liquid C phase. Dremin and coworkers [117] have developed a methodology to produce ultra-fine diamonds in a recovery chamber where the explosive charge is placed in the center and the products are recovered at the bottom. Explosive charges can be remotely loaded and detonated every 100 seconds. Titov [120], Liampkin, and Staver [119], and Petrov [121] have developed similar production capabilities.

Solid solutions of boron nitride and carbon as well as carbon nitrides present the possibility of very high hardness. Efforts at synthesizing these materials have been made by Yoshida and coworkers [115] as well as Sekine and co-workers [122, 123], and Sawaoka and coworkers in Japan. A new crystalline form of carbon, produced by shock compression of graphite to 65 GPa and 3700 K, has been reported by Hirai and Kondo [124].

The synthesis of compounds from chemical reaction, described in Section 8.7 (p. 219ff) can also be triggered by shock waves, and numerous compounds have been produced. A unique aspect of shock compression is that under some conditions, metastable phases can be produced. This is possible when the energy of reaction is sufficiently low so that the release of the shock pulse can, by virtue of its high rate, quench-in non-equilibrium phases from the shocked state. These reactions have been reviewed by Thadhani [125], Batsanov [126] and Meyers [129]. Batsanov's book [126] contains a complete account of the Russian research on shock chemistry. Vecchio et al. [139] studied the Nb—Si reaction between elemental Nb and Si. The reaction powders proceeds as:

$$Nb(s) + Si(l) \rightarrow NbSi_2(l)$$

There is a minimum energy termed by Krueger and Vreeland [132, 133] "threshold energy," at which the reaction is initiated. This energy corresponds to a pressure P.

$$E_{th} = \tfrac{1}{2}(P + P_0)(V_{00} - V) = 600 - 800\ \mathrm{J/g}$$

for the Nb—Si and Mo—Si reactions.

It was proposed by [130, 131] that this energy corresponds to the melting of a fraction of the silicon. The reaction proceeds as shown in Figure 8.14(b): $NbSi_2$ (in liquid form) is formed at the interface between solid niobium and liquid silicon and forms spherules. The sequence of events is depicted schematically in Figure 17.47. After the reaction, initially forming a liquid $NbSi_2$ layer (17.47(a)–(c)) between the two reactants, has proceeded sufficiently, the interfacial energy favors the formation of spherules with $\sim 1\ \mu m$ diameter (Figure 17.47(d)). These spherules then solidify (Figure 17.47(e)). Fresh reactive interface is created by the contraction of the reaction layer into a spherule (e). Subsequent spherules are formed adjacent to the first one (g). They exert forces on it, and contribute to the dissolution of the niobium surrounding it, eventually dislodging it from the interface (f). In such a manner, these spherules are formed and "float away" into the liquid silicon, as shown in Figure 8.14(b). These spherules are monocrystalline and faceted as shown from Figure 17.48(a). The mechanism described above is typical of the Nb—Si, Ti—Si, and Mo—Si systems. Different systems will exhibit different reaction mechanisms, dependent on the exothermicity of reaction, particle size, heat conduction, and other factors. Thadhani [125] provides a comprehensive treatment of the mechanisms involved. Two additional sources of information are the books by Graham [127] and Sawaoka and Horie [128]. The plastic deformation, superimposed on the shock deformation, favors the reaction further, and shear band regions in $MoSi_2$—Nb—Si mixtures subjected to shock compression at an energy level below E_{th} revealed reaction, as shown in Figure 17.48(b) [134].

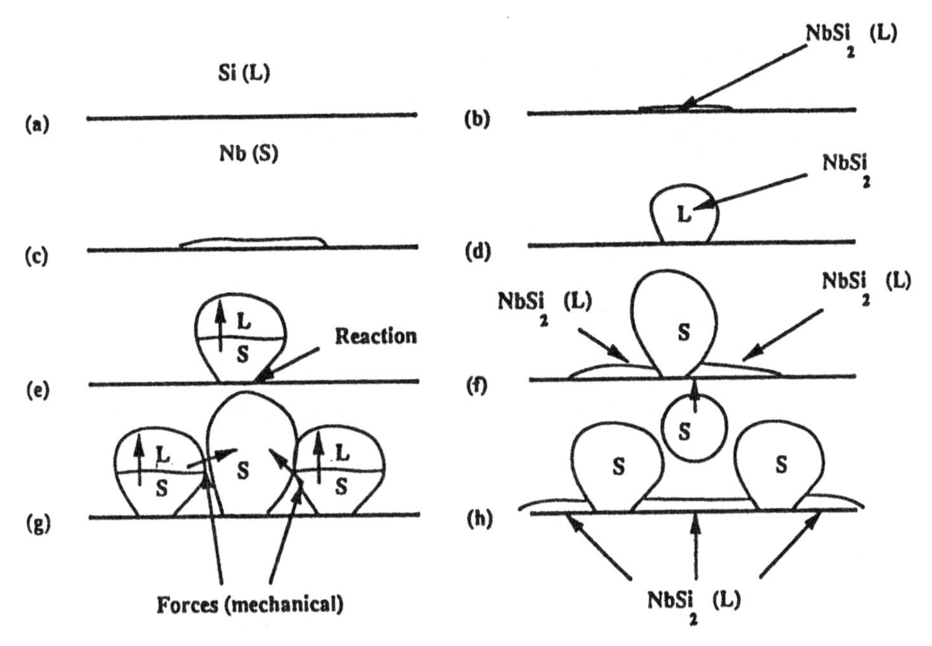

FIGURE 17.47 Sequence of events proposed by Meyers, Yu, and Vecchio [131] for the formation of $NbSi_2$ spherules from the reaction of Nb and Si powders under shock compression at an energy level of ~600–800 J/g. (a–c) nucleations and growth of thin layer; (d) interfacial energy producing spheroidization; (e) solidification of spherule due to slowing down of reaction; (f–h) formation of subsequent spherules and their ejection from interface due to dissolution and mechanical forces.

17.8 DYNAMIC EFFECTS IN GEOLOGICAL MATERIALS

Dynamic events are common on the surfaces of the earth and planets. The impacts of asteroids, meteorites, and comets on planetary surfaces produce craters and eject mass into space. The surface of the moon is a sequence of craters of different sizes, and the impact of space matter has been responsible for significant geological changes on the surface of the earth. The entire everglades area in southern Florida could have arisen from a large crater. Alvarez (the Berkeley Nobel laureate) and co-workers [135] proposed a theory for extinction of dinosaurs caused by a large meteorite impact in the Cretacious–Tertiary boundary. This large impact would have produced a large mass of ejecta that created a thick layer of dust around the earth. Evidence for this catastrophic event is produced by a high incidence of iridium and other noble metals, measured on geological strata corresponding to the Cretacious–Tertiary boundary. This iridium would have been present in the bolide and was ejected into the "dust cloud" settling later on the entire earth surface. This dust cloud would have shielded the surface of the earth from the radiative heating of the sun, resulting in significant drops of temperature that would alter the flora to

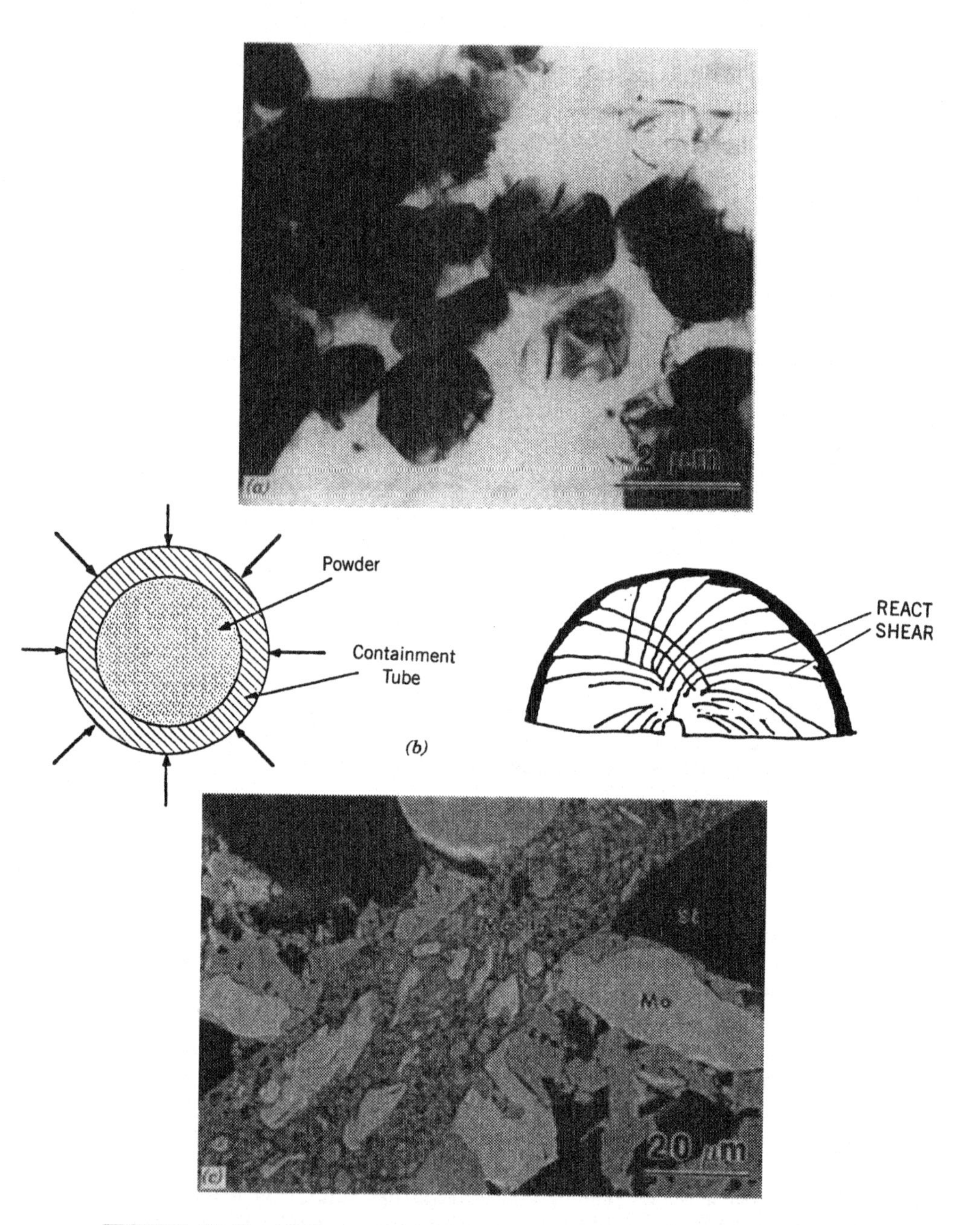

FIGURE 17.48 (a) Transmission electron micrograph of $NbSi_2$ spherules produced by shock synthesis (courtesy of K. S. Vecchio, UCSD); (b) Shear-band regions formed in shock consolidation of $MoSi_2$—MoSi reactive mixtures in cylindrical geometry—dark bands are reacted regions; (c) SEM of shear band showing three regions: white—Mo; black—Si; grey—$MoSi_2$ (back scattered image).

the extent that the food supply of large dinosaurs would be disrupted, leading to their extinction. Another possibility is that the impact would have produced chemical changes, such as large releases of CO_2 or NO from the impact with carbonate rocks. We will briefly discuss these physical and chemical changes below.

O'Keefe and Ahrens [136, 137] performed calculations in order to establish the required meteorite mass and velocity to produce an ejecta of sufficient mass to generate the "dust cloud." Melosh [138] is the author of a book entirely devoted to impact cratering. Hydrocode (see Chapter 6) calculations are needed to predict pressures, plastic deformation, fracturing, and ejecta mass and velocity in these impact events. As an illustration, Figure 17.49 shows the pressures generated, at different times (0.33τ, 0.99τ, 3.03τ, 6.11τ, where τ = projectile diameter/impact velocity). The calculations were conducted by O'Keefe and Ahrens [140, 141], for a 46.4-km-diameter iron projectile impacting gabbroic anorthosite (a geological material) at 15 km/s. Pressures as high as 420 GPa are observed at the impact interface. (Note that the length

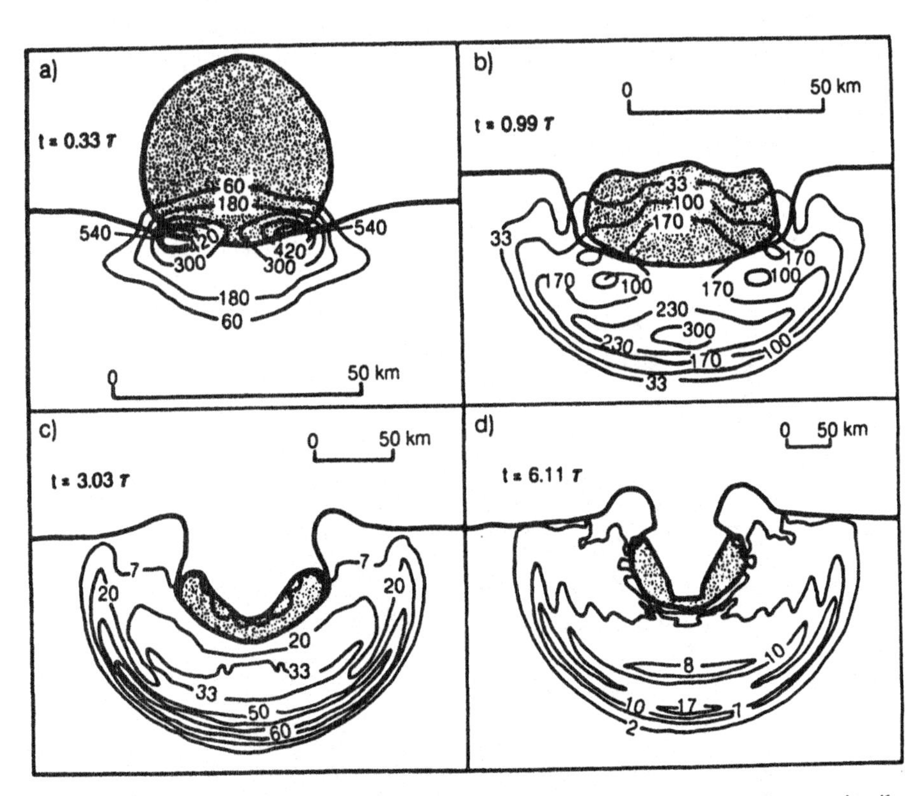

FIGURE 17.49 Pressures generated by impact of 46.4-km-diameter iron projectile against geological material at a velocity of 15 km/s; pressure contours are indicated in GPa; please note that length scale changes. Times are indicated in units of τ, where τ = projectile diameter/projectile velocity. (From O'Keefe and Ahrens [139].)

scales in Figure 17.49 were changed to accommodate the expanding pressure wave.) O'Keefe and Ahrens [136] concluded, from their calculations, that a bolide (asteroid, comet, or meteorite) with a diameter of 10 km impacting the earth at velocities between 15 and 45 km/s would produce an ejecta mass between 10 and 100 times the bolide mass. The formation of a global dust cloud would require particles with diameters less than 1 μm, and O'Keefe and Ahrens [136] calculated the total mass of these particles and estimated it to be 1–20 times the bolide mass. Figure 17.50 shows the effect of the ejecta; these calculations are due to O'Keefe and Ahrens [141]. They show two impact velocities for a projectile: (1) 15 km/s and (2) 45 km/s. The velocity and mass of ejecta increase drastically with impact velocity. Figure 17.50 shows that the ejecta velocity increases from 4 to 20 km/s. Ahrens [142] discusses this important point in the context of planetary accretion, that is, the effect of impacts on planetary mass. If the mass of the impactor is lower than the mass of the ejecta expelled at a velocity higher than the planetary escape velocity, there is a net decrease in mass as a result of the impact; this results in planetary erosion. If the mass of ejecta is lower than the mass of the impactor, planetary accretion takes place. It is estimated that dust cloud with a mass of 10^{16} g uniformly distributed over the earth's atmosphere would reduce photosynthesis by a factor of 10^5. A 10-km bolide possesses a mass of $\sim 10^{18}$ g, and therefore O'Keefe and Ahrens [136] concluded that the fine ejecta (<1 μm) mass was more than sufficient to produce an effective dust cloud.

The chemical changes produced by high-velocity impact are also very important. In Chapter 8 we studied chemical reactions induced by shock waves (Section 8.7). One possible chemical change with significant consequences to

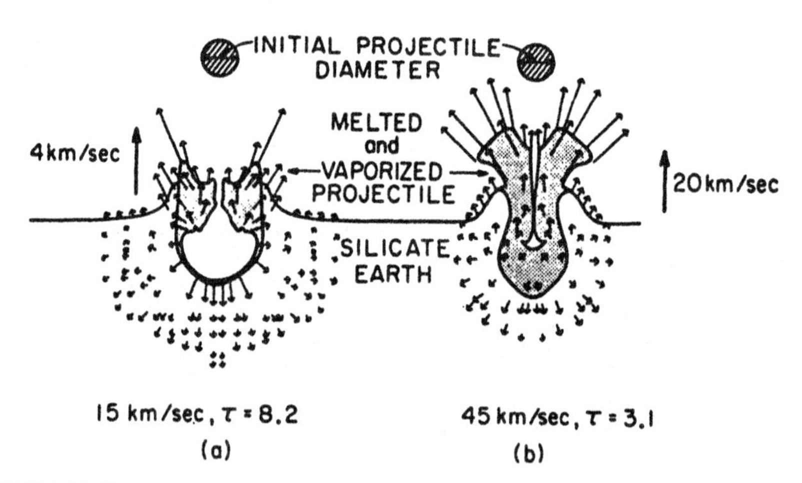

FIGURE 17.50 Particle velocity flow field (length of vector proportional to velocity) and molten region produced by impact of silicate projectile on silicate target at velocities of (a) 15 km/s and (b) 45 km/s. (Reprinted with permission from *Nature*, O'Keefe and Ahrens [104], Fig. 1, p. 124. Copyright 1989 Macmillan Magazines Limited.)

the global atmosphere, produced by the impact of the bolide during the Cretacious–Tertiary boundary, is the decomposition of carbonate rocks by the shock pressure. Lange and Ahrens [143] developed the following expression for the volume of vapor, V_v, generated from an impact:

$$\frac{V_v}{V_m} = 0.132 \left(\frac{\rho_m}{\rho_s}\right) \frac{U^2}{\Delta E} \tag{17.29}$$

where V_m is the volume of the impactor, ρ_m and ρ_s are the densities of impactor and planet surface, respectively, U is the impactor velocity, and ΔE is the energy of vaporization (per unit mass). For $CaCO_3$, it was estimated to be [144]

$$\Delta E_{CaCO_3} = 9.5 \times 109 \text{ erg g}^{-1}$$

Lange and Ahrens [143] conducted experiments in which they determined the amount of CO_2 produced from shock impact on calcite (calcium carbonate) rock. These results are seen in Figure 17.51. Volatilization starts at ~10 GPa,

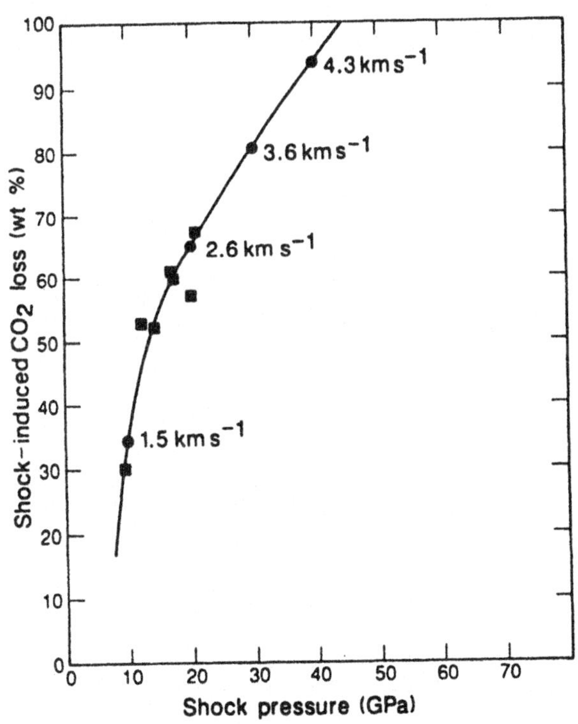

FIGURE 17.51 Fraction of CO_2 released from calcium carbonate as a function of shock pressure; corresponding impact velocities (silica impactor) indicated. (From Lange and Ahrens [103], Fig. 1, p. 412.)

and at 50 GPa there is full volatilization. The greenhouse effect (heating of the earth surface by increased CO_2 in the atmosphere) has been estimated to be given by

$$\Delta T = 10 \log \frac{m}{m_0} \tag{17.30}$$

where m is the CO_2 mass in the perturbed atmosphere and m_0 is the original CO_2 mass. Lange and Ahrens [143] estimated that a 10-km-diameter comet impacting the earth at 20 km/s (it can be shown from Fig. 17.51 that full volatilization is produced at this impact velocity) would produce a temperature increase $\Delta T = 10$ K. This would be sufficient to disrupt the biota and could account for the extinction of dinosaurs in the Cretacious–Tertiary boundary (65 Myears ago).

The collection of mineral specimens from impact craters has also yielded very valuable results, after analysis, and unique high-pressure phases (such as stichovite in SiO_2) have yielded direct evidence of the high pressures generated upon impact. Metallic meteorites (often, Fe–Ni) also exhibit deformation substructures (twins and dislocations) indicative of shock processes (see Chapter 14). The high pressures generated upon impact can lead to melting and vaporization of projectile and target. Figure 17.52 illustrates the pressure falloff from

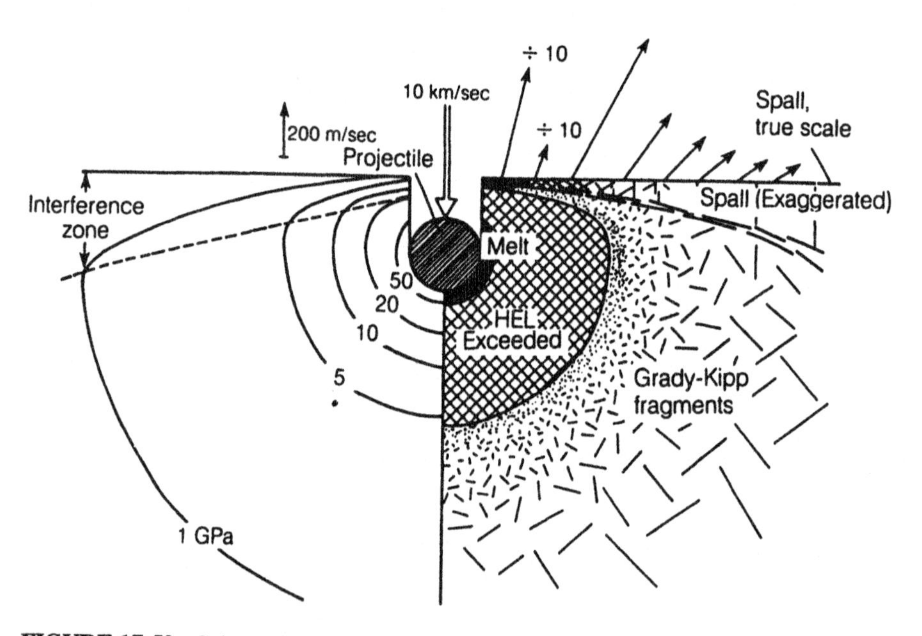

FIGURE 17.52 Schematic illustration of different effects caused in geological material as a function of distance from impact region. From Melosh [138], Fig. 5.4, p. 64. Reprinted with permission of the publisher.)

the impact point in an impact of 10 km/s against a target. Four regions are indicated, which show, in a schematic fashion, different processes occurring in the rock:

I. Melting
II. Region where pressure exceeds HEL
III. Grady–Kipp tensile failure region
IV. Spalling region

An additional geological application where high-pressure shock research has found a fertile ground is in the simulation of the environment in the core and mantle of the earth. It is possible to generate much higher pressures by impacts than the ones statically achieved. By obtaining, through shock wave experiments, the equations of state for the geological materials comprising the earth mantle and core, it is possible to enhance the understanding of the composition and pressure of these regions. The earth core, which extends from a depth of 2890 km to the center (at 6371 km), is subjected to pressures varying between 136 and 364 GPa (O'Keefe and Ahrens [141]). Shock experiments can access the pressure of 136 GPa and predict a density of 11 g/cm^3 for iron (the core material) at that pressure. The solid–liquid boundary in the core can equally well be calculated. A number of important pressure-induced effects, such as phase transformations in geological materials, have been investigated by shock techniques by Ahrens, Jeanloz, and colleagues [145, 146].

Mining and construction operations use, on a routine basis, blasting to fragment rock and move the earth. This constitutes a field of great technological importance, and we will provide here only a very preliminary view. Figure 17.52 shows a typical blasting setup. A hole is drilled in to the ground and an explosive is placed in it. The hole is sealed with the stemming (a grout), which ensures buildup of pressure in the hole after detonation. Blasting operations are usually conducted adjacent to a free face, and the fragment rock is pushed to the right. This material is then removed, and a new free face is then formed. The procedure can then be repeated. In blasting operations, the partitioning of the explosive energy between the initial shock pulse and the pressure produced by the gases generated by explosion is very important. Hard rock requires explosives with a high shock pressure, while in softer rock the energy should be provided mostly by the gas pressurization. Figure 17.53(b) shows the ground after the detonation of the explosive charge. Three regions are clearly distinguished: (1) a crushed zone, in the immediate vicinity of the borehole (it is produced by the compressive shock wave); (2) a region consisting of radial cracks, produced by the tensile tangential stresses behind the shock wave front; and (3) a slabbing region, produced by interaction with the free face (tensile reflection). Additionally, the high pressure in the borehole and in the cracks (when the detonation products penetrate the fractures) results in movement of the entire rock with additional fracturing. In field applications large arrays of boreholes are detonated at precisely calculated intervals in order to maximize

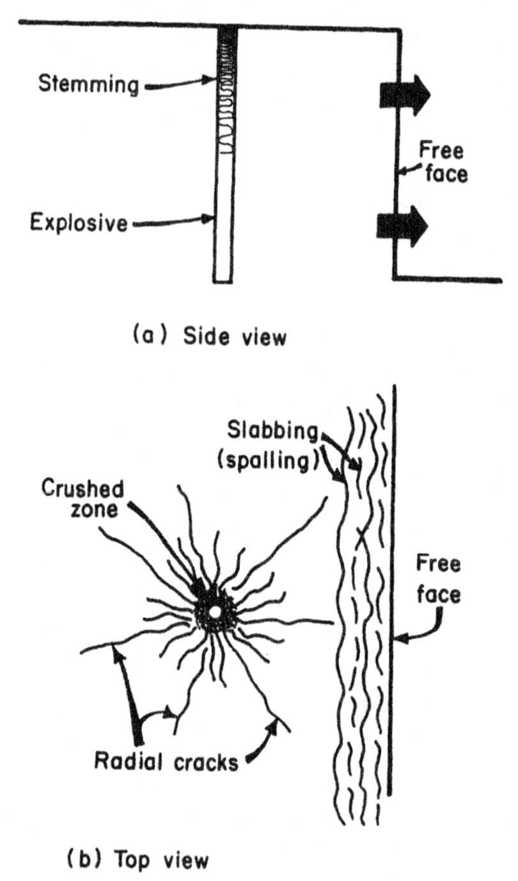

FIGURE 17.53 Blasting of rock for mining and construction operations: (a) side view prior to detonation of explosive; (b) top view after detonation.

rock breakage. Readers should consult books on blasting in order to obtain additional information (e.g., Dowding [147]). Aimone et al. [148] studied the effects of shock loading on the fragmentation of quartz monzonite rock and found that it is dependent on both pressure and pulse duration (see Chapter 16).

17.9 DYNAMIC EVENTS IN SPACE

A multitude of dynamic events take place in space, and the craters on the surface of the moon are an eloquent reminder of high-velocity impact and crater formation. These impacts are often accompanied by shock-induced phase transformations, and the high-pressure phase of silica, named after its discoverer, Russian Academician Stishov (Stishovite), is formed in areas surrounding large impact craters on earth. Meteorites impacting the earth often also bear the marks of impact and the high pressures.

Spacecraft are subjected to impact, and the velocities vary from 5 to 50 km/s. Special protection shields have been developed for them, and the "bumper" shields used by the Giotto probe sent into the Haley comet are an example. The small particles traveling at a velocity of 75 km/s are fragmented and vaporized during impact with the "shield" and a plume is formed in which the particles have a much lower velocity (although the mass is increased by the accretion of the bumper material mass) [149]. A variety of materials for the shield were tested, and a front plate of Al (1 mm thick) with a rear plate of Kevlar (13.5 mm thick) with a spacing between them (25 cm) presented optimum properties [149]. The frontal plate fragments the particle and partially vaporizes it; the rear plate "catches" the debris.

In March 1993, astronomers Gene and Carolyn Shoemaker and David Levy discovered a fragmented comet, consisting of spheres with 1–4 km in diameter in a pearl-necklace arrangement, in a trajectory around Jupiter. The fragmented comet consisting of approximately twenty chunks was bound to impact Jupiter on July 16–22, 1994. Several theories have been proposed for the effects of impact, and, as this book is being printed, the scientific community awaits anxiously the effects of the impact. Several major theories have been put forth for the sequence of events in impact, as reviewed by Time [150]. Large-scale computational codes are being used to predict the effects of impact, and the impact velocity will be of approximately 50 km/s. The comet fragmentation consists of a mixture of hydrogen and helium. The energy released by these sequential twenty impacts of 50 km/s is vastly superior to the one of the bolide impact in the tertiary-criterion boundary. The predictions vary, and the various hypotheses have been proposed:

a. A meteor shower, if the 1 km spheres disintegrate prior to penetration into Jupiter. The spheres might also crack into fewer fragments, in the Jupiter clouds.
b. Penetration into Jupiter, with the formation of an impact crater and a large ejecta mass. The theory of "soft catch" has been advanced by Ahrens; according to it, penetration of 300 km into Jupiter is feasible, with the formation of an ejecta cloud of over 2,000 km.

Example 17.6. The most powerful nuclear explosion on earth was carried out by the Russians and had the energy equivalent of 58 megatons of explosives. The impact of the Shoemaker-Levy comet with Jupiter would correspond to how many superbombs?

We will assume 20 spheres (1 km diameter) of ice impacting Jupiter at 50 km/s. We will only consider their kinetic energy and ignore subsequent fusion reactions.

Volume of each sphere:

$$V = \tfrac{4}{3}\pi r^3 = 5.23 \times 10^8 \text{ m}^3$$

Mass of each sphere (the density of ice is 10^3 kg/m^3):

$$M = V\rho = 5.23 \times 10^8 \times 10^3 = 5.23 \times 10^{11} \text{ kg}$$

Kinetic energy of twenty spheres:

$$E_c = 20 \times \tfrac{1}{2}Mv^2 = 10 \times 5.23 \times 10^{11} \times 2.5 \times 10^9$$
$$= 1.3 \times 10^{22} \text{ J}$$

From page 244, we can consider the chemical energy of explosive as 5 MJ/kg. Thus, the energy of the 58 megaton bomb is:

$$E_B = 58 \times 10^9 \times 5 \times 10^6 = 2.9 \times 10^{17} \text{ J}$$

The comet impact is equivalent to:

$$NR = \frac{1.3 \times 10^{22}}{2.9 \times 10^{17}} = 4.48 \times 10^4$$

Thus, it is equivalent to 44,800 supernukes. On this unauspicious note, we close the book.

REFERENCES

Shaped Charges

1. G. Birkhoff, D. P. MacDougall, E. M. Pugh, and G. I. Taylor, *J. Appl. Phys.*, **19** (1948), 563.
2. L. R. Carl, *Met.* Prog., **46** (1944), 102.
3. A. A. Deribas, in *Shock Waves and High-Strain-Rate Phenomena in Metals*, eds. M. A. Meyers and L. E. Murr, Plenum, New York, 1981, p. 915.
4. E. M. Pugh, R. J. Eichelberger, and N. Rostoker, *J. Appl. Phys.*, **23** (1952), 532.
5. S. K. Godunov, A. A. Deribas, and V. I. Mali, *Fizika Goreniya i Vzryva*, **11** (1975).
6. F. I. Grace, B. R. Scott, and S. K. Golaski, *Proc. 8th Intl. Symp. Ballistics*, ADPA, October 1984.
7. M. Raftenberg, U. S. Ballistic Research Laboratory, private communication, 1991.
8. F. I. Grace, in *Shock-Wave and High-Strain Rate Phenomena in Materials*, eds. M. A. Meyers, L. E. Murr, and K. P. Staudhammer, Dekker, New York, 1992.
9. P. C. Chou and J. Carleone, *J. Appl. Phys.*, **48** (1977), 4187.
10. P. C. Chou and W. J. Flis, *Propell. Expl. Pyrotech.*, **11** (1986), 99.

11. C. Fressengeas and A. Molinari, *Proc. Int. Conf. Mech. Prop. Materials at High Rates of Strain*, Inst. Phys. Conf. Ser. No. 102 (1989), 57.
12. D. E. Grady, *J. Impact Eng.*, **5** (1987), 285.
13. F. Jamet, "Methoden zur Untersuchung der Physikalischen Eigenshaften eines Hohlladungstrahles," Report CO 227/82, Institut St. Louis, France, December 1982.
14. A. H. Chokshi and M. A. Meyers, *Scripta Met.*, **24** (1990), 605.
15. M. A. Meyers, L. W. Meyer, J. Beatty, U. Andrade, K. S. Vecchio, and A. H. Chokshi, *Shock-Wave and High-Strain-Rate Phenomena in Materials*, Dekker, New York, 1992.
16. R. J. Eichelberger, in *Proc. 11th Intl. Ballistics Symposium*, American Defense Preparedness Assoc., 1989, p. 379.
17. M. Held, *Propell. Explos. Pyrotech.* **10** (1985), 125; **11** (1986), 170.
18. J. S. Reinehart and J. Pearson, *Behavior of Metals under Impulsive Loads*, ASM, OH, 1954.
19. G. Johnson and R. A. Stryk, *J. Impact Eng.*, **5** (1987), 411.

Penetrators

20. T. W. Wright, in *Computational Aspects of Penetration Mechanics*, eds. J. Chandra and J. E. Flaherty, Springer, 1983, p. 85.
21. W. Bruchey in *Computational Aspects of Penetration Mechanics*, eds. J. Chandra and J. E. Flaherty, Springer, 1983.
22. Backman and Goldsmith, *Intl. J. Eng. Sci.*, **16** (1978), 1.
23. J. A. Zukas, T. Nicholas, H. F. Swift, L. G. Greszczuk, and D. R. Curran, *Impact Dynamics*, Wiley, New York, 1982.
24. J. A. Zukas, ed., *High-Velocity Impact Dynamics*, Wiley, New York, 1990.
25. R. F. Bishop, R. Hill, and N. F. Mott, *Proc. Phys. Soc.*, **57** (1945), 147.
26. V. P. Alekseevskii, *Fizika Goreniya i Vzryva*, **2** (1966), 99.
27. A. Tate, *J. Mech. Phys. Sol.*, **15** (1967), 387; **17** (1969), 141.
28. D. R. Christman and J. W. Gehring, *J. Appl. Phys.*, **37** (1966), 1579.
29. A. Tate, *Intl. J. Impact Eng.*, **9** (1990), 327.
30. M. Ravid, S. R. Bodner, and I. Holoman, *Intl. J. Eng. Sci.*, **25** (1987), 473.
31. T. W. Wright and K. Frank, in *Impact: Effects of Fast Transient Loadings*, eds. W. J. Ammann, W. K. Liu, J. A. Studer, and T. Zimmermann, Balkema, 1988.

Armor

32. D. J. Viechnicki, A. A. Anctil, D. J. Papetti, and J. J. Prifty, "Lightweight Armor—A Status Report," U.S. Army MTL TR 89-8, 1989.
33. M. L. Wilkins, "Progress Report on Light Armor Program," Lawrence Livermore National Laboratory, 1969.
34. D. J. Viechnicki, M. L. Slavin, and M. I. Kliman, *Ceram. Bull.*, **70** (1991), 1035.

35. D. Shockey, A. H. Marchand, S. R. Skaggs, G. E. Cort, M. W. Birckett, and R. Parker, *Intl. J. Impact Eng.*, **9** (1990), 263.
36. J. Mescall, U.S. Materials Technology Laboratory, private communication, 1988.
37. J. Field, "Investigation of the Impact Performance of Various Glass and Ceramic Systems," Cavendish Laboratory, University of Cambridge, Contract DAJA 45-85-C-0021, European Research Office of the U.S. Army, 1988.
38. Hertz, *J. Math.* (*Grelle's J.*), **92** (1881).
39. B. Lawn and T. R. Wilshaw, *J. Mater. Sci.*, **10** (1975), 1049.
40. M. A. Meyers and C. L. Wittman, *Met. Trans.*, **21A** (1990), 3153.
41. D. Yaziv, G. Rosenberg, and Y. Partom, "Differential Ballistic Efficiency of Appliqué Armor," *Ninth Intl. Symp. on Ballistics, UK*, American Defense Preparedness Association, 1986.
42. M. Held, German Patent.

Explosive Welding

43. R. C. Carl, *Metal Prog.*, **46** (1944), 102.
44. J. S. Rinehart and J. Pearson, *Explosive Working of Metals*, MacMillan, New York, 1963.
45. A. A. Deribas, *The Physics of Explosive Hardening and Welding*, Nauka, Novosibirsk, USSR, 1972 (in Russian).
46. B. Crossland, *Explosive Welding of Metals and Its Application*, Clarendon, Oxford, 1982.
47. S. H. Carpenter, in *Shock Waves and High Strain Rate Phenomena in Metals*, eds. M. A. Meyers and L. E. Murr, Plenum, New York, 1981, p. 941.
48. H. El-Sobky, in *Explosive Welding, Forming, and Compaction*, ed. T. Z. Blazynski, Applied Science Publishers, England, 1983, p. 189.
49. M. D. Chadwick and P. W. Jackson, in *Explosive Welding, Forming, and Compaction*, ed. T. Z. Blazynski, Applied Science Publishers, England, 1983, p. 219.
50. T. Z. Blazynski, in *Explosive Welding, Forming, and Compaction*, ed. T. Z. Blazynski, Applied Science Publishers, England, 1983, p. 289.
51. N. V. Naumovich, A. I. Yadevich, and N. M. Chigrinova, in *Shock Waves for Industrial Applications*, ed. L. E. Murr, Noyes, Park Ridge, NJ, 1988, p. 170.
52. R. H. Wittman, *Proc. 2nd Intl. Symposium on the Use of Explosive Energy Manufacturing Metallic Materials*, Marian Sice Laxne, Czechoslovakia, 1973.
53. A. A. Deribas, V. A. Simonov, and I. D. Zakcharenko, *Proc. 5th Intl. Conf. on High Energy Rate Fabrication*, p. 4.1.1.
54. S. L. Wang, M. A. Meyers, and A. Szecket, *J. Mater. Sci.*, **23** (1988), 1786.
55. J. C. Lasheras and H. Choi, *J. Fluid Mech.*, **189** (1988), 53.
56. M. Hammerschmidt and H. Kreye, in *Shock Waves and High-Strain Rate Phenomena in Metals*, eds. M. A. Meyers and L. E. Murr, Plenum, New York, 1981, p. 961.

Shock Processing of Powders

57. O. Bergmann and N. F. Bailey, in *High Pressure Explosive Processing of Ceramics*, eds. R. A. Graham and A. B. Sawaoka, Trans. Tech Publications, Basel, Switzerland, 1987, p. 65.
58. S. Saito and A. B. Sawaoka, *Proc. 7th Intl. AIRAPT Conference, High Pressure Science and Technology*, Le Creusot, France, 1979, p. 541.
59. A. A. Deribas, Special Design Office of High-Rate Hydrodynamics, Siberian Division of the USSR Academy of Sciences, Novosibirsk, USSR, private communication, 1985.
60. W. H. Gourdin, *Progr. Mater. Sci.*, **30** and **39** (1986).
61. N. N. Thadhani, *Adv. Mater. Manufact. Proc.*, **3** (1988), 493.
62. R. Prümmer, *Explosivverdichtung Pulvriger Substanzen*, Springer, Berlin, 1987.
63. M. A. Meyers, L. H. Yu and N. N. Thadhani, in *Shock Waves for Industrial Applications*, ed. L. E. Murr, Noyes, Park Ridge, NJ, 1988, p. 265.
64. "Dynamic Compaction of Metal and Ceramic Powders," NMAB-394, National Materials Advisory Board, National Academy of Sciences, Washington, DC, 1983.
65. M. A. Meyers, L. E. Murr, and K. P. Staudhammer, eds., *Shock-Wave and High-Strain-Rate Phenomena in Materials*, Dekker, New York, 1992, p. 259.
66. L. E. Murr, K. P. Staudhammer, and M. A. Meyers, eds., *Metallurgical Applications of Shock-Wave and High-Strain-Rate-Phenomena*, Dekker, New York, 1986.
67. M. A. Meyers and S. L. Wang, *Acta Metall.*, **36** (1988), 925.
68. D. Raybould, in *Shock Waves and High-Strain-Rate Phenomena in Metals: Concepts and Applications*, eds. M. A. Meyers and L. E. Murr, Plenum, New York, 1981, pp. 895–911.
69. T. J. Ahrens, G. M. Bond, W. Yang, and G. Liu, in *Shock-Waves and High-Strain-Rate Phenomena in Materials*, eds. M. A. Meyers, L. E. Murr, and K. P. Staudhammer, Dekker, New York, 1992, p. 239.
70. W. J. Nellis et al., in *Shock-Wave and High-Strain-Rate Phenomena in Materials*, eds. M. A. Meyers, L. E. Murr, and K. P. Staudhammer, Dekker, New York, 1992, p. 795.
71. A. H. Mutz and T. Vreeland, in *Shock-Wave and High-Strain-Rate Phenomena in Materials*, eds. M. A. Meyers, L. E. Murr, and K. P. Staudhammer, Dekker, New York, 1992, pp. 323, 425.
72. A. Sawaoka and Y. Horie, in *Shock-Wave and High-Strain-Rate Phenomena in Materials*, eds. M. A. Meyers, L. E. Murr, and K. P. Staudhammer, Dekker, New York, 1992.
73. R. K. Gurney, "The Initial Velocities of Fragments from Bombs, Shells, and Grenades," BRL Report 405 Ballistic Research Laboratory, Aberdeen, MD, 1943.
74. M. A. Meyers, S. S. Shang, and K. Hokamoto, in *Applications of Shock Waves to Materials Science*, ed. A. B. Sawaoka, Springer, Tokyo 1993.
75. A. Ferreira, M. A. Meyers, S. N. Chang, N. N. Thadhani, and J. R. Kough, *Met. Trans.*, **22A** (1991), 685.

76. H. L. Coker, M. A. Meyers, and J. Wessels, *J. Mater. Sci.*, **26** (1991), 1277.

77. T. Akashi and A. Sawaoka, U.S. Patent 4,655,830, April 7, 1987.

78. F. R. Norwood, R. A. Graham, and A. Sawaoka, in *Shock Waves in Condensed Matter—1985*, ed. Y. M. Gupta, Plenum, New York, 1986, p. 837.

79. R. A. Graham and D. M. Webb, in *Shock Waves in Condensed Matter—1983*, eds. J. R. Asay, R. A. Graham, and G. K. Straub, Elsevier, Amsterdam, 1984, p. 211.

80. L. Davison, D. M. Webb, and R. A. Graham, in *Shock Waves in Condensed Matter—1981*, eds. W. J. Nellis, L. Seaman, and R. A. Graham, American Institute of Physics, New York, 1982, p. 67

81. R. A. Graham and D. M. Webb, cited in S. S. Shang et al. *J. Mater. Sci.*, **27** (1992), 5470.

82. S. L. Wang, M. A. Meyers, and R. A. Graham, "Shock Consolidation on IN-100 Nickel-Base Superalloy Powder," in *Shock Waves in Condensed Matter*, ed. Y. M. Gupta, Plenum, New York, 1986, p. 731.

83. G. E. Korth, J. E. Flinn, and R. C. Green, in *Metallurgical Applications of Shock-Wave and High-Strain-Rate Phenomena*, eds. L. E. Murr and K. P. Staudhammer, Dekker, New York, 1986, p. 129.

84. M. Yoshida, "Shock Consolidation of Pure Diamond Powders," Center for Explosives Technology Research, New Mexico Tech, Socorro, NM, October 1986.

85. L. H. Yu and M. A. Meyers, and T. C. Peng, *Mater. Sci. Eng.*, **A132** (1991), 257.

86. L. H. Yu and M. A. Meyers, *J. Mater. Sci.* **26** (1991), 601.

87. L. H. Yu, M. A. Meyers, and N. N. Thadhani, *J. Mater. Sci.* **5** (1990), 302.

88. M. A. Meyers, L. E. Murr, and B. B. Gupta, *J. Met.* **33** (1981), 21.

89. R. Graham, Sandia National Laboratories, private communication, 1991.

90. M. A. Meyers and H.-R. Pak, *J. Mater. Sci.*, **20** (1985), 2140.

91. A. Ferreira, M. A. Meyers, and N. N. Thadhani, *Met. Trans.*, **23A** (1992), 3251.

92. S. S. Shang, M. A. Meyers, and K. Hokamoto, *J. Mater. Sci.*, **27** (1992), 5470.

93. A. B. Schwarz, P. Kasiraj, T. Vreeland, Jr., and T. J. Ahrens, *Acta Met.*, **32** (1984), 1243.

94. M. M. Carroll and A. C. Holt, *J. Appl. Phys.*, **52** (1981), 2812.

95. M. M. Carroll, K. T. Kim, and V. F. Nesterenko, *J. Appl. Phys.*, **59** (1986).

96. V. F. Nesterenko, "High Rate Deformation of Heterogeneous Materials," Nauka, Novosibirsk, Russia, 1992, p. 17.

97. W. Tong and G. Ravichandran, *J. Appl. Phys.*, **74** (1993), 2425.

98. A. S. Helle, K. E. Easterling, and M. F. Ashby, *Acta Met.*, **33** (1985), 2163.

99. R. L. Williamson, *J. Appl. Phys.*, **68** (1990), 1287.

100. D. J. Benson and W. J. Nellis, in "Shock Compression of Condensed Matter—1993," *Am. Inst. Phys.*, 1994, in press.

101. M. A. Meyers, D. J. Benson, and S. S. Shang, in "Shock Compression of Condensed Matter—1993," *Am. Inst. Phys.*, **1994,** in press.

102. M. A. Meyers, S. S. Shang, and Hokamoto, in "Shock Waves in Materials Science," ed. A. B. Sawaoka, Springer, Tokyo, 1992, p. 145.

103. A. Ferreira and M. A. Meyers in "Shock-Wave and High-Strain-Rate Phenomena in Materials," eds. M. A. Meyers, L. E. Murr, and K. P. Staudhammer, M. Dekker, New York, 1991, p. 361.
104. K. Hokamoto, S. S. Shang, L. H. Yu, and M. A. Meyers, ibid, p. 453.
105. T. J. Ahrens, G. M. Bond, W. Yang, and G. Liu, ibid., p. 339.
106. H. Kunishige, Y. Horie, and A. B. Sawaoka, ibid., p. 353.
107. K. Kondo and S. Sawai, *J. Am. Cer. Soc.* **72** (1990) 1983.
108. S. Sawai and K. Kondo, *J. Am. Cer. Soc.* **72** (1990) 2428.
109. K. Kondo, in "Shock Compression of Condensed Matter," eds. S. C. Schmidt, R. D. Dick, J. W. Forbes, and D. G. Tasker, North Holland, Netherlands, 1992, p. 609.
110. A. H. Mutz and T. Vreeland Jr. in "Shock-Wave and High-Strain-Rate Phenomena in Materials," eds. M. A. Meyers, L. E. Murr, and K. P. Staudhammer, M. Dekker, NY, 1992, p. 425.
111. G. E. Korth, J. E. Flinn, and R. C. Green, in "Metallurgical Applications of Shock-Wave and High-Strain-Rate Phenomena," eds. L. E. Murr, K. P. Staudhammer, and M. A. Meyers, M. Dekker, NY, 1986, p. 129.
112. O. R. Bergmann in "Shock Waves in Condensed Matter," eds. J. R. Asay, R. A. Graham, and G. K. Straub, North Holland, Amsterdam, 1984, p. 429.
113. M. Yoshida and N. N. Thadhani, in "Shock Compression of Condensed Matter 1991," eds. S. C. Schmidt, R. D. Dick, J. W. Forbes, and D. G. Tasker, Elsevier, (1992) p. 585.
114. S. Fujihara, K. Narita, Y. Saito, K. Tatsumoto, S. Fujiwara, M. Yoshida, K. Aoki, Y. Kakudate, S. Usuba, and H. Yamawaki, in "Shock Waves," ed. K. Takayama, Springer, Berlin, (1992) p. 367.
115. M. Yoshida and S. Fujiwara, Proc. Symp. on Shock Synthesis of Materials, eds. N. N. Thadhani and E. S. Chen, Georgia Tech, May 24–26, 1994, U.S. Army Research Office.
116. G. A. Adadurov, Z. G. Aliev, L. O. Atoo-myan, T. V. Babina, Yu. G. Borodsko, Q. N. Breusov, A. N. Dremin, A. K. Muranevich, and S. V. Pershin, *Sov. Phys. Dokl.* **12** (1967) 173.
117. A. N. Dremin and O. N. Breusov, in "Shock Waves in Materials Science," ed. A. B. Sawaoka, Springer, Tokyo, Japan, 1993, p. 17.
118. A. N. Dremin and Staver, eds., Proc. 5th All Union Meeting on Detonation, Krasnoyarsk, Russia (in Russian) 1991.
119. A. I. Lyampkin, E. A. Petrov, A. P. Ershov, G. V. Sakovich, A. M. Staver, and V. M. Titov, *Sov. Phys. Dokl.* **33** (1988) 705.
120. V. M. Titov, "Proc. 2nd. Intl. Symposium on Intense Dynamic Loading and its Effects—1992," Sichuan U. Press, China, 1992, p. 851, Chengdu.
121. E. A. Petrov, G. V. Sakovitch, and P. M. Brylyekov, in "Shock-Wave and High-Strain-Rate Phenomena in Materials," eds. M. A. Meyers, L. E. Murr, and K. P. Staudhammer, M. Dekker, NY, 1992, p. 483.
122. T. Sekine, H. Kanda, Y. Bando, M. Yokoyama, and K. Hojou, *J. Mater. Sci. Lett.* **9** (1990) 1386.
123. T. Sekine and T. Sato, *J. Appl. Phys.* **74** (1993) 2440.
124. H. Hirai and K. Kondo, *Proc. Jap. Acad.*, **67** (1991) 22.

125. N. N. Thadhani, *Prog. Mater. Sci.* **37** (1992) 117.
126. S. S. Batsanov, "Effects of Explosives in Materials," Springer, NY, 1994.
127. R. A. Graham, "Solids under High-Pressure Shock Compression," Springer, New York, 1993.
128. Y. Horie, and A. B. Sawaoka, "Shock Compression Chemistry of Materials," Terra, Tokyo, 1993.
129. L. H. Yu and M. A. Meyers, in "Shock-Wave and High-Strain-Rate Phenomena in Materials," eds. M. A. Meyers, L. E. Murr, and K. P. Staudhammer, M. Dekker, 1992, p. 303.
130. K. S. Vecchio, L. H. Yu, and M. A. Meyers, *Acta Metall. Mater.* **42** (1994) 70.
131. M. A. Meyers, L. H. Yu, and K. S. Vecchio, *Acta Metall. Mater.* **42** (1994) 715.
132. B. R. Krueger, A. H. Mutz, and T. Vreeland, Jr., *J. Appl. Phys.* **70** (1991) 5362.
133. B. R. Krueger and T. Vreeland, Jr., in "Shock-Wave and High-Strain-Rate Phenomena in Materials," eds. M. A. Meyers, L. E. Murr, and K. P. Staudhammer, M. Dekker, NY, 1992, p. 241.
134. M. A. Meyers, L. H. Yu and M. S. Hsu, unpublished results (1994).
135. L. E. Alvarez, W. Alvarez, F. Asaro, and H. V. Michel, *Science*, **208** (1980), 1095.
136. J. D. O'Keefe and T. J. Ahrens, *Nature*, **298** (1982), 123.
137. J. D. O'Keefe and T. J. Ahrens, *Proc. Conf. on Large Body Impacts and Terrestrial Evolution*, Sp. Paper No. 190, Geological Society of America, 1982, p. 103.
138. H. J. Melosh, *Impact Cratering*, Oxford University Press, New York, 1989.
139. J. D. O'Keefe and T. J. Ahrens, Proc. Conf. on *Large Body Impacts and Terrestrial Evolution*, Sp. Paper No. 190, Geological Society of America, 1982, p. 103.
140. J. D. O'Keefe and T. J. Ahrens, *Proc. 6th Lunar Sci. Conf.* (1975), p. 2831.
141. J. D. O'Keefe and T. J. Ahrens, *Phys. Earth Plan. Int.* **16** (1978), 341.
142. T. J. Ahrens, in *Shock Waves in Condensed Matter*, ed. Y. M. Gupta, Plenum, New York, 1986, p. 571.
143. M. A. Lange and T. J. Ahrens, *J. Earth Plan. Sci. Lett.*, **77** (1986), 409.
144. J. D. O'Keefe and T. J. Ahrens, *Nature*, **338** (1989), 247.
145. R. Jeanloz and T. J. Ahrens, *Geophys. J. R. Astron.* Soc., **62** (1980), 505.
146. R. Jeanloz, T. J. Ahrens, H. K. Mao, and P. M. Bell, *Science*, **266** (1979), 829.
147. C. H. Dowding, *Blast Vibration Monitoring and Control*, Prentice-Hall, Englewood Cliffs, NJ, 1985.
148. C. T. Aimone, M. A. Meyers, and N. Mojtabai, in *Rock Mechanics in Productivity and Protection*, eds. C. H. Dowding and M. M. Singh, SME-AIME, New York, 1984, p. 979.
149. E. Schneider and A. Stilp, *J. Impact Eng.* **5** (1987) 561.
150. Time, 143, No. 21 (1994) pp. 60–61.

AUTHOR INDEX

Numbers underlined refer to page where complete reference is given.

SUBJECT INDEX

CPSIA information can be obtained at www.ICGtesting.com
Printed in the USA
BVOW06*0305270816

460300BV00005B/32/P

9 780471 582625